A CENTURY OF
AMERICAN SCULPTURE
THE ROMAN BRONZE WORKS FOUNDRY

Lucy D. Rosenfeld

Schiffer Publishing Ltd
4880 Lower Valley Road, Atglen, PA 19310 USA

Library of Congress Cataloging-in-Publication Data

Rosenfeld, Lucy D., 1939-
A century of American sculpture: the Roman Bronze Works Foundry/by Lucy D. Rosenfeld.
p. cm.
ISBN 0-7643-1519-6
1. Roman Bronze Works Foundry--History. 2. Foundries--New York (State) 3. Bronze founding--New York (State) 4. Bronze sculpture, American--20th century. I. Title.
NB1230 .R68 2002
730'.973'0904--dc21
2001006596

Designed by "Sue"
Type set in Dutch801 Rm BT

ISBN: 0-7643-1519-6
Printed in China
1 2 3 4

Published by Schiffer Publishing Ltd.
4880 Lower Valley Road
Atglen, PA 19310
Phone: (610) 593-1777; Fax: (610) 593-2002
E-mail: Schifferbk@aol.com
Please visit our web site catalog at **www.schifferbooks.com**
We are always looking for people to write books on new and related subjects. If you have an idea for a book please contact us at the above address.

This book may be purchased from the publisher.
Include $3.95 for shipping.
Please try your bookstore first.
You may write for a free catalog.

In Europe, Schiffer books are distributed by
Bushwood Books
6 Marksbury Ave.
Kew Gardens
Surrey TW9 4JF England
Phone: 44 (0) 20 8392-8585; Fax: 44 (0) 20 8392-9876
E-mail: Bushwd@aol.com
Free postage in the U.K., Europe; air mail at cost.

CONTENTS

Acknowledgments **5**
Preface **7**

I The Role of Roman Bronze Works **8**
II. The Process of Casting **14**
III. Monuments and Portrait Commemoratives:
Honoring America and Americans in the Early Twentieth Century **21**
Daniel Chester French **22**
Augustus Saint-Gaudens **29**
John Quincy Adams Ward **36**
Karl Bitter **37**
Herbert Adams **38**
Alexander Stirling Calder **40**
Gutzon Borglum **42**
Robert Ingersoll Aitken **45**
Jo Davidson **48**
Rudolph Evans **53**
Charles Keck **54**
Charles Grafly **55**
William Ordway Partridge **57**
John Massey Rhind **58**
Adolph Alexander Weinman **59**
Margaret French Cresson **62**
Bryant Baker, George Grey Barnard, John Flanagan, Augustus Lukeman, Philip Martiny, Samuel Murray, Charles Henry Niehaus, Bela Lyon Pratt, Ettore Ximenes and others **64**

IV. Cowboys, Indians, and Buffalo:
The Western Image and Sculpture of Wildlife **69**
Frederic Remington **70**
Charles Russell **79**
Solon Borglum **88**
Sally Farnham **93**
Charles Humphriss **96**
Cyrus Dallin **98**
Hermon Atkins MacNeil **101**
Adolph Alexander Weinman **105**
Mahonri Young **107**
Alexander Proctor **109**
Anna Hyatt Huntington **113**
Edwin W. Deming **120**
James Clark **122**
Henry Merwin Shrady **125**
Albert Laessle **129**
Frederic Roth **131**
Other Sculptors of Western and Wildlife Images **134**
Carl E. Akeley, Antoine Louis Barye, Isidore Jules Bonheur, Eli Harvey, Louis Potter, Charles Schreyvogel, Charles Umlauf

V. Modernism in American Sculpture:
New Modes of Expression From 1913 .... **139**
Alexander Archipenko **140**
Elie Nadelman **143**
Gaston Lachaise **145**
Jacques Lipchitz **150**
Max Weber **151**
Isamu Noguchi **153**
Robert Laurent **155**
Paul Manship **156**
Lee Lawrie **165**
William Zorach **167**
Hugo Robus **169**
Alexander Calder **172**
Reuben Nakian **180**
Other Modernist Sculptors **183**
John Storrs, Max Beckmann, Mary Callery

VI. Within the Realist Framework:
Genre, Classicism, Allegory, and the Exotic **185**
John Rogers **186**
Thomas Eakins **187**
Bessie Potter Vonnoh **190**
Mahonri Young **199**
Isidore Konti **201**
Frederic William MacMonnies **204**
Paul Troubetskoy **209**
Robert Tait McKenzie **211**
Gertrude Vanderbilt Whitney **216**
James Earle Fraser **217**
Edward McCartan **219**
Albin Polasek **222**
John Gregory **224**
Harriet Frishmuth **226**
Josef Mario Korbel **229**
Malvina Hoffman **234**
Anthony deFrancisi **242**
Gleb Derujinsky **246**
Paul Jennewein **248**
René Chambellan **250**
Alan Clark **252**
Avard Fairbanks **255**
Wheeler Williams **257**
Joseph Kiselewski **259**
Archimedes Giacomantonio **261**
OTHER REALIST SCULPTORS **264**
Chester Beach, Edward Berge, Gustav Bohland, Joseph Coletti, Pompeo Coppini, Hunt Diederich, Abstenia St. Leger Eberle, Max Kalish, Leo Lentelli, Henry Linder, Evelyn Longman, Boris Lovett-Lorski, Nathaniel Kaz, Eleanor Mellon, Attilio Piccirilli, Joseph Pollia, Brenda Putnam and others

VII. To the Present:
R.B.W.'s Current Sculptors **279**
Marvin Bell, Sandro Chia, Dante V. de Florio, Arturo di Modica, Leonda Finke, Brian Hunt, Beatrice Jaberg, Bruno Lucchesi, Lawrence M. Ludke, Gregory Perillo, Alfredo Sasso, Adolf Sehring, Rhoda Sherbell, Eleanor Stillman, Bismarck Victoria

Photo Credits **287**
Index of Artists **288**

ACKNOWLEDGMENTS

I would like to acknowledge with grateful thanks the help of many people who guided my research, helped me find rare photos, or contributed their expertise to this project. Thanks go particularly to Gwen Pier of the National Sculpture Society, as well as to Ann Lebleu of Hirschl and Adler Galleries, to Robert Kashey of the Shepherd Gallery, and to Walter Maibaum, all of New York. Robin Salmon of Brookgreen Gardens in South Carolina, Wanda Styka at Chesterwood in Massachusetts, and Charles Robertson of the National Museum of American Art in Washington D.C. all contributed their expertise to this project. Librarians Deborah Bigelow and Theresa Wyman were always helpful, and I could not have completed the picture collection without the tireless aid of Robyn Stoltzfus. Philip Schiavo of Roman Bronze Works provided me with an invaluable collection of photographs and other material, without which this book would have been impossible to write. And finally I would like to thank my special and always most faithful reader and supportive critic, my husband Peter. To all, many thanks!

PREFACE

In an era of so much artistic change, it is rare to find a part of the art world that has survived for a full century. Roman Bronze Works has seen the 20th century's sculpture come and go, has produced thousands of castings, has dealt with hundreds of artists, galleries, collectors, and museums. A reading of its ledgers is like a visit to other eras—when people like Remington and Russell, Archipenko and Manship, Nadelman and Calder stopped by in Corona, Queens to deliver their works for casting. Interspersed with these hundreds of artists' names are those of a glamorous era of collecting: John D. Rockefeller, Cecil B. DeMille, Durand-Ruel.

As a long-time writer on art, when I was asked to do this book I could not resist the challenge of discovering the works that R.B.W. cast and of trying to make sense of the various styles and types of sculpture that characterized the last century. For where better to visualize the stylistic progression of America's monuments and statues than through the foundry that cast the works of virtually every important sculptor in the nation? This one hundred year long parade of names and sculptures in the ledgers is carefully dated, with works titled: there are war monuments and cowboys, nude figures—both realistic and abstract, muscled steel workers and exotic Asian dancers, giant spheres and works described simply as "Abstract"—in short a sampling of each artistic movement and aspect of American taste.

In addition to finding pictures of the works and biographies of the sculptors—many once famous and now long forgotten—I have also found another almost forgotten subject: the virulence of the battle between sculptural realism and modernism that characterized the first half of the century. For large sculpture, unlike painting, required commissions, and in many cases, civic subject matter. The battle for the mind and aesthetic taste and backing of the art establishment has been long and furious, rife with political intrigue. (In fact, I found in one archive the transcript of one sculptor's testimony against his modernist colleagues at the House Un-American Activities Committee in the 1950s.)

Another controversial element of this subject is the disposition of the maquettes or "sketches" left by the artists at the foundry over many years. This is an area that has long been debated; the related roles of sculptor, founder, dealer, and collector have been a source of conflict and even lawsuits in the history of practically every foundry both here and abroad, and R.B.W. has had its share of controversy. My focus has been entirely on the art itself, and I have not attempted to write a history of R.B.W.'s business dealings or copyright issues, or those of any gallery or collector. I have neither legal expertise nor access to the materials involved.

My focus is on the works themselves. Slightly over one hundred artists are included. I have reproduced only those works by each artist that are documented as having been cast at R.B.W. (Some sculptors used several different foundries—occasionally even for the same piece.) It should also be noted that the statues reproduced here are by no means the only ones cast by each artist at Roman Bronze Works, but are merely a sampling of the sculptor's work. Needless to say, there are numerous copies of these works in a variety of museums and private collections—some made by R.B.W., and others cast by different foundries.

I have begun the book with rare pictures from R.B.W.'s own archives, showing the procedure for lost-wax casting. Following are hundreds of pictures of works cast at the foundry. Please note that dates given for works almost always refer to the R.B.W. ledger dates when the work was first cast there. When pieces were cast over and over, as so many were, I have given the earliest mention. Some of the dates are necessarily guess-work; discrepancies in dates of casting may often be the result of many notations in the ledgers that merely refer to "statue" or "figure," or other unspecific listing.

I hope that my readers will enjoy this visit to the world of American sculpture, both as an aesthetic experience, and as a way of looking at the conflicting issues of the art world through a century of change.

I. THE ROLE OF ROMAN BRONZE WORKS

Throughout Roman Bronze Work's century of operation, it has seen many of the great sculptors of America bring their models for casting. The nation's patriotic monuments and statuary, its ornate public doors and small private fountains, its bronze cowboys and Indians, its nudes and its first abstract sculptures—all have been part of this foundry's extraordinary history. As the nation's cities expanded, its skyscrapers built, and its monuments constructed, bronze sculpture—both architectural and free-standing—became a necessity for civic pride and the furthering of national ideals. R.B.W.'s history reflects a nation celebrating progress and nationalism, heroism and the ideals of democracy, as well as conflicting national aesthetic sensibilities. In studying the sculpture of the period we can see the many diverse strains of American life in the twentieth century in three-dimensional form: the Beaux-Arts and neo-classical monuments in European style, the national war heroes, the home-grown celebrations of American life and belief, the first efforts at modernism.

When R.B.W. opened its doors at the turn of the twentieth century, it was entering a volatile world, as a collaborator with America's best-known sculptors—both conservative and modernist, native-born and newly arrived, well established or anti-establishment and avant-garde. As R.B.W.'s hand-written ledgers show, Roman Bronze Works became the premier lost-wax casting foundry in the United States in the first half of the twentieth century, and, the foundry of choice of the nation's major sculptors.

A foundry has an unusual place in the art world, for the sculptor in bronze or other metals joins into a kind of partnership with it. Unlike artists in paint or other two-dimensional media, the sculptor in metal will never see his or her work in its full size and desired media without the help of the foundry. When the sculptor brings a small, fragile plaster model to the foundry for casting, he or she takes a leap of faith: will the full sized, bronze or aluminum work capture the essence of that small cast? Will the foundry work with the artist to imitate exactly, or to modify as wished? And, on the other hand, will the foundry recognize the limits of its own role and not modify the work on its own?

Throughout the history of metal casting of sculpture—from ancient times in fact—the foundry's role has been a complex one, and it still is today. The bonds between artist and foundry are at best warm and collaborative, at worst a cause for argument and lawsuits. The expanded market for privately owned sculpture, as well as the vast growth of museum collections, has meant that sometimes dozens of castings are made from one plaster model. The ownership of the copies can be in doubt, as can the authenticity of the castings. Sculptors frequently leave their plaster casts at a foundry for years and years, waiting for requests for a new edition. Foundries, for their part, sometimes cast copies at the request of galleries or collectors instead of the artist, while the sculptor may have taken a copy of the same work to another foundry for production. Unlike their European counterparts, American foundries never had contractual arrangements with their clients. Plaster models were often moved from place to place. These confusions are endemic to the American foundry business, and Roman Bronze Works has had its share of them.

Riccardo Bertelli (1870-1955) arrived in the United States from Genoa in 1895. He had been trained both in science and in art, and had a degree in chemical engineering. During his first visit to America, he was persuaded by artists and sculptors that there was a need for a new foundry that would cast sculpture in the historic lost-wax method. With an obscure sculptor and founder named Giuseppe Moretti, he founded Roman Bronze Works in 1897 in Manhattan (making it the oldest working foundry in America). Its name was registered in New York in 1900, when it moved to Greenpoint, Brooklyn under the sole direction of Bertelli. The new founder made a trip to his native country to bring back artisans skilled in the traditional lost-wax process. Among them were the foundry superintendent, John Zappula and Philip Schiavo, the latter's brother-in-law and assistant. The Schiavo family would continue to be involved in the organization throughout the century. A wealthy silk importer named Celestino Piva became a backer. The foundry remained in Brooklyn until 1928.

The first foundry in the United States to devote itself exclusively to lost-wax casting, R.B.W. made casting the most intricate bronze sculpture possible by allowing for greater clarity of detail and more complex textures. The lost-wax method also invited the collaboration of the sculptor during the process of casting. Using the thin wax model, the artist could change one casting of a work from the next. R.B.W. led the way in lost-wax casting in America, transferring the technique from Europe to America. Bertelli visited many sculptors' studios to acquaint them with the process. His was a charming and sympathetic personality, and he was generous enough to lend money and extend credit to sculptors. Close relationships developed between the founder and many different sculptors. Some spent time living in his home; others used the foundry as their own studios.

Among the first artists to be convinced by the method and Bertelli's commitment to casting fine sculpture was Frederick Remington. By 1901 the sculptor who mytholo-

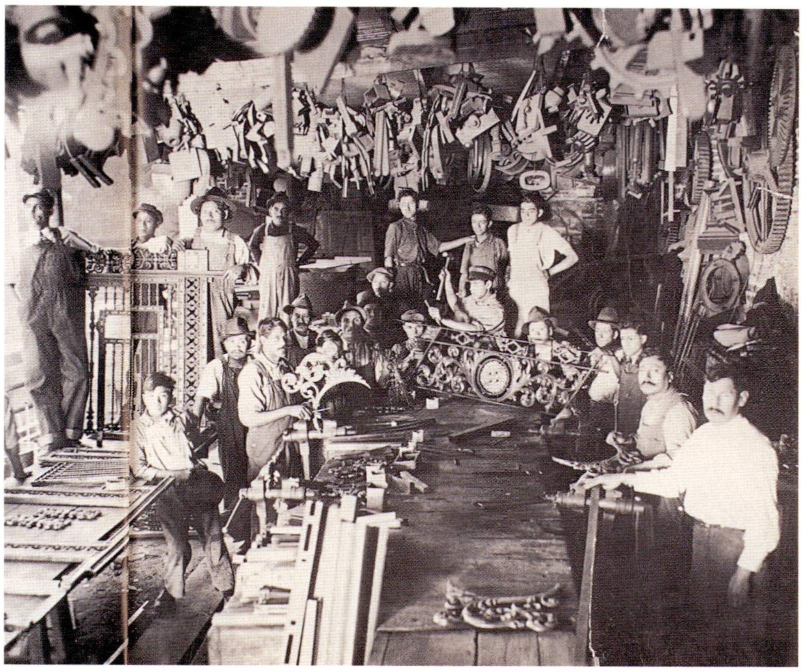

I-1. Workers in the first Roman Bronze Works in the pattern shop in Greenpoint, Brooklyn, circa 1901.

gized the American West with his cowboys and Indians and bucking broncos, had left behind the sand-casting method and taken his business to R.B.W., both transferring models from his previous foundry and bringing new works in for casting. The collaboration between Bertelli and Remington was close and personal and it lasted until the artist's death in 1909. Together they explored new methods, used the lost-wax casting technique to alter small aspects of the sculptor's work, and found different ways of supporting bronze sculpture on its base.

Another close relationship developed with Tiffany & Company, which had sold Remington bronzes with great success. (Remington's "Bronco Buster" was the most popular American bronze statuette of the nineteenth and early twentieth centuries.). Bertelli had seen Remington's statues at Tiffany's as early as 1895. While working with R.B.W. Remington continued to sell his bronzes through Tiffany's; the artist kept the store supplied with casts that Bertelli made, though they had a hard time keeping his statuettes in

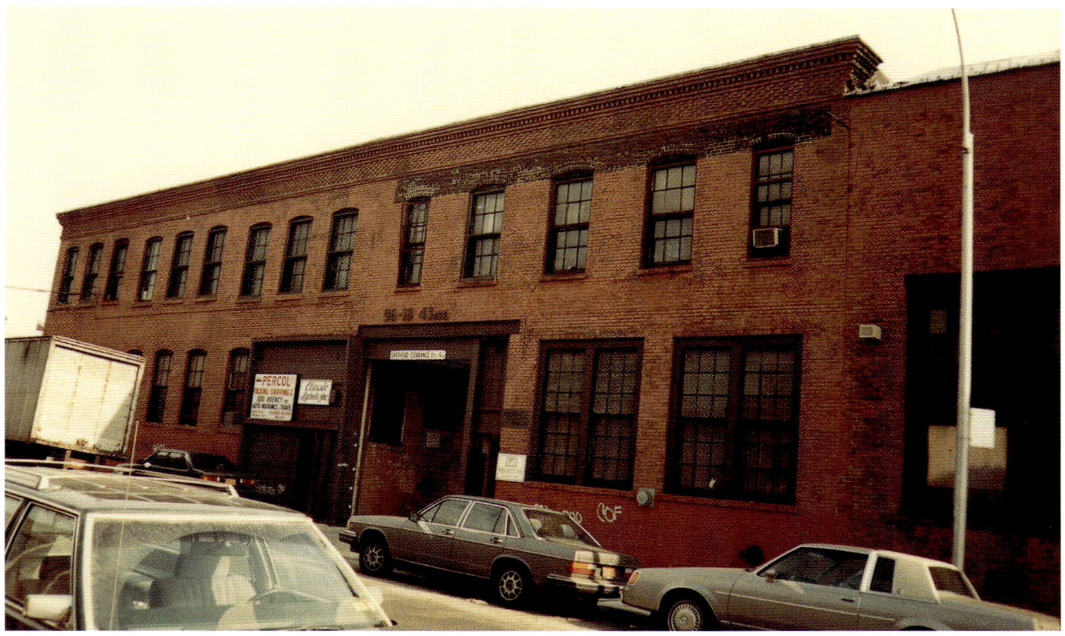

I-2 & I-3. Exterior views of the Corona foundry building.

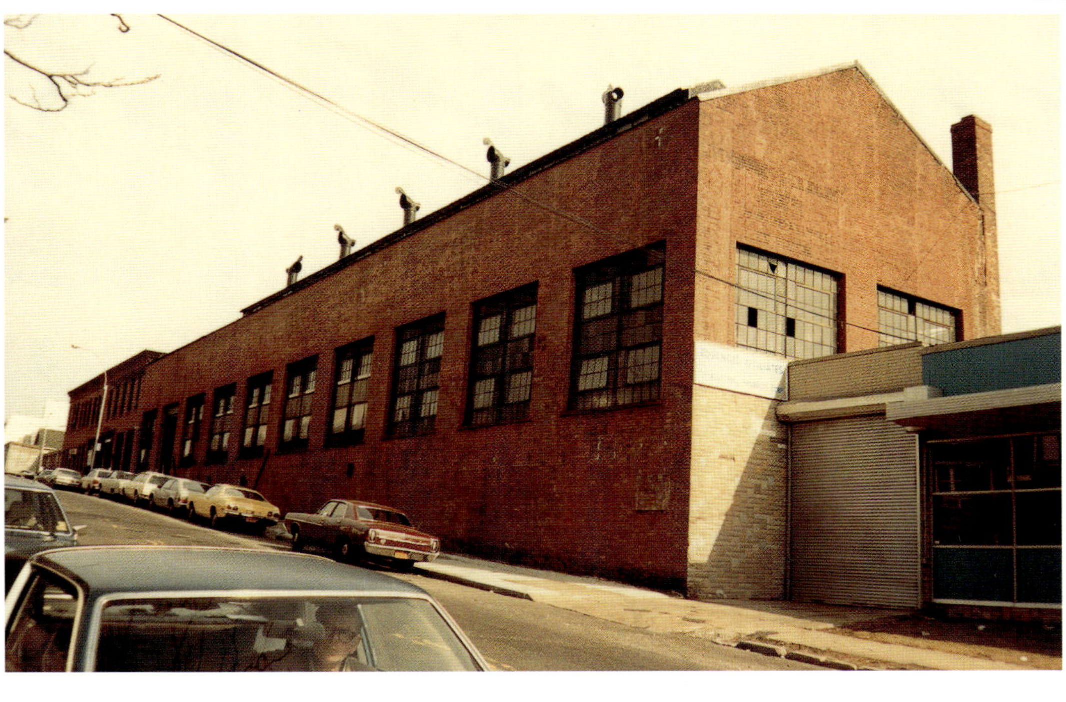

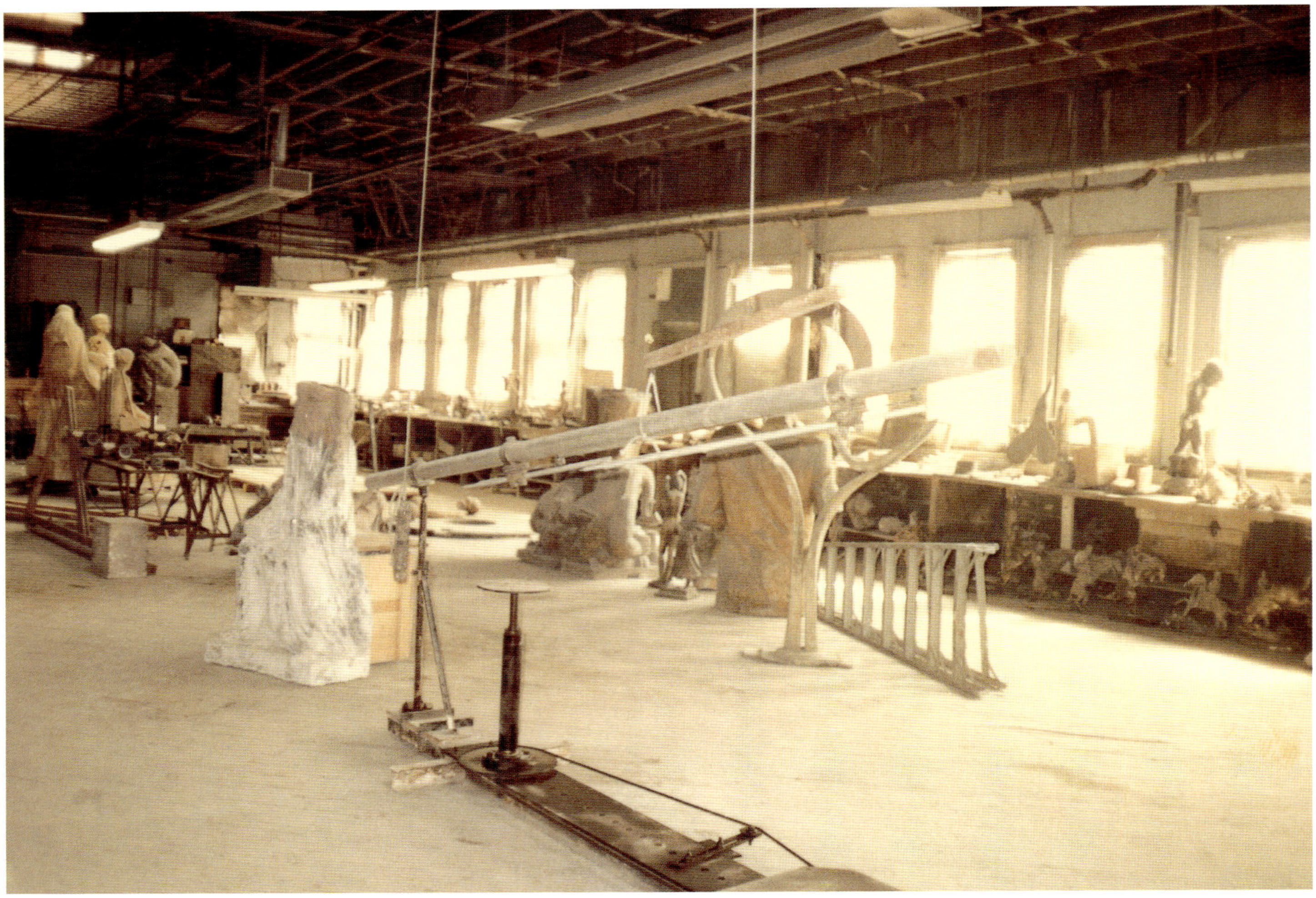

I-4. The enlarging studio in the Corona foundry.

I-5. Storage basement of the Corona building, where shelves hold both sculptural maquettes and foundry supplies.

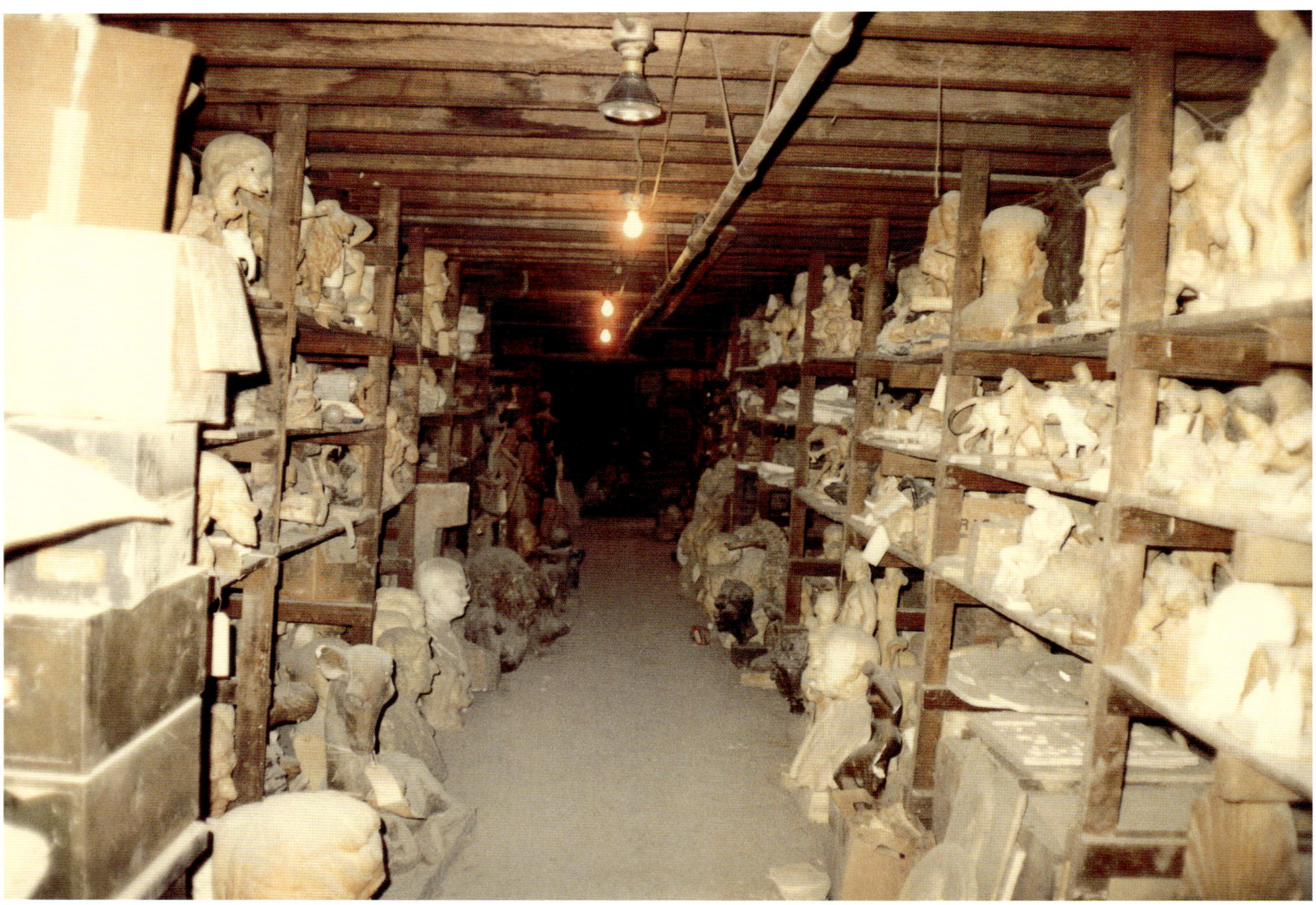

stock. In addition to this relationship, Bertelli began casting the bronze bases for the lamps and sculpture of Tiffany Studios.

Other major sculptors, including Augustus Saint-Gaudens, Daniel Chester French, and Henry Merwin Shrady, discovered the process and the foundry, casting major bronze monuments there between 1901 and 1912. R.B.W. also acquired models of several well-known European sculptures, including those of Rodin. After a number of successful years, R.B.W., like other American foundries, faced a steep decline with the beginning of World War I and the difficulty of obtaining metals. But Roman Bronze Works not only survived, it prospered as a huge number of sculptors—both professional and amateur—cast their works there. Bertelli (and the owners of R.B.W. who followed him) typically acquired one or more cast from each original edition of sculpture at the foundry. If not, they often subsequently purchased a cast, in order to complete a comprehensive collection of all of the bronzes cast at the foundry.

In 1928 R.B.W. became a subsidiary of General Bronze Corporation of New York, a manufacturer of architectural pieces. The foundry then moved to Corona, Queens to the old Tiffany Studios bronze facility. When the Tiffany Studio went bankrupt in 1932 Roman Bronze Works took over the entire historic complex that had housed the glass and metal operations. Among the objects produced by R.B.W. for Tiffany were lamp bases, bronze sculptures, window frames for its decorative glass, and such small but essential items as the whip carried by their popular Remington cowboy statues.

At the height of its operations, from the early 1900s to World War II, Roman Bronze was the foundry of choice of most of the major sculptors in the United States. The 106 sculptors whose work is included in this book were among dozens of less well-known or amateur clients, as well as architects and designers, to use the foundry. Roman Bronze Works even maintained its own Manhattan gallery on 56th Street between 1930 and 1949.

R.B.W. worked extensively on architectural sculpture with Stanford White, Cass Gilbert, Horace Trumbauer, and McKim, Mead, and White, as well as many other architects. Among the many architectural pieces R.B.W. cast were door jambs, floor registers, window casings, and lamps. Many were in the Neo-Classical and Art Deco styles popu-

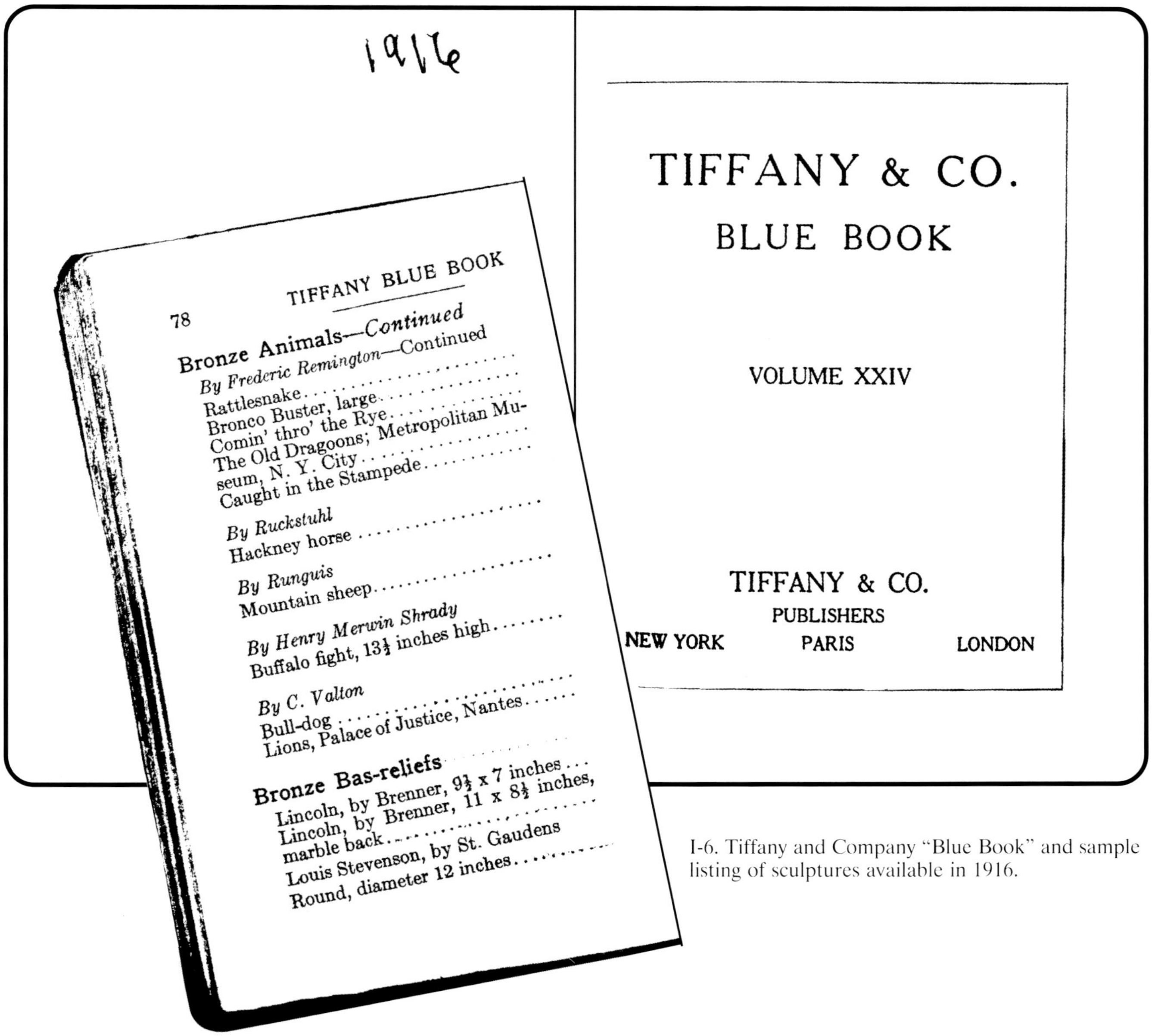
1916

78 TIFFANY BLUE BOOK

Bronze Animals—*Continued*

By Frederic Remington—Continued
Rattlesnake
Bronco Buster, large
Comin' thro' the Rye
The Old Dragoons; Metropolitan Museum, N. Y. City
Caught in the Stampede

By Ruckstuhl
Hackney horse

By Runguis
Mountain sheep

By Henry Merwin Shrady
Buffalo fight, 13½ inches high

By C. Valton
Bull-dog
Lions, Palace of Justice, Nantes

Bronze Bas-reliefs
Lincoln, by Brenner, 9½ x 7 inches
Lincoln, by Brenner, 11 x 8½ inches, marble back
Louis Stevenson, by St. Gaudens
Round, diameter 12 inches

TIFFANY & CO.
BLUE BOOK

VOLUME XXIV

TIFFANY & CO.
PUBLISHERS
NEW YORK PARIS LONDON

I-6. Tiffany and Company "Blue Book" and sample listing of sculptures available in 1916.

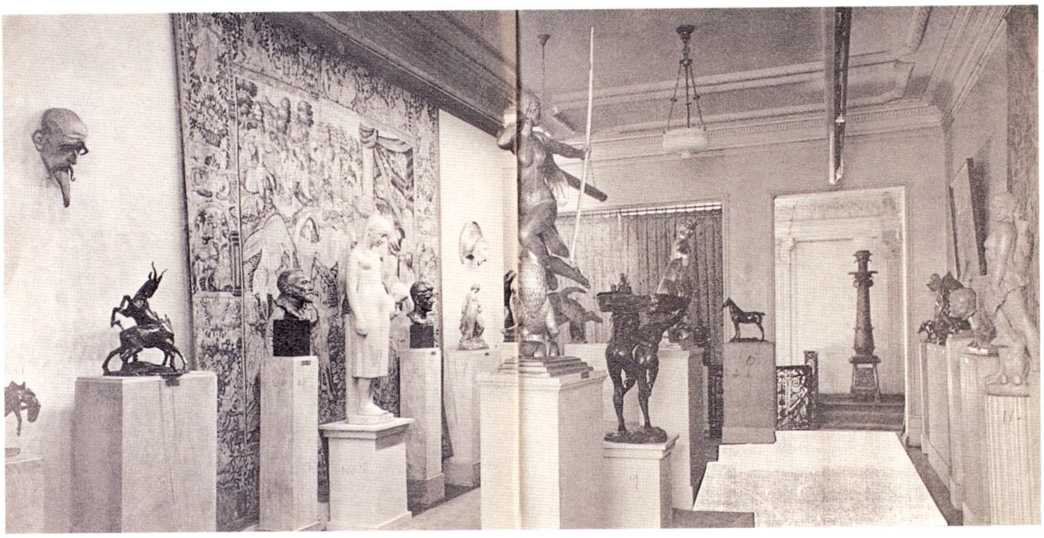

I-7. Roman Bronze Works had its own 56th Street gallery in New York between 1930 and 1949.

I-8. John P. Schiavo with his son, Philip Schiavo, in the gallery.

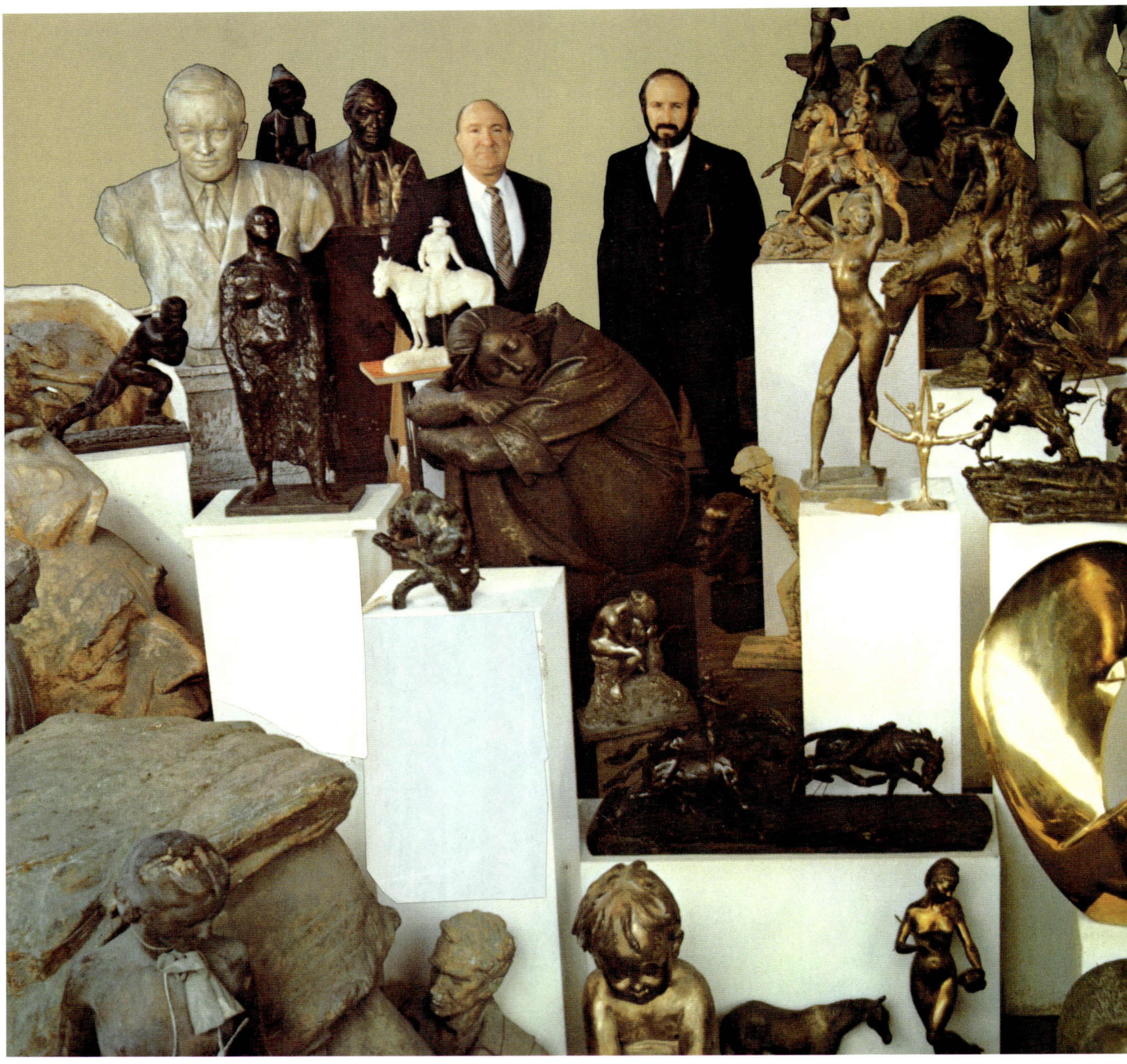

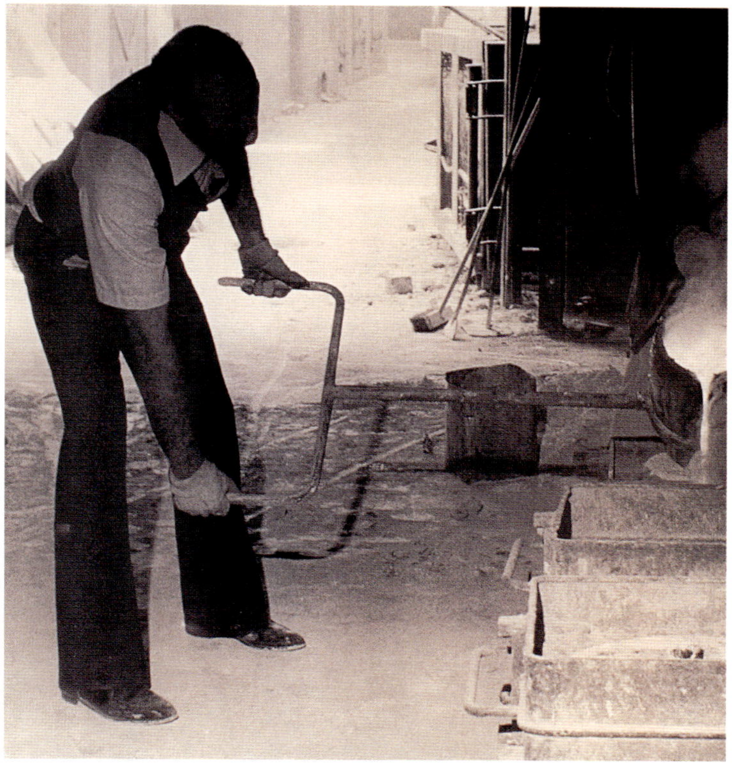

I-9. Philip Schiavo, who took a hands-on approach to running the foundry, pours molten metal.

lar in the first half of the century. At the archives for the foundry (now at the Amon Carter Museum in Fort Worth, Texas) there are seven hundred ninety-one architectural drawings and blueprints for decorative elements and fixtures for ten different buildings, all constructed between 1931 and 1941. Among them were the United States Supreme Court Building, the Municipal Airport Building at La Guardia, and Cincinnati's Union Terminal. Another notable client was the designer Gustav Stickley.

Sculptors and architects were not the only clients of R.B.W. Roman Bronze also cast works directly for museums, ranging from the American Museum of Natural History to the Metropolitan Museum of Art, to the Whitney Museum of American Art. Other major clients included numerous New York galleries, among them Grace Borgenicht, Hammer, Knoedler and Company, Grand Central, Iolas, David Findlay, Kraushaar, Milch, Curt Valentin, and Viviano. In addition, many collectors and philanthropists brought works for casting: names found in the ledgers include Enrico Caruso, John D. Rockefeller, Louis Mayer, Mrs. Andrew Carnegie, R.H. Macy, B. Altman, Harry Guggenheim, Adam Gimbel, Adolph Lewissohn, Anna Mahler, Walter Chrysler, Charles Scribner, Cecil B. De Mille, and Billy Rose.

The association with General Bronze Corporation lasted for twenty years, during which the foundry produced virtually all of the sculpture for Rockefeller Center, numerous national monuments, and a number of sculptures for the W.P.A., in addition to its usual complement of private commissions. But World War II, and the shortage of available metals for casting, forced a temporary halt to production in the early 1940s.

In 1948 the ownership of R.B.W. changed once again. The foundry had become a corporate entity rather than a family business run by the magnetic personality of Bertelli, and the Schiavo family, who had been associated with the firm since 1901, decided to purchase it. But the foundry had some trouble regaining its prominent position in the art world in the post-war period. While the original owner had encouraged sculptors of both conservative and avant-garde bent, the new owners favored representational sculpture at a time when modernism was finally being widely accepted in America. In addition, a new emphasis on the process of carving directly into stone or wood had replaced casting in bronze for many sculptors. And by the 1950s many younger sculptors were experimenting with non-objective and nonrepresentational art and new media, and Roman Bronze Works—and indeed, bronze casting itself— was increasingly seen as too conservative for them. More traditional sculptors, however, continued to use the foundry and its business in monuments and smaller statuary went on as it always had.

In 1976 Philip Schiavo, the current owner of R.B.W., took over the foundry from his uncle. He has recently moved the business to its new location in Copiague, Long Island, where he is renovating the former torpedo factory he purchased; as we go to press much of the new site is already operational. Though this book is primarily a study of the history of the extraordinary range of American sculpture that passed through the foundry between 1900 and 1960, a sampling of the work of some of R.B.W.'s most recent clients appears at the end.

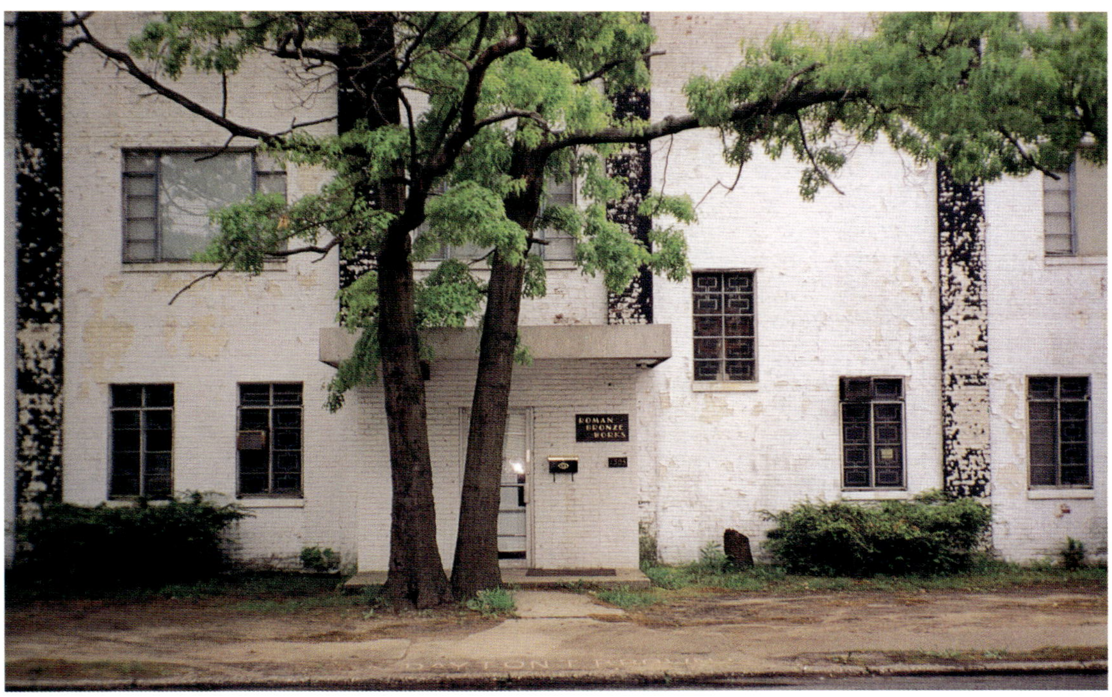

I-10. The new foundry building in Copiague, Long Island.

II. THE PROCESS OF CASTING

Bronze has long been American sculptors' metal of choice for its versatility, tensile strength, and permanence. And the lost-wax system truly revolutionized the process of casting American bronze sculpture. Unlike the sand-casting method that in the nineteenth century was widely used in the United States (in which pieces of the sculpture were cast separately and then rejoined), the lost-wax system allowed sculpture to be cast in one piece. It also offered the artist a chance to work on the piece during the process of casting. Its flexible mold was the key to these changes. A brief look at the process of lost-wax casting may help to explain how the system works.

The sculptor arrives at the foundry with a plaster model of the original clay work. The first step is to make a mold from the sculptor's model. The process begins when clay is put over the original plaster model in preparation for a "mother" mold. This flexible mold has a rubbery, gelatin lining, which can be easily peeled away. The interior of the mold is an exact negative of the original sculpture.

Hot melted liquid wax is brushed into the insides of the mold, exactly reproducing the plaster model. (Great care is taken to cover every crevice.) Sheets of wax are then placed in it to make the wax stronger and to make uniform the thickness of the bronze that will eventually fill the same area. (Bronze can be cast up to an inch in thickness, but in small statues is generally about three-eighths of an inch thick.)

When the flexible rubber mold is removed, it leaves a wax positive, a hollow replica, which can still be altered or retouched by the sculptor. (Thus, every wax model changed by the sculptor becomes an "original.")

The "gates" come next; they are the channels through which the molten metal is poured into the mold. The wax model is covered with a silica plaster material and gated or vented with tubes, in order to create a mold for the final bronze casting. The gates form a kind of circulatory system so that the metal will reach all parts of the mold, as well as creating ducts or vents for the release of gases.

The silica-covered molds are put in a hand-built oven, in preparation for the wax melt-out. As the mold is fired, the wax melts out or is "lost." The process of heating can take two or three days until the wax has completely melted out.

Bronze, which is a combination of 80 or 90% copper and other metals including primarily zinc, lead, and tin, has a lower melting point than copper alone. It is highly versatile in every way, enabling the artist to use a great variety of shapes, surfaces, and tones. It is heated separately in a graphite crucible until it is around 1005 degrees Celsius.

When the molten metal is poured into the plaster investment mold, filling the space left by the wax, the metal sculpture is created. The moment in which the molten metal fills the mold was described by one observer: "The top of the mold appears to be a pale submarine monster with glowing fiery eyes and mouth...it is impossible even in the most prosaic technical description not to give a sense of the beauty and excitement of this relatively brief moment in the process of bronze casting."

After the bronze cools for several days, the plaster investment is chipped away to expose the raw bronze casting. The core is raked out. The sculpture is cleaned in a bath of nitric acid. In the final finishing touches, a die grinder is used to remove the vents, and the holes that remain are filled with bronze. If the sculpture was cast in several pieces, they are welded together at this point. The artist may also choose to chisel away parts, or to file or polish one part and not another. Some contemporary sculptors, who have made a very rough mold, use this moment for the creative changing of the shape.

The sculpture is then hand finished (or chased), if desired. Among the tools used are buffers with brushes and steel wool. Either the artist or the foundry does the surface finishing. While the natural color of the bronze is a rich gold tone (reflecting the copper in the metal), many artists prefer a darker hue. Liquid chemicals are applied either by fire or a cold solution to create a finished bronze patina that appeals to the artist; the copper content enables various chemicals to change the color from a deep brown to a bluish or greenish tone. Occasionally the statue is gilded with gold leaf or inlaid with details made of other metals.

As these historic photographs from Roman Bronze Works archives show, the process was a complex and lengthy one. Most finished works at the foundry were identified with the words "Roman Bronze Works" or just the initials "R.B.W." on the bottom of the piece, but a few sculptors preferred not to have the foundry's mark on their work. (Like so many other aspects of the casting process, this decision has made identifying the origins of certain bronzes more difficult.)

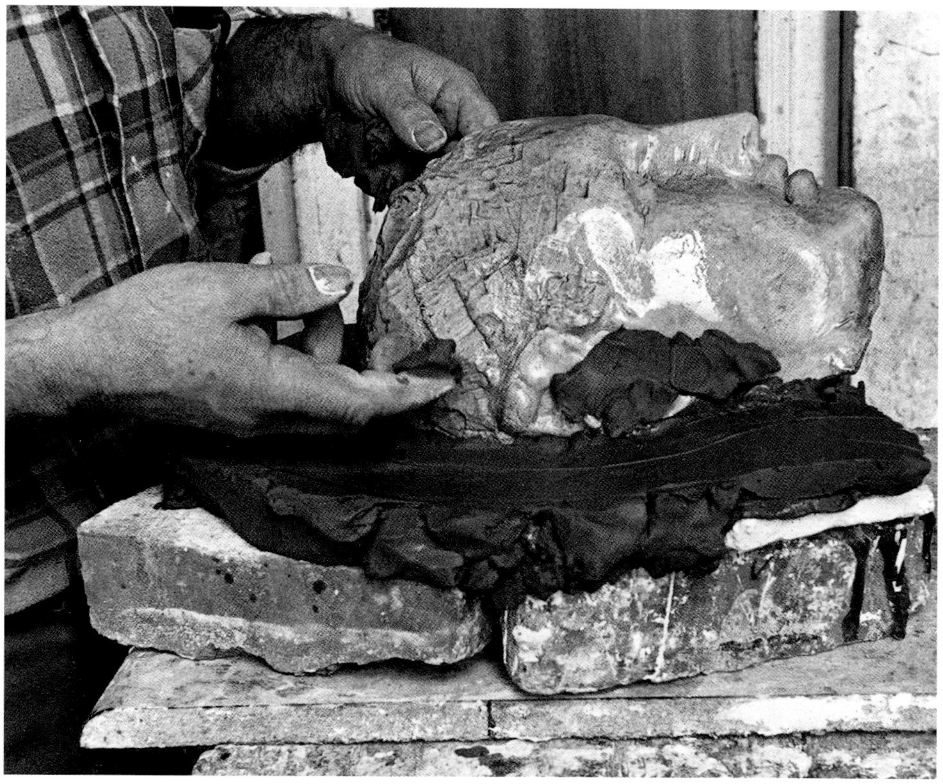

II-1. Clay is put over the original plaster model in preparation for the mother mold.

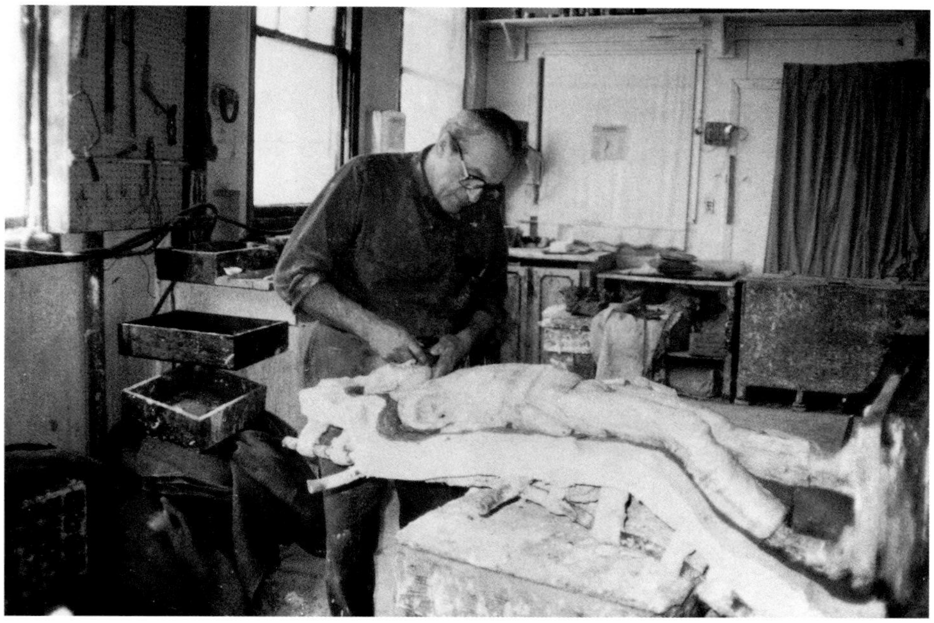

II-2. A plaster shell is made by craftsman Stanley Spring.

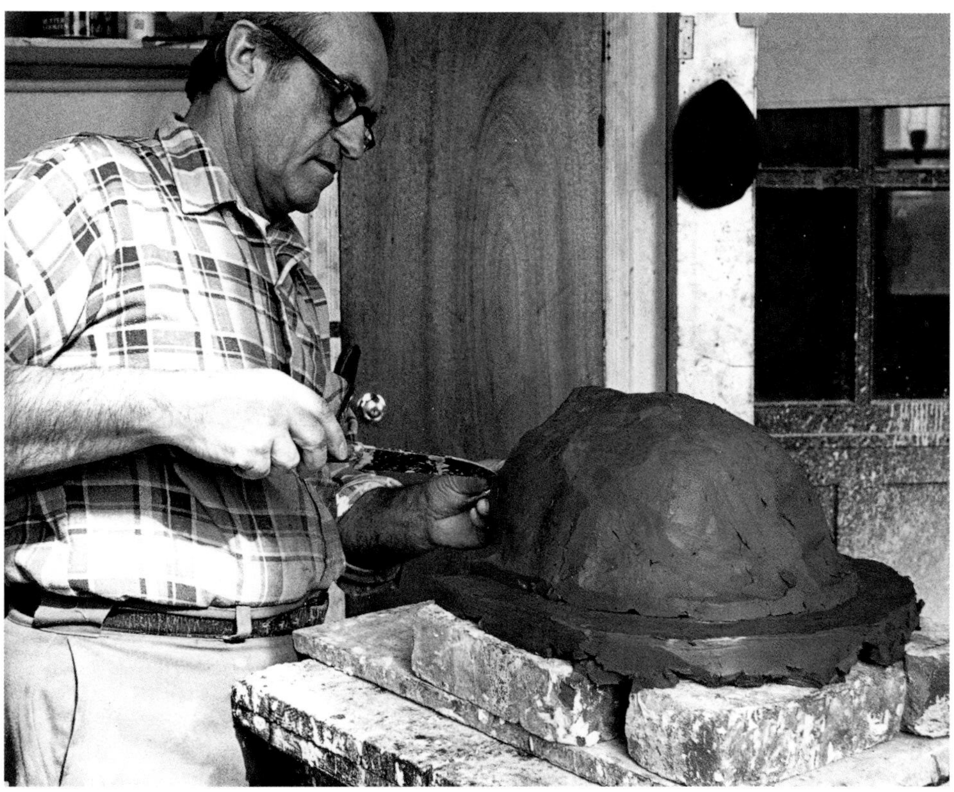

II-3. The plaster model is covered with clay, in preparation for the rubber.

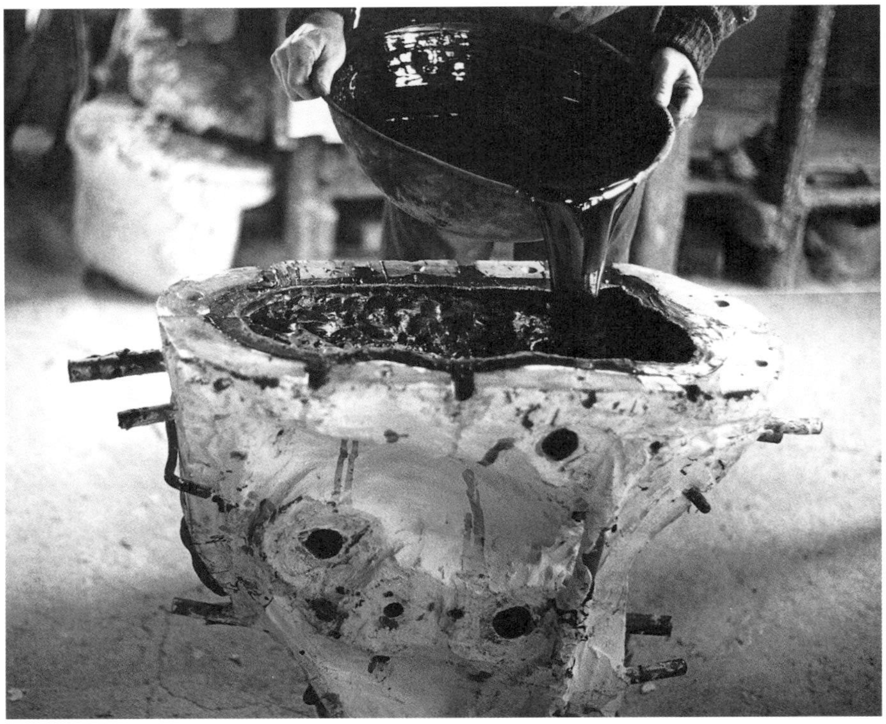

II-4. Hot melted wax is poured into the rubber mold.

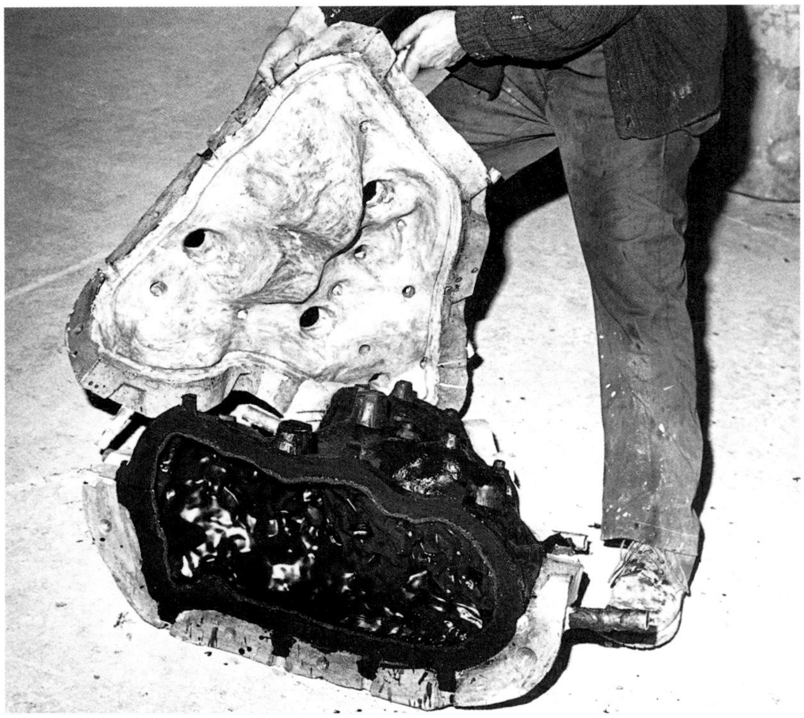

II-5. The completed rubber mold is opened with wax inside.

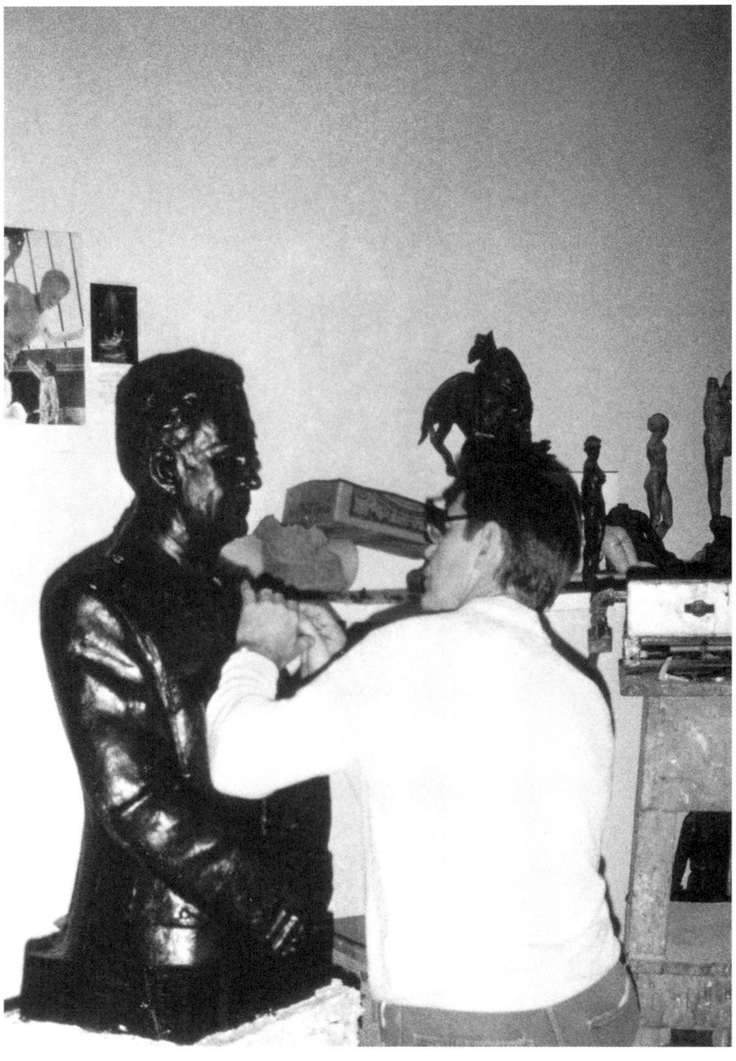

II-6. The wax is retouched.

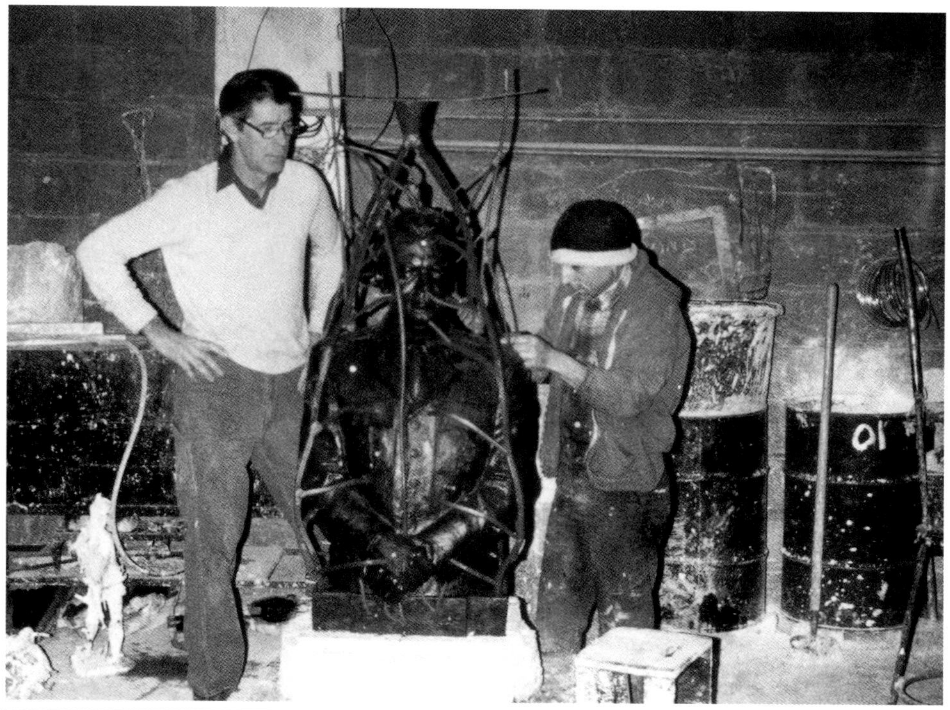

II-7. "Gating," or putting vents and sprues onto the wax. (a sprue in metal casting is the hole through which melted metal is poured.)

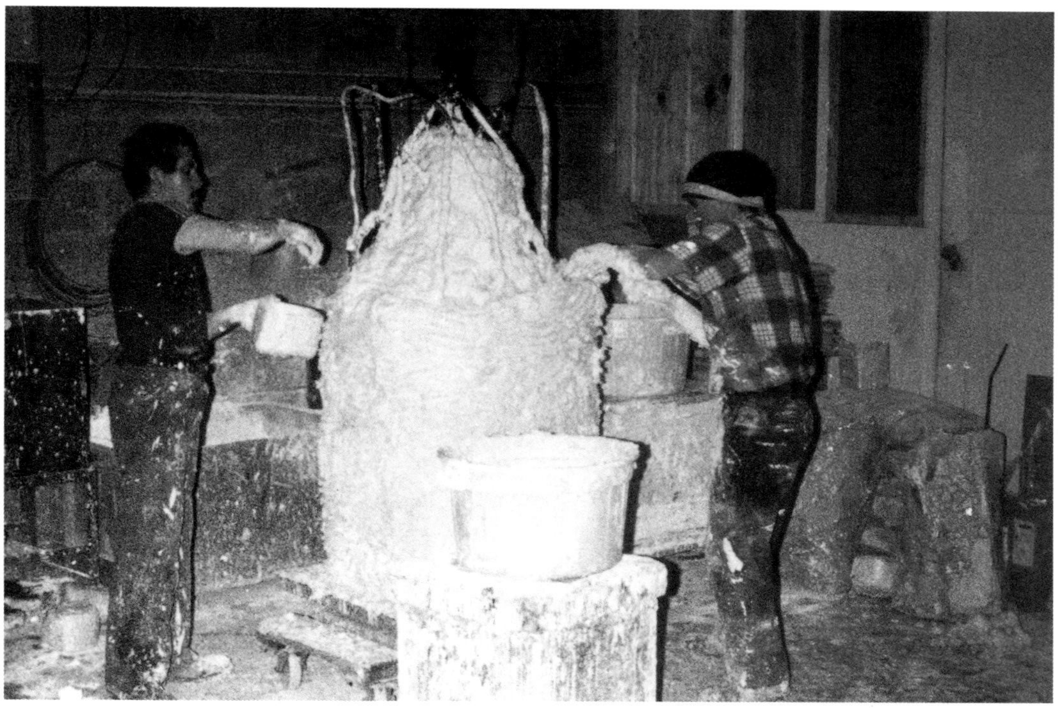

II-8. The wax is invested in a silica plaster mold.

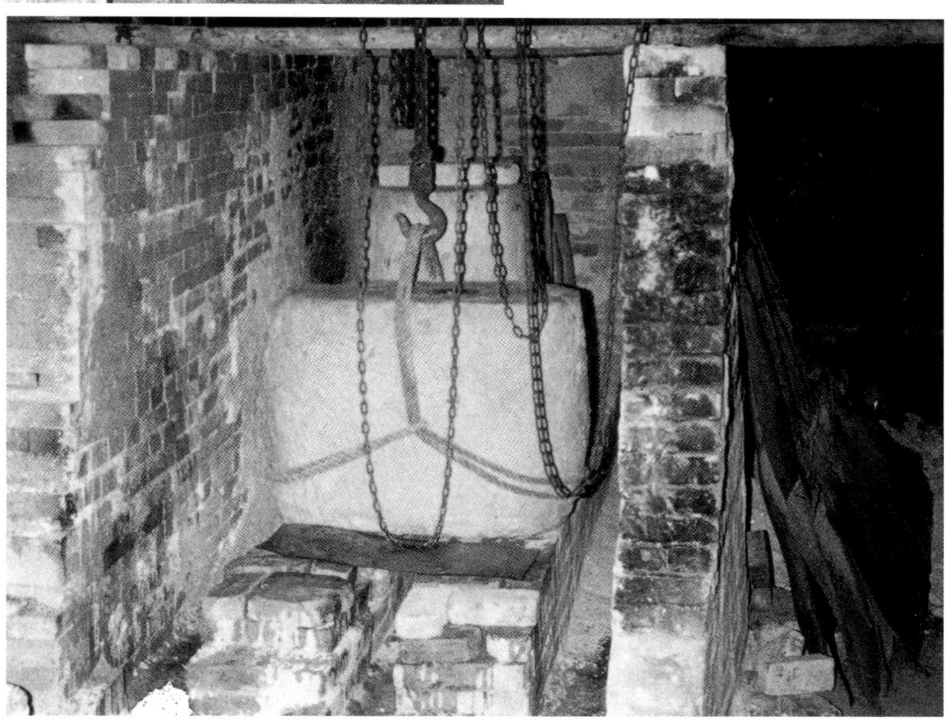

II-9. The molds are put in the hand-built oven, in preparation for the wax melt-out.

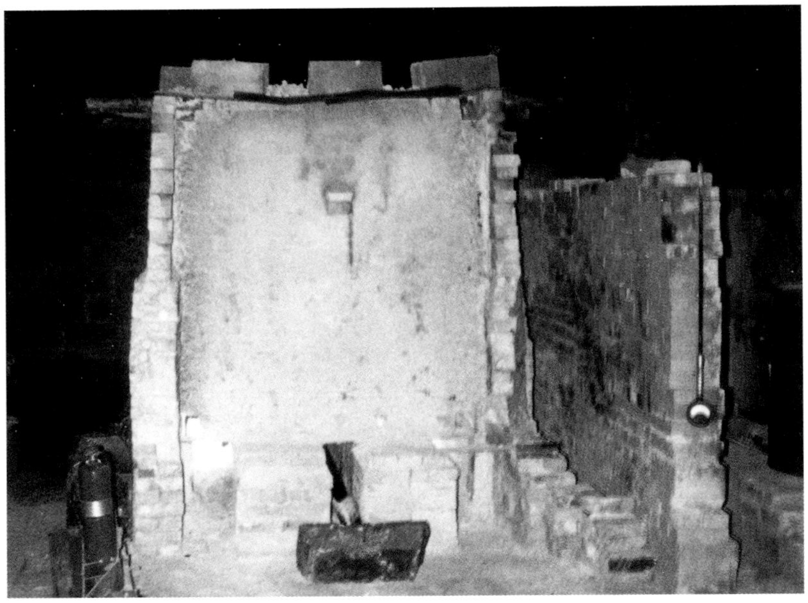

II-10. The molds are cooked for the wax melt-out.

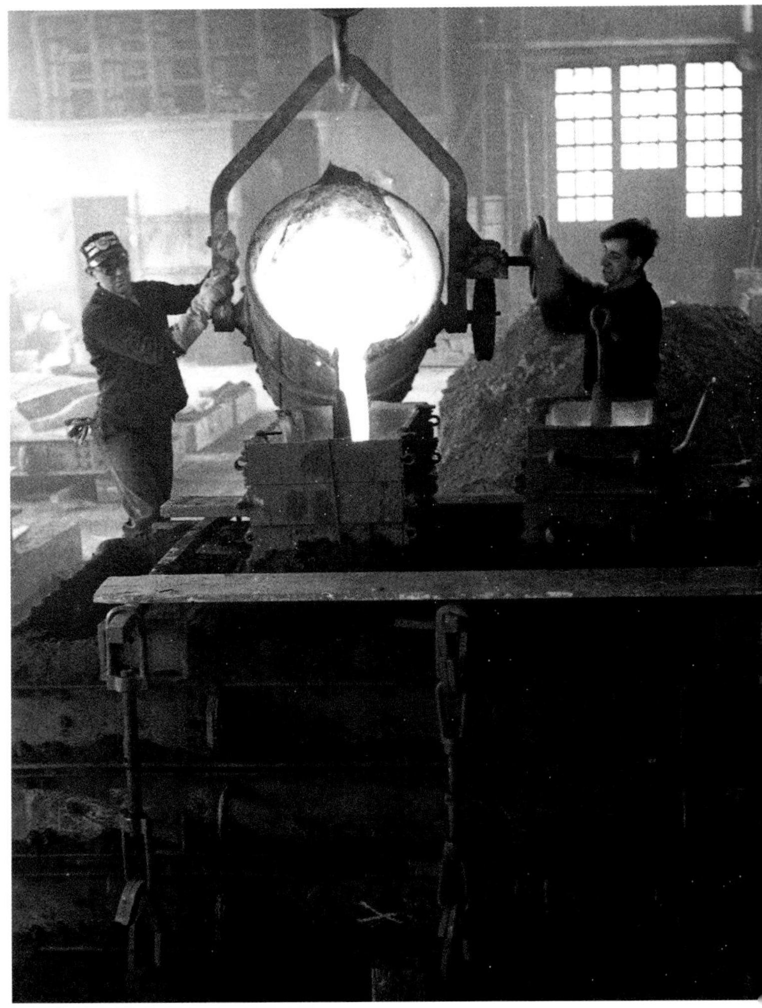

II-11. Cleaning metal prior to pouring.

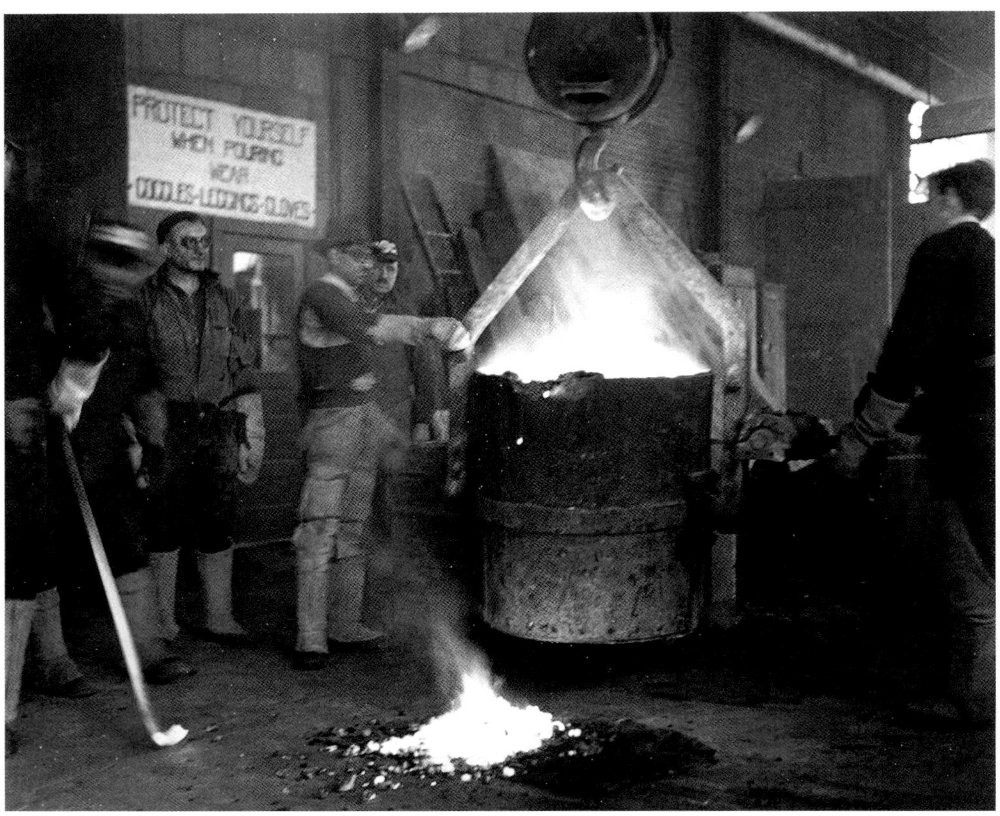

II-12. Pouring metal into the plaster investment mold.

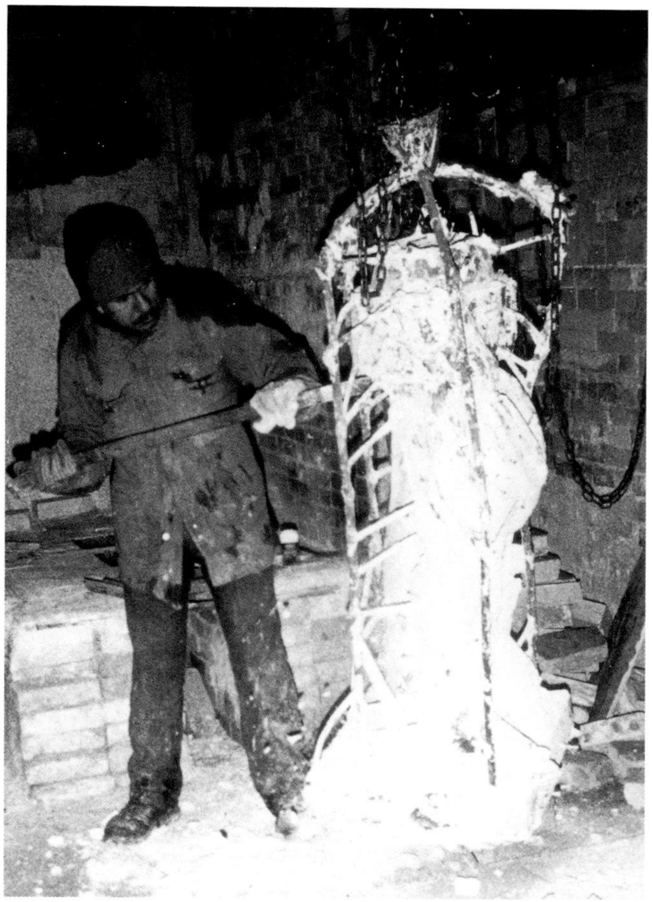

II-13. After the metal has been poured into the mold, the plaster investment is chipped away to expose raw bronze casting.

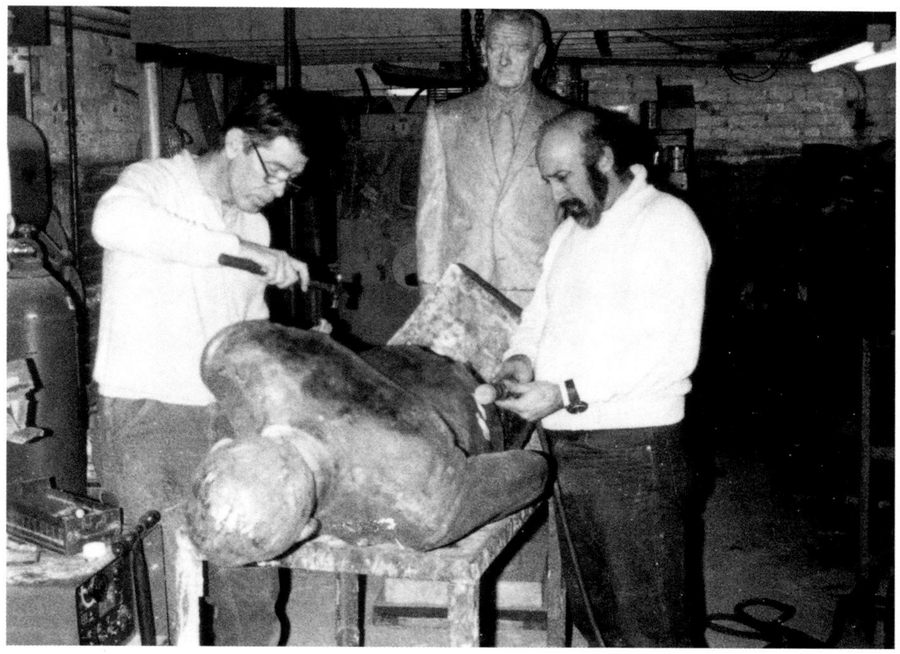

II-14. Philip Schiavo and sculptor Larry Ludke work on the rough casting; Ludke is chasing the surface while Schiavo grinds it.

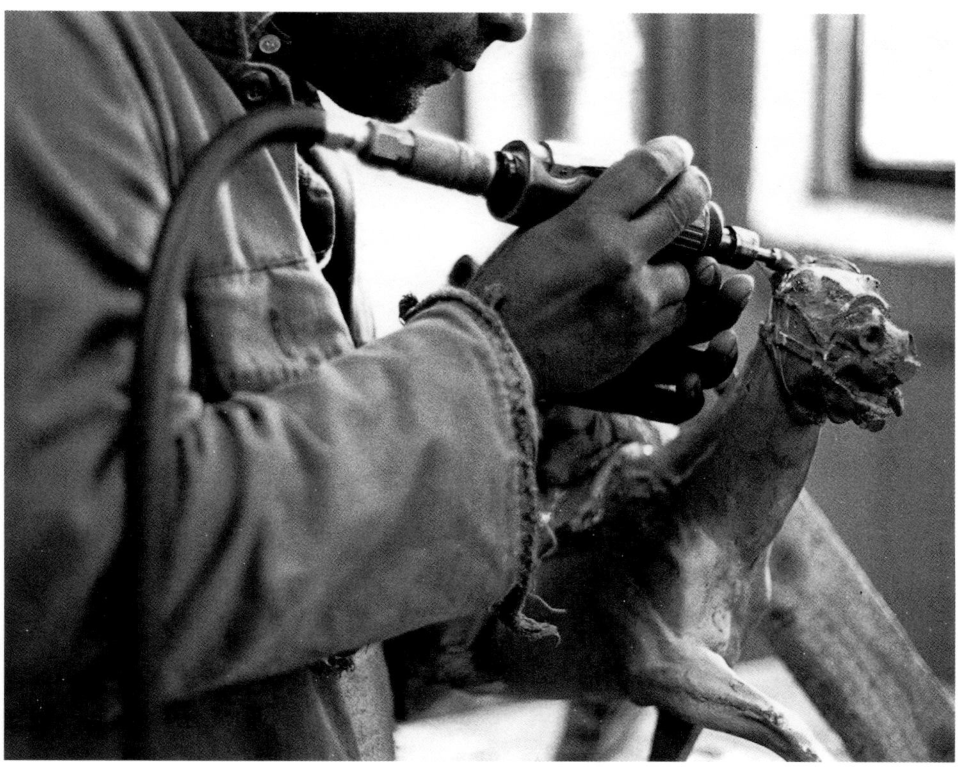

II-15. Final finishing of the bronze; craftsman Brian Ramnarine uses a die grinder to take the vents off.

Right:
II-16. Worker applies liquid chemicals heated by a torch flame to the finished bronze patina.

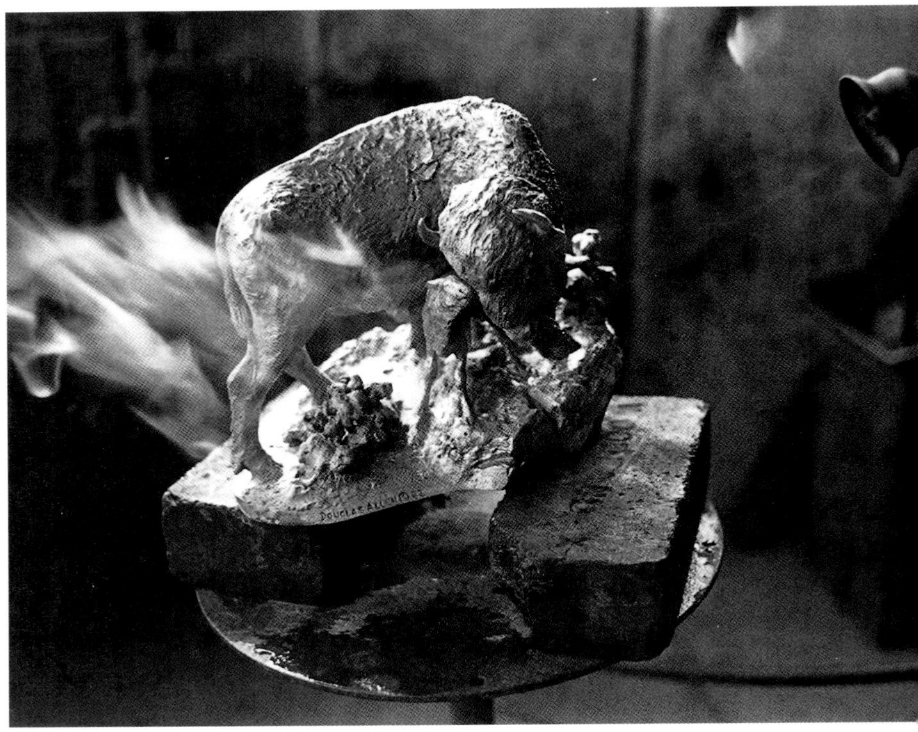

Below:
II-17. The ceramic shell method is used for producing multiple series or editions, such as plates. Used between 1979 and 1984, the process was designed and installed by Gerard Libero at R.B.W.

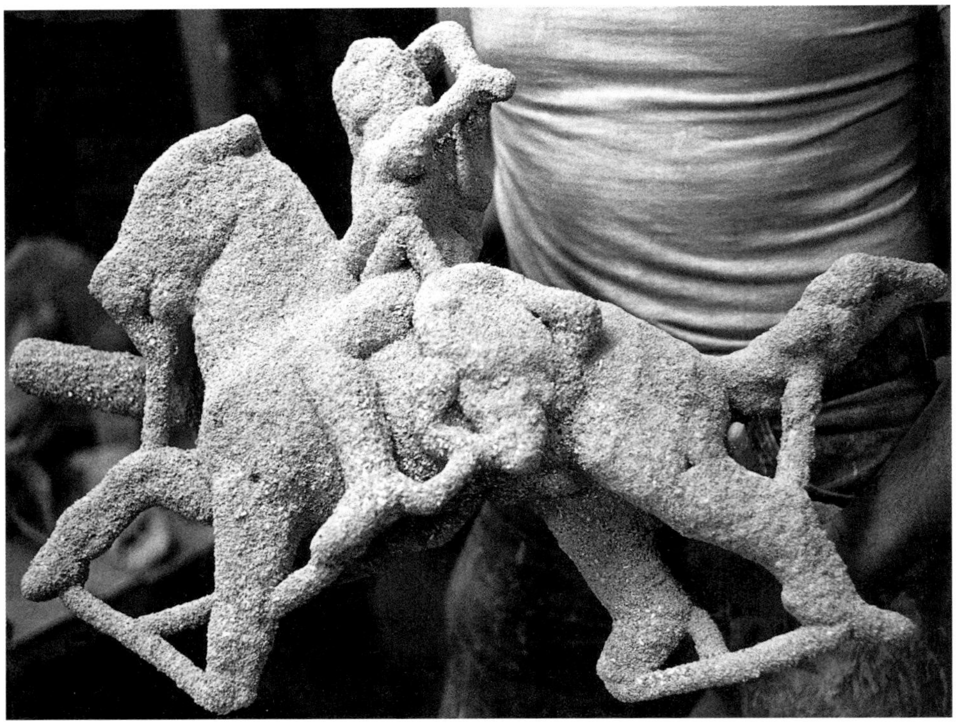

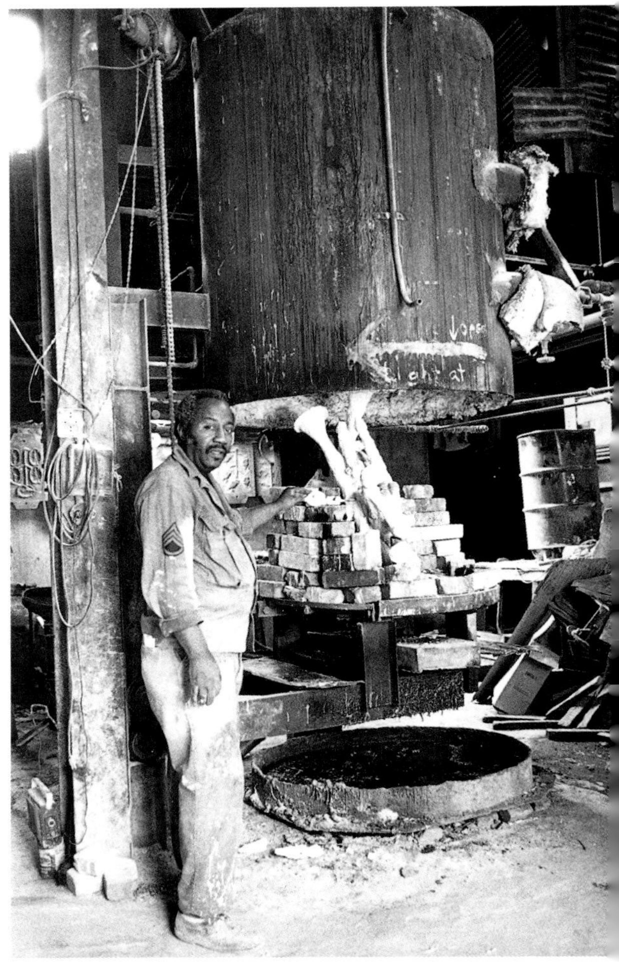

II-18. Craftsman LeRoy Gooden at the ceramic shell melt-out.

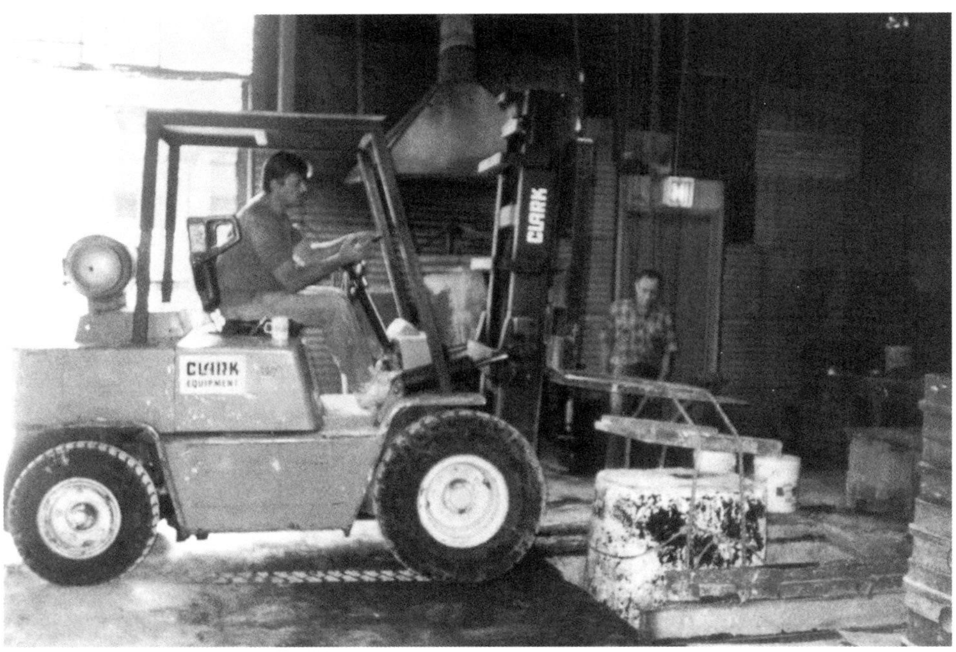

Left:
II-19. Craftsmen Ed Duke (left) and Salvatore Feruggio removing the cast mold from the large casting port.

III. MONUMENTS and PORTRAIT COMMEMORATIVES: HONORING AMERICA AND AMERICANS IN THE EARLY 20th CENTURY

When Roman Bronze Works opened its foundry at the turn of the last century, it was responding to an exciting moment in the nation's history and taste. The growing need for civic monuments and commemoratives matched the booming economy, growing cities, and sense of national pride and patriotism. Every city wanted its monuments, every new skyscraper its decorative elements, every courthouse its symbolic sculptures representing civic virtues. The nation's great leaders needed to be honored, and its war heroes and ordinary soldiers commemorated. Portrait heads and commemorative statues of noted people were wanted to decorate the buildings of government, the theater lobby, the university campus, even the newly elegant private homes and gardens of the wealthy. It was an era of commemoration.

For the first time in the nation's history, patrons—in the guise of civic governments, architects, and the well-to-do—made monumental sculpture as a profession viable. The making of statues (for that is how sculpture was known) became a public art, with the taste of the patrons paramount. Sculptors worked only on commission (unlike painters). Eccentricities and individual flights of fancy had little place in the sculptor's commissions. Instead, the artist became the purveyor of civic ideals in styles acceptable to his or her patrons.

Just before 1900 the Library of Congress commissioned a vast collection of statues and memorials, bringing together many American sculptors. And a series of expositions, such as Chicago's Columbian Exposition of 1895, the Pan-American Exposition in Buffalo in 1901, the St. Louis Exposition in 1904, and the San Francisco Panama-Pacific International Exposition of 1915 gave sculptors a national public showing. Awards and prizes were given out for sculpture at all of these expositions, providing sculptors with added visibility and renown. The National Sculpture Society was organized after the Columbian Exposition "to spread the knowledge of good sculpture; foster the taste for, and encourage the production of, ideal sculpture for the household and museums...and to promote the decoration of public and other buildings, squares and parks with sculpture of a high class."

As America became a proud, more cohesive nation, the haphazard state of municipal sculpture of nineteenth century America had given way to a more dignified and monumental taste, mirroring that of the great European cities. American artists went abroad to study as a matter of course. Politicians and "taste makers" made pleasure trips to Europe where many encountered for the first time the monument-filled squares of each elegant metropolis. Along with the steady influx of foreign artists, a civic taste developed; it was for European-style (particularly French) Beaux-Arts elegance and dignity.

Combined with the traditional American preference for recognizable and natural-looking art, the monuments and commemoratives that its sculptors produced (and Roman Bronze Works cast) were strongly realistic. Many were Neo-Classical, drawing on the Gods of ancient Greece and Rome to propound the nation's ideals and civic virtues. Trained in the studios of French and Italian academies, the country's sculptors were for the most part grounded in traditions of the past. Public taste (and the prizes and medals that were awarded) firmly favored these styles of pictorial realism, with archaic, Neo-Classical, medieval Gothic, or Renaissance overtones.

Only a few American artists reached beyond traditional, academic realism. Many of them were influenced by Auguste Rodin, the recognized leader of French sculptors. Daniel Chester French, who became the dean of America's sculptors, brought a strong sense of individualism to his work. Several other American sculptors, recognizing in Rodin and his colleagues in France a more emotional way of expressing themselves than the academic style they had been taught, traveled to Paris to study with the great artist. Augustus Saint-Gaudens, a master of American monumental sculptors, led the way. Rodin, with his strongly impressionistic—but nonetheless classical— approach thus became a major influence on such American sculptors as Saint-Gaudens, Gutzon Borglum, and Rudolph Evans.

In addition to the teachings of Rodin, there were several other strong influences on American sculptors. One was the new spirit of progress and invention at the turn of the century; several major monument sculptors, including Karl Bitter, Robert Aitken, and Alexander Stirling Calder, captured the spirit of the new era by using the symbols of progress and electricity in their works. The recognition of individual achievement brought a new emphasis on personality in portraiture; a few artists became adept at conveying personality, like Jo Davidson, who sculpted striking portraits of the world's important personages. In the first decade of the twentieth century there were occasional hints of Art-Nouveau decorative design.

In addition to commemorative monuments and portraits, many of these early 20th century artists produced decorations for the buildings that were going up in cities across the nation. Architectural sculpture became a viable and important way to make a living for sculptors, as well as an outlet for creative design.

Before 1913 and the Armory Show that introduced modernism in New York, barely a hint of non-realistic sculpture appeared in America. However, the first three decades of the century saw subtle changes in monument design. Though the classical influence of Saint-Gaudens and his many disciples continued, some sculptors, including A.S. Calder, George Grey Barnard, and Adolph Weinman, began to fuse the rhythmic, formal lines of modernism into their (still recognizable) commemorative works. But most continued to work in what had become an American idiom: realistic portraits of great leaders, commemorations of doughboys and famous battles, and allegorical figures symbolizing national ideals and civic responsibilities. (Of the forty-five sculptors exhibiting at the San Francisco exposition in 1915 only four took modernist directions.) A few American sculptors, like Paul Manship, developed their own styles of realism; he was an early advocate of a decorative, symbolic style that, though drawn from archaic Greek sculpture, was nonetheless new-looking, schematic, and graceful. Manship and his more modernist colleagues will be discussed later.

New York was already the center of the American art world (as it would remain throughout the twentieth century). Even though art schools flourished in Philadelphia, Boston, Cincinnati, and Chicago, most artists headed abroad for study, and then to New York for their careers. Roman Bronze Works was perfectly placed to produce the sculpture created there. And the emphasis on bronze rather than marble in many of the new works gave the foundry continuous work. Though monuments were commissioned from all over the country many were nonetheless sculpted and cast in New York. And of course New York, itself, was a major market for commemorative monuments. The city's Department of Parks was among the many civic clients of R.B.W.

The following selection of monumental artists suggests that Roman Bronze Works was the foundry of choice for many of the prominent sculptors of the first three decades of the century. Ledgers include the names of virtually every important monument maker of the time, though some cast only a few works there. Although it would be impossible to include visual images of all of their major works, this representative collection gives a broad overview both of the artistic taste and the civic sentiments of an era.

DANIEL CHESTER FRENCH (1850-1931)

Daniel Chester French has become a legendary figure in American sculpture. With Augustus Saint-Gaudens he established the Beaux-Arts style of heroic realism in American monuments, contributing perhaps the best- known statue in the nation—his seated Abraham Lincoln at the Lincoln Memorial in Washington. During a lifetime of impressive monument making, French created lasting representations that have become indelible parts of the nation's images, among them "The Minute Man" in Concord, Massachusetts, the "Alma Mater" at Columbia University, his symbolic "Four Continents" for the New York Customs House, "Disarmament," and his "Standing Lincoln." A number of his most important works were cast at Roman Bronze Works.

French was born in Exeter, New Hampshire, and raised in Concord, Massachusetts, where he was encouraged by Louisa May Alcott, and made a portrait of Ralph Waldo Emerson (III-1). He studied briefly with John Quincy Adams Ward, and in 1873 received his first commission, for "The Minute Man," sculpted when he was just twenty-three. In 1874 he went to Italy—he was one of the last Americans to study abroad there instead of in Paris—and worked with Thomas Ball. Two years later he returned to America, settling in Washington. He went to Paris to study in 1886 and on his return a year later opened a studio in New York.

French's personal fame and his circle of artists and intellectuals made his New York studio into something of a cultural center. At the height of his career, internationally known and able to live in grand style, he decided to create a perfect working and living environment for himself and his family in the country. His noted summer studio and home at Chesterwood in Stockbridge, Massachusetts, a testament to his illustrious career, is now open to the public as a National Historic Site.

Though schooled in the traditional style of classical realism, French's work became more idealistic and heroic than purely factual. Combining life-like naturalism and a powerful, almost mythical grandeur, as well as images that represented the lofty civic and patriotic ideals of the nation, French's sculpture was widely influential in setting the tone for American monuments for the next generations. His works commemorated national events, graced public spaces, and ornamented buildings across the country.

Among his most majestic allegorical works with female figures were his "Spirit of Life" (III-2), "Draped Female Figure" (III-3), "Woman in Drapery and Helmet and Breastplate" (III-4), and his classical images "Diana" (III-5) and "Narcissa" (III-6). The grandly winged angel, softly draped but powerful, appeared in several of his most moving works, such as World War I monuments "Victory: The First Division War Memorial" (III-7), and "Benediction" for St. Miehl, France (III-8).

In addition to his many noted war memorials, including "Victory" and "Disarmament" (III-9) and the Exeter War Memorial (III-10), French made a number of images of national leaders, including an equestrian statuette of George Washington (III-11). His great Lincoln Memorial statue was not his only image of that president; he also created an "Abraham Lincoln Standing" (III-12) and a Lincoln portrait bust (III-13). French also sculpted figures from literature, such as Rip Van Winkle (III-14), and an image of an ideal college student for Princeton University (III-15).

His association with Roman Bronze Works began in 1902 and he continued casting there until his death in 1931; his daughter Margaret Cresson cast many of her father's works (as well as her own) there as well.

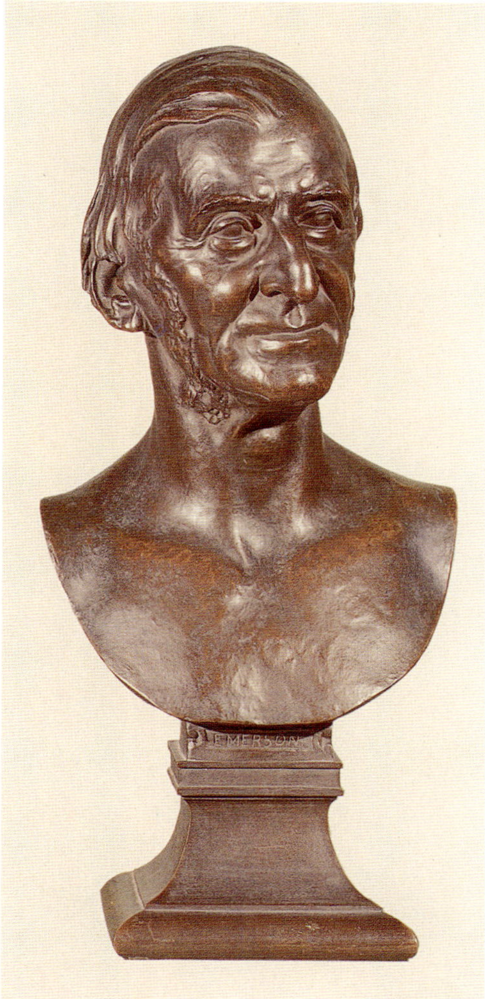

III-1. "Ralph Waldo Emerson" by Daniel Chester French. Dated 1879; cast 1901. (Emerson was so taken with this likeness that he remarked "That is the face I shave.")

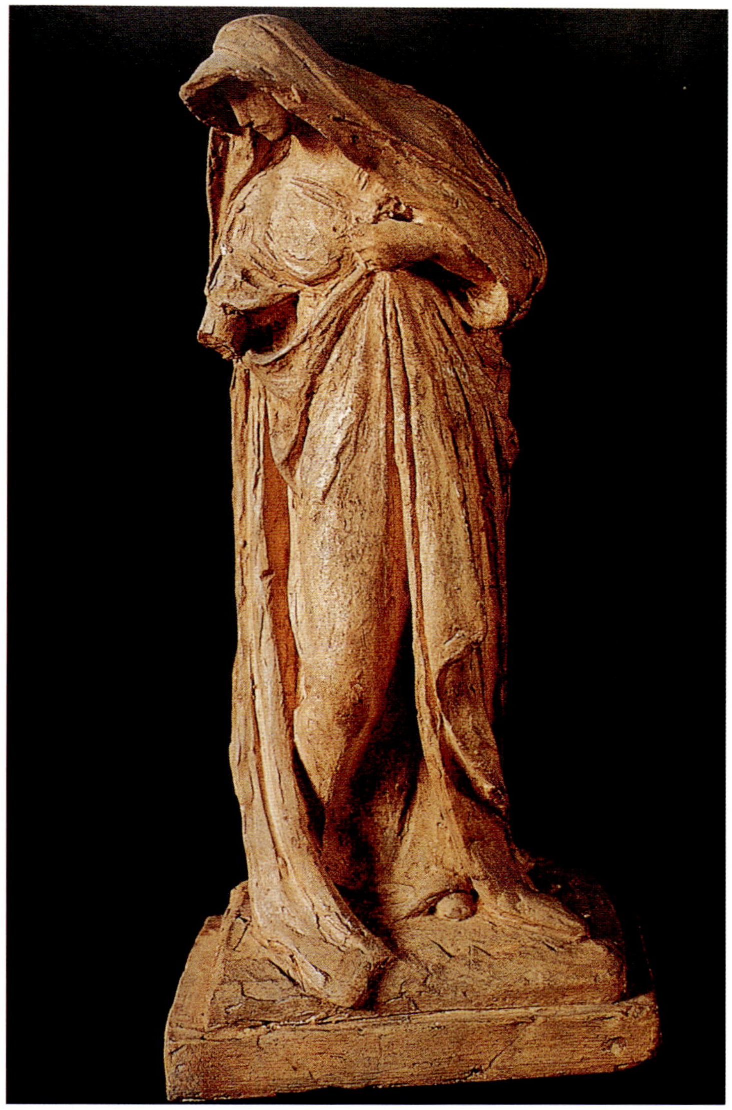

III-3. "Draped Female Figure" (plaster cast) by Daniel Chester French.

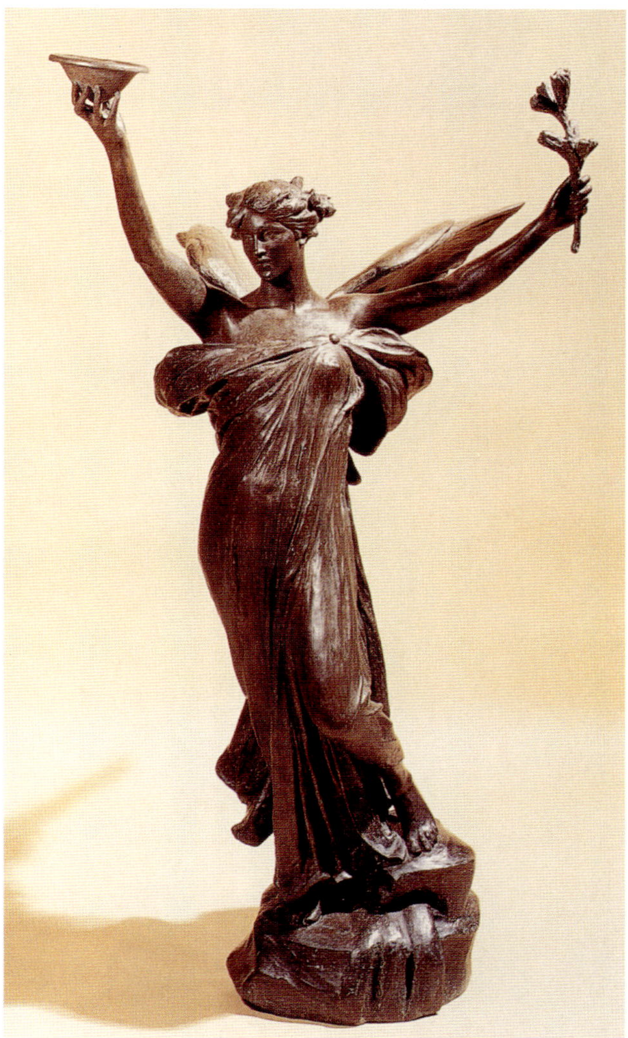

III-2. "The Spirit of Life" by Daniel Chester French. Dated 1914. This was the central figure in a memorial to Spencer Trask, a New York businessman; it was erected at Saratoga Springs, New York.

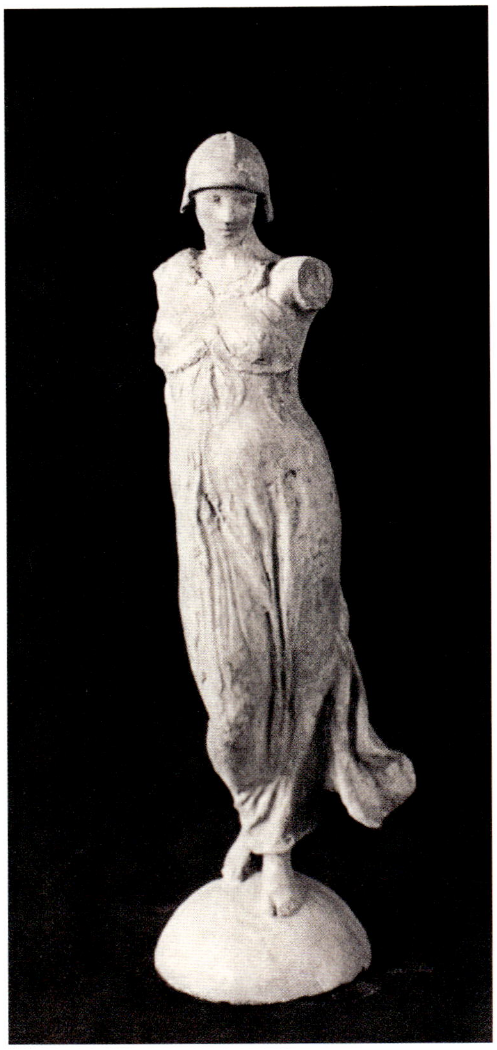

III-4. "Woman in Drapery, Helmet, and Breastplate" (plaster cast) by Daniel Chester French. 1921.

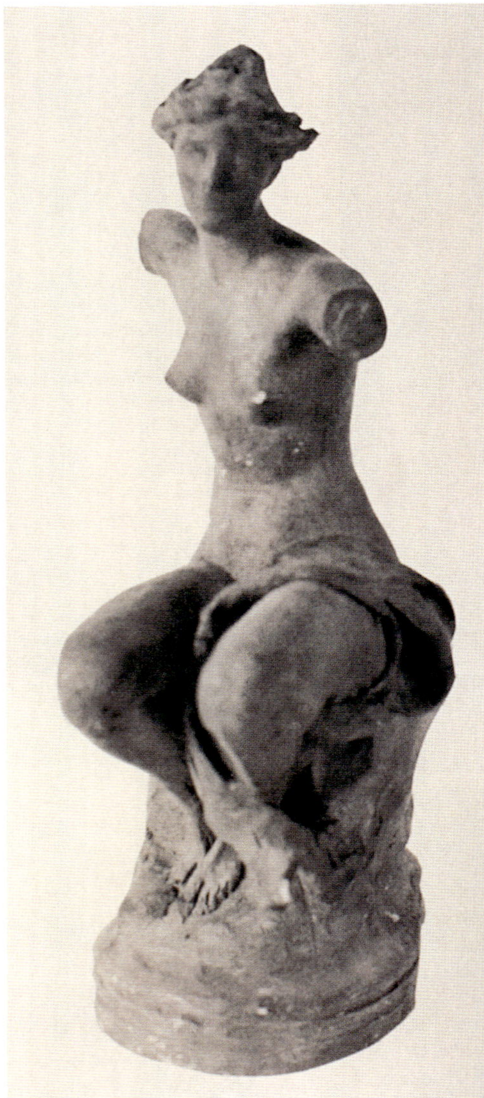

III-5. "Diana" (plaster cast) by Daniel Chester French

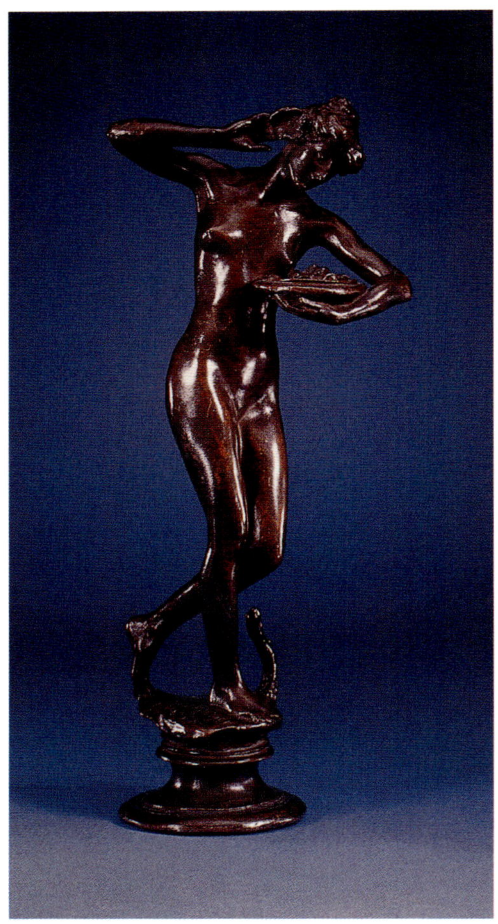

III-6. "Narcissa, (or "Vanity") by Daniel Chester French. 1901.

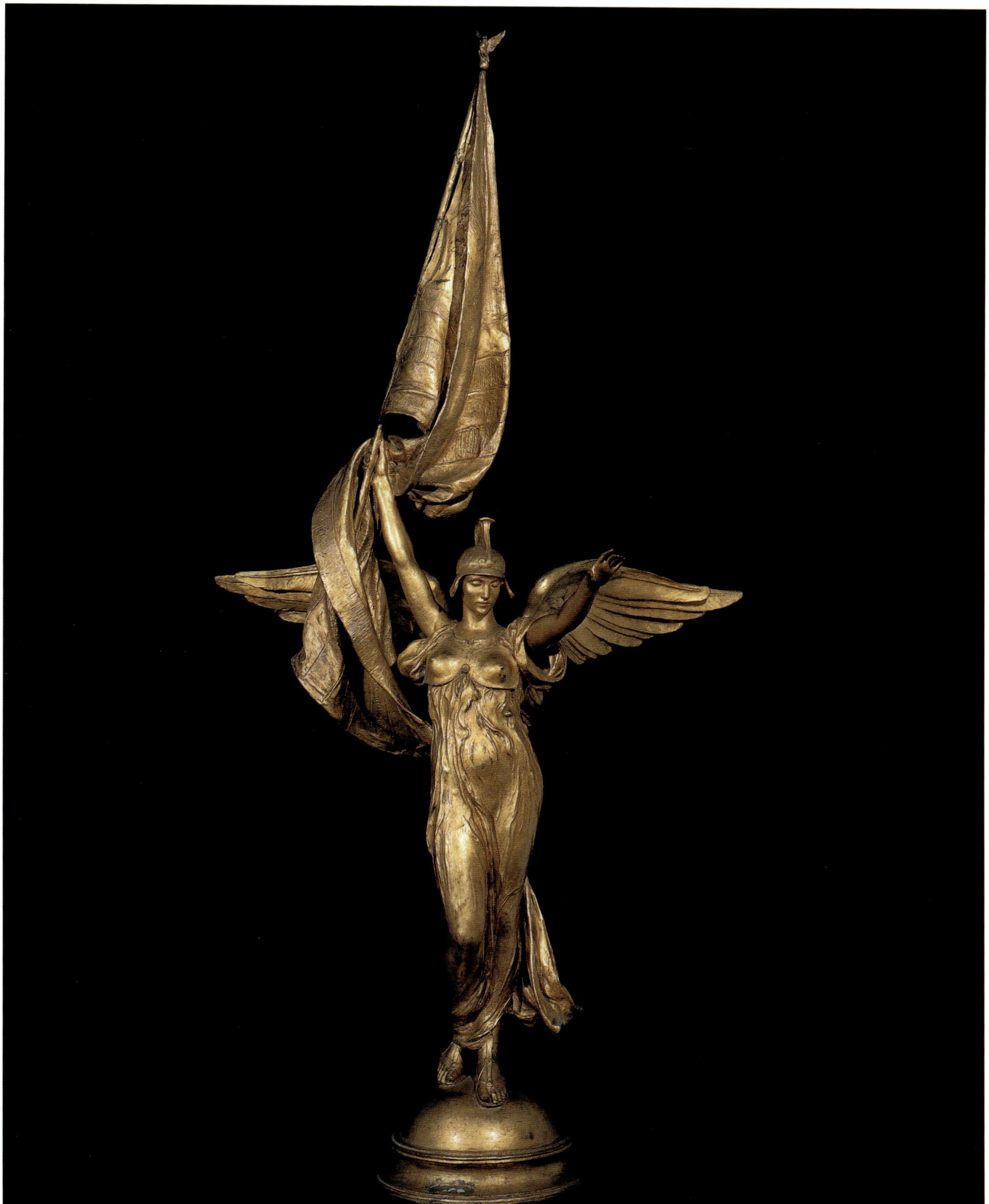

III-7. "Victory" for the First Division Memorial, Washington, D.C. by Daniel Chester French. Gilded bronze model. R.B.W. 1921.

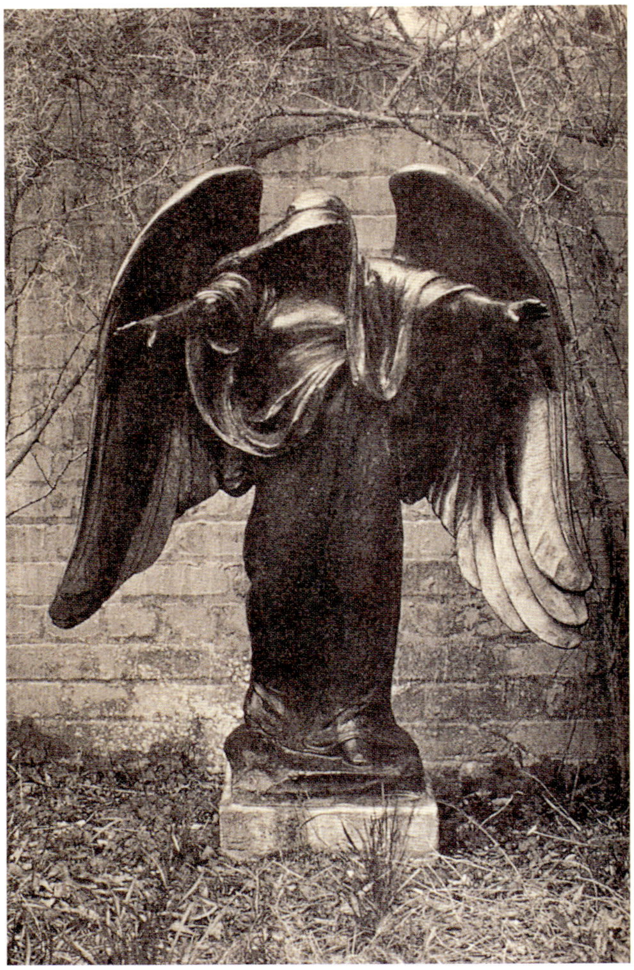

III-8. "Benediction" World War I Memorial for St. Miehl, France by Daniel Chester French. 1928.

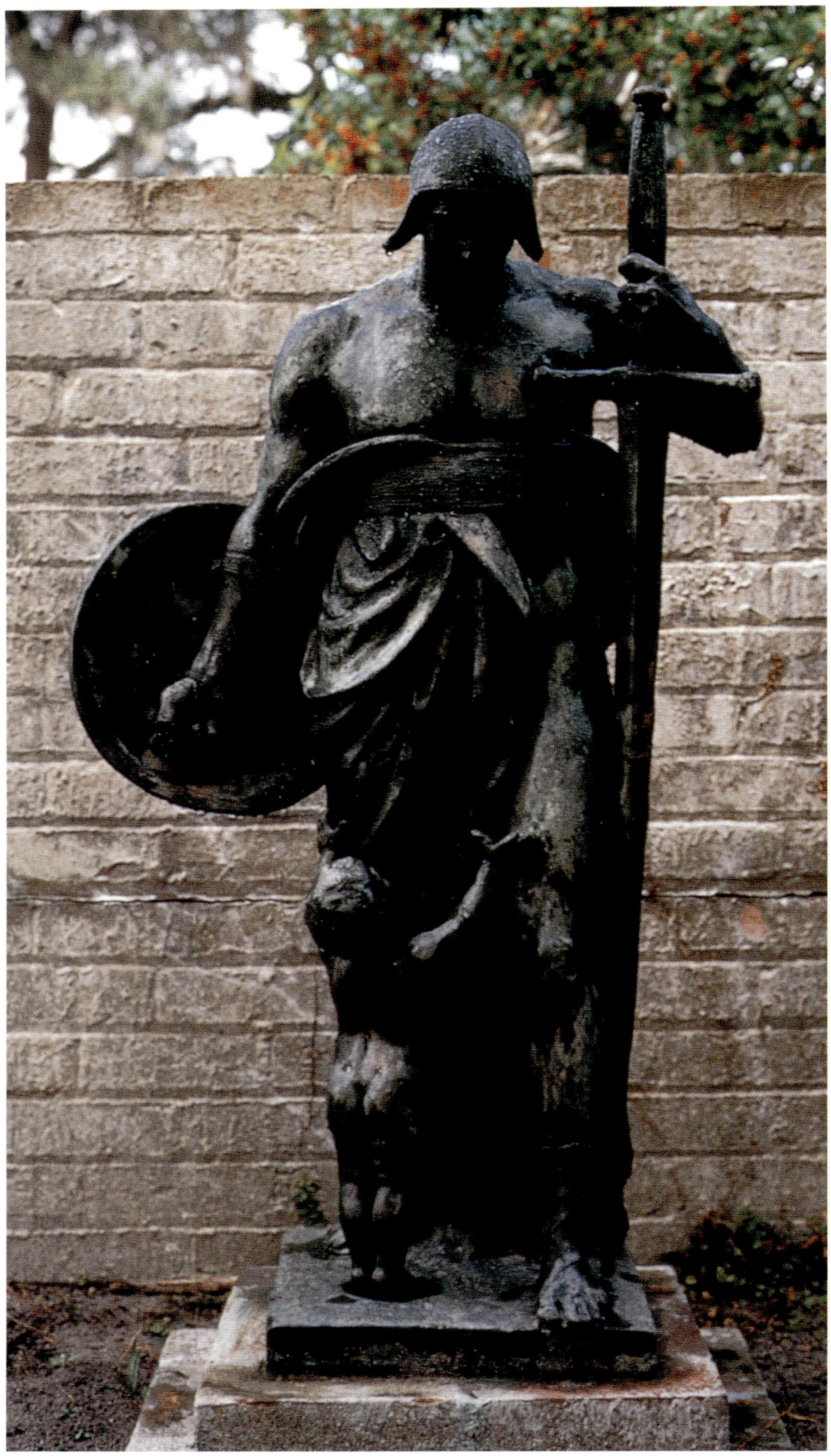

III-9. "Disarmament" by Daniel Chester French. (Part of 1919 Victory Arch in New York.) R.B.W. 1930.

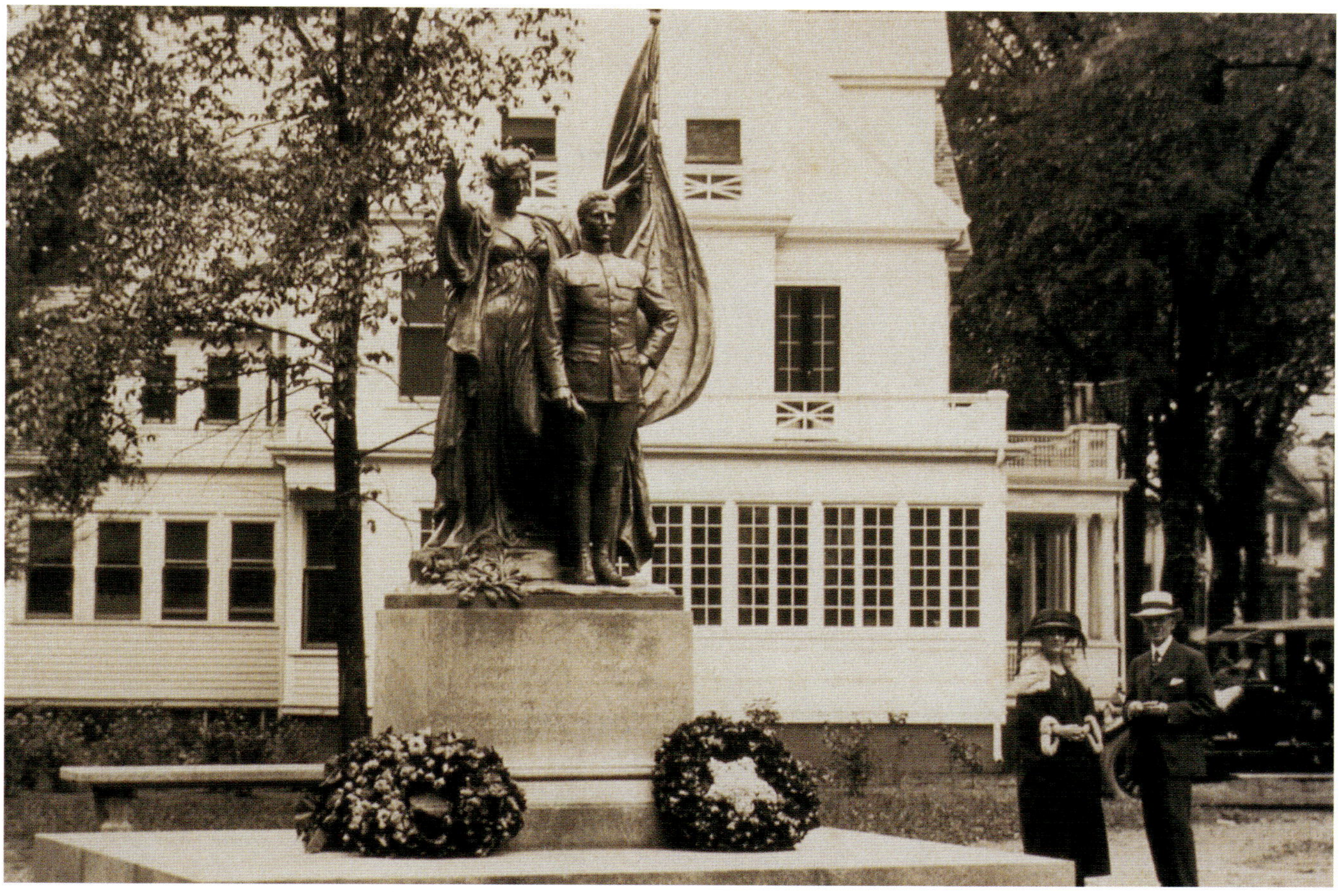

III-10. "Exeter War Memorial" by Daniel Chester French. 1922. Historic photo: Dedication Day, July 4, 1922.

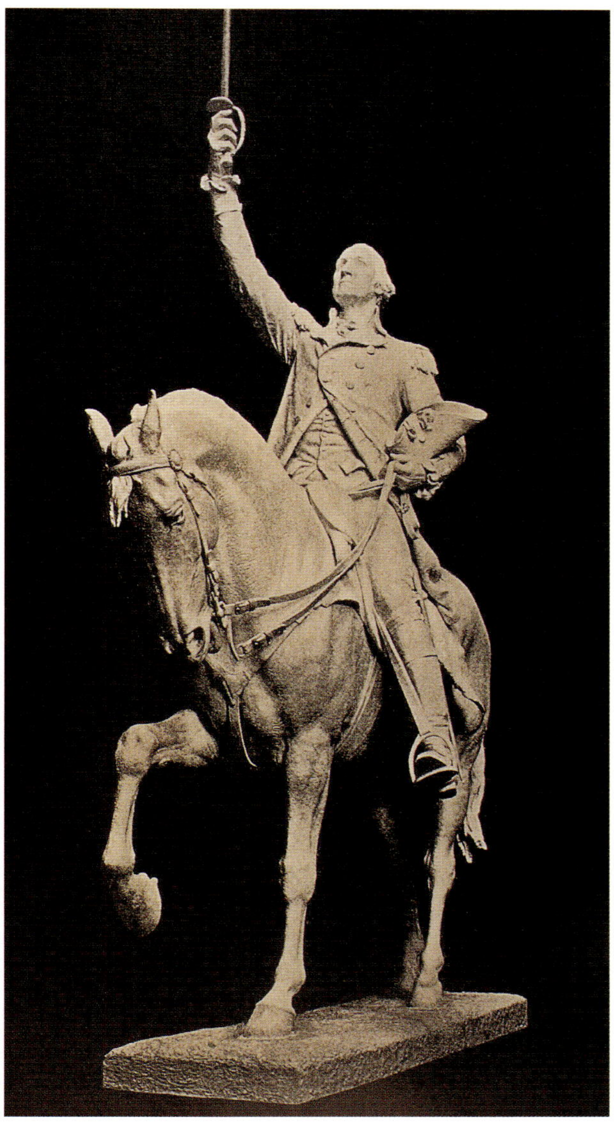

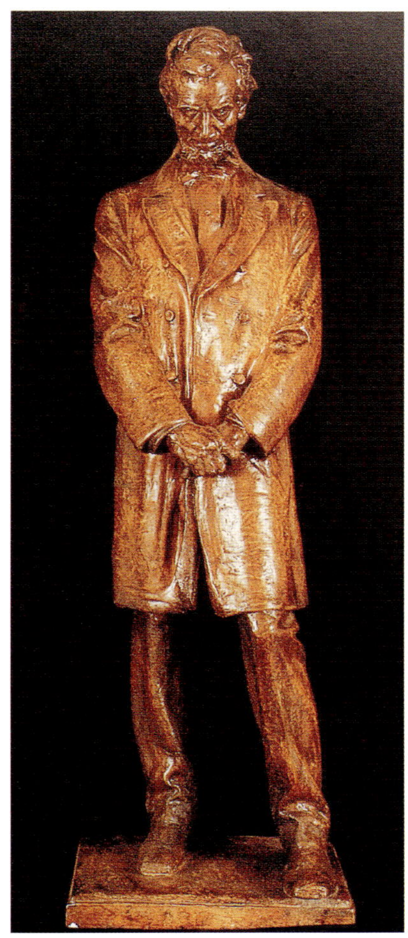

Left:
III-11. "General Washington," equestrian statuette, by Daniel Chester French. Original statue for France was unveiled in 1900; large replicas were made in 1901 and 1904, and small models thereafter. This sculpture was a collaboration with Edward Potter.

Right:
III-12. "Abraham Lincoln Standing" (plaster cast) by Daniel Chester French. Lincoln is shown before delivering the Gettysburg Address. Working model completed 1911.

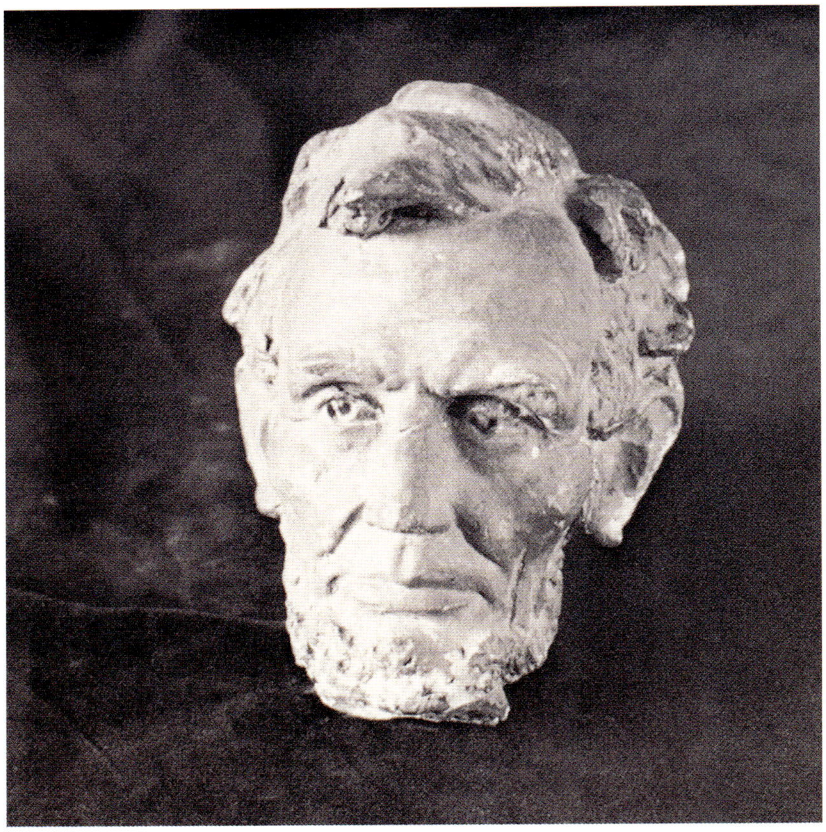

III-13. "Abe Lincoln" (dark patina plaster cast) by Daniel Chester French

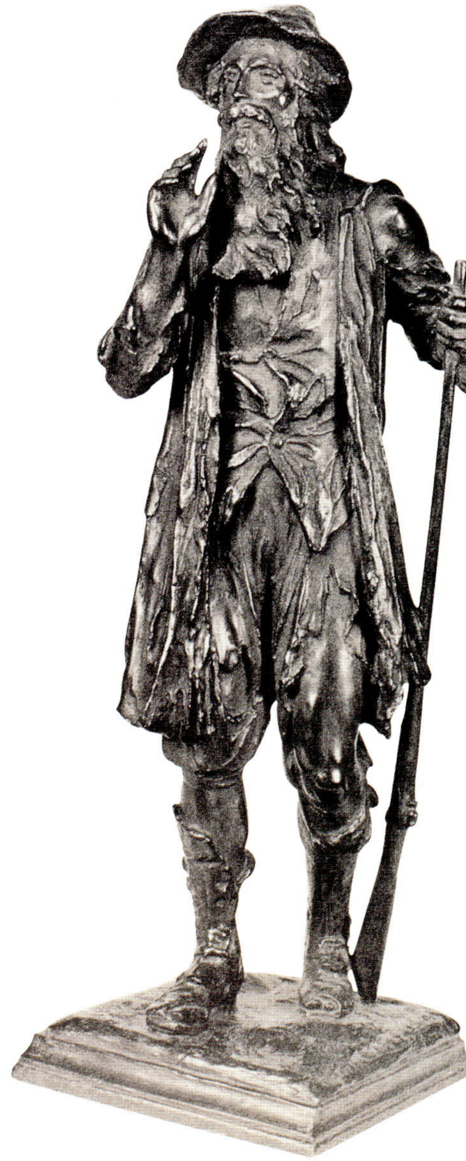

III-14. "Rip Van Winkle" by Daniel Chester French. 1925.

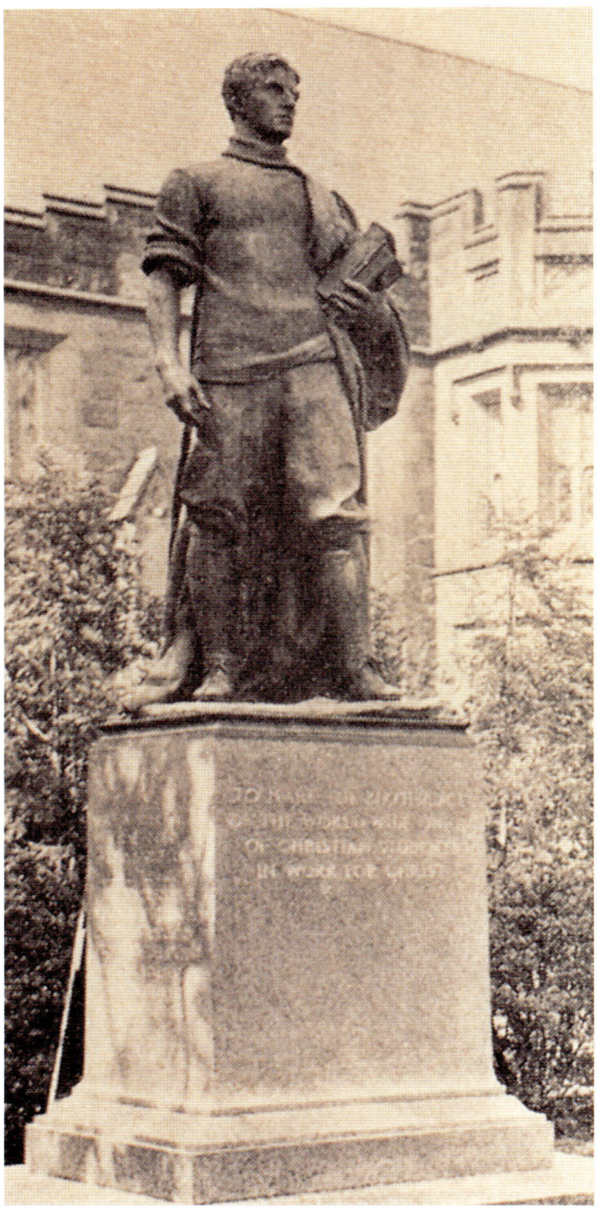

III-15. "College Youth—a Christian Student" for Princeton University by Daniel Chester French. 1913.

AUGUSTUS SAINT-GAUDENS (1848-1907)

Augustus Saint-Gaudens was one of the most important sculptors of the final decades of the nineteenth century; he was considered the father of "the American Renaissance." His monuments had a dynamic, graceful, and dramatic quality never before seen in American sculpture. Examples of his work can be seen throughout the country. A leading figure in establishing the Beaux-Arts style in America, he also pioneered the integration of sculpture with architecture and landscape in America, working with leading figures such as Stanford White, John La Farge, and Charles McKim.

Saint-Gaudens was born in Dublin and arrived as an infant in the United States. At nineteen he went to Paris to study at the Ecole des Beaux-Arts. His first commissions followed; in 1881 a monument to Civil War hero Admiral Farragut brought him immediate acclaim, including a gold grand prix at the Paris Exposition in 1897 for his Sherman Monument (III-16). The strongly modeled head from the Sherman Monument, which he called "Nike-Eirene" or

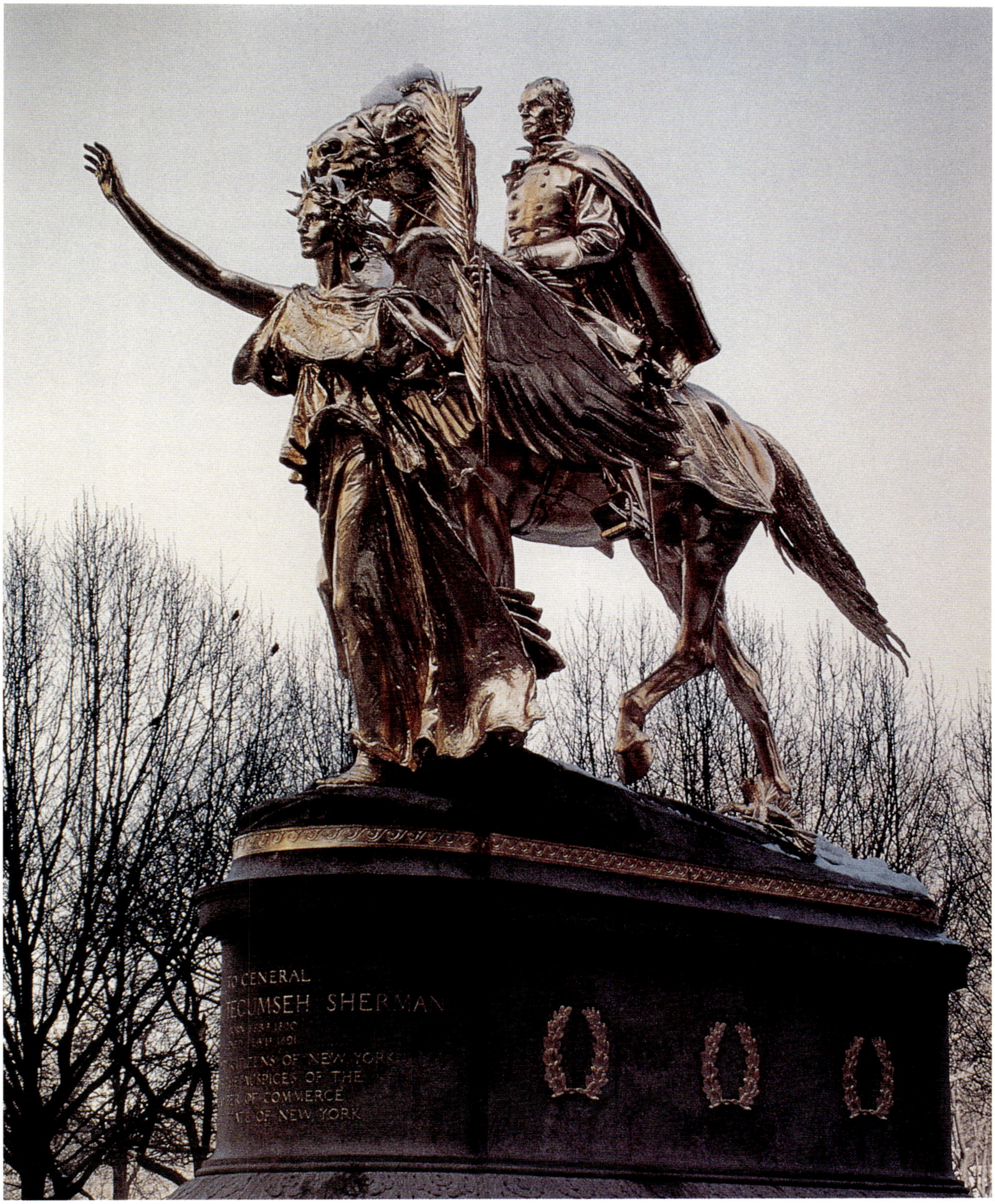

III-16. "General William Tecumseh Sherman Monument," at the Grand Army Plaza, New York, by Augustus Saint-Gaudens. 1897-1903.

(Nikh-eiphnh") (III-17) was cast separately too. He worked on numerous commissions thereafter, as well as teaching at the Art Students' League in New York and tutoring generations of young sculptors. He was an artistic advisor to the Columbian Exposition in 1893, a participant at the American Academy in Rome, and a figure of prominence throughout the art world. In 1885 he first came to Cornish, New Hampshire and from 1892 he summered there at his own estate, "Aspet," where many assistants helped him produce his major monuments. Some of his large monuments were reproduced in small versions and sold by Tiffany's. Numerous other artists followed Saint-Gaudens to create an art colony at Cornish. Today his extraordinary house and studios are open to the public as a National Historic Site.

Saint-Gaudens is remembered primarily for his strong civic monuments. His ability to combine realism and allegory was a departure from the more academic realism then in fashion in American sculpture. Among his major monuments was his "Shaw Memorial (III-18 and 19), with its sensitive portrayal of a black regiment in the Civil War. (Sections of the memorial were cast separately too. See III-20). Other monumental works included his two images of Abraham Lincoln: his "Bust of Lincoln" (III-21), his "Abraham Lincoln: The Head of State" (III-22) for Grant Park in Chicago, and his well-known "Pilgrim" (also called "The Puritan") (III-23).

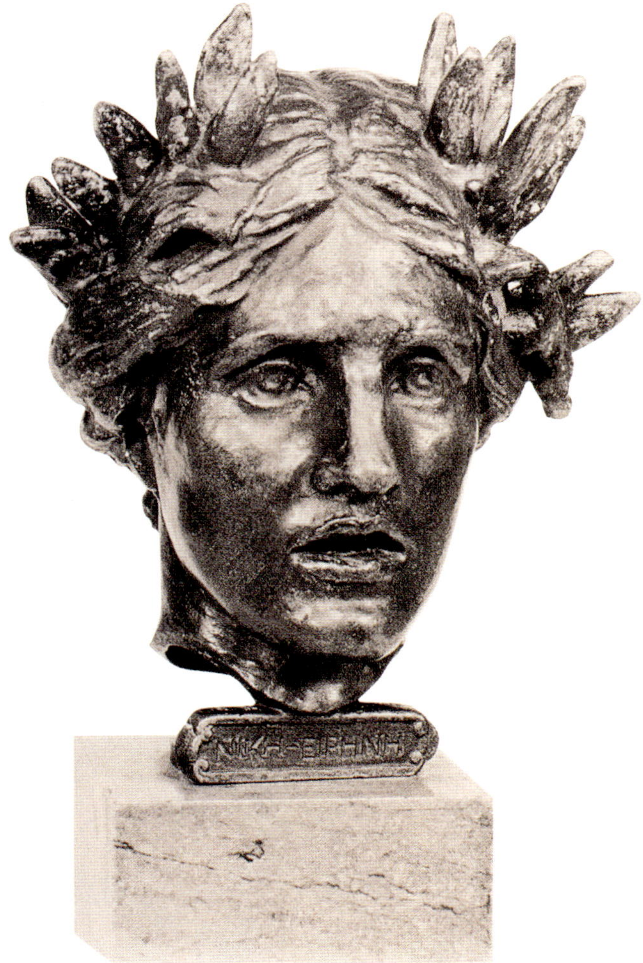

III-17. "Nike-Eirene" (or"Nikh-eiphnh") by Augustus Saint-Gaudens. The title means Victory-Peace. This head from the Sherman monument was cast separately circa 1906. It is said to be a likeness of a friend named Alice Butler.

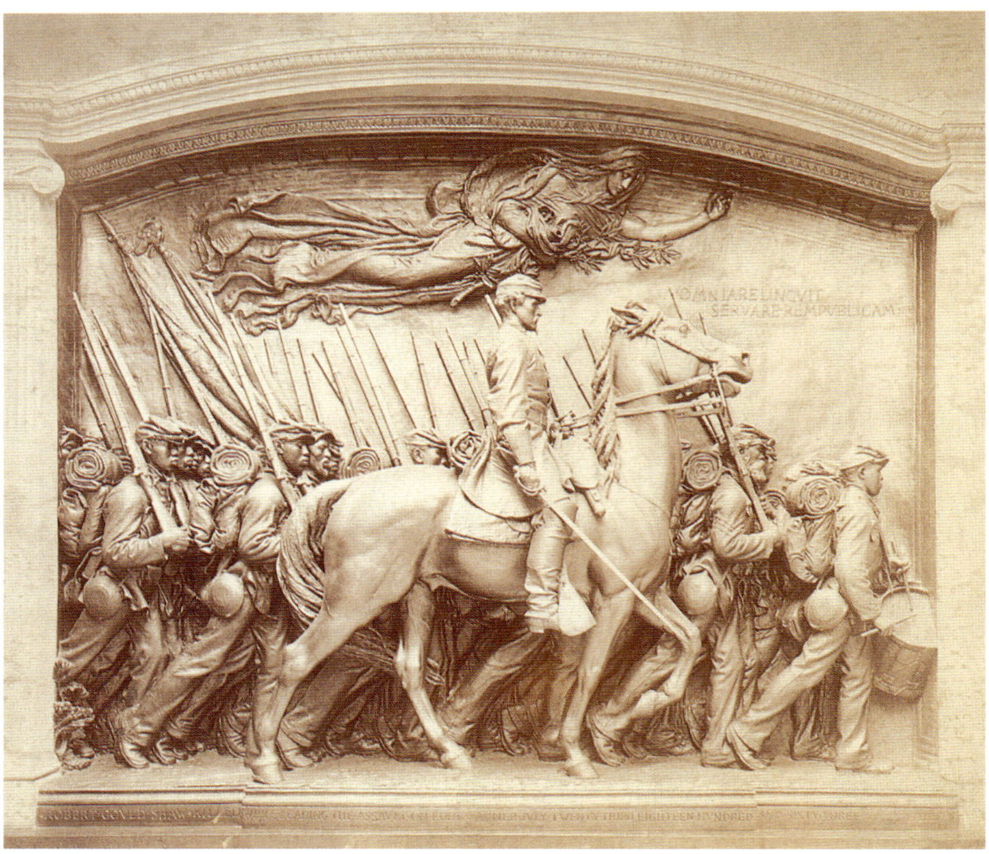

III-18. "Shaw Memorial" by Augustus Saint-Gaudens. 1900.

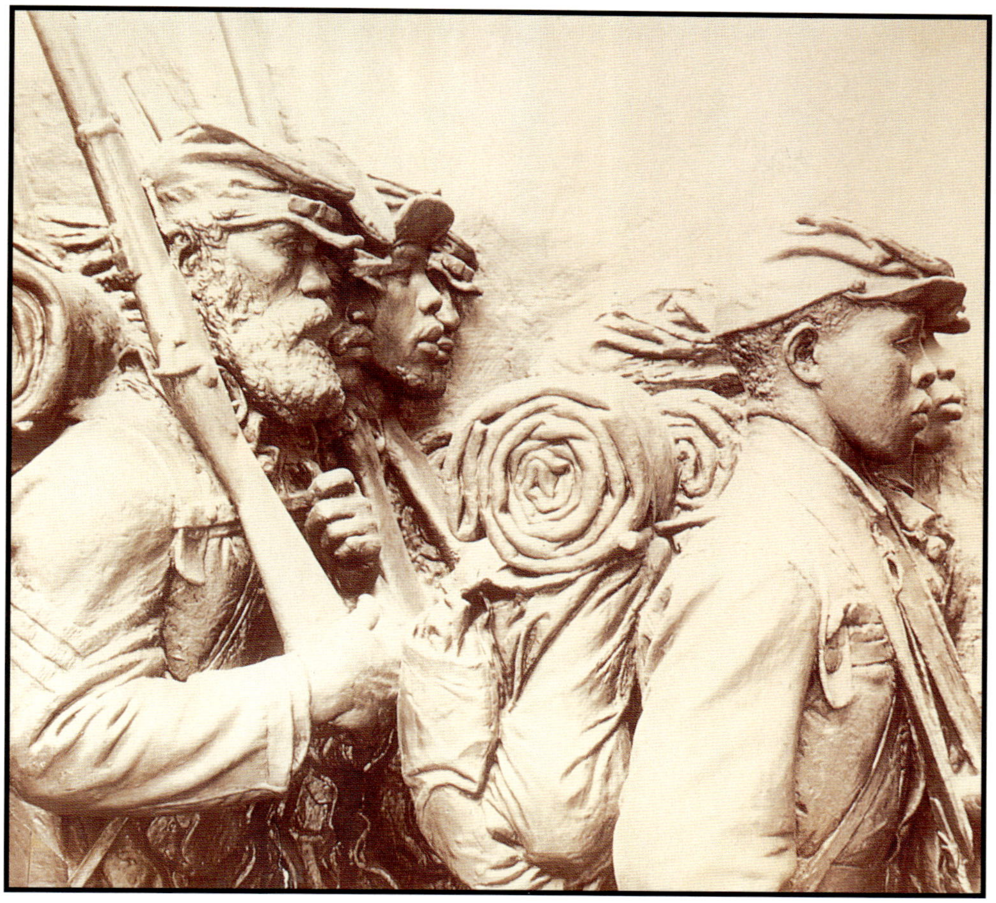

III-19. Detail of the "Shaw Memorial" showing African American soldiers.

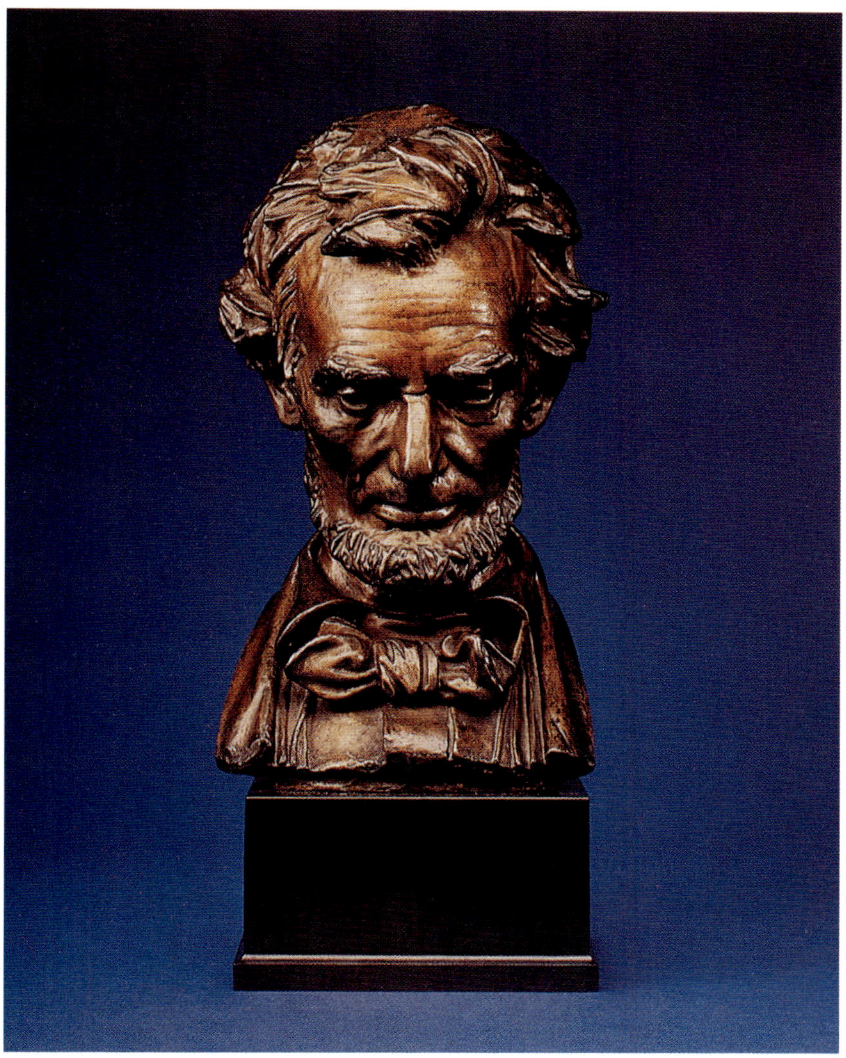

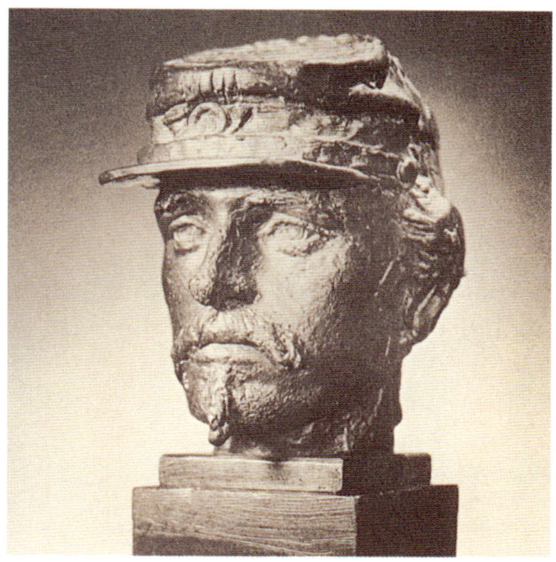

III-20. Replica of head of Robert Gould Shaw from the "Shaw Memorial" by Augustus Saint-Gaudens. 1908.

III-21. "Bust of Lincoln" by Augustus Saint-Gaudens

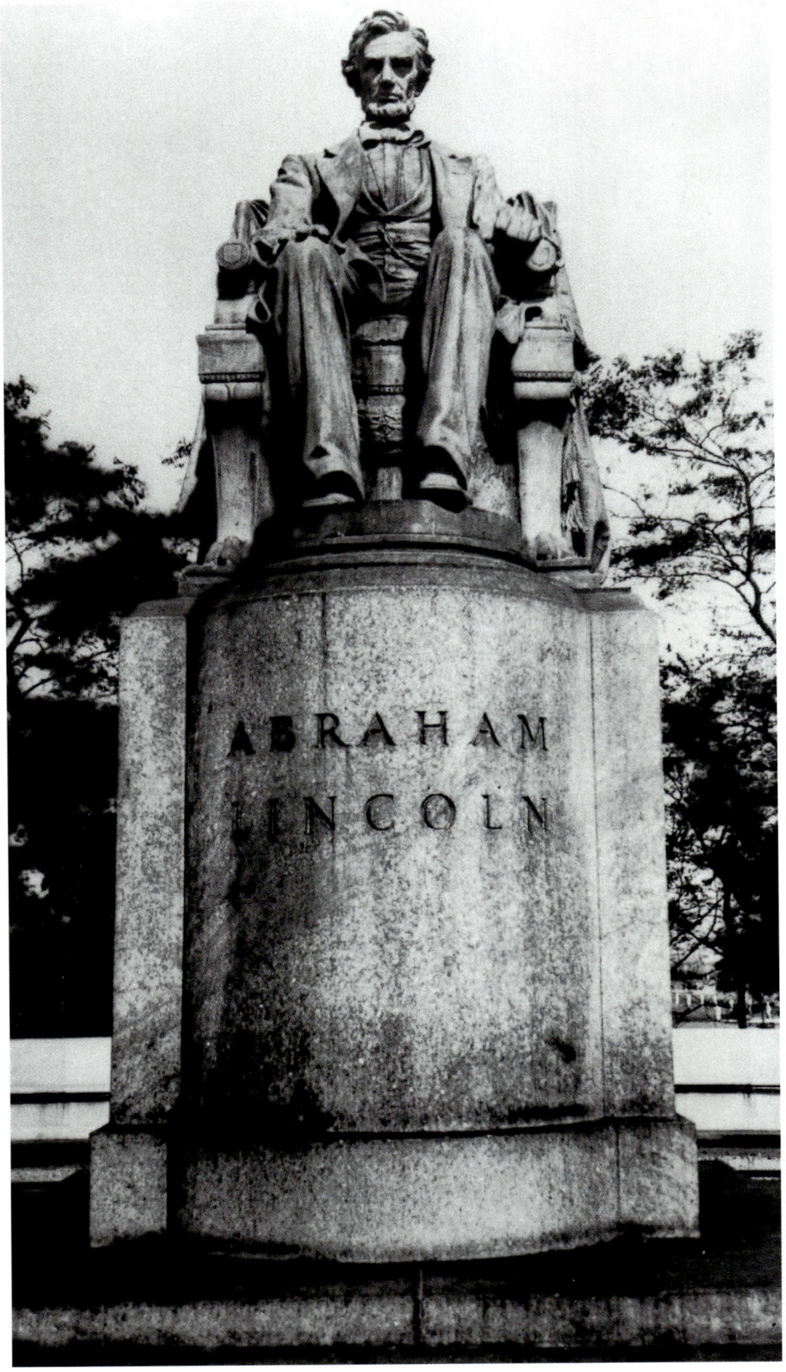

III-22. "Abraham Lincoln, The Head of State," Grant Park, Chicago, by Augustus Saint-Gaudens. Original c. 1887. R.B.W. 1905.

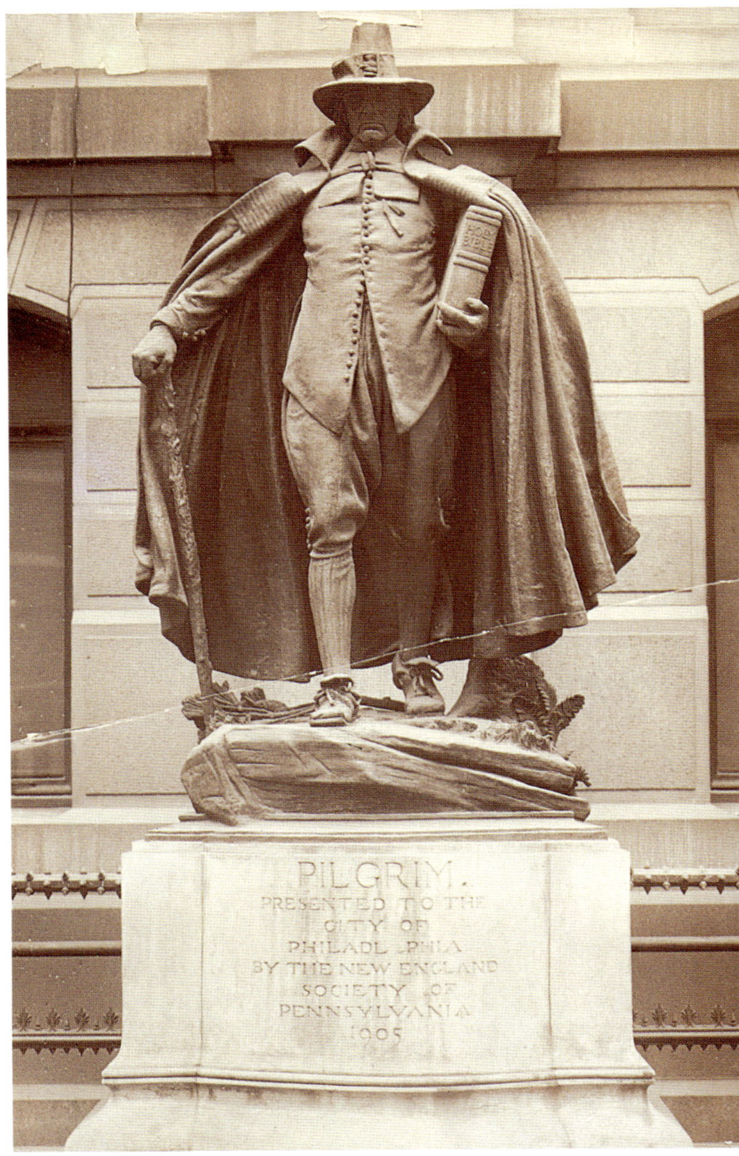

III-23. "The Pilgrim" (also called "The Puritan"), Philadelphia, by Augustus Saint-Gaudens. Originally cast in 1889. R.B.W. 1905.

In a more allegorical vein were his two versions of Diana; the original caused a furor when it was erected at the top of Madison Square Garden (III-24), but its subsequent popularity led to many small replicas being cast by R.B.W. and sold by Tiffany and Company (III-25).

Saint-Gaudens sculpted numerous commemorative statues of leading figures of his time. Among them were his low-relief sculpture of Robert Louis Stevenson (III-26), the Reverend Whitney Bellows Memorial (III-27), a memorial to "Cornelius Vanderbilt" (III-28), the Charles Stewart Parnell Monument for Dublin (III-29), and a plaque dedicated to the "Butler Children" (III-30). Saint-Gaudens also produced a number of medals, including a twenty-dollar gold piece for the United States Government (III-31). Numerous copies of his works were cast during and after his lifetime; Roman Bronze Works began reproducing Saint-Gaudens' works at his own and his heirs' requests at the beginning of the twentieth century.

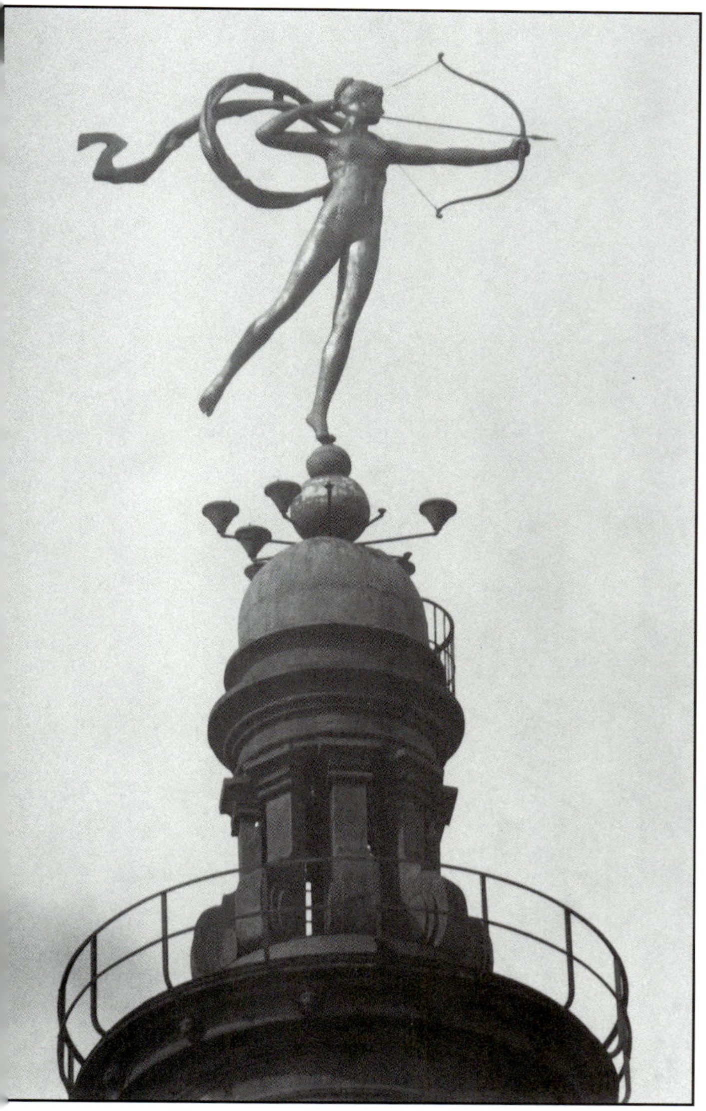

III-24. "Diana" (historic photo of statue atop Madison Square Garden) by Augustus Saint-Gaudens. 1891.

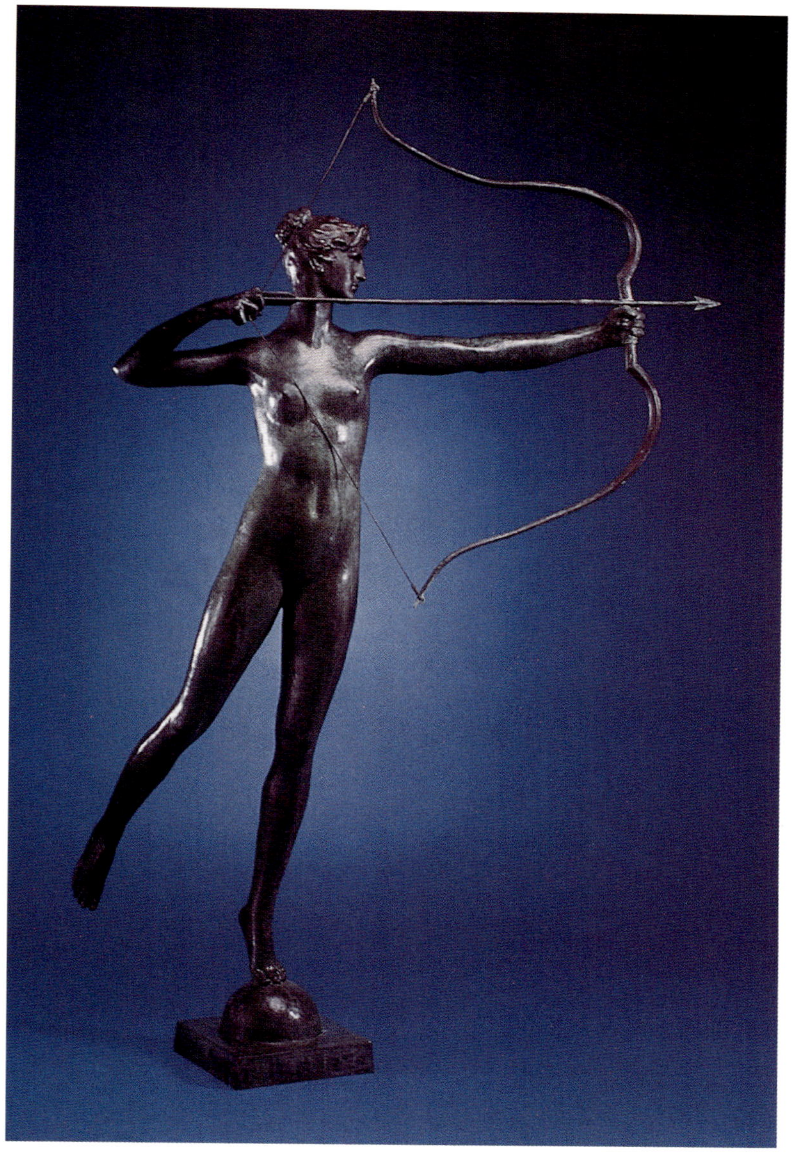

III-25. "Diana" (replica) by Augustus Saint-Gaudens. This is a second version, originally cast in 1892.

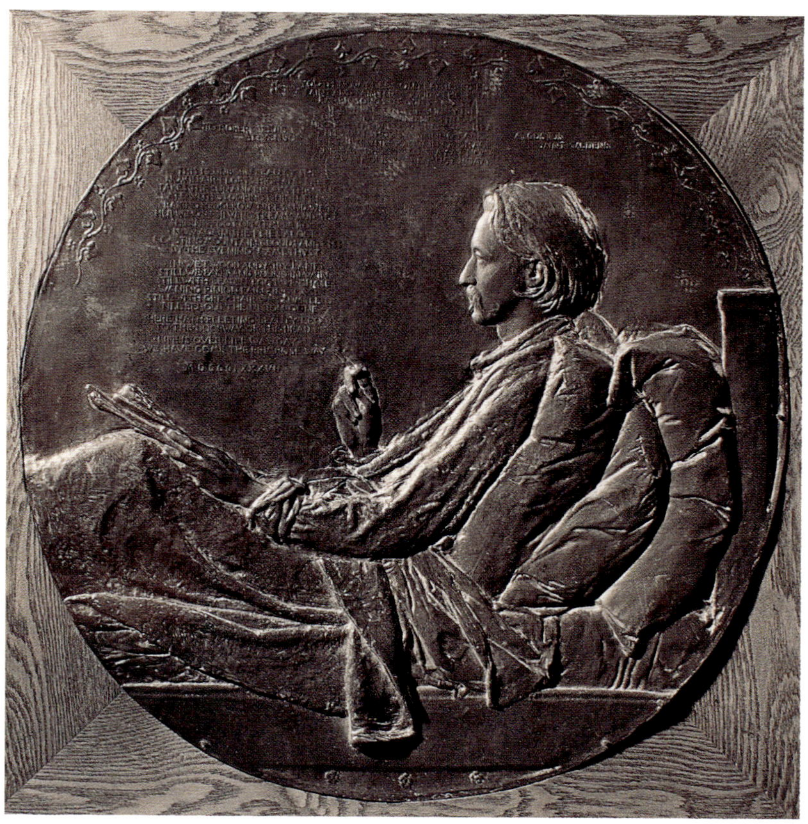

III-26. "Robert Louis Stevenson" (relief portrait) by Augustus Saint-Gaudens. Original casting 1887.

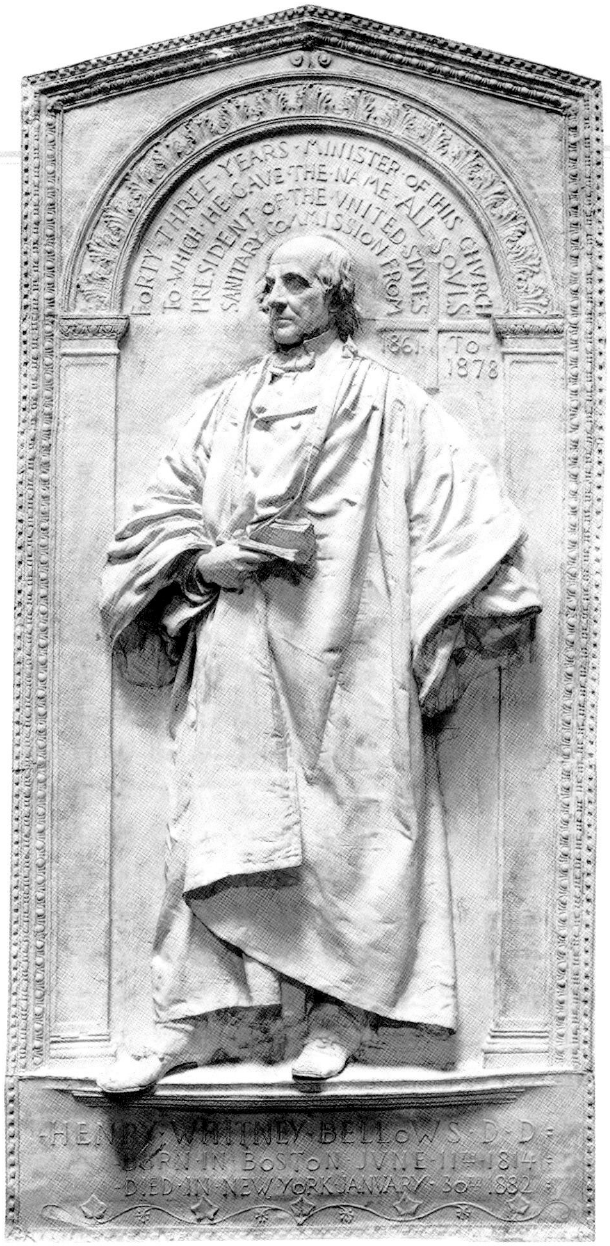

III-27. "Reverend Whitney Bellows Memorial" by Augustus Saint-Gaudens. Originally modeled c. 1885.

Below:
III-28. "Cornelius Vanderbilt Memorial" by Augustus Saint-Gaudens. Original casting 1880.

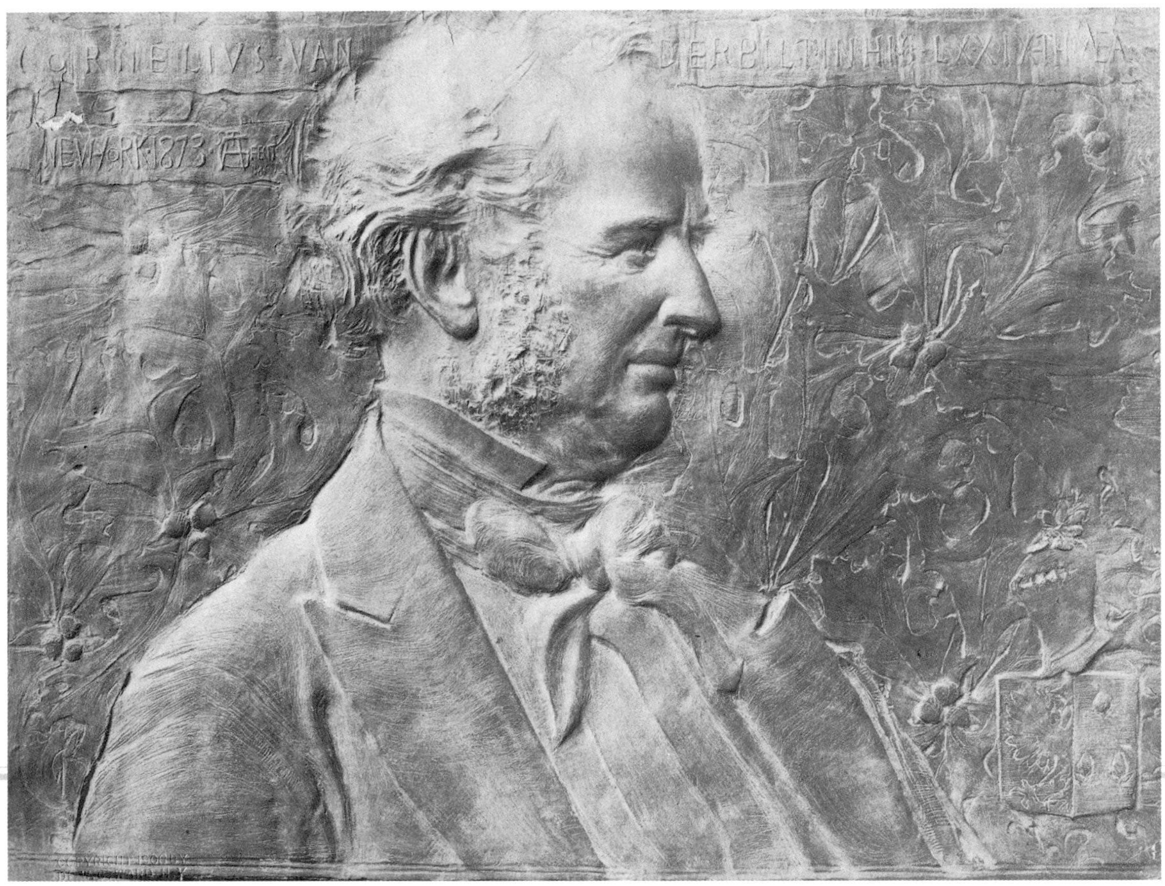

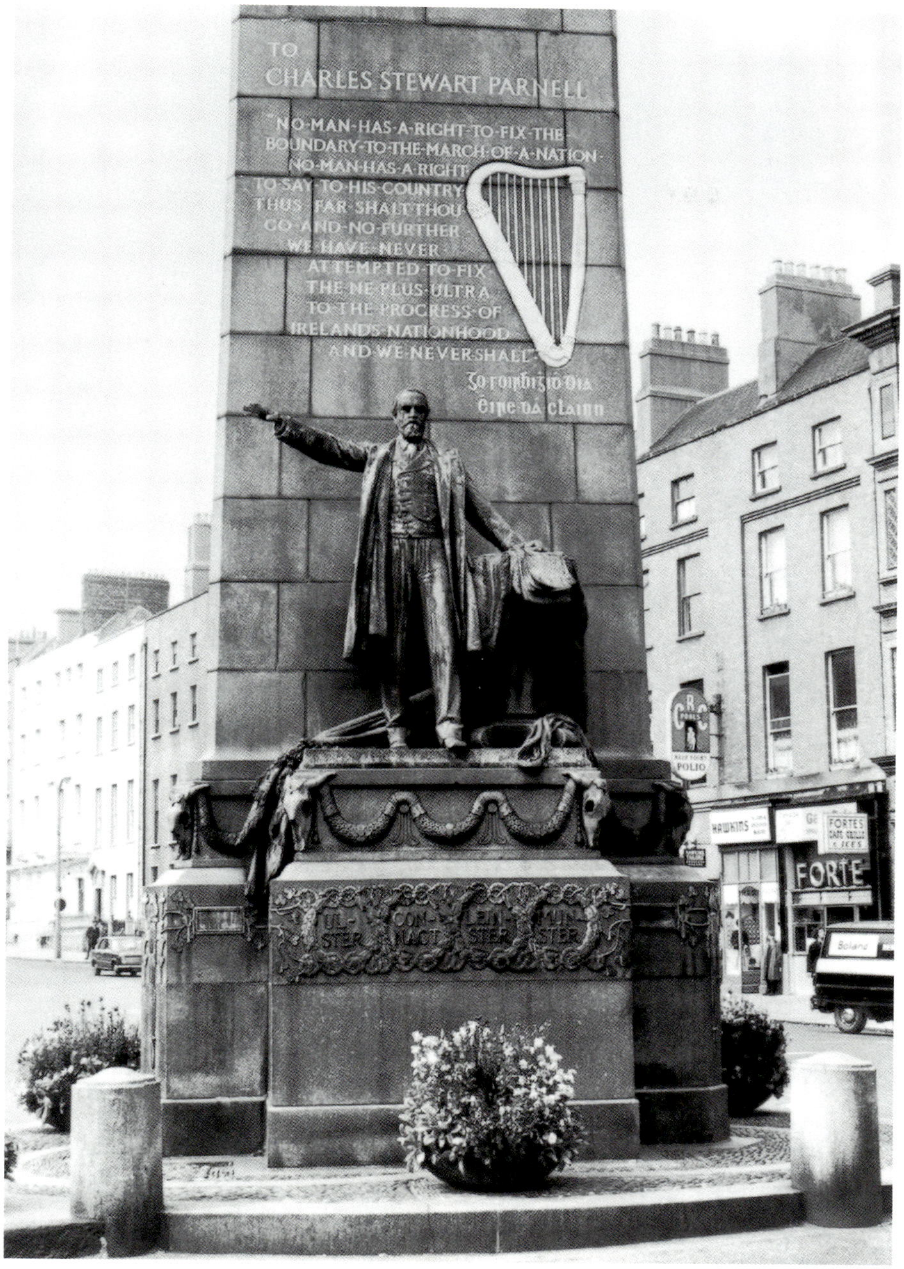

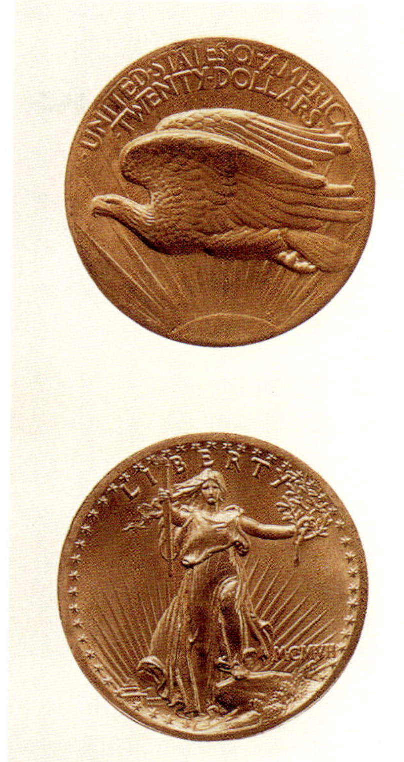

III-31. Twenty dollar gold piece by Augustus Saint-Gaudens. 1907.

III-29. "Charles Stewart Parnell Monument," Dublin, by Augustus Saint-Gaudens. Modeled 1907; unveiled in 1911.

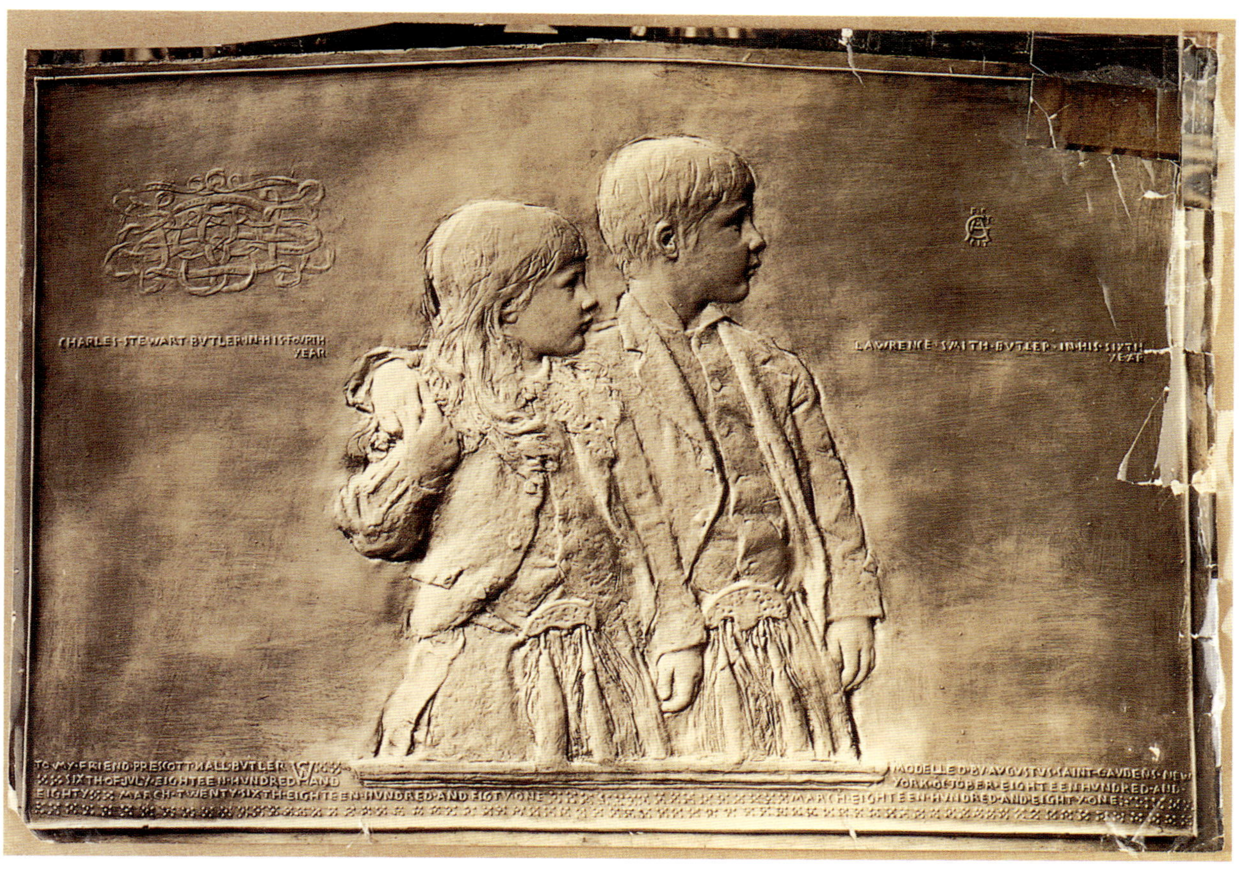

III-30. "The Children of Prescott Hall Butler," plaque by Augustus Saint-Gaudens. Original c. 1881.

JOHN QUINCY ADAMS WARD (1830-1910)

John Quincy Adams Ward was one of the nation's most revered sculptors of monuments and commemorative portraits. Though he was already an elder statesman of sculpture when Roman Bronze Works opened its foundry, they cast several of his many important monuments. Born in Ohio, Ward was a student of the renowned Henry Kirke Brown in New York between 1850 and 1857. A founding member of the National Academy of Design, and then its president in 1874, Ward had reputation and prestige matched only by Daniel Chester French and Augustus Saint-Gaudens in the late 19th century. He was seldom without a commission. He also made a number of architectural sculptures, including the pedimental figures at the New York Stock Exchange.

Ward's work can be seen throughout the country. In Washington D.C. he is represented by the Garfield Monument, and the Poetry figure in the Rotunda at the Library of Congress, among other monuments. In New York major works include the "Indian Hunter" in Central Park—in 1864 it was the first statue placed there—and his noted equestrian of George Washington in Union Square. Ward's portrait figures were noted for their dignity and straightforward naturalism.

At R.B.W. he cast statues of Henry Ward Beecher (originally modeled around 1891) (III-32), and the Ball Monument, of which a plaster sketch model is seen in (III-33).

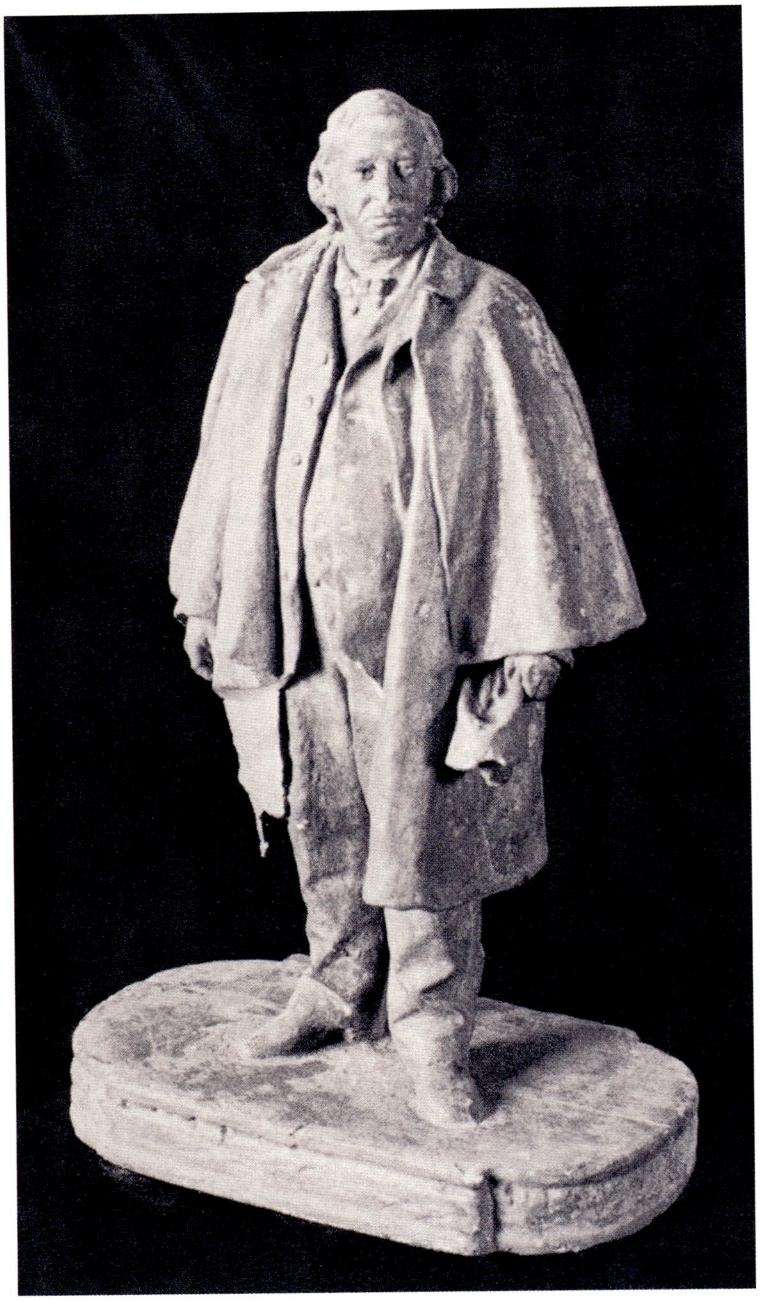

III-32. "Henry Ward Beecher," (plaster cast) by John Quincy Adams Ward. Original c. 1891. This served as a model for the nine-foot statue in Cadman Plaza, Brooklyn.

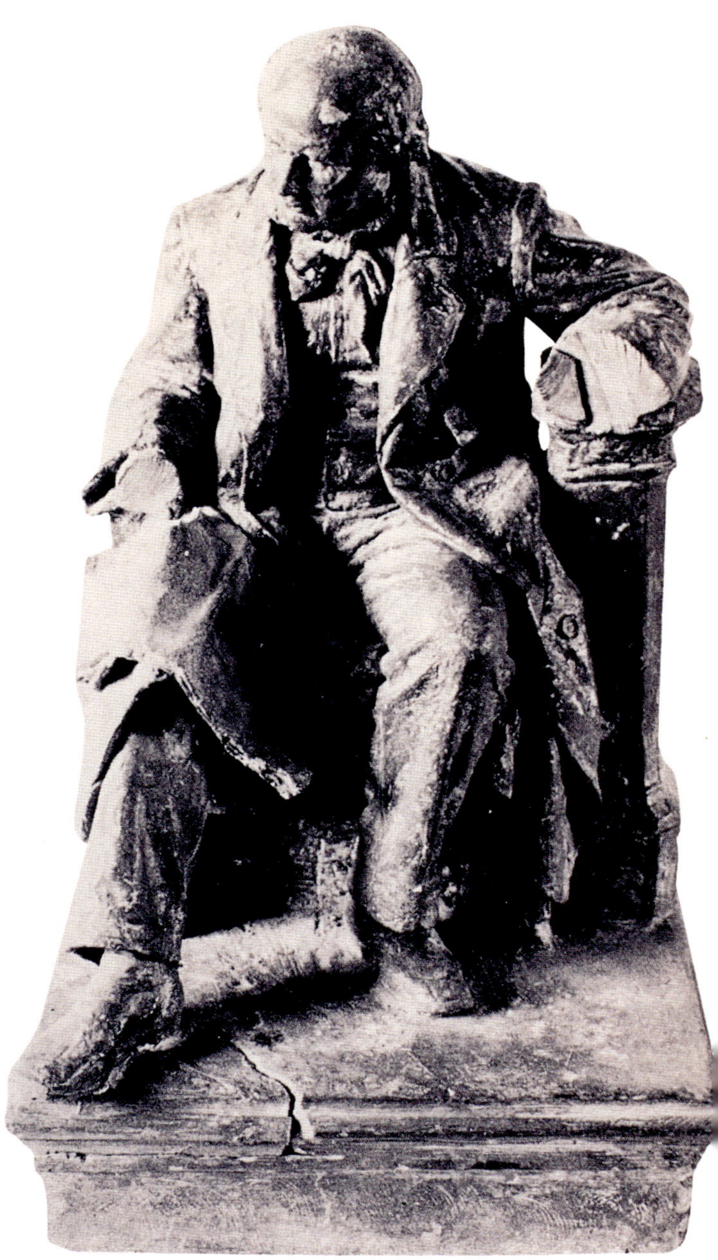

III-33. "Ball Monument" (plaster sketch model) by John Quincy Adams Ward.

KARL BITTER (1867-1915)

Karl Bitter was a major sculptor and an important force in the art world of the early twentieth century. In a period of tremendous emphasis on building and decorative sculpture to ornament it, Bitter organized several vast sculptural programs, as well as creating numerous monuments and statues. Among the sculptural venues that Bitter arranged and produced works for were the Pan-American Exposition at Buffalo in 1901, the Louisiana Purchase Exposition in St. Louis in 1904, and the Exposition in San Francisco in 1915. He was a guiding force in the National Sculpture Society.

Bitter was born and trained in Vienna. He arrived in New York in 1888, where he obtained work for the leading architect Richard Morris Hunt. Hunt gave him the chance to create bronze gates for New York's Trinity Church, sculptural decorations for the Vanderbilt mansion, and ornaments for the Administration Building at the Columbian Exposition in 1893 (widely considered a seminal moment for American sculpture.) His architectural ornaments were well known, appearing on both major civic and private buildings. By 1900 Bitter had become a noted sculptor of civic monuments and public figures.

His style was forceful and often allegorical (he created robed figures that were personifications of the arts, for example). He wrote that "sculpture must give inspiration to the mind and assist the reason...(having) a conception which step by step and link by link, should lead the receptive mind to grasp one big idea and ignite a fire of true and lasting enthusiasm." Among his major monuments were a memorial work to Carl Schurz (III-34), a monument to the Louisiana Purchase (III-35) and a statue of Thomas Jefferson (III-36) at the University of Virginia (unveiled the day after his death). His memorial "Faded Flowers" (III-37) is an example of his more decorative work.

Bitter's sculptures, including a "Diana" in 1911, were cast at Roman Bronze Works between 1911 and 1915.

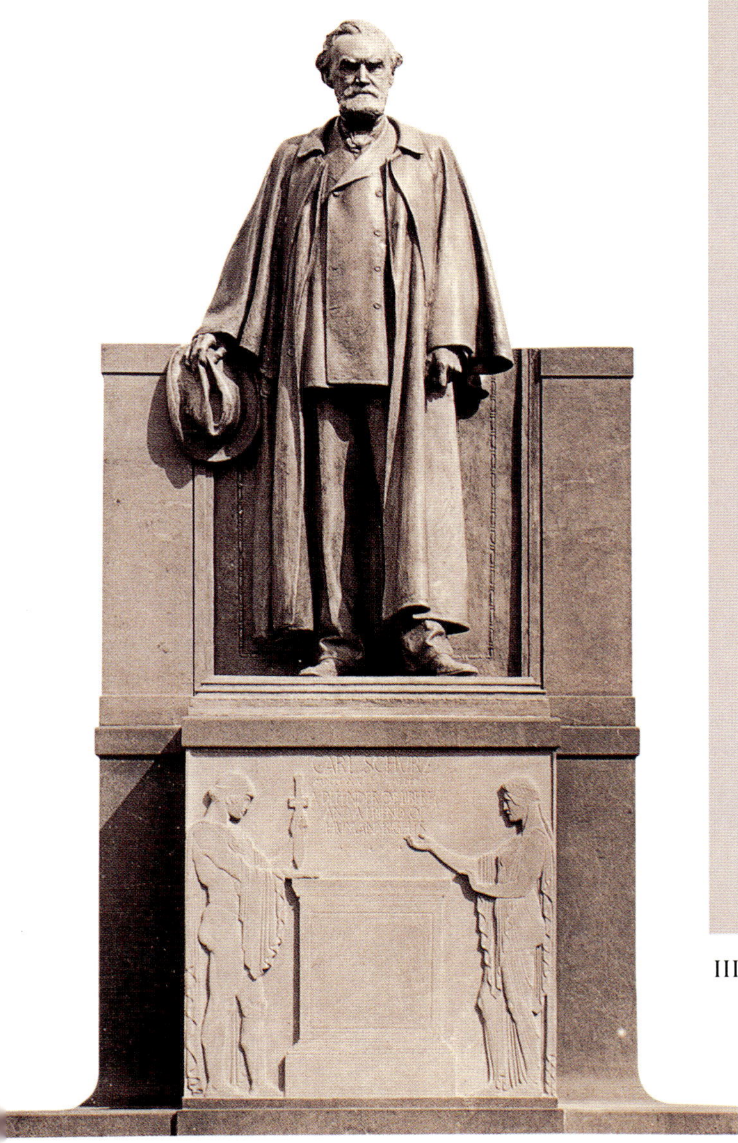

III-34. "Carl Schurz Memorial," Morningside Park, New York by Karl Bitter. 1909; erected 1913.

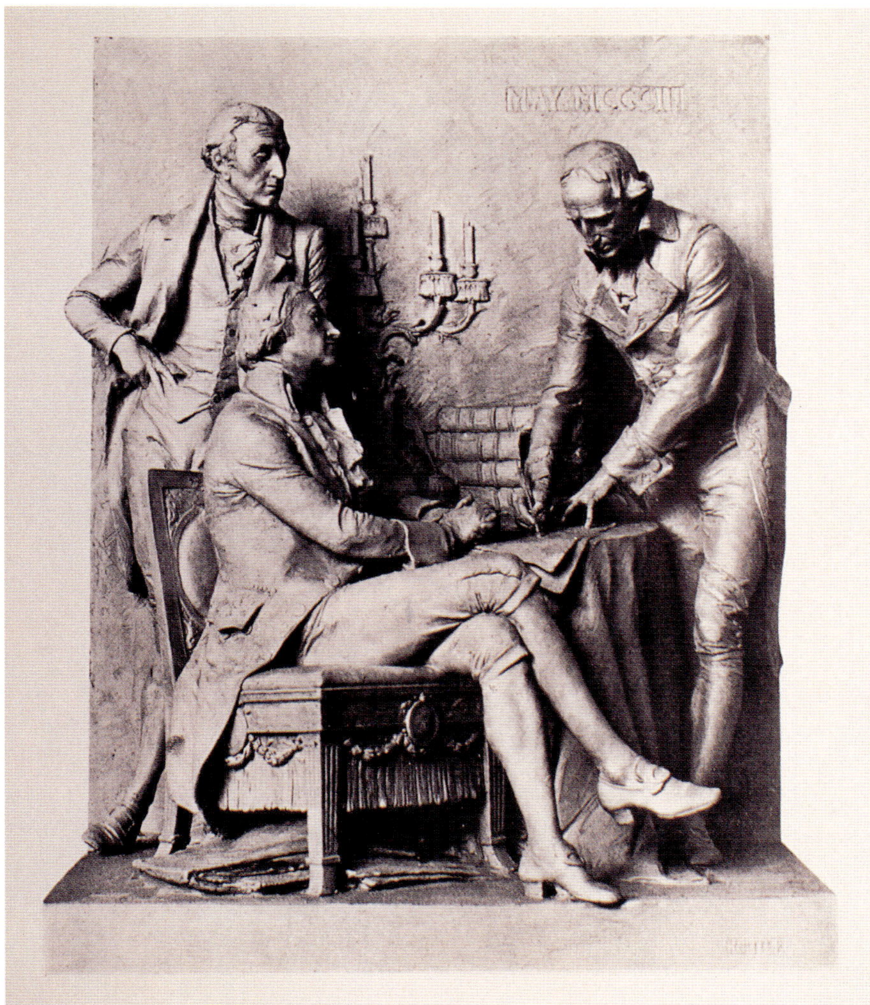

III-35. "Monument to the Louisiana Purchase," by Karl Bitter. 1904.

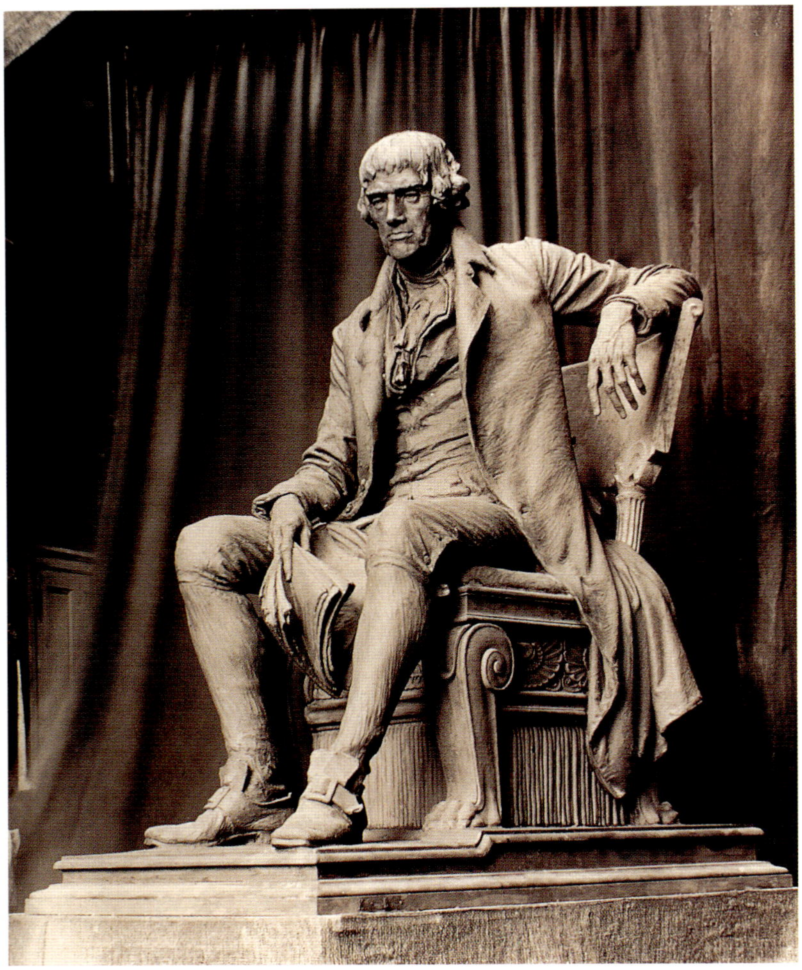

III-36. "Thomas Jefferson" by Karl Bitter. Unveiled 1915.

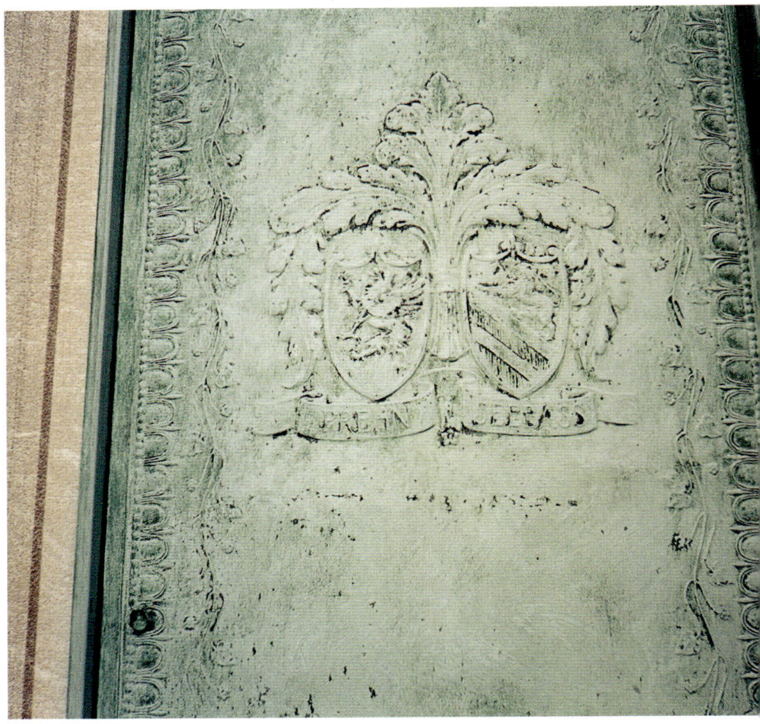

III-37. "Faded Flowers" Cedar Lawn Cemetery, Paterson, New Jersey, by Karl Bitter. 1911-12. Also produced in a small version by R.B.W.

HERBERT ADAMS (1858-1945)

Herbert Adams was a prominent sculptor from the 1890s through the first quarter of the twentieth century. After his education in Massachusetts, he spent five years studying in Paris, where he produced his first notable portrait busts and sculpted figures. In 1890 he became an instructor at Pratt Institute, but his many commissions made him give up teaching by 1898.

Adams's career brought him many important public commissions, including sculpture for the Columbian Exposition in 1893, architectural figures, and numerous busts for the Congressional Library. He became known for his attractive busts of women and his ideal female nudes (III-38).

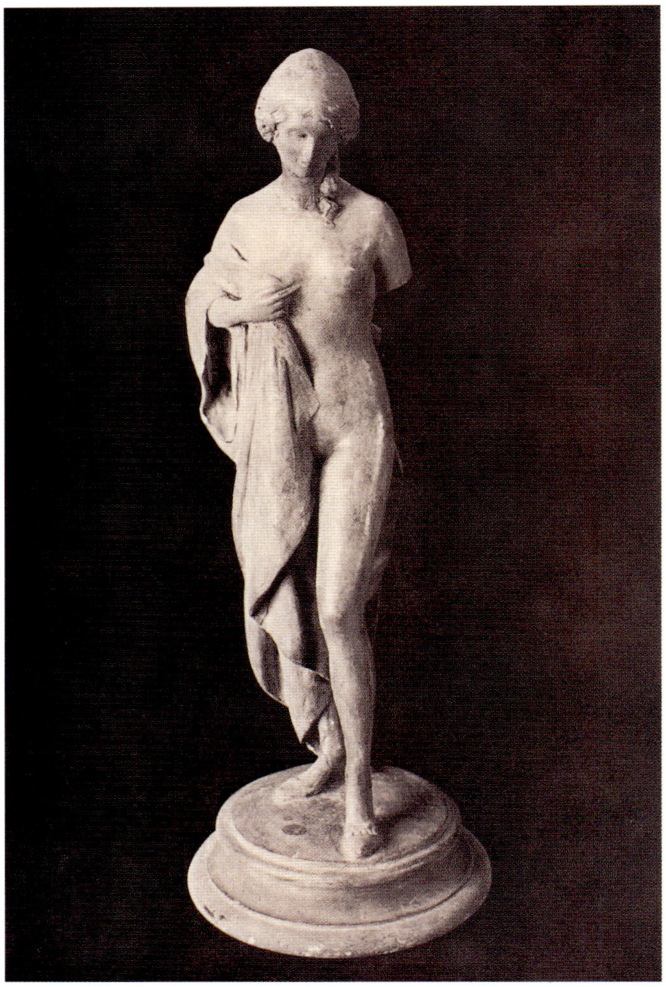

III-38. "Standing Nude Figure" (plaster cast) by Herbert Adams. 1938.

His works included many strong statues of important Americans, a number of sets of decorative bronze doors (including those at St. Bartholomew's Church in New York), and several War Memorials honoring World War I soldiers. His sculpture, though traditional in style, was noted for a certain flair, as can be seen in his allegorical statue of a soldier called "Enlightenment" (III-39). He occasionally colored his works.

In addition to "Enlightenment," a number of Adams' small decorative sculptures, such as "Infant Burbank" (III-40) were cast at Roman Bronze Works. Additional listings in their ledgers between 1914 and 1938 are described as "colonial soldiers, a standing nude, cupids, and a fish fountain."

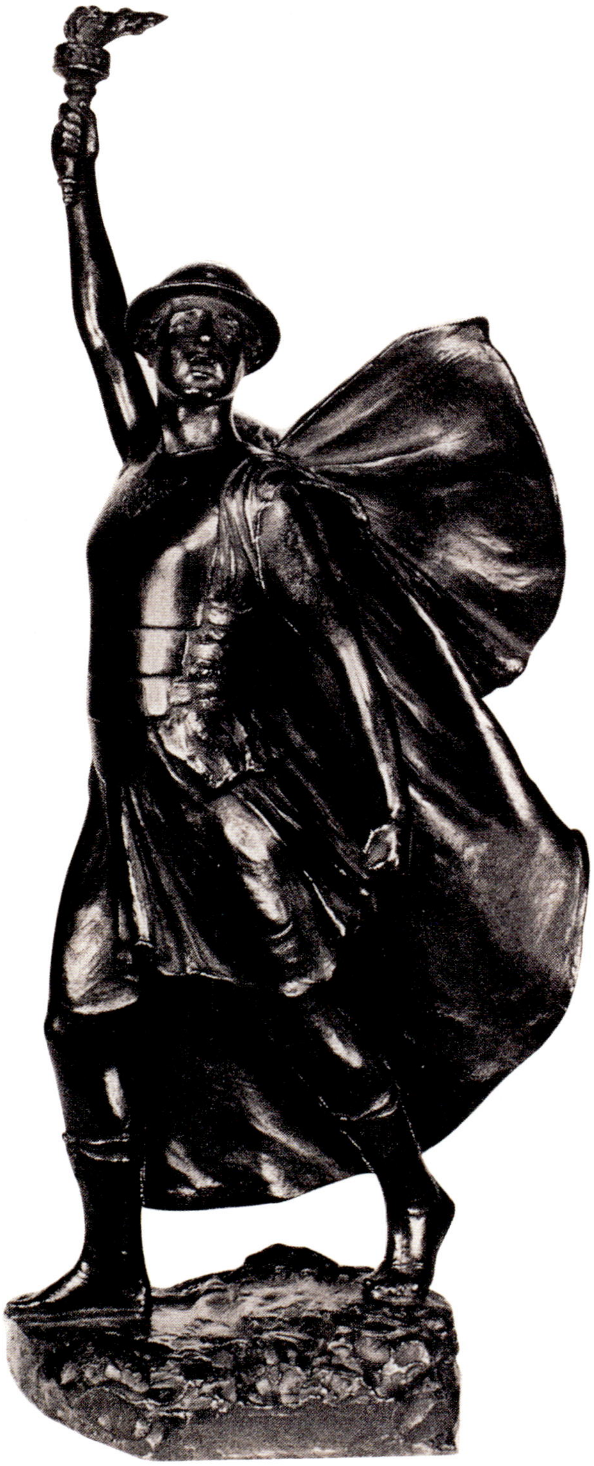

III-39. "Enlightenment" by Herbert Adams. c. 1918.

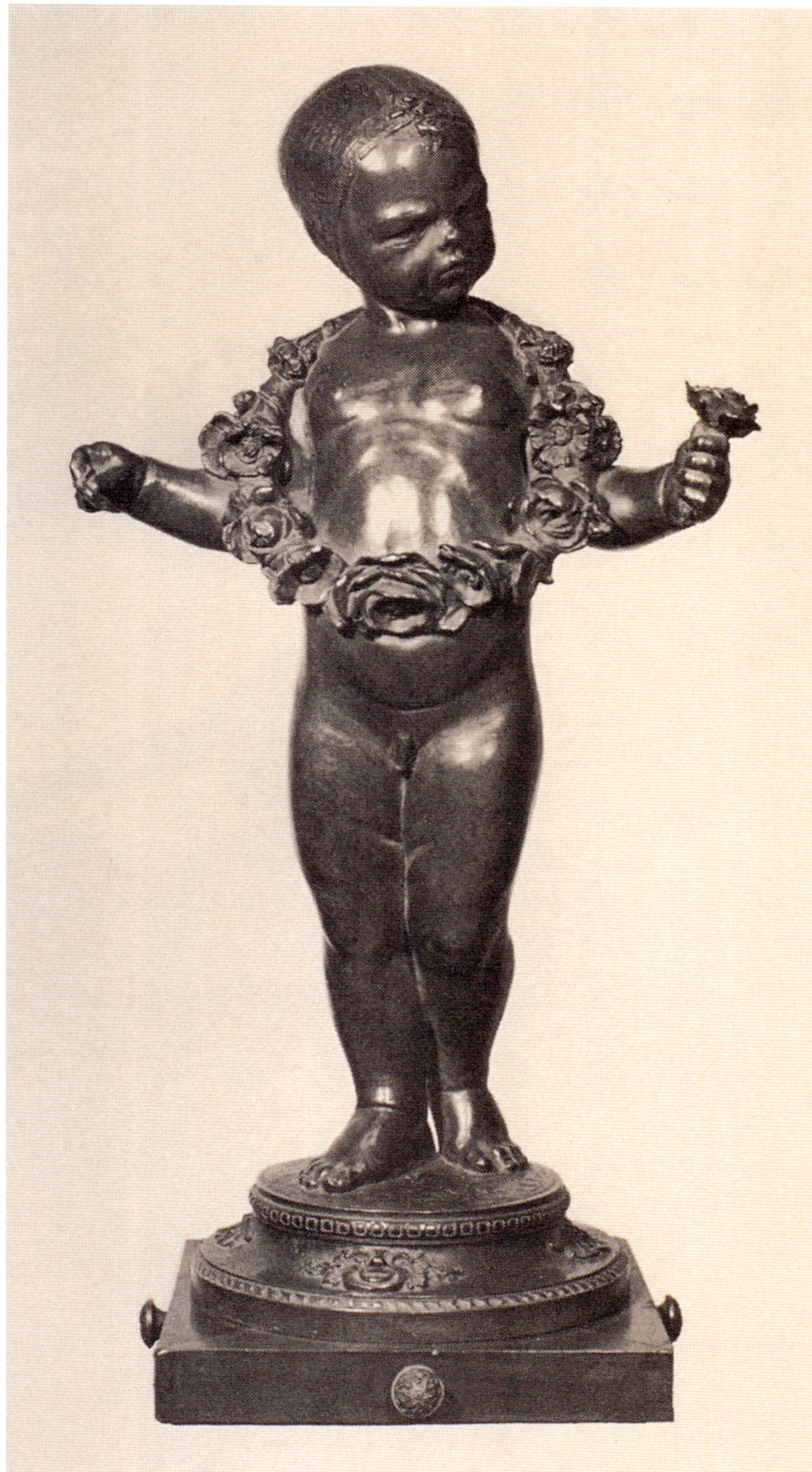

III-40. "Infant Burbank" by Herbert Adams. 1905.

ALEXANDER STIRLING CALDER (1870-1945)

Alexander Stirling Calder was a leading creator of American monuments in the Beaux-Arts tradition, although he moved toward a more modernist style later in his life. (He was the father of Alexander Calder, the noted sculptor of mobiles.) In addition to many civic statues and memorials for both his native city of Philadelphia and for New York, he produced a variety of sculptures later in his life that attempted to combine a traditional approach with some of the stylistic influences of modern art.

In 1885 Alexander Stirling Calder became a student at the Pennsylvania Academy of Fine Arts, where he studied with realist painters Thomas Eakins and Thomas Anshutz. In 1890 he went to Paris to study at the Académie Julian and the Ecole des Beaux-Arts. By 1892 he had returned to the United States, and several years later became an instructor at the Philadelphia Museum School of Industrial Arts. Like so many sculptors of his generation, Calder's career as a noted artist began at one of the large national exhibitions: he won a medal at the St. Louis Exposition in 1904. In 1910 he moved to New York, becoming a teacher at the National Academy of Design and the Art Students' League.

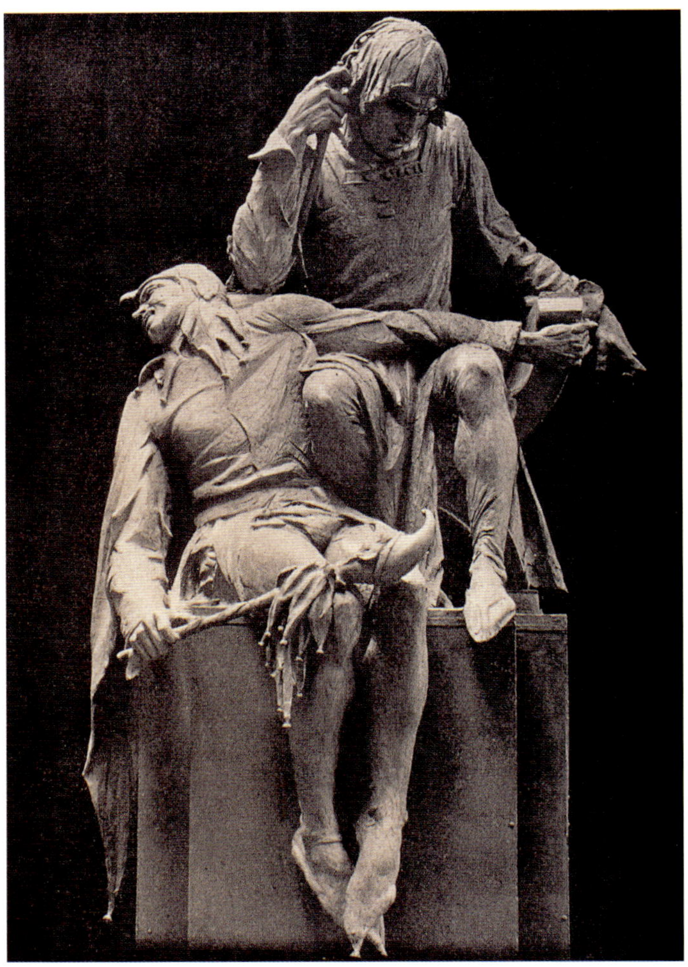

III-41. "Tragedy and Comedy," Shakespeare Memorial, Philadelphia, by Alexander Stirling Calder. 1928.

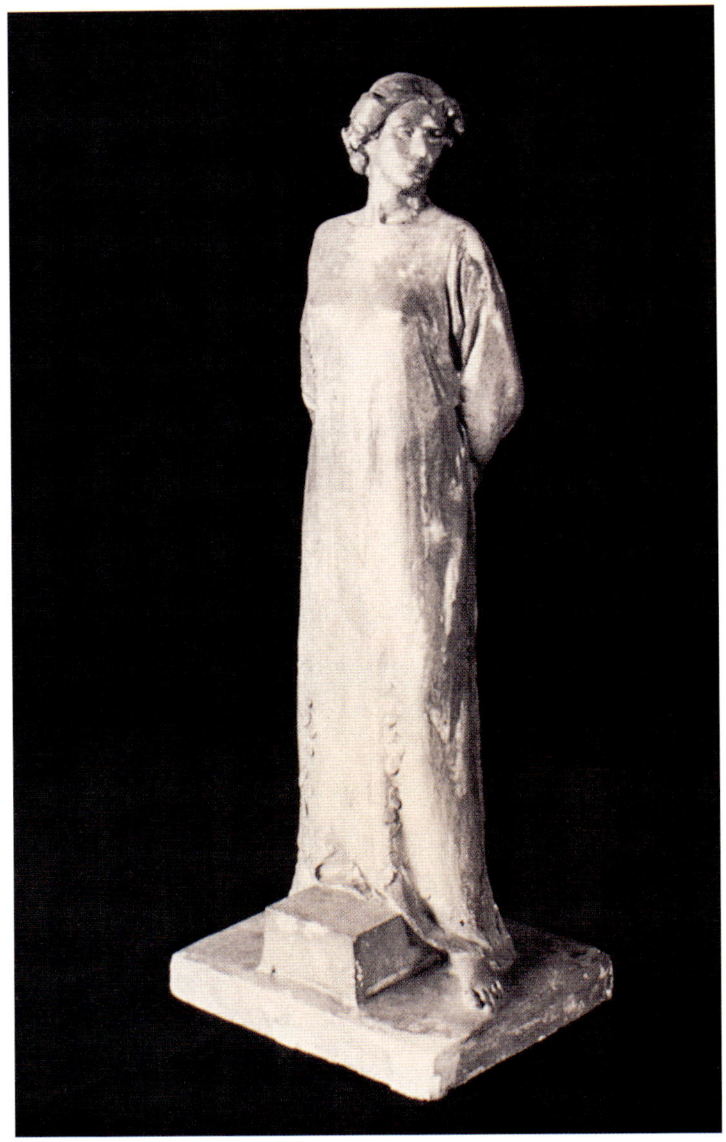

III-42. "Young Girl" by Alexander Stirling Calder. 1927.

He soon had many commissions and recognition from various civic organizations. With Karl Bitter he directed the sculpture program for the San Francisco Exposition in 1915; his "Fountain of Energy" was a resounding success with its twentieth century theme of progress and science. Other monuments included a statue of George Washington for Washington Square in New York, the Swann Memorial Fountain for Philadelphia, and "Tragedy and Comedy" (III-41). Smaller works included his "Young Girl" (III-42) and his "Pioneer Woman" (III-43), and portraits of Lincoln and Edison.

Unlike many of his contemporaries, who were wedded to thoroughly academic concepts of art, by 1920 Calder had interested himself in abstract ideas of form. His synthesis of classical ideas with a somewhat more modernist approach can be seen in his "Girl Scratching her Heel" (III-44) and his "Reclining Nude" (III-45).

Calder first cast sculptures at Roman Bronze Works in 1910; he continued to use the foundry for his brass castings to the end of his life, and several works were cast posthumously in the early 1950s.

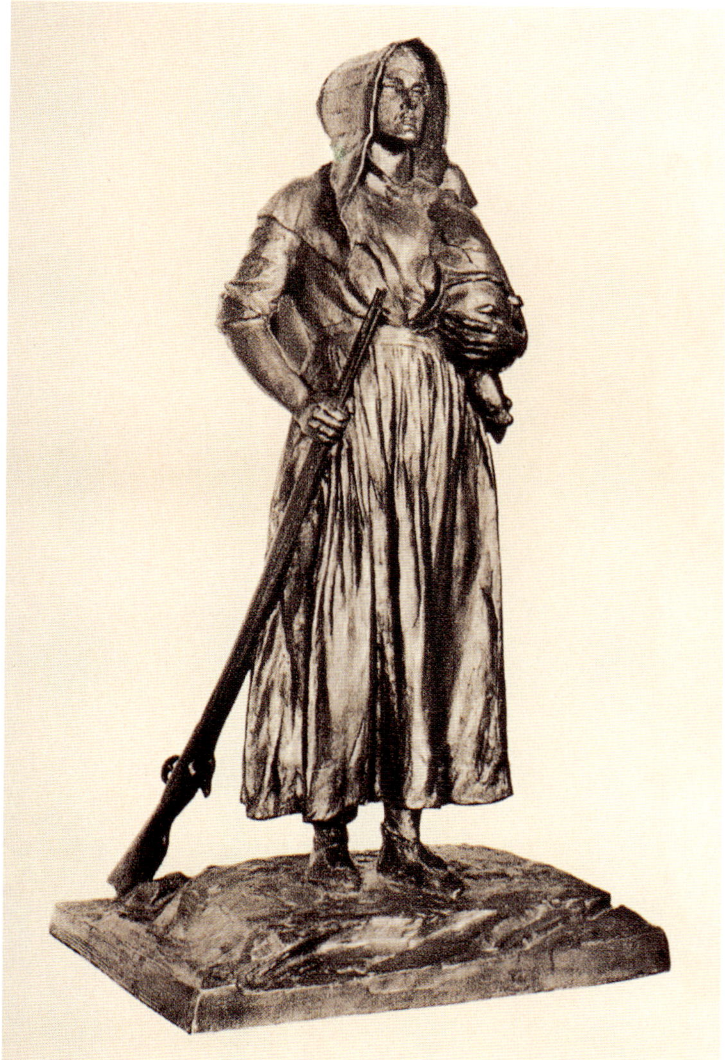

III-43. "Pioneer Woman" by Alexander Stirling Calder. 1927

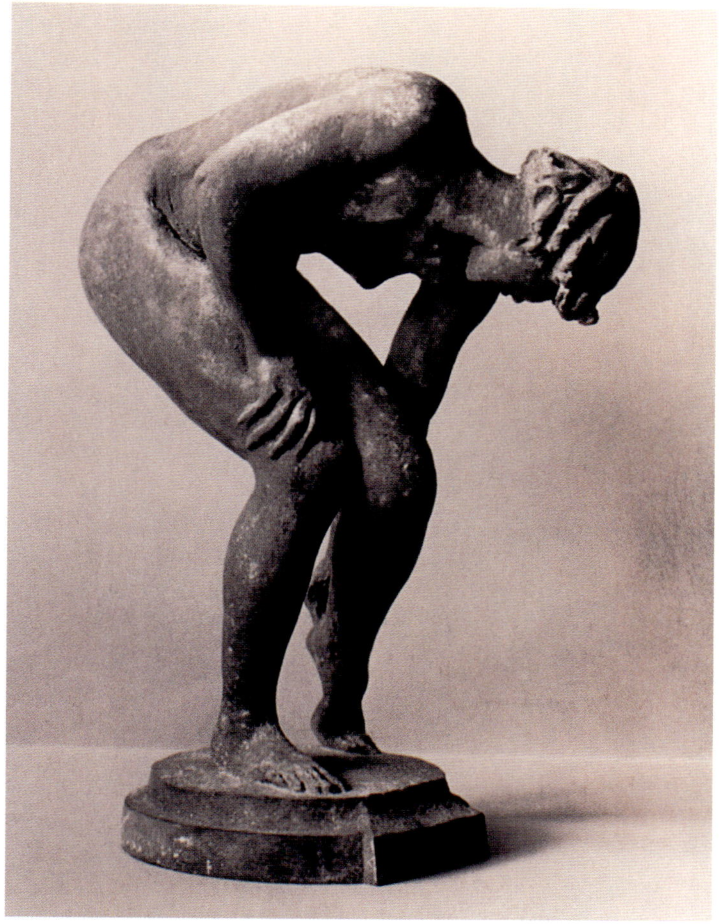

III-44. "Girl Scratching Her Heel" by Alexander Stirling Calder. c.1920.

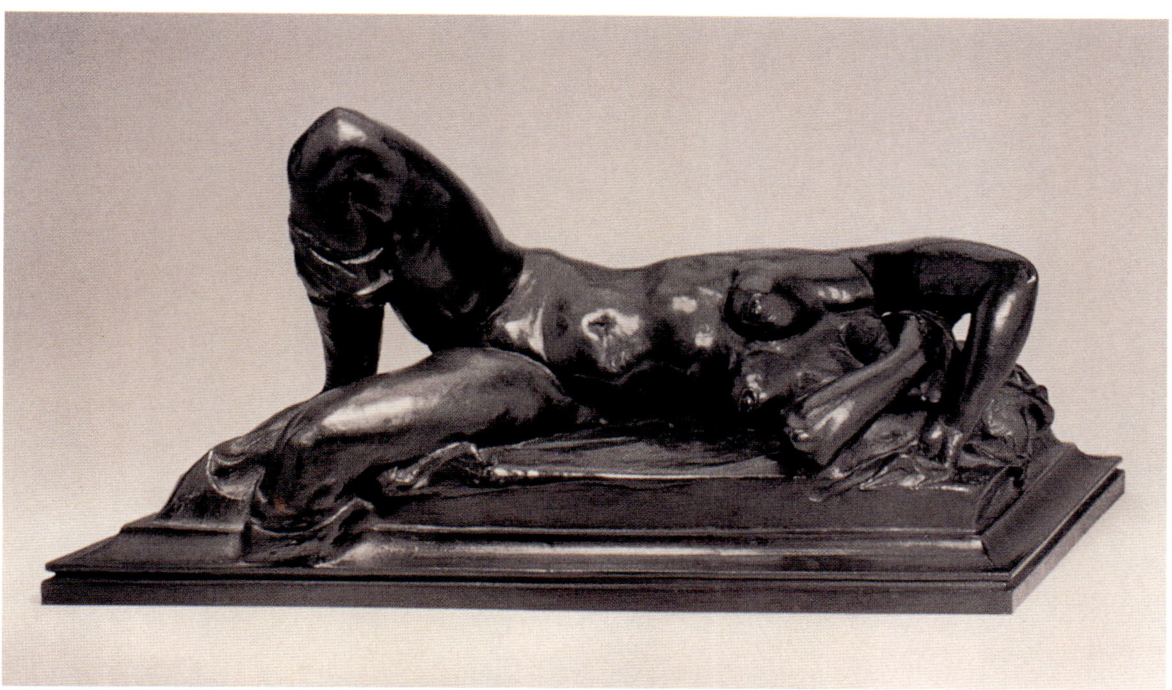

III-45. "Reclining Female Nude" by Alexander Stirling Calder.

GUTZON BORGLUM
(1867-1941)

Gutzon Borglum, was one of America's best known sculptors for his massive carving of four presidents' heads at Mount Rushmore. He was one of two brothers, children of Danish pioneers, to achieve renown in American sculpture. He was born in Nebraska, and apprenticed as a young man to a lithographer and fresco painter before attending art school in San Francisco. He studied painting in Paris but was influenced by Rodin's work; while there he was made a member of the Société des Beaux-Arts. A life-long fascination with Western images led him to making small Western-themed sculptures, such as "Apaches Pursued (III-46). They were well received and he exhibited an example at the Columbian Exposition in Chicago in 1893. By 1902 he showed seven works at the St. Louis Exposition, including a notable statuette of John Ruskin (III-47), and won the gold medal for his huge and impressive "Mares of Diomedes" (a massive mythical work featuring wild horses).

In New York, where he settled, he produced statues of the twelve apostles for the Cathedral of St. John the Divine, and in 1908 a head of Abraham Lincoln for the Library of Congress. He created more than 170 works of sculpture, including many portraits of famous figures, moving

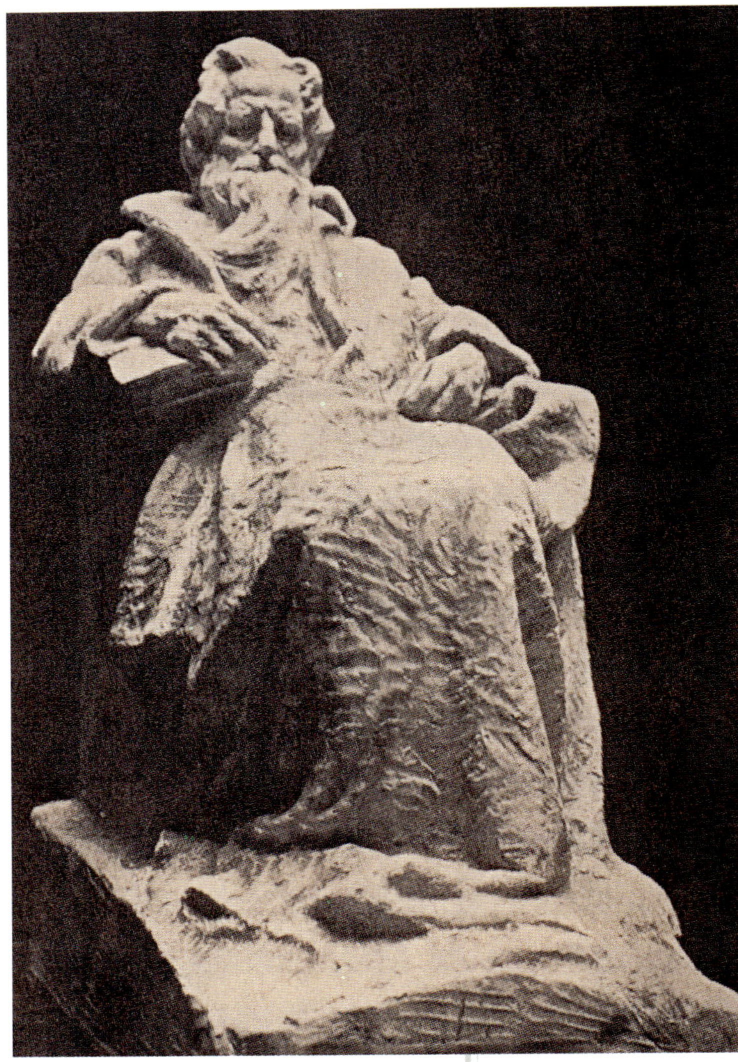

III-47. "Ruskin" by Gutzon Borglum. Modeled 1903.

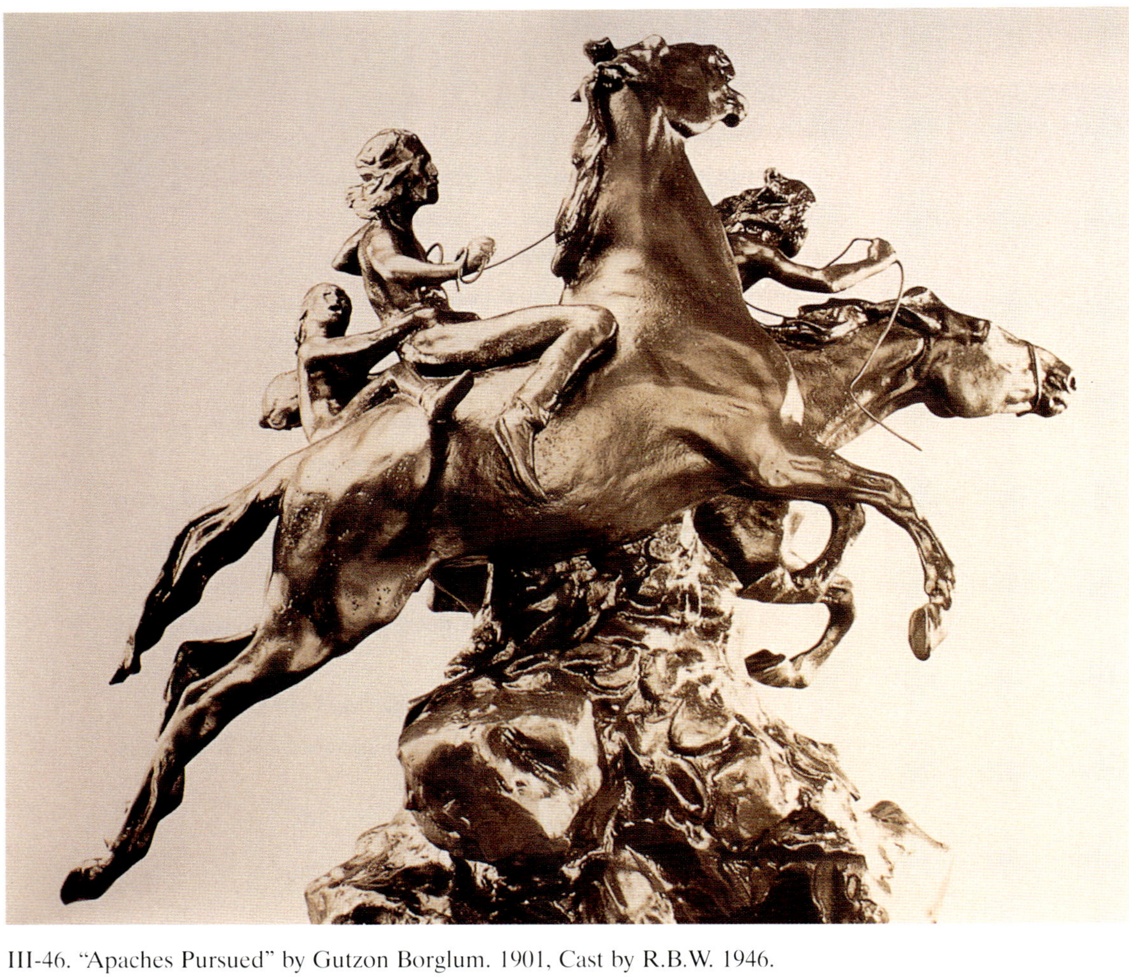

III-46. "Apaches Pursued" by Gutzon Borglum. 1901, Cast by R.B.W. 1946.

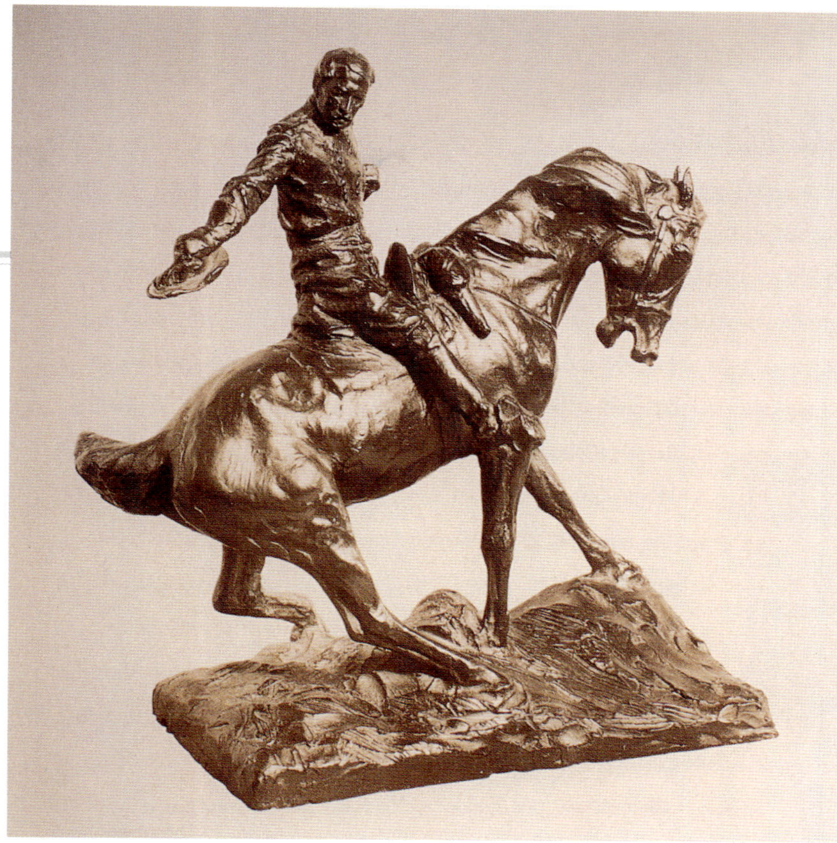

III-48. "General Philip Sheridan" (bronze model) by Gutzon Borglum. 1913.

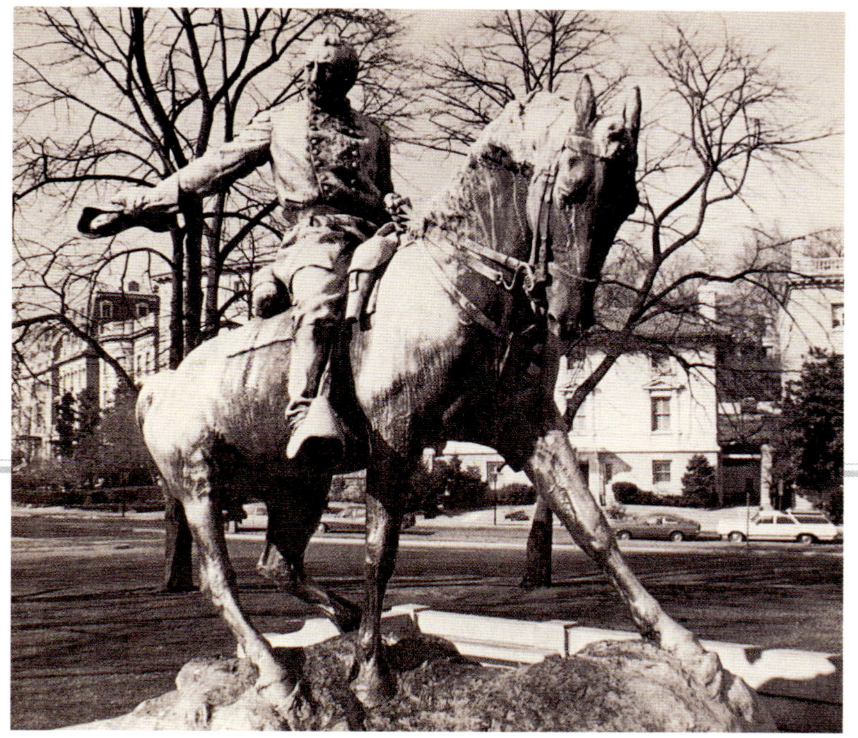

III-49. "General Philip Sheridan," Washington, D.C., by Gutzon Borglum. 1908.

from purely Western subjects to more national subjects, including General Sheridan (III-48 and 49), Theodore Roosevelt (III-50) and Robert E, Lee, and a massive composition of figures called "Wars of America" in Newark. Between 1927 and 1939 his bold signature carvings on Mount Rushmore were created, with his son Lincoln finishing the project after his father's death.

Just as Borglum's major works were distinguished by their colossal size and complexity, many of his smaller statues also bore his signature dynamic composition, with large forms and rough-hewn surfaces. He believed that American art should have American subject matter, and should idealize American achievement; he complained bitterly about the artistic influences of classicism and other historic forms, describing the nation's art and architecture as "ruined by the trinkets and gewgaws plagiarized from the Middle Ages."

His relationship with Roman Bronze Works began in 1902, when he moved to New York, and continued until 1916. During that period he cast a variety of portrait busts, including statues of Homer and the Emperor Nero (III-51), as well as several major equestrian works.

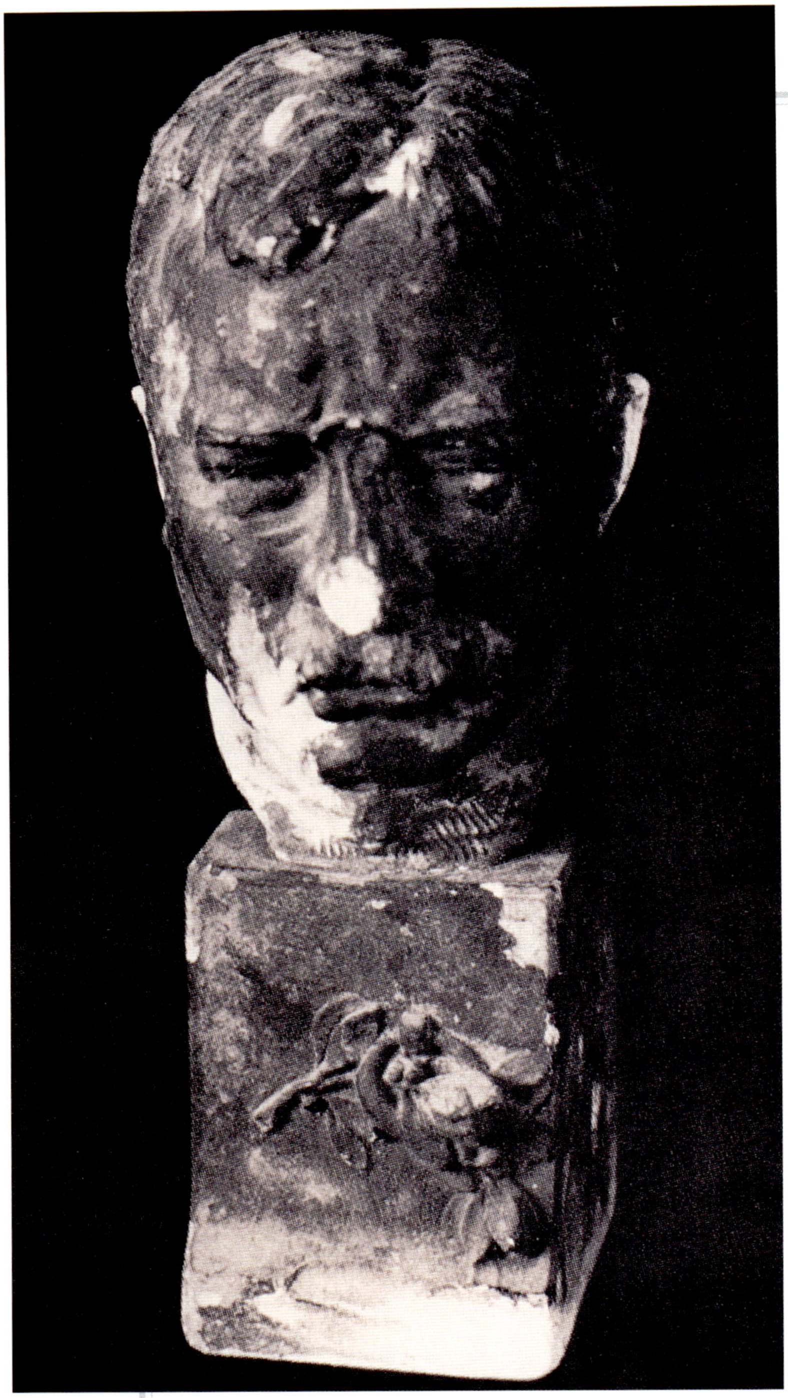

III-50. "Theodore Roosevelt" (plaster cast) by Gutzon Borglum.

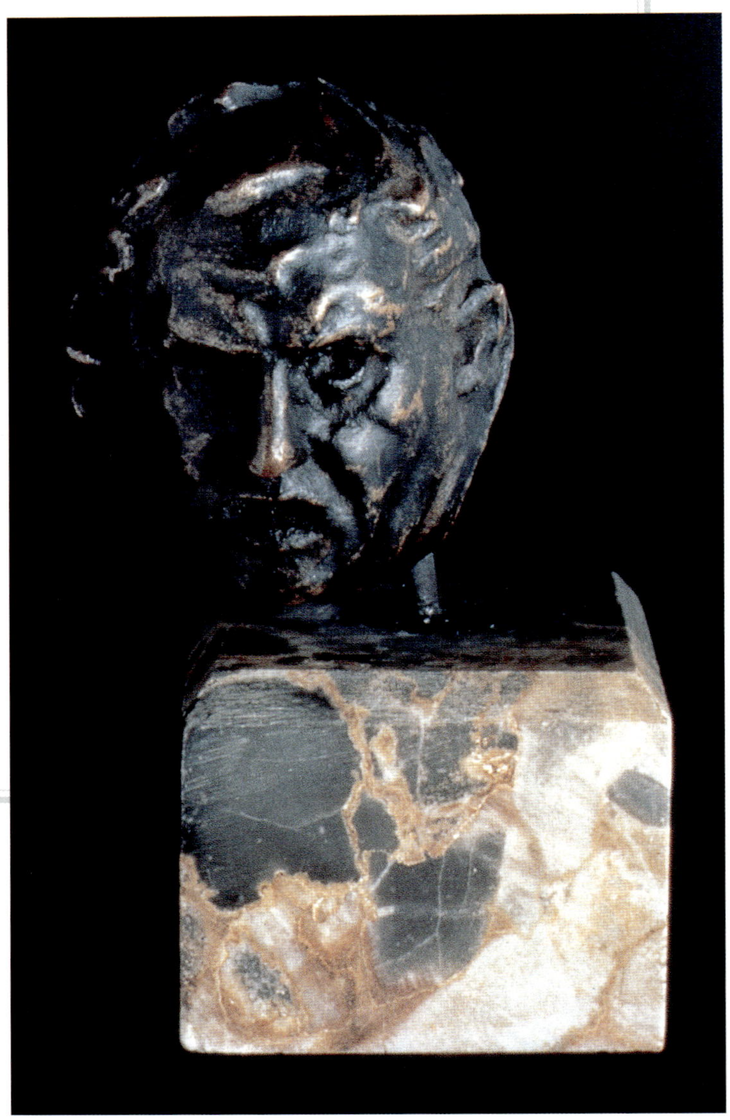

III-51. "Head of Nero" by Gutzon Borglum. 1903.

ROBERT I. AITKEN (1878-1949)

Robert Aitken was a leading figure in representational sculpture in America. But he employed both modern devices and contemporary ideas in his work, making him a representative of the new century and new thought. His reputation for originality was created at the San Francisco Exposition in 1915 when he produced a notable work called "Fountain of the Earth." It used such innovative devices as electric illumination and jets of steam, as well as presenting themes ranging from Darwinism to "sexual love and physical parenthood without enlightenment."

Aitken was born in San Francisco, spent three years in Paris just after the turn of the century, and returned to the United States to become an instructor at the Art Students League and the National Academy of Design in New York. Among his major commissions were sculpture for the American Telephone and Telegraph Company building in New York and a large monument to the labor leader Samuel Gompers (III-52).

He sculpted numerous busts of famous people, including American statesmen such as Thomas Jefferson (III-53) and Henry Clay (III-54), and artists and writers including Edgar Allen Poe (III-55), Robert Burns (III-56), and painter George Bellows (III-57). His "Babe in Arms" (III-58) shows more formalist influences, while he also made such whimsical works as "Pan" (III-59) and "Xoros" (III-60) and a plaque of female heads entitled "Hear No Evil, Speak No Evil" (III-61). Other works listed in R.B.W.'s ledgers between 1910 and 1929 include "mermaids, a dancing faun, lizards, and a standing Diana."

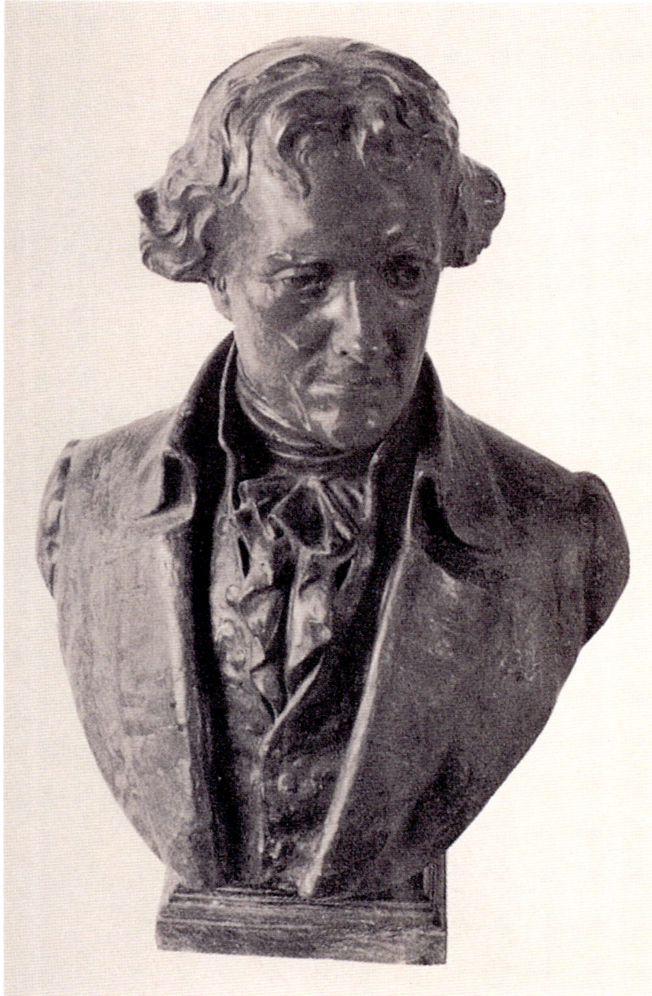

III-53. "Bust of Thomas Jefferson" by Robert Aitken. 1930.

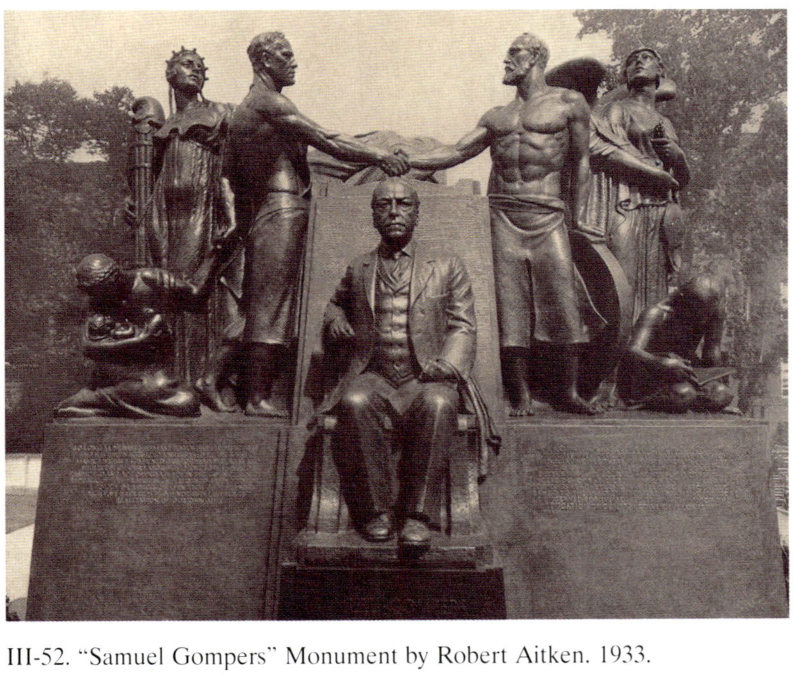

III-52. "Samuel Gompers" Monument by Robert Aitken. 1933.

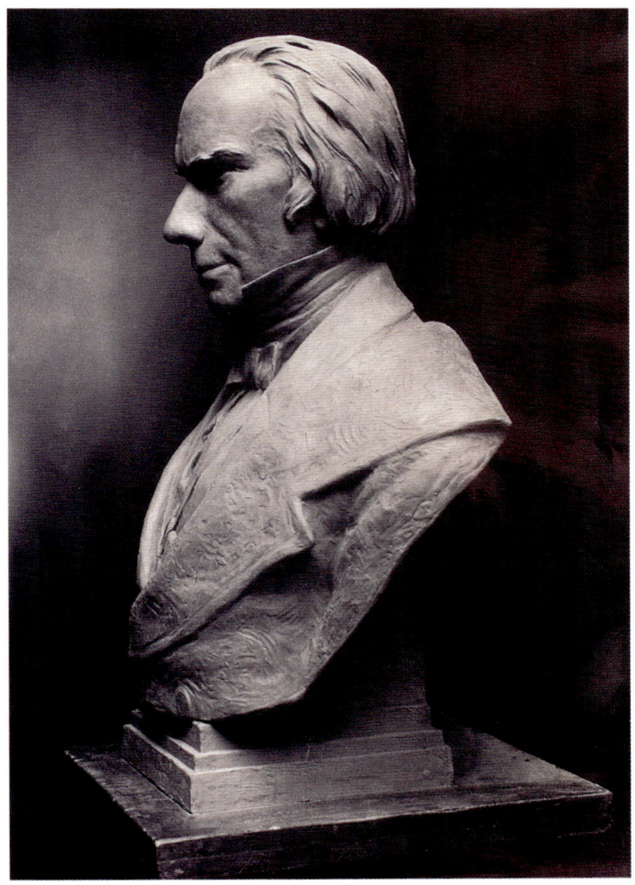

III-54. "Bust of Henry Clay" by Robert Aitken. 1929.

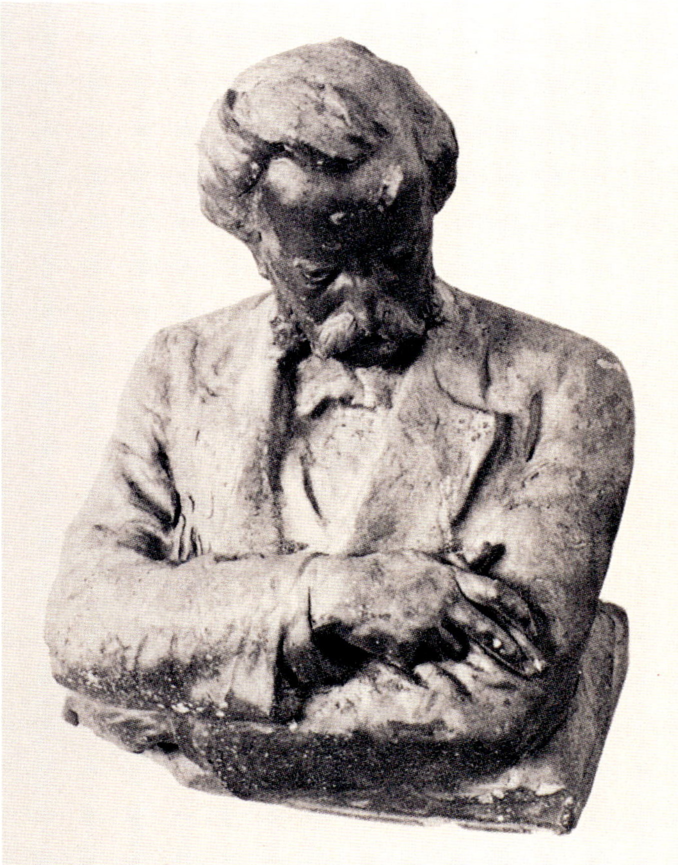

III-55. "Bust of Edgar Allen Poe" (plaster) by Robert Aitken.

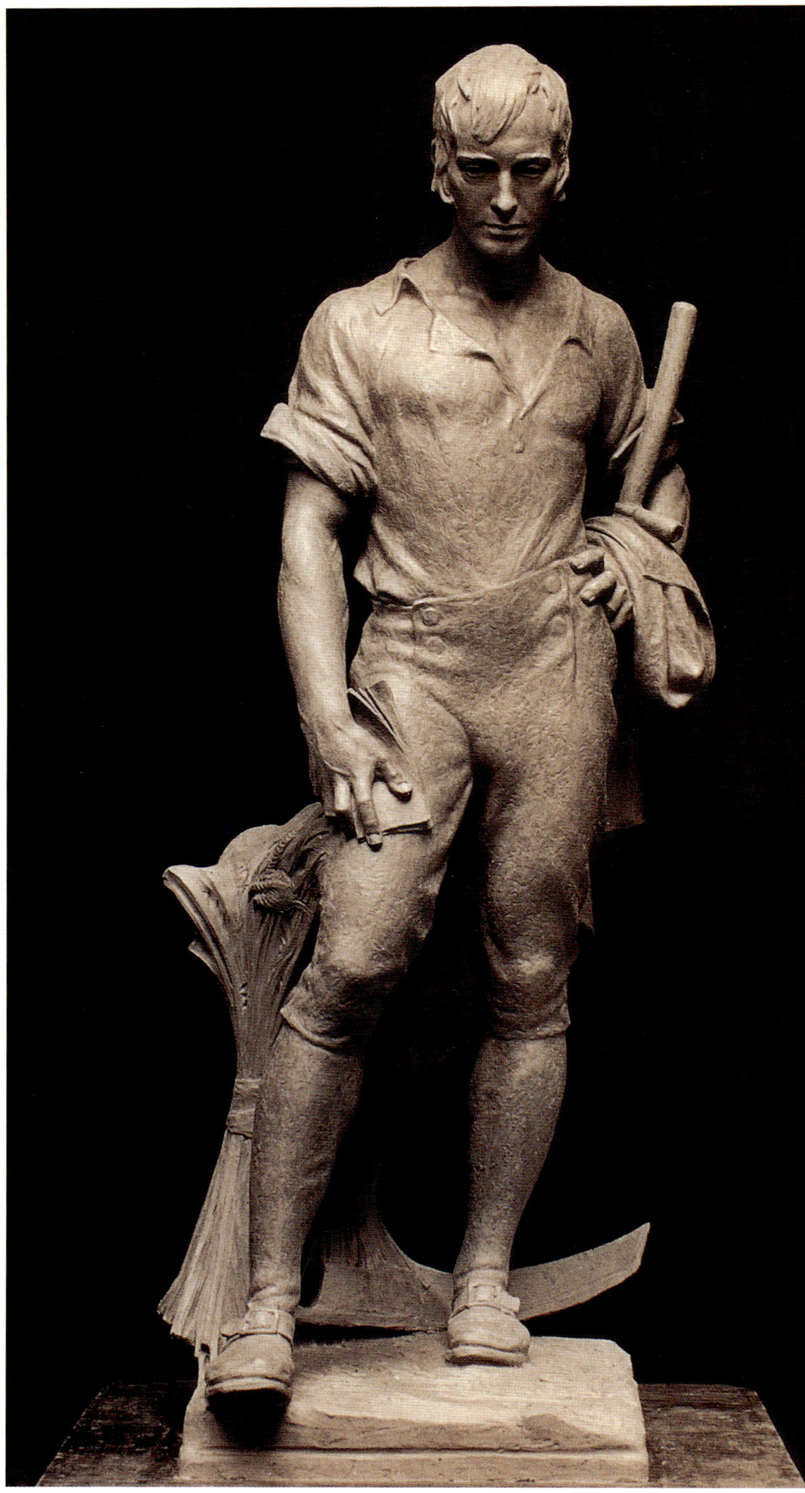

III-56. "Heroic Bronze Statue of Robert Burns," Washington University, St. Louis, by Robert Aitken. c. 1920.

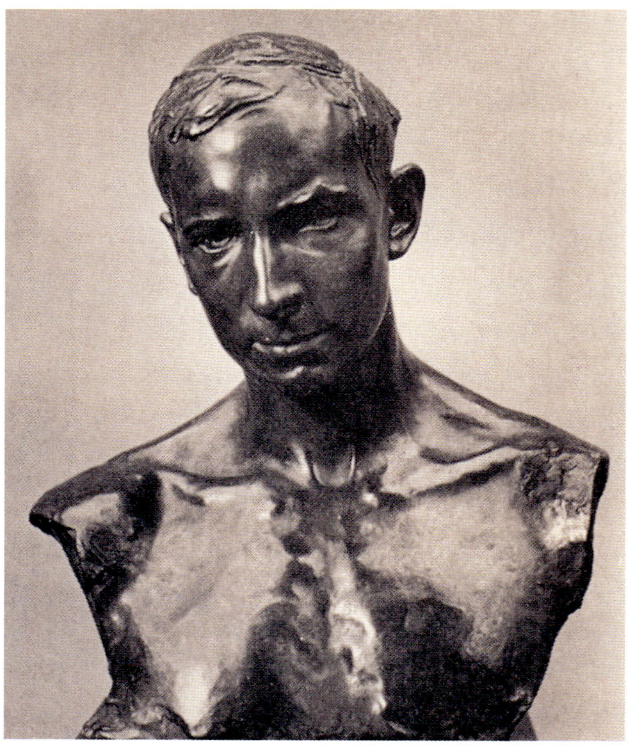

III-57. "Portrait Bust of George Bellows" by Robert Aitken. Cast posthumously by R.B.W. from original model in 1951.

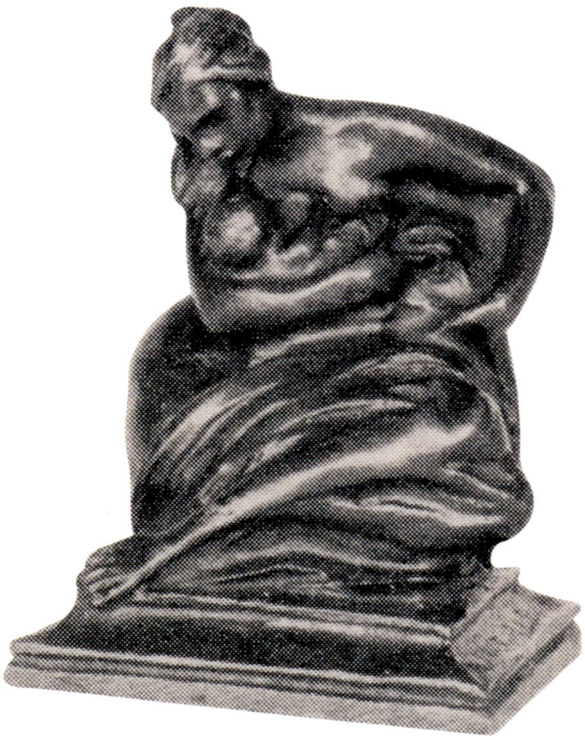

III-58. "Babe in Arms" by Robert Aitken. c. 1930.

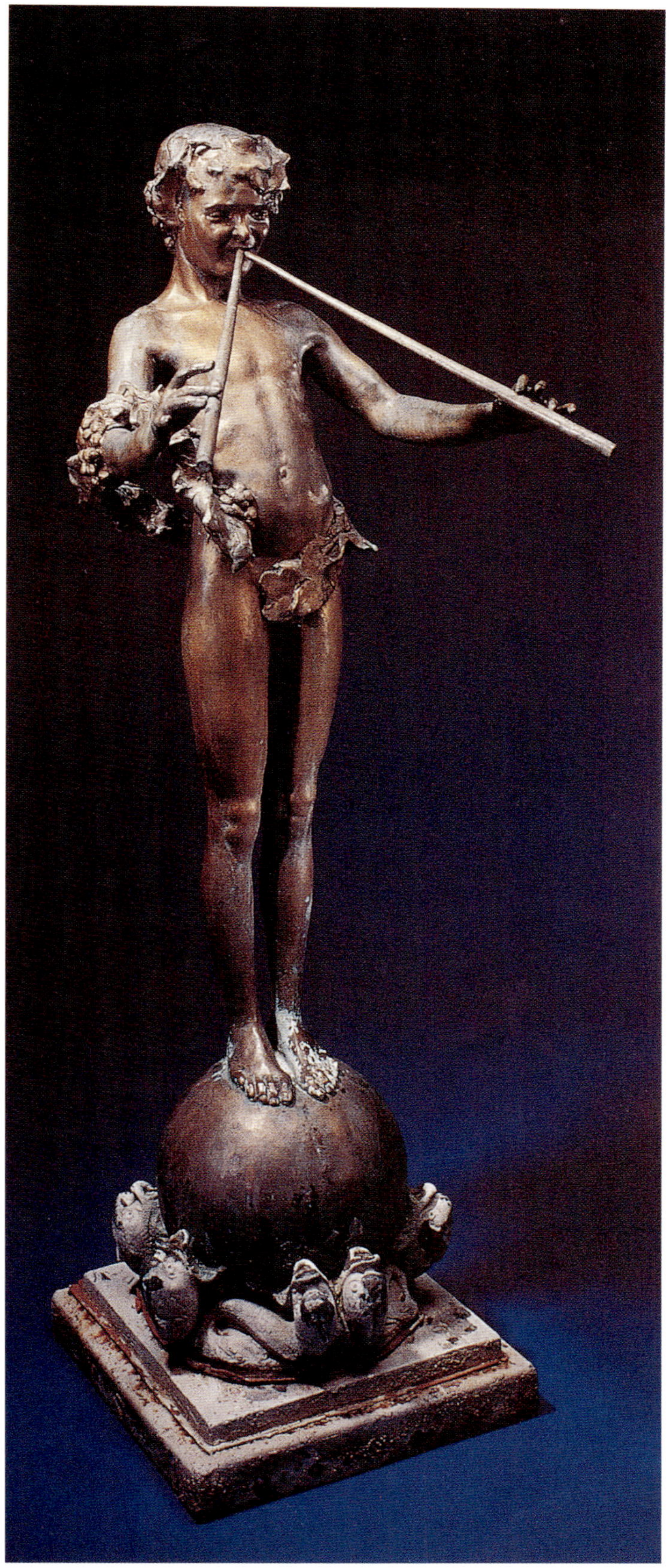

III-59. "Pan, a Bronze Fountain Figure" by Robert Aitken. Installed c. 1916 at Claypool Hotel, Indianapolis, later sold to private collector.

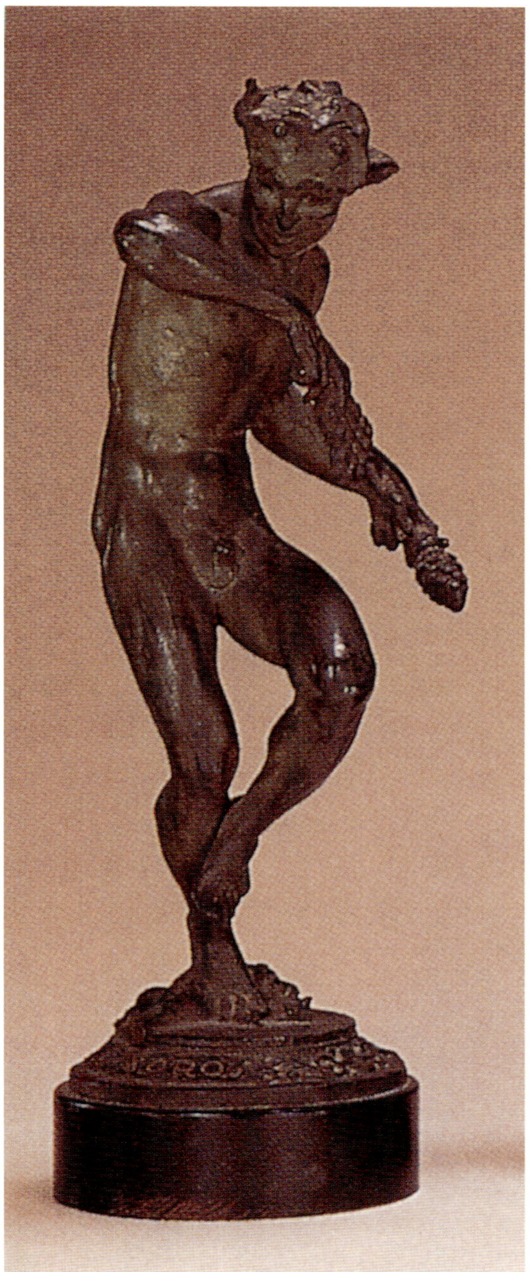

III-60. "Xoros" by Robert Aitken.

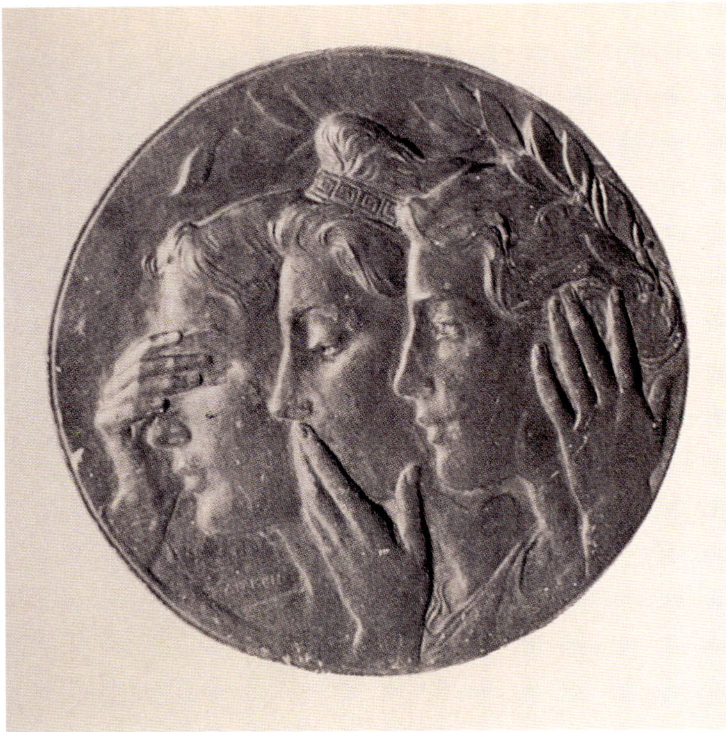

III-61. "See No Evil, Speak No Evil, Hear No Evil" plaque by Robert Aitken.

JO DAVIDSON (1883-1952)

Jo Davidson was a noted sculptor of portrait busts, making images of an extraordinary number of the well-known figures of his time. His subjects ranged from Gertrude Stein to Ghandi, from Franklin Roosevelt to Einstein. In modeling portraits from life, he had the opportunity to get to know each of his sitters—a vast panoply of the major rulers, business tycoons, thinkers, artists, and writers of the first half of the twentieth century. He was always interested in personality and greatness; his ability to capture the spirit as well as the appearance of his subjects made him a sought-after portraitist. "Portraiture became an obsession," he wrote. "Meeting and knowing people meant becoming acquainted with their thinking." His autobiography, "Between Sittings," describes the numerous encounters and conversations he had with his illustrious subjects.

Born in New York, Davidson went to Paris to study in 1907, and it was there that he discovered his gift for sculptural portraiture, though he also made other pieces. He sent seven sculptures and ten drawings to the Armory Show in 1913, and was soon sculpting such figures as President Wilson and a number of other World War I leaders. Other portraits followed in his busy career in France and America. Among his sitters were a series of American presidents, great writers ranging from James Joyce to H.G. Wells; ten South American presidents, and numerous business tycoons (Vanderbilt, Rockefeller, and Pulitzer); as well as noted Americans ranging from Clarence Darrow to Mayor La Guardia to Albert Einstein (numbers III-62 to III-71). Occasionally he made full figures, as in his statue of Will Rogers (III-72) and his "Pioneer Woman" (III-73).

Davidson's busts–some 200 pieces–were noted for their strength and sculptural vitality, as well as for capturing the sitter's innate personality; having recorded the faces of many of the century's most illustrious figures he became known as a "plastic historian." Davidson cast many of his portrait busts at Roman Bronze Works, beginning in 1906 and continuing into the 1940s.

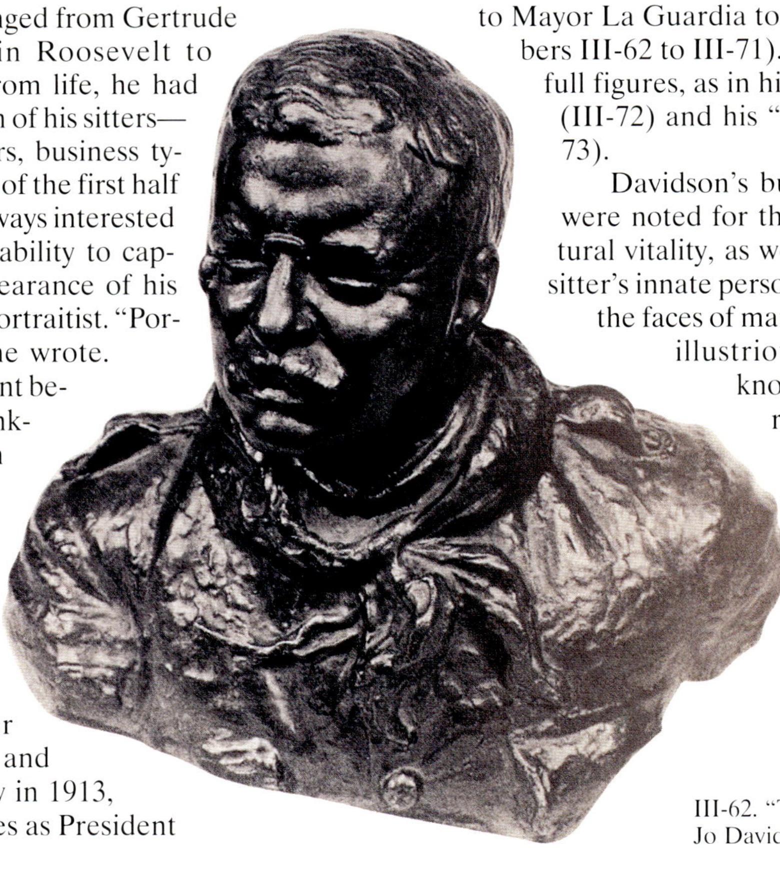

III-62. "Theodore Roosevelt" by Jo Davidson. c. 1909.

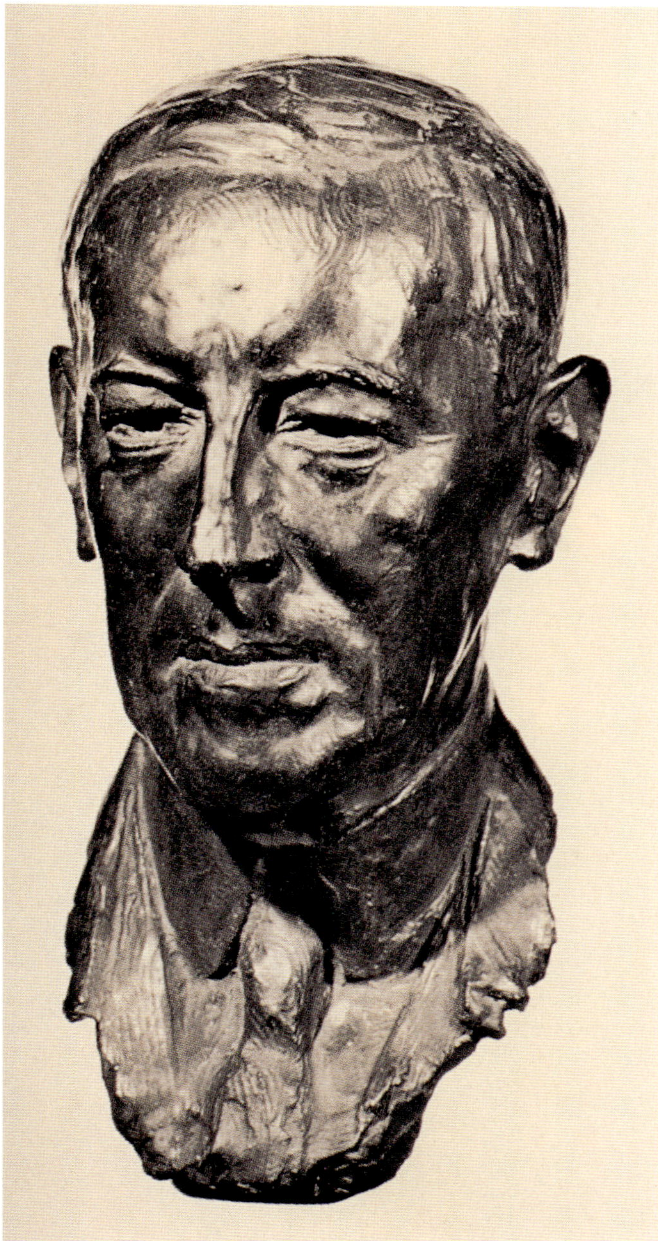

III-63. "Woodrow Wilson" by Jo Davidson. c. 1918.

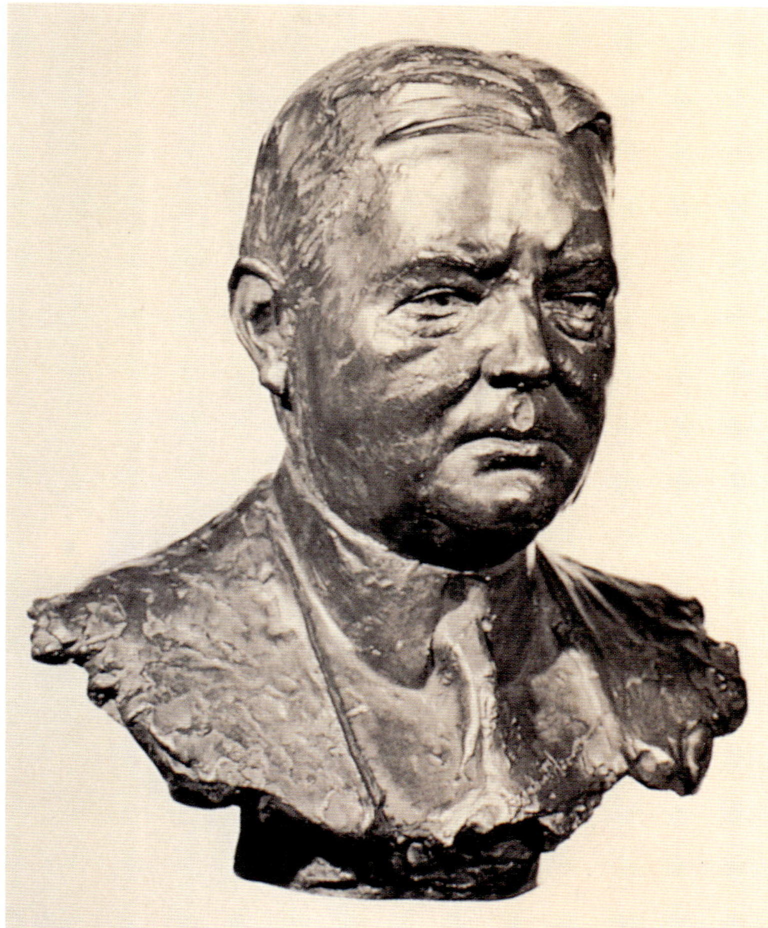

III-65. "Herbert Hoover" by Jo Davidson. 1929.

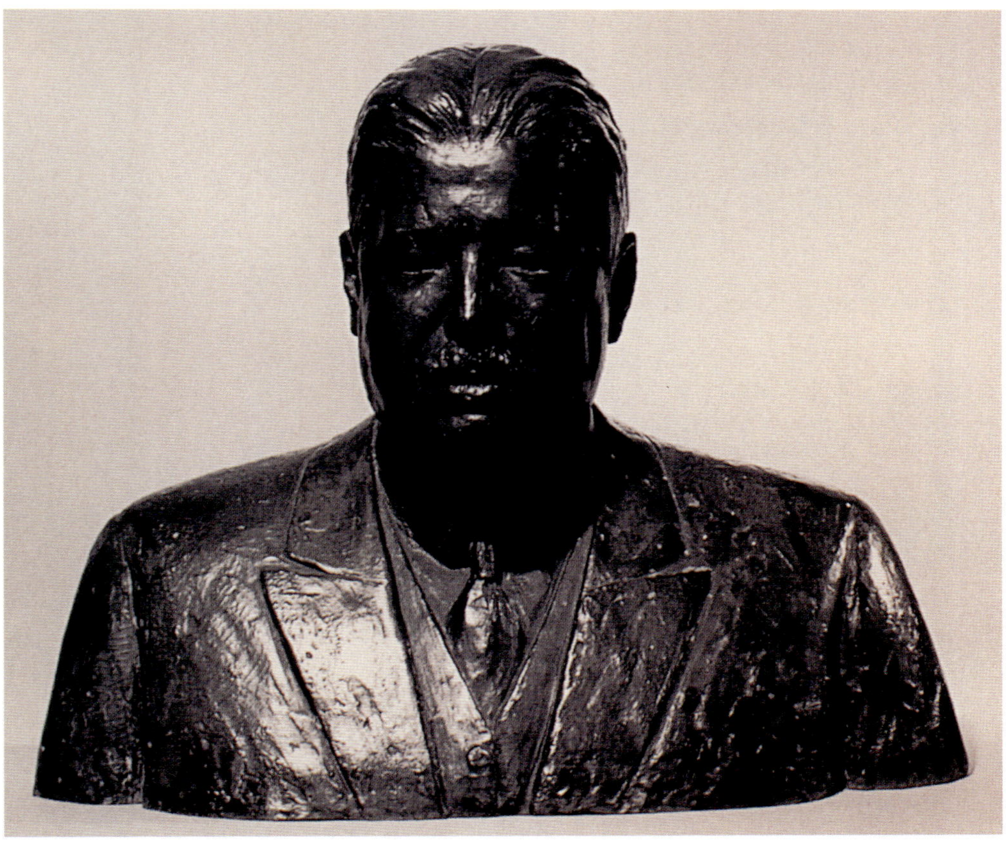

III-64. "Cornelius Vanderbilt" by Jo Davidson. 1925.

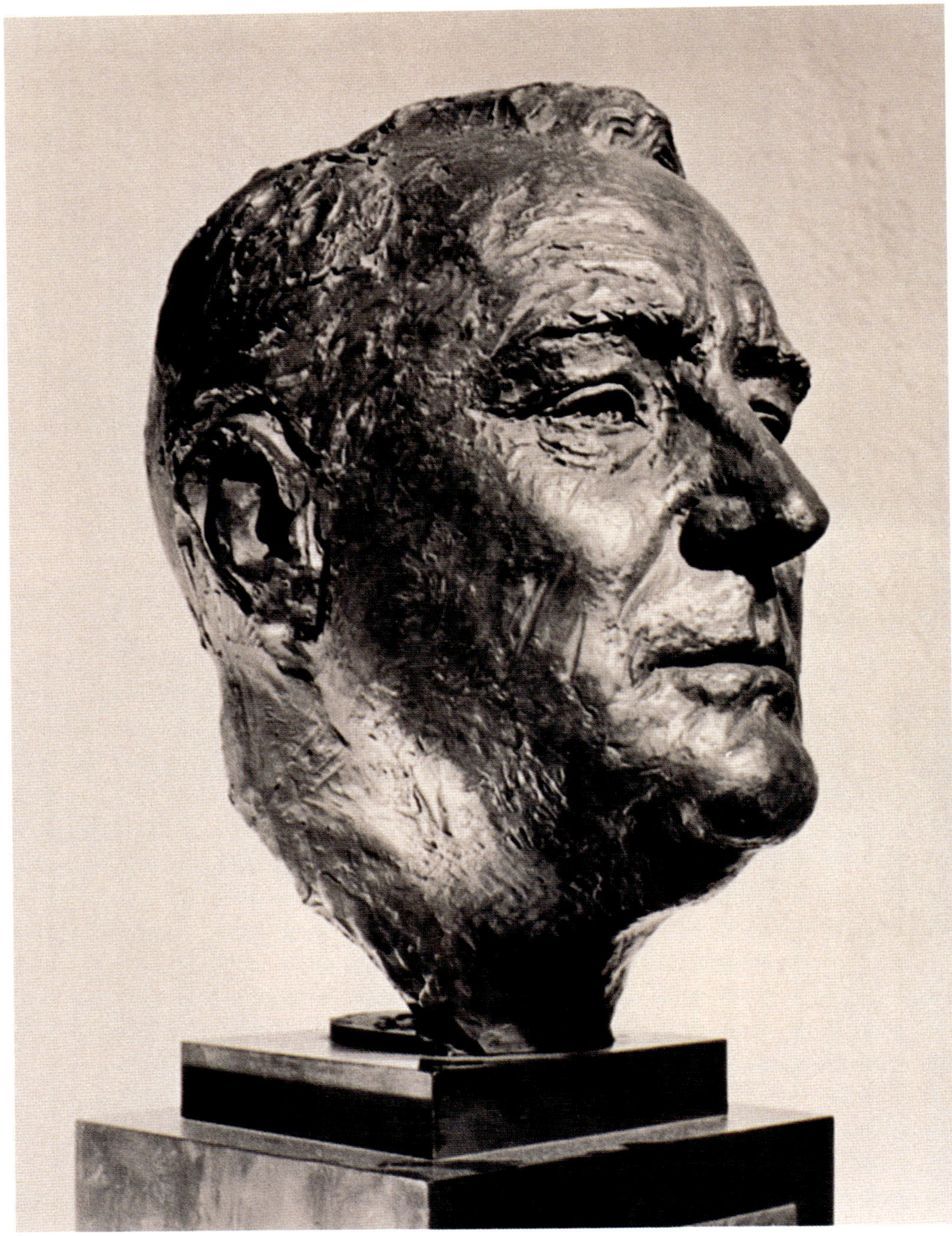

III-66. "Franklin Delano Roosevelt" by Jo Davidson. 1934.

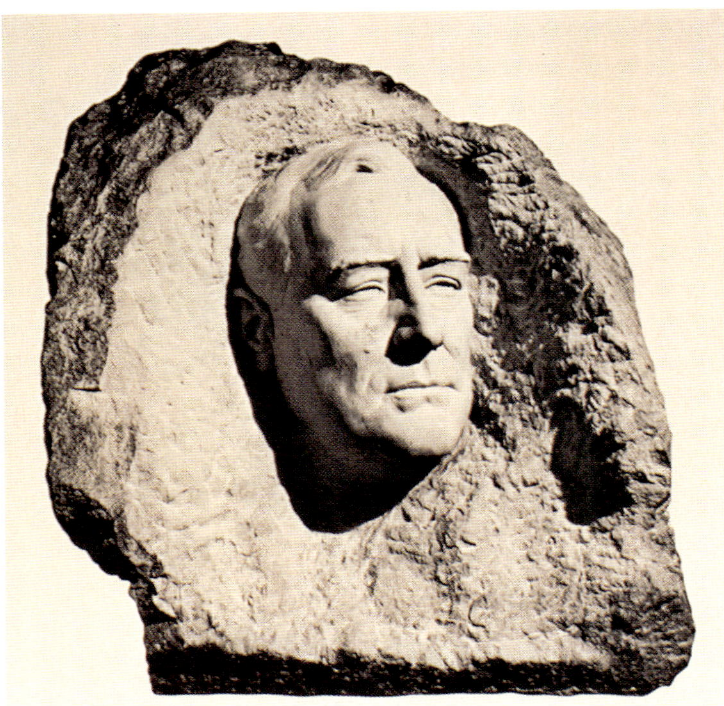

III-67. "Franklin Delano Roosevelt" (in stone setting) by Jo Davidson. 1951.

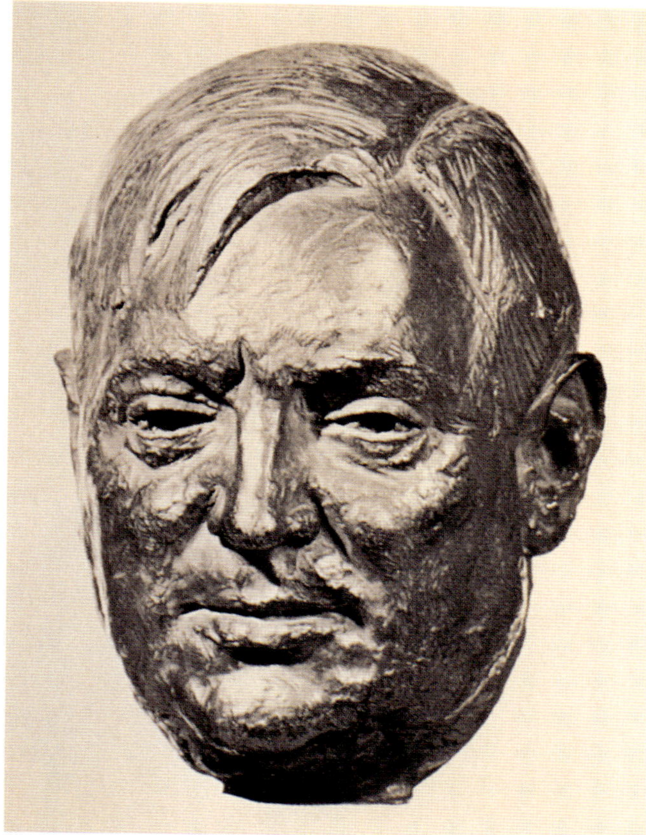

III-69. "Fiorello La Guardia" by Jo Davidson. c. 1933.

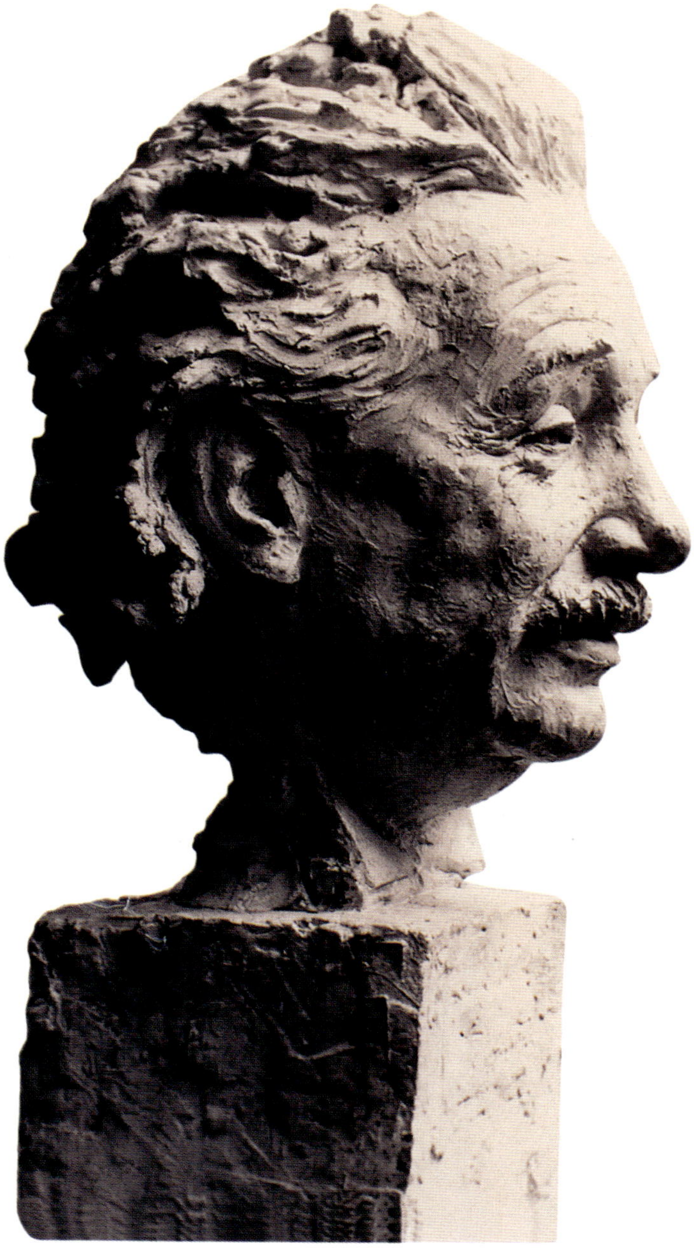

III-68. "Albert Einstein" by Jo Davidson. 1934.

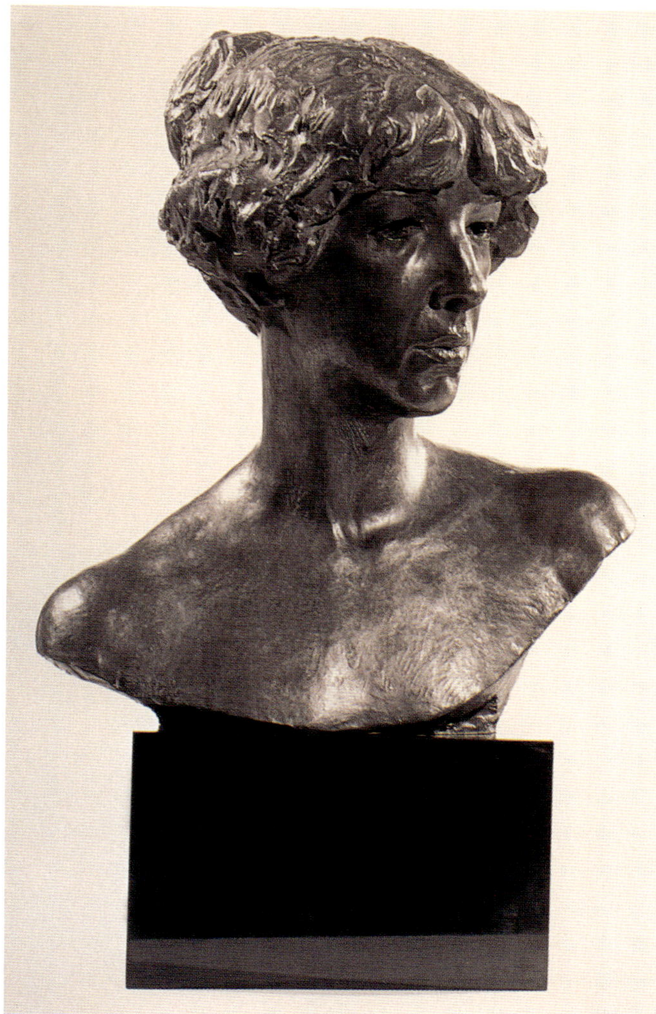

III-70. "Gertrude Vanderbilt Whitney" by Jo Davidson. 1917.

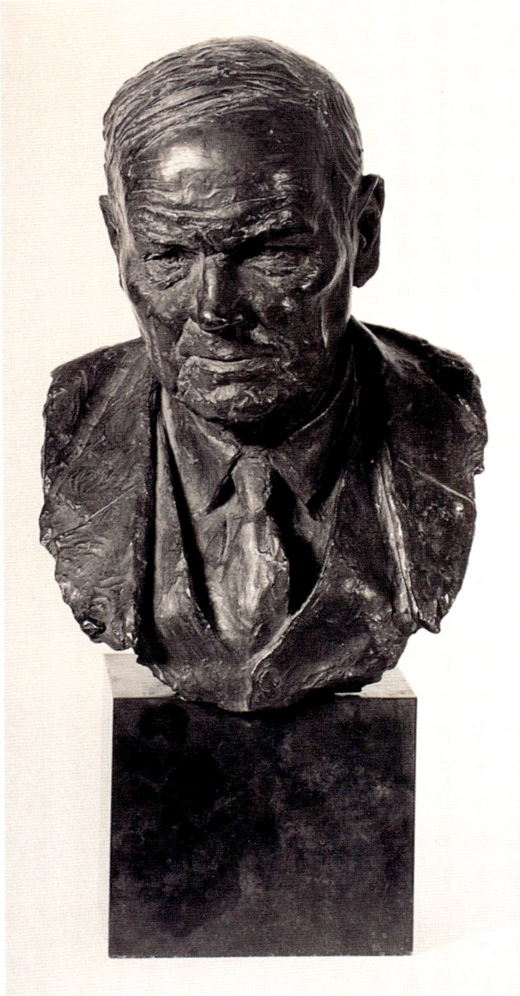

III-71. "Clarence Darrow" by Jo Davidson. 1929.

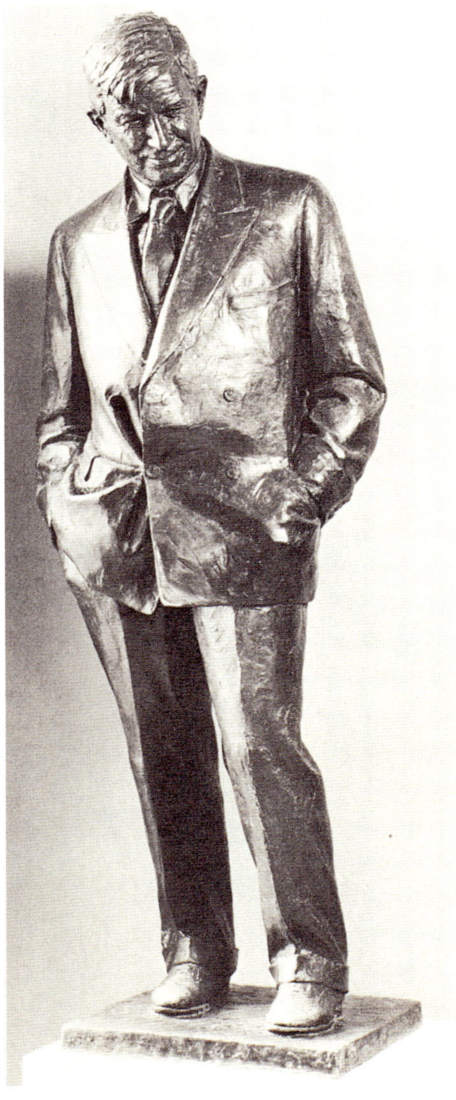

III-72. "Will Rogers," U.S. Capitol, by Jo Davidson. 1938-39.

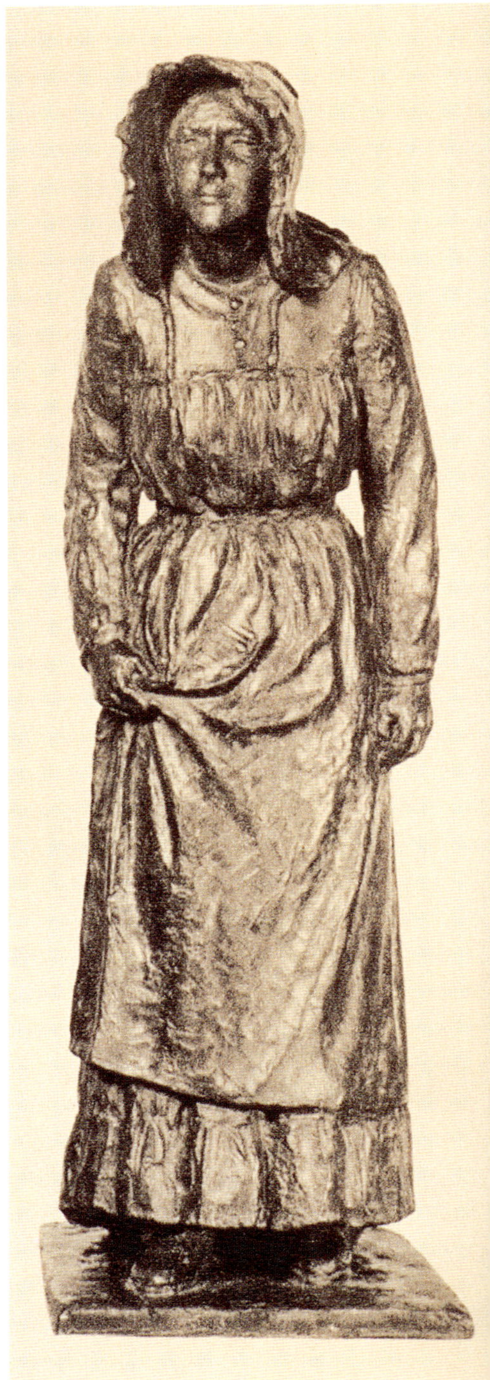

III-73. "Pioneer Woman" by Jo Davidson. 1927.

RUDOLPH EVANS
(1878-1959)

Best known for his heroic bronze statue of Thomas Jefferson at the Jefferson Memorial in Washington (III-74), Rudolph Evans also made a number of other important monuments including a portrait bust of Jefferson (III-75). Evans' idealized portraits of important Americans included images of Robert E. Lee (III-76), and Henry Wadsworth Longfellow. Some of his many portrait busts of notable Americans are in the Hall of Fame at New York University. He made a number of statues and busts of children.

Evans was born and died in Washington. In the United States he studied at the Corcoran School of Art and the Art Student's League. In Paris he was a student of several notable sculptors, including Rodin. Like several of his American colleagues who were students of Rodin, he reached for a certain immediacy of expression in his work. He won a bronze medal in the Paris Exposition in 1914, and the Watrous Gold Medal at the National Academy of Design in 1919. He won a competition to design the Jefferson statue for the Memorial in 1941.

Evans brought sculptures to Roman Bronze Works for casting between 1928 and the 1950s.

III-74. "Thomas Jefferson," Jefferson Memorial, Washington, D.C. by Rudolph Evans. 1938-1943; dedicated on the 200th anniversary of Jefferson's birth.

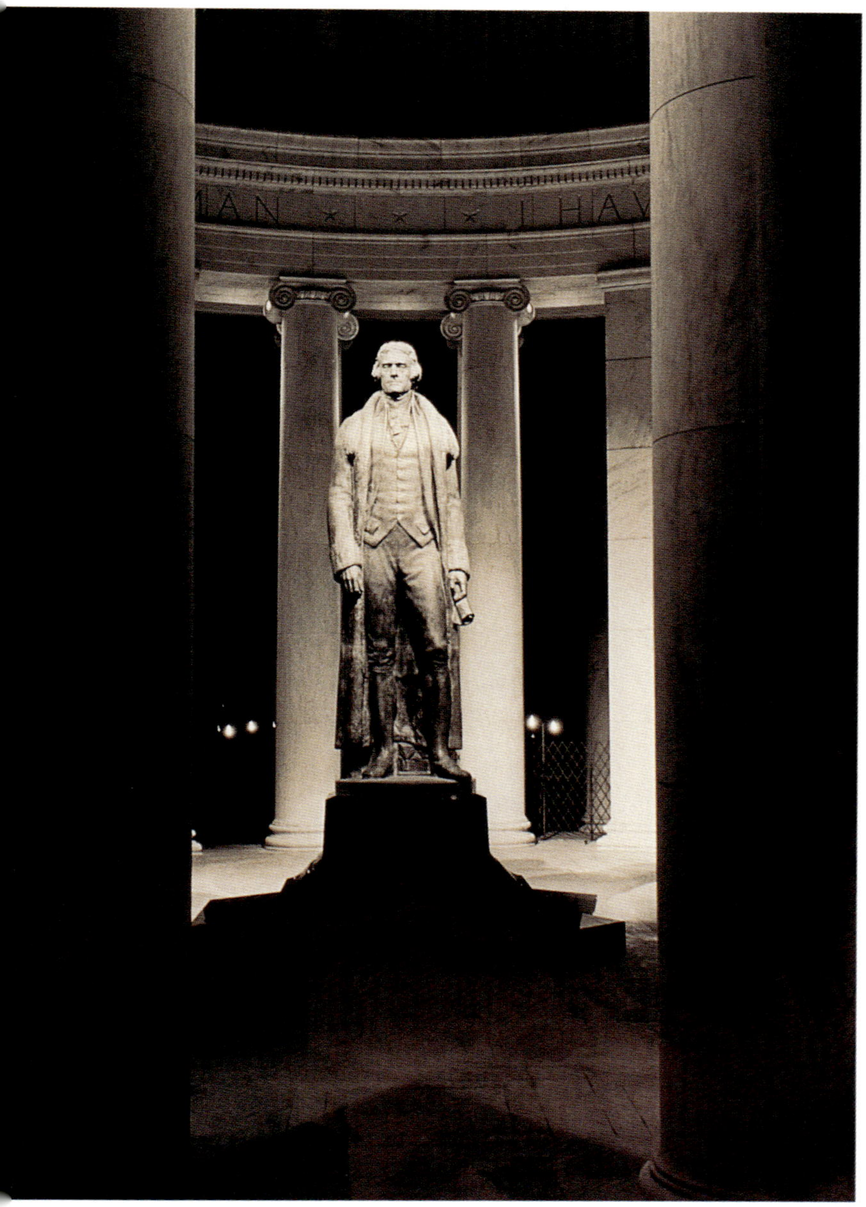

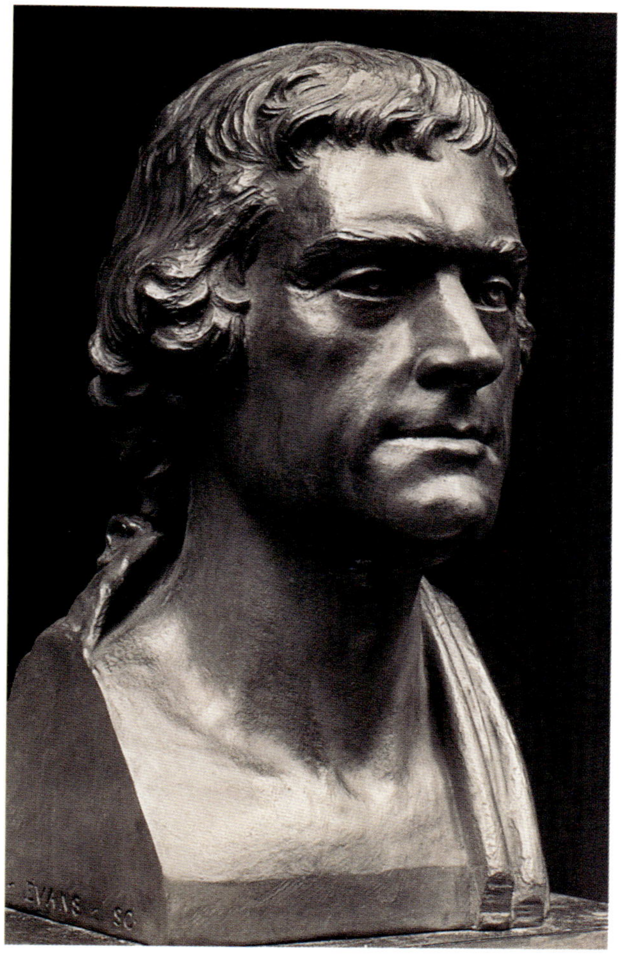

III-75. "Bust of Jefferson" by Rudolph Evans

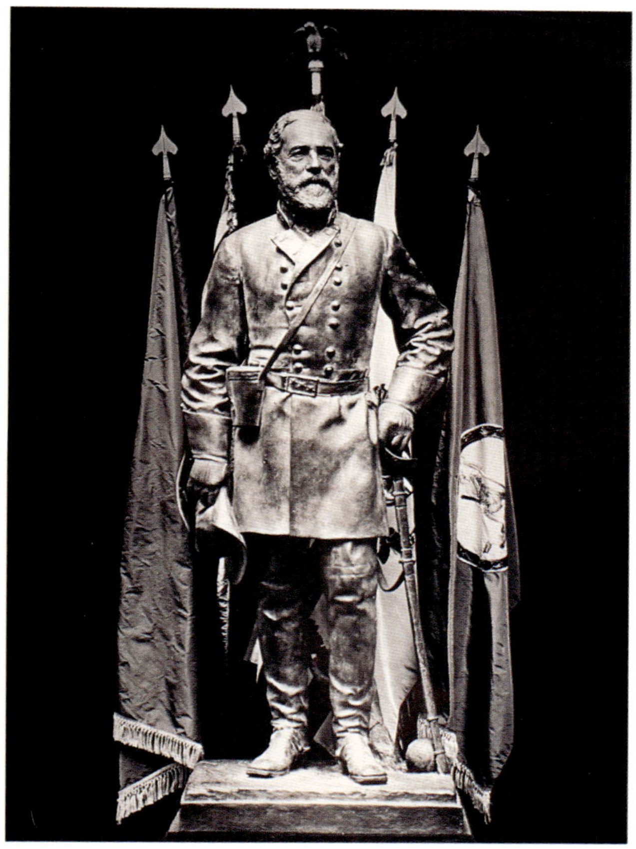

III-76. "Robert E. Lee," in the Capitol, Richmond, Virginia, by Rudolph Evans

CHARLES KECK
(1875-1951)

Charles Keck is noted for his many likenesses of Abraham Lincoln (III-77 and 78) and other major commemorative portraits of notable Americans. His historical monuments and portrait busts in bronze and marble were highly regarded and he received many commissions. He made a portrait bust of President Truman (III-79), now in the United States Capitol in Washington. New York City sculptures include images of Governor Alfred E. Smith, Father Duffy (in Times Square), and imposing allegorical figures for Columbia University. Other important works included statues of Stonewall Jackson, Booker T. Washington, and Andrew Jackson. In addition to his statues, he designed architectural sculpture and medals.

Keck was born in New York, and was a student of Philip Martiny and of Augustus Saint-Gaudens, with whom he worked from 1893 to 1898. He continued his studies in Europe, including a stay at the American Academy in Rome and travels in Florence, Italy, Greece, and Paris. In 1904 he opened a studio in New York, receiving an early commission for a figure representing "The Genius of Islam" for the façade of the Brooklyn Museum. The recipient of a number of awards, he was a member of the National Academy of Design, the American Numismatic Society, the Architectural League of New York, and the National Sculpture Society.

Keck first cast works at R.B.W. in 1906, and continued to cast there until his death in 1951.

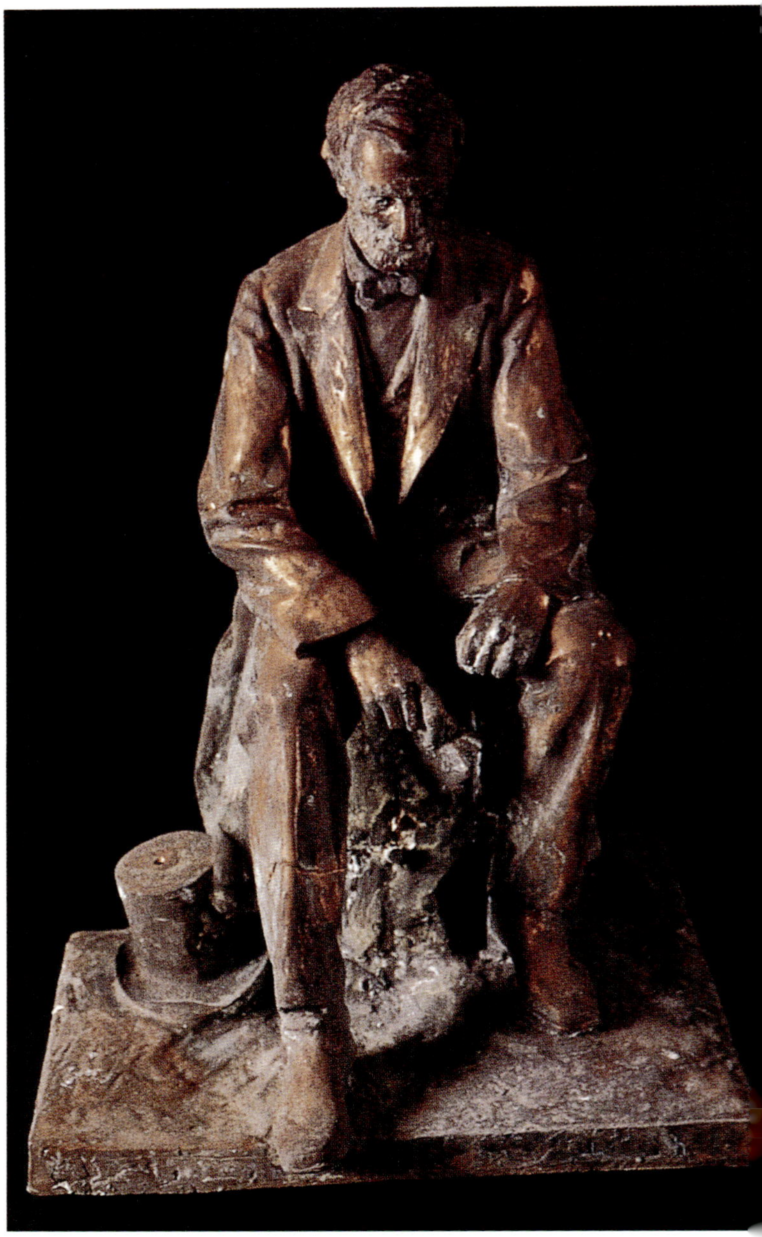

III-77. "Young Lincoln" by Charles Keck. This is a sketch for the large seated Lincoln in the Chicago Library. Cast c. 1950 by R.B.W.

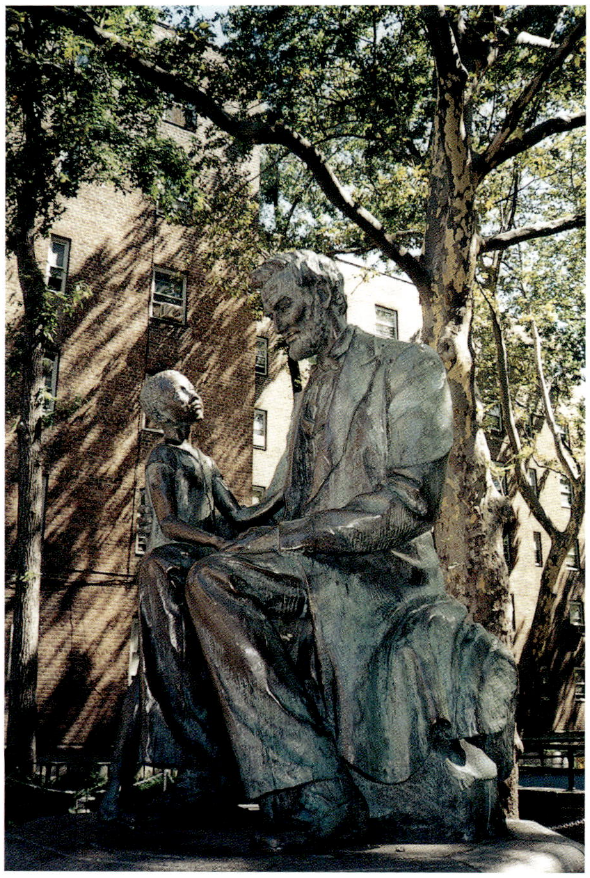

III-78. "Abraham Lincoln and Child," New York City by Charles Keck. 1948.

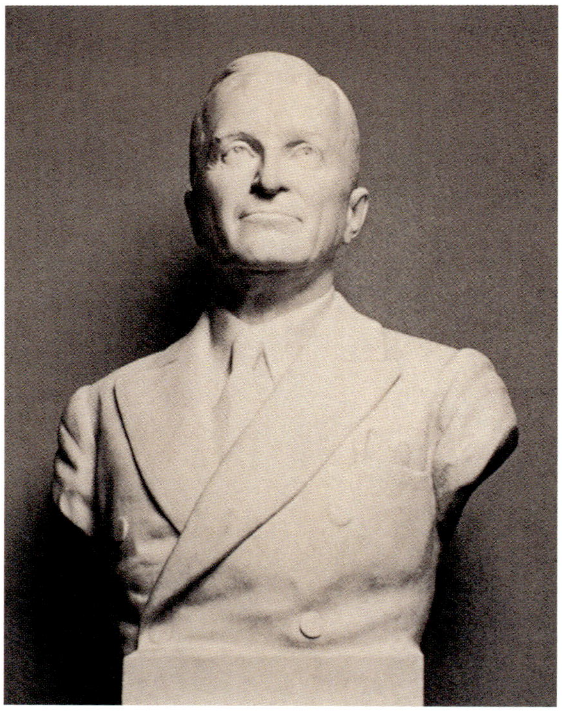

III-79. "Bust of Truman as Vice-President," U.S. Capitol, by Charles Keck. 1961.

CHARLES GRAFLY
(1862-1929)

Noted for his likenesses of other artists, as well as several large monuments, Charles Grafly was the recipient of many awards. His best known work was "Vulture of War" (III-80), modeled in Paris in 1896, and widely recast and exhibited in the United States. Among his portrait subjects were Thomas Anshutz (III-81), and Childe Hassam. He also made a portrait bust of Hermann Kotschmar (III-82), and a bust (now in the Hall of Fame for Great Americans at New York University) of John Paul Jones (III-83). Major monuments include the Pioneer Mother (III-84) and the General Meade Memorial in Washington.

Charles Grafly was born and died in Philadelphia. After apprenticeship to a stonecutter, he studied in Paris at the Académie Julian and the Ecole des Beaux-Arts from 1888 to 1892. In that year he returned to the United States, opened a studio in Philadelphia, and became a teacher at the Pennsylvania Academy of Fine Arts and Drexel Institute; he was known in the 1920s as the leading teacher of sculpture of his time. As fellow sculptor Lorado Taft commented in 1925, "Grafly still rules the destinies of young sculptors. To be known as America's best teacher is a sufficiently proud distinction, but to be recognized as our greatest master of portraiture is likewise Mr. Grafly's due." He exhibited in and won prizes at expositions in Paris, Charleston, Buffalo, and Buenos Aires. Grafly was a member of the National Sculpture Society and the National Institute of Arts and Letters.

Grafly cast works at R.B.W. beginning in 1907 and continuing until the end of his life.

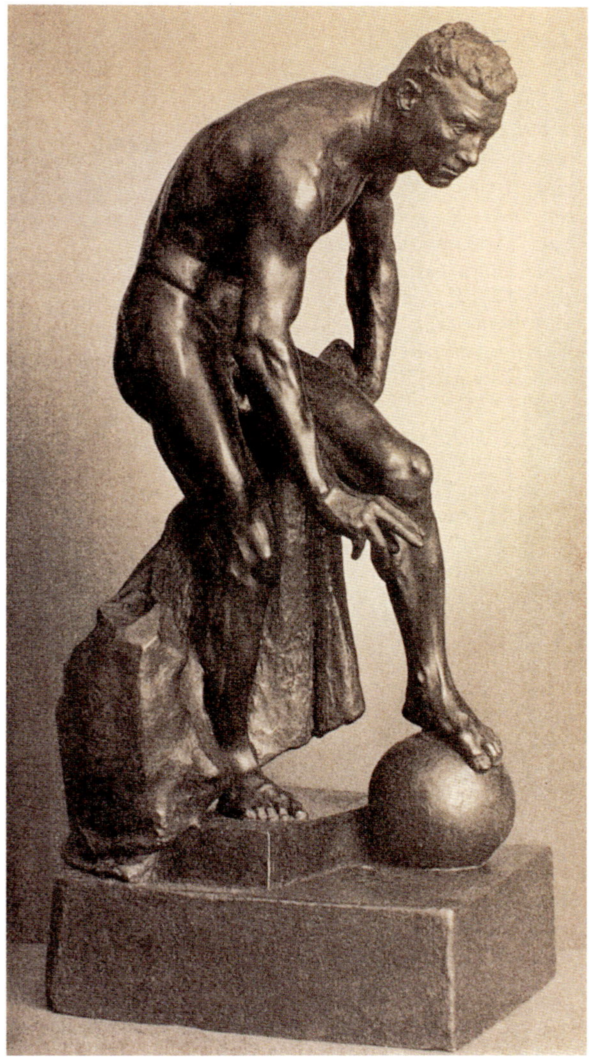

III-80. "Vulture of War" by Charles Grafly. Modeled 1896; R.B.W. 1930.

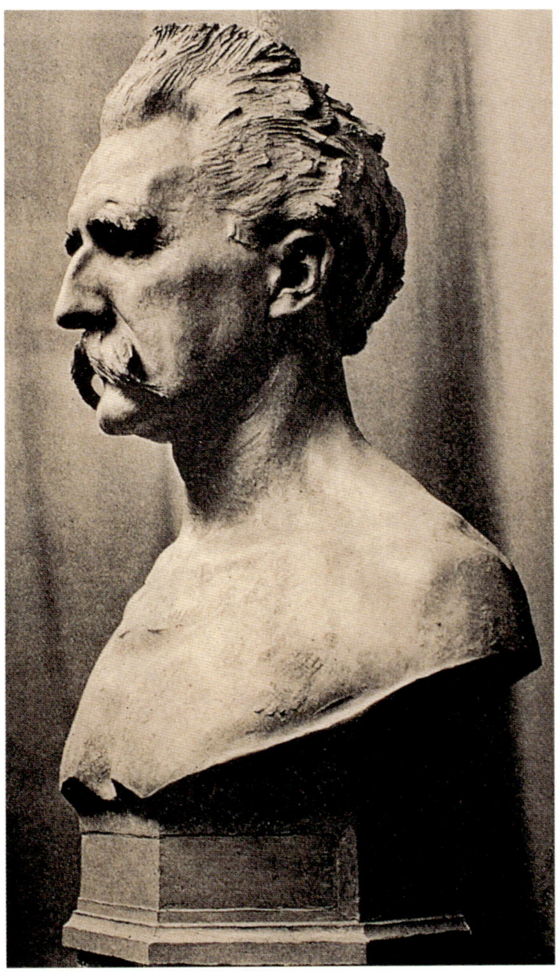

Left:
III-82. "Bust of Hermann Kotszchmar," Portland, Maine by Charles Grafly. 1912.

Right:
III-81. "Thomas P. Anshutz" by Charles Grafly. 1913.

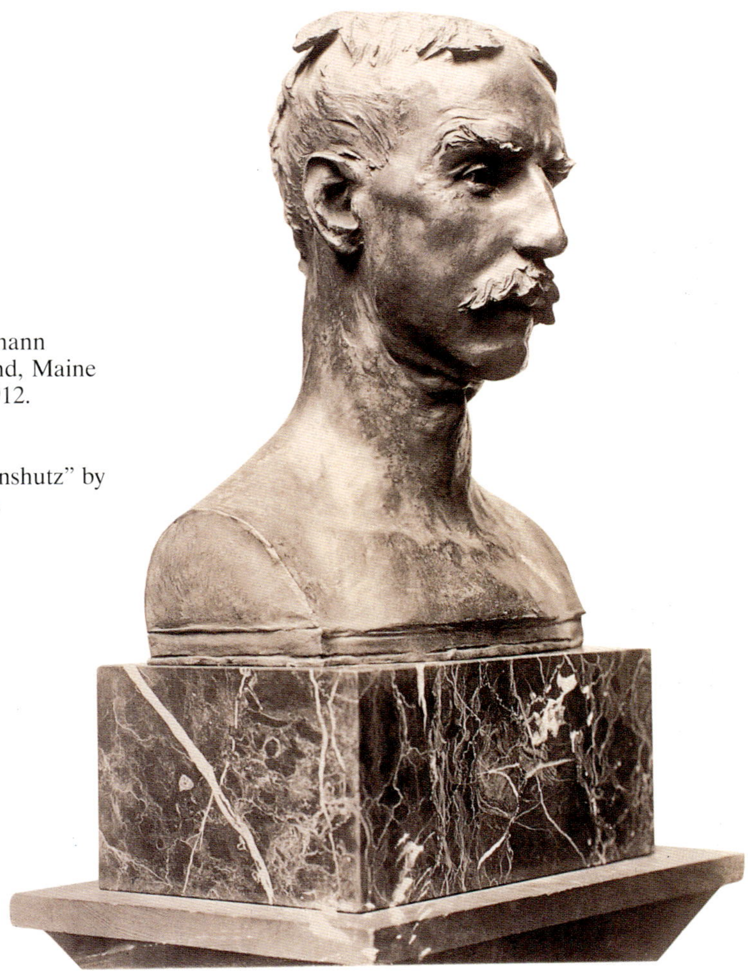

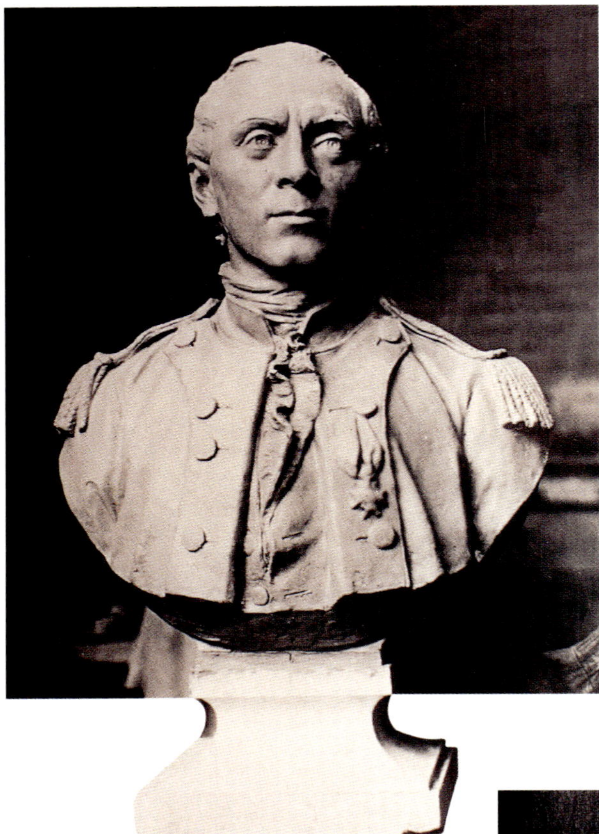

III-83. "John Paul Jones," Hall of Fame for Great Americans at New York University, by Charles Grafly. 1916.

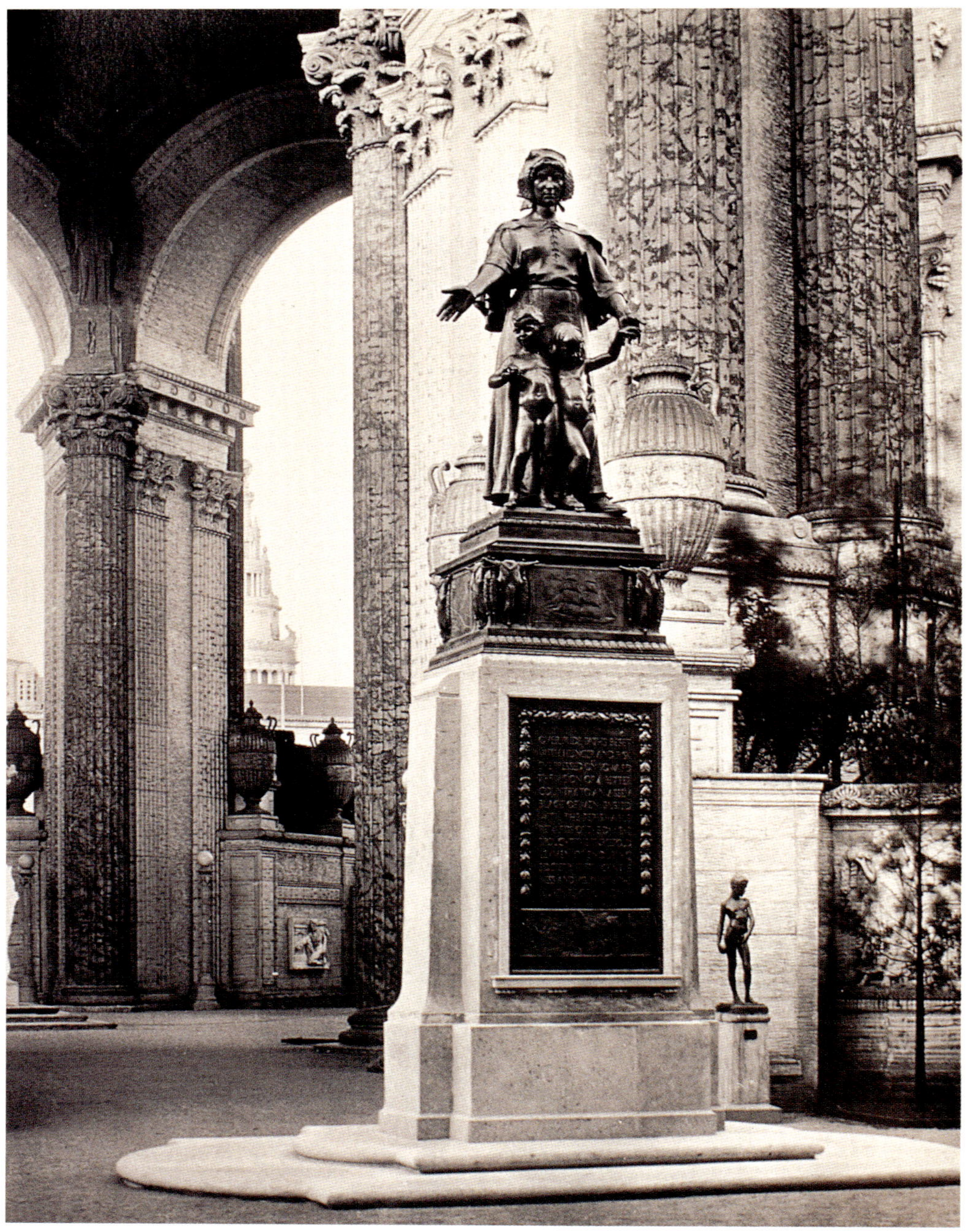

III-84. "Pioneer Mother" by Charles Grafly. 1915.

WILLIAM ORDWAY PARTRIDGE (1861-1930)

William Ordway Partridge was an important creator of monuments and commemorative statuary, particularly in New York City. He was noted for his images of Alexander Hamilton and Thomas Jefferson, as well as for numerous portrait busts. Among his New York works was a Pieta at St. Patrick's Cathedral and a tomb for Joseph Pulitzer at Woodlawn Cemetery.

Partridge published a number of articles on American sculpture, and several books of poetry. He taught or lectured at several universities. He had a more international view of art than many of his colleagues and his sculpture was occasionally spontaneously impressionistic, with only partially realized forms and rough surfaces.

Born in Paris of American parents, he was educated in New York and in Rome and Florence. His career began with the exhibition of several portrait busts at the Columbian Exposition in 1893; he soon had commissions for an equestrian statue of General Grant, and for statues of Hamilton and Jefferson (III-85), and Governor Samuel J. Tilden of New York (III-86). He made numerous portrait busts of notable figures, including Beethoven, Shelley, Richard Wagner, Horace Greeley, and Alfred Lord Tennyson (III-87).

The accidental destruction of all of his plaster casts (by having coal dumped on them), and increasingly ill health made his last years difficult. Partridge cast his bronzes at Roman Bronze Works between 1912 and 1918.

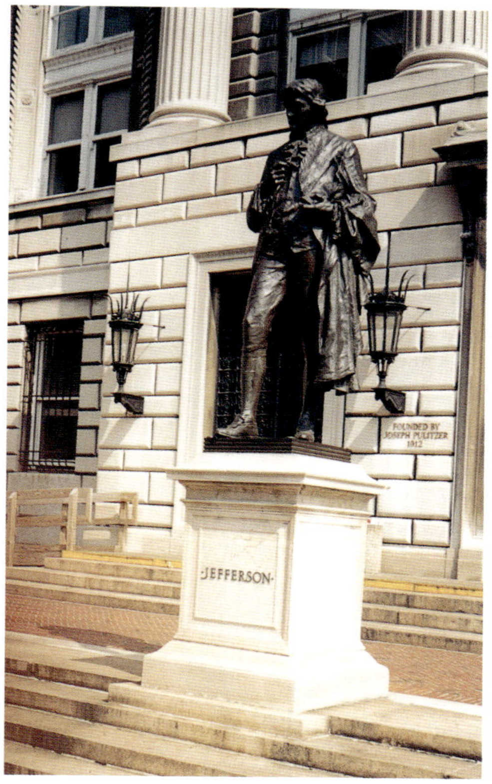

III-85. "Thomas Jefferson," Columbia University, by William Ordway Partridge. 1914.

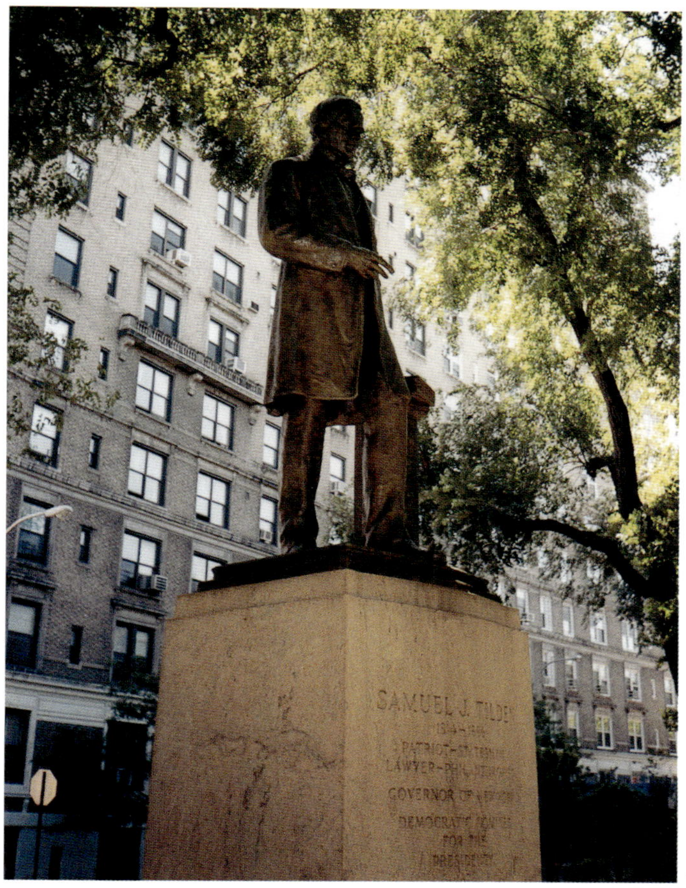

III-86. "Samuel J. Tilden," New York City, by William Ordway Partridge. 1926.

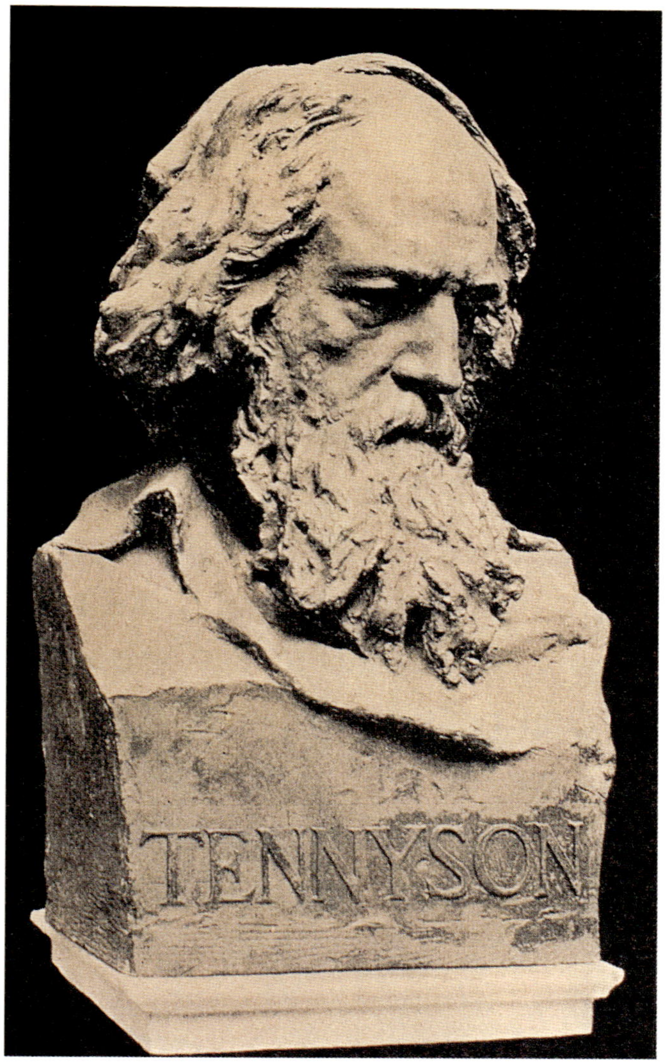

III-87. "Bust of Alfred Lord Tennyson" by William Ordway Partridge. 1913.

JOHN MASSEY RHIND (1860-1936)

John Massey Rhind, a third generation sculptor and a native of Scotland, came to the United States in 1889, having already studied in England and France. His ability as an architectural sculptor brought him numerous commissions in his first decade in New York when the Beaux-Arts style increasingly incorporated sculptural ornamentation. Rhind (along with Karl Bitter and Charles Henry Niehaus) was a winner of a competition to design the new doors of the Trinity Church. His success at creating the sculptural figures there led to more commissions for architectural designs.

But he also excelled at commemorative portraits. Following the unveiling of his impressive statue of John C. Calhoun in the 1890s, he was asked to make many portraits of notable Americans; among them were images of George Washington, Theodore Roosevelt (III-88), William McKinley (III-89), Pieter Stuyvesant (III-90), Elihu Root, and William Howard Taft.

His output was vast; he also created a variety of monuments, allegorical statues, and even a well-known Indian figure, "On the Lookout" (III-91). Though Rhind lived until 1936, his work retained its Beaux-Arts style of academic naturalism that rooted it firmly in tradition.

At R.B.W., Rhind cast many works ranging from portrait busts (Carnegie, Burns, Ives, Hanna, Taft) to allegorical figures representing "Authority" and "Wisdom," to a sarcophagus. Rhind's name appears in the ledgers between 1903 and 1930.

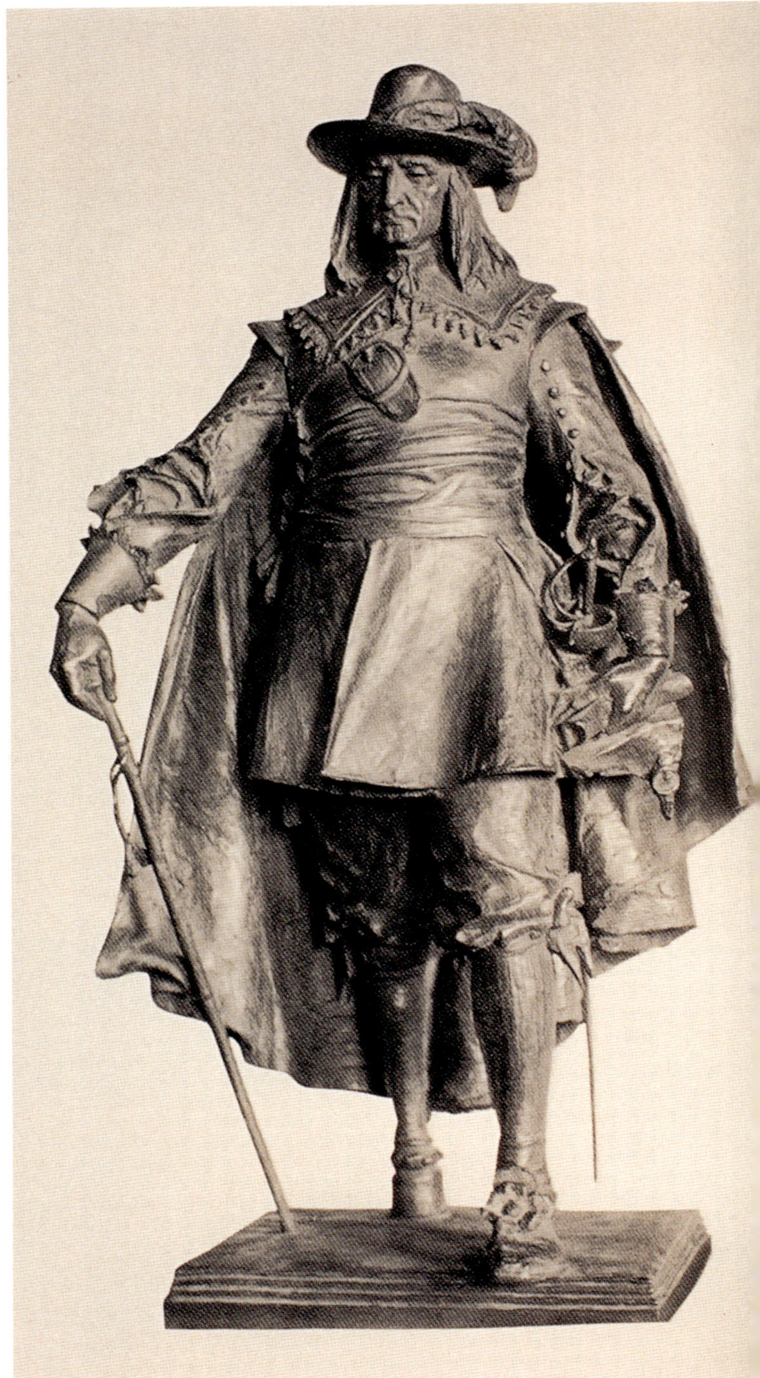

III-90. "Pieter Stuyvesant" by John Massey Rhind. 1913.

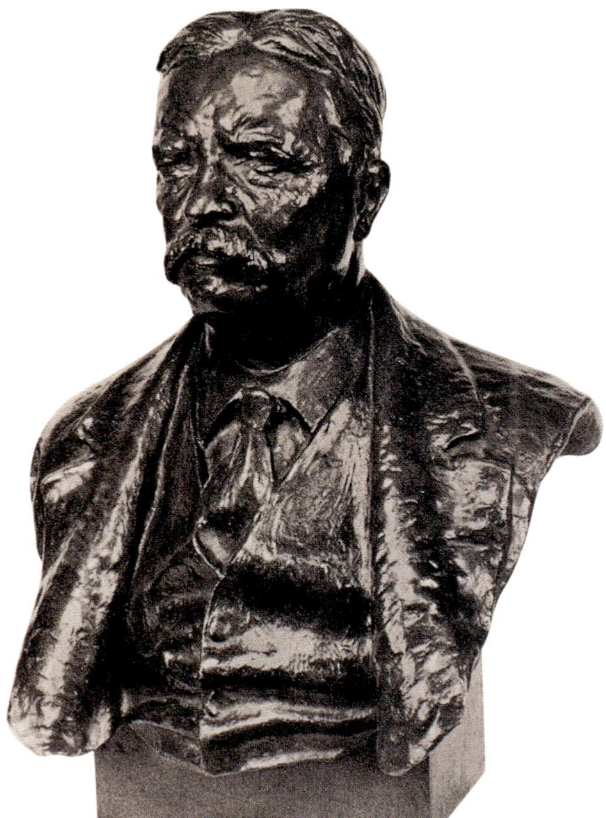

III-88. "President Teddy Roosevelt" by John Massey Rhind. c. 1915.

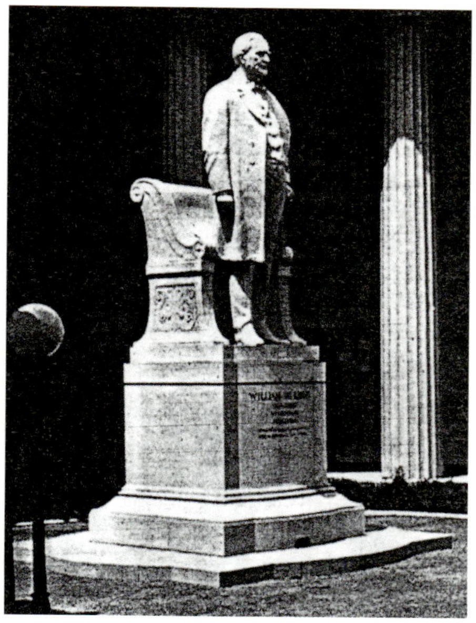

III-89. "William McKinley," Niles, Ohio by John Massey Rhind. 1916.

ADOLPH ALEXANDER WEINMAN (1870-1952)

Adolph Alexander Weinman was a creator of monuments, architectural reliefs, and medals, as well as a notable sculptor of Indian subjects (see page 105). As a portrait artist, Weinman made several different images of President Lincoln, including his "Seated Abraham Lincoln" (III-92). He also made portrait statues of many other well known individuals, including Alexander Hamilton, the

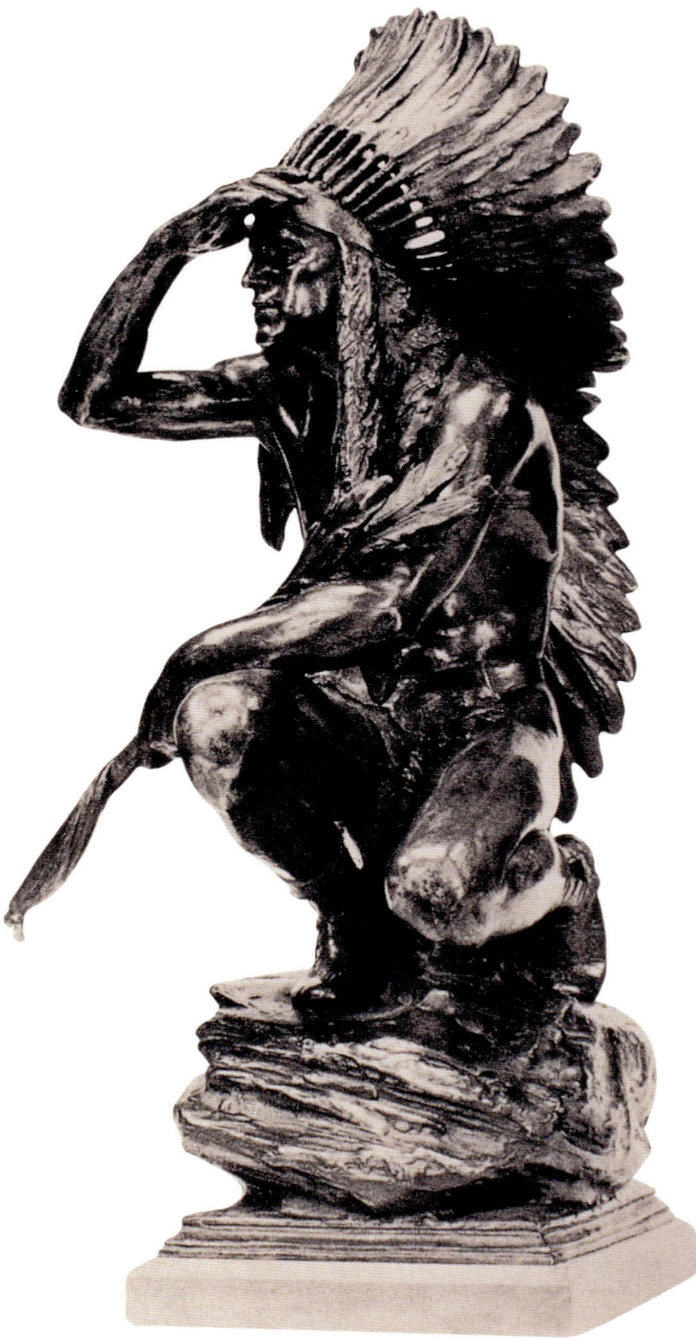

III-91. "Lookout" by John Massey Rhind. 1919.

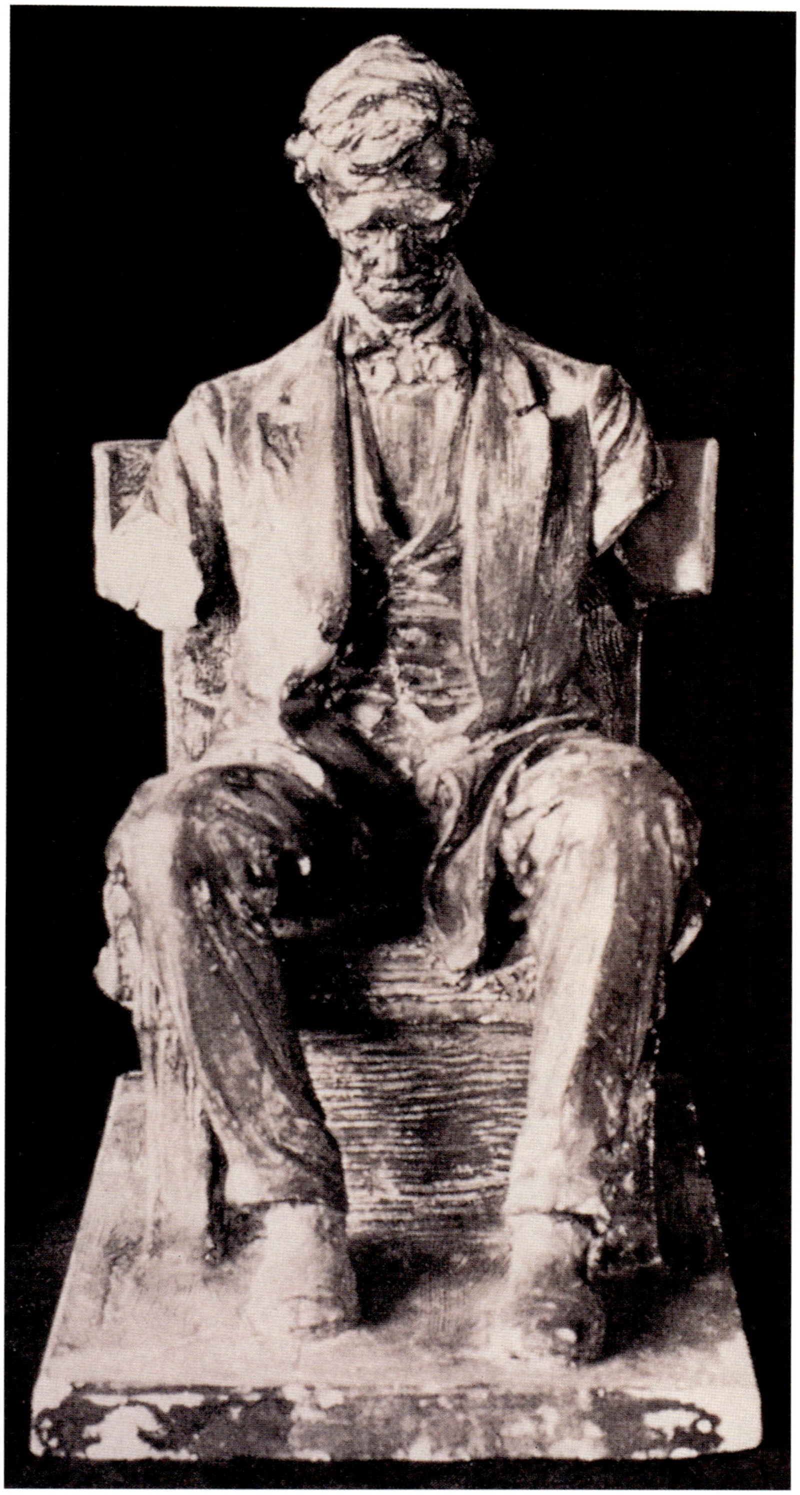

III-92. "Seated Lincoln" (plaster cast) by Adolph Weinman. 1912-1914.

Governor of New York DeWitt Clinton (III-93), and Pennsylvania Railroad President Samuel Rea (III-94), all in New York, as well as several memorials to generals and governors and Civil War heroes. His work as a monument and portrait artist was noted for its straight-forward realism and careful modeling; his Indian images show a more personal and empathetic approach.

Weinman was born in Germany, arriving in America in 1880. He was apprenticed to a carver of wood and ivory, and enrolled as a student at Cooper Union. He became an assistant to Philip Martiny, accompanying him to Chicago's Columbia Exposition in 1893, and assisted a number of leading sculptors, including Daniel Chester French and Augustus Saint-Gaudens. By 1894 he had embarked on his own career with his first Indian sculpture for the St. Louis Exposition. Weinman's architectural reliefs brought him fame and many commissions; among his commissions for reliefs or pediments were important buildings in New York and Washington (including the U.S. Archives and the Post Office Building), and the state capitols of Wisconsin and Missouri.

He also made a variety of symbolic and idealistic works, including "Narcissus" (III-95), "The Duet" (III-96), and companion pieces "Descending Night" (III-97) and "Rising Day" (III-98). Other allegorical works listed in R.B.W.'s ledgers include "Heroic Courage," "Head of Victory," and a "Winged Figure." His bronzes, including "Putting the Shot" (or "Ball Pitcher") (III-99) were cast at Roman Bronze Works between 1902 and the 1950s.

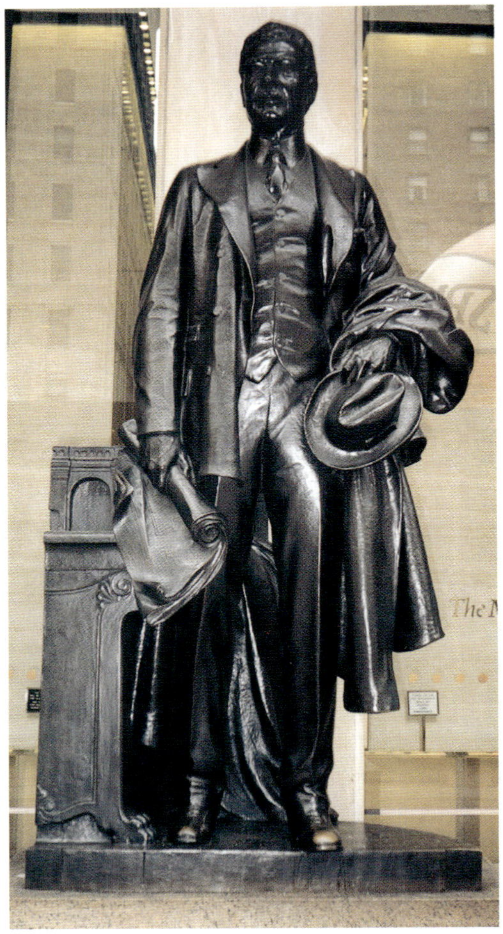

III-94. "Samuel Rea" by Adolph Weinman. c. 1910.

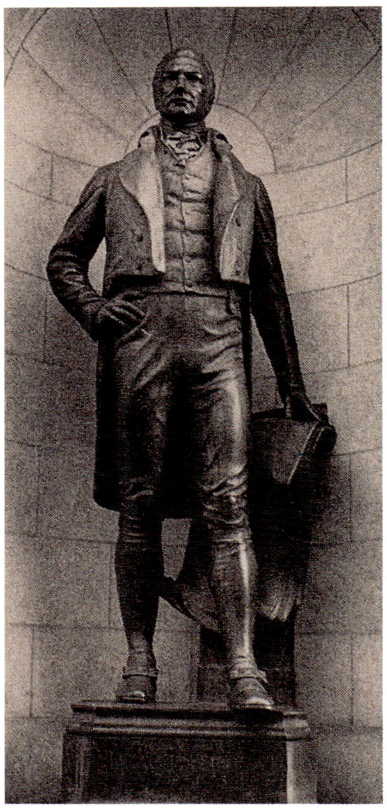

III-93. "DeWitt Clinton" by Adolph Weinman.1941. This statue was cast in nickel silver.

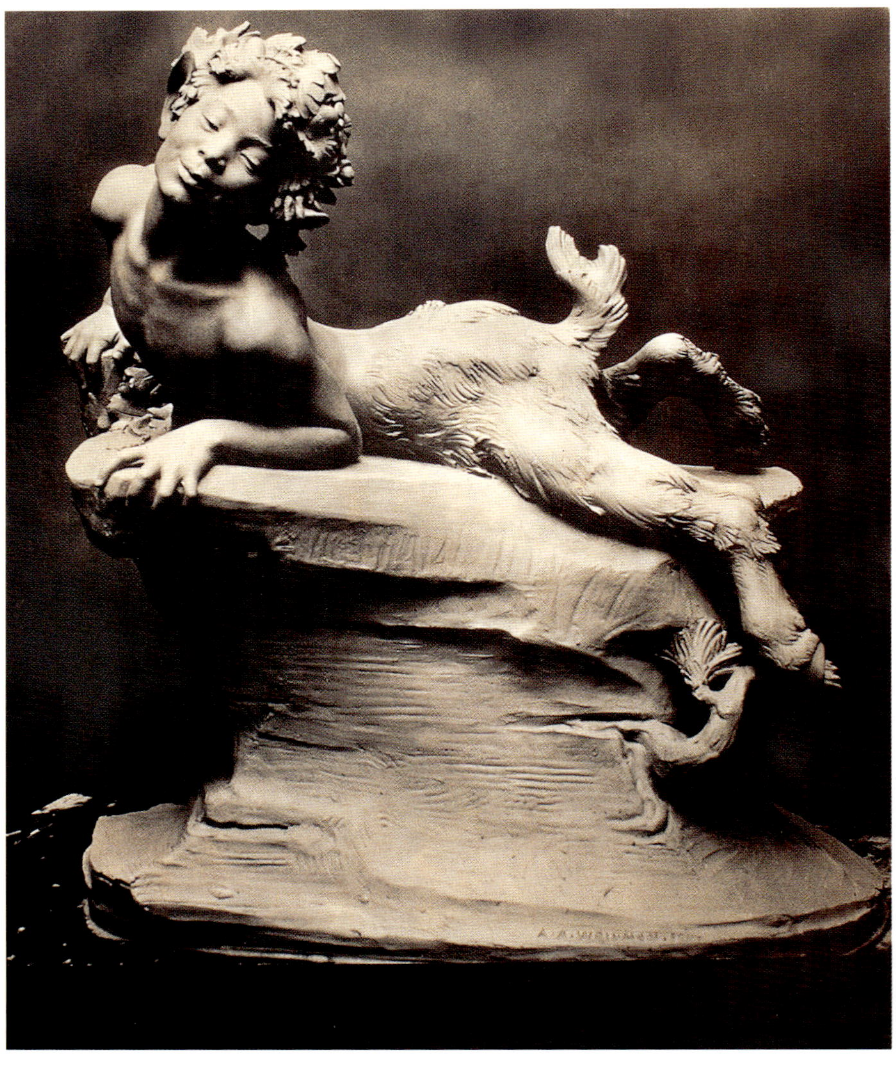

III-95. "Narcissus," a garden pool figure by Adolph Weinman. 1923.

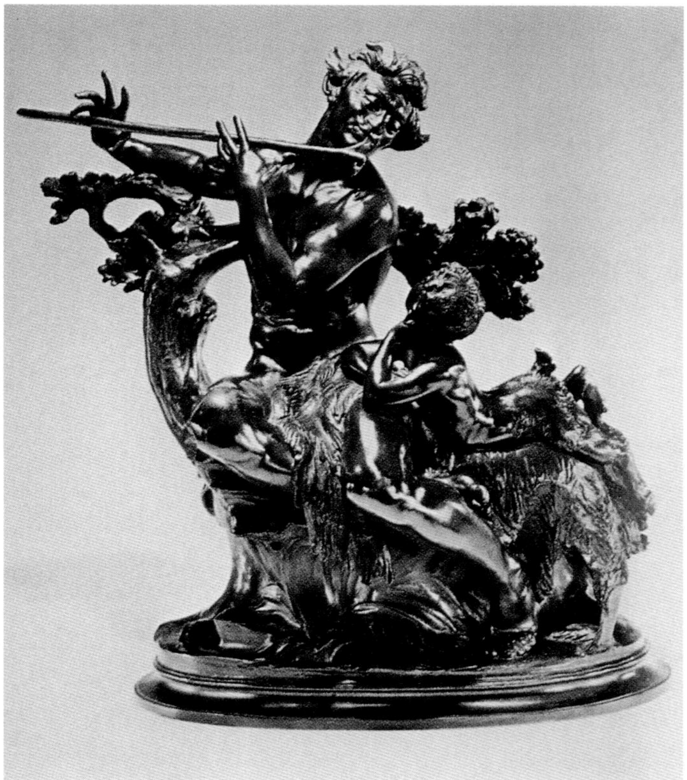

III-96. "The Duet" by Adolph Weinman. 1924.

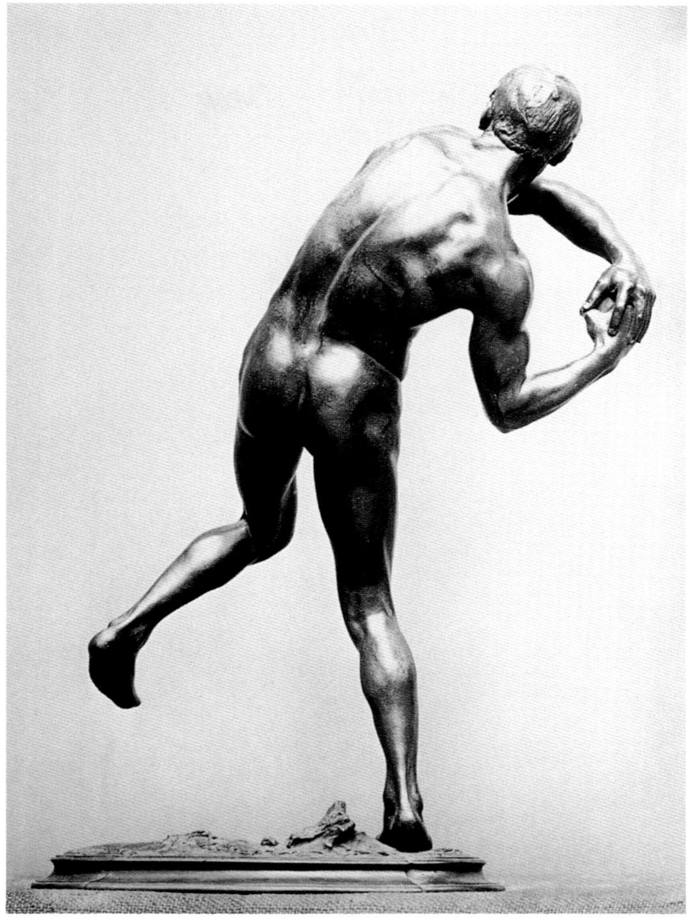

III-99. "Ball Pitcher" (or "Putting the Shot") by Adolph Weinman. c. 1910.

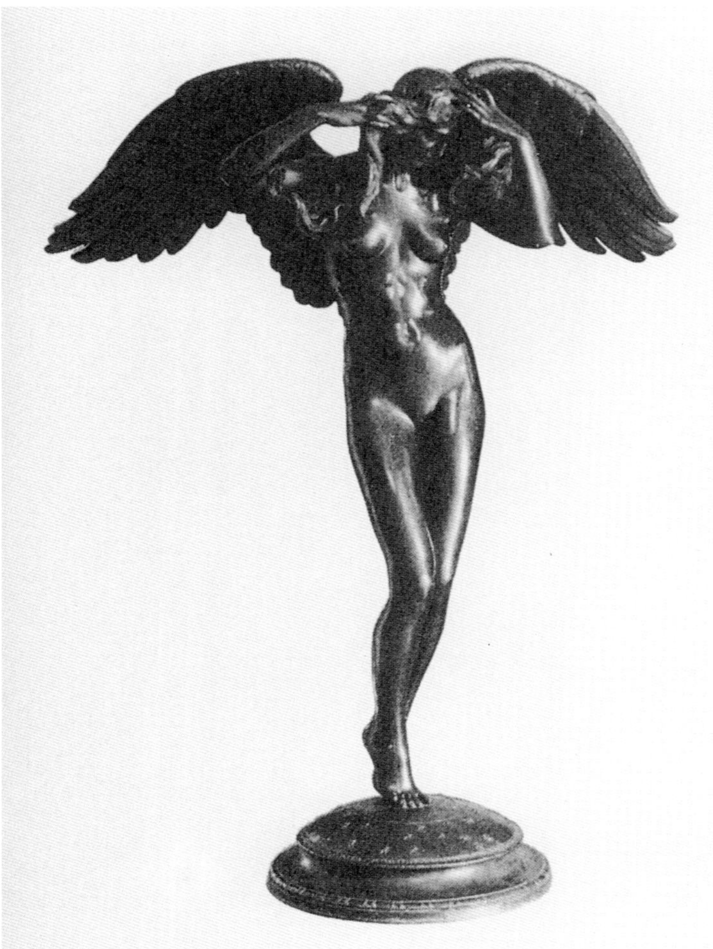

III-97. "Descending Night" (also called "The Setting Sun") by Adolph Weinman. With "The Rising Day" designed for a fountain at the Panama Pacific International Exposition in 1915.

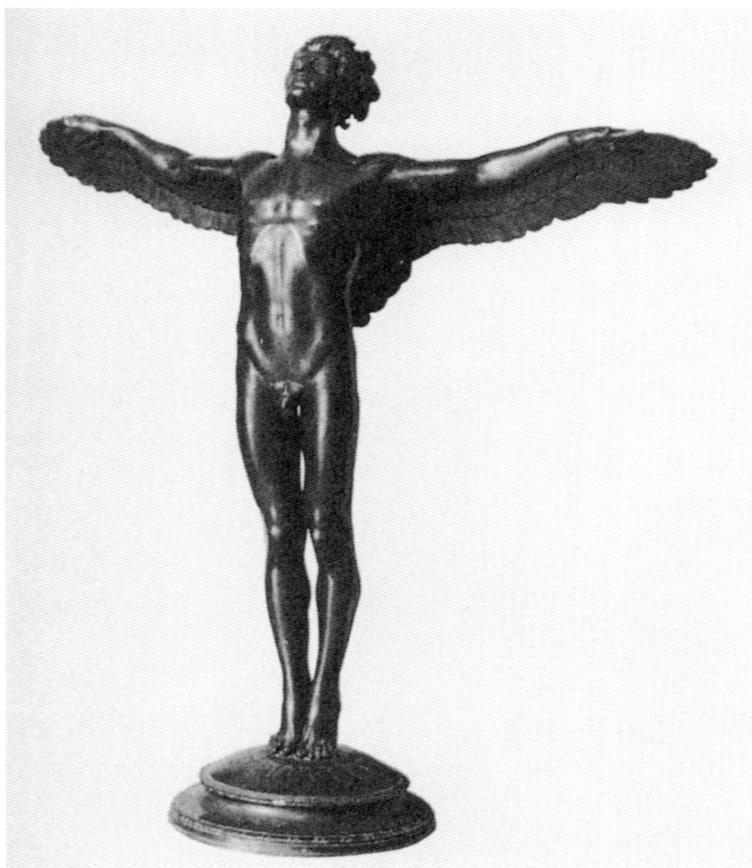

III-98. "Rising Day" by Adolph Weinman. 1915

MARGARET CRESSON (1889-1973)

The daughter of Daniel Chester French, Margaret Cresson was a successful sculptor of portrait heads who used Roman Bronze Works to cast both her own and her father's works. She lived and worked at Chesterwood, her father's notable estate and studio in Massachusetts for most of her life, eventually making it into a permanent memorial to French.

After studying with her father, and at the New York School of Applied Design for Women, as well as with Abstenia Eberle, she went to Sicily where she met and married an American diplomat. On her return to America, she took up portrait sculpture in marble and bronze, working in a style of realistic naturalism. She also spent much of her time on her father's career, writing two books and numerous articles, in addition to having his works recast.

Cresson was a contributor to many magazines and exhibited her portrait heads at the Art Institute of Chicago, the National Academy of Design, the World's Fair of 1939, and the Whitney Museum in 1940. She won many prizes and commissions. Her bust for a tomb of the Unknown Soldier of World War I (III-100) was much admired. Among her portrait subjects were heads of D.C. French, James Monroe, Olga Mestrovic (the wife of the sculptor) (III-101), George G. Merrill (III-102), and Dr. Carl Voegtlin (III-103).

Cresson dealt with Roman Bronze Works, both on her own and her father's behalf, between 1928 and 1957.

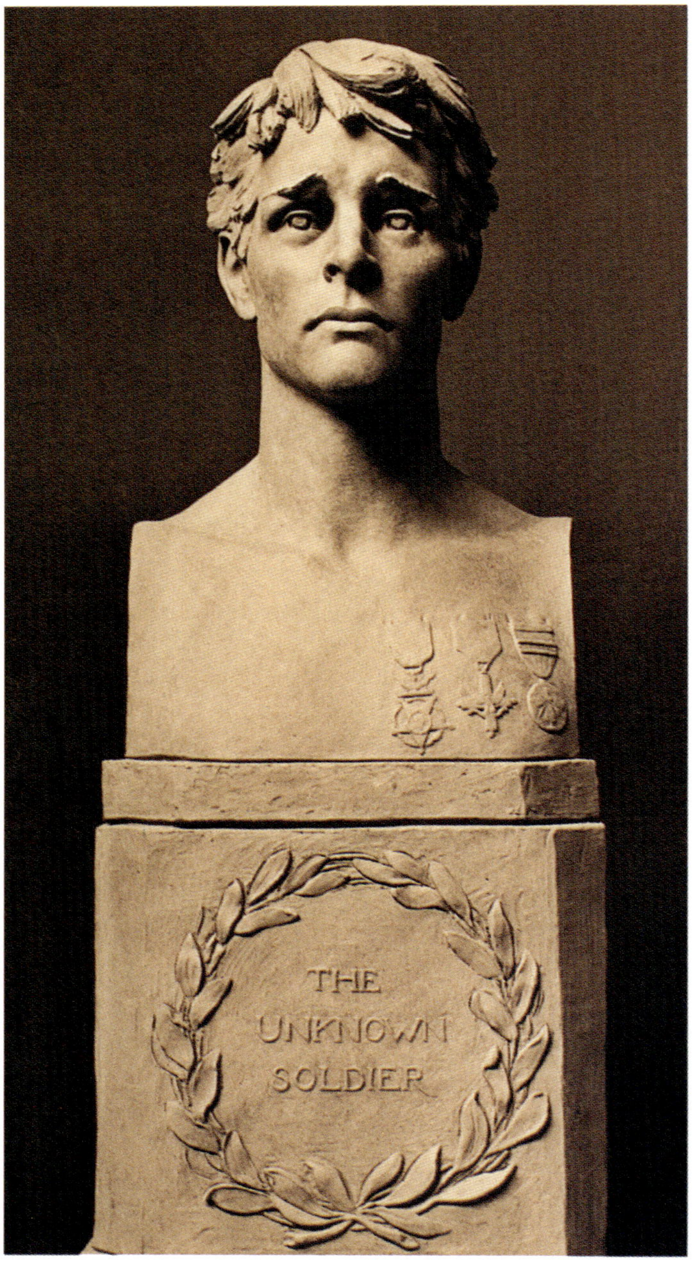

III-100. "Unknown Soldier" by Margaret Cresson. 1941.

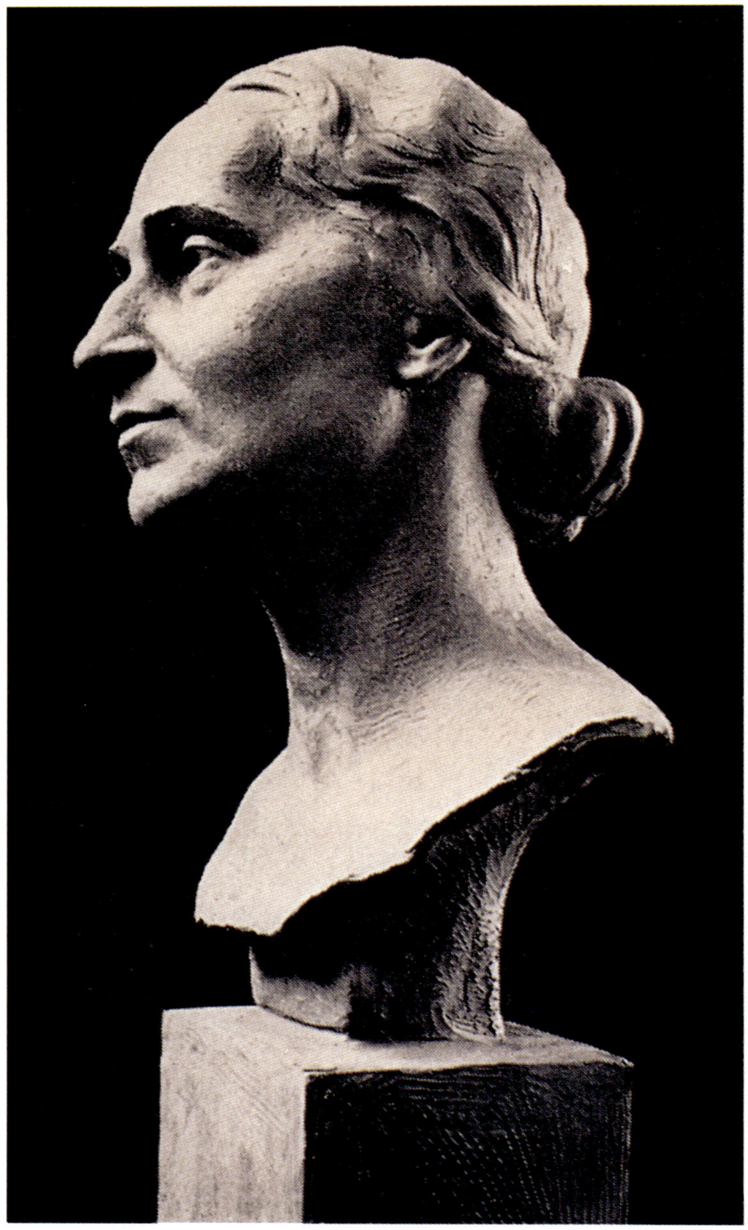

III-101. "Bust of Olga Mestrovic" by Margaret Cresson

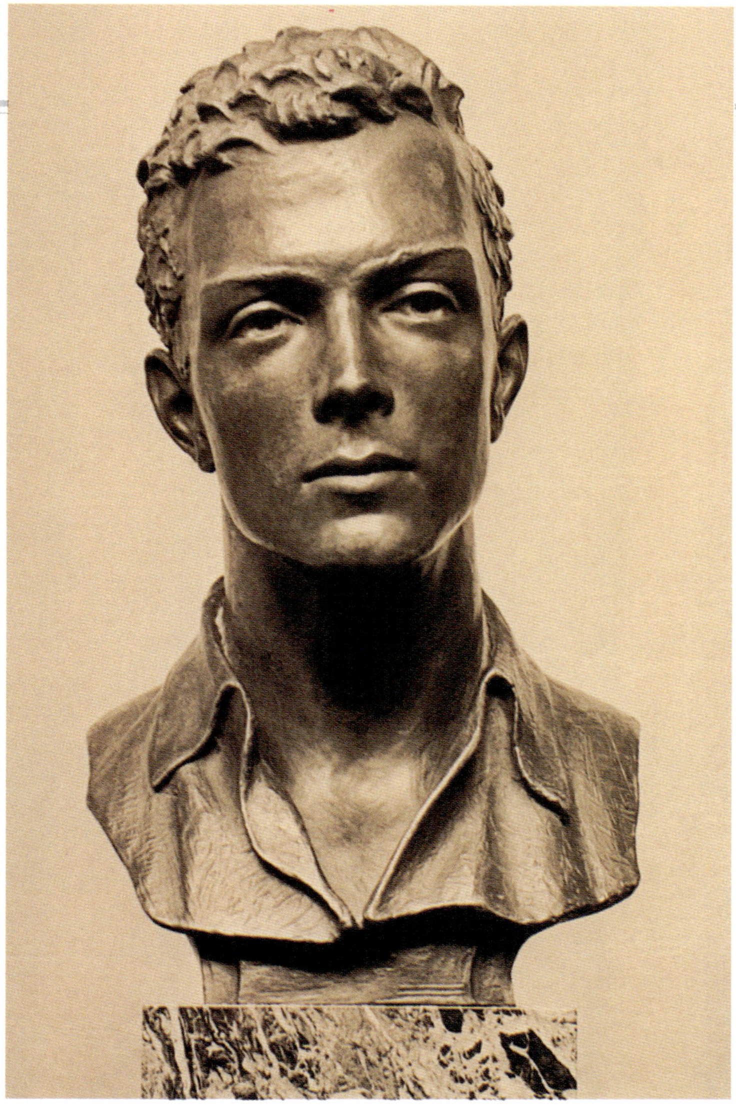

III-102. "Bust of George C. Merrill" by Margaret Cresson. 1938.

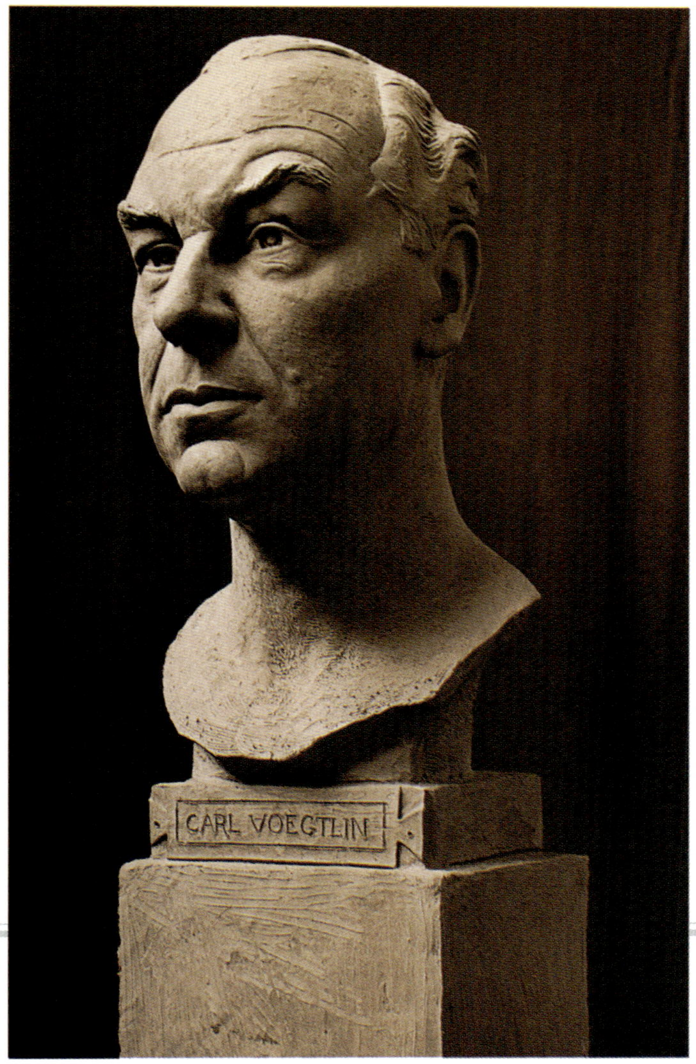

III-103. "Bust of Dr. Carl Voegtlin" by Margaret Cresson. 1940.

OTHER EARLY MONUMENT AND PORTRAIT SCULPTORS:

BRYANT BAKER, GEORGE GREY BARNARD, JOHN FLANAGAN, AUGUSTUS LUKEMAN, PHILIP MARTINY, SAMUEL MURRAY, CHARLES HENRY NIEHAUS, BELA LYON PRATT, ETTORE XIMENES

A number of noted sculptors of monuments and portrait commemoratives cast only a few of their statues at Roman Bronze Works:

III-104. "Pioneer Woman" (plaster cast) by Bryant Baker. 1927.

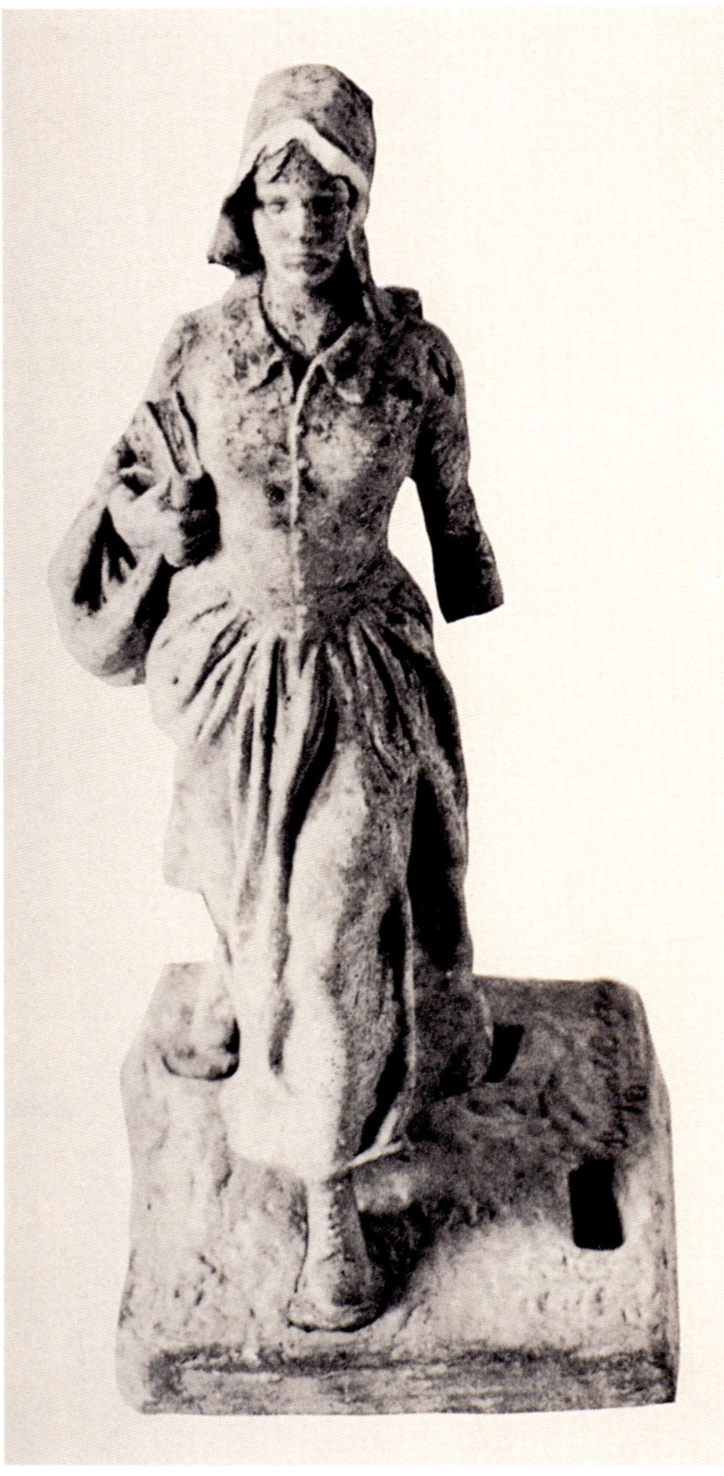

BRYANT BAKER (1881-1970)

Bryant Baker, who was born in England, was a well-known maker of monuments and statues of prominent people in the United States. He is represented in numerous American museums, and in the Capitol Building in Washington, among many other sites. At Roman Bronze Works he cast several portrait statues, including one of Grover Cleveland, as well as one of his best known works, "Pioneer Woman" (shown here in plaster cast (III-104) and as a sixteen foot sculpture (III-105).

III-105. "Pioneer Woman," Ponca City, Oklahoma. Unveiled 1930. This statue was selected in a competition in which 750,000 people voted.

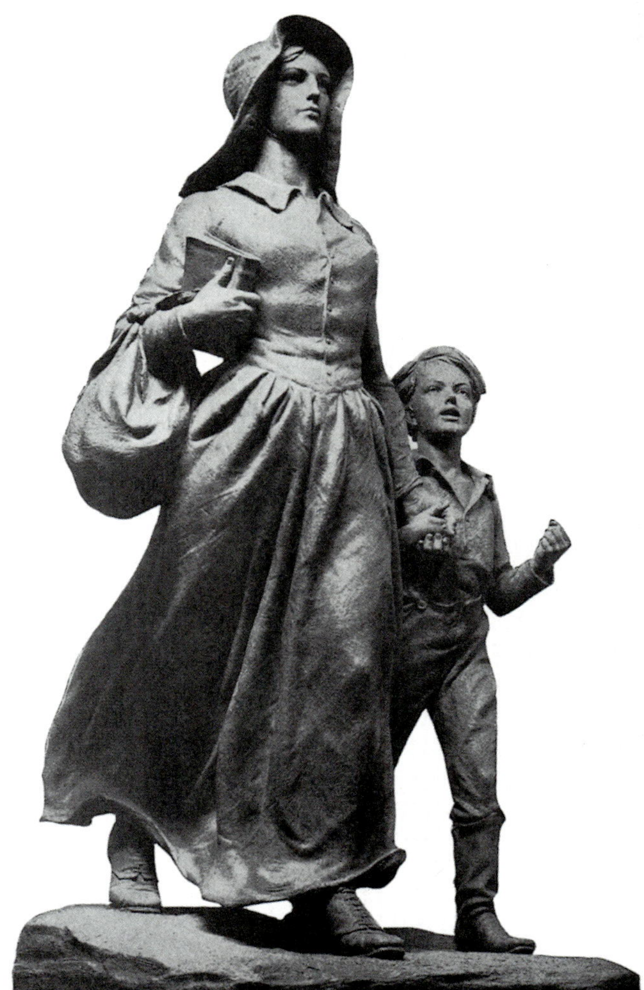

GEORGE GREY BARNARD (1863-1938)

George Grey Barnard was a noted monument maker in a somewhat more modernist style than that of his contemporaries; his career was marked by hostility from the conservative wing of the sculpture world. He is particularly noted for his images of Abraham Lincoln, and for his "Struggle of the Two Natures in Man." Among the works he cast at R.B.W. was a plaque honoring Lindbergh, "Spirit of St. Louis" (III-106).

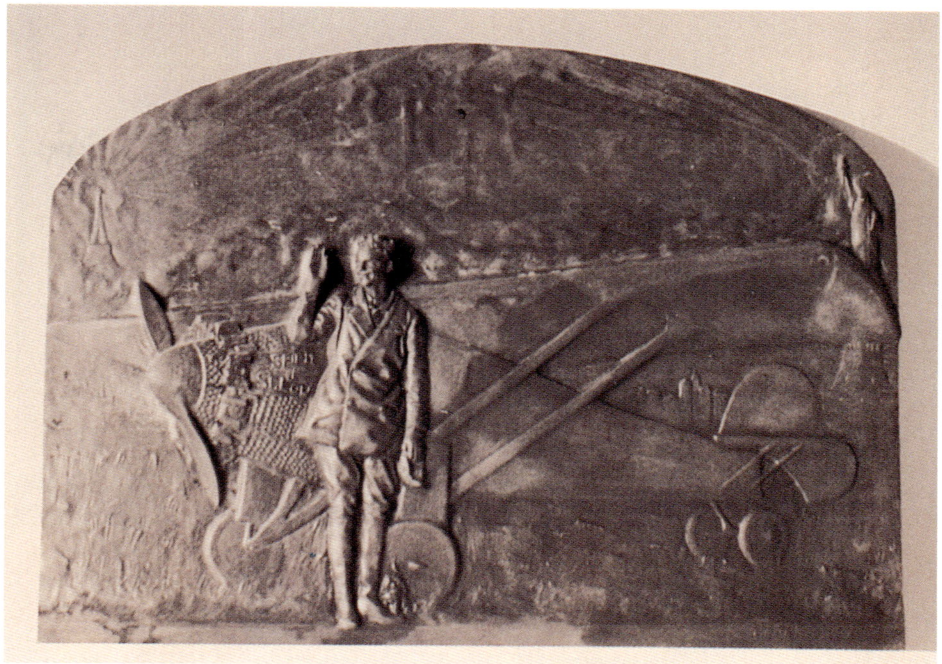

III-106"The Spirit of Saint Louis" by George Grey Barnard. This bronze plaque celebrated Charles Lindberg. 1927.

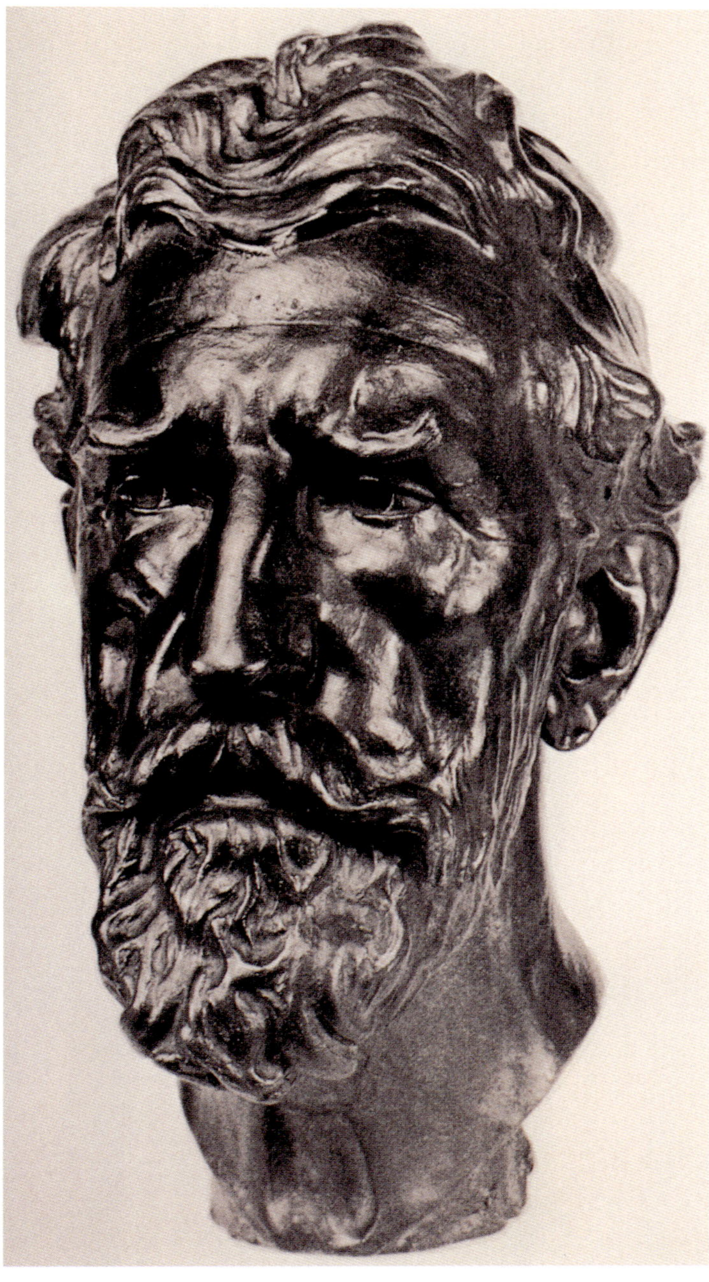

JOHN FLANAGAN (1865-1952)

John Flanagan, though best known for his commemorative medals (and the monumental clock at the Library of Congress), also made a number of portrait bronze busts. Among them was the portrait of fellow sculptor (and his teacher) Augustus Saint-Gaudens (III-107), cast in 1924.

III-107. "Head of Augustus Saint-Gaudens" by John Flanagan. Begun in 1905, completed in 1924.

AUGUSTUS LUKEMAN (1870-1935)

Augustus Lukeman made many portrait commemoratives of well-known American figures, including Daniel Boone, Bishop Alma White, Jefferson Davis, and President McKinley. In 1926, he made a portrait of Francis Asbury (III-108), for whom Asbury Park, New Jersey is named. His charming memorial to philanthropists Ida and Isador Straus (III-109) is in New York.

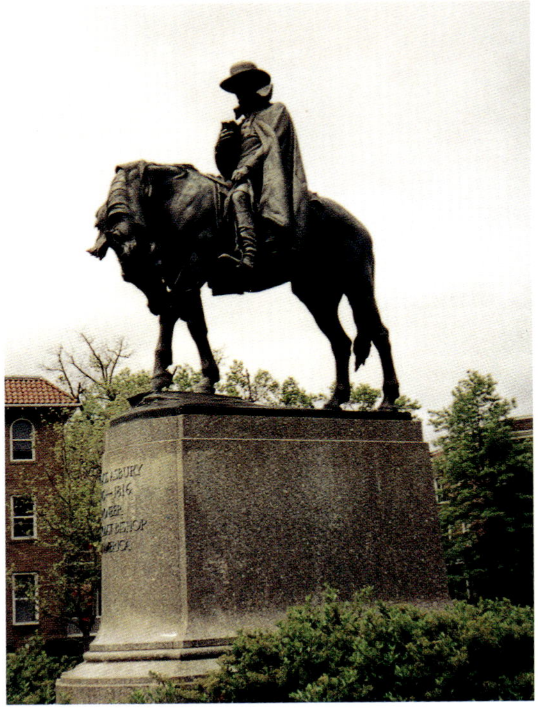

III-108. "Monument to Francis Asbury," Washington, D.C., by Augustus Lukeman. 1926.

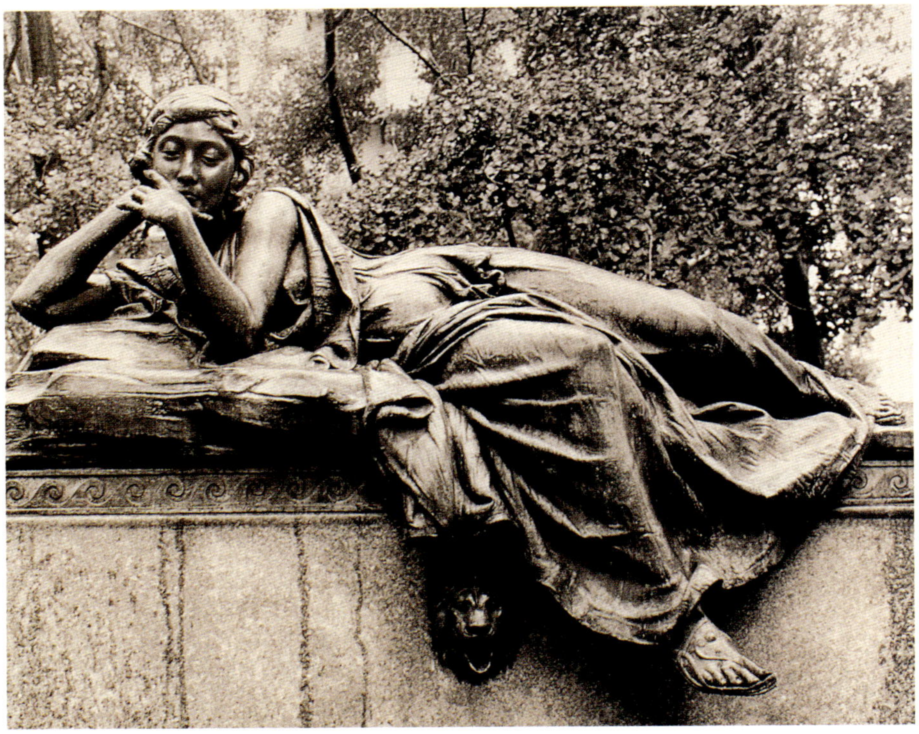

III-109. "The Straus Memorial," New York City, by Augustus Lukeman. 1913-14.

PHILIP MARTINY (1858-1917)

Philip Martiny, who was a colleague of Karl Bitter in several large sculptural projects, was a maker of monuments and architectural ornamentation. Among the memorial statues he made was New York's "Soldiers and Sailors Monument" (III-110), which was cast at Roman Bronze Works.

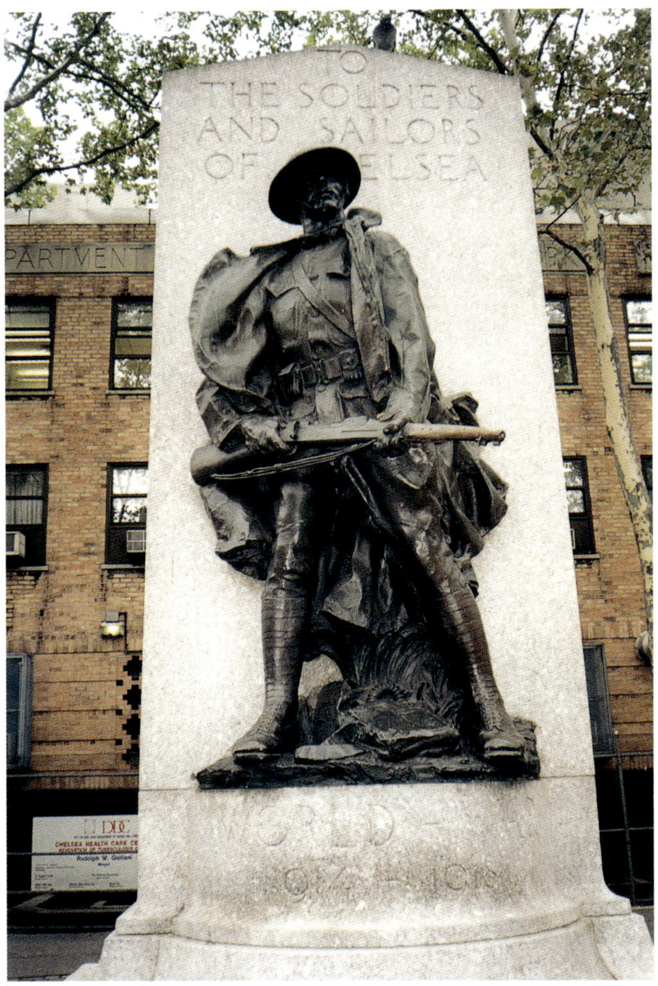

III-110. "Soldiers and Sailors Monument" (also called "Chelsea Park Memorial,") New York City, by Philip Martiny. c. 1920.

SAMUEL MURRAY (1870-1941)

Samuel Murray was a lifelong resident of Philadelphia, where he was a protégé of painter Thomas Eakins, sharing a studio with him and creating a bronze statue of him (III-111) that was cast by R.B.W. in 1907. He also cast a statue of Senator Boies Penrose, a bust of Whitman, a boxer, and a monument to peace at R.B.W.

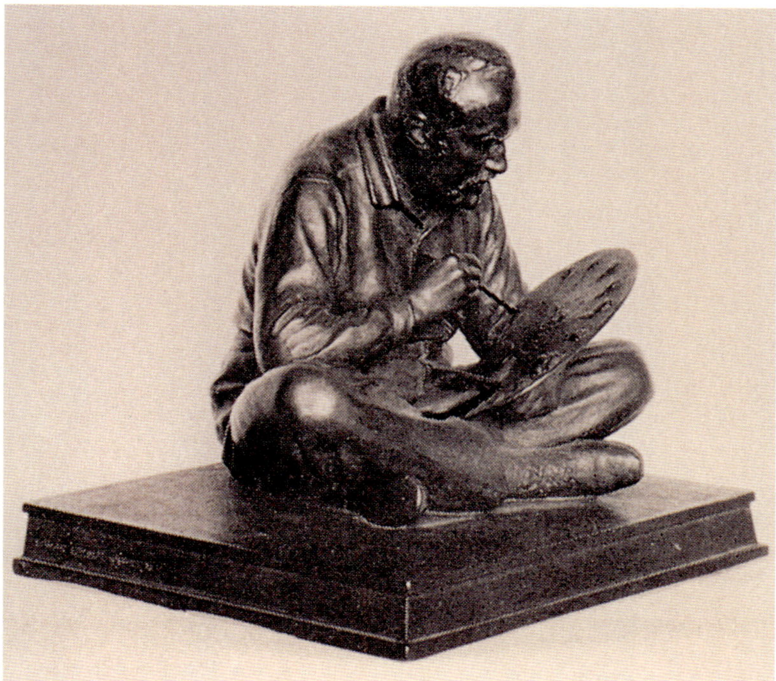

III-111"Thomas Eakins" by Samuel Murray. 1907.

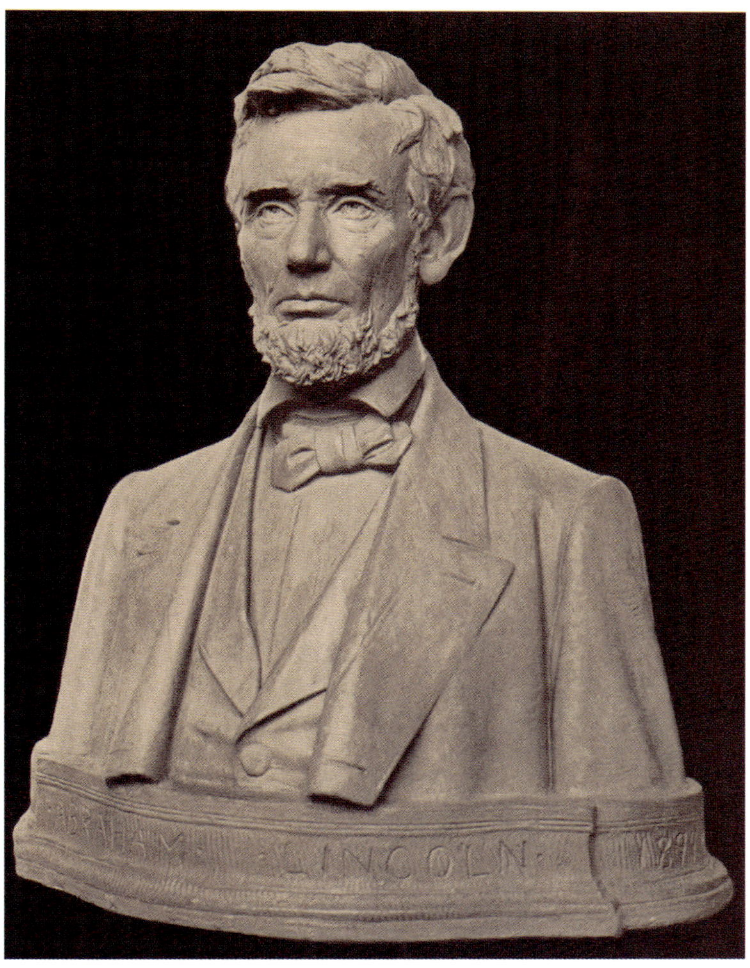

CHARLES HENRY NIEHAUS (1855-1935)

Charles Henry Niehaus was noted for large monuments, portrait busts and some architectural sculpture. At R.B.W. he cast a portrait bust of Abraham Lincoln (III-112) and a monument to Dr. Edward McDowell. (III-113).

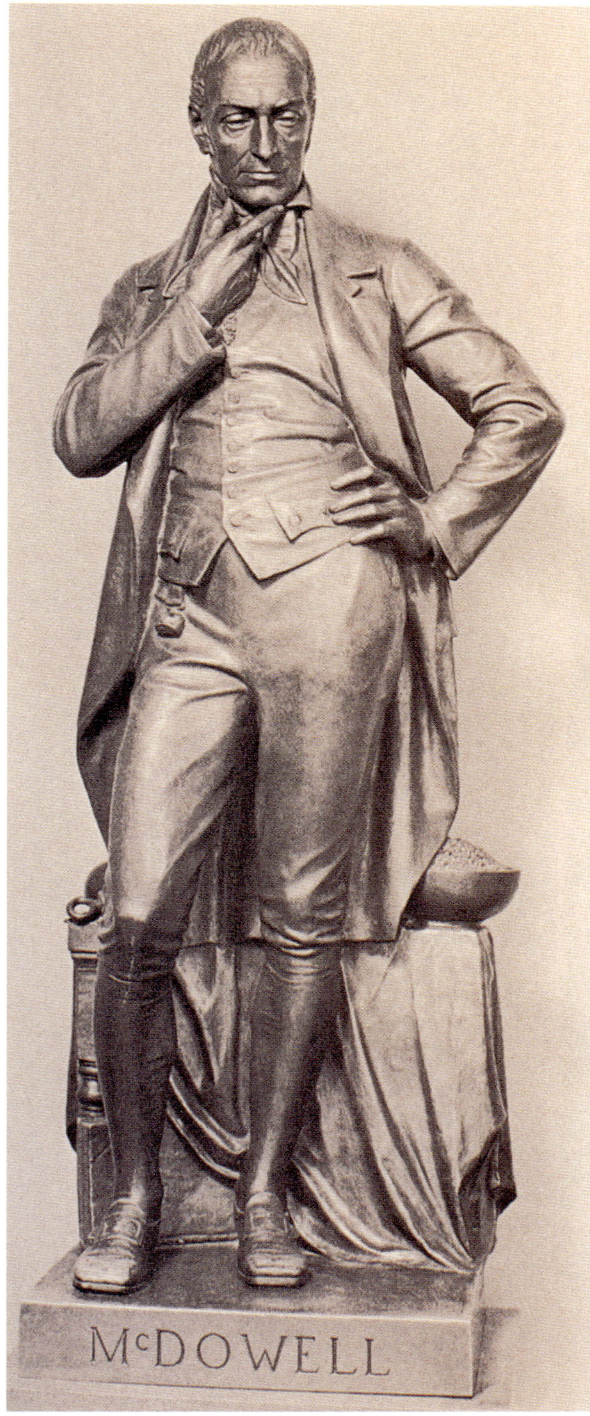

III-113. "Dr. Ephraim McDowell" by Charles Henry Niehaus. 1928.

III-112. "Abraham Lincoln" by Charles Henry Niehaus. First cast in 1897.

BELA LYON PRATT (1867-1917)

Bela Lyon Pratt was a Boston-based sculptor of monuments and portrait busts, including many for the Library of Congress. One of his best known sculptures was cast at Roman Bronze Works: "Captain Nathan Hale" (on the campus of Yale University and in Washington, D.C.) (III-114). Also listed in R.B.W.'s ledgers are works by Pratt portraying polo players, nudes, nymphs, and cupids.

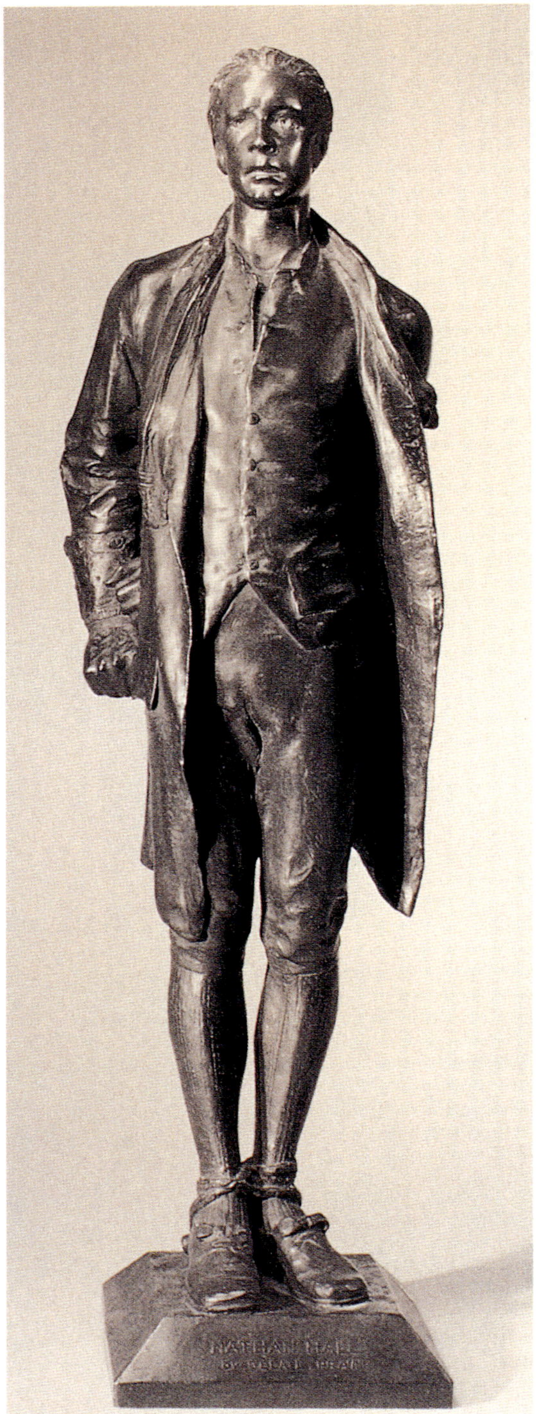

III-114. "Captain Nathan Hale" by Bela Lyon Pratt. 1913.

ETTORE XIMENES (1855-1926)

Ettore Ximenes created a number of statues of notable Italians, including a memorial to Dante in New York, and a monument to the explorer, Giovanni da Verrazzano (III-115) which was erected in 1909.

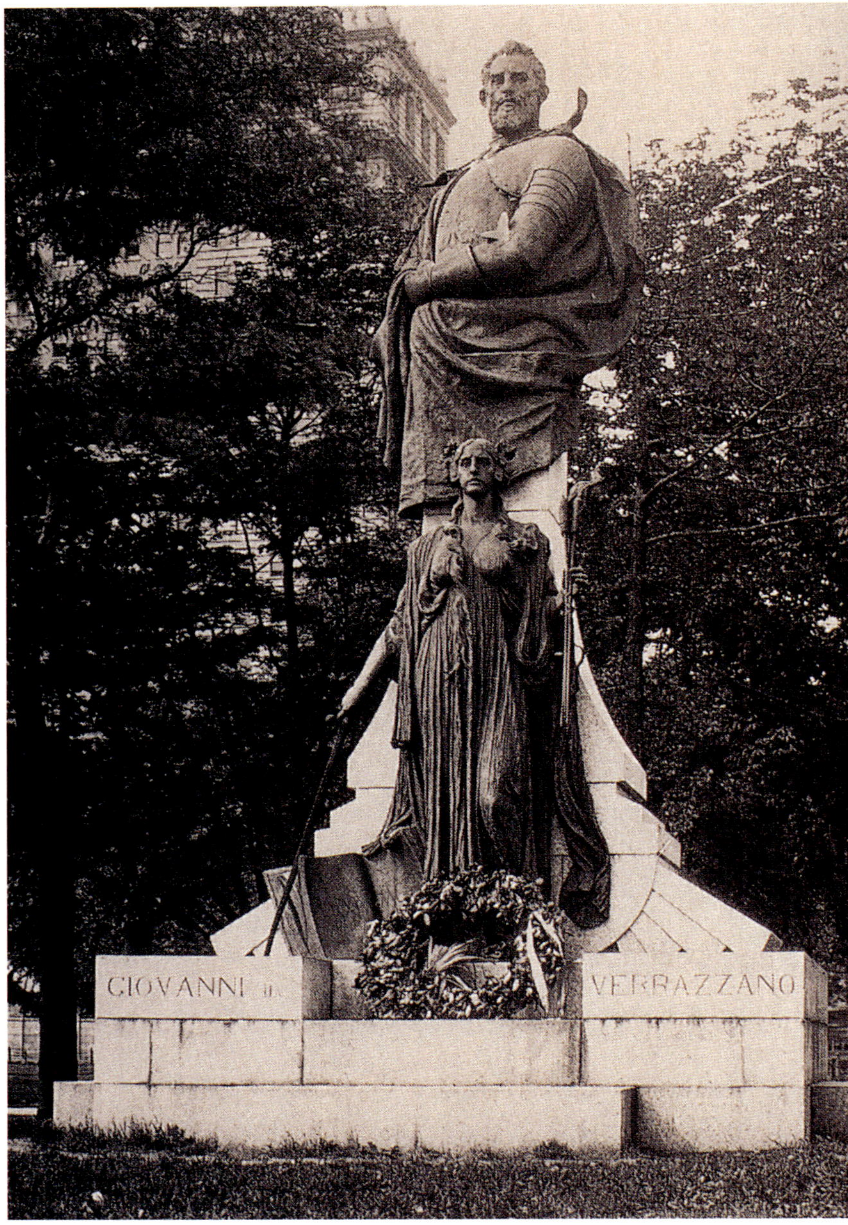

III-115. "Giovanni da Verrazzano," New York City, by Ettore Ximenes. 1909.

Other sculptors of monuments and portraits who cast an occasional work at R.B.W. include Jonathan Scott Hartley (a monument to inventor John Ericsson in New York, a bust of Chaucer, and a "Pan"), Henry Hering ("Indiana World War Memorial"), Lorado Taft ("Lincoln" in Chicago), Oscar J.W. Hansen ("Winged Figure" at the Hoover Dam), Trygve Hammer ("Bust of Amundsen") and architect and sculptor John Russell Pope ("Garvan Memorial" in Philadelphia). McKim, Mead and White cast New York's Lafayette Memorial Library sculptures at R.B.W.

IV. COWBOYS, INDIANS, and BUFFALO: THE WESTERN IMAGE AND THE SCULPTURE OF WILDLIFE

From the grand canvases of panoramic Western scenery and exotic Indian encampments came a new American mythology that still continues today. Artists and writers created "the Wild West" for an audience back home that longed for scenes of action and adventure. Dynamic pictures of Indian wars, thundering stampedes, carefree cowboys, and men surviving in a dangerous and dramatic natural setting appealed to the armchair-bound all over the country. These images of the west have been memorialized by the art of the late 19th and early 20th centuries—and they continue today in cinematic form. In fact, despite their realistic technique and recognizable subjects, this kind of art could be described as a form of romanticism, since it idealized certain pictorial images and gave a glorified vision of reality. Like so much of 19th century American art it was based on a nostalgic vision of a simpler time and place, when civilization had not yet destroyed the simple freedoms and natural wonders of the West. For a population increasingly urban and faced with complex social issues, the West pictured by these artists seemed carefree and thrilling, as well as beautiful.

No artist did more to popularize Western images than Frederick Remington, who was a painter, illustrator, and sculptor in bronze. An Easterner, Remington went west, worked as a cowhand, lived among native tribes, and became an illustrator for the newspapers before devoting himself fully to painting and sculpture. Remington's action-packed images of Western culture brought about an entire genre in American art; for example, his success with a bronze sculpture of the bucking bronco (of which some 200 copies were cast) led many other artists to head west.

Among the other sculptors who worked in the Western genre were Charles Russell, Cyrus Dallin, Solon Borglum, and Sally Farnham. Their contributions to the genre were much in the style of Remington. In fact, all of the artists shared an outlook: they hoped to record and preserve a way of life that was fast disappearing, and to portray American Indians in a dignified and sympathetic way. Remington spoke for all of them: "The West," he wrote, "is no longer the West of picturesque stirring events. Romance and adventure have been beaten down in the rush of civilization." It was their aim to recreate the West they remembered in romantic images that would be available to households all over the country.

Though the reality of Western life might change, artists could still give viewers a recognizable (if glorified) picture of a nostalgic and familiar theme. Like Western movies, the images could be repeated with small changes numerous times, for the very existence of this type of art depended on the recreation of a familiar collection of characters and events. Remington's ability to portray both character and motion made his familiar images noteworthy; many other artists of Western imagery peopled their works with similar stock types with varying degrees of success. Some works achieved lasting fame, creating certain American icons that were later seen over and over in Western movies.

The artists who portrayed wildlife were similarly engaged. They too saw a disappearing world of buffalo, wild horses and wolves. Like the "animaliers" of 19th century Paris who had made animal sculpture into a fashionable style, they turned their attention to life-like sculptures of indigenous animals ranging from giant buffalo to graceful deer. Increasingly popular in both monumental and smaller forms, these animal sculptures were successful throughout the country as images representative of a beautiful America, filled with roaming animals. Like their European counterparts, other animal sculptors turned to more exotic beasts, visiting zoos and portraying lions and tigers. Some artists worked for the museums of natural history, creating panoramic images of wildlife in a pristine environment. Among the most noteworthy of animal sculptors were Anna Hyatt Huntington and Alexander Proctor. Like the sculptors of Western images, they too were attempting to portray a kind of natural beauty, untrammeled by civilization.

It was hardly surprising that art based on such nostalgic and heroic images was so successful at this time of great economic and social change, with a fading frontier, vanishing natural wildlife, and a belated understanding of the plight of the native American. As various other styles of art swept Europe, and eventually came to America, the artist of Western genre and wildlife continued undeterred. Theirs was a celebration of a way of life that carried a strong message, as well as being a type of artistic expression, and they had no intention of seeking other pictorial means or styles.

Roman Bronze Works was the foundry of choice for the casting of bronzes of both Western and animal subjects. The close personal relationship that developed between Riccardo Bertelli and Frederick Remington, and the increased interest and technical development in the lost-wax casting process, led more and more sculptors to bring their works to the foundry. As can be seen in the following pages, the major figures in the history of American Western and animal sculpture were regular customers at Roman Bronze Works.

FREDERIC REMINGTON (1861-1909)

Frederic Remington is recognized as the artist who brought Western images to the American nation, and thus inspired a lasting fascination with the "wild" west, cowboys, Indians and bucking broncos. He was followed by numerous other artists, and eventually movie makers, in creating a picture story of mythical proportions that became the American "Western." His work—both paintings and sculpture—constitutes the most extensive and accurate picture of what became known as a glamorous period of American history, when rough and rugged individuals confronted a wilderness.

Remington's horses and cowboys, Indians and frontiersmen appear in his work with startling, action-packed realism. His ability to picture people and animals at the very moment of dramatic action gave his art the excitement and naturalism that made him a popular success. His horses rear, his figures lean precipitously, and in his sculptures—dozens of them—he creates the sensation of spontaneous movement and vitality. Though many artists followed in his footsteps, few matched the colorful and dramatic action of his work.

Remington was born in Canton, New York, and was raised in nearby Ogdensburg (where there is now a museum devoted to his life and art). After attending Yale, he went west, where he learned the trade of a cowboy. He became an itinerant cowhand, riding cattle trails from Mexico to Canada, sketching as he went. His realization that the way of life he was pursuing was fast disappearing, led him to return to New York in 1885 to chronicle in art the glamorous life of the Western frontier. His success with paintings and sculpture was rapid, and he soon became one of the best known and most highly paid illustrators in the world.

R.B.W.'s founder, Ricardo Bertelli, had seen Remington's "Bronco Buster" in 1895; some forty copies were originally cast by the Henry Bonnard Foundry and the Tiffany Company was displaying exact replicas of the statue for sale. When Remington moved to Bertelli's foundries in 1901 to use the lost-wax process, he discovered he could tinker slightly with the statue during the casting process. Soon a number of different versions of "Bronco Buster" appeared (numbers IV-1 to IV-4); some 275 were cast at R.B.W. (For example, in some there are "leather" chaps, in others wooly ones, and the whip is frequently in different positions.)

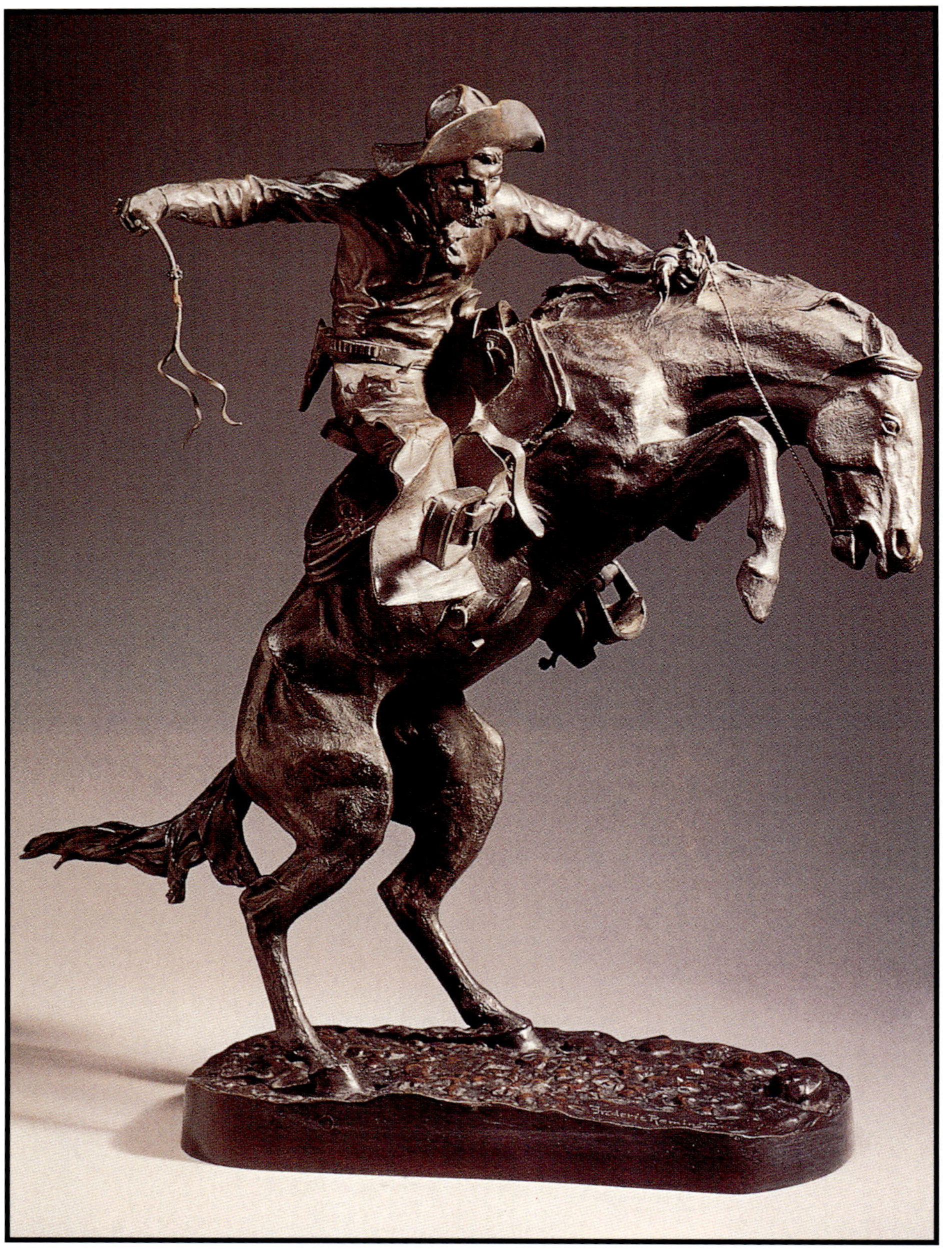

This and opposite page:
IV-1 to IV-4. Four versions of "Bronco Buster" by Frederic Remington. First cast in 1895, and recast numerous times with slight changes at R.B.W.

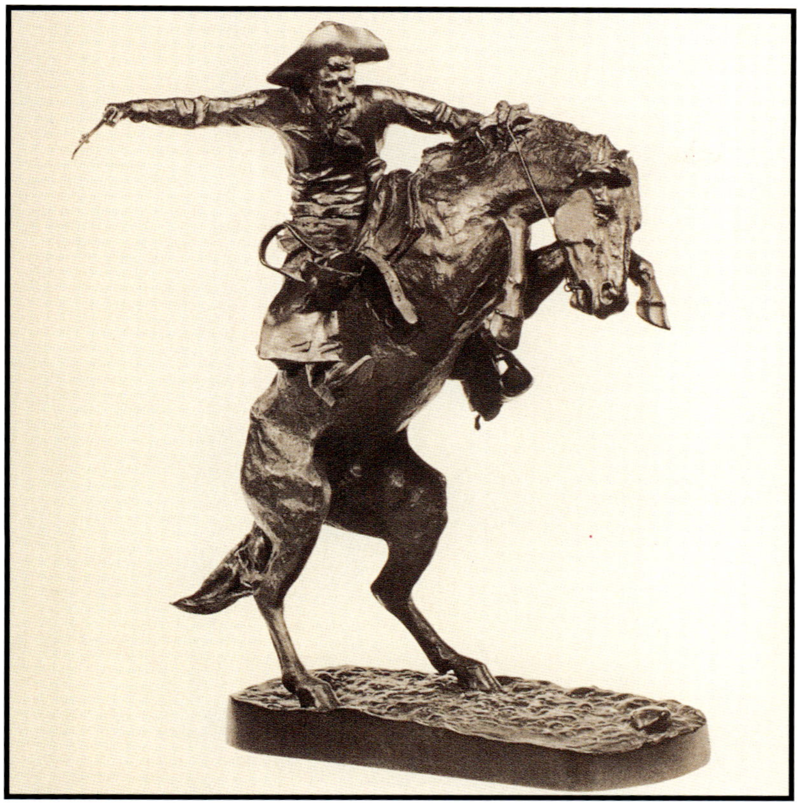

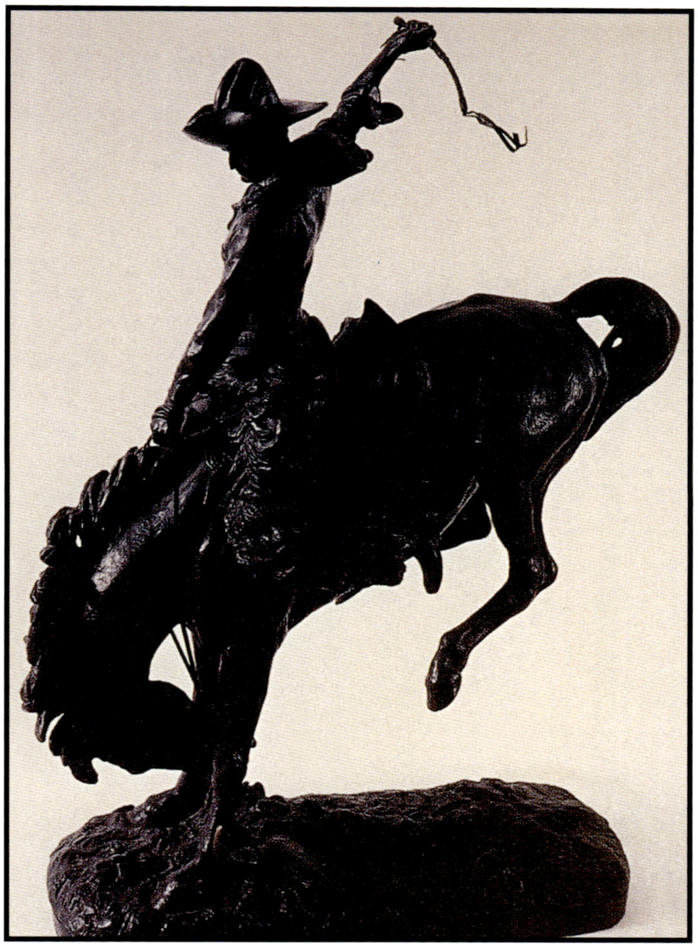

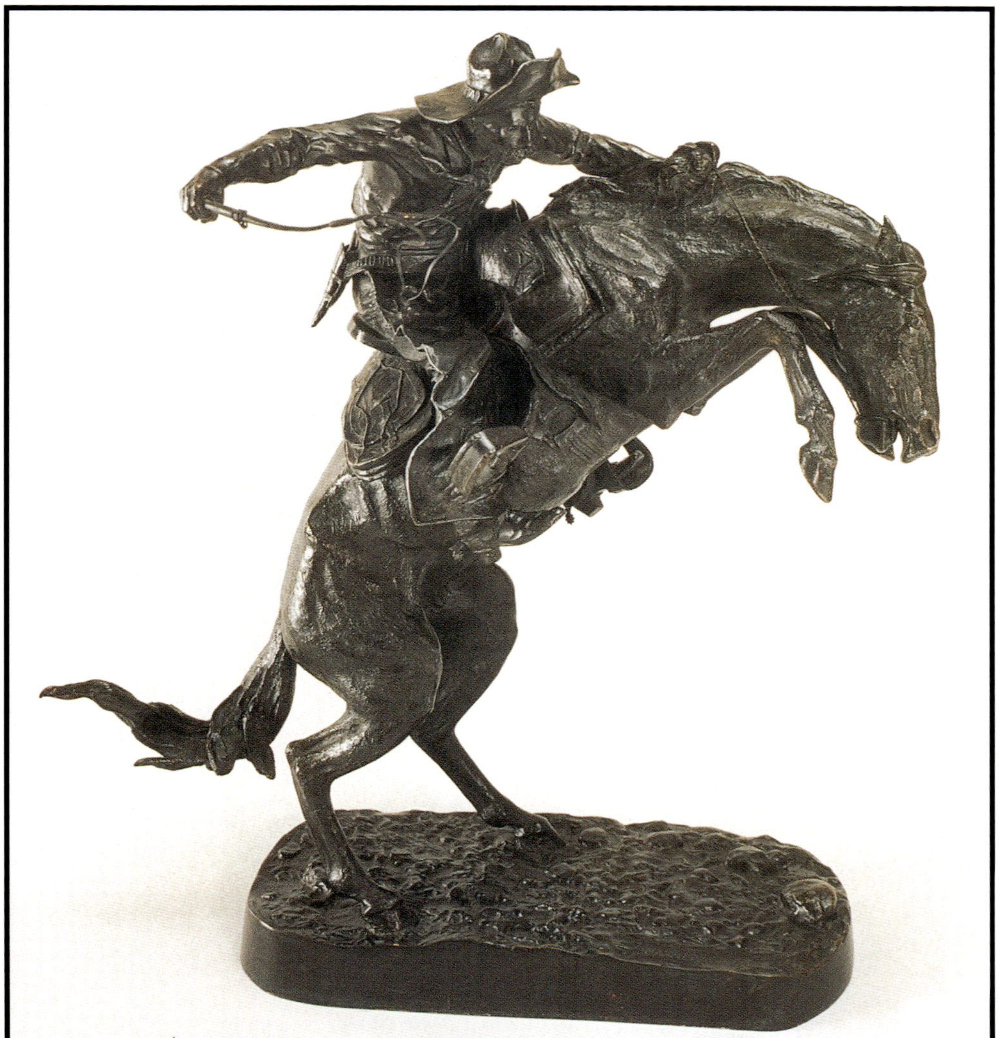

Remington was among the first American artists to embrace the lost-wax method of casting. Soon after he brought his models to Riccardo Bertelli, other sculptors followed. Between 1901 and his death in 1909 he worked closely with Bertelli, exploring methods for altering the sculpture during the casting process, and for supporting the work on its base. (His sculptures of rearing horses that rested only on two legs were examples of new technology in the supporting base.) When Remington died at the age of forty-eight, at the height of his artistic powers, he left an entire vision of a vanishing time and way of life: numerous paintings, drawings, and bronze sculptures.

Many of the bronzes were cast at Roman Bronze Works between 1901 and 1909, and posthumously thereafter. Remington's sculpture and painting represent the Western genre in most major collections of Americana. These sculptures, as can be seen here, picture a panoply of characters and events, almost all of them drawn from his Western experiences (numbers IV-5-through IV-22).

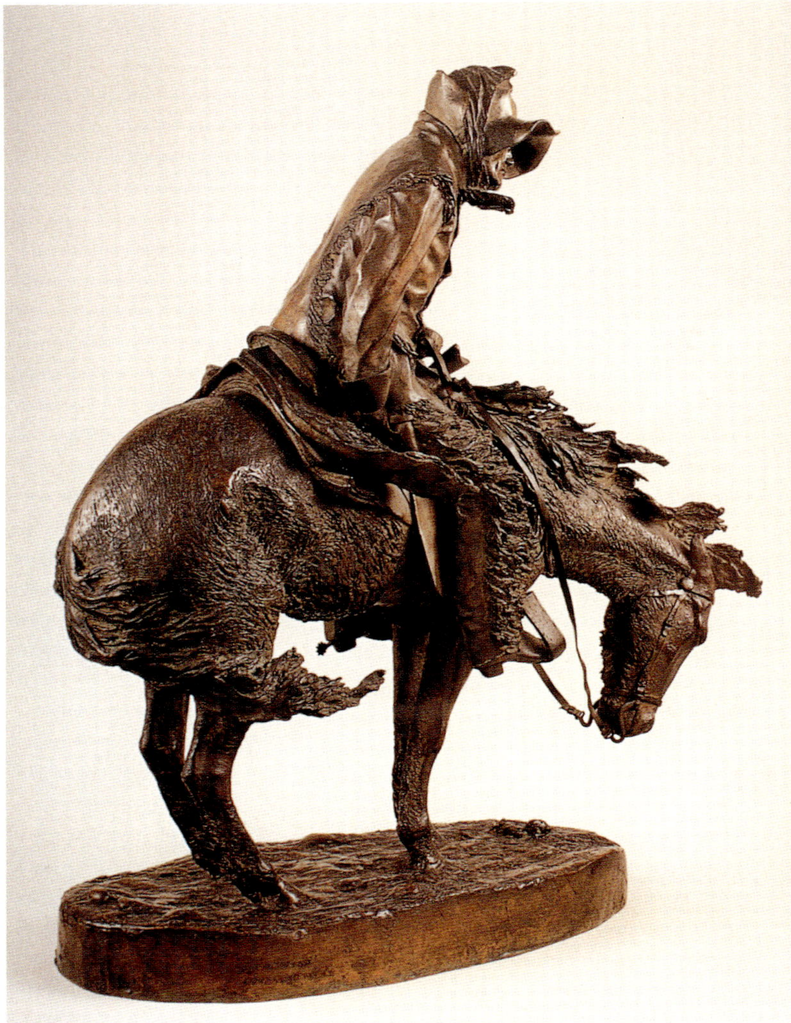

IV-6. "The Norther" by Frederic Remington. 1900.

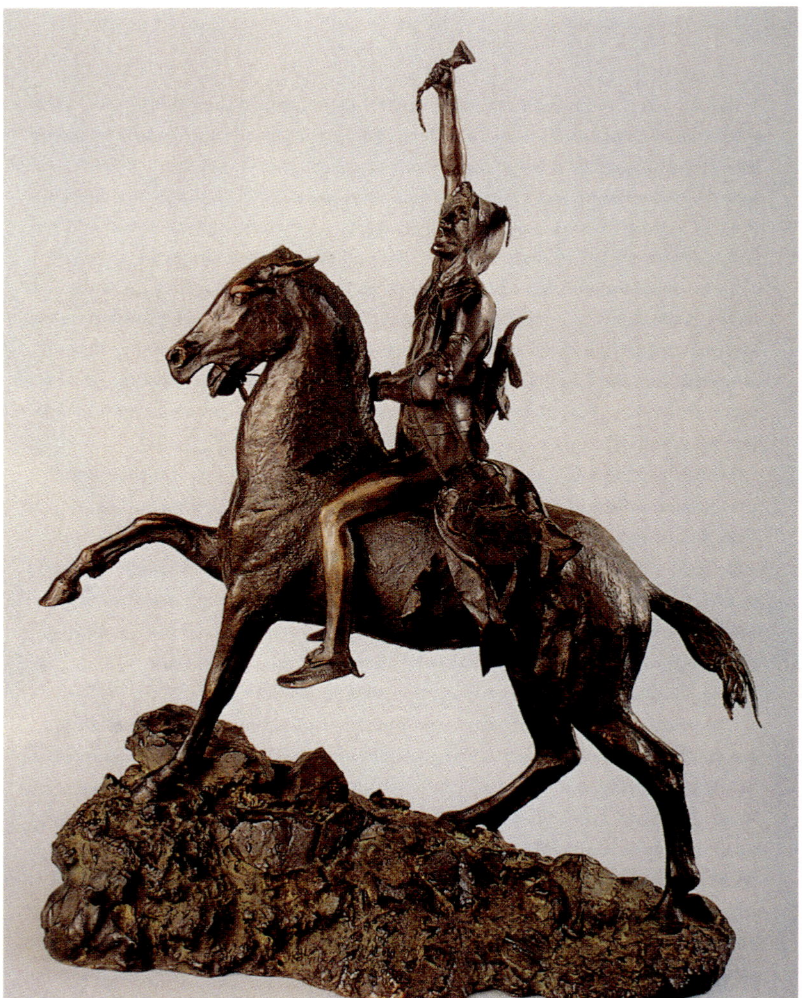

IV-5. "The Scalp" (also called "The Triumph") by Frederic Remington. First cast in 1898. Transferred to R.B.W. where 76 castings were made.

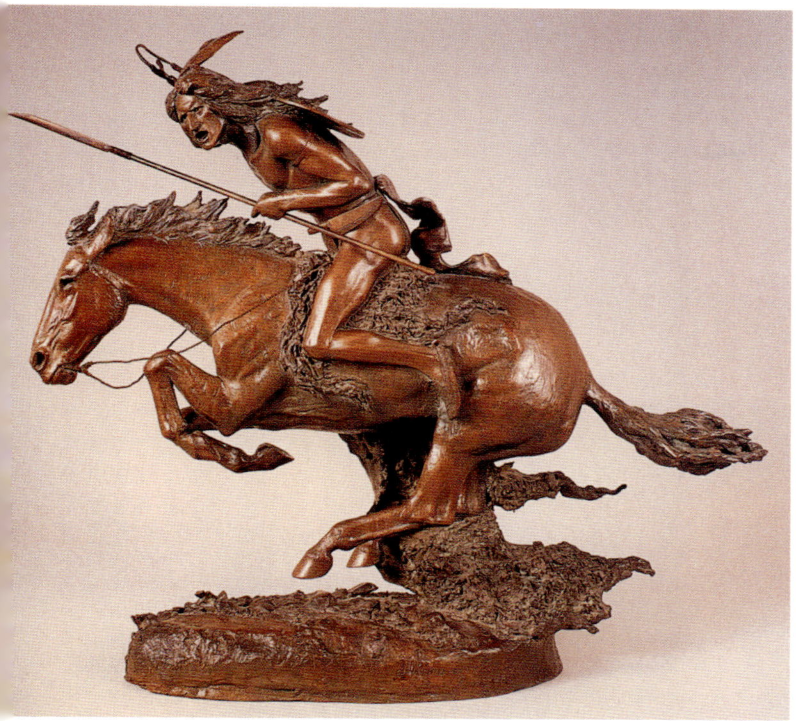

IV-7. "The Cheyenne" by Frederic Remington. 1901. This was the first Remington bronze produced at R.B.W. There were some 87 castings.

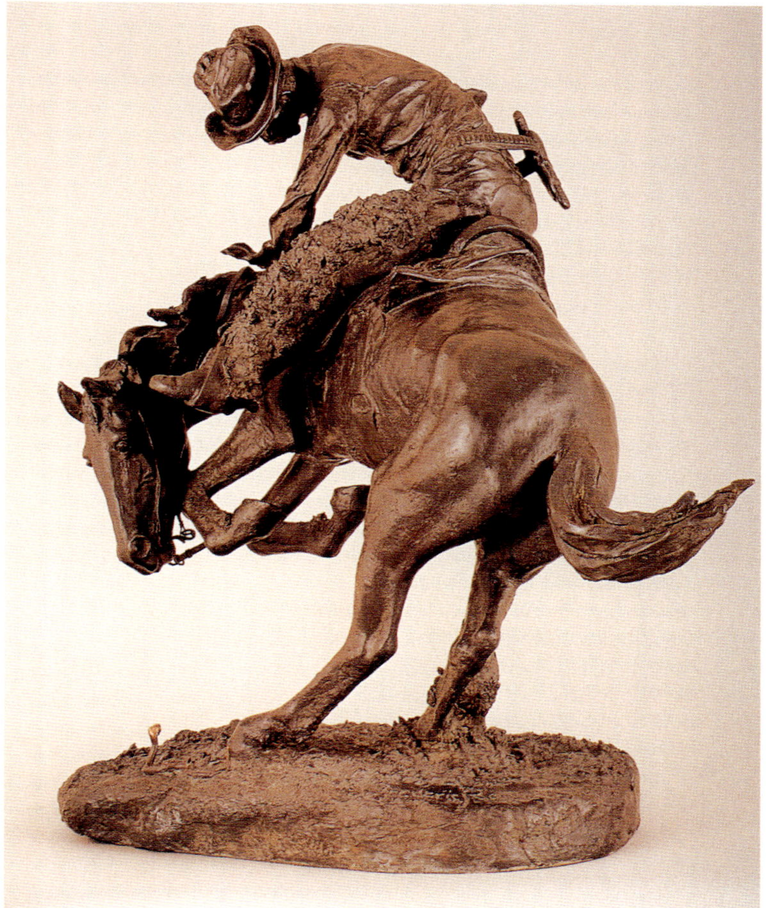

IV-8. "The Rattlesnake" (also called "Snake in the Path") by Frederic Remington. 1905.

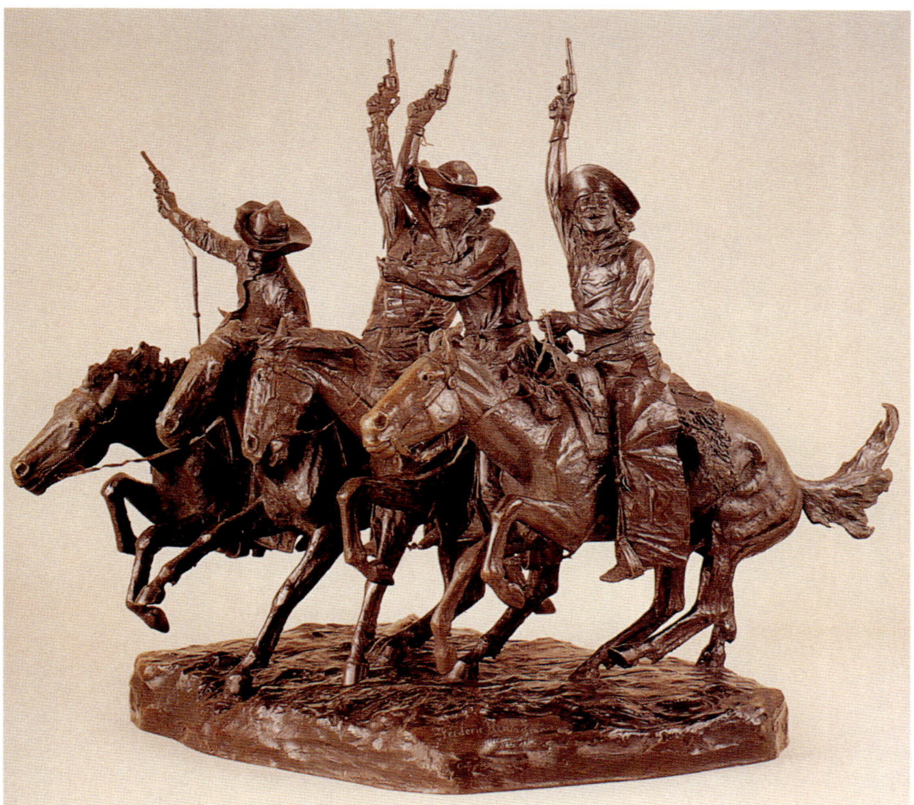

IV-9. "Coming Through the Rye" by Frederic Remington. 1902.

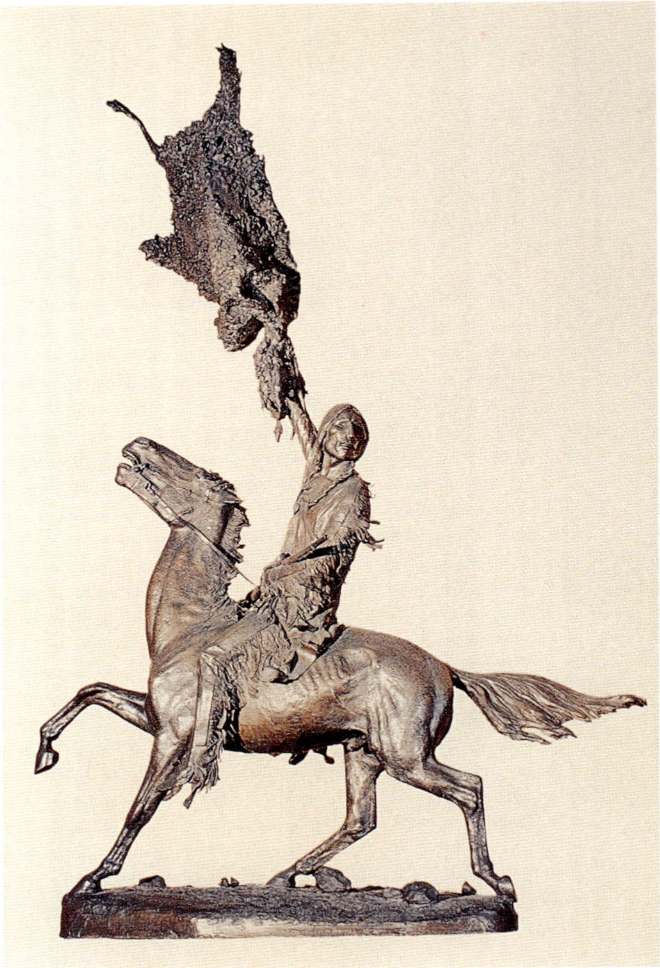

IV-10. "The Buffalo Signal" by Frederic Remington. 1902.

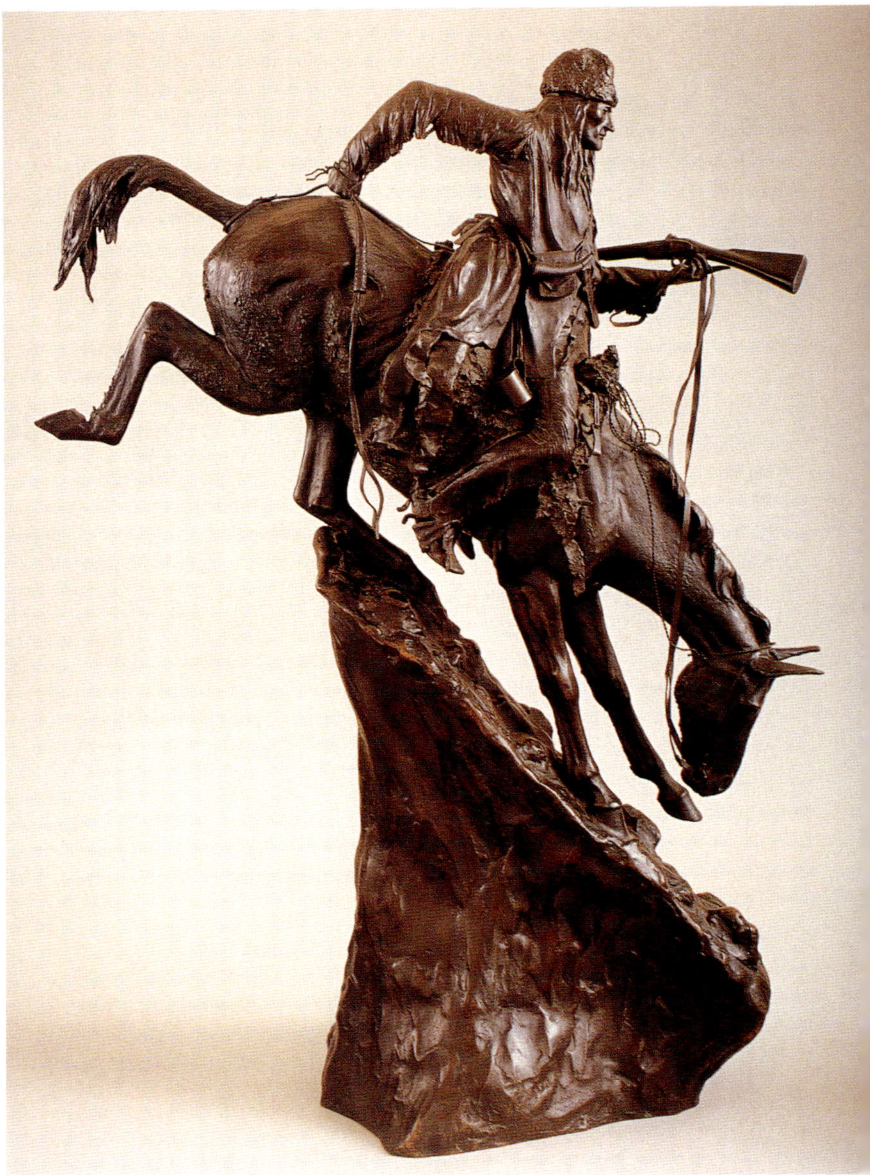

IV-11. "The Mountain Man" by Frederic Remington. 1903. There are approximately 81 casts of this work.

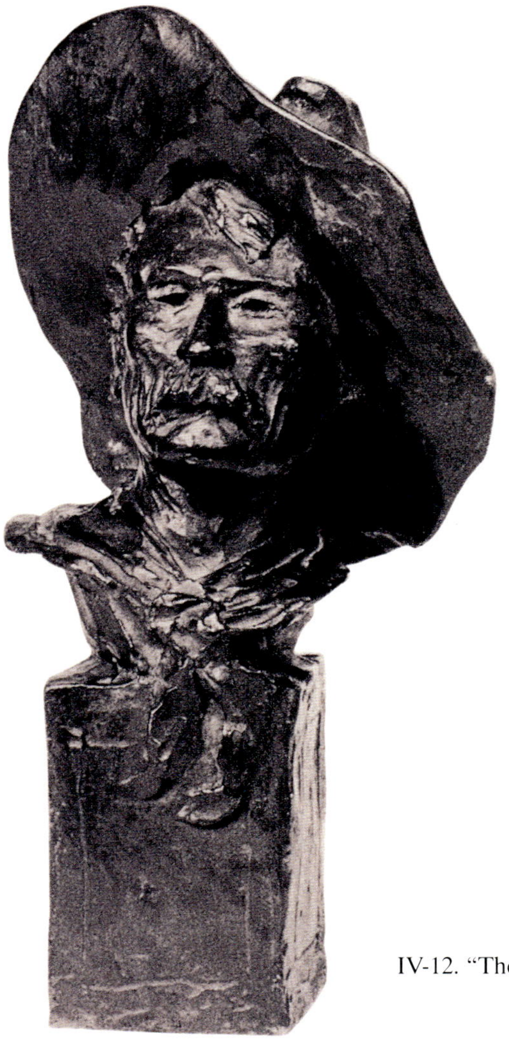

IV-12. "The Sergeant" by Frederic Remington. 1904.

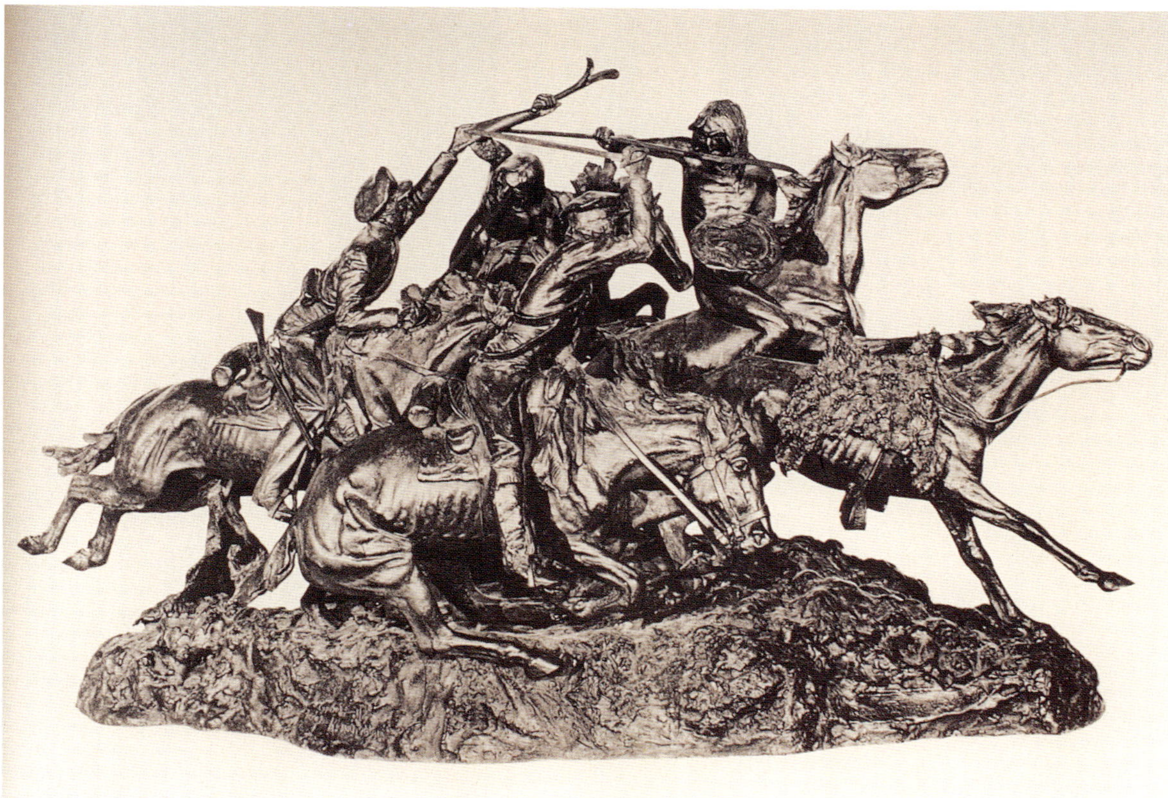

IV-13. "Dragoons—1850" by Frederic Remington. 1905.

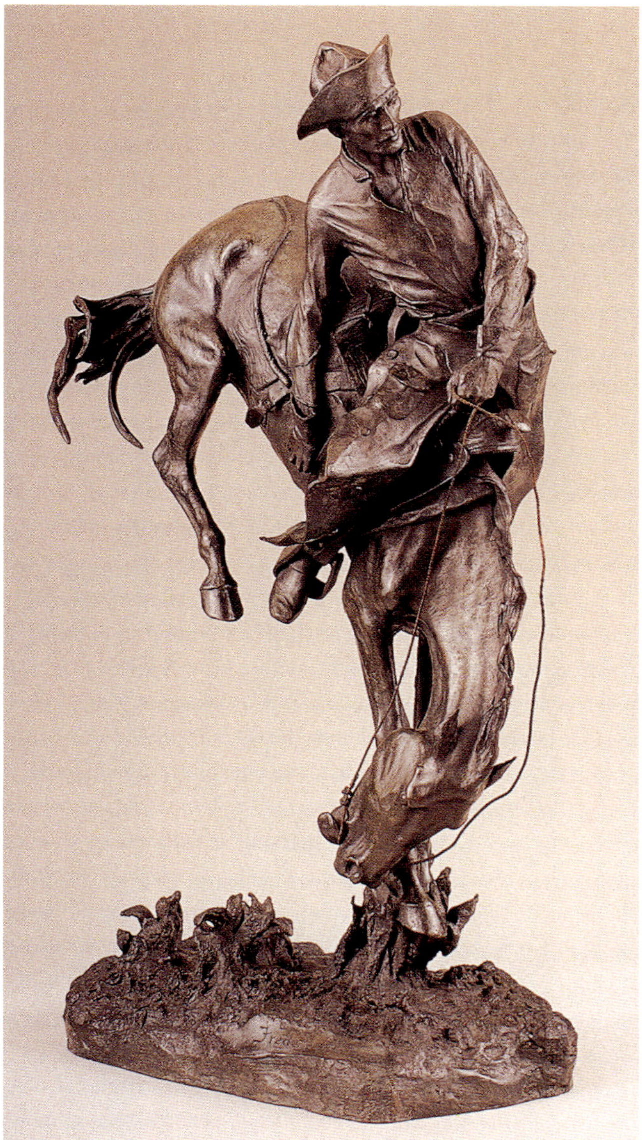

IV-14. "The Outlaw" by Frederic Remington. 1906. There are approximately 59 casts of this work.

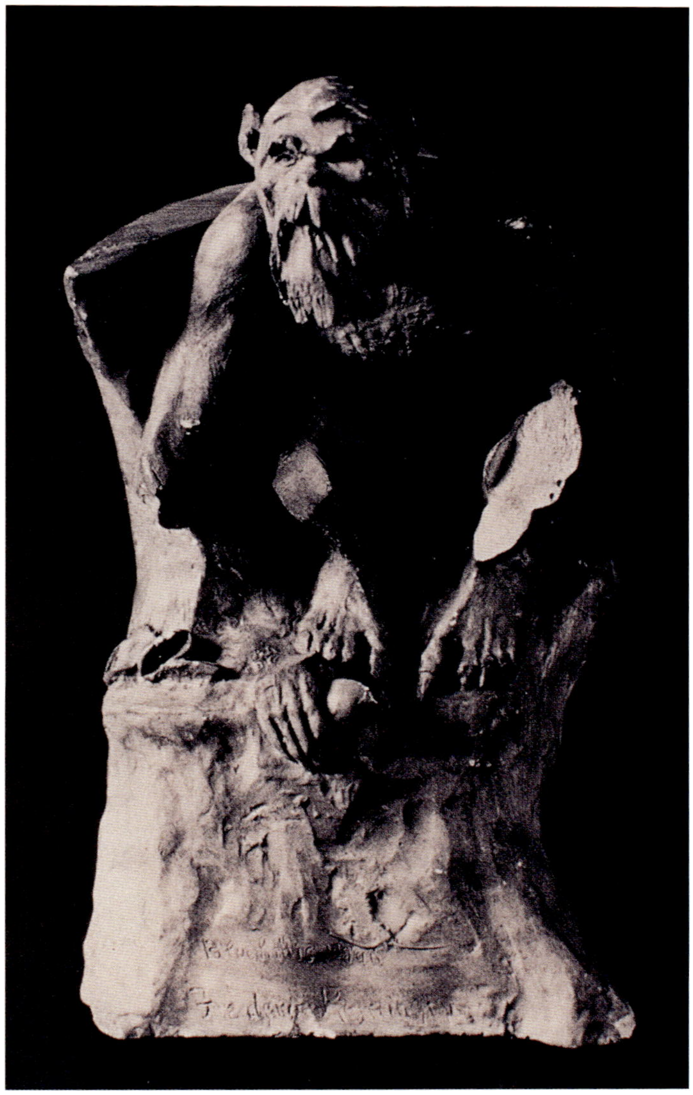

IV-15. "Paleolithic Man" by Frederic Remington. 1906. There are approximately 14 casts of this sculpture.

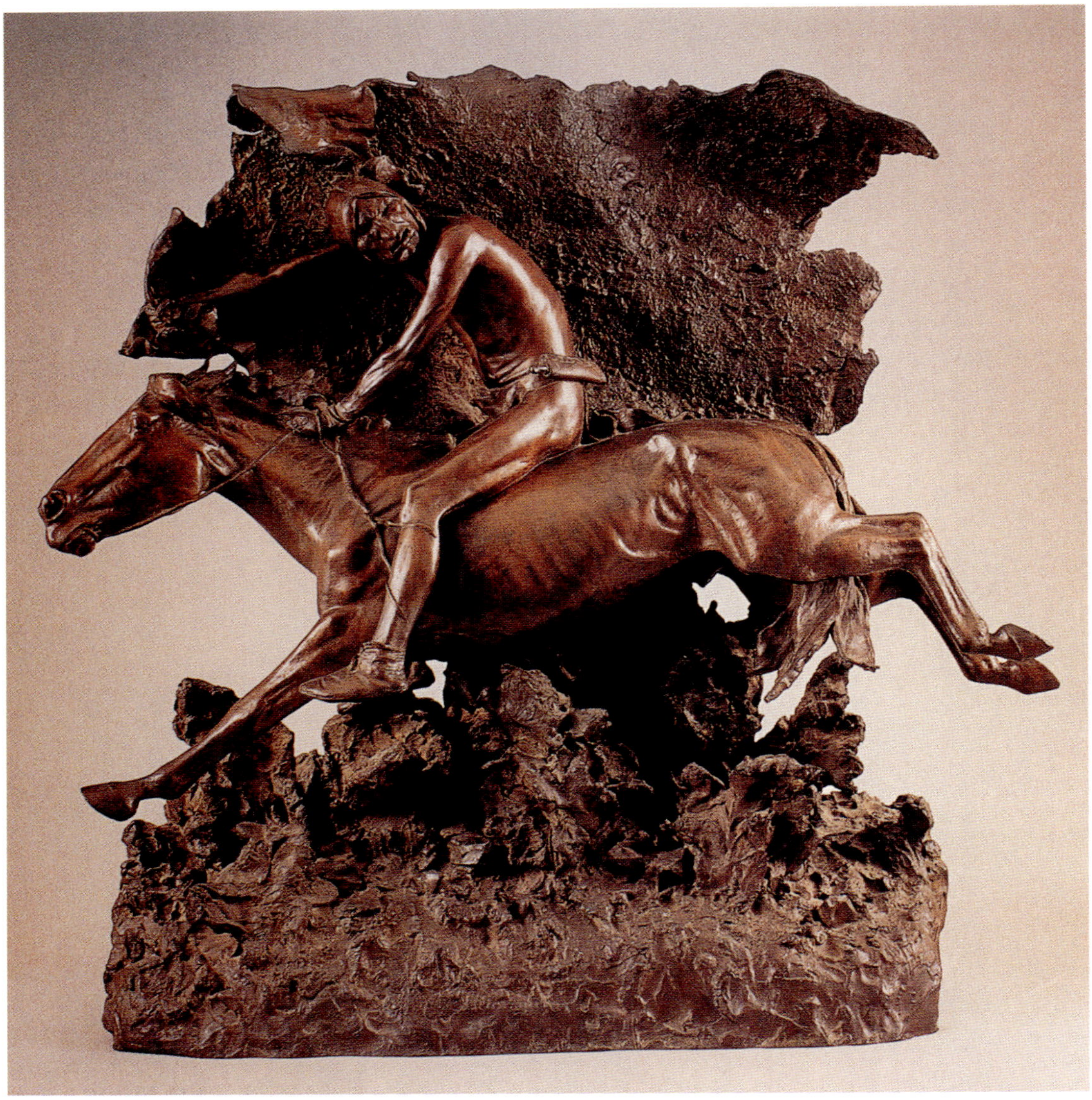

IV-16. "The Horse Thief" by Frederic Remington. 1907.

Below:
IV-18. "The Savage" by Frederic Remington. 1908.

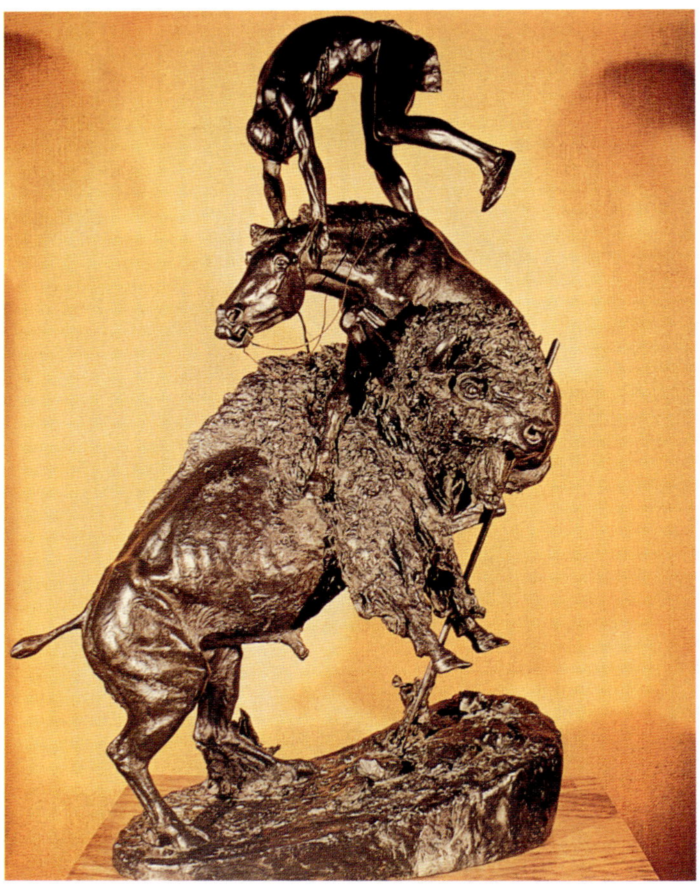

IV-17. "The Buffalo Horse" by Frederic Remington. 1907.

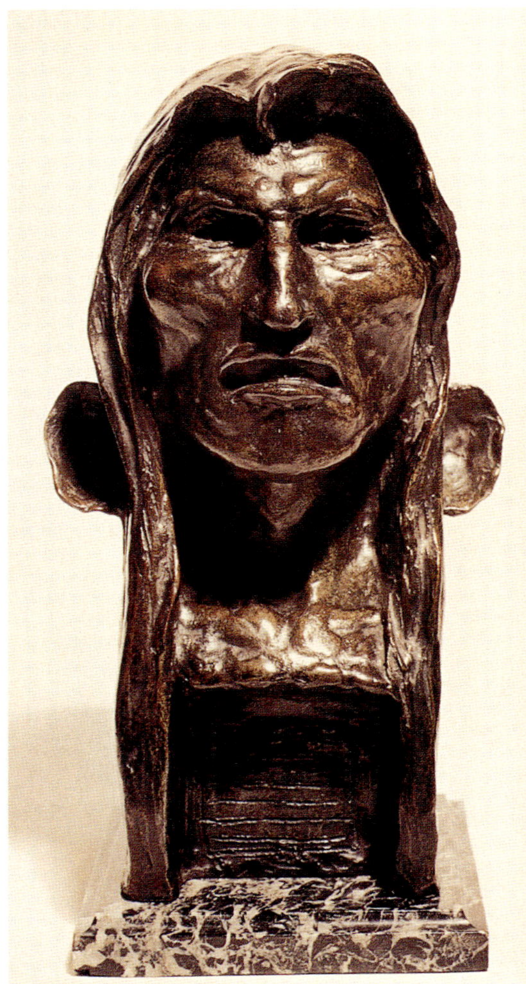

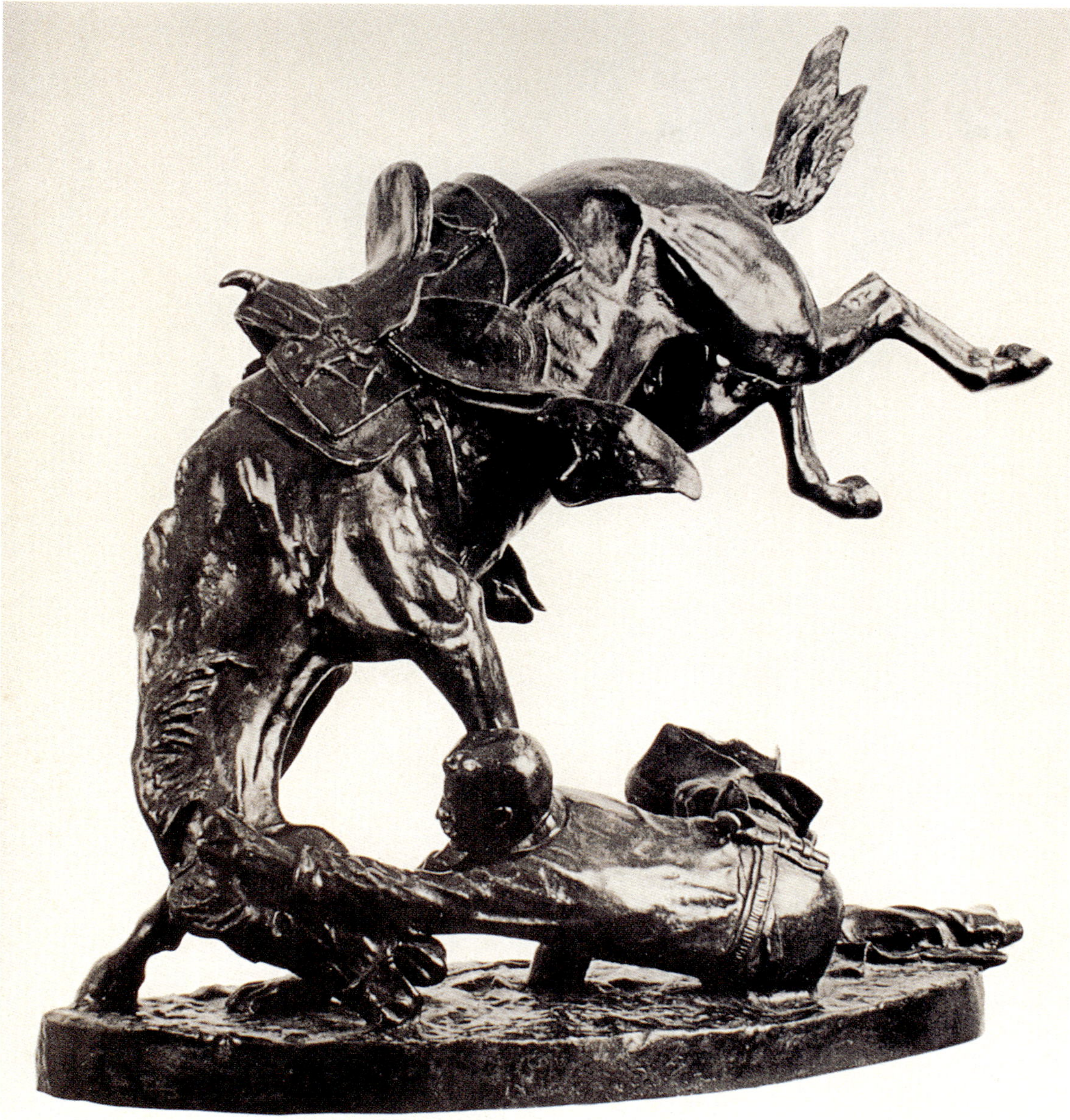

IV-19. "The Wicked Pony" (also called "The Fallen Rider") by Frederic Remington. Originally cast in 1898 and transferred to R.B.W. where it was cast seventeen times.

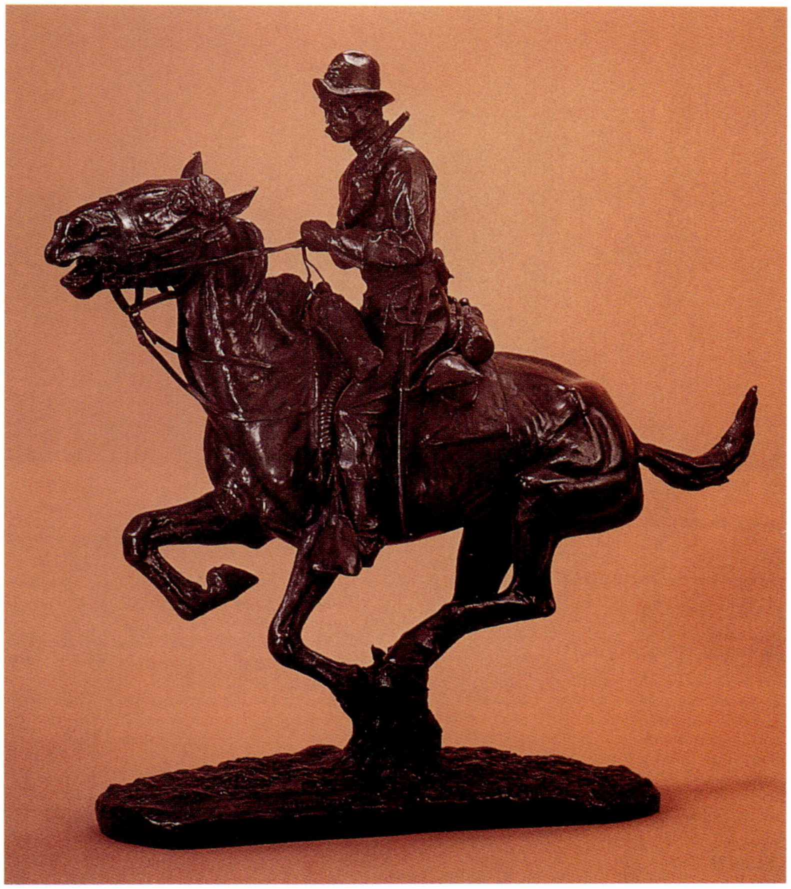

IV-20. "Trooper of the Plains" by Frederick Remington. R.B.W. 1909.

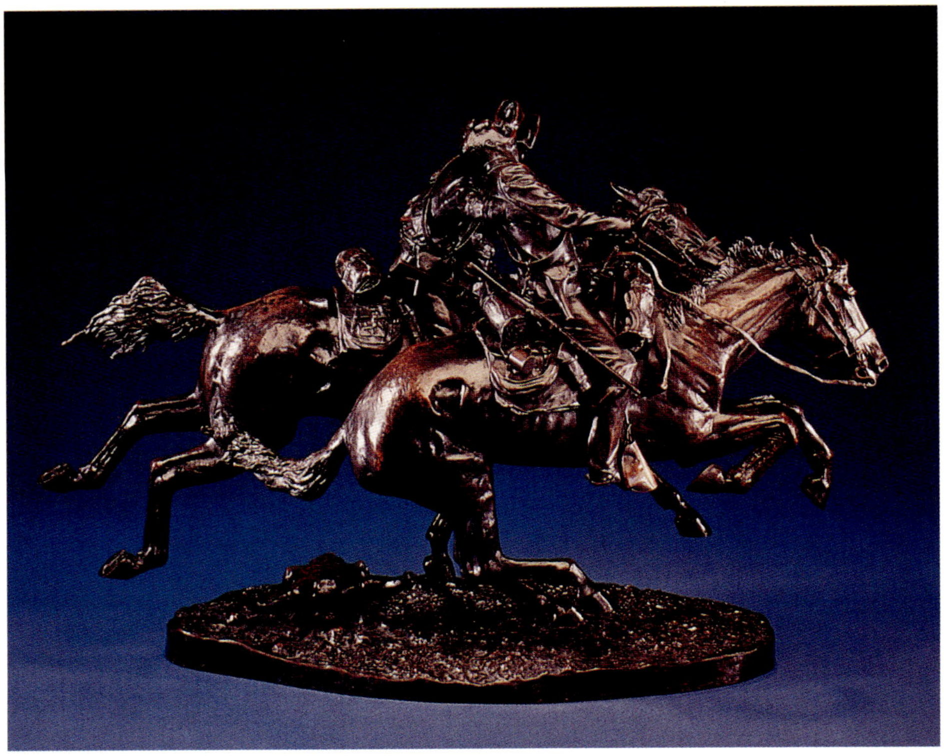

IV-21. "The Wounded Bunkie" by Frederic Remington. First cast in 1896.

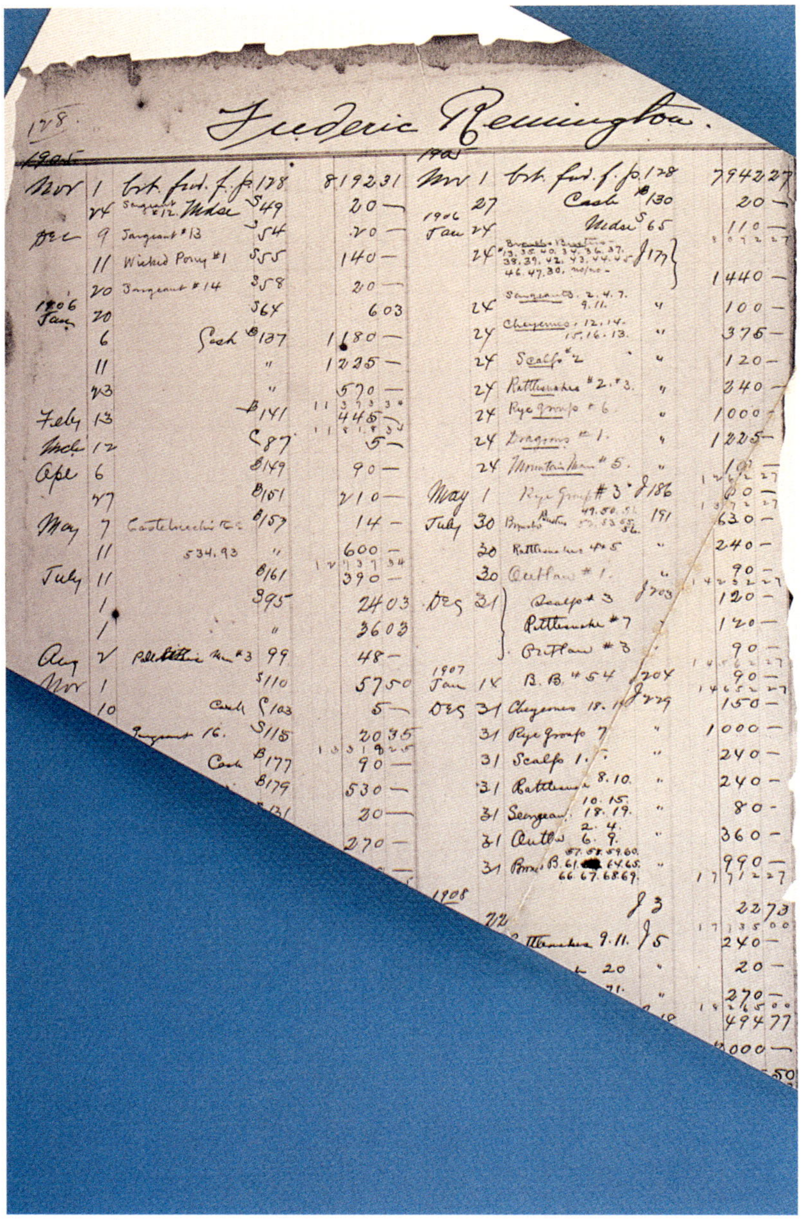

IV-22. Page of R.B.W. ledger for Frederick Remington. 1905.

CHARLES MARION RUSSELL (1865-1926)

Charles Russell, a self-taught artist of Western life, was known for his small bronze statuary (some fifty-three of them were cast during his lifetime), as well as for paintings of frontier life. Russell carried a small wad of beeswax around with him, so that he was able to model on the spot the bucking broncos, drunken cowboys, night herders, and buffalo, wolves, and other wildlife that he encountered in a life on the range. Working as a painter, a sculptor, and an illustrator, Russell achieved success as an artist who could picture the mythology of the West with life-like accuracy.

Russell was born in Missouri and as a young man went to Montana Territory, where he lived with a trapper for two years and then became a cowhand for the following eleven. In 1893 he received his first major commission, and from that time on became a full-time artist of Western images. Most of his sculptural work was cast in small editions of about twenty-four.

Less interested in the heroic exploits of cowboys and Indians than Remington, Russell was an early conservationist, who cared deeply about the plight of native Americans and deplored the destruction of natural wildlife. He saw in his art a chance to show harmony between man and nature, as he saw the West before "civilization" overtook it. Many of his sculptural images were efforts to show the public back home what natural beauty and harmony existed in the "old" West.

In 1903 he visited New York, where he modeled his first statuette to be cast in bronze: "Smoking Up" (IV-23). (It pictured a drunken cowboy shooting up the town as he rides.) Following its success he made many more such small bronzes of cowboy and pioneer life (numbers IV- 24 to IV-27), and of native Americans (IVs-28 through 37). He also cast a number of images of bears, buffalo, antelopes, wolves, steer, and mountain goats (numbers IV-38 to IV-44).

Though Russell used a California foundry as well, the majority of his works were eventually cast at Roman Bronze Works between 1903 and 1915, and posthumously. (Among those cast at R.B.W., but not pictured here, are: "The Lunch Hour," "Mountain Sheep," "Nature's Cattle," "American Cattle," "Weapons of the Weak," "Bear with the Jug," "Meat for Wild Men," "Buffalo Lying Down," "Jim Bridger," "Watcher of the Plains," "Lone Wolf," "The Robe Flesher," "The Texas Steer," "Smoking with the Spirit of the Buffalo," and a pair of buffalo bookends.) Altogether, Russell made more than forty different sculptures with Western themes. Museums throughout the country have Russell's Western bronzes in their collections. An example of R.B.W.'s ledger for Russell can be seen in (IV-45).

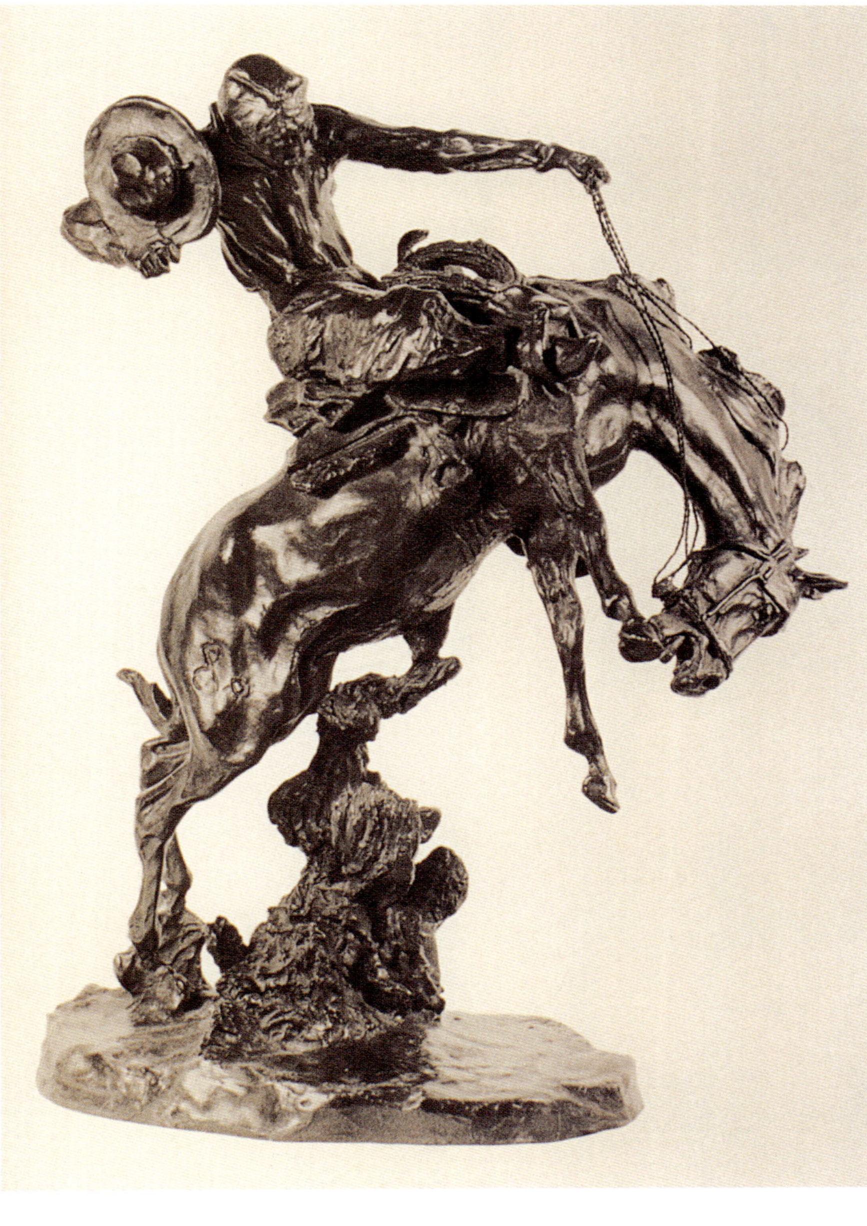

IV-23. "Smoking Up" by Charles Marion Russell. 1903.

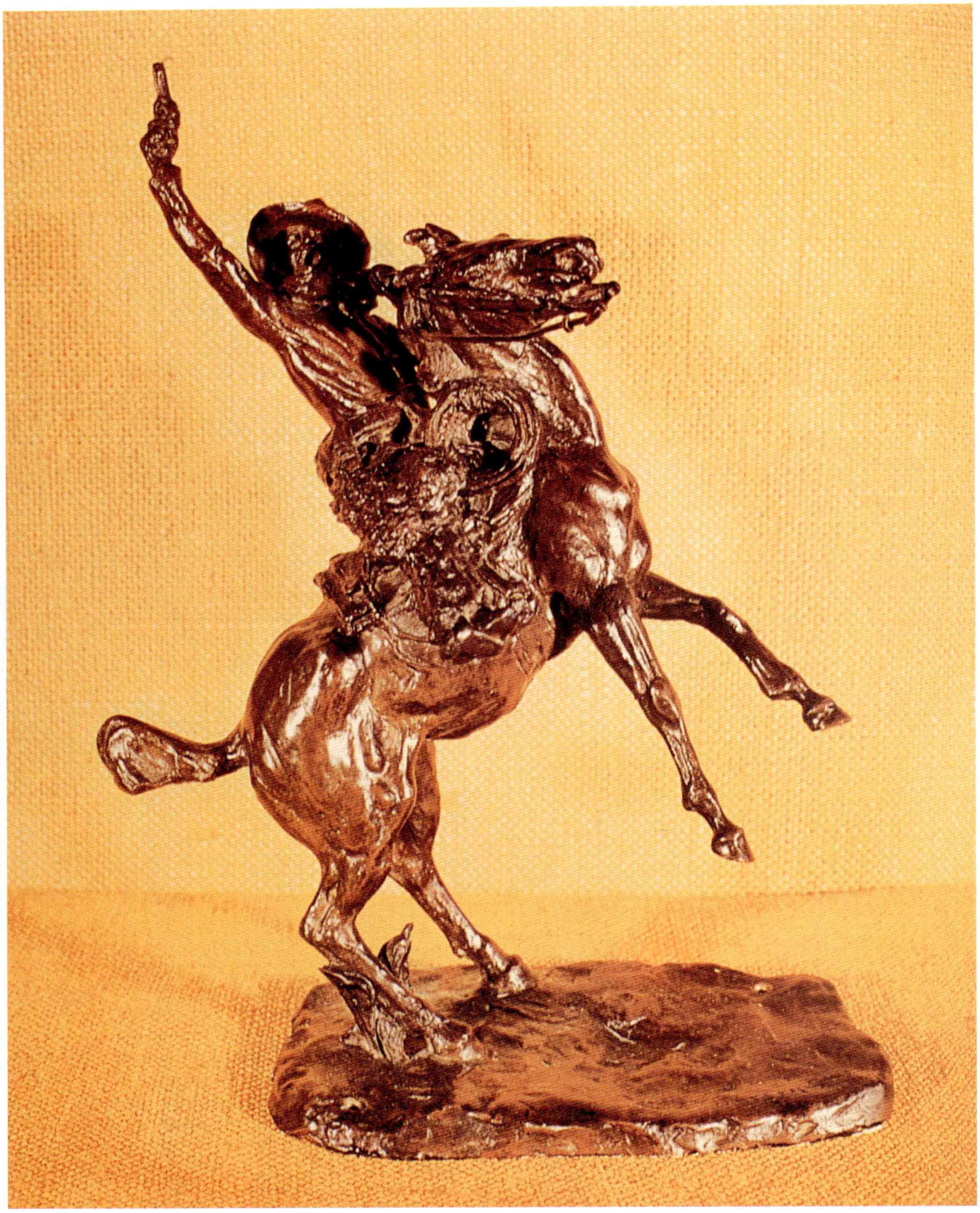

IV-24. "The Range Father" by Charles Marion Russell. 1903.

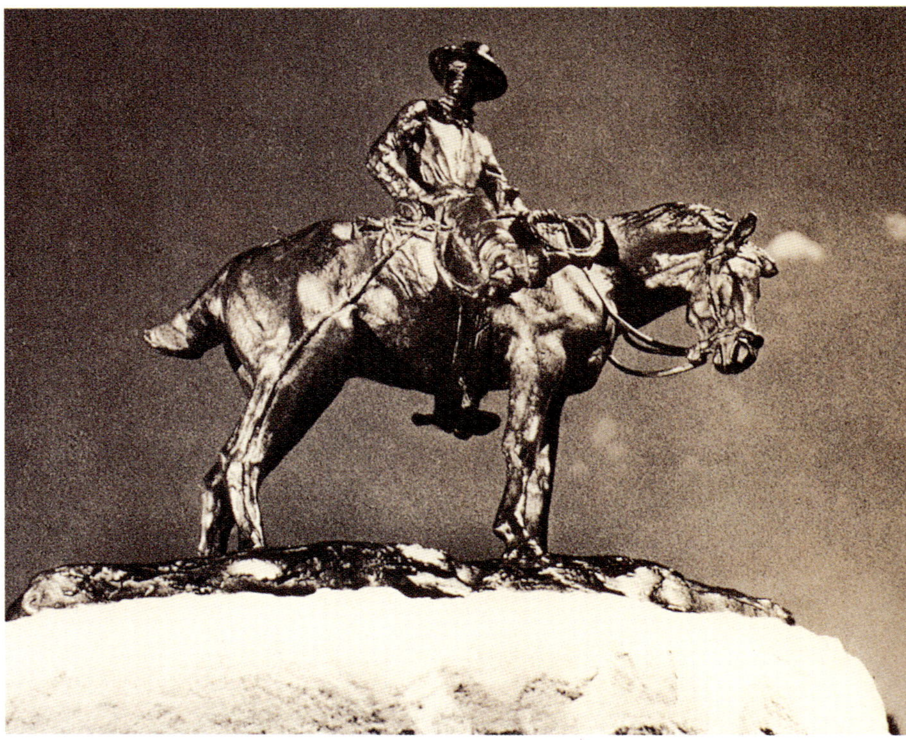

IV-25. "Night Herder (Self Portrait)" by Charles Marion Russell. 1925.

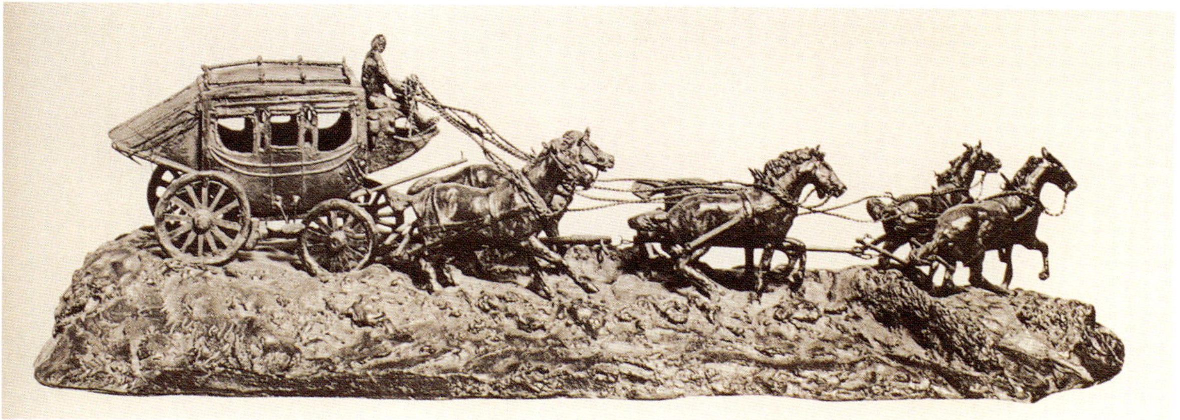

IV-26. "It Ain't No Lady's Job" by Charles Marion Russell. 1926. R.B.W. 1962.

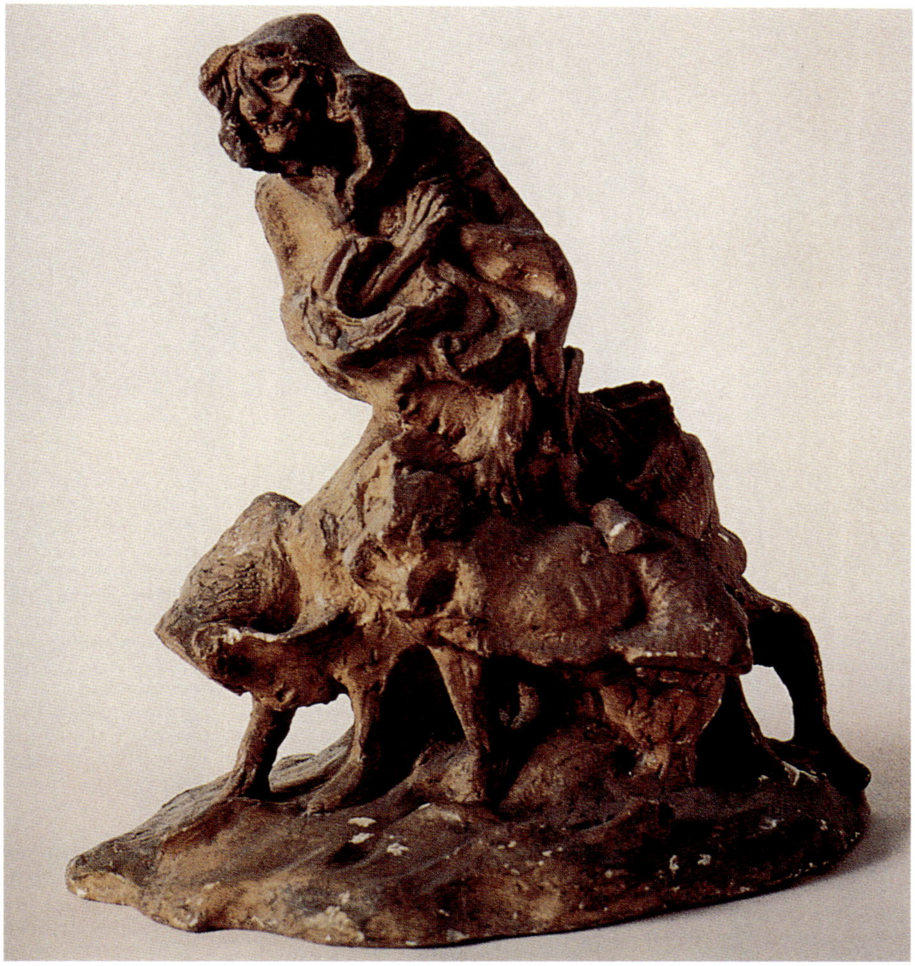

IV-27. "Spirit of Winter" by Charles Marion Russell. c. 1926.

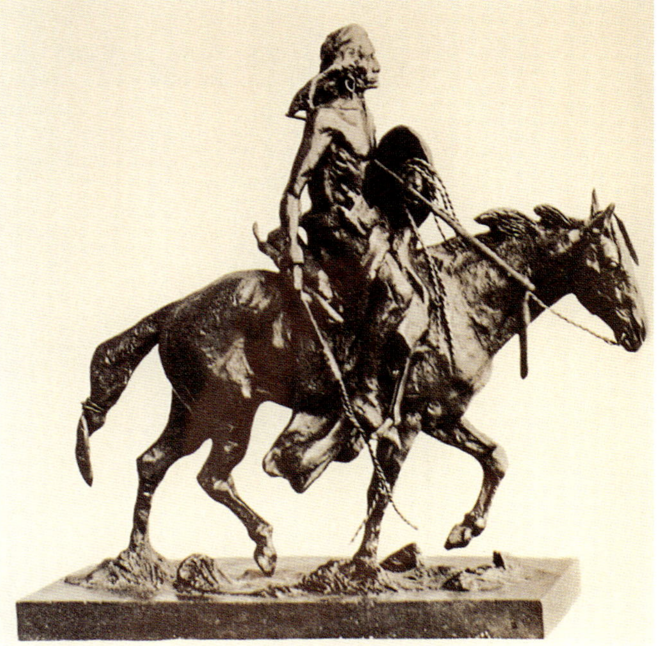

IV-28. "Blackfoot War Chief" by Charles Marion Russell. c. 1900.

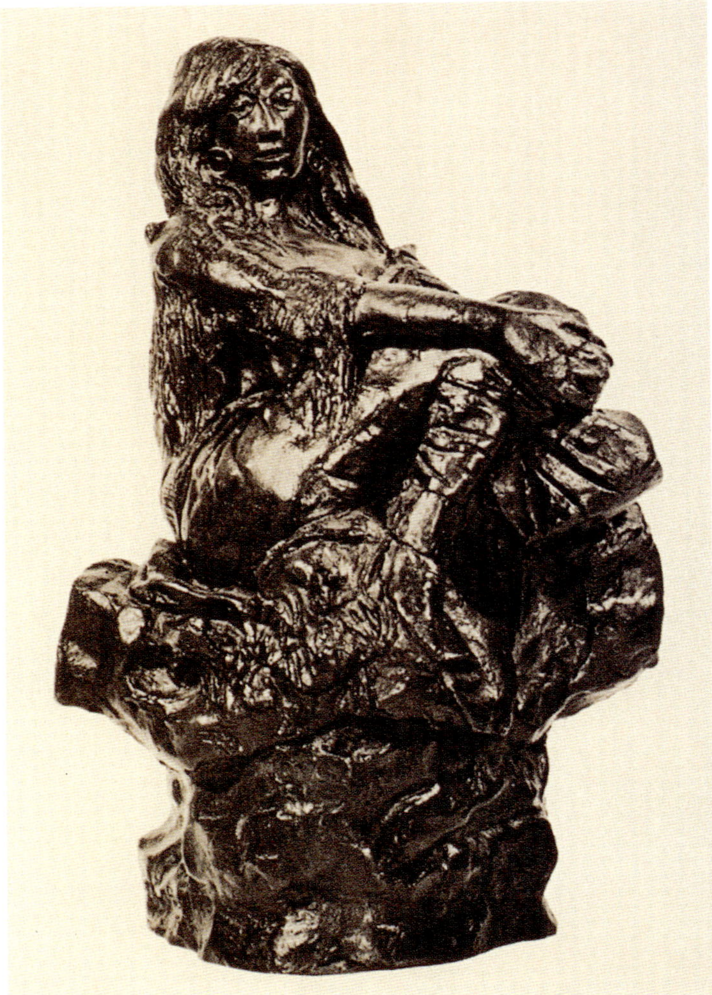

IV-29. "Piegan Girl" by Charles Marion Russell. 1902.

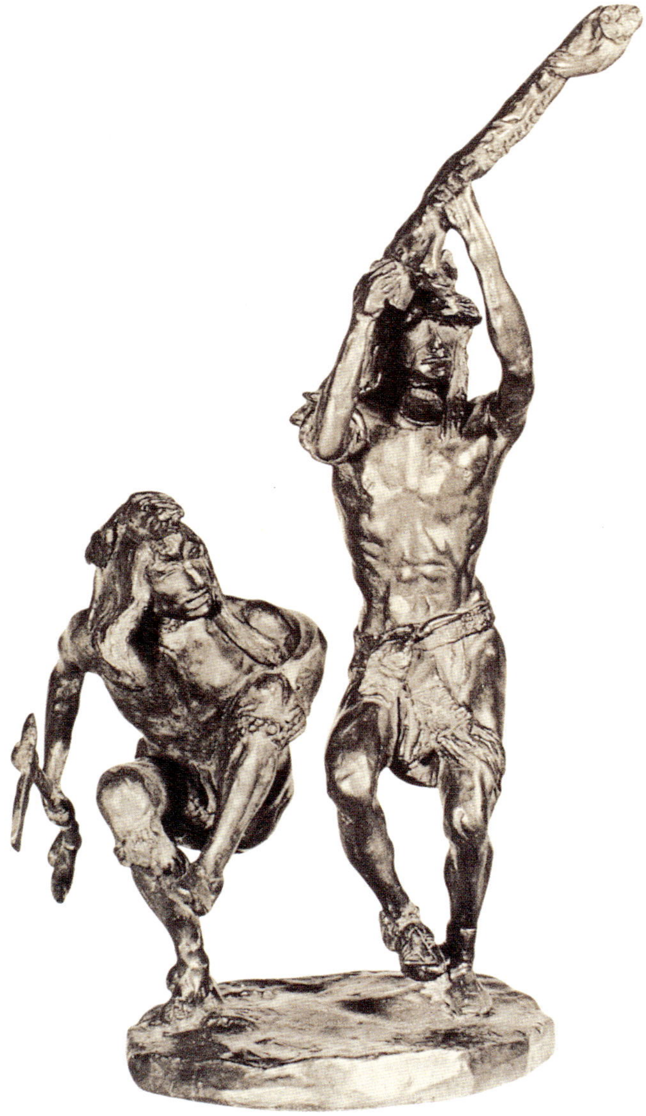

IV-30. "Scalp Dance" by Charles Marion Russell. c. 1904

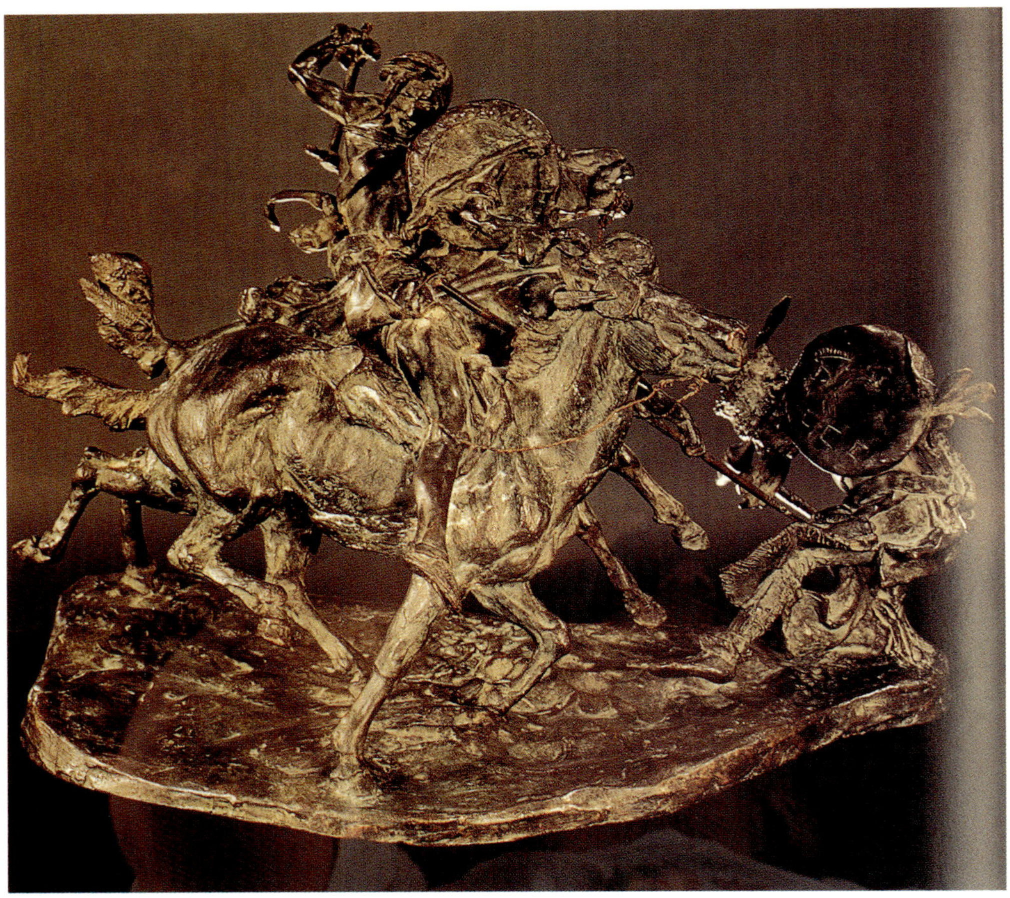

IV-31. "Counting Coup" by Charles Marion Russell. 1904.

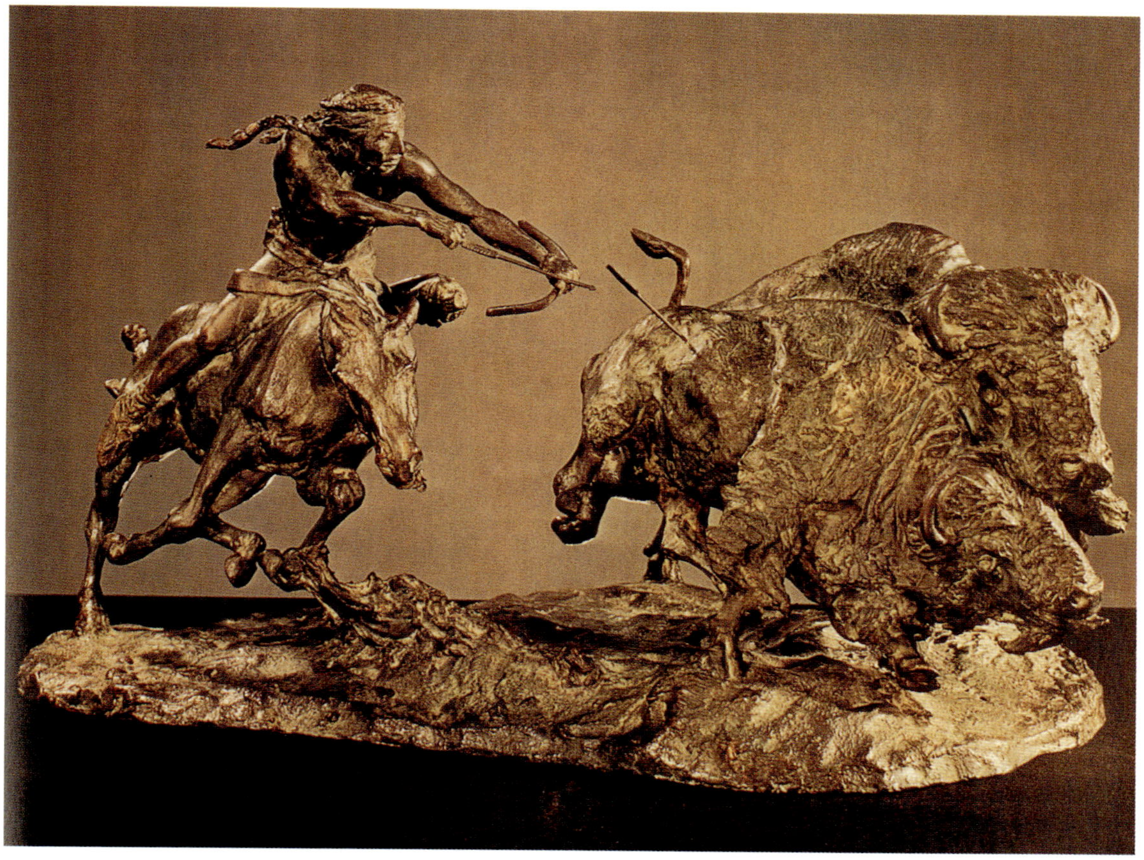

IV-32. "The Buffalo Hunt" by Charles Marion Russell. 1904-5.

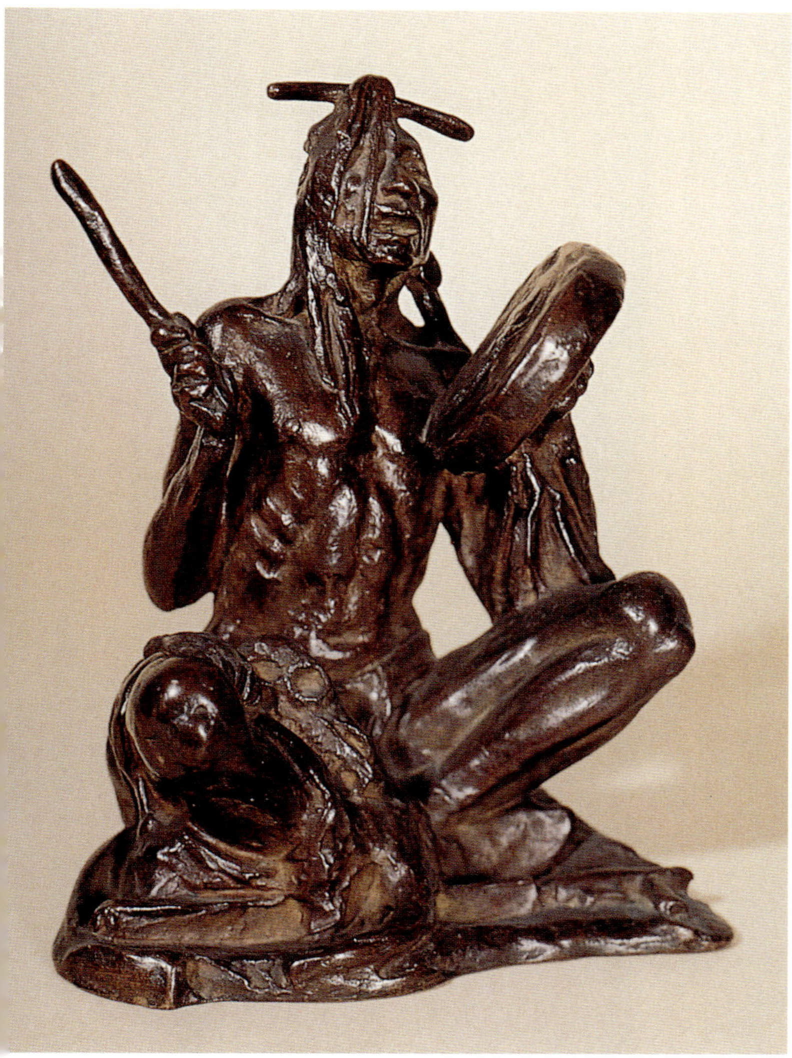

IV-33. "The Medicine Man" by Charles Marion Russell. c. 1920.

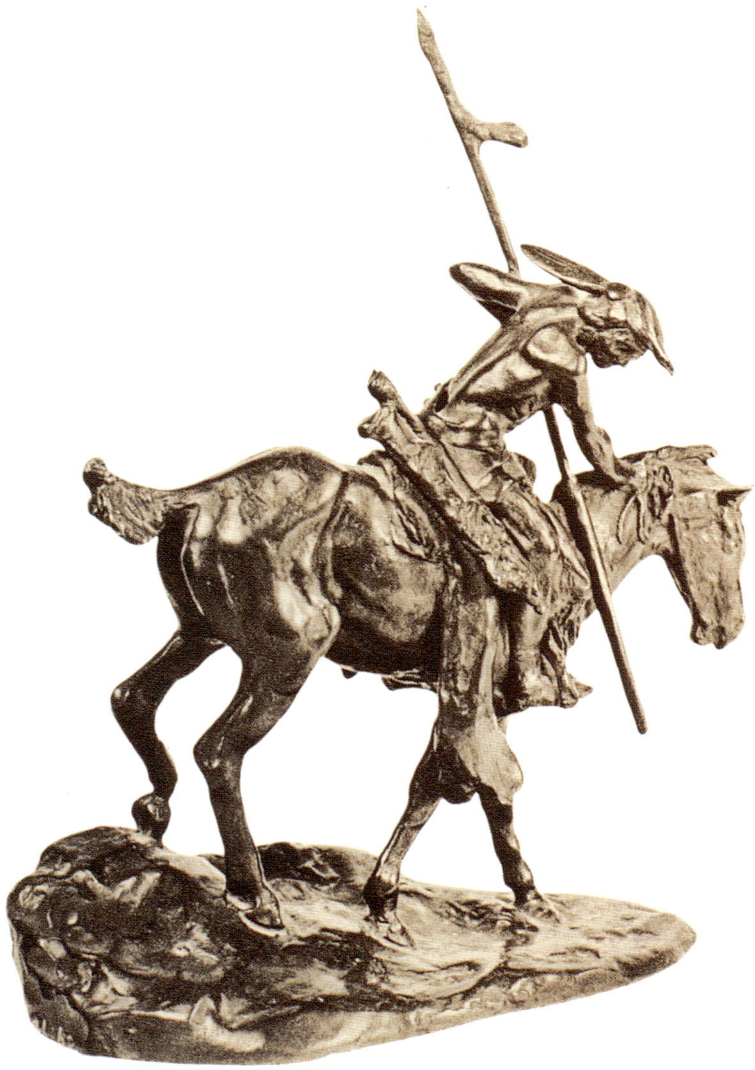

IV-34. "The Enemy's Tracks" by Charles Marion Russell. c. 1920.

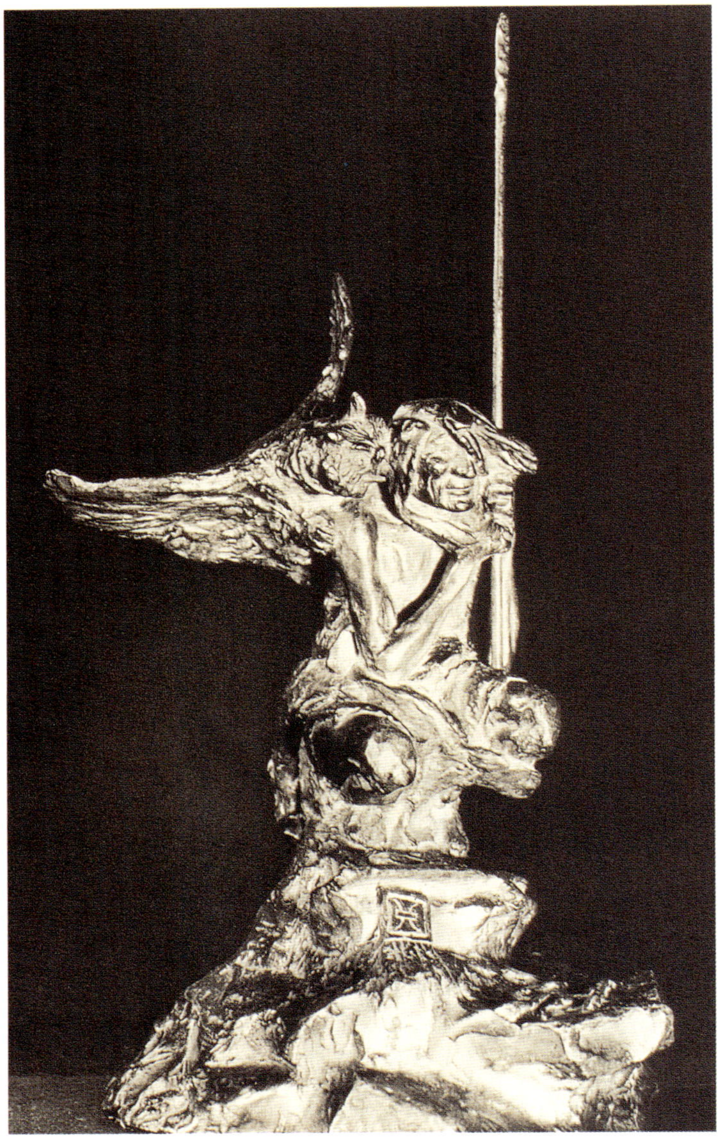

Left:
IV-35. "Secrets of the Night" by Charles Marion Russell. c. 1926.

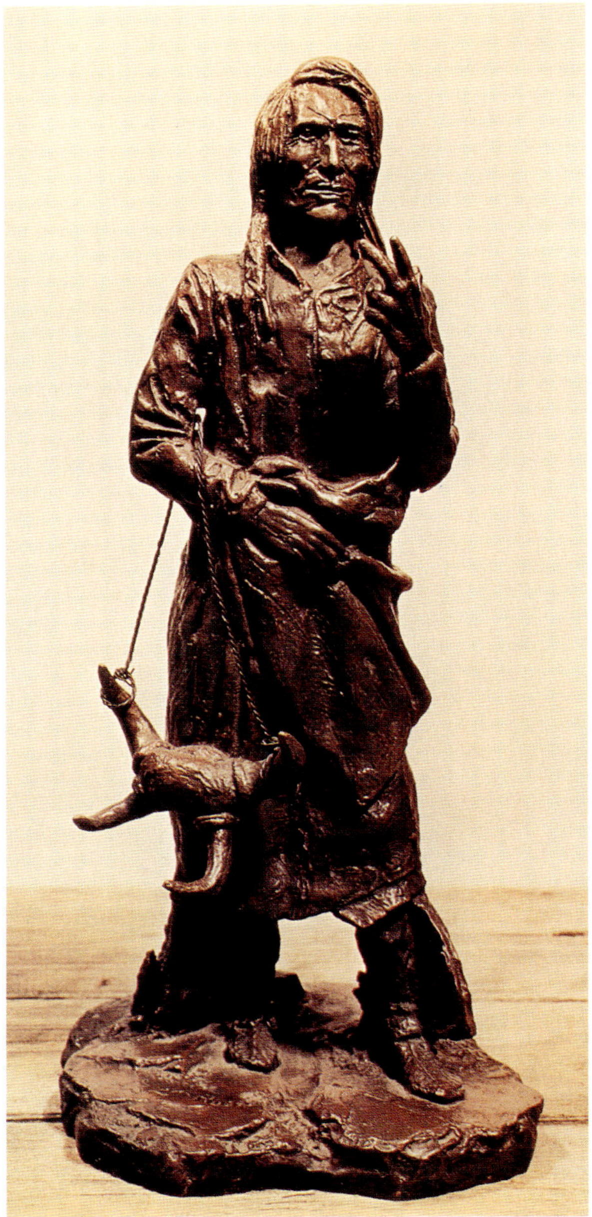

IV-36. "In the White Man's World" by Charles Marion Russell. 1947.

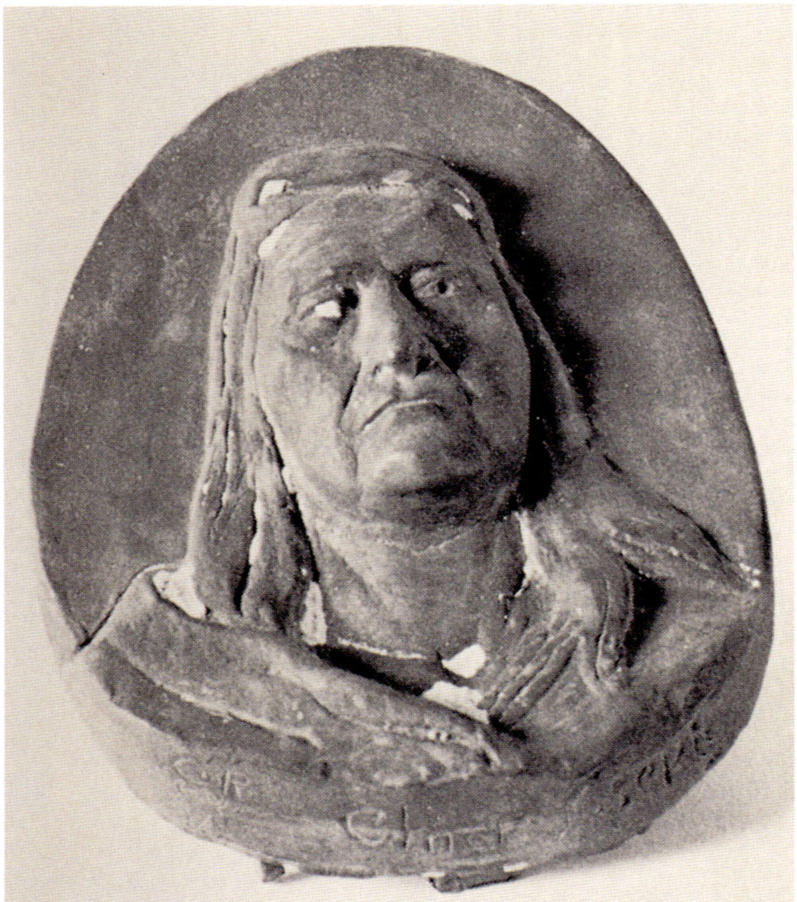

Left:
IV-37. "Chief Joseph" (plaque) by Charles Marion Russell.

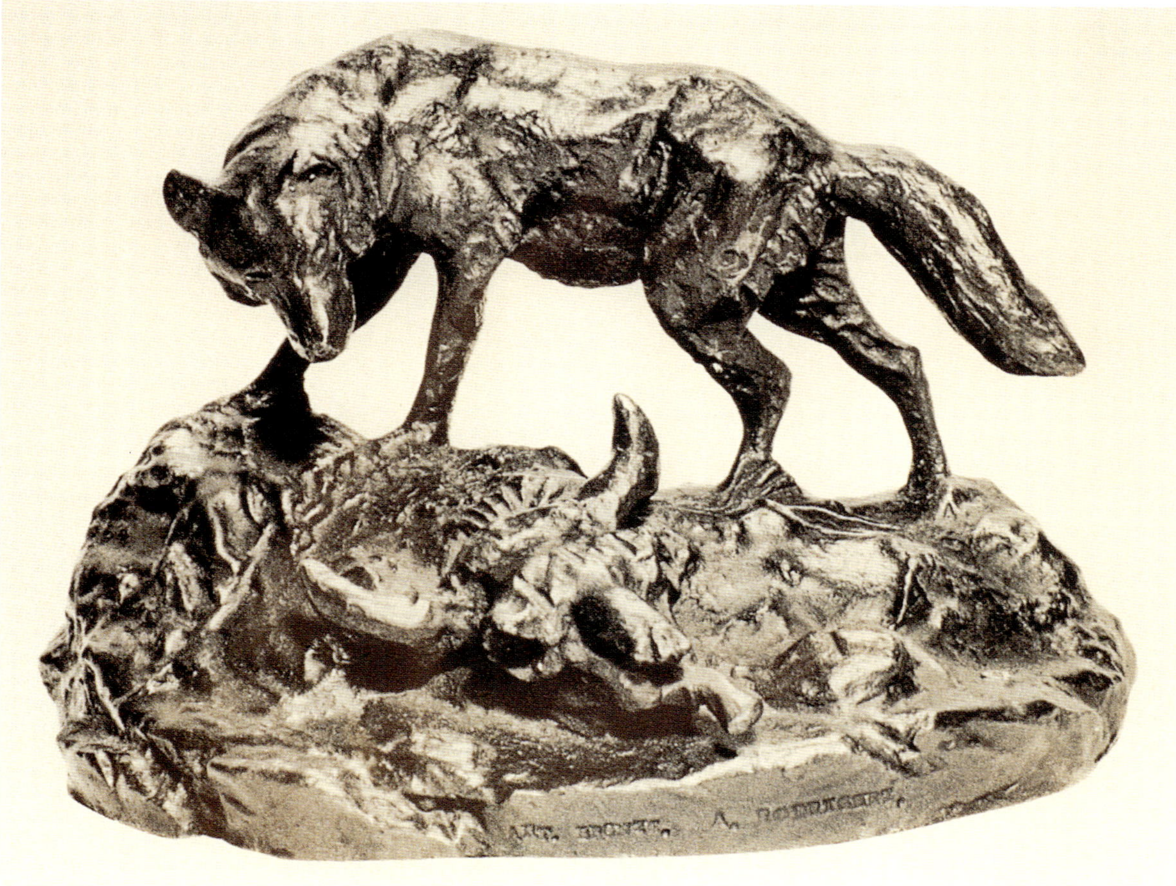

IV-38. "The Fallen Monarch" by Charles Marion Russell. c. 1909.

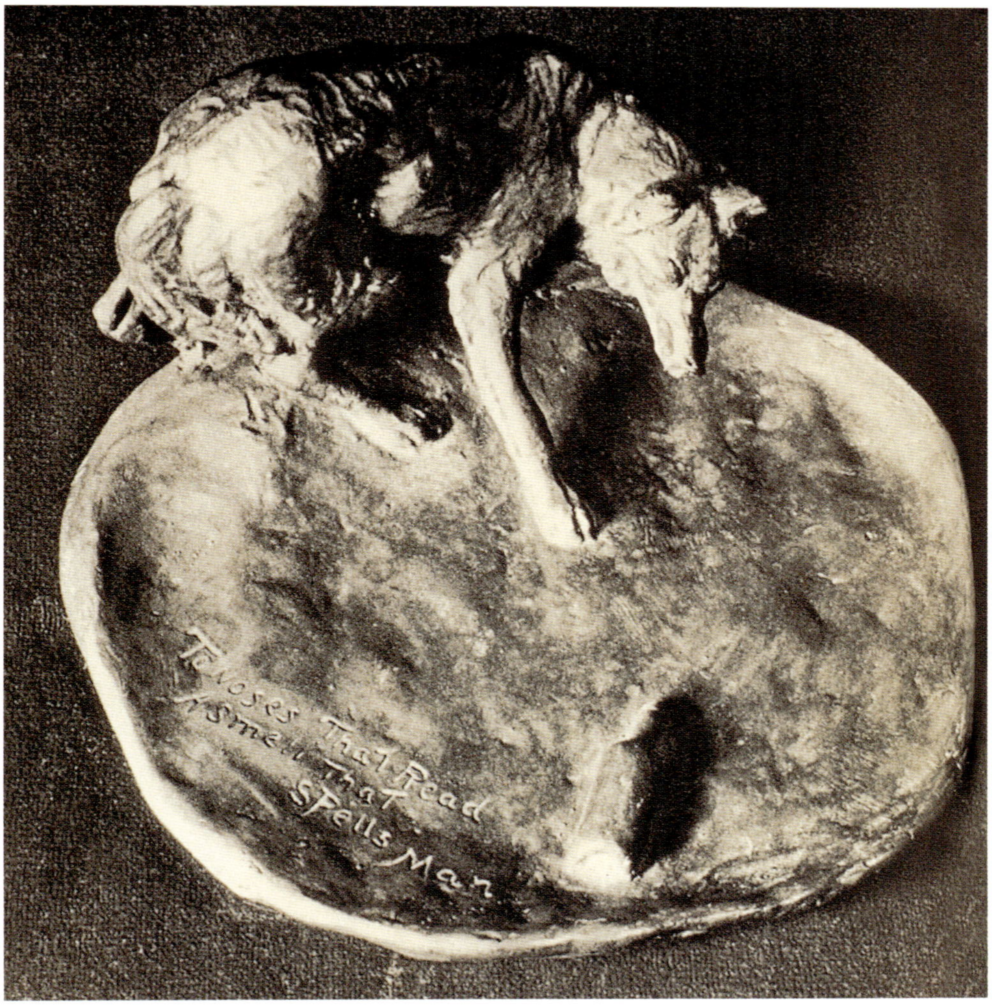

IV-39. "To Noses That Read, a Smell That Spells Man" by Charles Marion Russell. c. 1920.

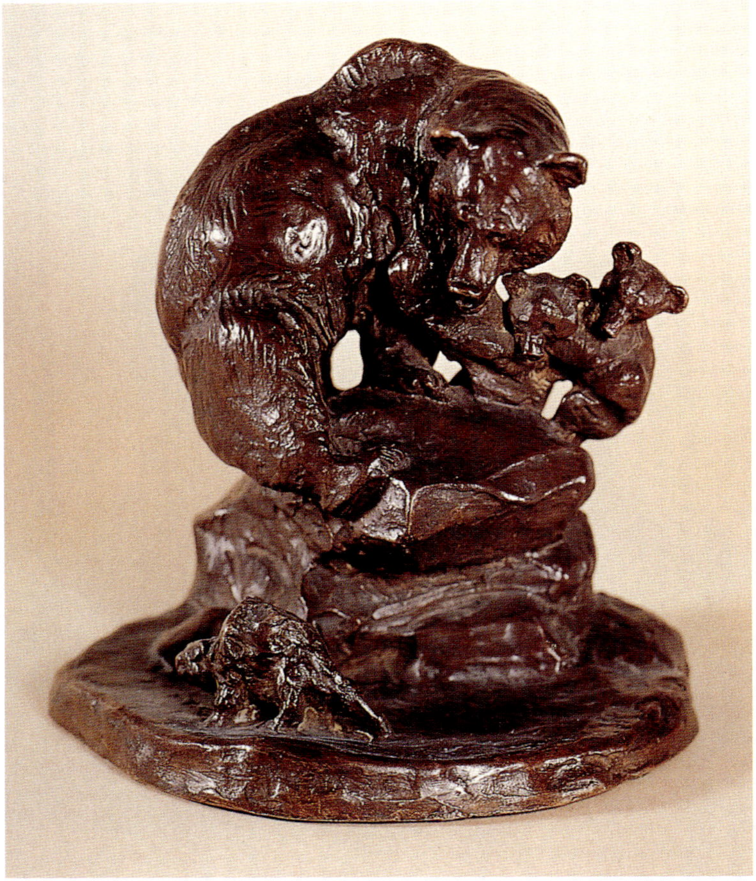

IV-40. "Weapons of the Weak" by Charles Marion Russell. Modeled 1921; cast 1925.

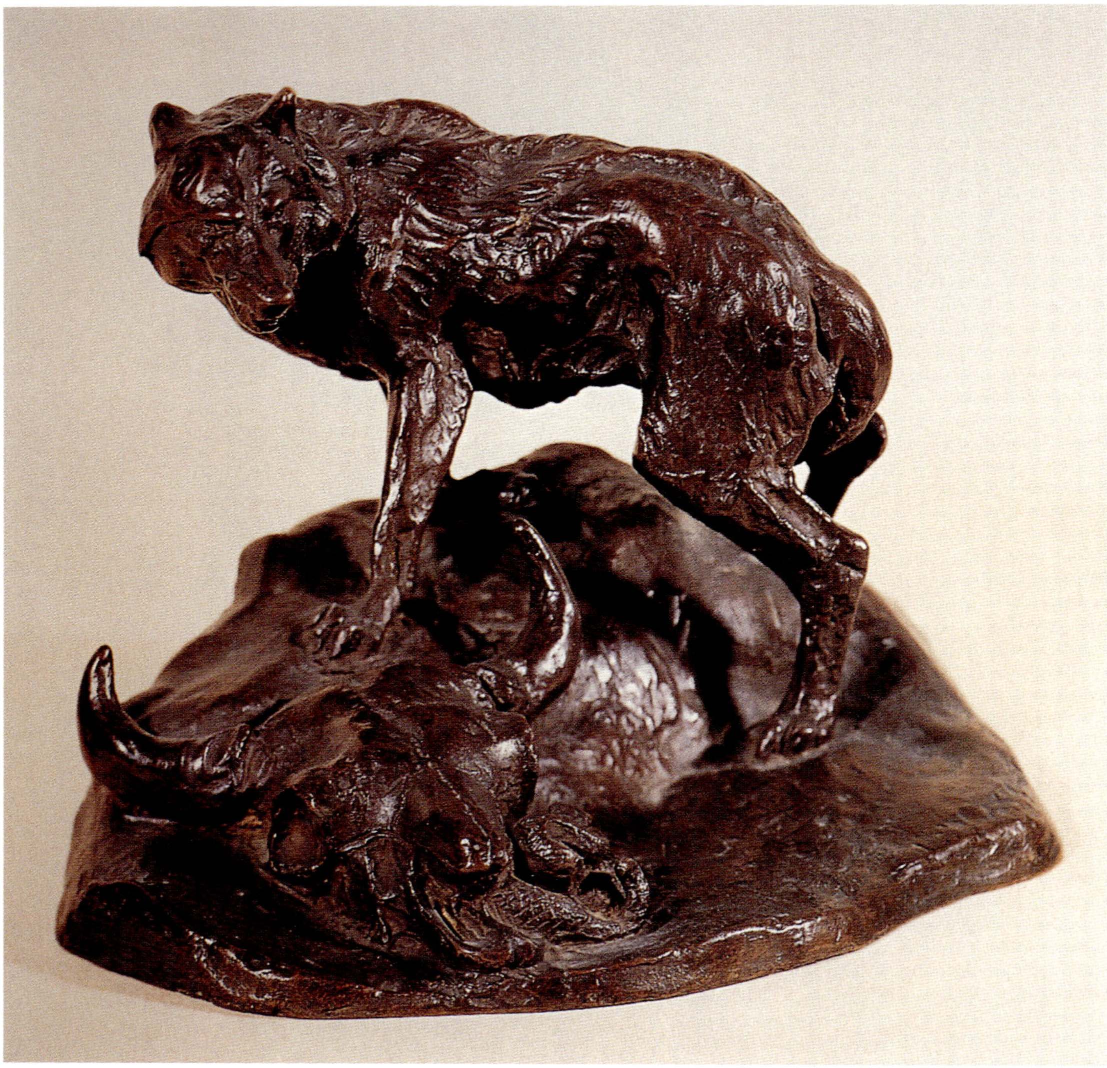

IV-41. "An Enemy That Warns" by Charles Marion Russell. 1921.

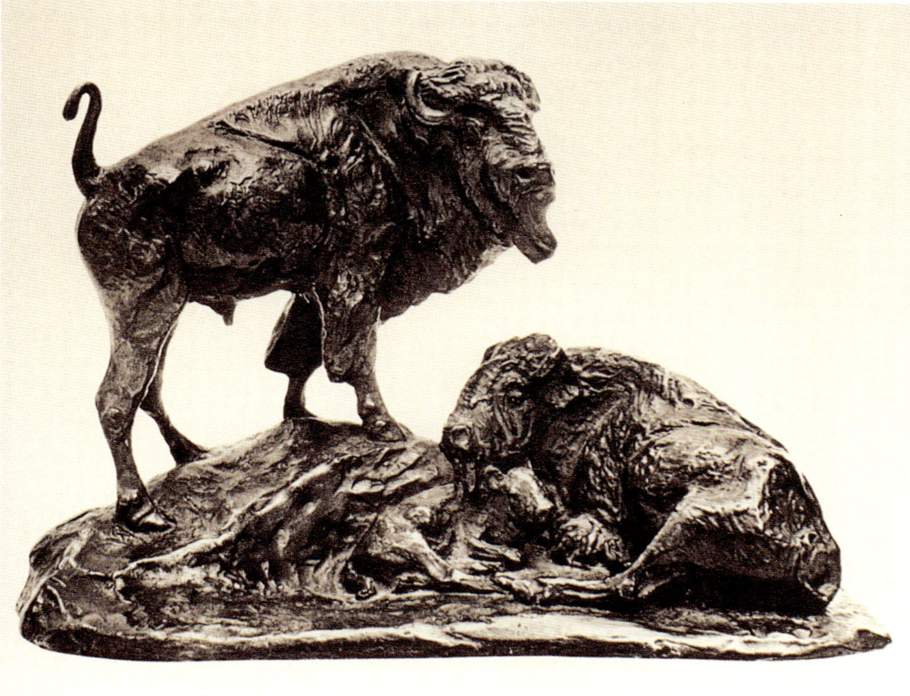

IV-42. "The Buffalo Family" by Charles Marion Russell. c. 1925.

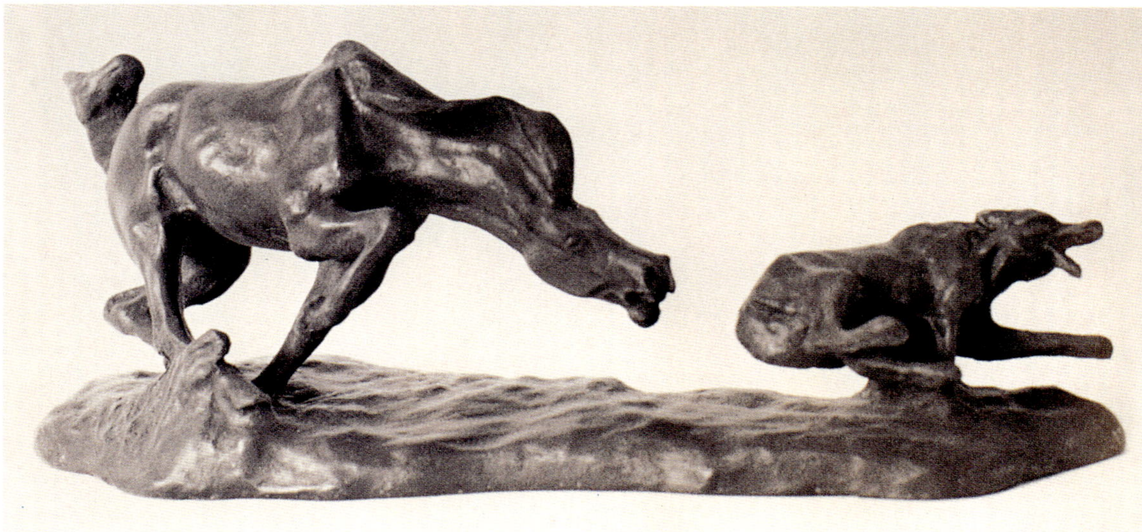

IV-43. "The Range Father" by Charles Marion Russell. c. 1926.

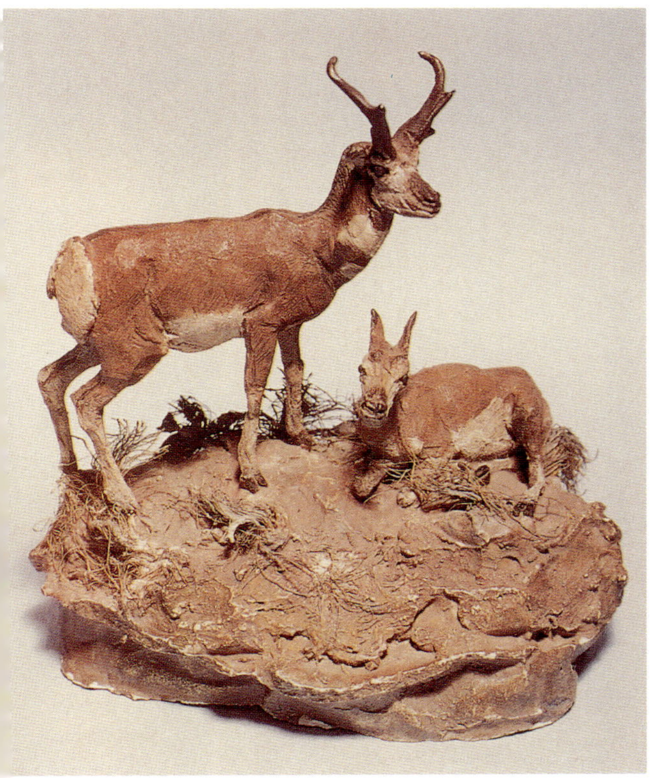

IV-44. "Nature's People" by Charles Marion Russell. Modeled 1905; cast at R.B.W. 1968.

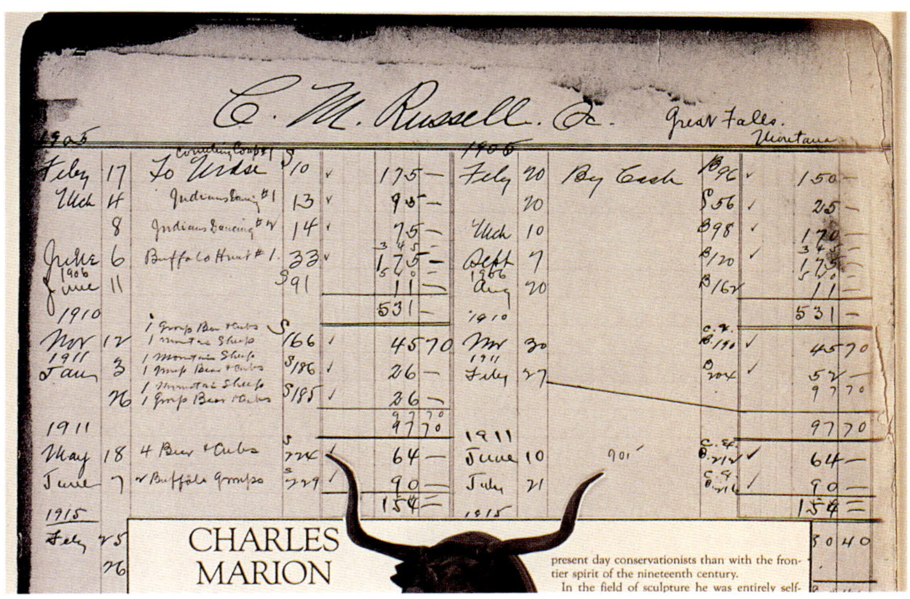

IV-45. An example of a Roman Bronze Works ledger page for Charles Marion Russell, 1905.

SOLON BORGLUM (1868-1922)

The younger brother of Gutzon Borglum, Solon Borglum became one of the nation's first, and most prolific sculptors of Western images. As a youth he worked on a cattle ranch and became fascinated by Plains Indians and Western scenes. The primitive "Wild West" became a lifelong theme for his art, and he produced numerous action-filled sculptures depicting cowboys, Indians, and horses, as well as several equestrian monuments.

Borglum, encouraged by his brother with whom he studied in California, attended the Cincinnati Art Academy, and in 1897 traveled for further study to the Académie Julian in Paris. By the next year he was already exhibiting Western images. In 1899 his "On the Border of White Man's Land" (IV-46) won a medal at the Paris Exposition. He settled in New York in 1901, and exhibited widely, including his well-known "Blizzard" (IV-47) and "Sioux Indian Buffalo Dance" (IV-48) at the St. Louis Exposition in 1904. His monument to one of the Rough Riders, "Captain Bucky O'Neill" (IV-49) was one of his best known works.

Borglum produced dozens of natural-looking sculptures that conveyed the vitality of Western life with a Rodin-influenced combination of realism and blurred outlines that effectively conveyed action and dynamism. His themes ranged from bucking broncos, to careful representations of native American life, to romantic images of night on the prairie, such as his "Night Hawking" (IV-50). An expert in the realistic portrayal of anatomy, Borglum wrote a book on comparative anatomy that was posthumously published. His thorough knowledge of anatomy was evident in his many studies of horses and cattle, among them "The Bucking Bronco" (IV-51), "New Born" (IV-52), "Horse Tamed" (IV-53), "Cowboy at Rest" (IV-54), "Intelligent Bronco" (IV-55), "On the Trail" (IV-56), "The Bronco Buster" (IV-57), and "Evening" (IV-58). In addition to his Western bronzes, he cast an equestrian statue of George Washington at the Roman Bronze Works foundry.

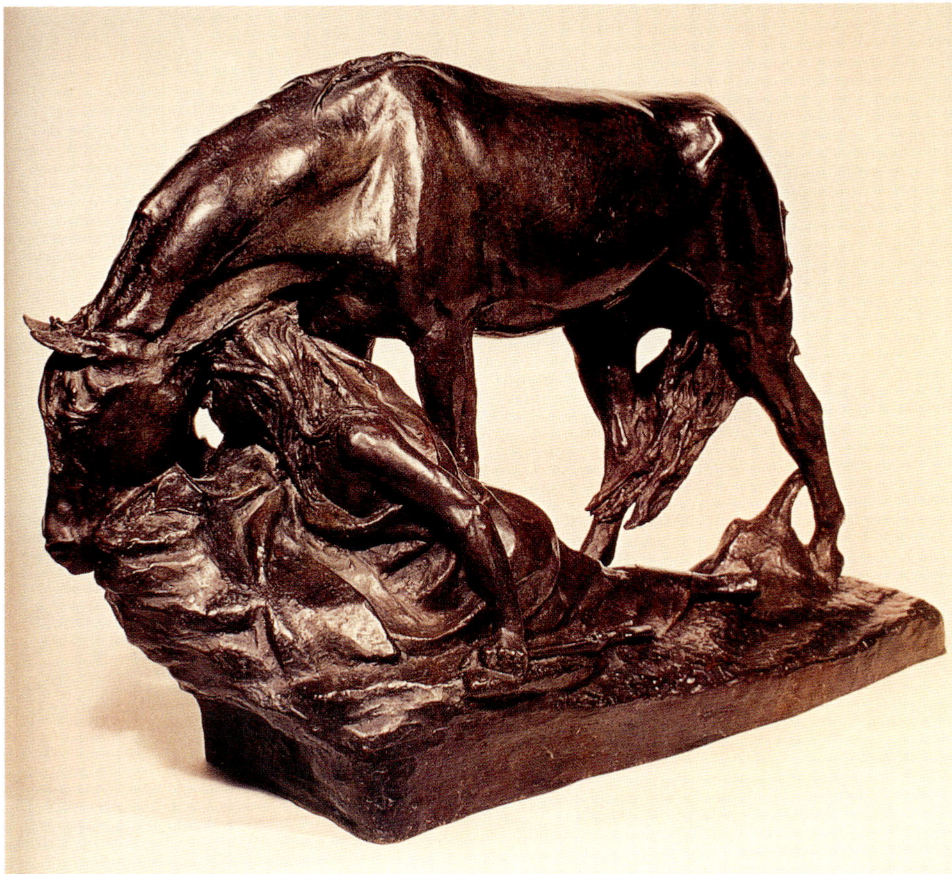

IV-46. "On the Border of the White Man's Land" by Solon Borglum. Modeled 1899, R.B.W. 1906.

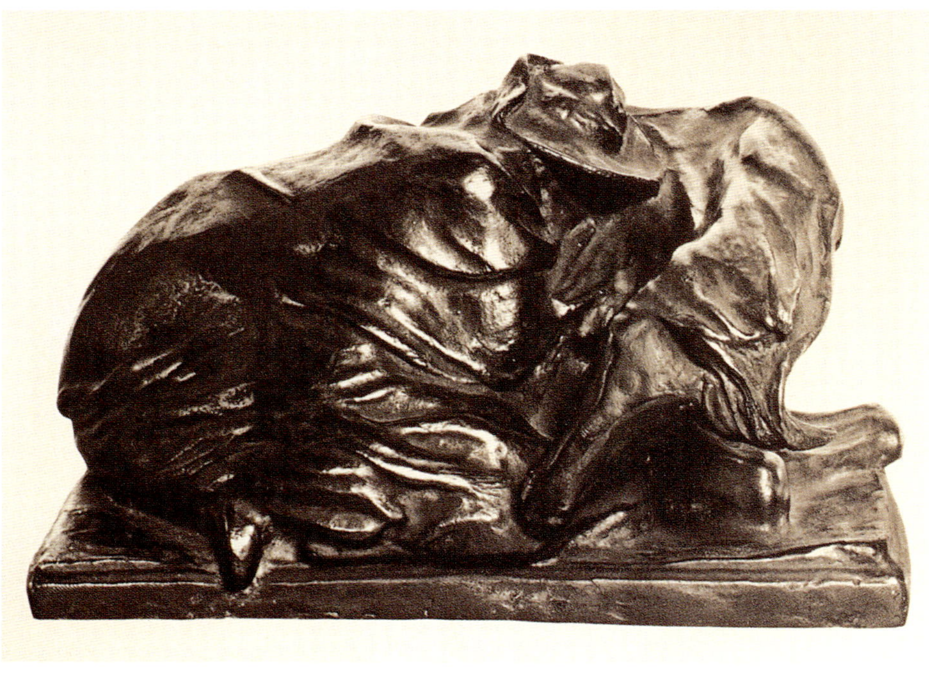

IV-47. "Blizzard" by Solon Borglum. 1900. This was one of the nine works portraying Western life that Borglum showed at the St. Louis Exposition in 1904.

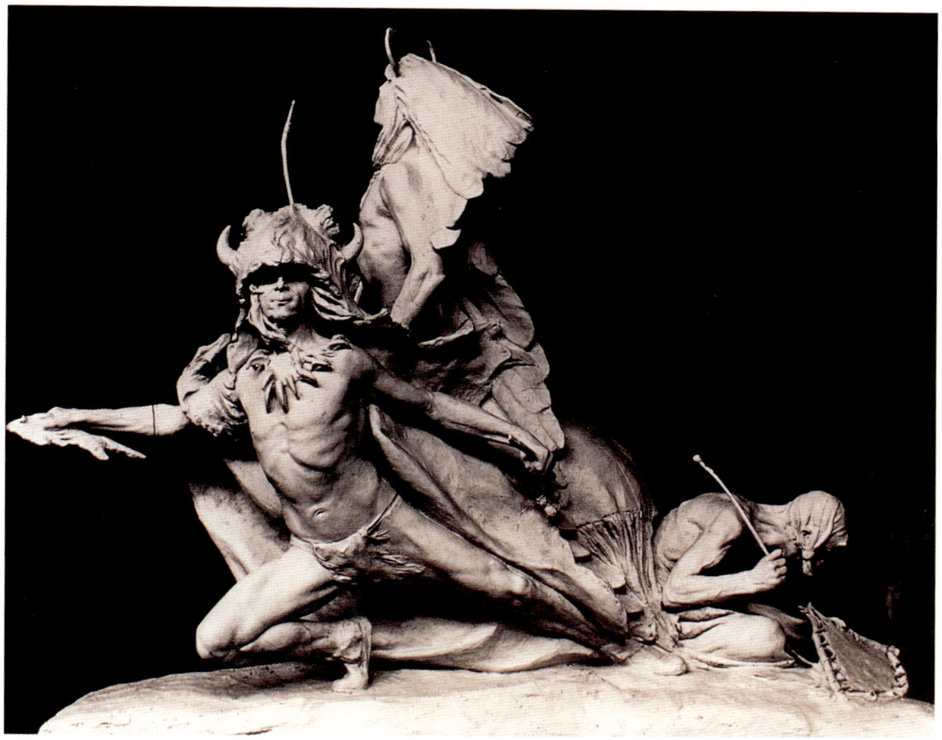

IV-48. "Sioux Indian Buffalo Dance" (plaster cast) by Solon Borglum. This was one of nine sculptures he exhibited at the St. Louis Exposition in 1904.

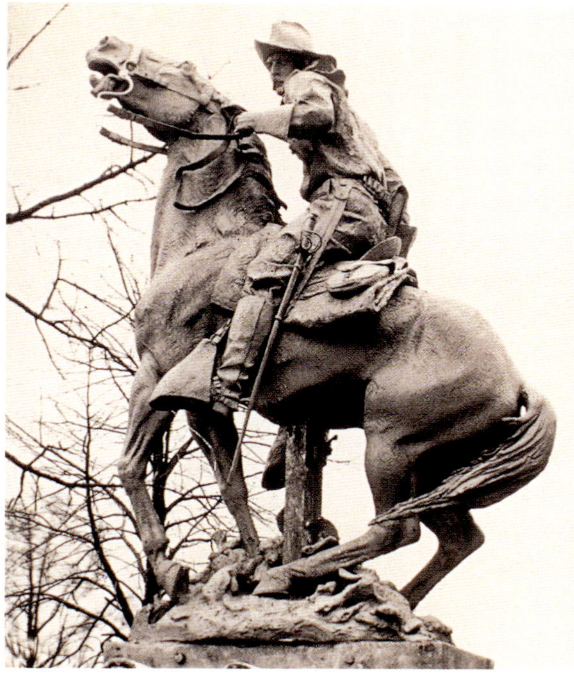

IV-49. "Captain Bucky O'Neill" by Solon Borglum. 1907.

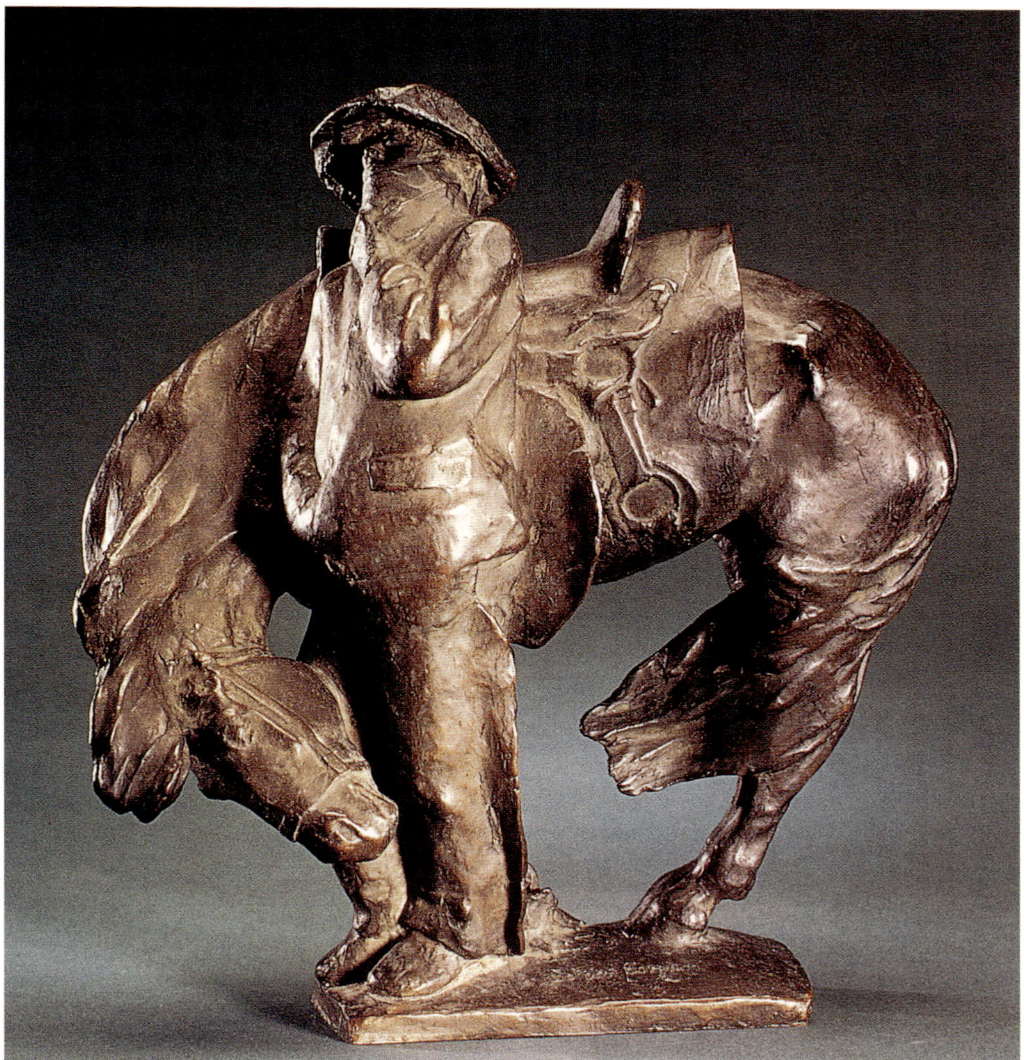

IV-50. "Night Hawking" by Solon Borglum. 1900.

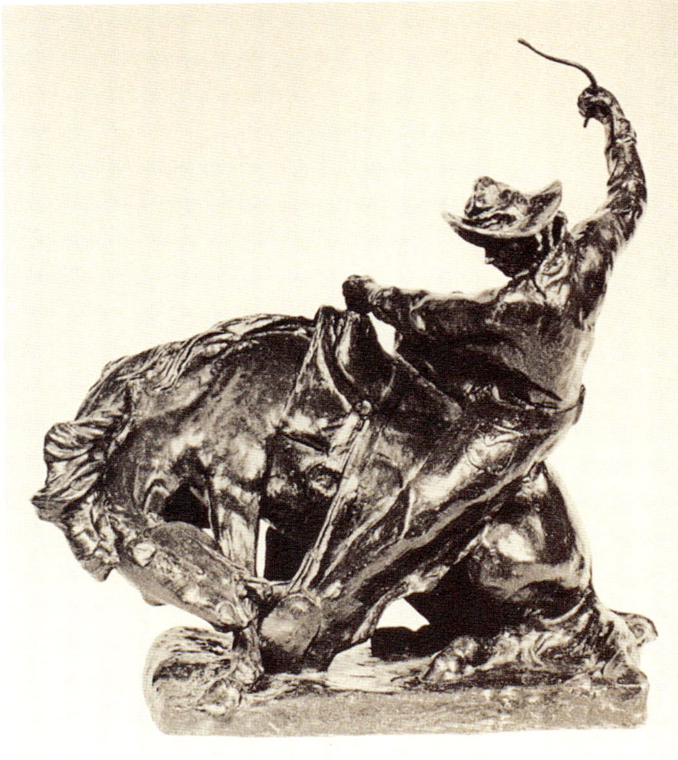

IV-51. "The Bucking Bronco" by Solon Borglum. Modeled 1898. R.B.W. 1914.

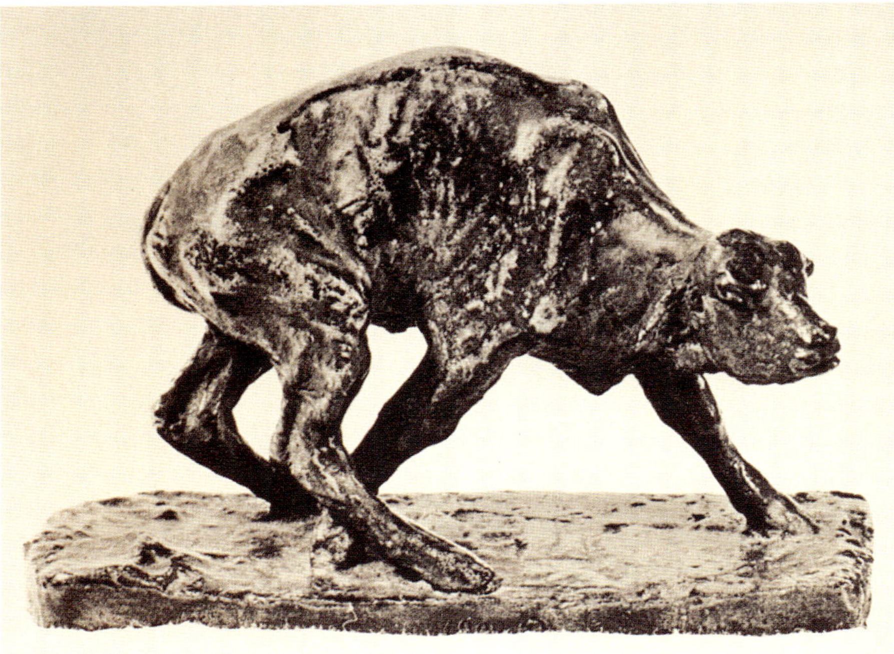

IV-52. "New Born" by Solon Borglum. 1900.

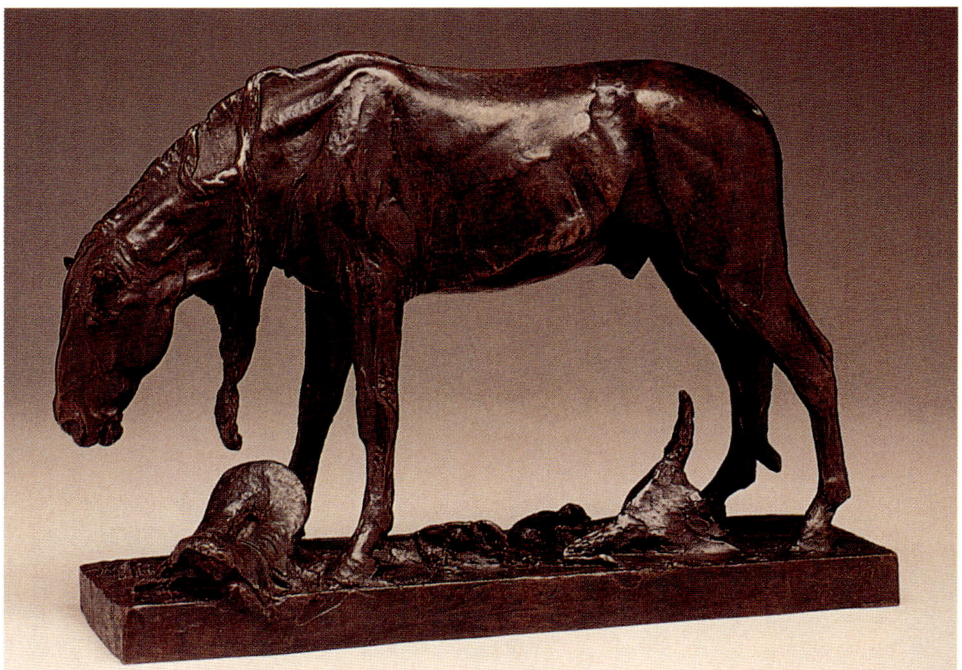

IV-53. "Horse Tamed" by Solon Borglum. 1915.

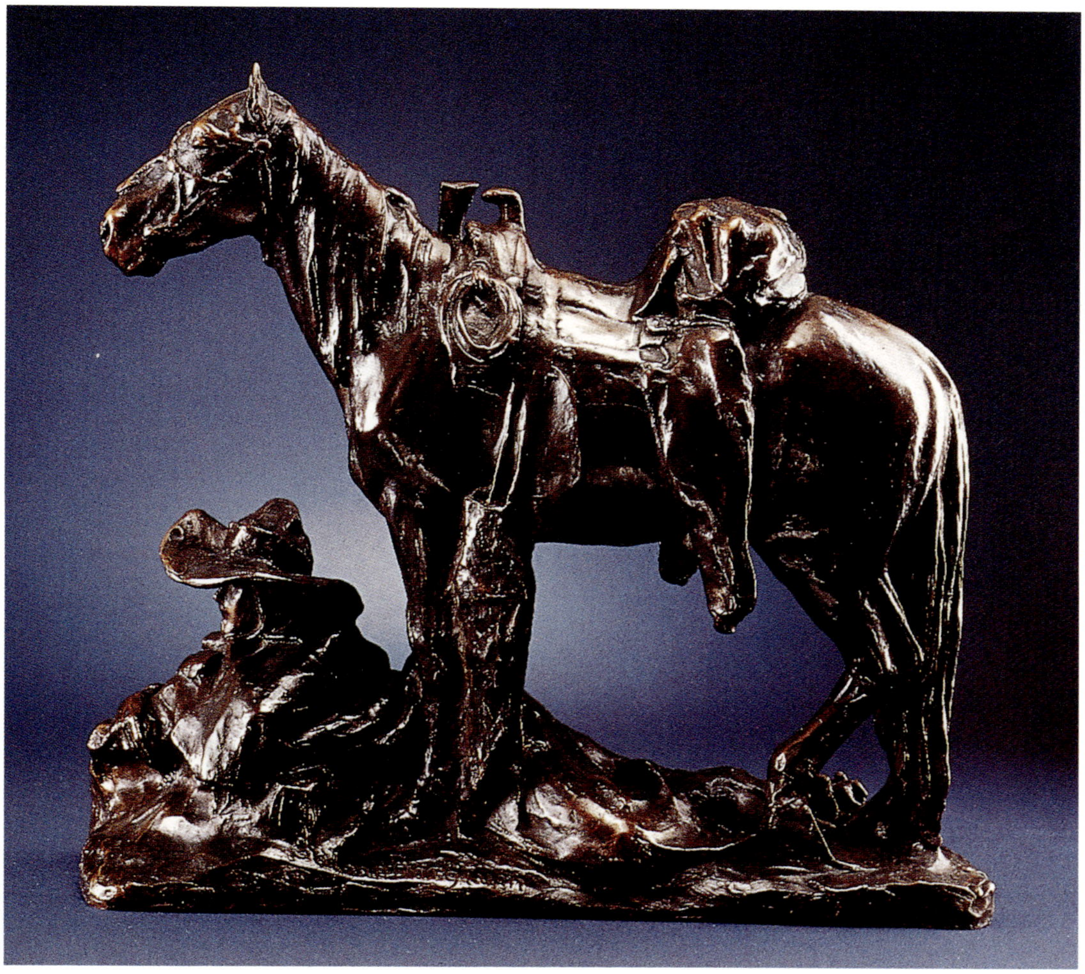

IV-54. "Cowboy at Rest" by Solon Borglum. 1906.

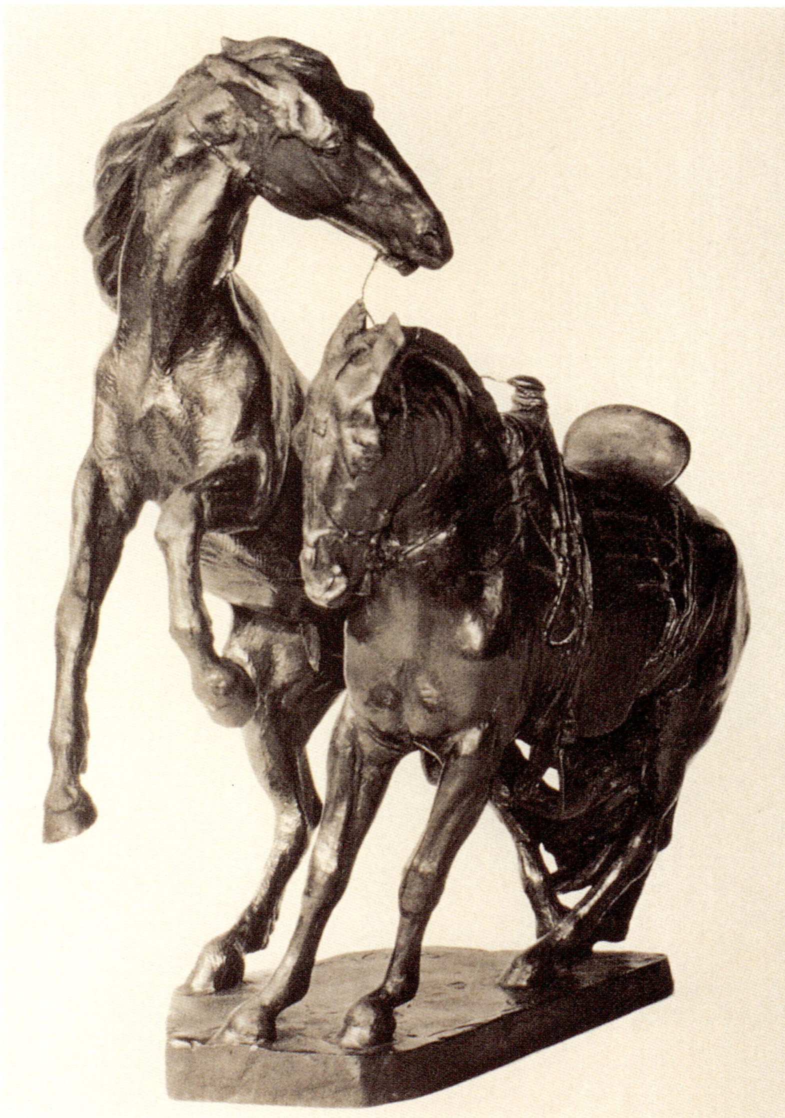

IV-55. "Intelligent Bronco" by Solon Borglum.

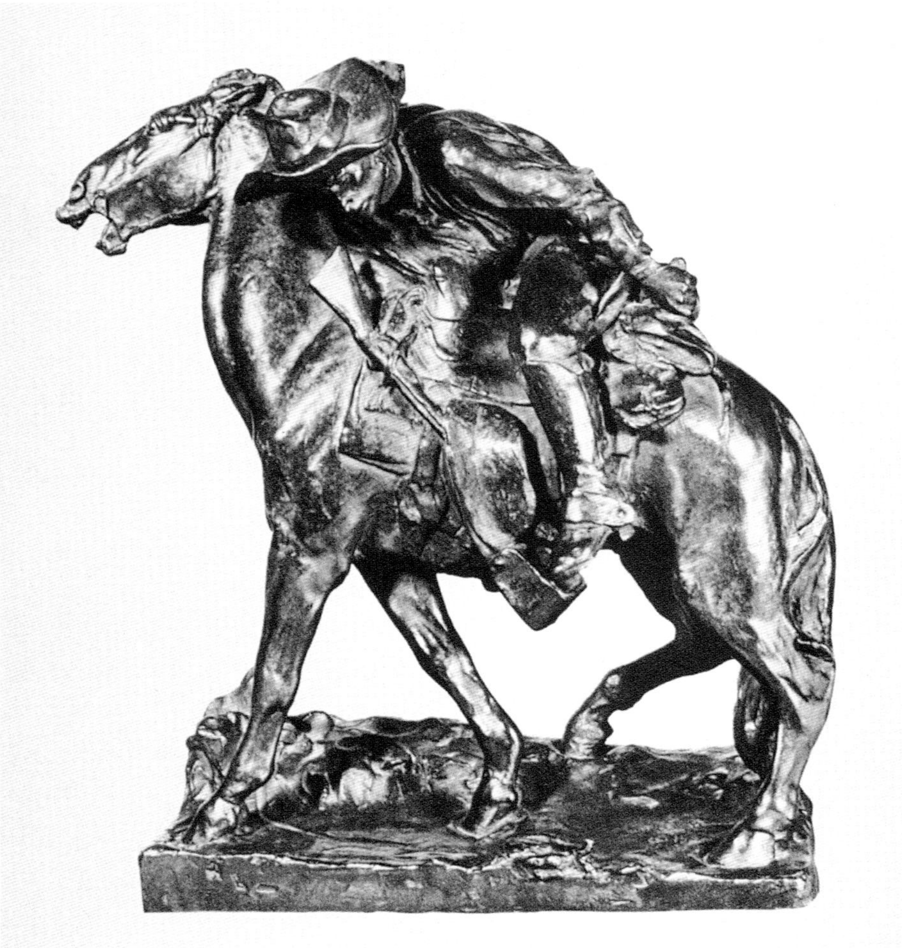

IV-56. "On the Trail" by Solon Borglum. 1900.

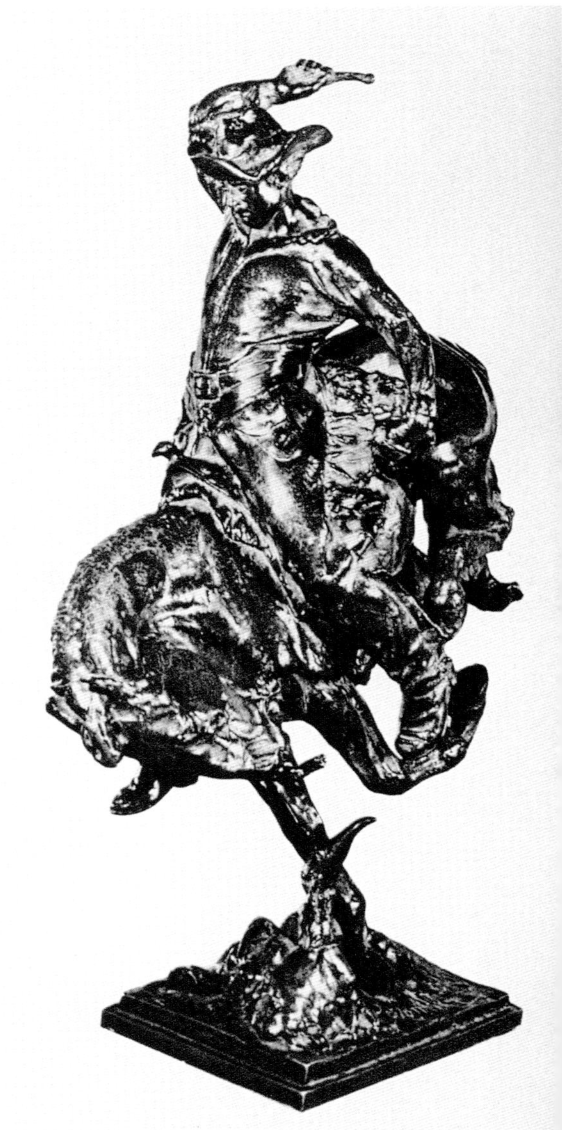

IV-57. "The Bronco Buster" (or "One in a Thousand") by Solon Borglum. 1901.

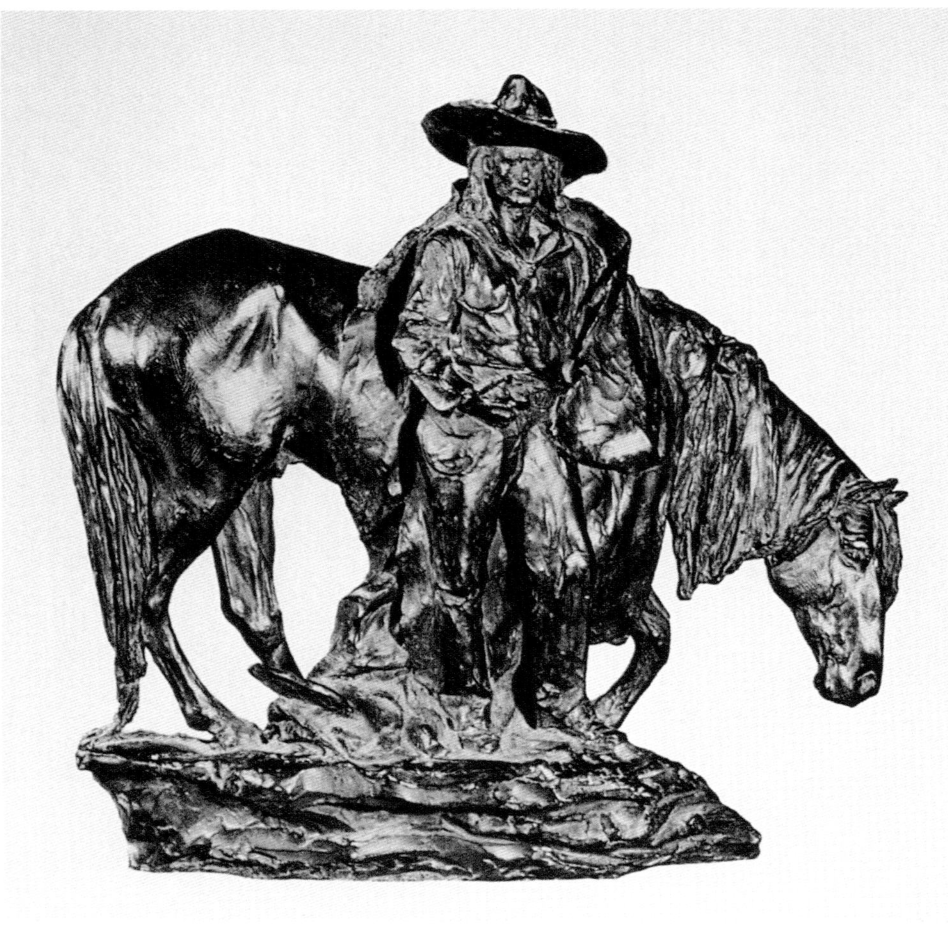

IV-58. "Evening" by Solon Borglum. c. 1905.

SALLY JAMES FARNHAM (1876-1943)

A protégé and friend of Frederic Remington, Sally James Farnham was strongly influenced by his style, but she also became a sculptor of Western images in her own right. Her horses and equestrian statues were well-regarded for their naturalism, and though her images lacked Remington's dramatic power, she had her own career as a maker of cowboy and rider statuary. After Remington's death Farnham was entrusted with the task of completing his unfinished work.

Farnham was born in Ogdensburg, New York, where Remington lived. She began her career in 1901 when she took a small work to Remington for criticism. (He reportedly said of her first effort "ugly as hell, but full of ginger.")Though she never had formal training, her work was critiqued by several leading sculptors of animals and Western themes, including Henry Merwin Shrady, Augustus Lukeman, and Frederic Roth.

Farnham's largest commission was the result of a competition for an equestrian of Simon Bolivar (IV-60); it was placed at an entrance to Central Park in New York in 1921. Her most popular work, often recast, was "Payday Going to Town" (IV-59) (which bore striking similarity to Remington's "Coming Through the Rye.") Her equestrian image of Will Rogers on his pony, Soapsuds, (IV-61), was made in 1938; it was one of her many images of well-known politicians (including a statue of President Herbert Hoover) and artists. She also made occasional sculptures of family subjects, such as her "Three Children" (IV-62).

Other Western images—all cast at Roman Bronze Works from 1900 until her death—include "Horse and Rider" (IV-63), "The Sun Fisher" (IV-64), and "Scratchin 'Im" (IV-65).

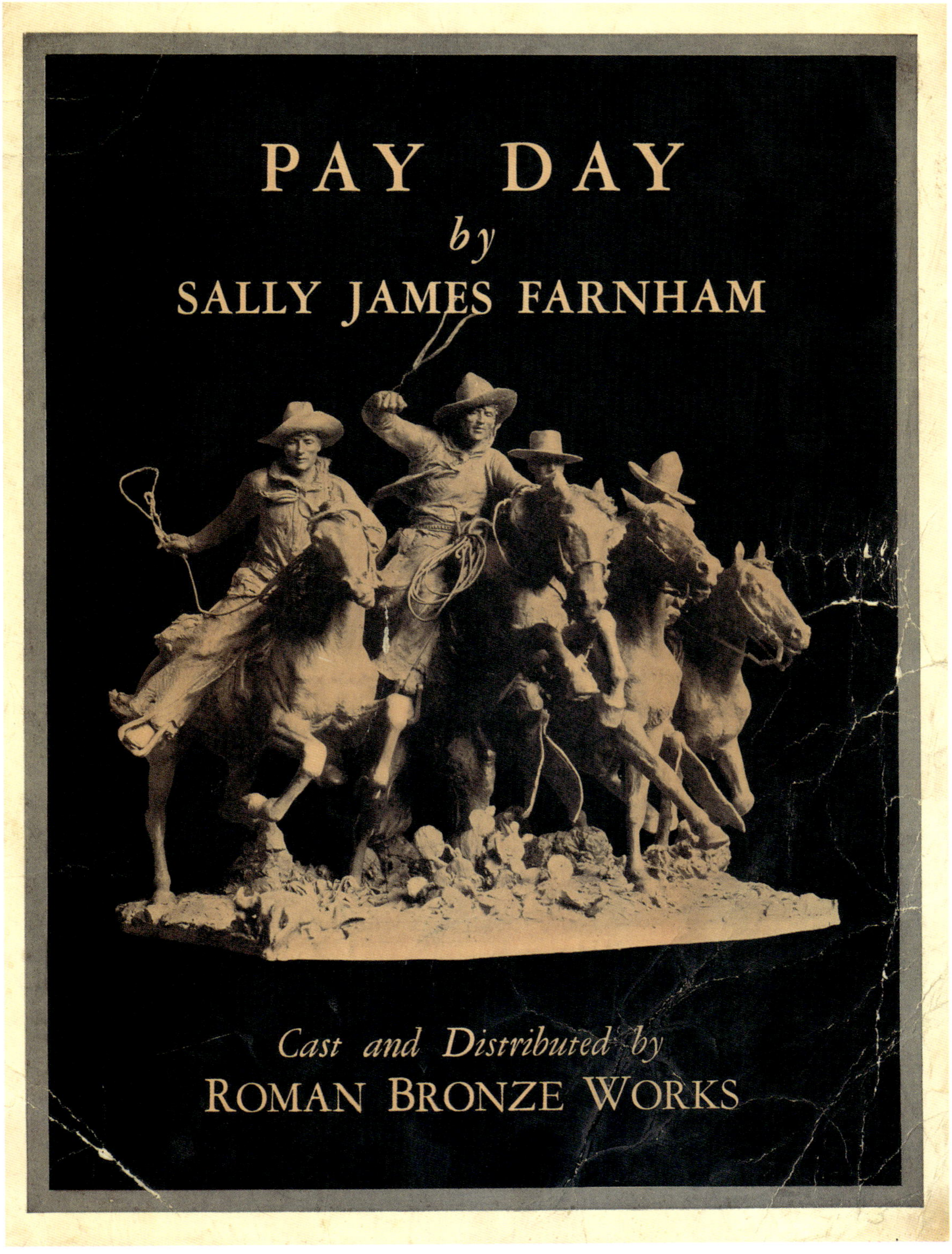

IV-59. "Payday" (as offered for sale by R.B.W.) by Sally James Farnham. 1931.

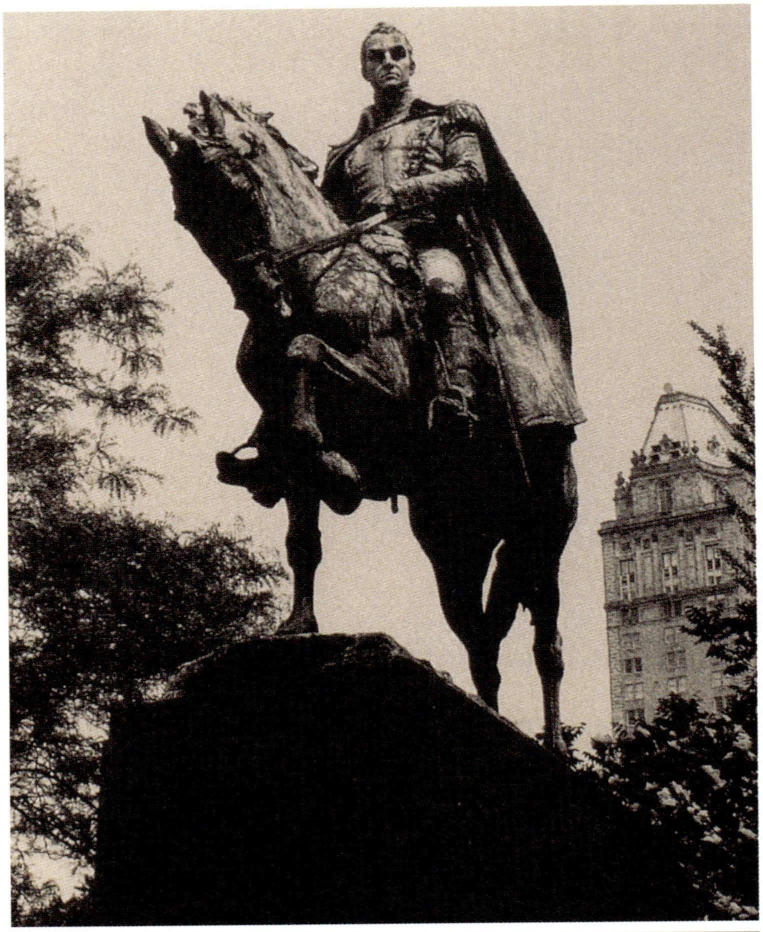

IV-60. Statue of Simon Bolivar, New York City, by Sally James Farnham. 1919.

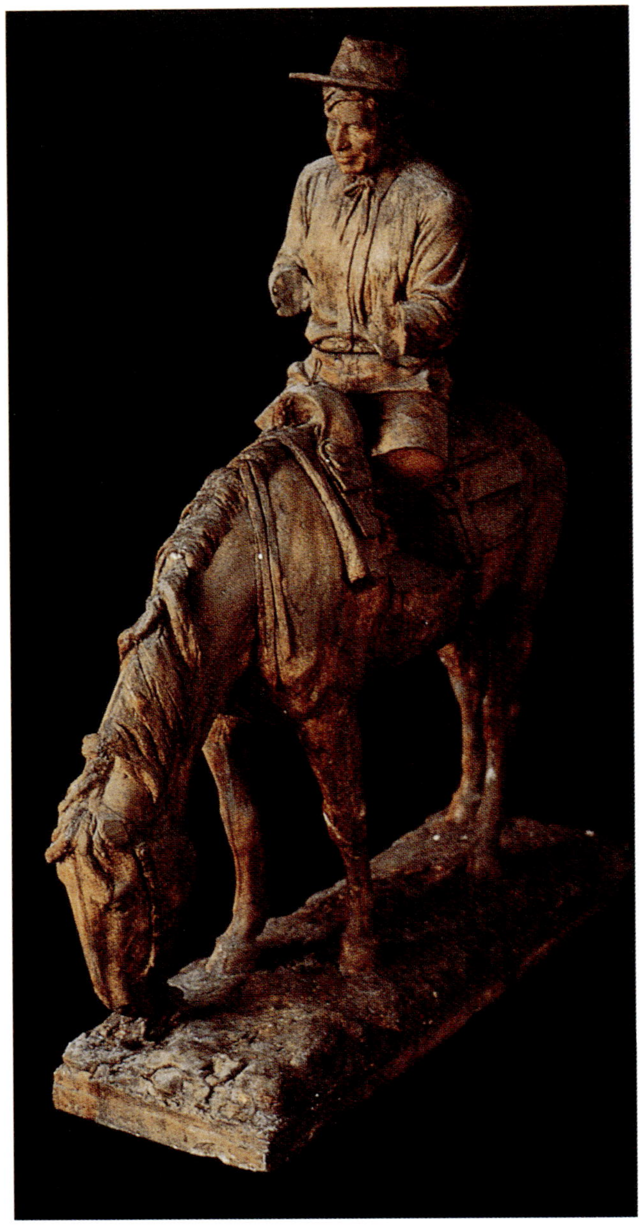

IV-61. "Will Rogers" (plaster cast) by Sally James Farnham. 1938.

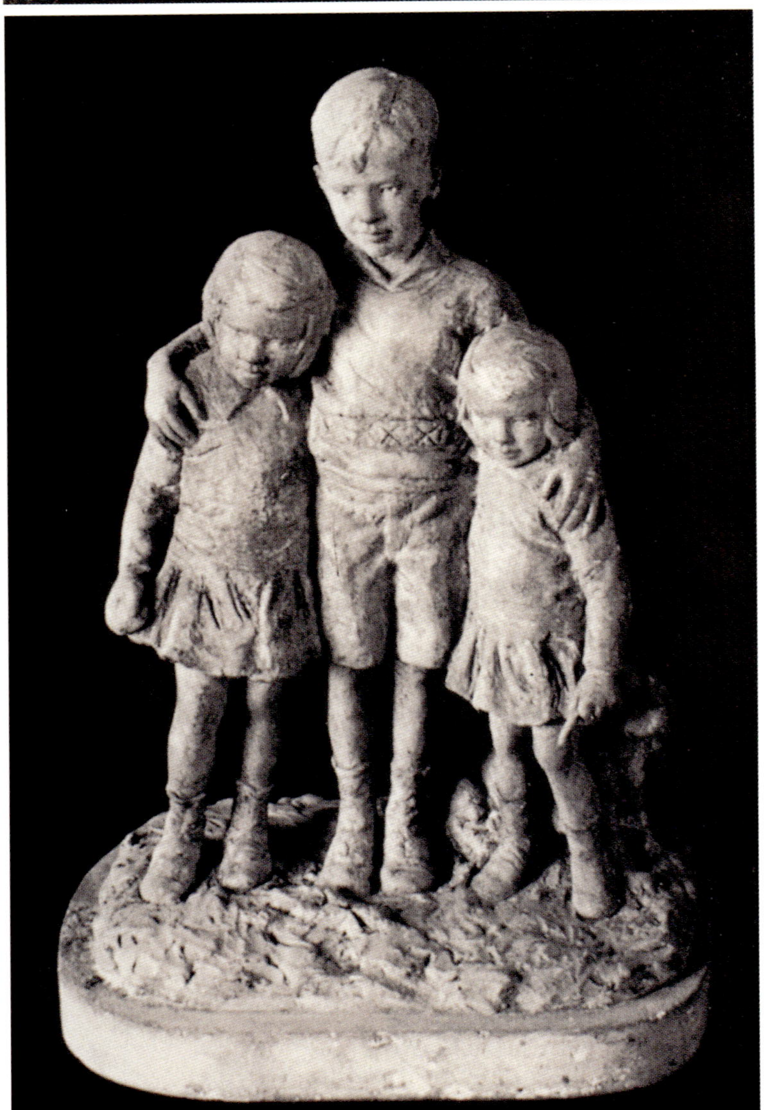

IV-62. "Three Children" (plaster cast) by Sally James Farnham. 1931.

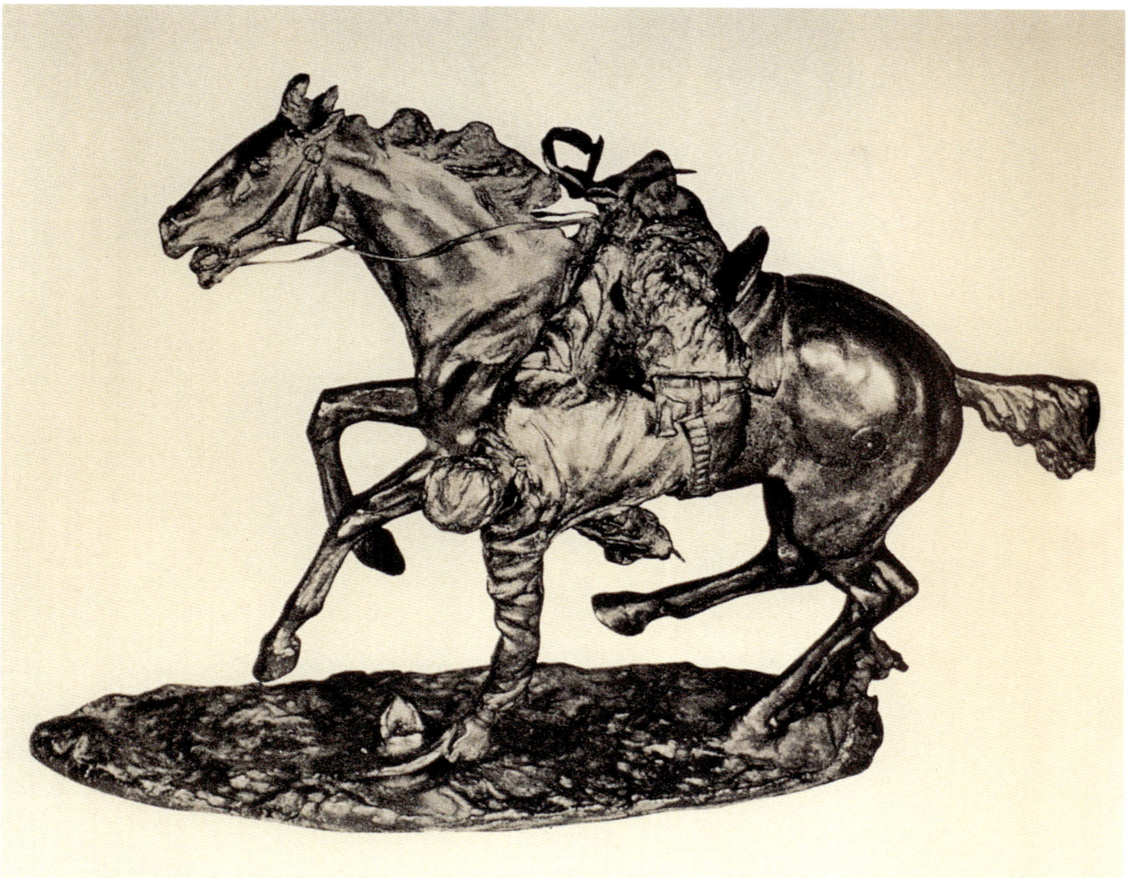

IV-63. "Horse and Rider" by Sally James Farnham. c. 1900.

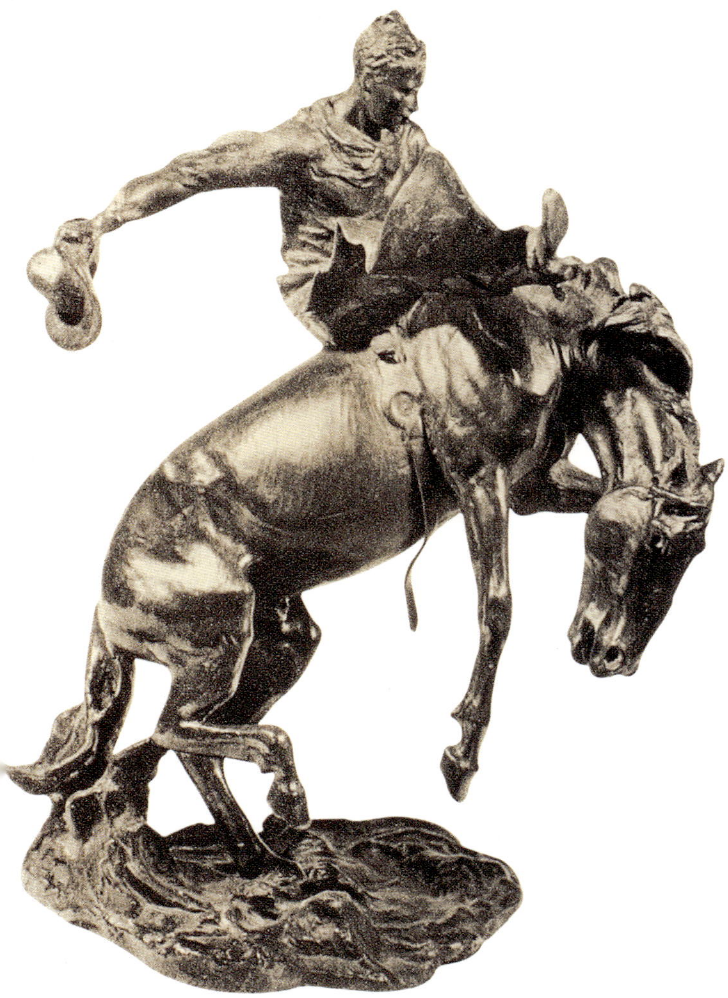

IV-64. "The Sun Fisher" by Sally James Farnham. c. 1920.

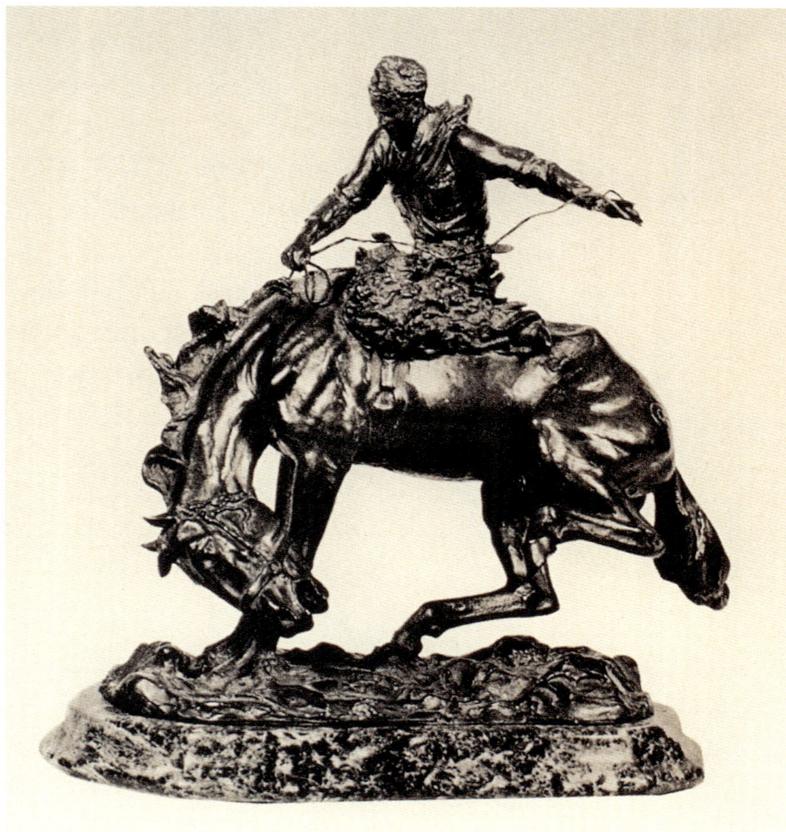

IV-65. "Scratchin' Im" by Sally James Farnham.

CHARLES HUMPHRISS (1867-1934)

Charles Humphriss was a specialist in sculpting American Indian subjects, particularly themes of Indian worship. These dignified sculptural images portrayed the native American as a man of peace and piety, turning to the Great Spirit in his adversity. Humphriss also made a number of cowboy and horse statues, as well as sculpture with wide-ranging subject matter: among his works were images of Lincoln, King Arthur, clowns, soldiers, and a Lotus-eater. But it is for his sculpture of America's West that he is best remembered.

Humphriss was born in England and settled in New York, though his career was primarily concerned with Western images. He won several awards at the Panama-Pacific Exposition. Among his sculptures depicting life in the "Old West" were some forty different works cast at Roman Bronze Works between 1905 and the end of his life. They included "Indian Appeal" (IV-66), "Indian Chief" (IV-67), "Appeal to the Great Spirit" (IV-68), "Cowboy on a Bucking Horse" (IV-69), "Indian Chief with Feathered Headdress" (IV-70), and "Old Soldier" (IV-71).

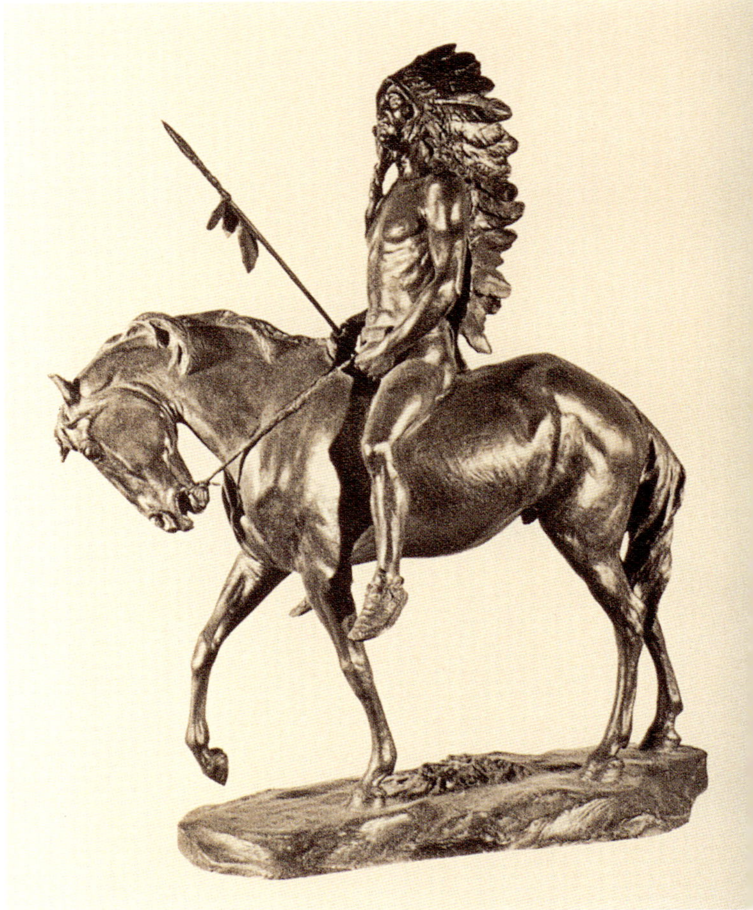

IV-67. "Indian Chief" by Charles Humphriss. c. 1905.

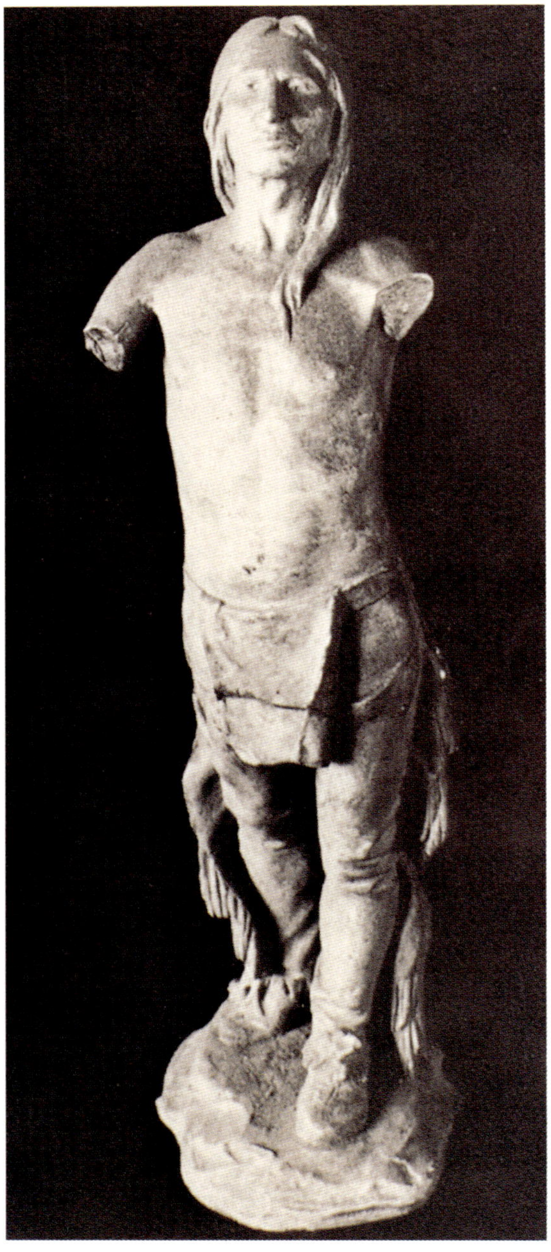

IV-66. "Indian Appeal" (plaster cast) by Charles Humphriss. 1929-31.

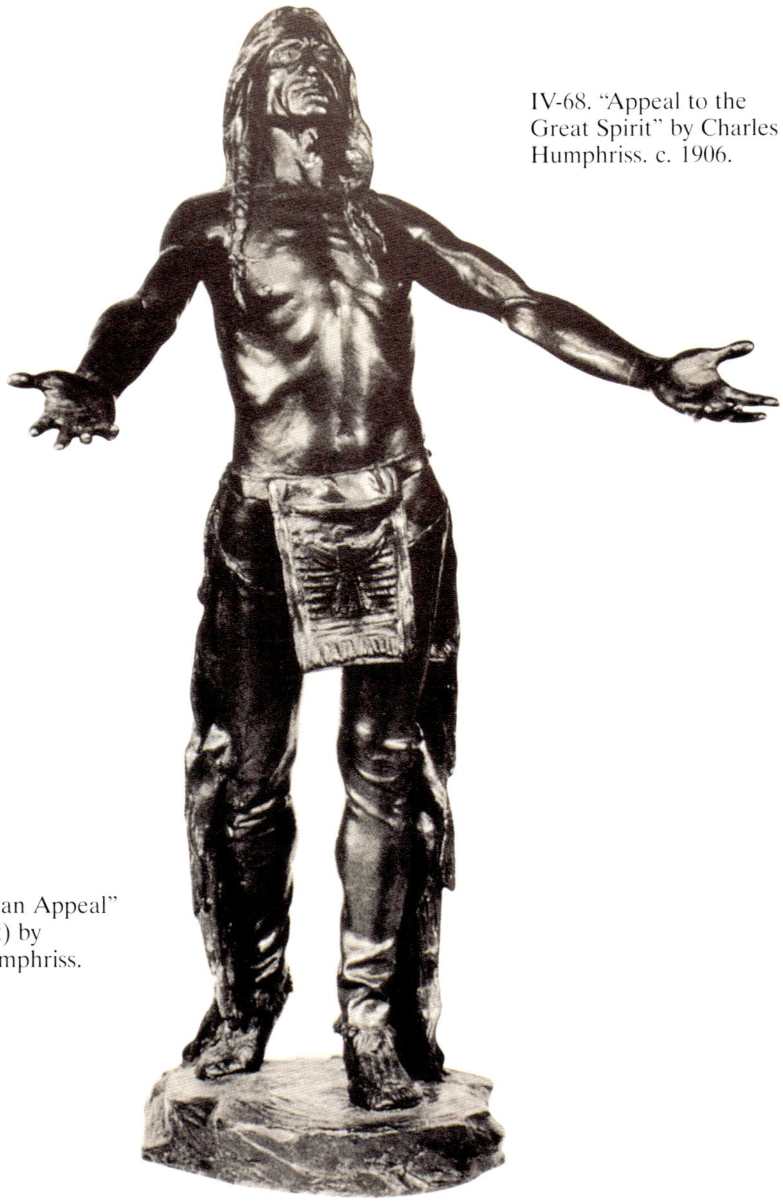

IV-68. "Appeal to the Great Spirit" by Charles Humphriss. c. 1906.

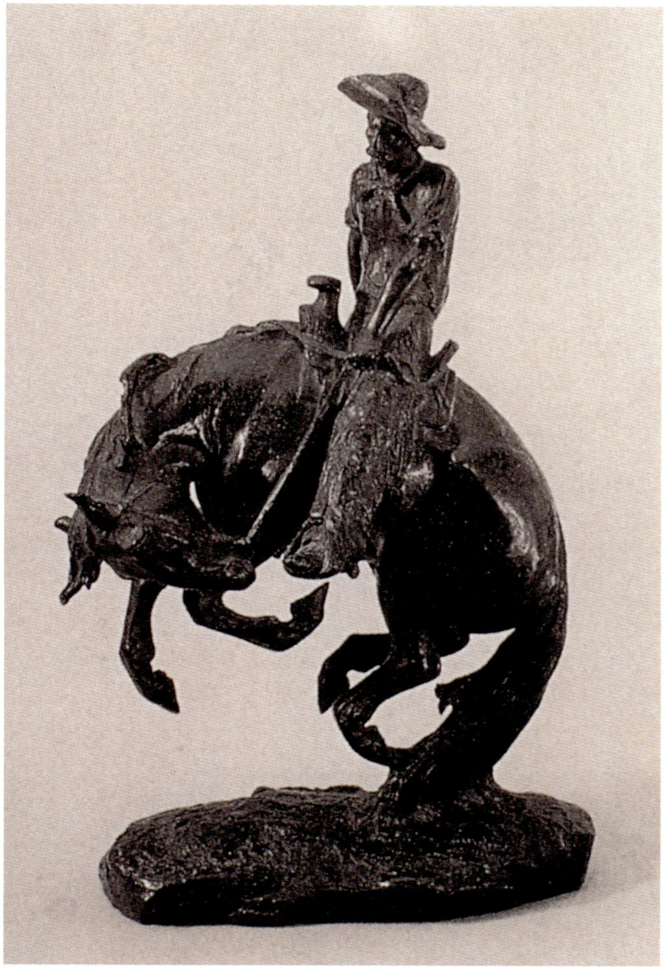

IV-69. "Cowboy on Bucking Horse" by Charles Humphriss. 1905.

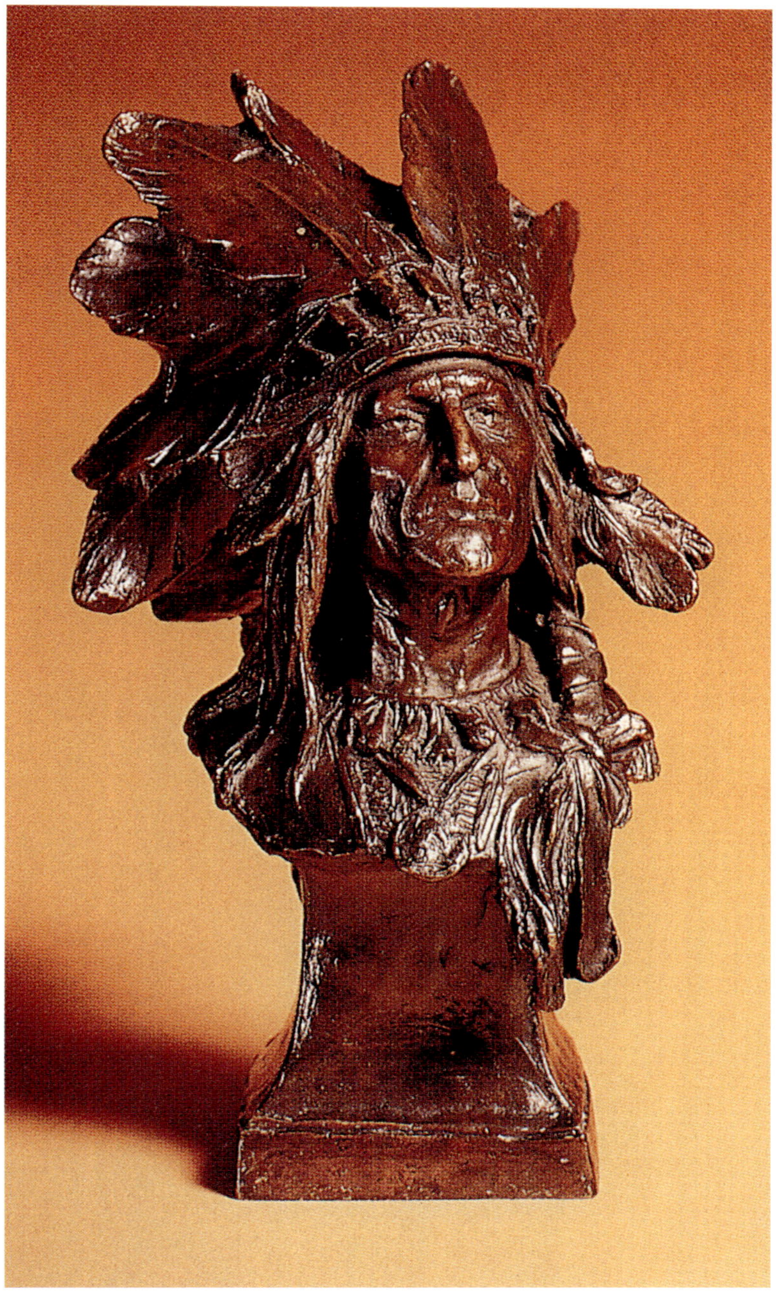

IV-70. "Indian Chief with Feathered Headdress" by Charles Humphriss. c. 1908.

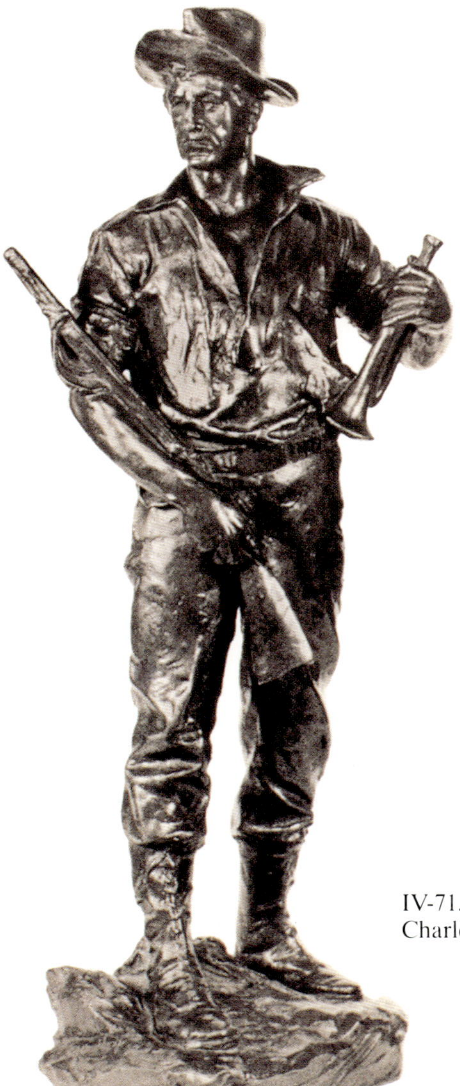

IV-71. "Old Soldier" by Charles Humphriss. 1905.

CYRUS E. DALLIN (1861-1944)

A leading sculptor of images of the American West's native Americans, Dallin was born in Utah of pioneer parents and was well-acquainted with Ute Indians at an early age. It was at the mine where his father worked that he first modeled with native clay. Dallin devoted his entire career to depicting his subject with empathy and grave dignity, showing many different aspects of native American life: "The Scout" (IV-72), "The Archery Lesson (IV-73), and a pow-wow consideration of "War or Peace" (IV-74).

He was trained in Boston and Paris, where he made his first equestrian Indian; several years later, also in Paris he produced "Medicine Man" (IV-75), which won a prize at the Paris Exposition in 1900 and became one of his best known works. Another Dallin equestrian sculpture, "The Protest" (IV-76) was exhibited at the St. Louis Exposition in 1904. While continuing to fulfill many major commissions, the sculptor taught for forty years at the Massachusetts Normal Art School.

Dallin's careful depiction of Indian life and strong naturalism brought a sense of authenticity, dignity, and monumentality to the genre. His outdoor monuments can be seen in Chicago, Salt Lake City, Kansas City, and in front of the Museum of Fine Arts in Boston, as well as at many other sites. (He also made statues of non-Indian subjects; among them are war memorials, a cavalryman, pioneer women, and an image of Paul Revere). But Dallin is most remembered for endowing the American Indian image (in both his art and his writings) with nobility and dramatic dignity, as well as a certain sadness, as in his famous "Appeal to the Great Spirit" (IV-77).

At Roman Bronze Works, Dallin cast a number of Indian images, beginning in 1912. Some of these were small models of larger outdoor monuments.

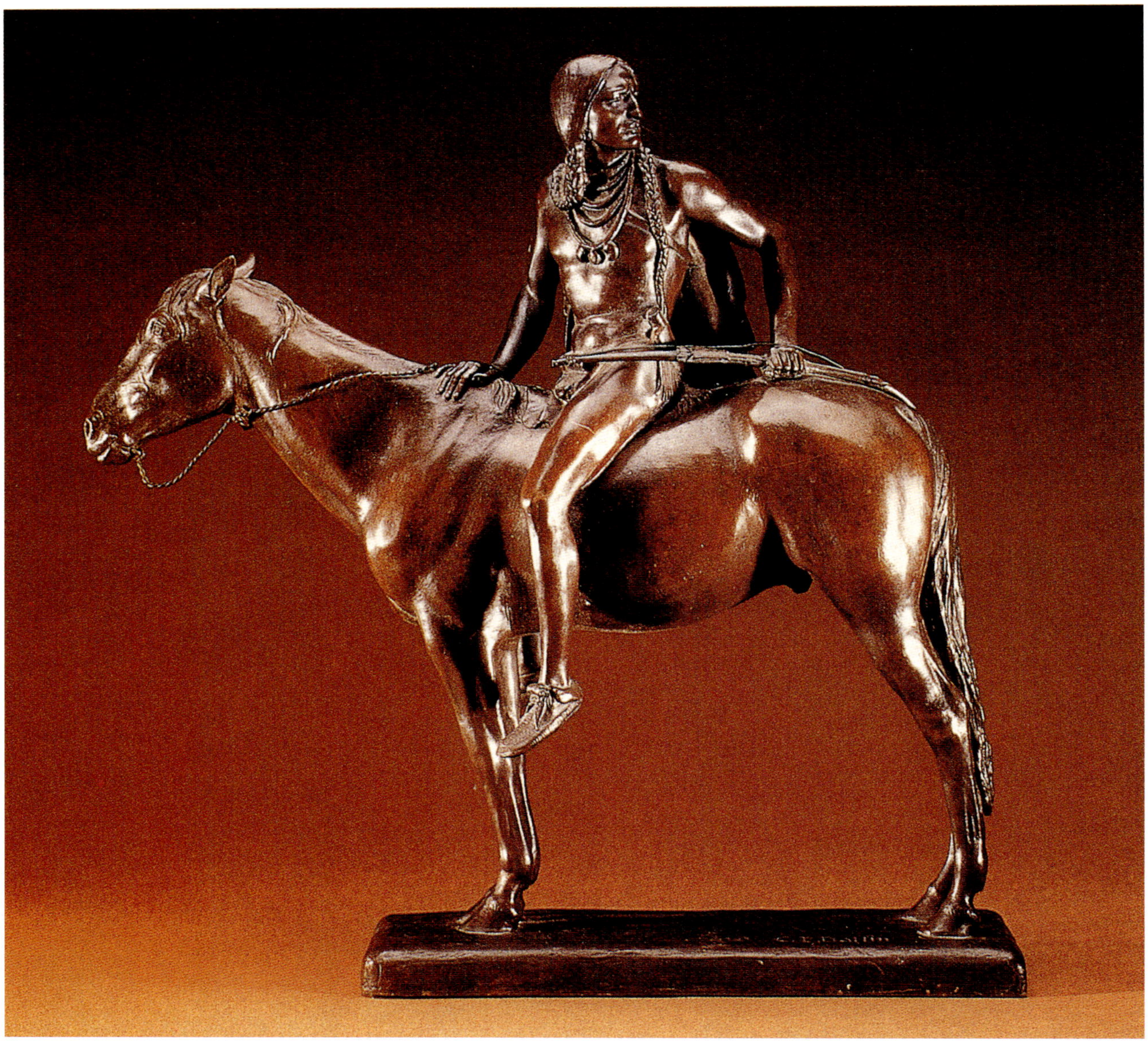

IV-72. "The Scout" by Cyrus Dallin. 1902.

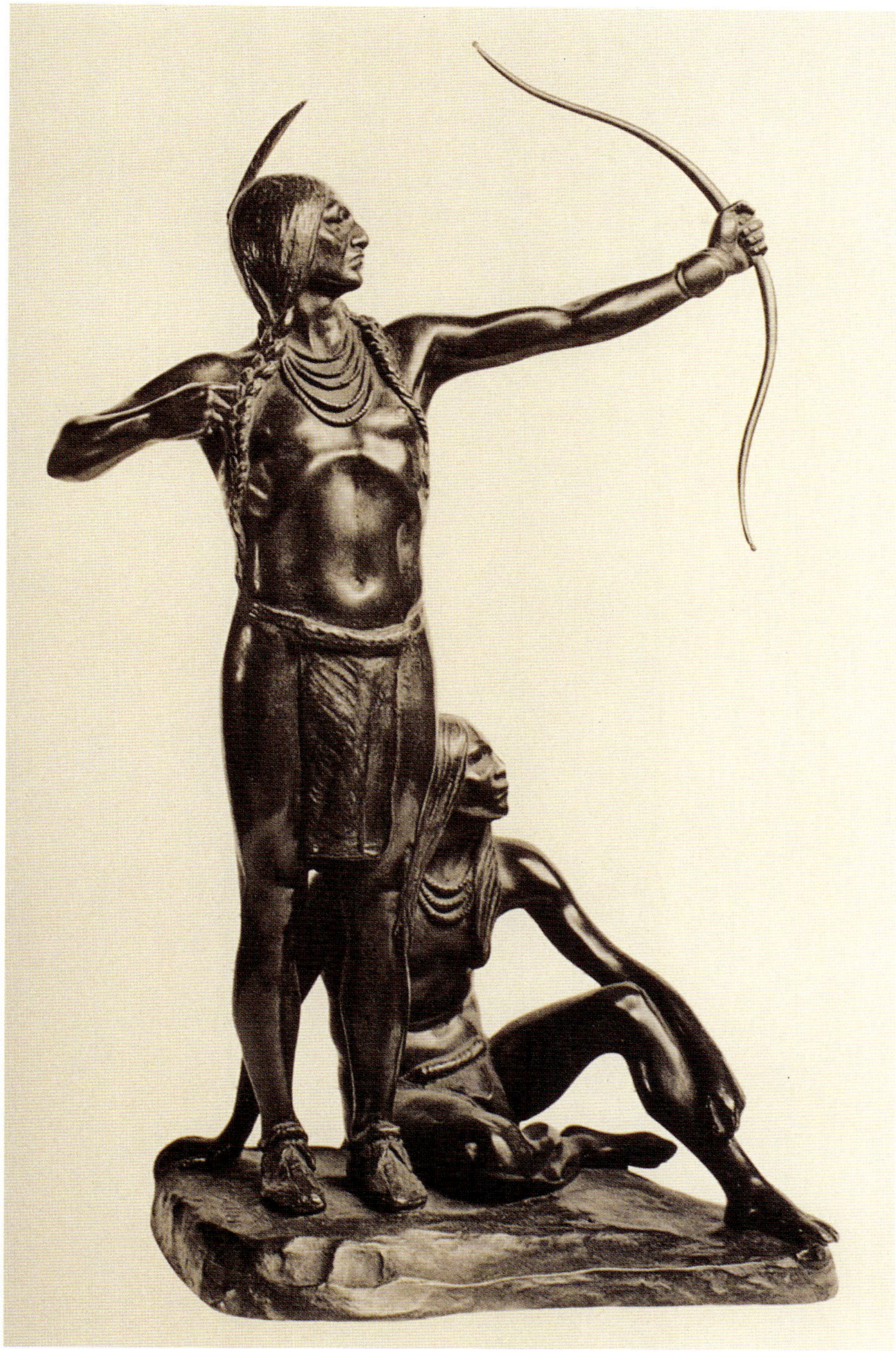

IV-73. "Archery Lesson" by Cyrus Dallin. 1907.

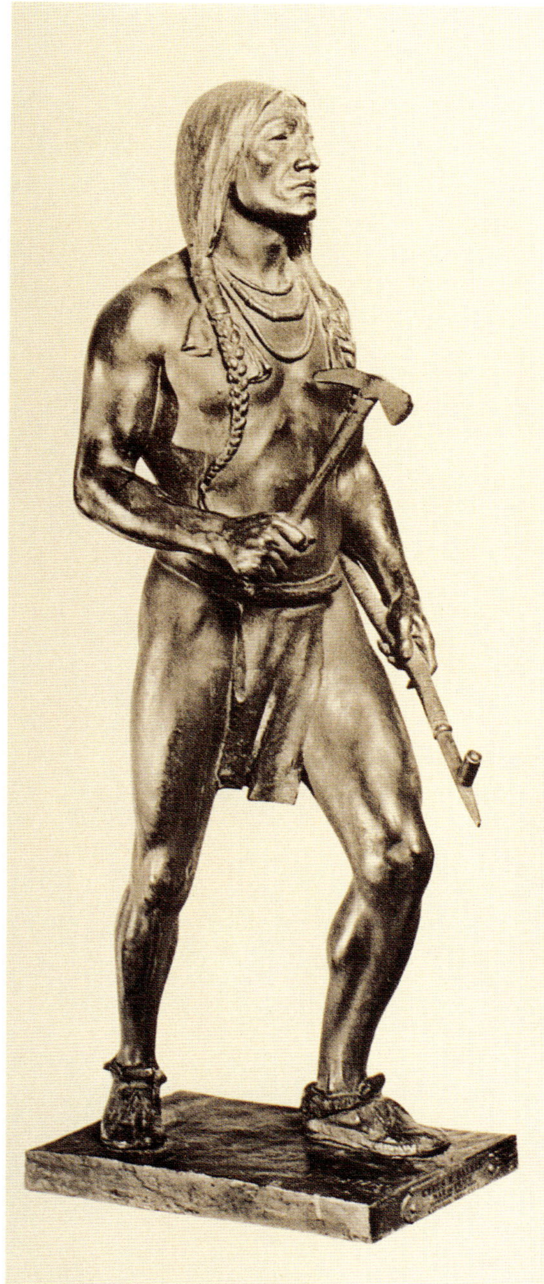

IV-74. "War or Peace" by Cyrus Dallin. c. 1907.

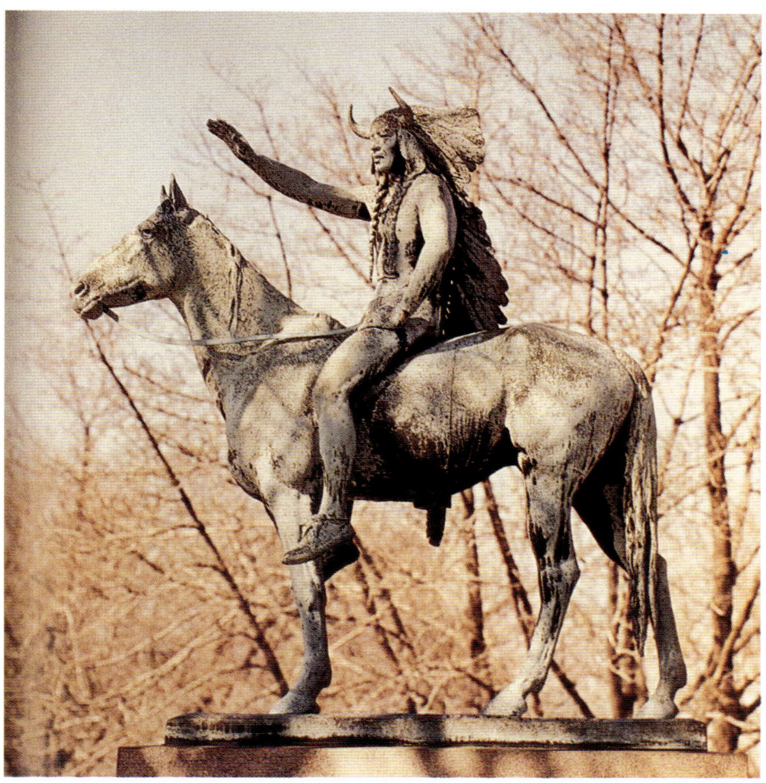

IV-75. "The Medicine Man" by Cyrus Dallin. 1903.

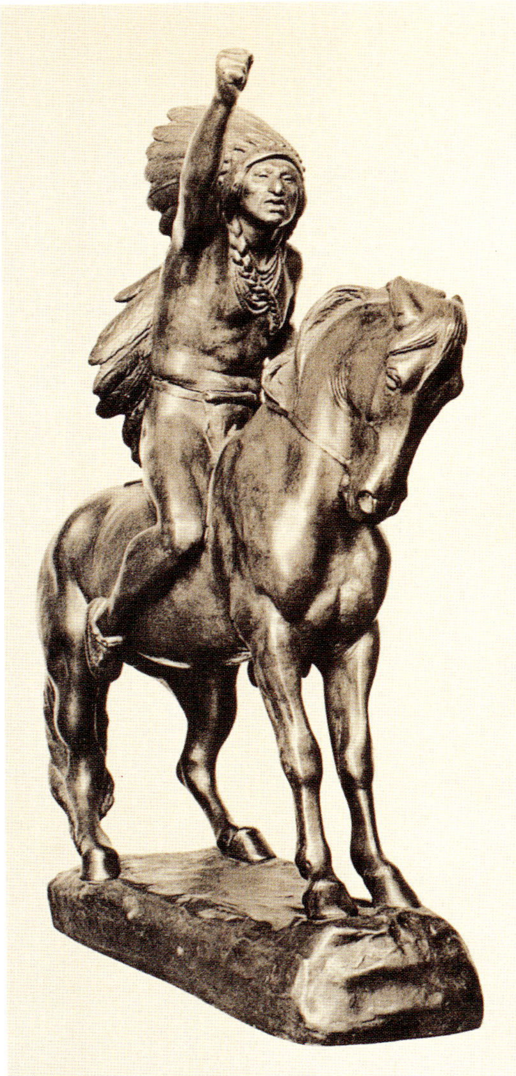

IV-76. "The Protest" by Cyrus Dallin. 1904.

Below:
IV-77. "Appeal to the Great Spirit" by Cyrus Dallin. 1909.

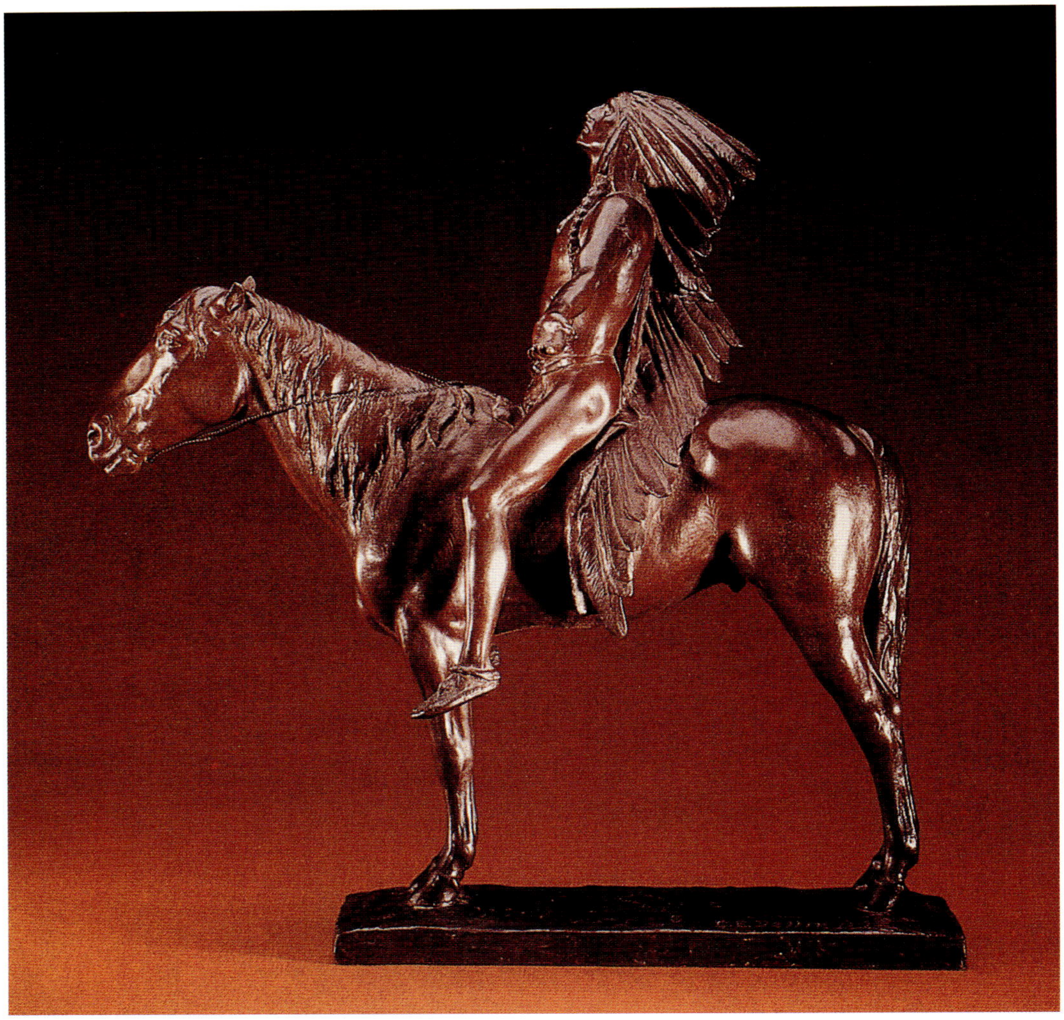

HERMON ATKINS MacNEIL (1866-1947)

Hermon Atkins MacNeil is best known for his depictions of Indian life, though he also made a number of portrait statues of Abraham Lincoln and various other images drawn from American history. His interest in Western life was reportedly fostered by the visit of Buffalo Bill Cody to the Columbian Exposition in Chicago when MacNeil was a young man; it was there that he modeled his first Indian subject.

MacNeil was born in Massachusetts, educated at the Normal School of Art in Boston, and became an instructor at Cornell University. In 1888 he went to study in Paris under Falguière, the noted modeler of animal statuary at the Ecole des Beaux-Arts, and at the Académie Julian. He began his career in the United States in 1891 as an architectural sculptor for the Columbian Exposition in Chicago. He then made several trips to the West to study Indian life, as well as spending a year as a student in Rome. His career as an architectural sculptor took him to the major expositions in America and included such commissions as the sculptural pediment for the United States Supreme Court Building in Washington. As a monument maker he received a number of important commissions, such as the Ezra Cornell statue for Cornell University, "Washington as Commander in Chief," and the McKinley Memorial. He became president of the National Sculpture Society and received several notable awards.

Many of MacNeil's smaller statues, (mostly made before 1910) were devoted to Indian themes; he drew on Indian folklore for his dramatic, rather impressionistic style. Many of these works, including his most famous work "The Sun Vow" (IV-78), "Return of the Snakes"(IV-79), "A Chief of the Multnomah Tribe" (IV-80), and "A Primitive Chant" (IV-81), were cast at Roman Bronze Works. Works with reference to American history included his unusually powerful "The Slave Block" (IV-82) and a portrait of Abraham Lincoln (IV-83). Other small pieces cast there included "Returning to the Snakes" (IV-84) and a "Bronze Fountain Figure of a Nymph" (IV-85). MacNeil cast at R.B.W. between 1902 and 1940.

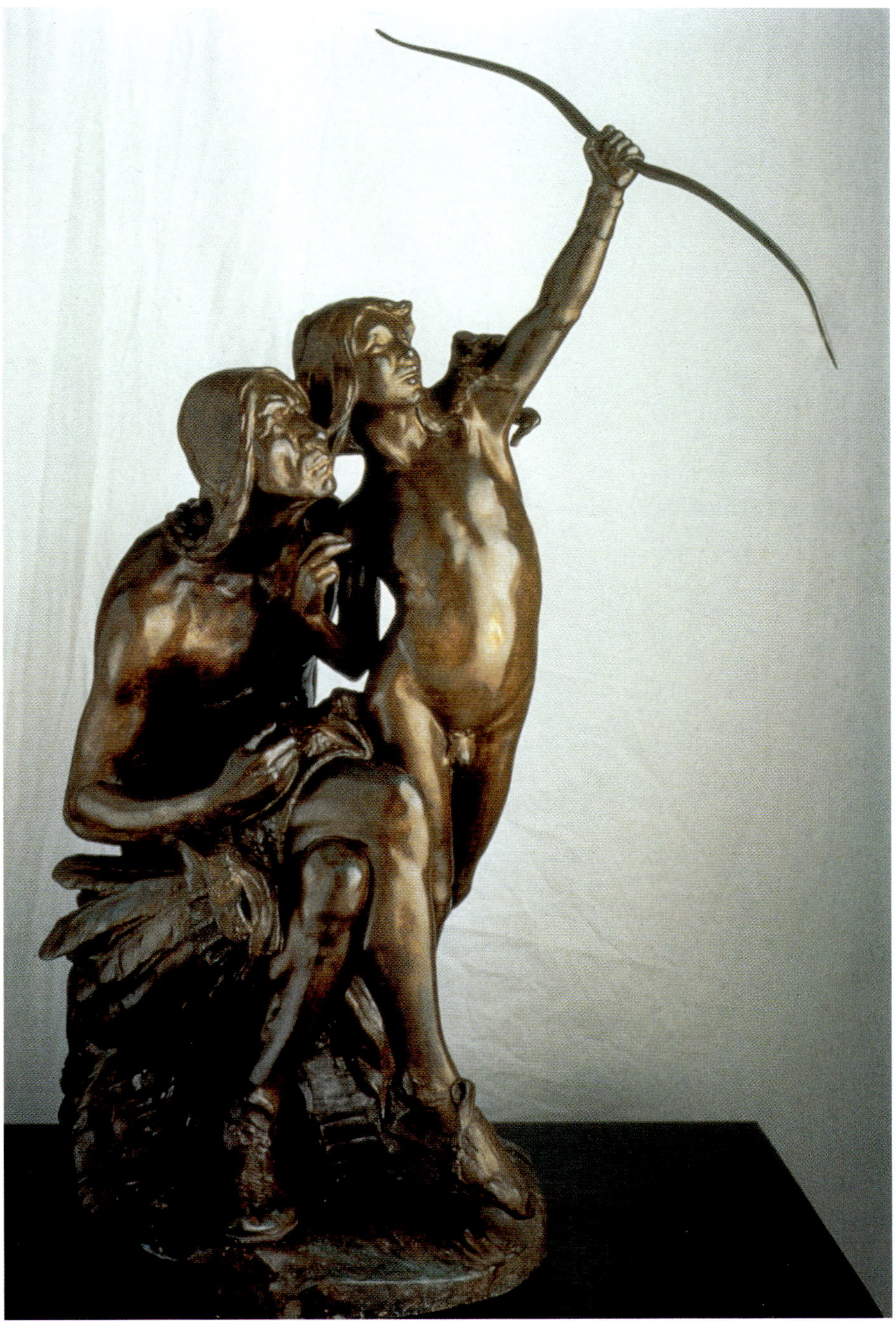

IV-78. "Sun Vow" by Hermon Atkins MacNeil. Modeled 1899. R.B.W. 1905.

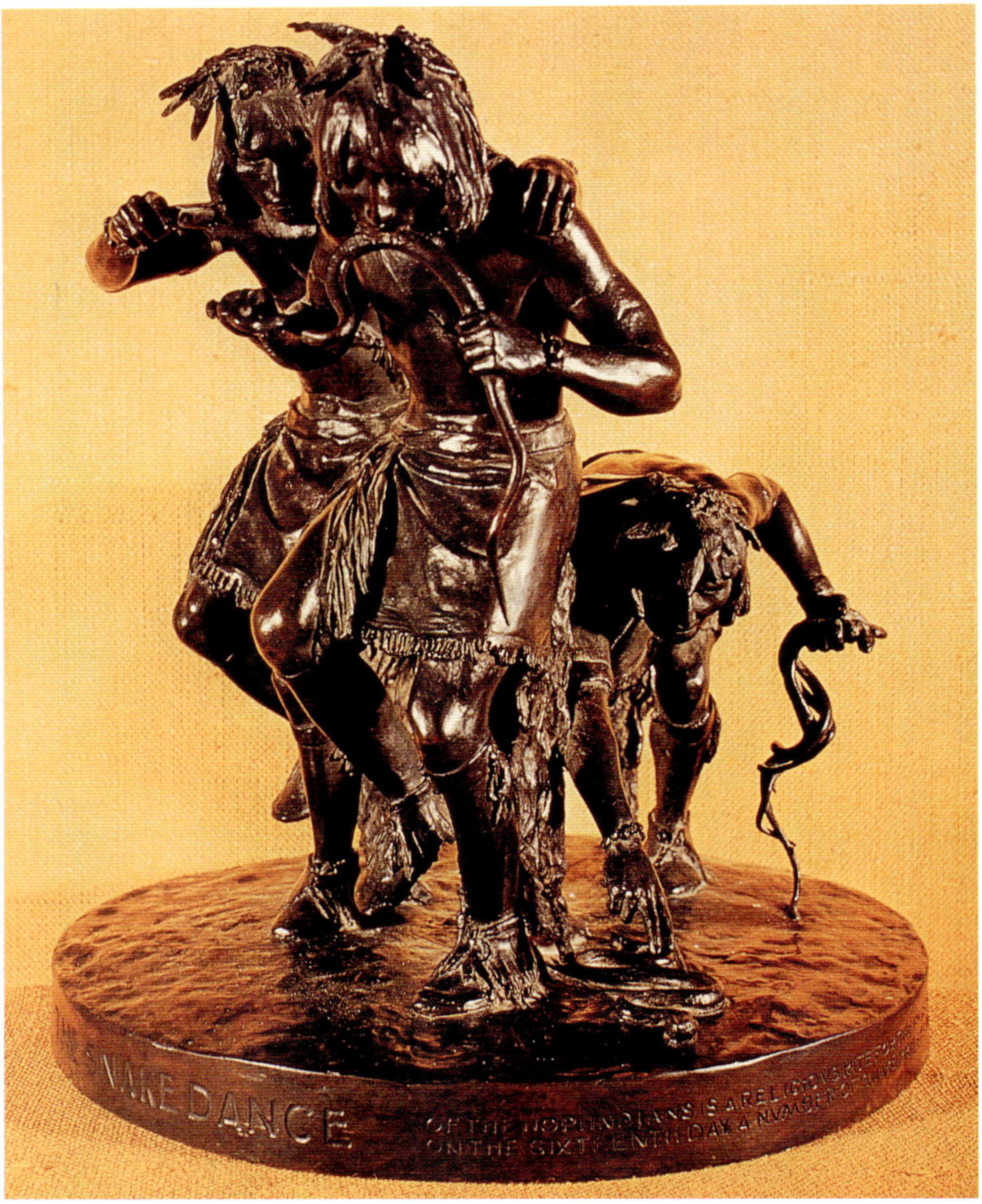

IV-79. "Return of the Snakes" by Hermon Atkins MacNeil. Modeled c. 1896.

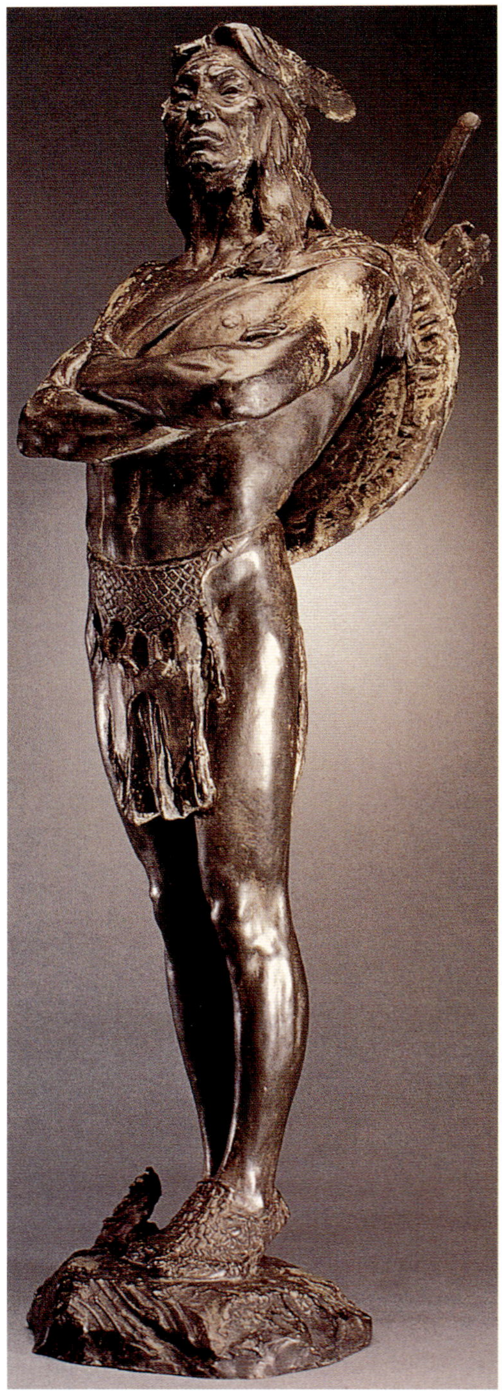

IV-80. "Chief of the Multnomah Tribe" by Hermon Atkins MacNeil. 1903.

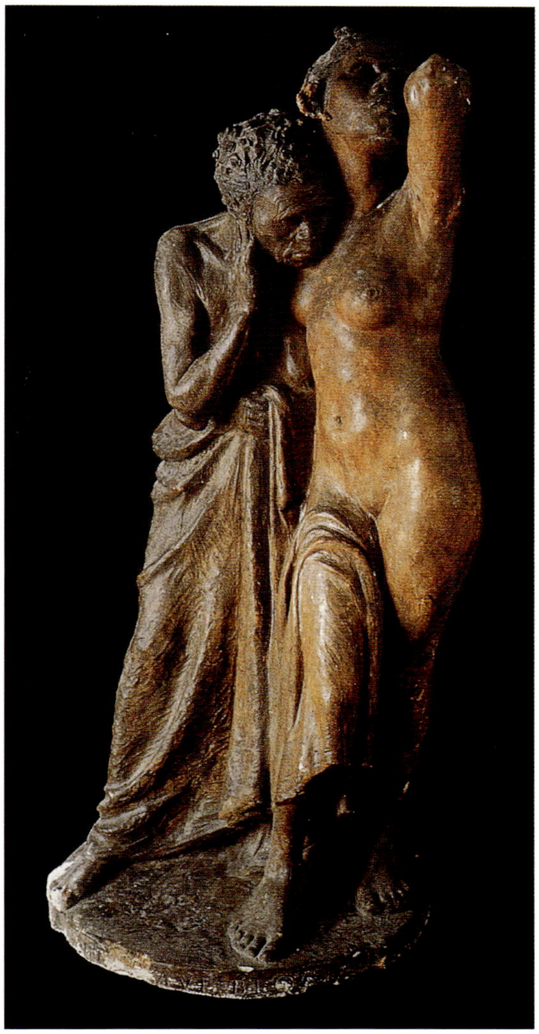

Left and below:
IV-82. "The Slave Block," (plaster cast and statue) by Hermon Atkins MacNeil. Plaster dated 1937; statue cast posthumously.

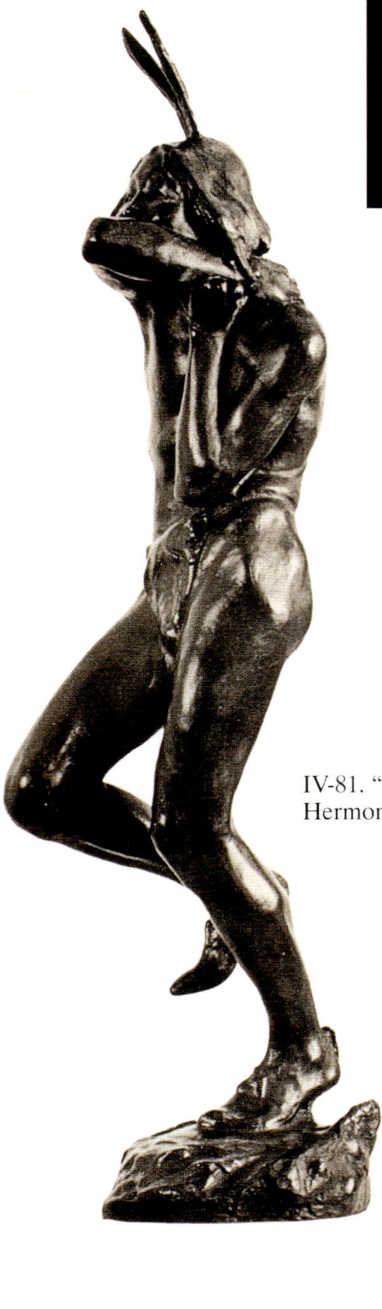

IV-81. "Primitive Chant" by Hermon Atkins MacNeil. 1906.

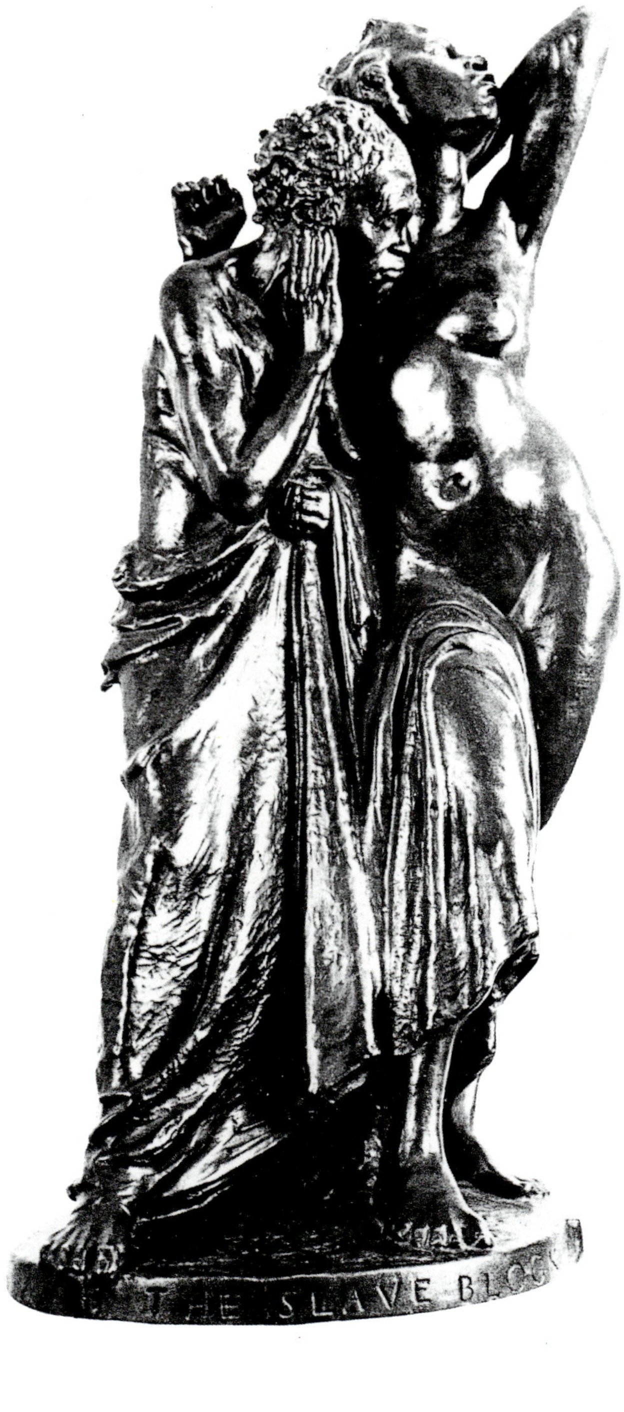

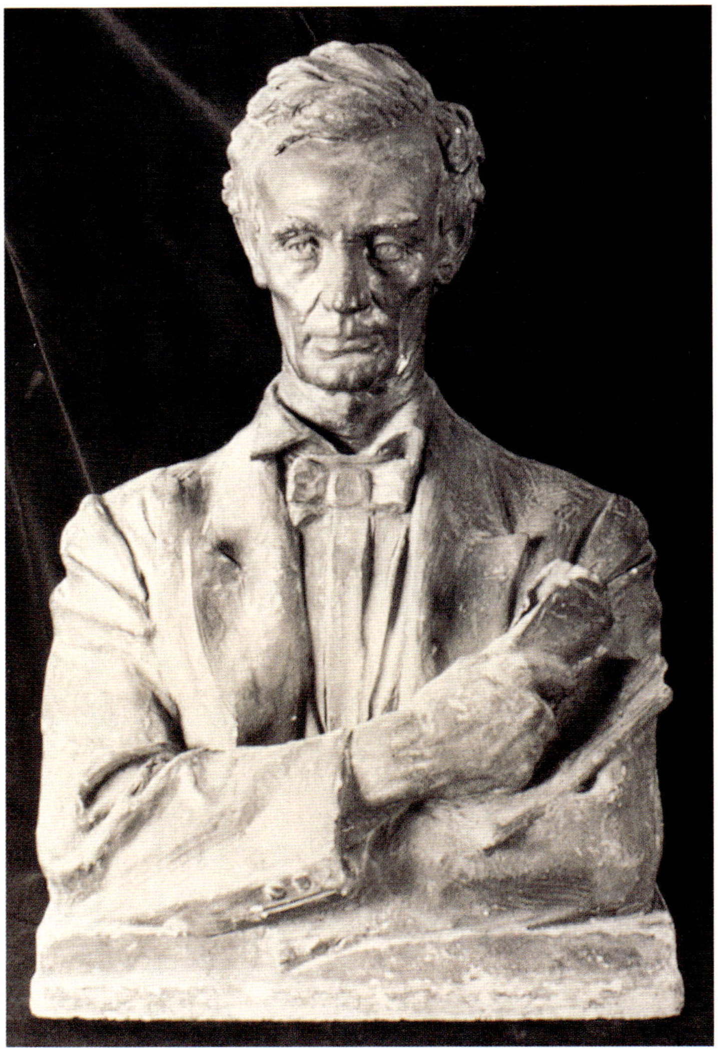

IV-83. "Abe Lincoln" (plaster cast) by Hermon Atkins MacNeil. 1915.

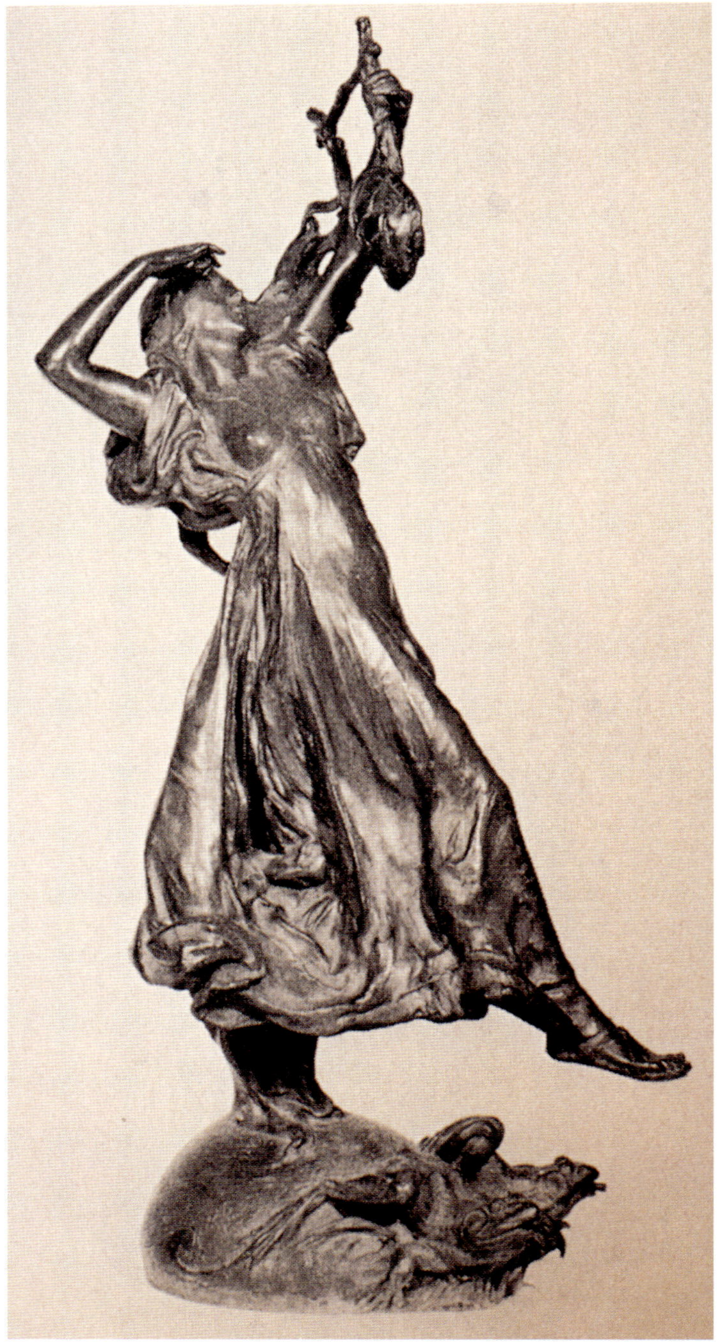

IV-85. "Bronze Fountain Figure of a Nymph" by Hermon Atkins MacNeil.

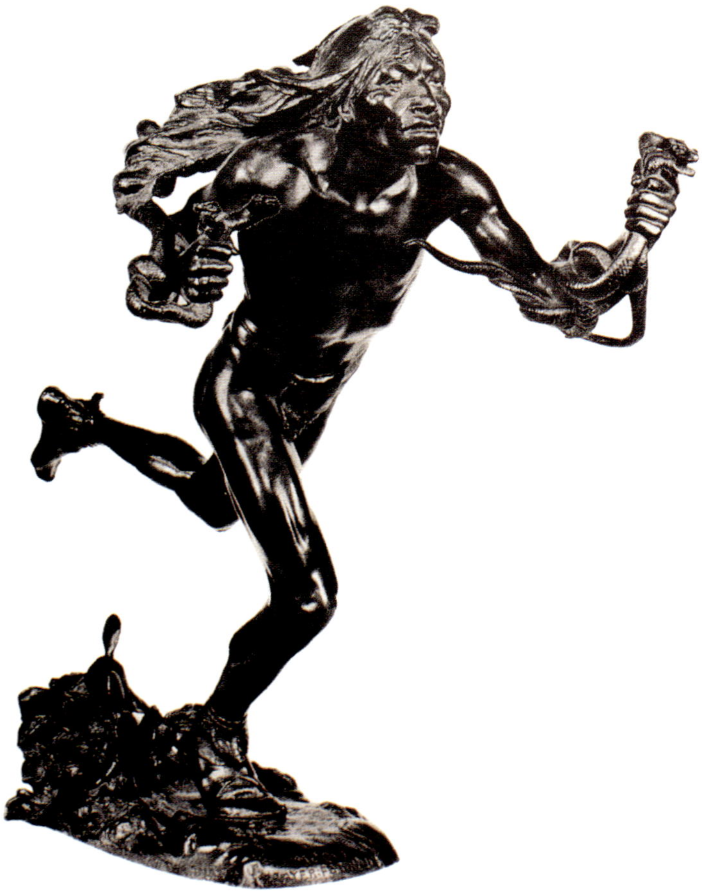

IV-84. "Returning to the Snakes" by Hermon Atkins MacNeil. Modeled c. 1897.

ADOLPH ALEXANDER WEINMAN (1870-1952)

Though Adolph Weinman was known as a sculptor of monuments (see page 59) he also created a number of memorable images of native Americans, whom he portrayed with uncommon delicacy and dignity. His Indian portrait busts treat each subject as an individual; Weinman was, in fact, an eloquent spokesman for the cause of Indian rights. "The tragic decline of a noble race," he wrote in 1904, "has touched a sympathetic chord in the minds of many writers and artists...I have endeavored to symbolize the passing of an heroic race, the North American Indian."

Among his notable sculptures are his portrait bronzes "Womboli" (IV-86), "Chief Blackbird-Ogalalla Sioux" (IV-87 and 88), and "Chief Flatiron" (IV-89). Particularly revealing of his viewpoint was his "Destiny of the Red Man" (IV-90). All of these bronzes were cast at Roman Bronze Works between 1902 and the 1950s.

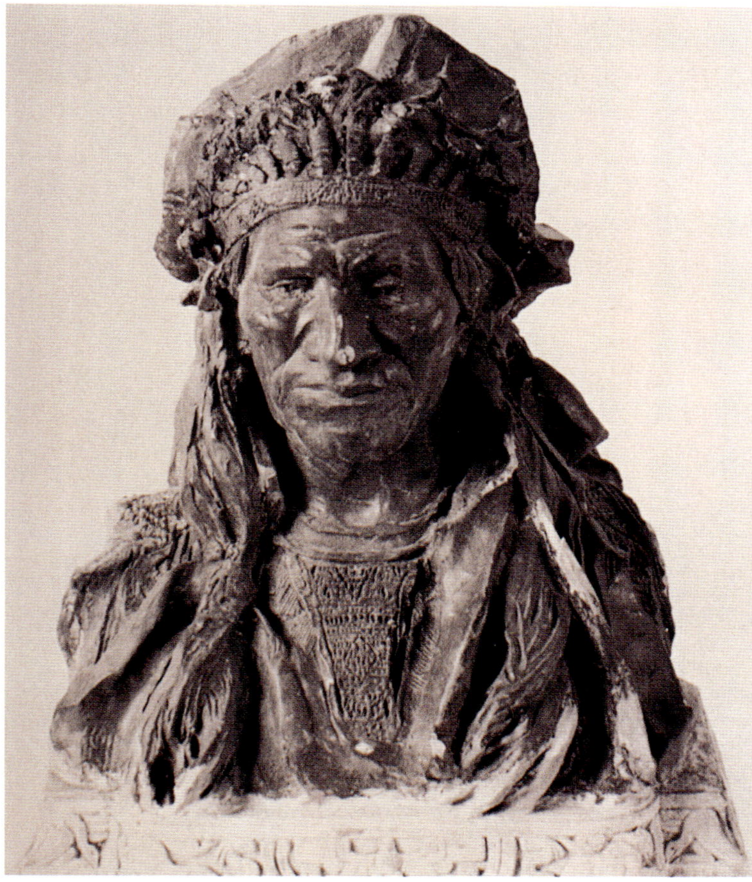

IV-87. "Chief Blackbird—Ogalalla Sioux" (plaster cast) by Adolph Alexander Weinman. Modeled 1902. Cast 1907.

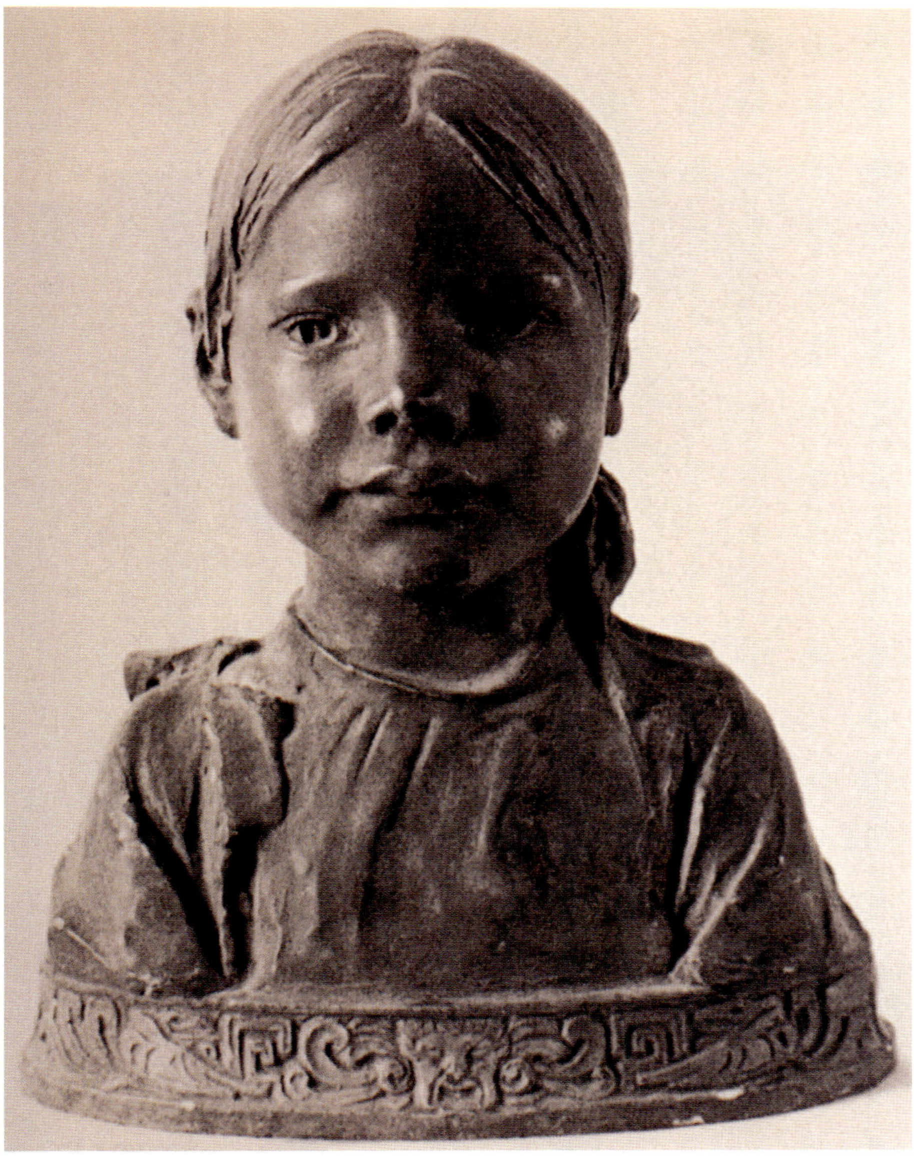

IV-86. "Womboli" (plaster cast) by Adolph Alexander Weinman. Modeled 1902. Cast 1907.

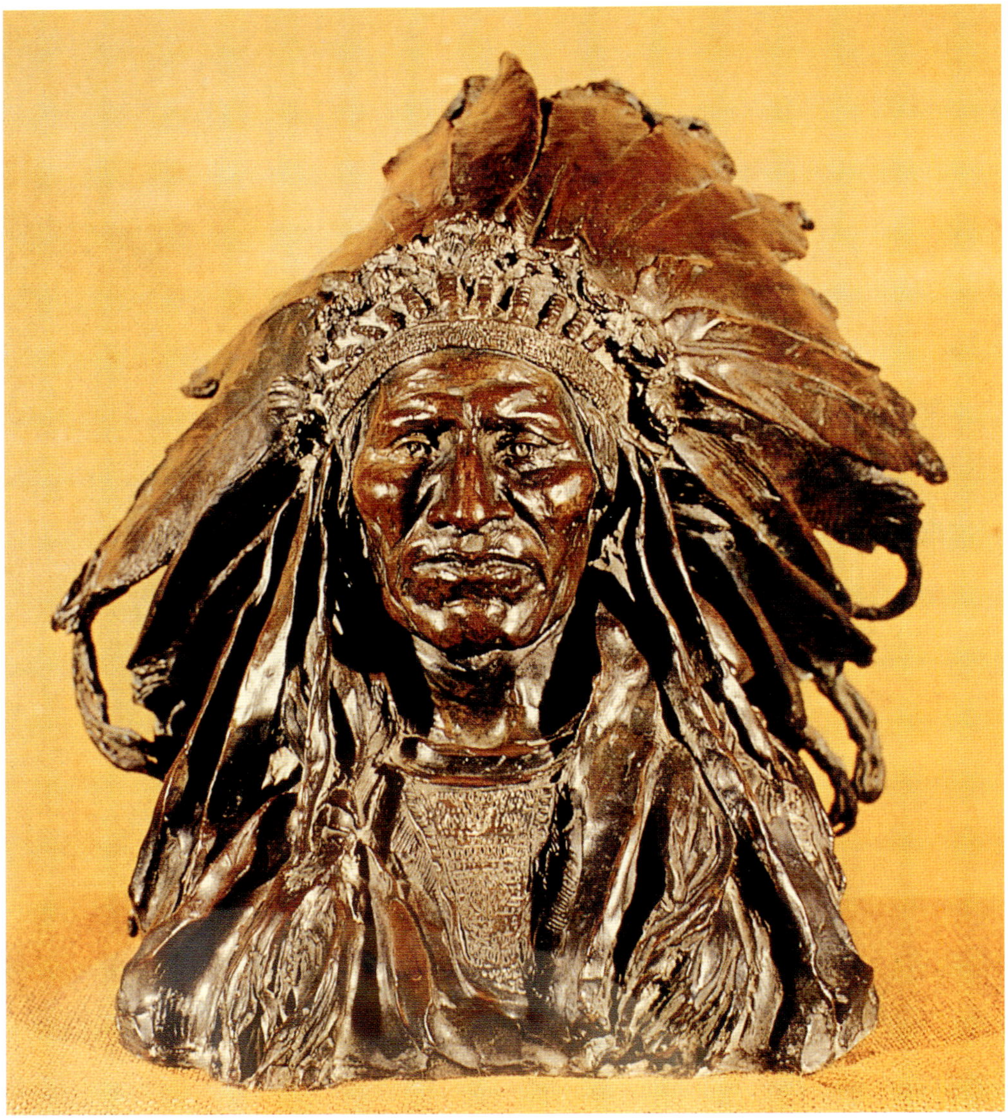

IV-88. "Chief Blackbird" by Adolph Alexander Weinman. Modeled 1903. Cast 1907.

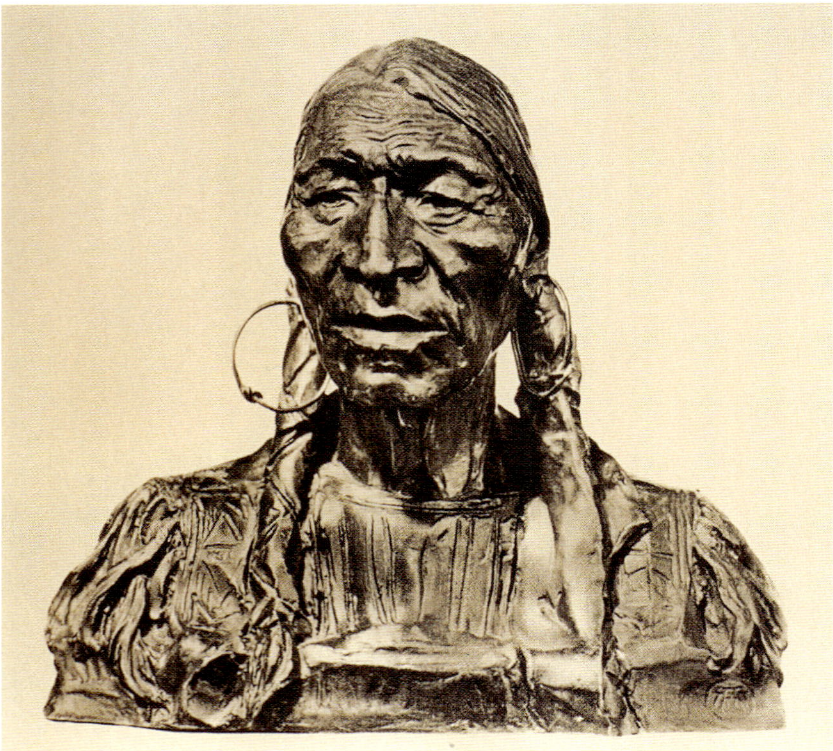

IV-89. "Chief Flatiron" by Adolph Alexander Weinman. c. 1903.

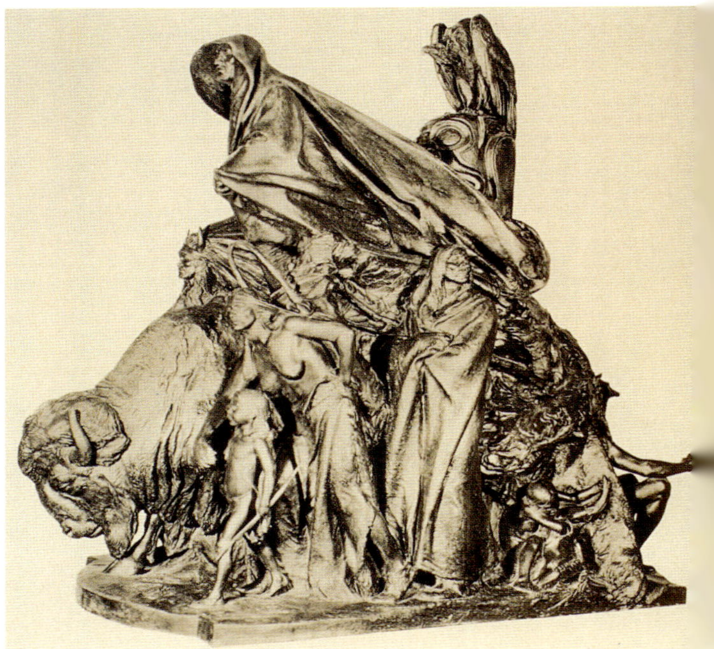

IV-90. "Destiny of the Red Man" by Adolph Alexander Weinman. 1903. Cast 1947.

MAHONRI YOUNG (1877-1957)

Mahonri Young was a sculptor both of genre figures of working men (see page 199) and of Western images. His sculptures of cowboys, pioneers, pony express riders, and frontiersmen were influenced by his Western upbringing and several trips to Indian territories.

Young, the grandson of Mormon leader Brigham Young, was born in Utah and had early training as an illustrator. A visit by Cryus Dallin to Salt Lake City to work on his Pioneer Monument influenced Young to go to New York; he arrived in 1899. He studied there at the Art Student's League, and in Paris at the Académie Julian. An early sculpture called "Bovet Arthur–A Laborer" won him a major prize from the National Academy of Design. Young made a number of trips to the West and became proficient at modeling Indian and Western images. He made Apache, Hopi, and Navajo groups for the American Museum of Natural History. In 1913 he helped to organize the Armory Show, in which he had six of his own works. Young's recognition as a sculptor followed with his "Sea Gull Monument" in Salt Lake City (where he also created a monument to the Mormons), and with several important prizes for sculpture.

Among Young's favorite images were those inspired by the "Wild West;" several of them, including "Cowboy and Horse" (IV-91), "Pioneer Woman" (IV-92), "Rolling His Own" (IV-93), "Lunch in the Desert" (IV-94), and "Pony Express" (IV-95) were cast at Roman Bronze Works (as were a number of his other genre pieces). Young's association with R.B.W. began in 1910 and continued into the 1950s.

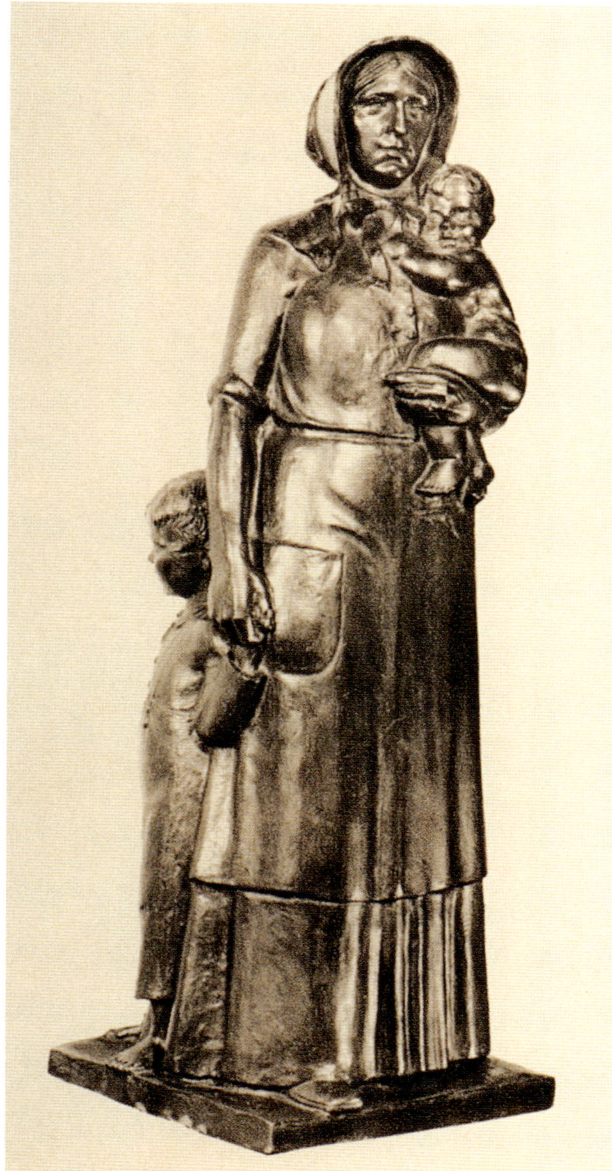

IV-92. "Pioneer Woman" by Mahonri Young. 1927.

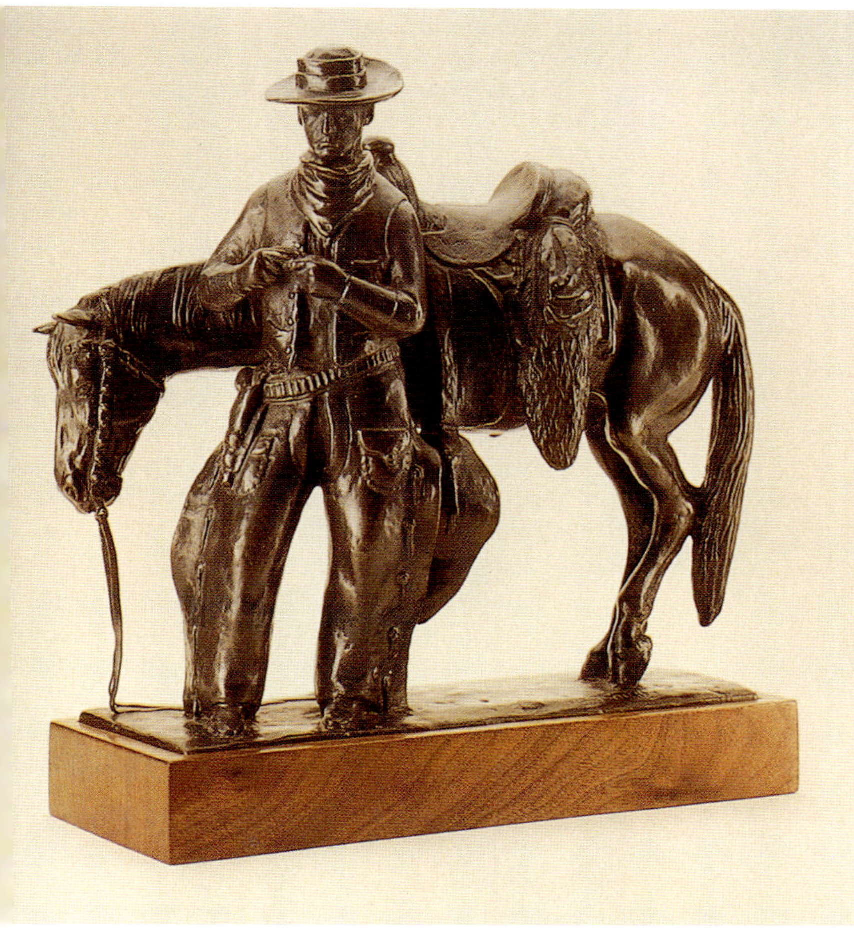

IV-91. "Cowboy and Horse" by Mahonri Young. c. 1924.

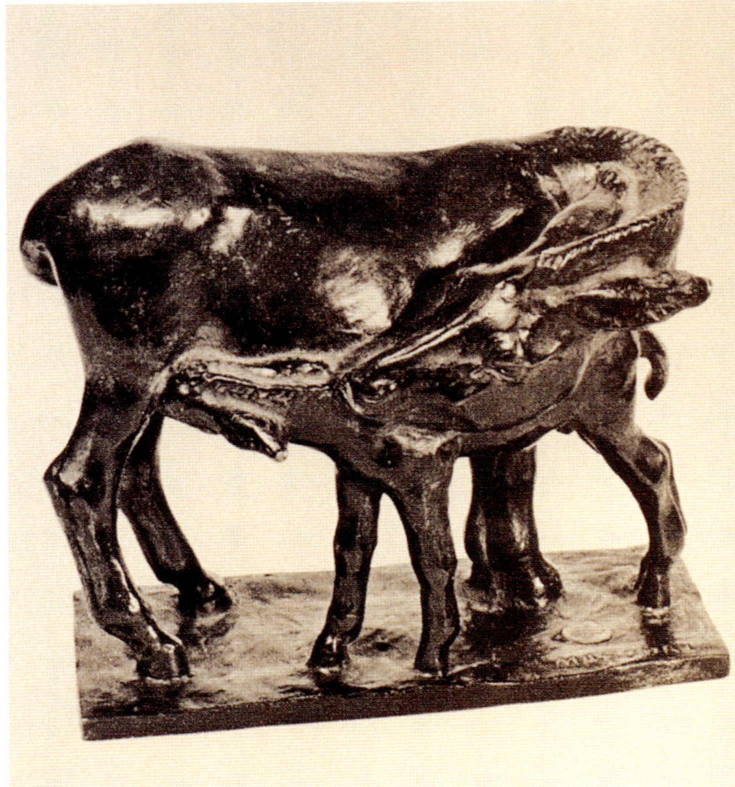

IV-94. "Lunch in the Desert" by Mahonri Young. c. 1930.

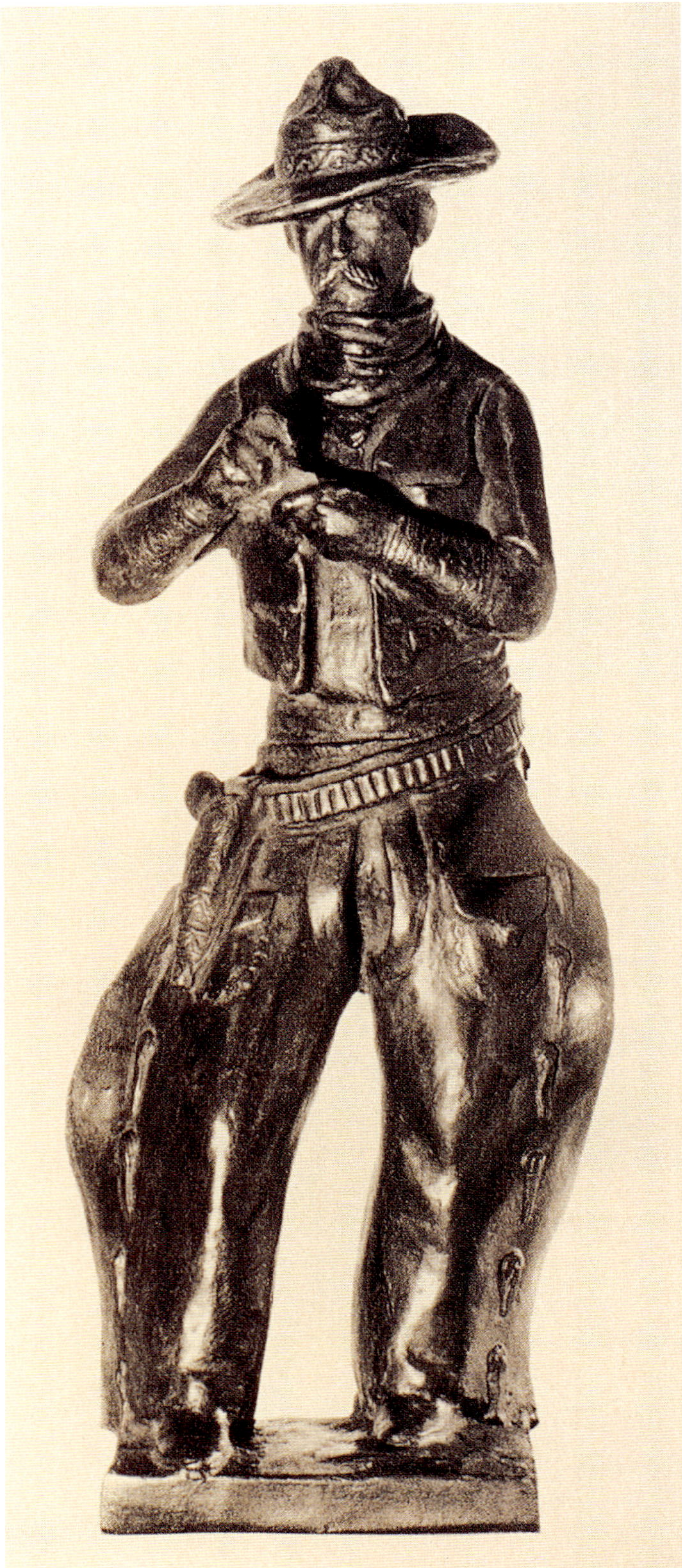

IV-93. "Rolling His Own" by Mahonri Young. c. 1928.

IV-95. "Pony Express" by Mahonri Young. c. 1932.

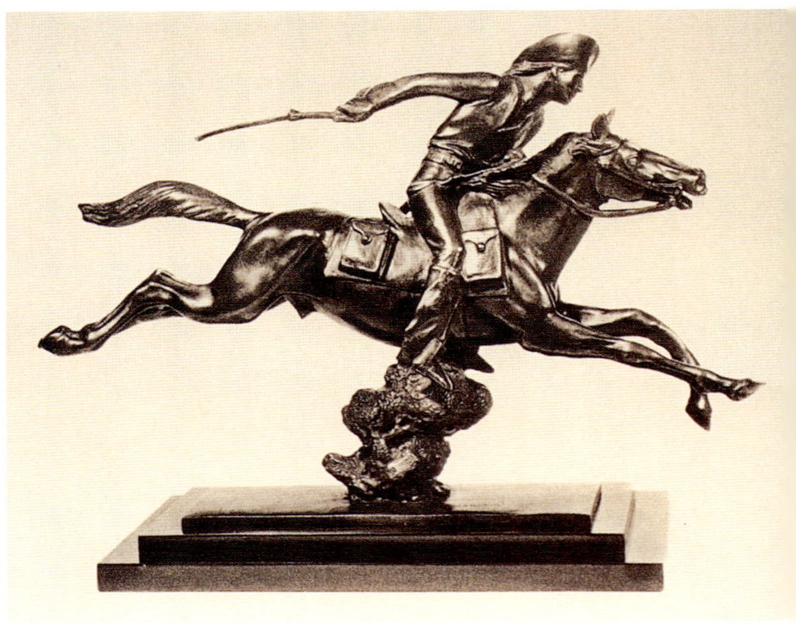

ALEXANDER PHIMINSTER PROCTOR (1860-1950)

Alexander Proctor was a sculptor of Western images, life-size animal statuary, and several well-known monuments. Proctor was born in Canada, and spent much of his youth in the American frontier and wilderness where images of Western life and wild animals intrigued him.

He began his art education in Colorado, becoming an accomplished engraver. In 1885 he studied in New York and in 1893 modeled thirty-seven animal figures, a cowboy, and an Indian for the Columbian Exposition in Chicago. In Paris, where he worked with Saint-Gaudens and took part in the Exposition Universelle of 1900, he won a gold medal for his "Goddess of Liberty" and his first notable animal sculpture—the pair of panthers now in Prospect Park, Brooklyn. He also modeled the horses for Saint-Gauden's Sherman and Logan Memorials. His animal sculptures were noted for their accuracy, with lean, dramatic shapes and naturalistic tension.

On his return from France, Proctor began an illustrious career, receiving many commissions. He was a sculptor-in-residence at the American Academy in Rome, and won several major awards. He executed a number of large works with Western themes, such as "On the War Trail" (IV-96) and "Buckaroo" (IV-97) (for the Denver Civic Center), "The Indian Warrior" (IV-98), and "Pursued" (IV-99). He also made smaller Western bronzes like his "Bust of a Cowboy" (IV-100).

Major animal monuments included the lions at the McKinley Monument in Buffalo, and the tigers at Princeton University, as well as the "Buffalo" (IV-101) in Washington, (also seen in small bronze version: IV-102), "Stalking Panther" (IV-103), and "Moose" (IV-104). His "Bear Cub and Rabbit" (IV-105) (also called "The Surprise") was cast many times. His association with Roman Bronze Works began in 1909 and continued through 1935.

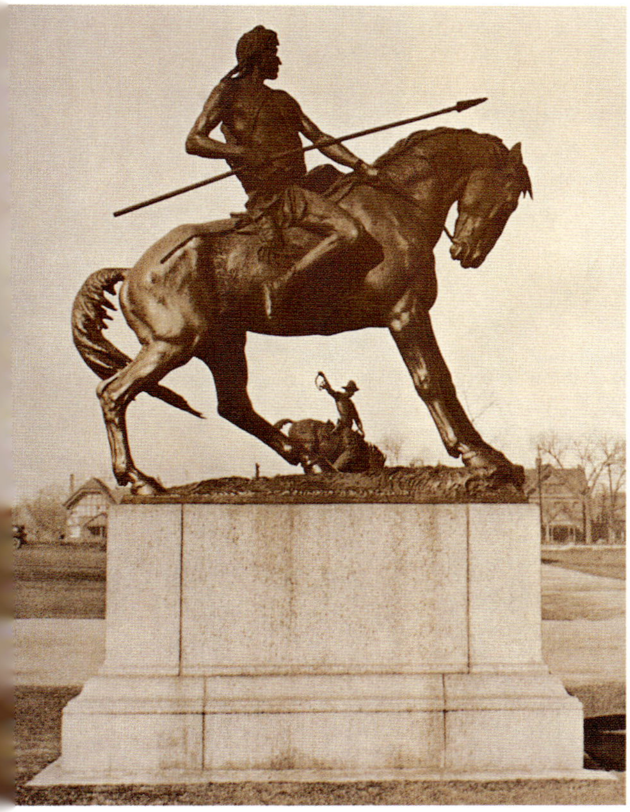

IV-96. "On the War Trail," Denver, Colorado, by Alexander Phiminster Proctor. c. 1930.

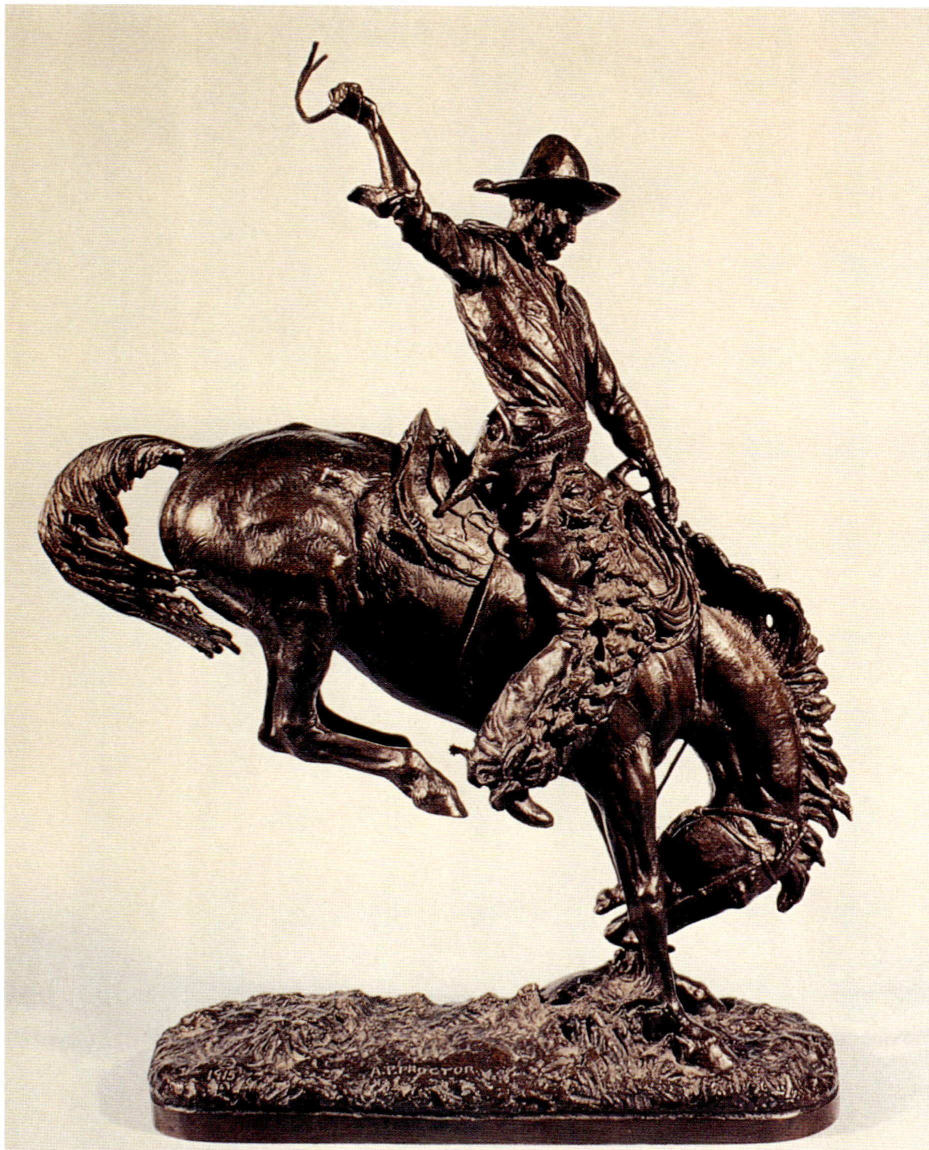

IV-97. "The Buckaroo" by Alexander Phiminster Proctor. 1915.

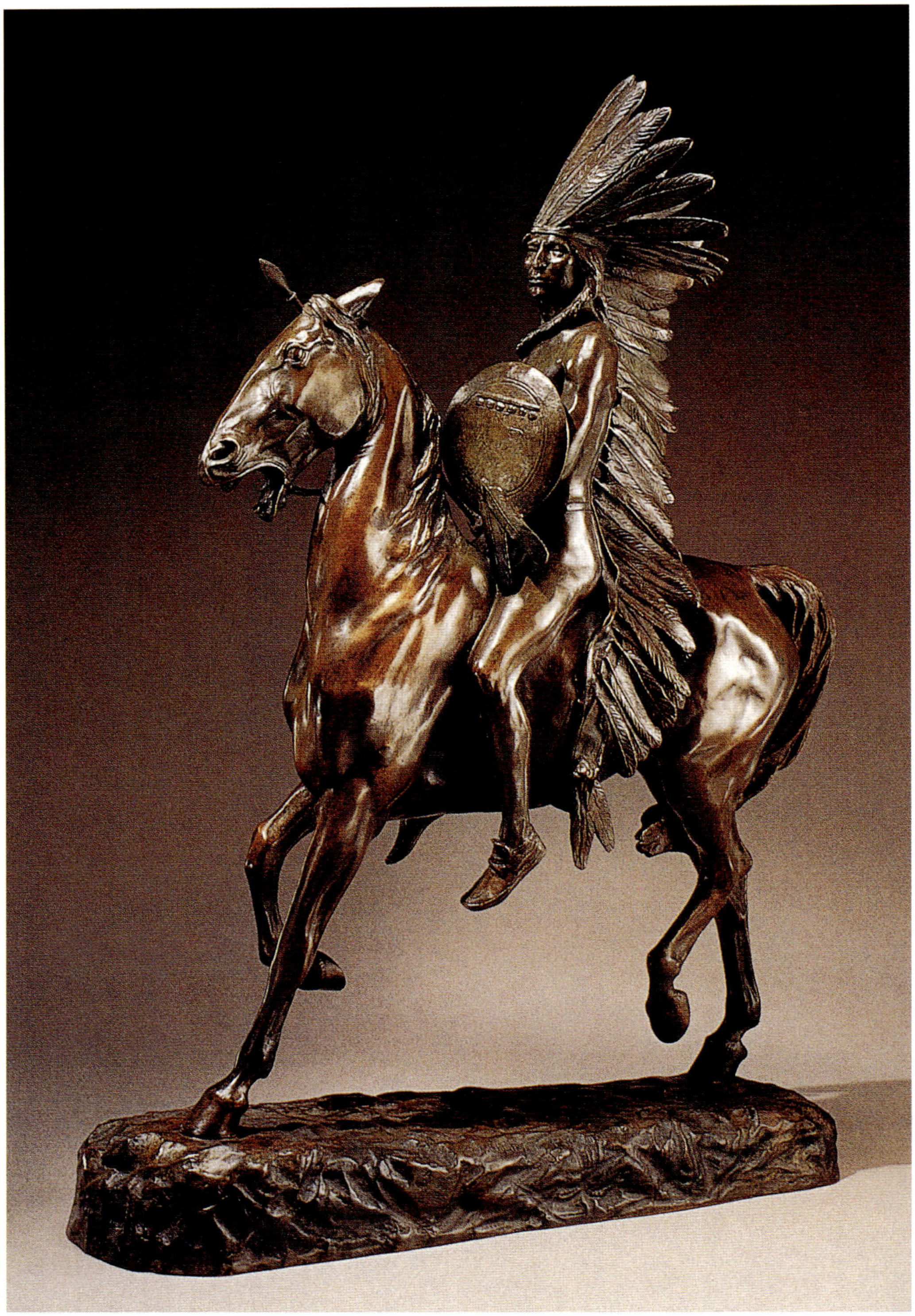

IV-98. "The Indian Warrior" by Alexander Phiminster Proctor. 1915.

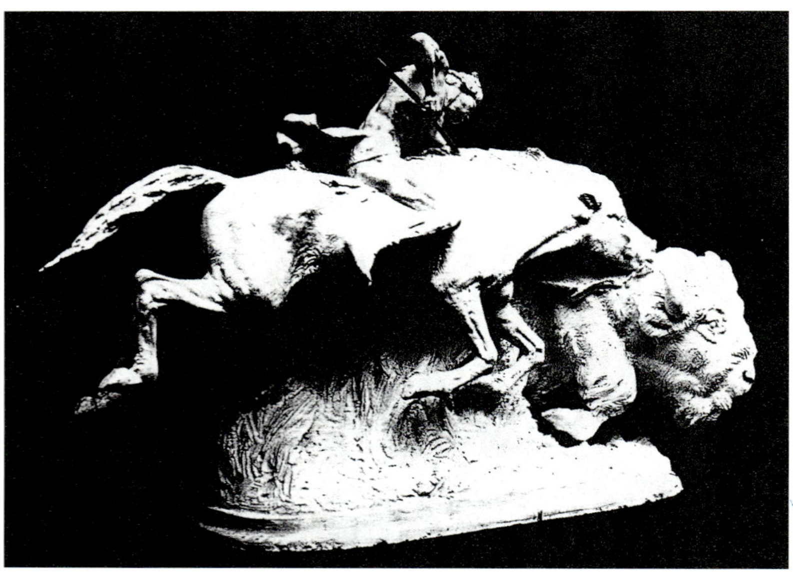

IV-99. "Pursued" or "Indian Pursuing Buffalo" by Alexander Phiminster Proctor. 1916.

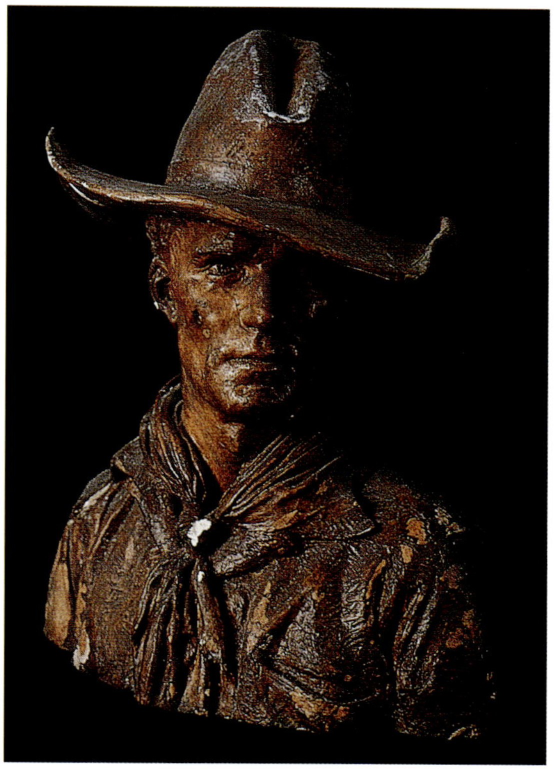

IV-100. "Bust of a Cowboy" (plaster cast) by Alexander Phiminster Proctor. c. 1934.

IV-101. "Buffalo," Washington, D.C., by Alexander Phiminster Proctor. 1914.

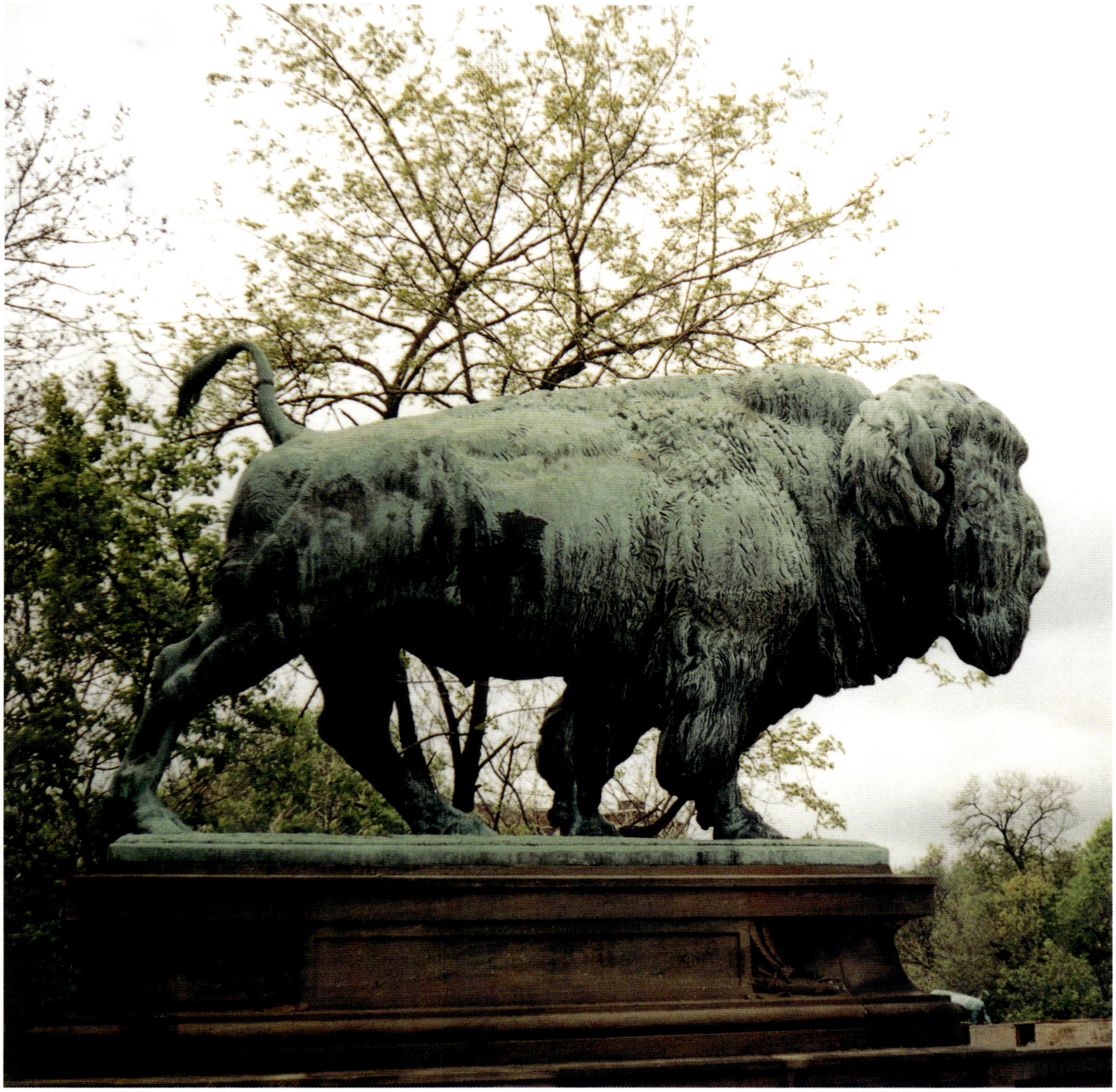

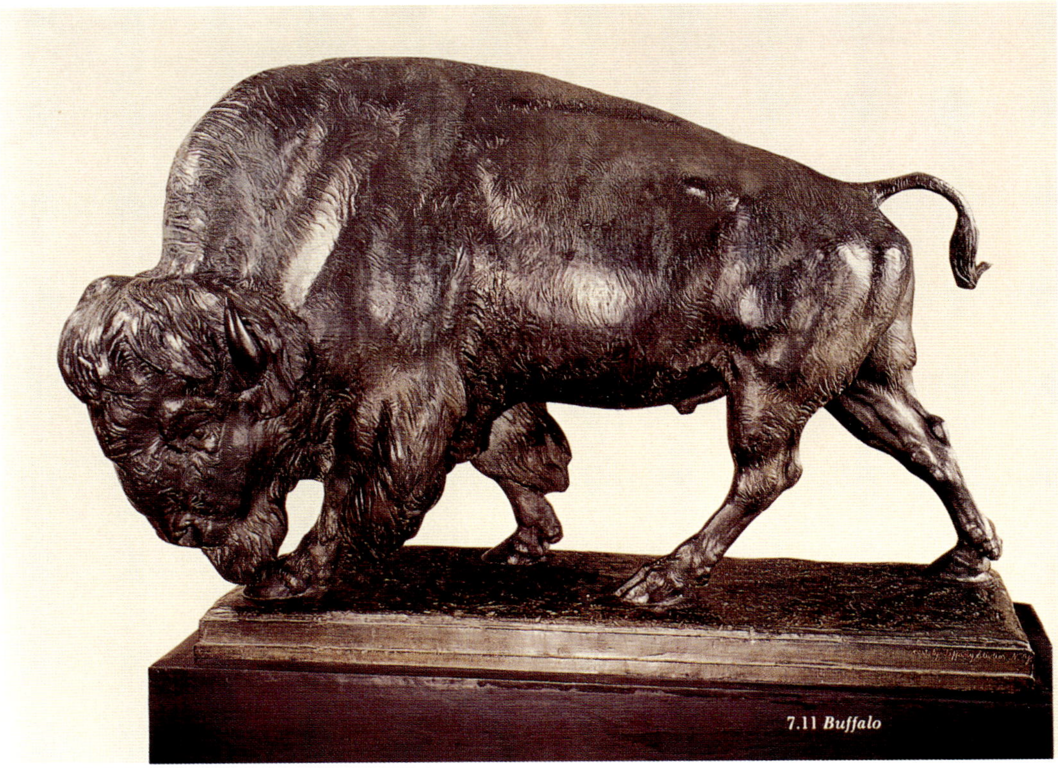

IV-102. "Buffalo" (statuette) by Alexander Phiminster Proctor. 1911.

Below:
IV-103. "Stalking Panther" by Alexander Phiminster Proctor. Cast 1904. Presented to President Theodore Roosevelt by the members of his cabinet in 1908.

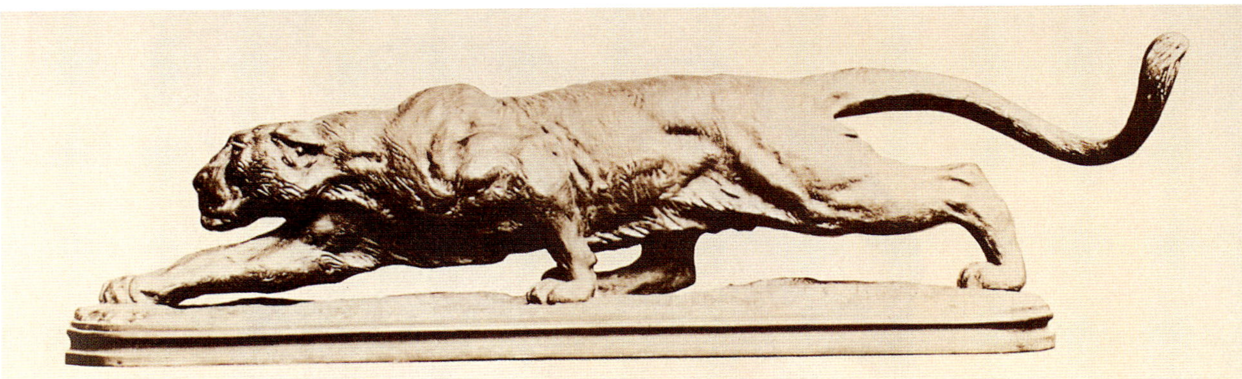

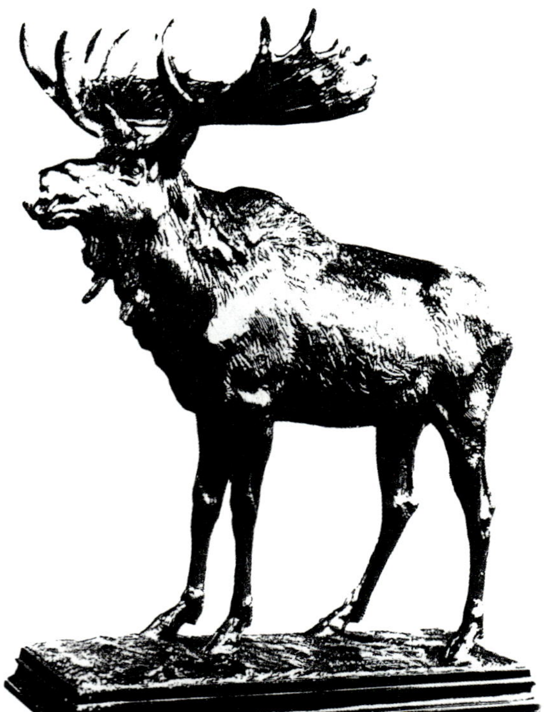

IV-104. "Moose" by Alexander Phiminster Proctor. Cast 1913. This small bronze was a reworking of the life-size moose Proctor made for the World's Columbian Exposition in 1893.

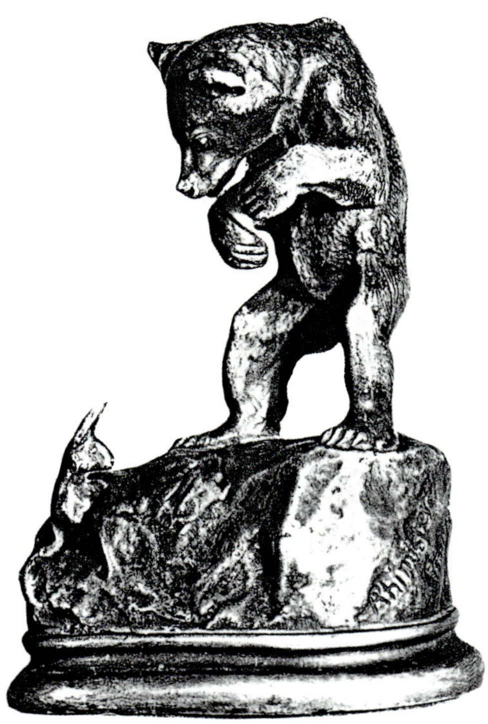

IV-105. "Bear Cub and Rabbit" (also called "Surprise") by Alexander Phiminster Proctor. Originally modeled 1885-88; first cast by R.B.W. 1914.

ANNA HYATT HUNTINGTON (1876-1967)

Anna Hyatt Huntington had a distinguished career as a sculptor of animals and equestrian monuments in the first half of the twentieth century, a time few women artists could make their way in America. She received numerous public commissions and executed many smaller works, with an emphasis on animal sculptures. She was the first American woman to make a heroic equestrian monument, and the first woman sculptor to be admitted to the Academy of Arts and Letters.

Born in Cambridge, Massachusetts, she studied with Henry Kitson and had her first one woman show of forty animal sculptures at the age of twenty-four. In 1902 she went to New York where she studied at the Art Students League and with both Gutzon Borglum and Hermon Atkins McNeil. She also worked in France and Italy. By the time she was thirty she had made her well-known 1915 equestrian of Joan of Arc (numbers IV-106 and IV-107) for which she was decorated with the French Legion of Honor. (The statue can be seen on Riverside Drive in New York, as well as in Blois, France, and in Gloucester, Massachusetts.) Her work became known for its realism, its meticulous detail, and its sympathetic understanding of animal life.

After her marriage to Archer Huntington, a wealthy patron of the arts and a poet, she also became involved with the creation of several museums. Among them were the Hispanic Society in New York, where several of her works grace the plaza, and Brookgreen Gardens in South Carolina, which was to become a major outdoor site for American sculpture. She was a member of the National Sculpture Society, the National Academy of Design, and many other organizations, many of which recognized her with awards.

Her success continued with many commissions (for which she received several major prizes). Like her statue of Joan of Arc, many of her larger works included beautifully modeled horses; among them were "El Cid Compeador" (IV-108), "The Torchbearers" (IV-109), and "Winter Noon" (IV-110). Also featuring horses though in smaller form were her bronzes "Riders to the Sea" (IV-111), and "Young Jackson" (IV-112), a study for a larger bronze equestrian statue.

Huntington was well known for her smaller, meticulously naturalistic bronze animal sculptures. Most of these were cast at R.B.W. They include her "Polar Bear and Cubs" (IV-113), her greyhound, "Speedy" (IV-114), "Mother and Sheep" (IV-115), "Horse's Head" (IV-116), "Donkey Braying" (IV-117), "Fawn Running" (IV-118), "Red Doe and Fawn" (IV-119), "Goats Butting" (IV-120), "Rolling Bear" (IV-121), and "Mongoose and Cobra" (IV-122). Huntington cast at R.B.W. from 1905 throughout her lifetime.

Although she continued to work in an academic tradition of realism, she was one of the earliest sculptors to experiment with cast aluminum; she had a touring show of twenty aluminum sculptures in 1940s. In 1952 the National Arts Club included sixty of her works in a retrospective. Huntington worked steadily, with an enormous output, until she was ninety-one, when she made her last equestrian statue.

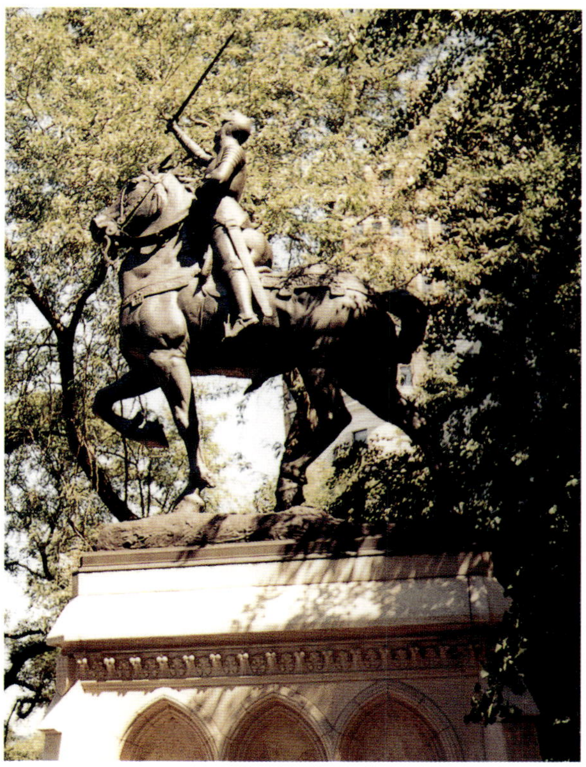

IV-106. "Joan of Arc," New York City, by Anna Hyatt Huntington. 1915.

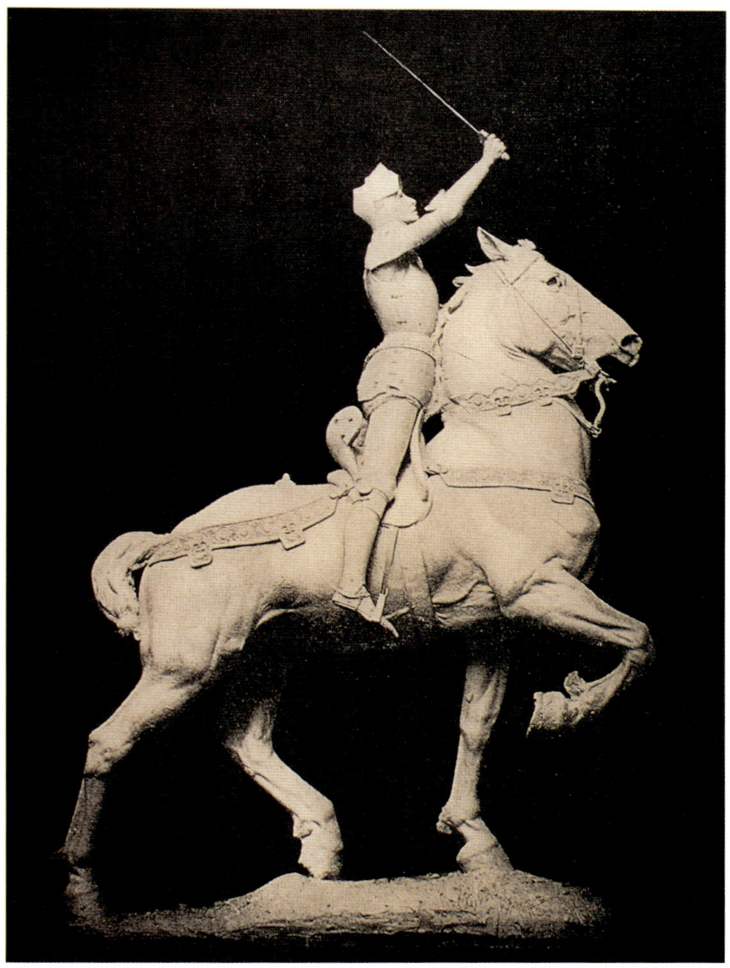

Right:
IV-107. "Joan of Arc" by Anna Hyatt Huntington. There are several versions of this statue in the United States, Canada, and France.

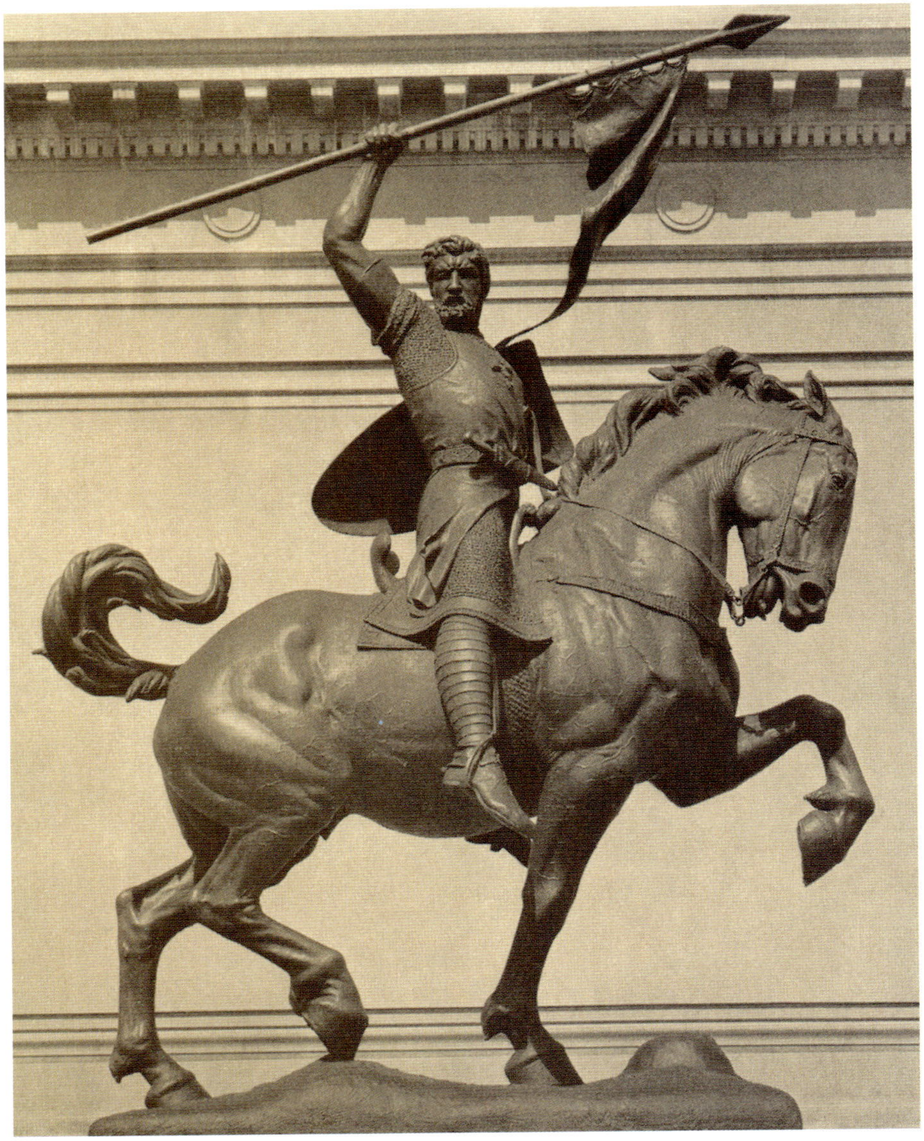

IV-108. "El Cid Compeador," New York City, by Anna Hyatt Huntington. 1927.

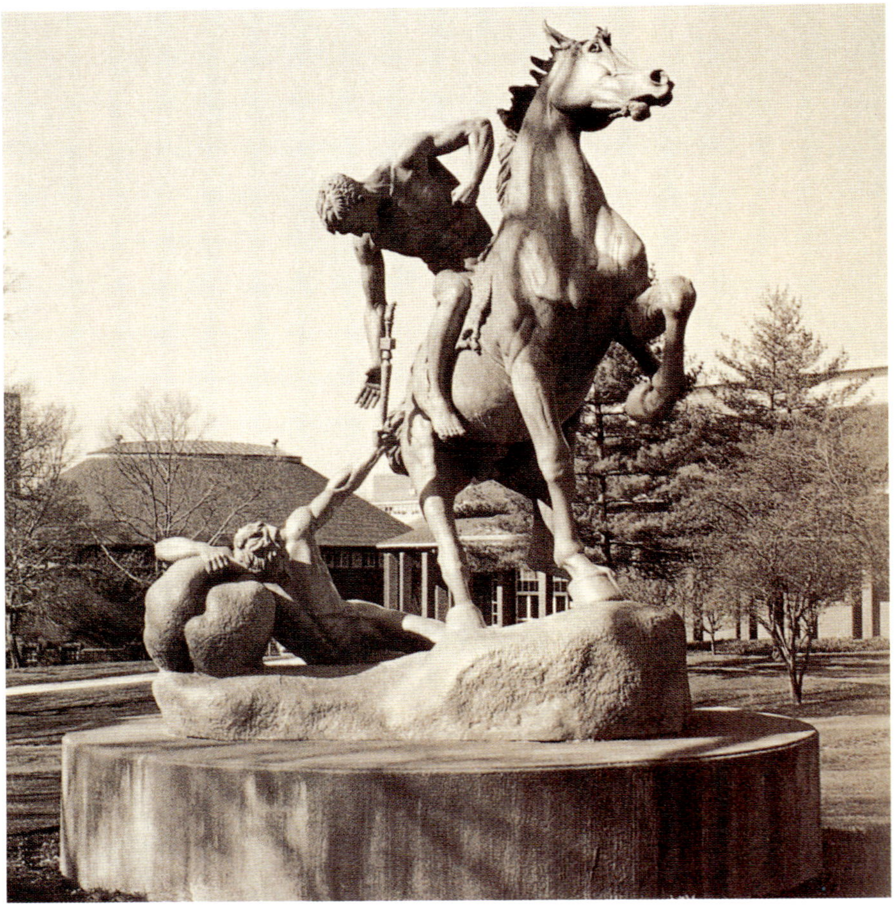

IV-109. "The Torch Bearers," Hoboken, New Jersey, by Anna Hyatt Huntington. 1953.

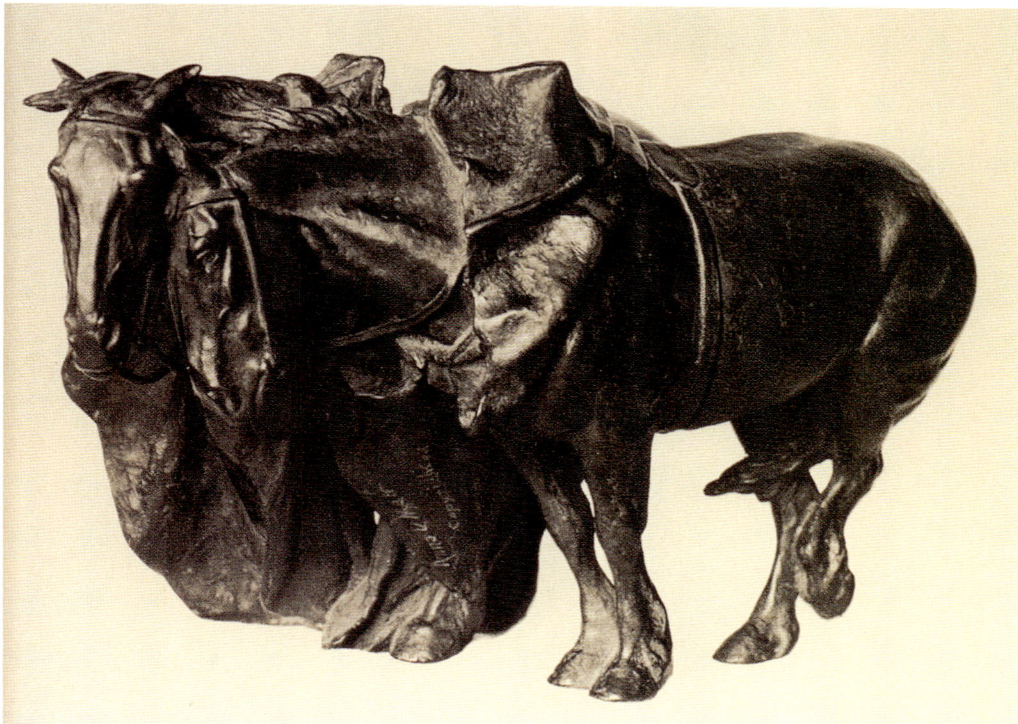

IV-110. "Winter Noon" by Anna Hyatt Huntington. c. 1902.

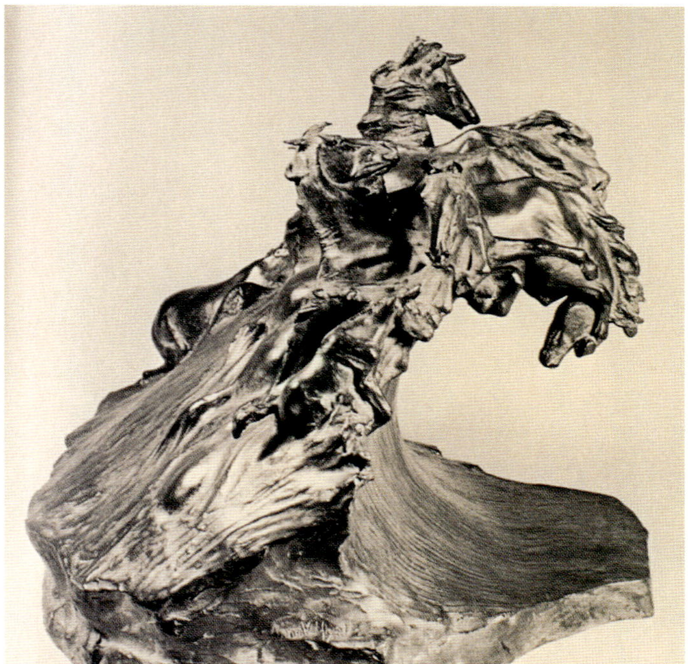

IV-111. "Riders to the Sea" by Anna Hyatt Huntington. c. 1912.

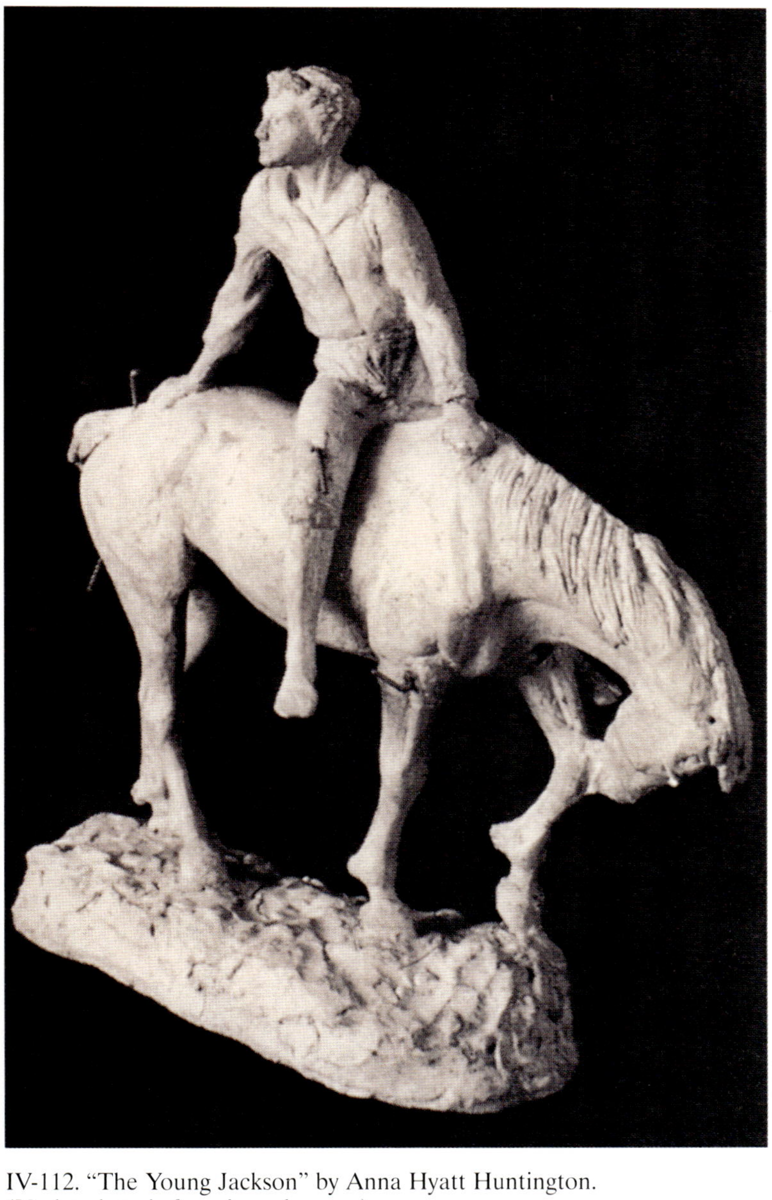

IV-112. "The Young Jackson" by Anna Hyatt Huntington. (Undated study for a large bronze.)

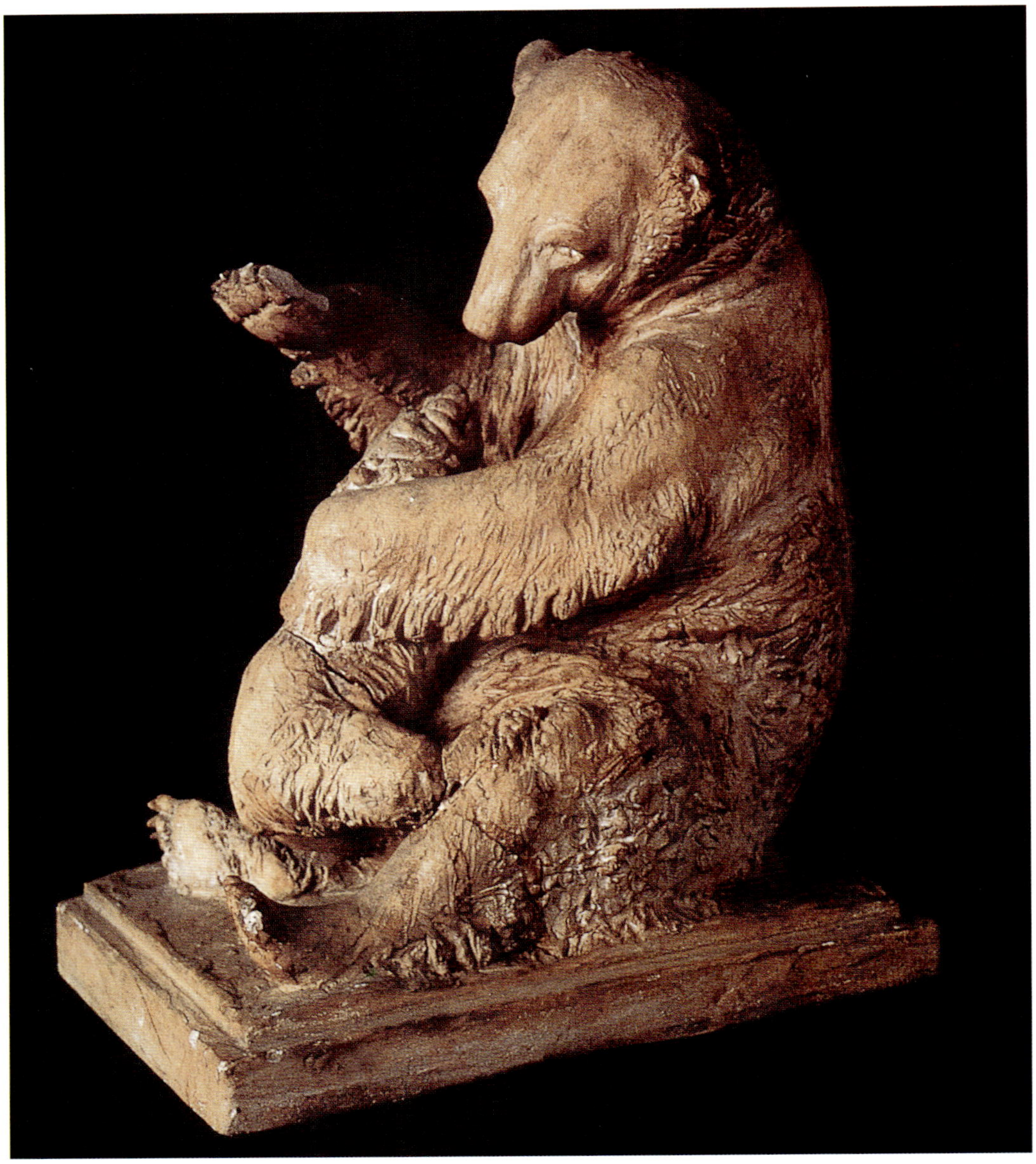

IV-113. "Polar Bear and Cub" (plaster cast) by Anna Hyatt Huntington. 1954.

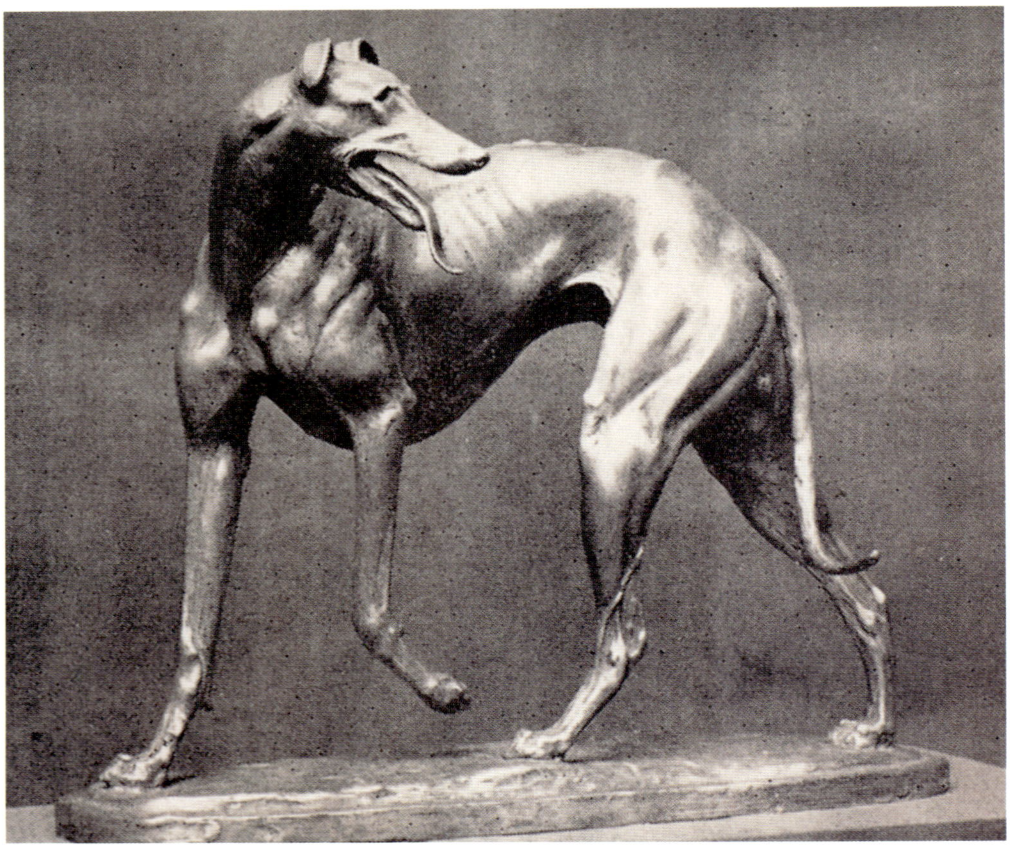

IV-114. "Speedy" by Anna Hyatt Huntington. c. 1936.

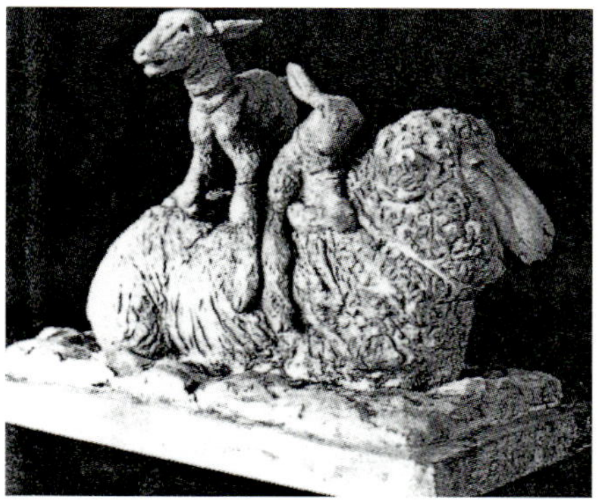

IV-115. "Mother and Sheep" (plaster cast) by Anna Hyatt Huntington. 1958.

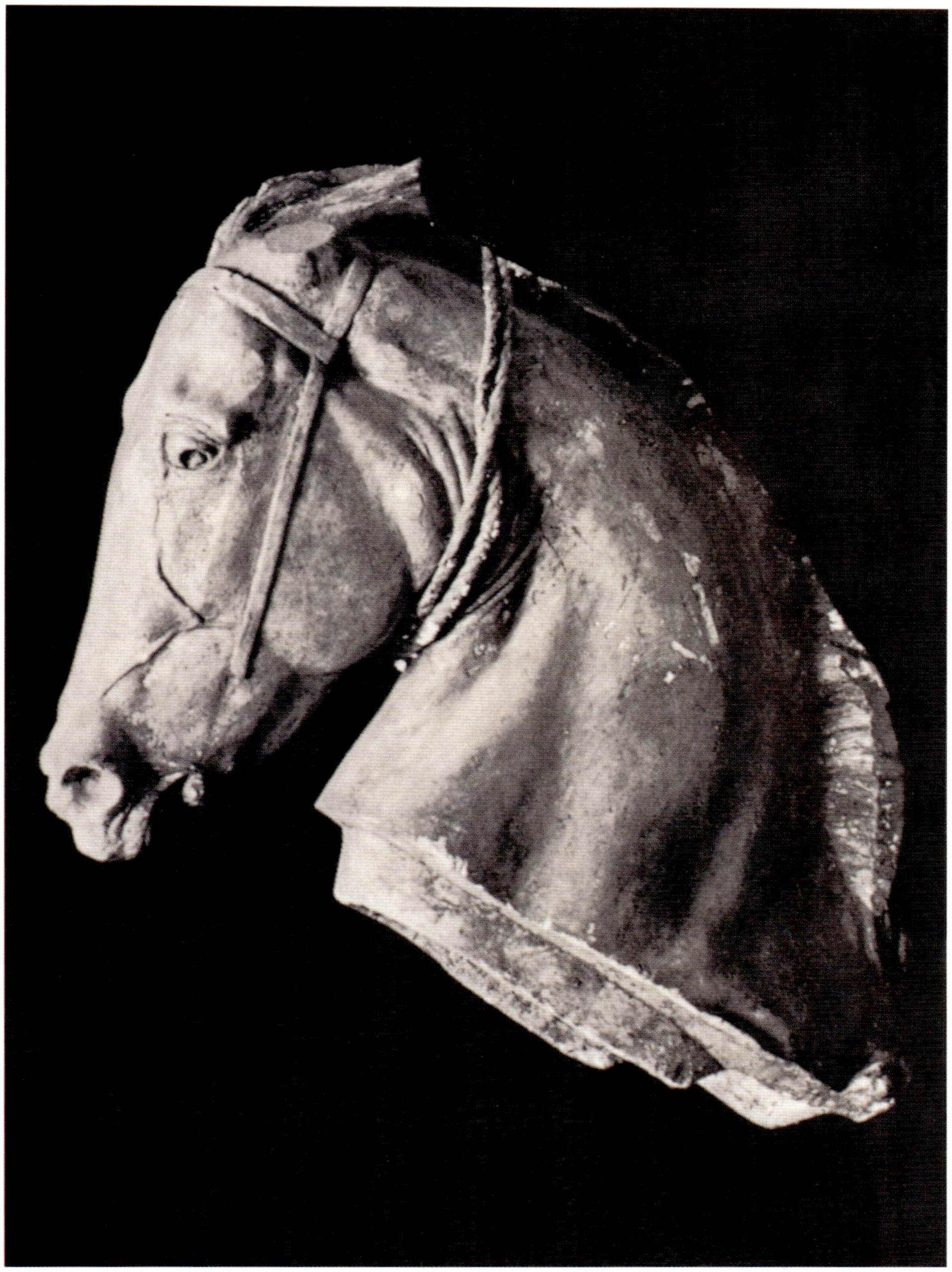

IV-116. "Horse's Head" by Anna Hyatt Huntington. 1937.

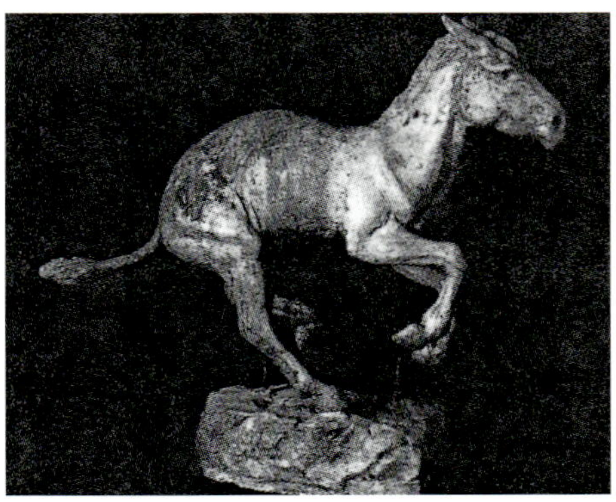

IV-117. "Donkey Braying" (plaster cast) by Anna Hyatt Huntington.

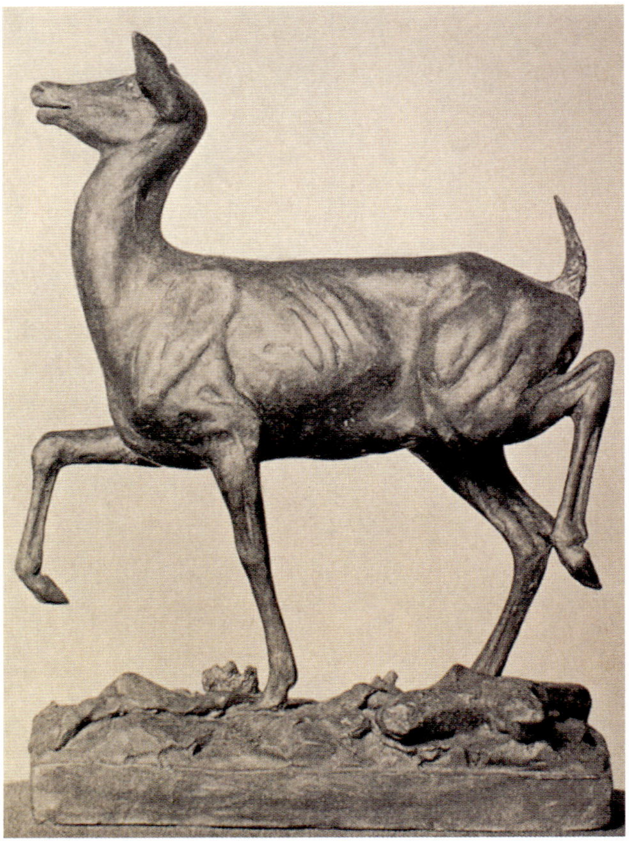

IV-118. "Fawn Running" by Anna Hyatt Huntington. c. 1952.

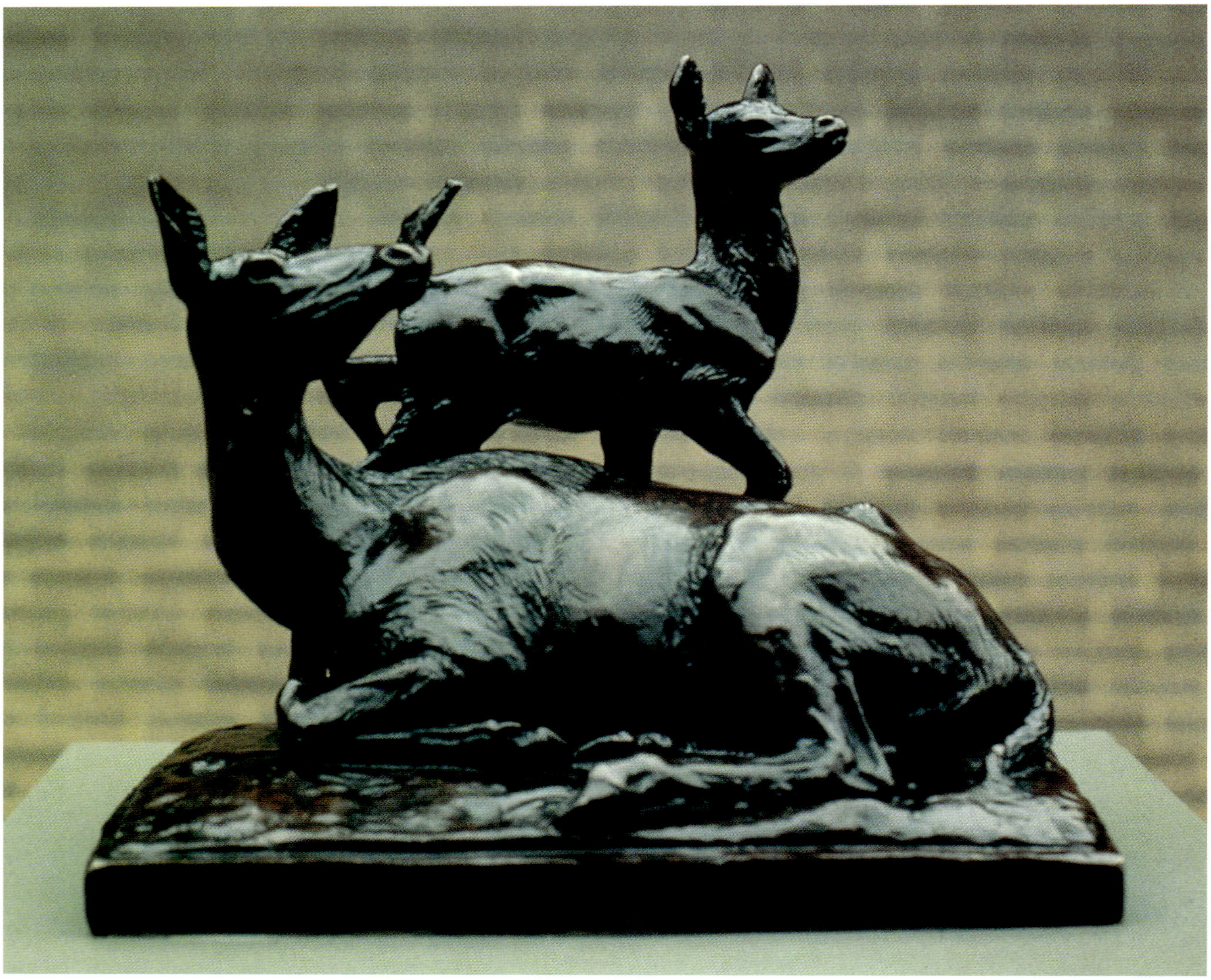

IV-119. "Red Deer and Fawn" by Anna Hyatt Huntington. 1933.

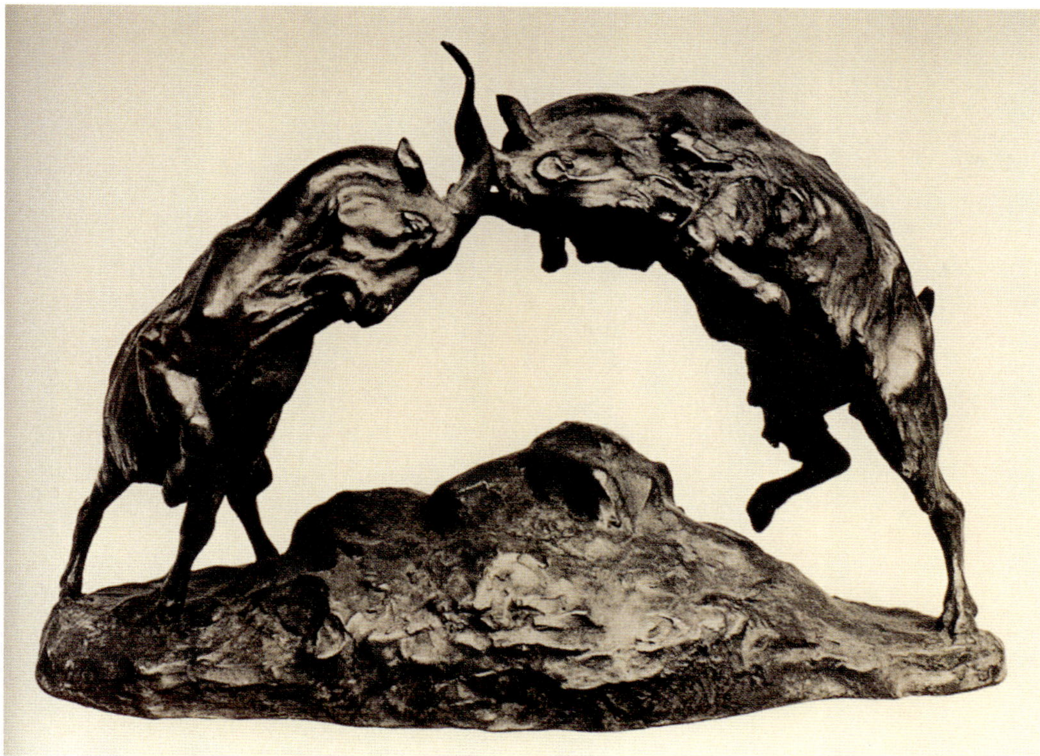

IV-120. "Goats Butting" by Anna Hyatt Huntington. 1904.

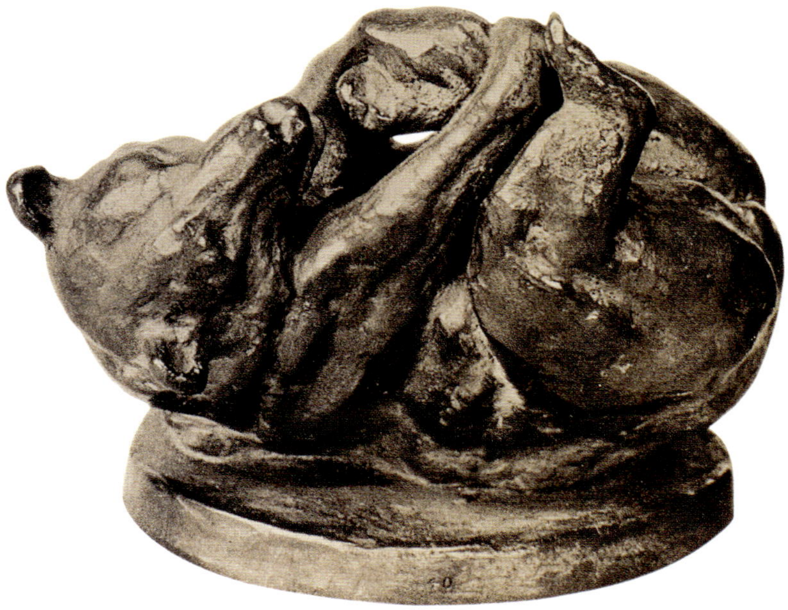

IV-121. "Rolling Bear" by Anna Hyatt Huntington. Before 1914.

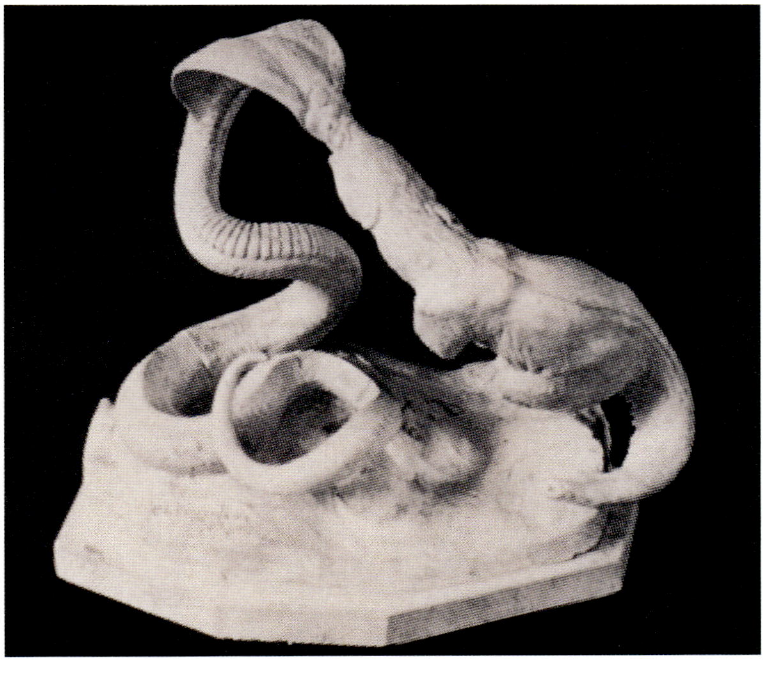

IV-122. "Mongoose and Cobra" (plaster cast) inscribed "Hold On Rikki-tikki-tavi," by Anna Hyatt Huntington. 1955.

EDWIN W. DEMING (1860-1942)

Though primarily known as a painter, Edwin Deming also made a significant number of bronze animal sculptures. He was interested in Western and Indian culture, having grown up on the prairie frontier. He was born in Ohio, but as a young man he made his way to Indian Territory in Illinois, where he studied and sketched the Ponca, Osage, Oto, and Pawnee Indians. Though he was briefly a student at New York's Art Students League and at the Académie Julian in Paris, Deming was primarily self-taught, and spent most of his life observing and drawing Indians and natural life of the Far West.

Deming's major contributions to art showing Western life were his realistic murals depicting Indian life, with a particular emphasis on spiritual and mystical aspects, for the American Museum of Natural History. His careful attention to detail is also evident in his animal sculptures of wildlife found on the Plains.

His small bronzes, made primarily between 1905 and 1910, include his best-known sculptures "The Fight" (IV-123) depicting a mountain lion and bear locked in combat, and "Mutual Surprise" (IV-124); additional animal sculptures include his "Bears in Combat" (IV-125) and "Buffalo" (IV-126). These works were all cast at Roman Bronze Works between 1906 and 1917.

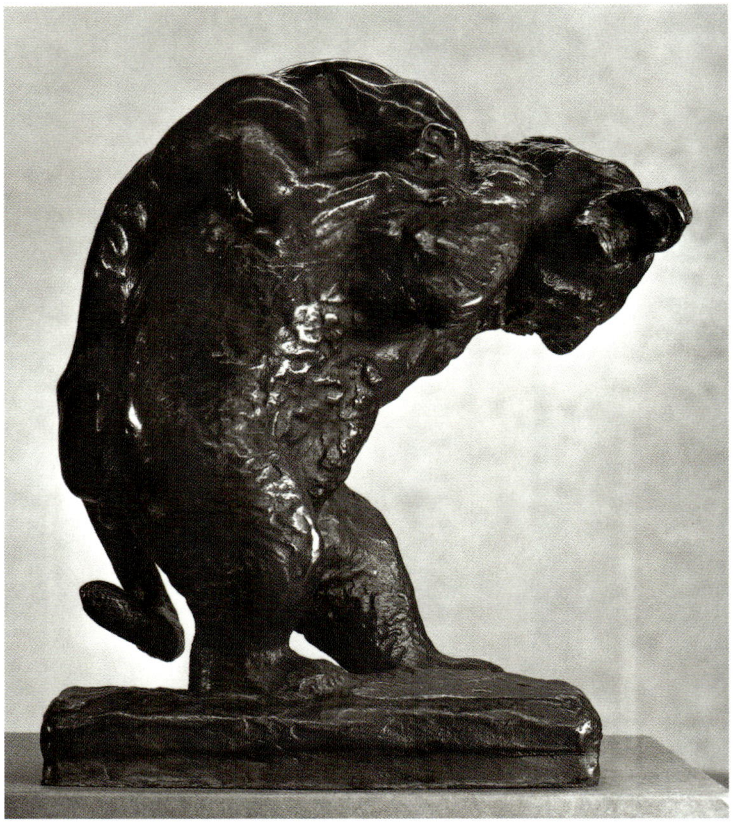

IV-123. "The Fight" by Edwin W. Deming. 1906.

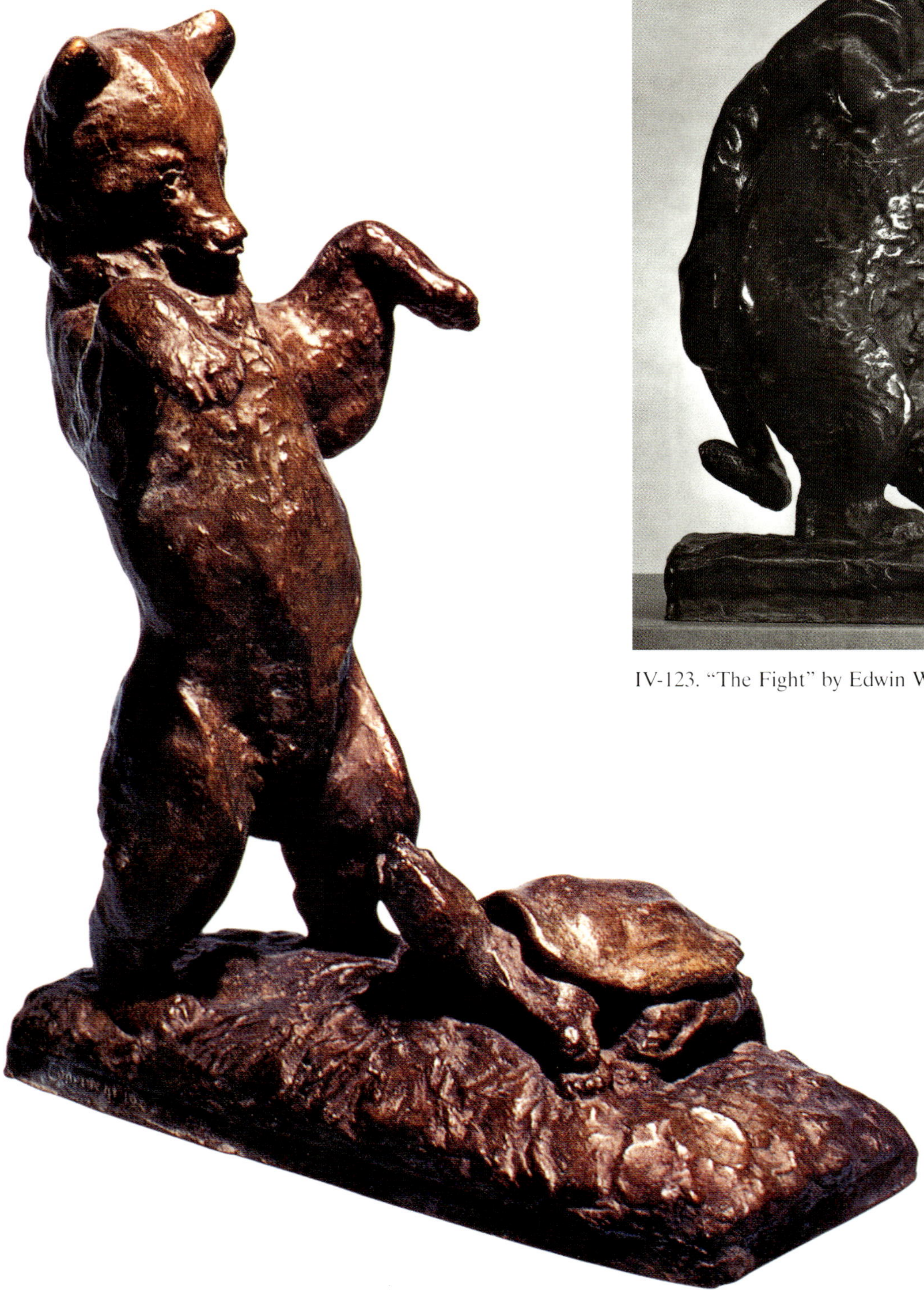

IV-124. "Mutual Surprise" by Edwin W. Deming. 1907.

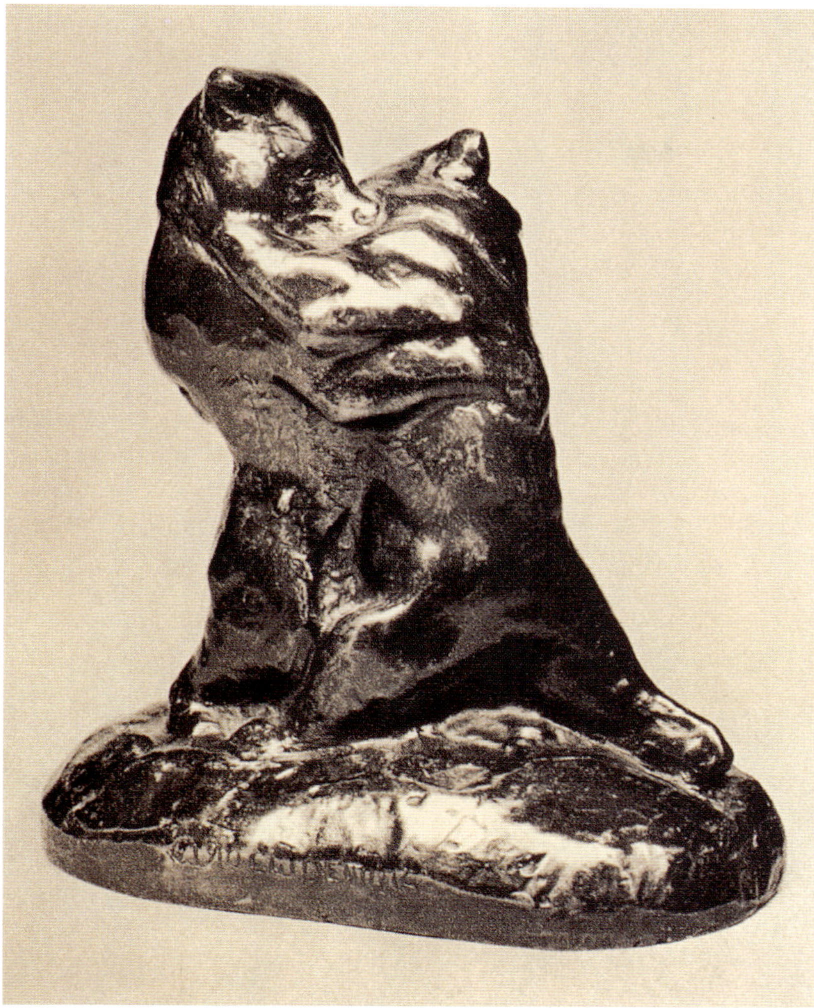

IV-125. "Bears in Combat" by Edwin W. Deming. 1916.

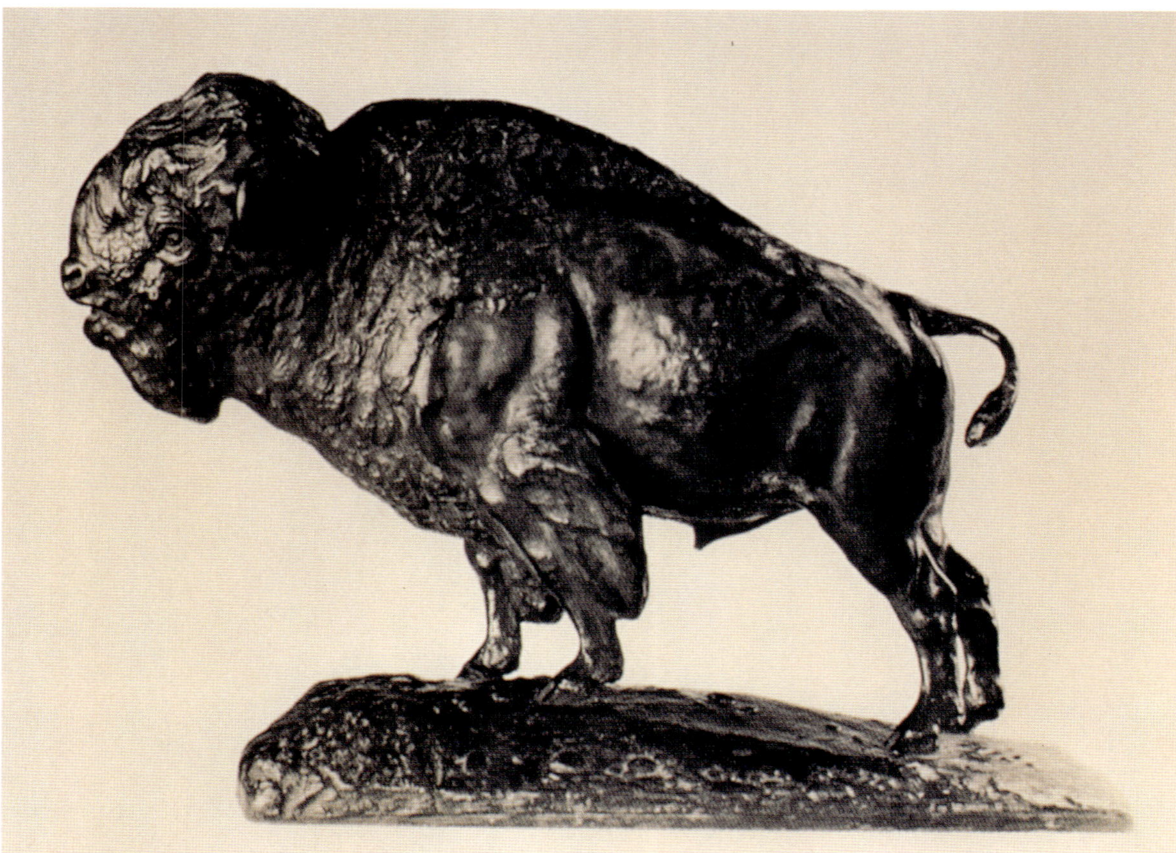

IV-126. "Buffalo" by Edwin W. Deming. 1908.

JAMES LIPPITT CLARK (1883-1969)

James Clark was a leading naturalist sculptor, whose images of animals can be seen throughout the American Museum of Natural History. Clark spent almost fifty years on the staff of the Museum and was responsible for the creation of the halls devoted to Asian, North American, and African mammals. His careful attention to the anatomy of the animals and the environments around them showed both his scientific and artistic expertise.

Born in Providence, Clark attended the Rhode Island School of Design. He began his career making small animal sculptures for the Gorham Silver Company. He started work at the Museum in 1902; he worked there until 1949. His travels to study and sketch animals in their native habitats took him to the American West, Africa, Mongolia, and numerous other places around the world. He also produced the first film of African wildlife and wrote many books on his travels.

Before working on the taxidermy and settings for the displays at the Museum, Clark made smaller models of the animals, many of which were cast in bronze (see numbers IV-127 to IV-132), and are in museum collections throughout the country. His first castings at Roman Bronze Works were in 1904 and he continued to use the foundry through the 1950s.

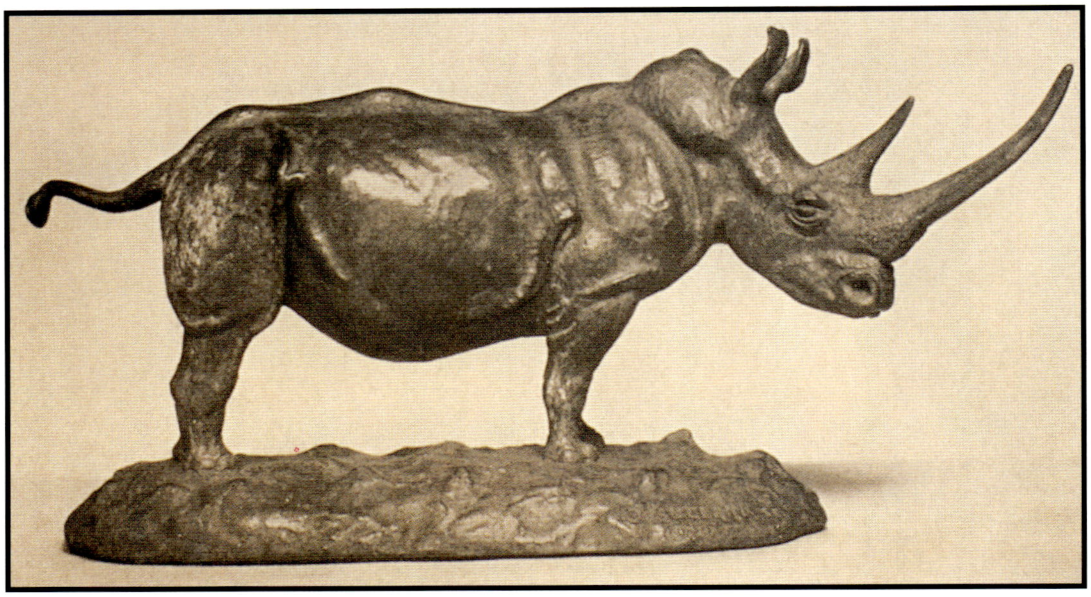

IV-127. "African White Rhino" by James L. Clark. 1913.

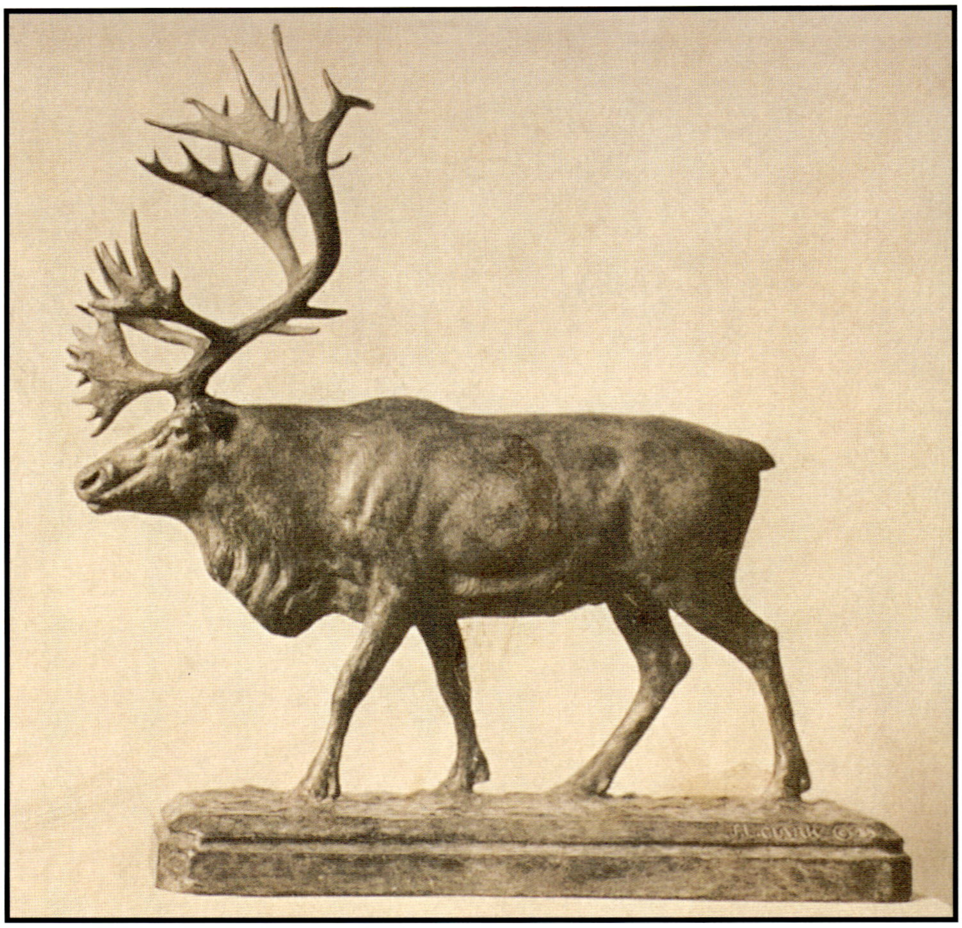

IV-128. "Osborn Caribou" by James L. Clark. c. 1913.

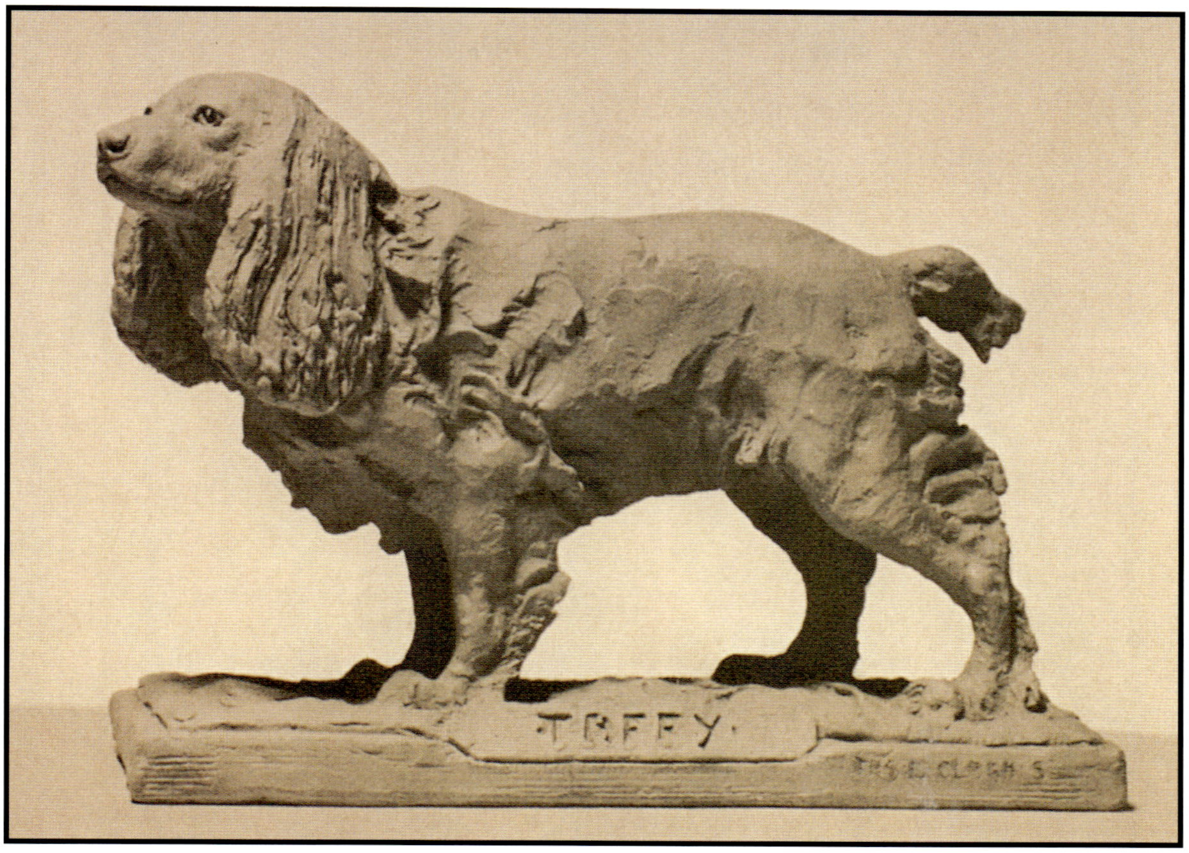

IV-129. "Taffy, a Portrait" by James L. Clark. 1913.

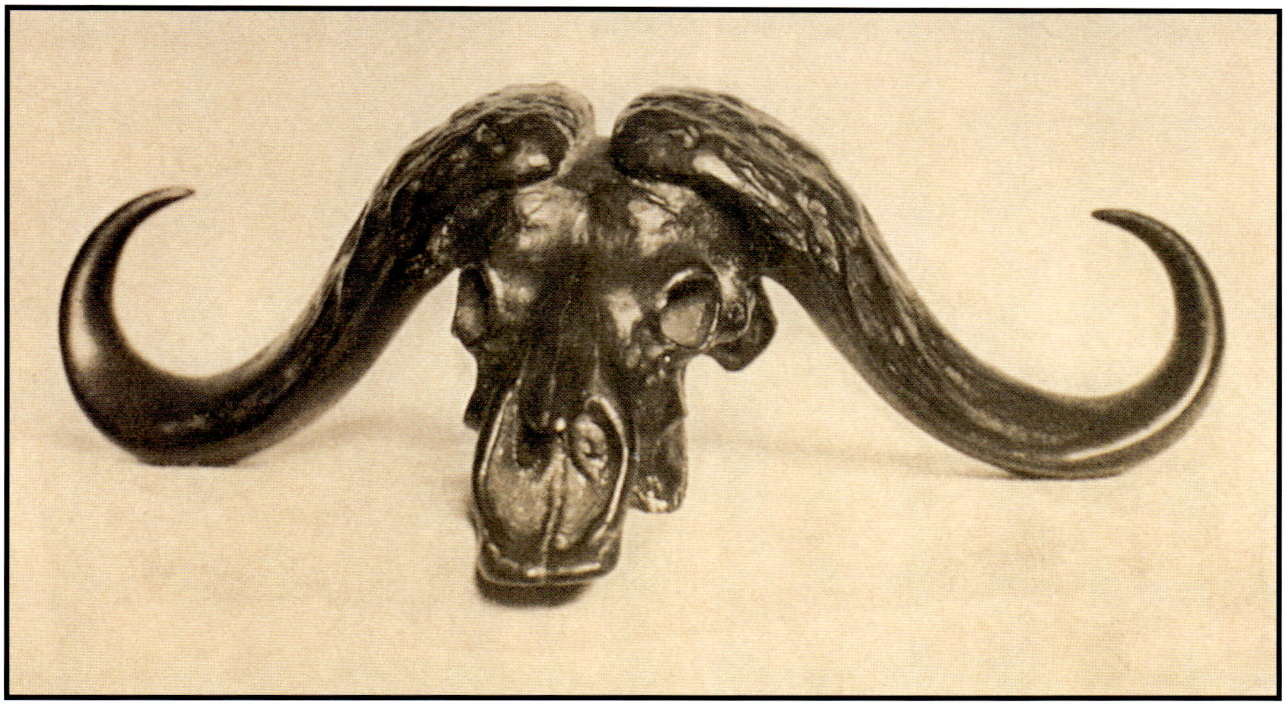

IV-130. "African Buffalo Horns" by James L. Clark. c. 1913.

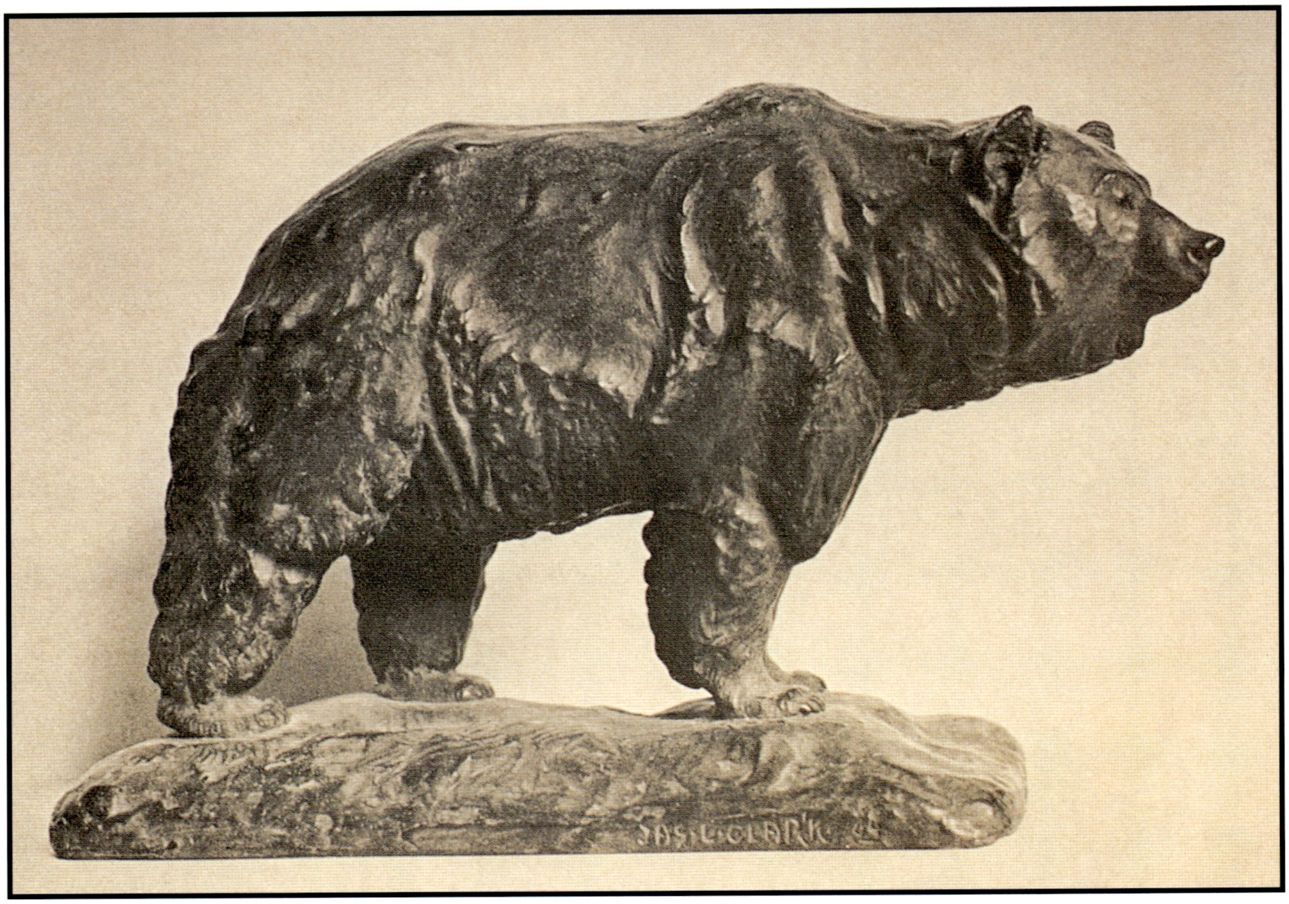

IV-131. "Alaskan Kodiak Bear" by James L. Clark. R.B.W. 1955.

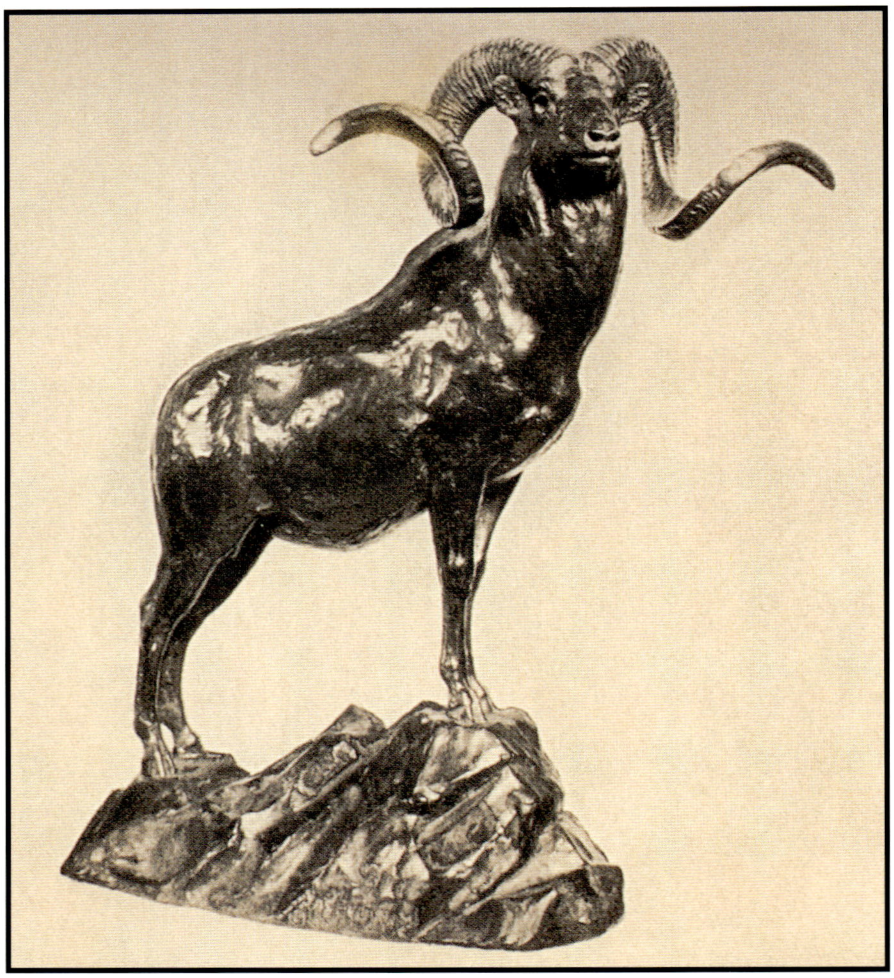

IV-132. "Ovis Poli" (The Marco Polo Sheep of Central Asia) by James L. Clark. R.B.W. 1954.

HENRY MERWIN SHRADY (1871-1922)

Henry Merwin Shrady was a self-taught sculptor who spent almost his entire career working on the Appomatux Memorial Monument to General Ulysses S. Grant in Washington (IV-133); it occupied twenty-two of his twenty-four years as a sculptor (and was unveiled two weeks after his death). This monumental undertaking included several groups of horses and soldiers, among them a cavalry charge (IV-134) and marching soldiers (IV-135), all cast at Roman Bronze Works. He also made several other equestrian monuments, including memorials to George Washington at Valley Forge, seen here in a small bronze replica (IV-136) (like the Grant Memorial, he won the commission in a contest), and General Robert E. Lee.

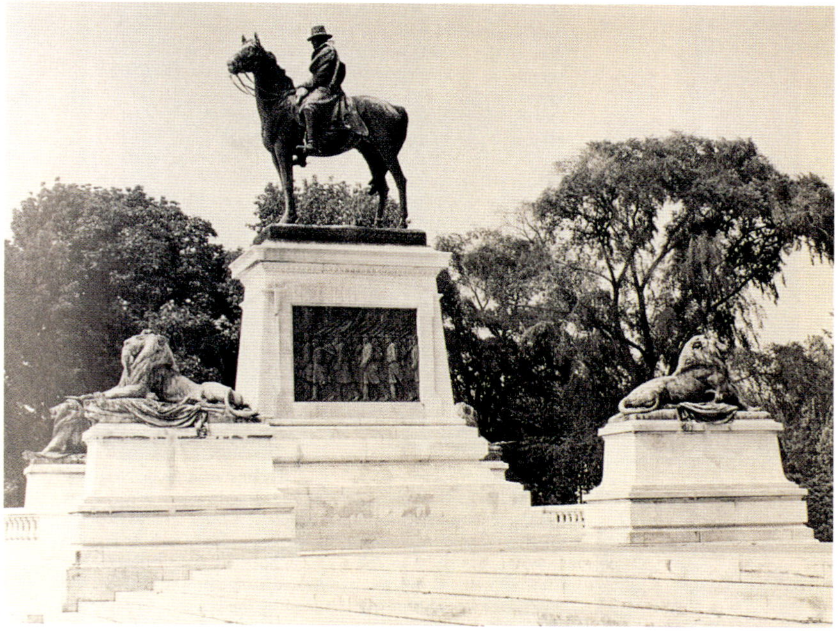

IV-133. "General Ulysses S. Grant Memorial," Washington, D.C., by Henry Merwin Shrady. Parts cast in 1911, 1912, 1916. Completed 1922.

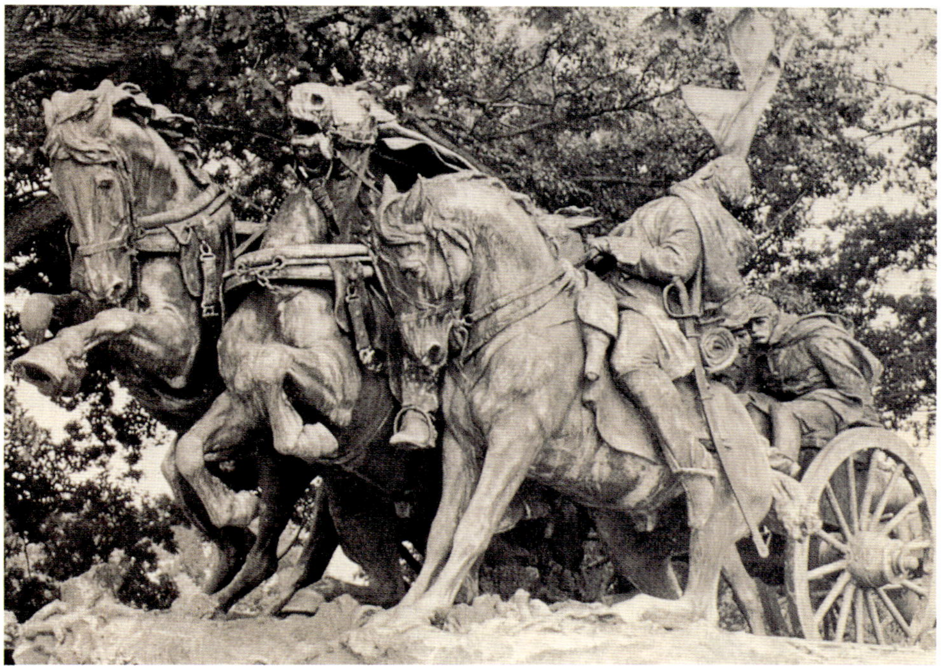

IV-134. Detail "General Ulysses S. Grant Memorial: The Artillery." By Henry Merwin Shrady. 1922.

Born in New York, Shrady worked in business until, while recuperating from an attack of typhoid fever, he began visiting the zoo to sketch. He began his new career making small animal models for the Gorham Silver Company. Before the massive undertaking of the Grant Monument, he devoted himself to very realistic, carefully detailed bronze animal sculpture, including many Western images; these smaller works were prized by collectors.

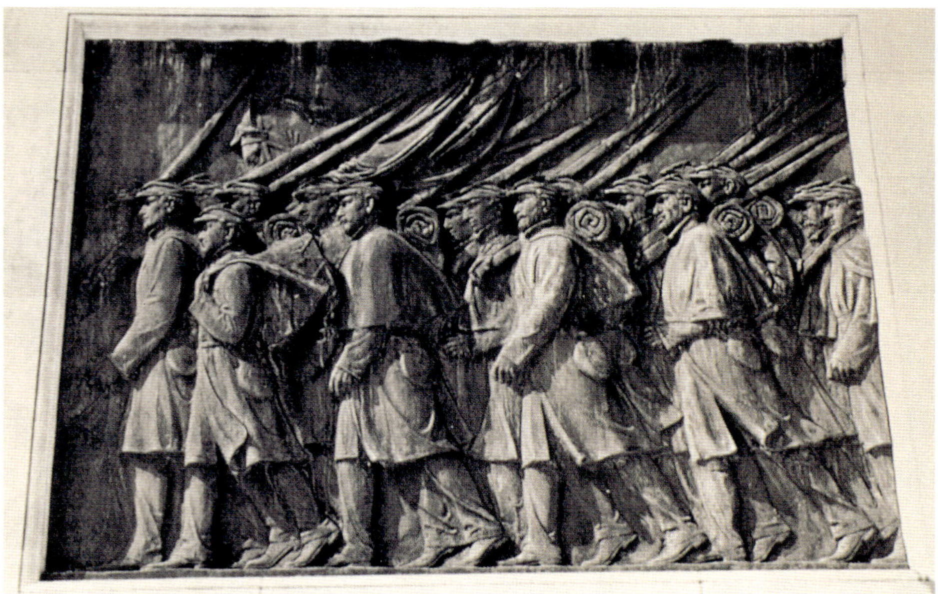

IV-135. Detail "General Ulysses S. Grant memorial: The Infantry" by Henry Merwin Shrady. 1922.

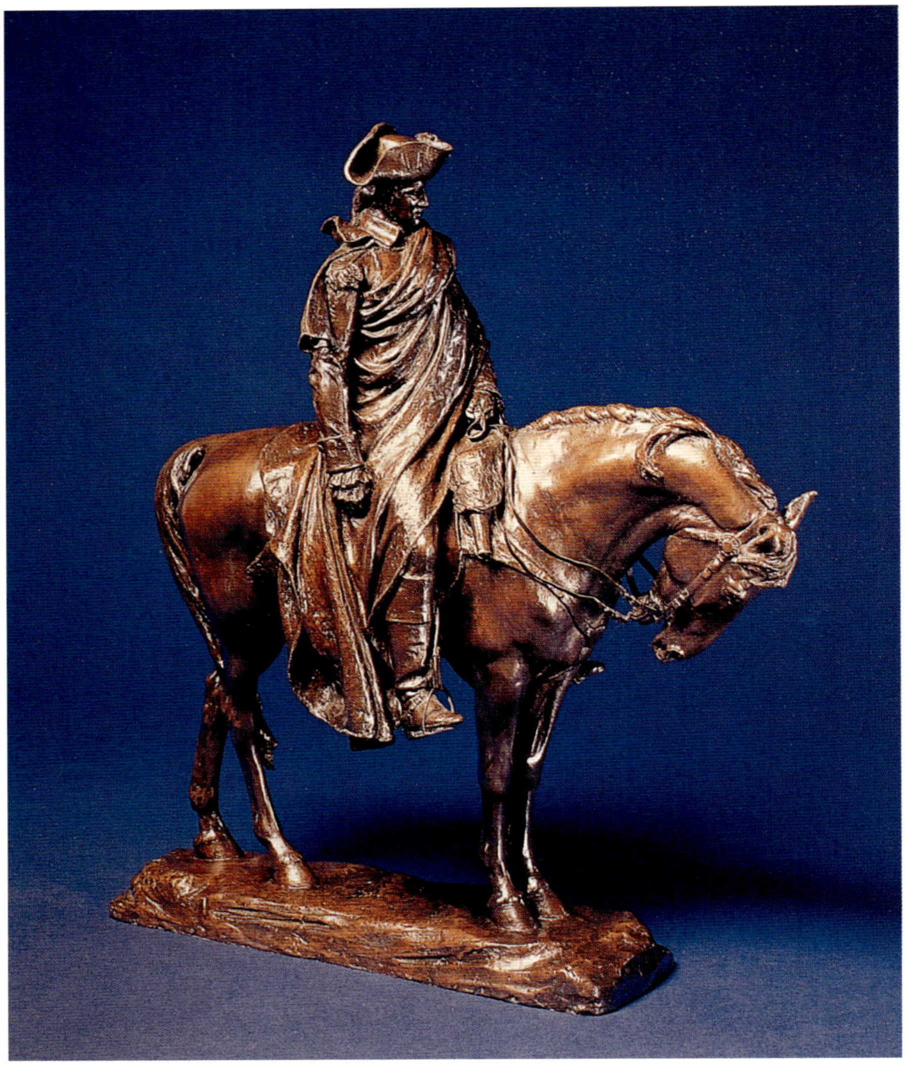

IV-136. "George Washington at Valley Forge" (statuette) by Henry Merwin Shrady. 1904.

He used his own zoo sketches to create a small bull moose (IV-137) and an elk buffalo (IV-138). Sculptor Karl Bitter encouraged the young sculptor to enlarge them, and they were exhibited at the 1901 Pan-American Exposition in Buffalo. Eight of Shrady's animal monuments decorated the bridges spanning the Fair's canals. Today his giant "Elk Buffalo" (IV-139) is a well-known landmark in Washington. Shrady went on to create additional Western images, including "The Empty Saddle" (IV-140) and "Horse Head" (IV-141). Shrady also made several portrait heads, including one of General Grant (IV-142). Roman Bronze Works cast his bronzes between 1905 and the 1950s.

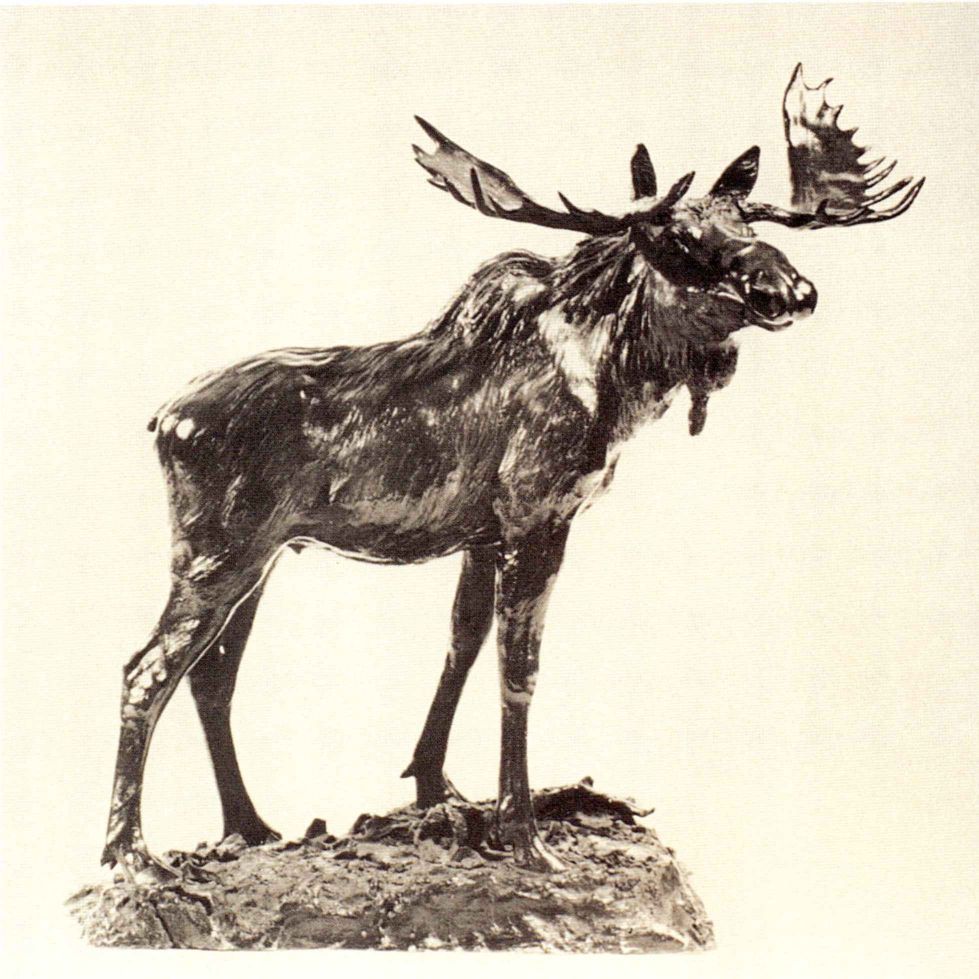

IV-137. "Bull Moose" by Henry Merwin Shrady. 1900.

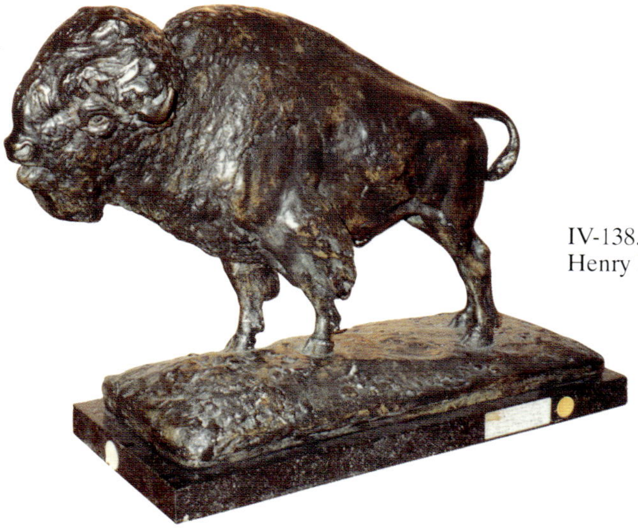

IV-138. "Buffalo" (statuette) by Henry Merwin Shrady. c. 1900.

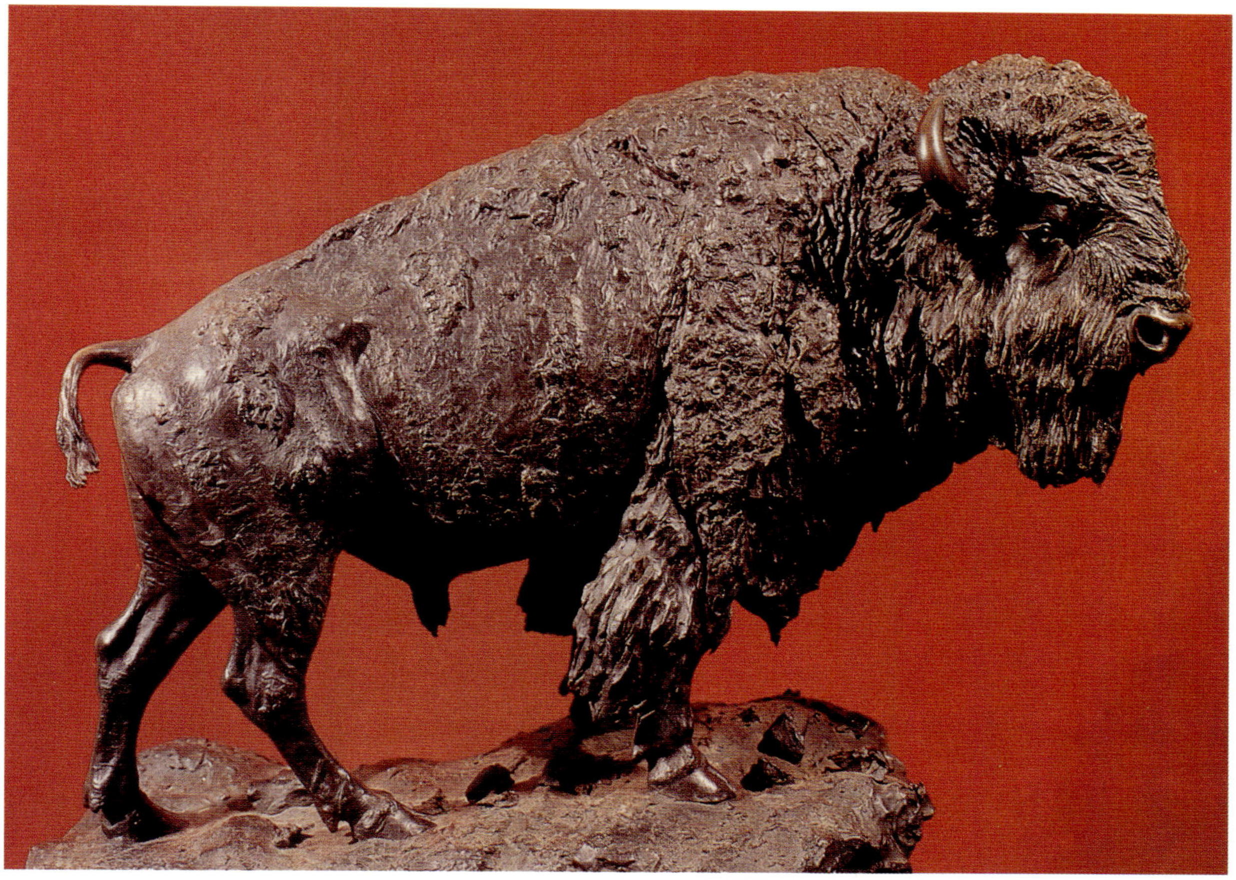

IV-139. "Elk Buffalo: Monarch of the Plains" by Henry Merwin Shrady. 1901.

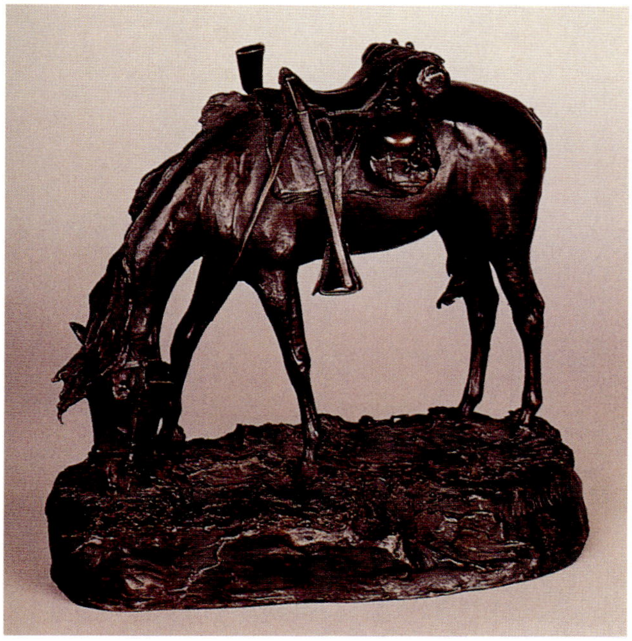

IV-140. "The Empty Saddle" by Henry Merwin Shrady. 1900.

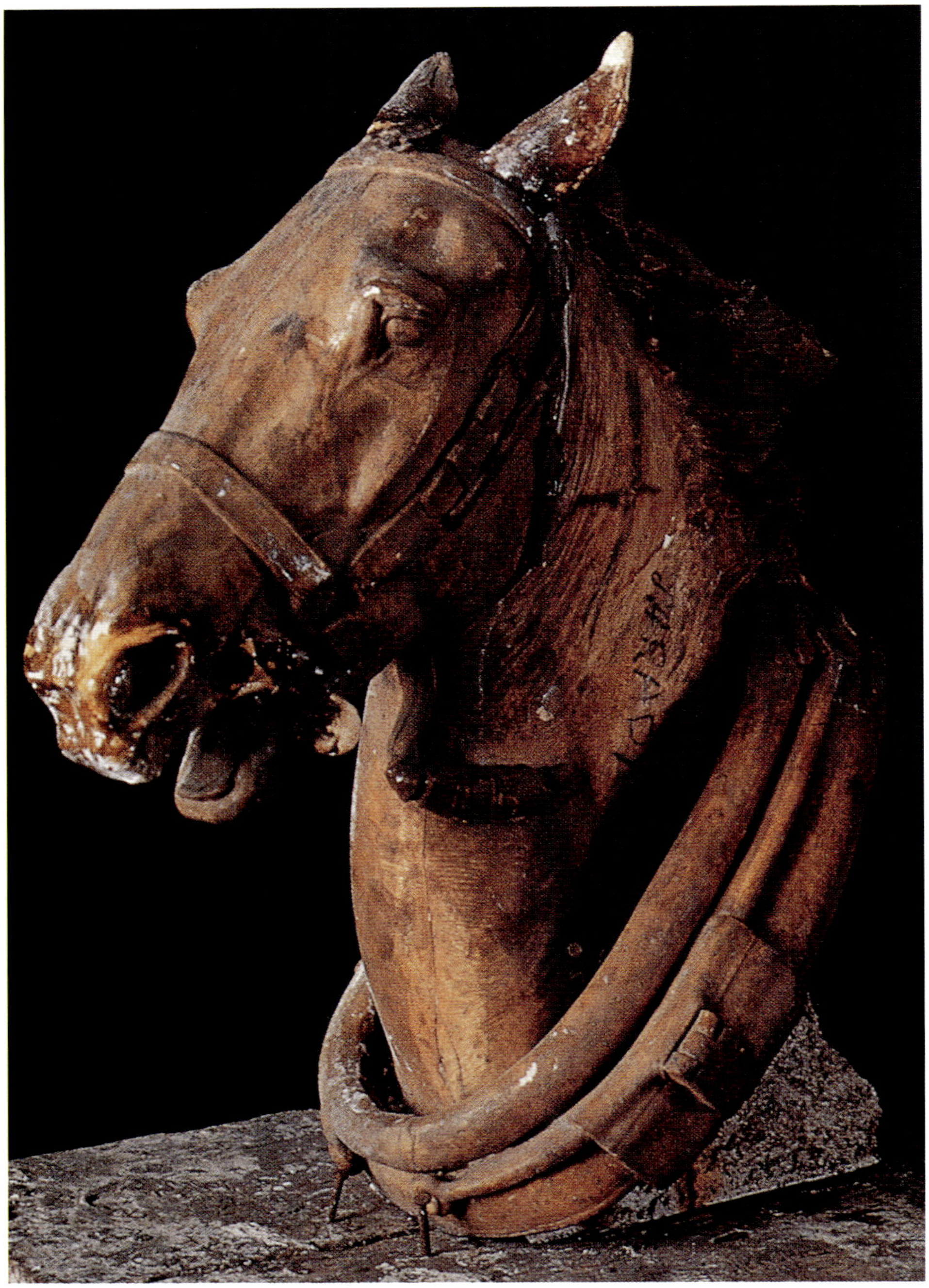

IV-141. "Horse's Head" by Henry Merwin Shrady. (plaster study for the Grant Memorial). 1902.

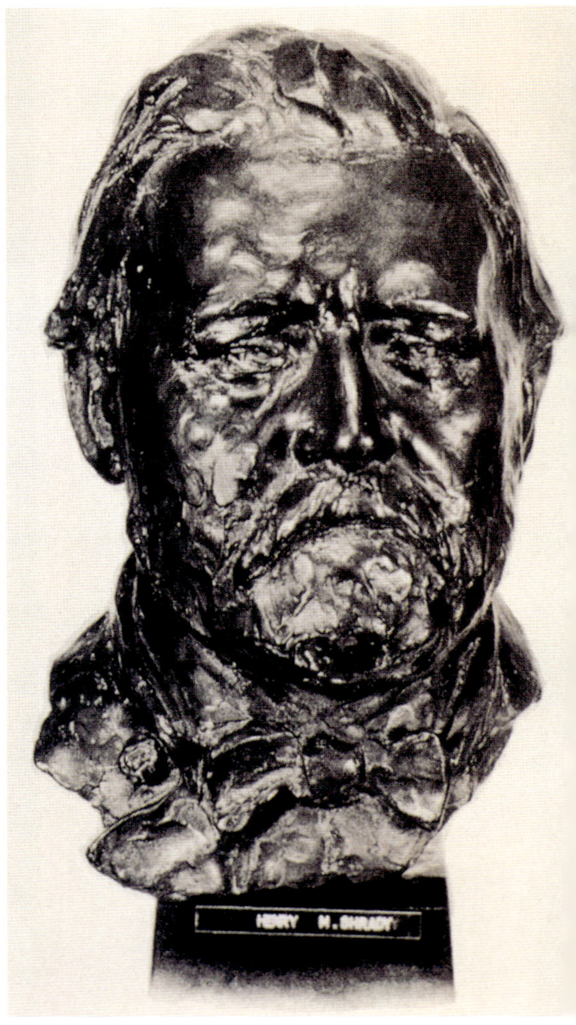

IV-142. "Bust of Ulysses S. Grant" by Henry Merwin Shrady. c. 1915.

ALBERT LAESSLE (1877-1945)

Although Albert Laessle also made monuments, portrait busts, and medals, he is best known for his animal sculptures, specializing in smaller creatures. His execution was so realistic of animals like turtles, crabs, and birds that he was occasionally thought to have cast the real thing, but it was his meticulous study and detailed realization that made his sculptures so lifelike. He also adopted a careful finish, like that of Japanese craftsmen.

Born in Philadelphia, he was a student of Thomas Anshutz and Charles Grafly at the Pennsylvania Academy of Fine Arts. From 1904 to 1907 he studied in Paris. He was a teacher at the Pennsylvania Academy from 1921 to 1939. Among his many awards was a gold medal at the Panama-Pacific Exposition in 1915. His sculptures were placed in many public venues, particularly in Philadelphia. He was a member of the National Academy of Design, the National Sculpture Society, and the National Society of Arts and Letters.

A number of Laessle's animal sculptures were cast at Roman Bronze Works between 1904 and continuing after his death until the 1950s. The foundry's ledgers list a turkey, pigeon, heron, lizard, toad, king crab, bantam rooster, polar bear, Billy goat, and katydids, among some forty castings. Seen here are his eagle "Victory" (IV-143) "Grouse" (IV-144), "Penguins" (IV-145) and his much loved "Dancing Goat" (IV-146).

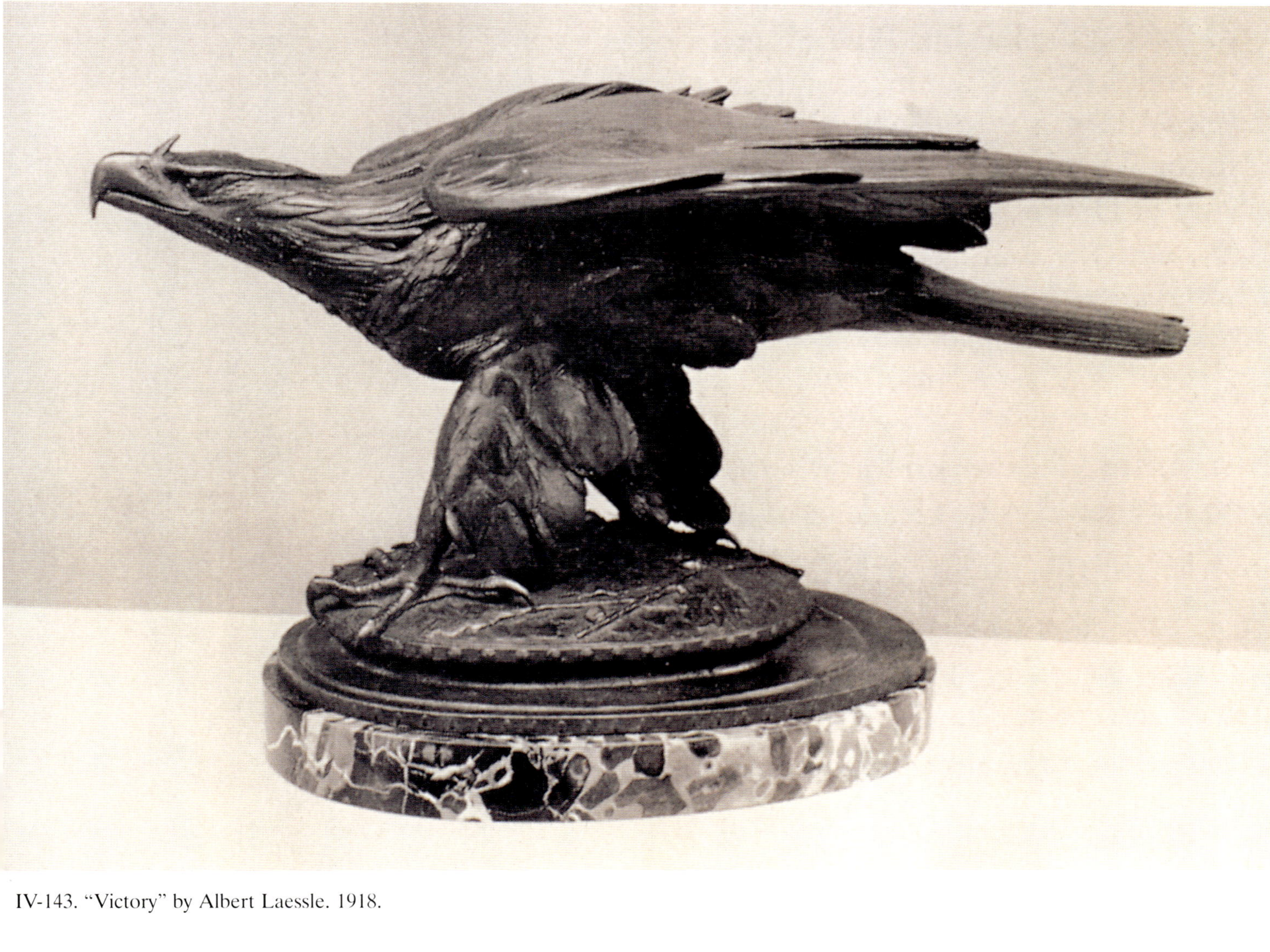

IV-143. "Victory" by Albert Laessle. 1918.

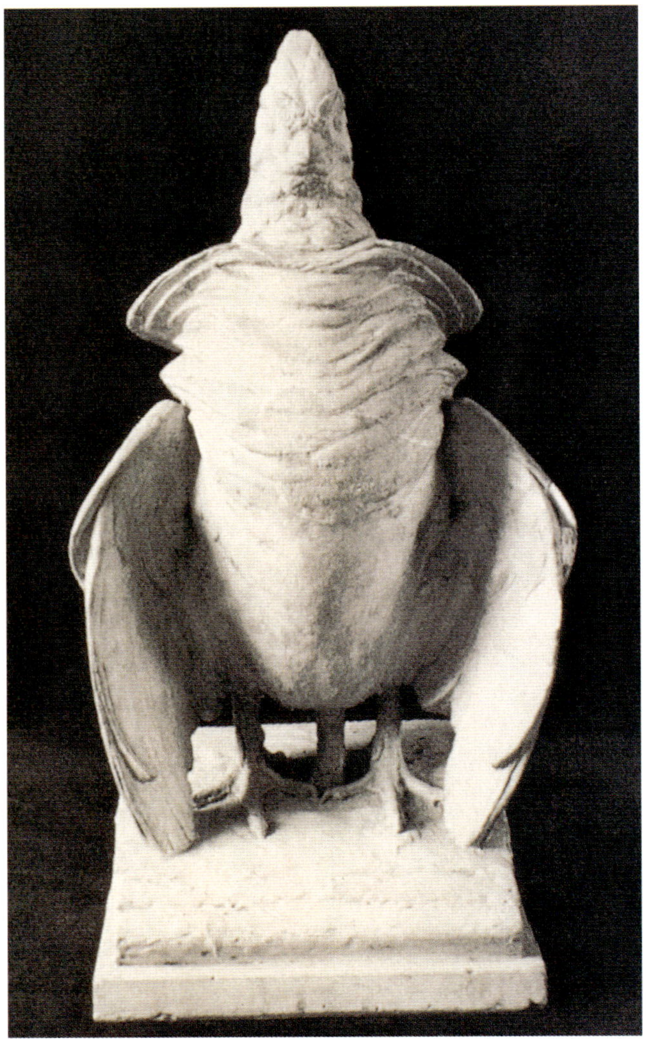

IV-144. "Grouse" (plaster cast) by Albert Laessle. R.B.W. 1952.

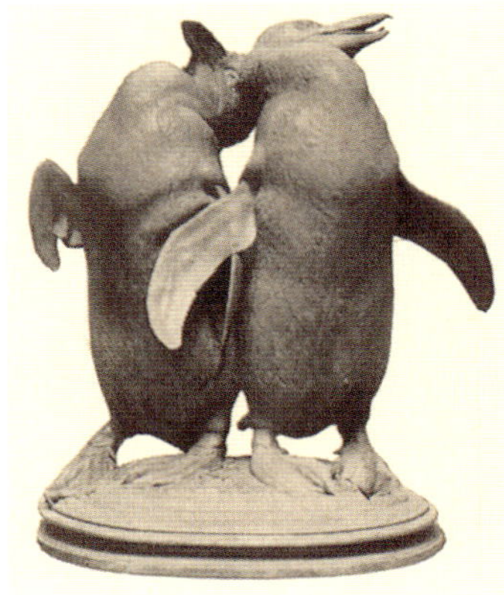

IV-145. "Penguins" by Albert Laessle. 1918.

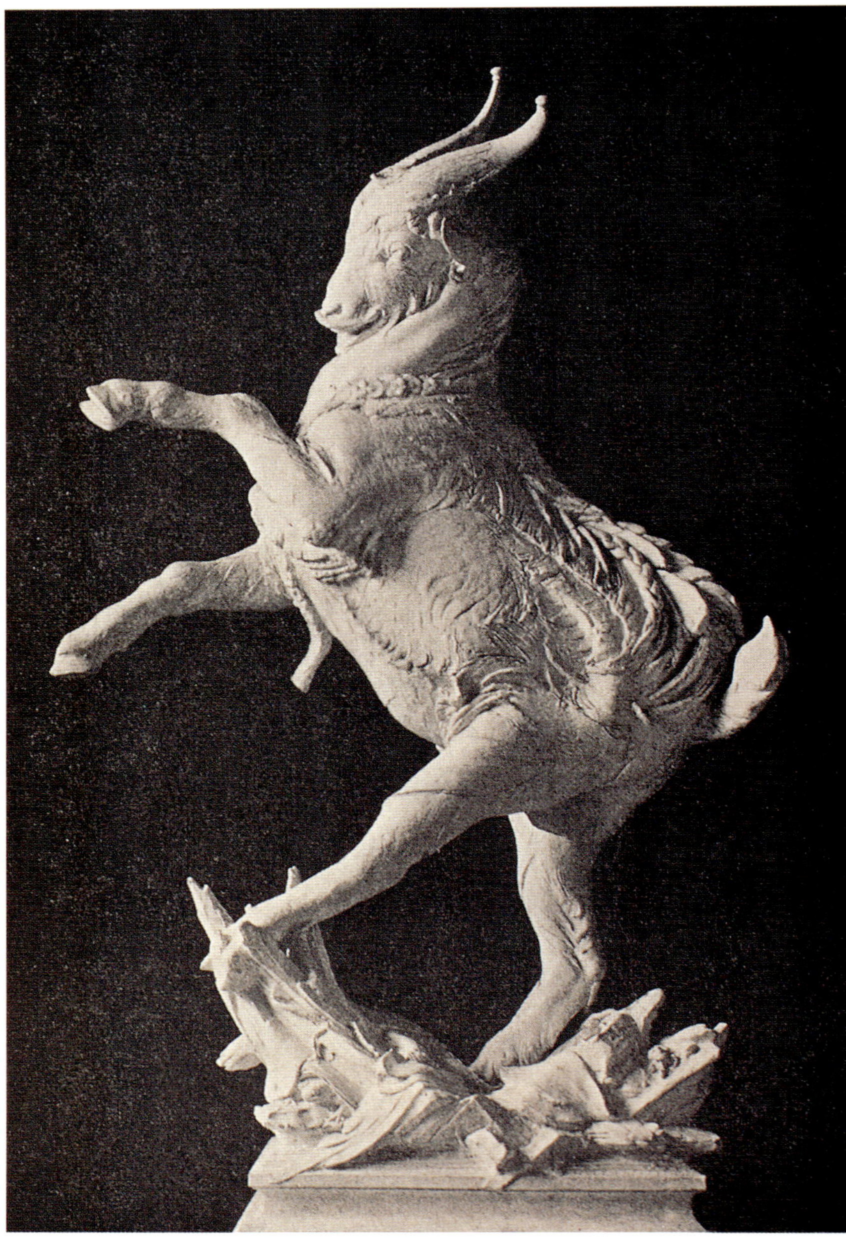

IV-146. "Dancing Goat" by Albert Laessle. c. 1927.

FREDERIC ROTH
(1872-1944)

Frederic Roth was one of the best known animal sculptors of his time. His works can be seen both in museum collections and in many public venues; the "Columbia Lion," a landmark at Columbia University, "Balto""(IV-147) in Central Park, and his "Dancing Goat" (IV-148) and "Dancing Bear" (IV-149) at the Central Park Zoo are great favorites in New York. His work was both accurate and charming, having a gentle sense of humor and a fluid sense of movement.

Roth was born in Brooklyn, New York, and studied sculpture in Berlin and Vienna. By 1903 he had received recognition for his "Roman Chariot Race" for the Pan-American Exposition in Buffalo. He also designed architectural sculpture for subsequent expositions, and for the W.P.A. was chief sculptor for the Department of Parks in New York. He was president of the National Sculpture Society, and a member of the National Institute of Arts and Letters.

His small bronzes of animals, mostly made between 1905 and 1909 show his interest in depicting animals in motion; he often chose to depict the animals at moments of particular charm. In addition to the monuments cited above, he cast many other sculptures at Roman Bronze Works between 1909 and 1940; among them were a "Pig Tied to a Stake and Scratching" (IV-150), two performing elephants (IV-151) and (IV-152), a polar bear (IV-153), and a drinking rhino (IV-154).

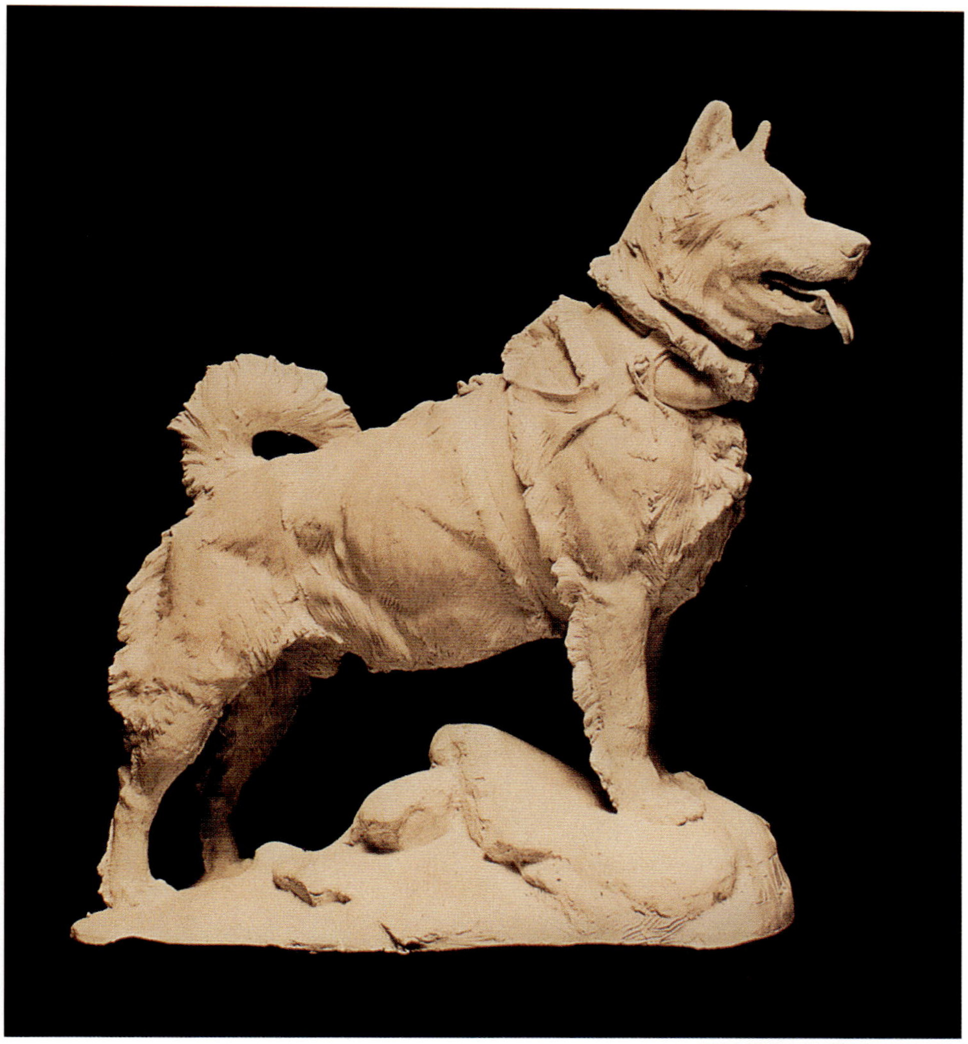

IV-147. "Balto" in Central Park, New York City, by Frederic Roth. 1925.

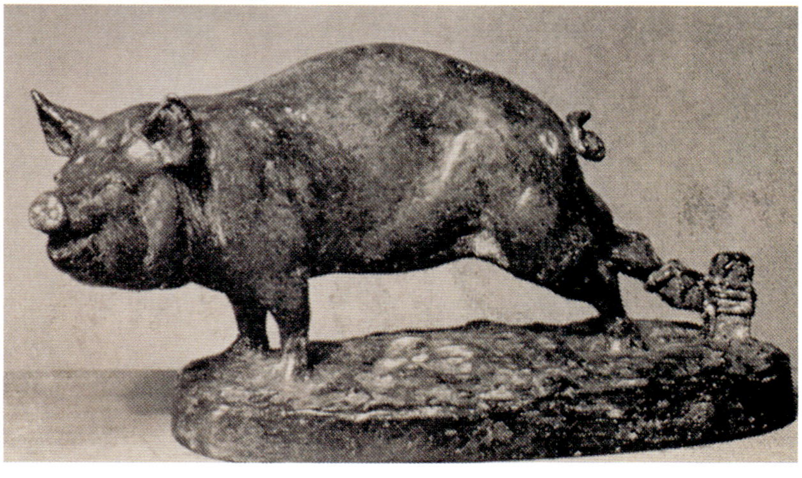

IV-150. "Pig Tied to a Stake and Scratching" by Frederick Roth. 1903.

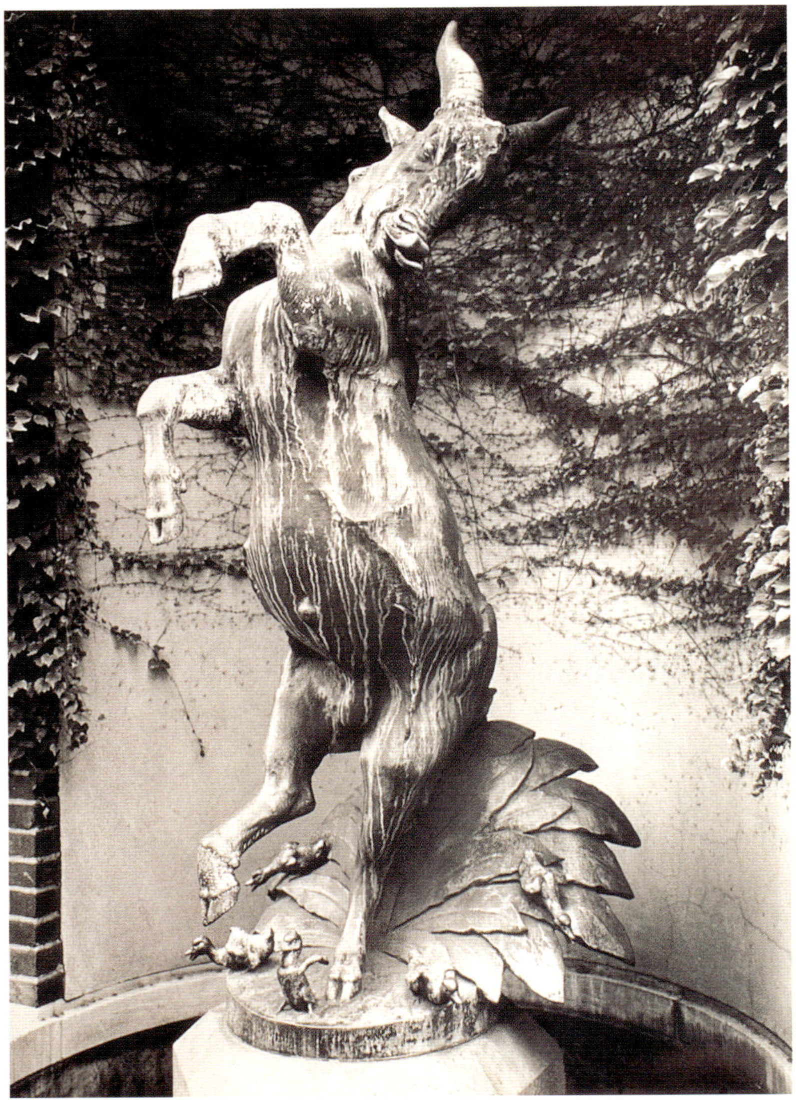

IV-148. "Dancing Goat" in Central Park Zoo, New York City, by Frederick Roth. c. 1935.

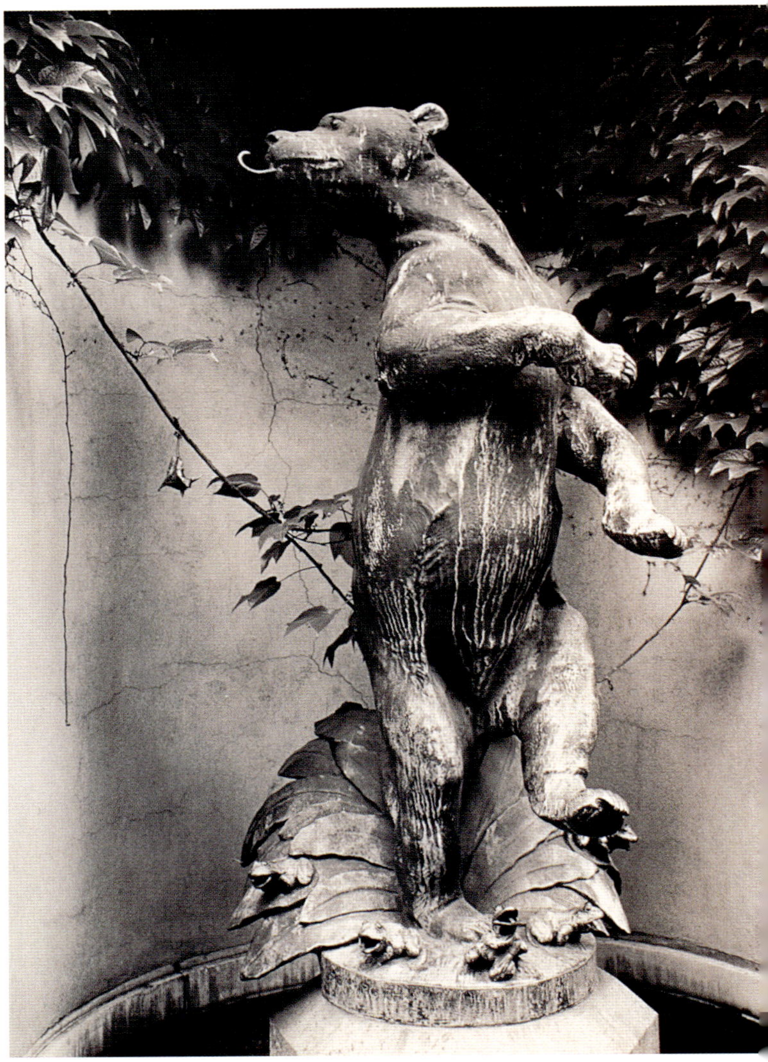

IV-149. "Dancing Bear" in Central Park Zoo, New York City, by Frederick Roth. c. 1935.

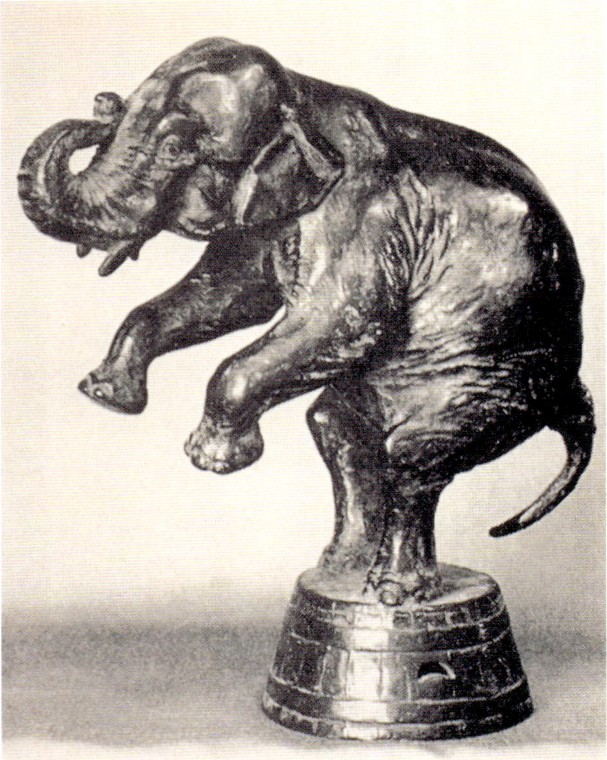

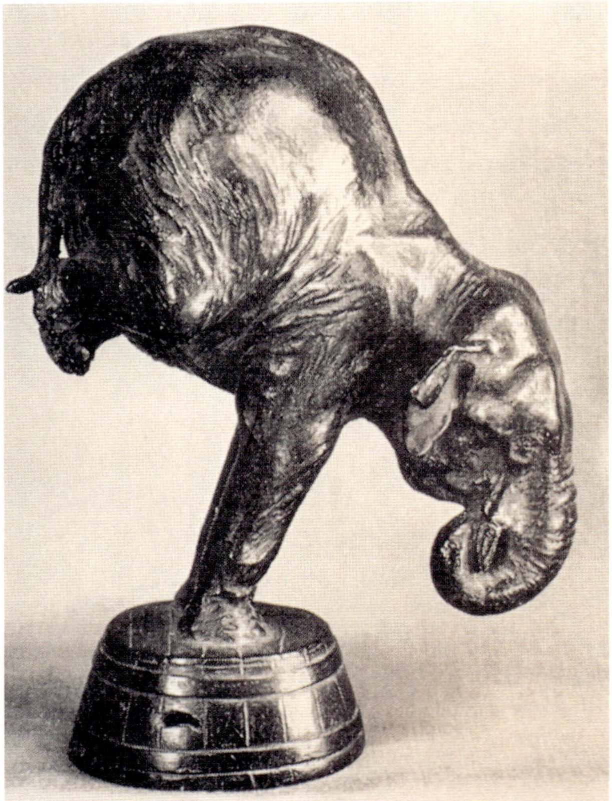

IV-151 (left) & IV-152 (right). "Performing Elephants" by Frederick Roth. 1902.

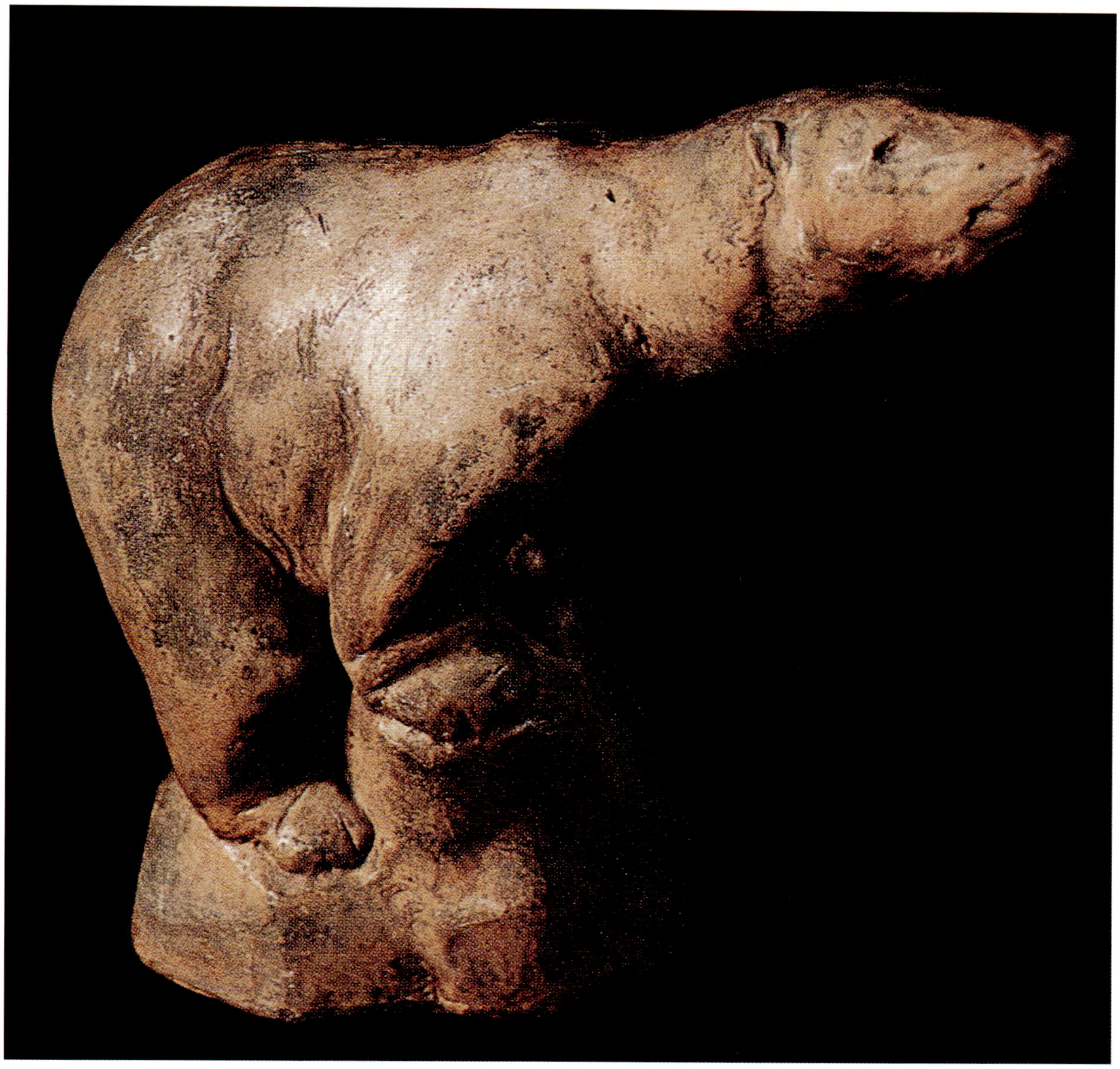

IV-153. "Polar Bear" (plaster maquette used for large polar bear sculpture at Bowdoin College, Maine) by Frederic Roth. c. 1929.

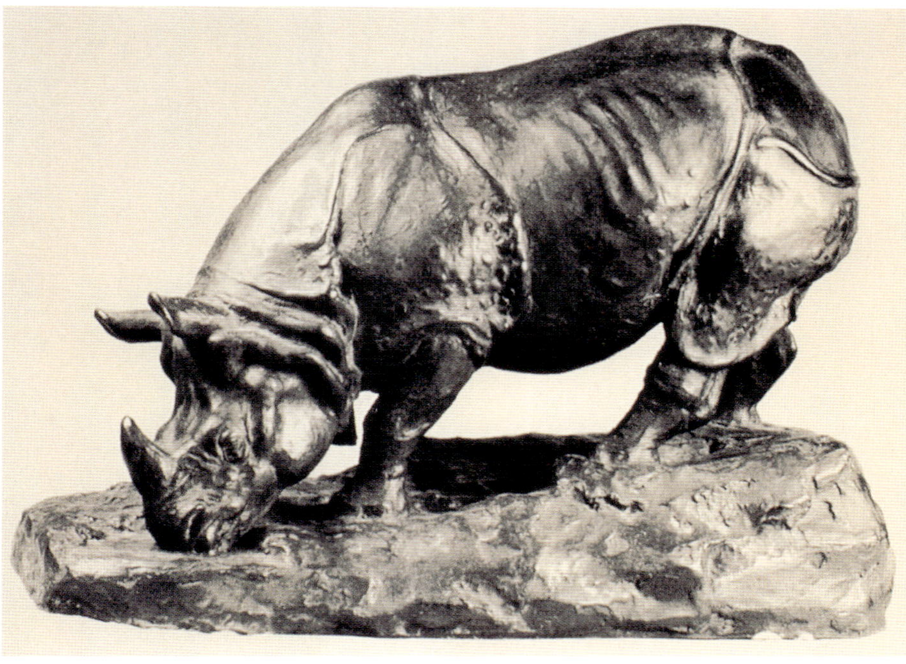

IV-154. "Drinking Rhino" by Frederic Roth. 1904.

OTHER SCULPTORS OF WESTERN AND WILDLIFE IMAGES

CARL E. AKEKEY, ANTOINE LOUIS BARYE, ISIDORE JULES BONHEUR, ELI HARVEY, LOUIS PAUL JONAS, LOUIS McCLELLAN POTTER, CHARLES SCHREYVOGEL, CHARLES UMLAUF

These notable sculptors of wildlife and Western images cast only a few pieces at Roman Bronze Works.

CARL E. AKELEY
(1864-1926)

Carl E. Akeley was a noted animal sculptor; his work is in the American Museum of Natural History. Among his works cast at R.B.W. was his study of the elephant "Patriarch" (IV-155).

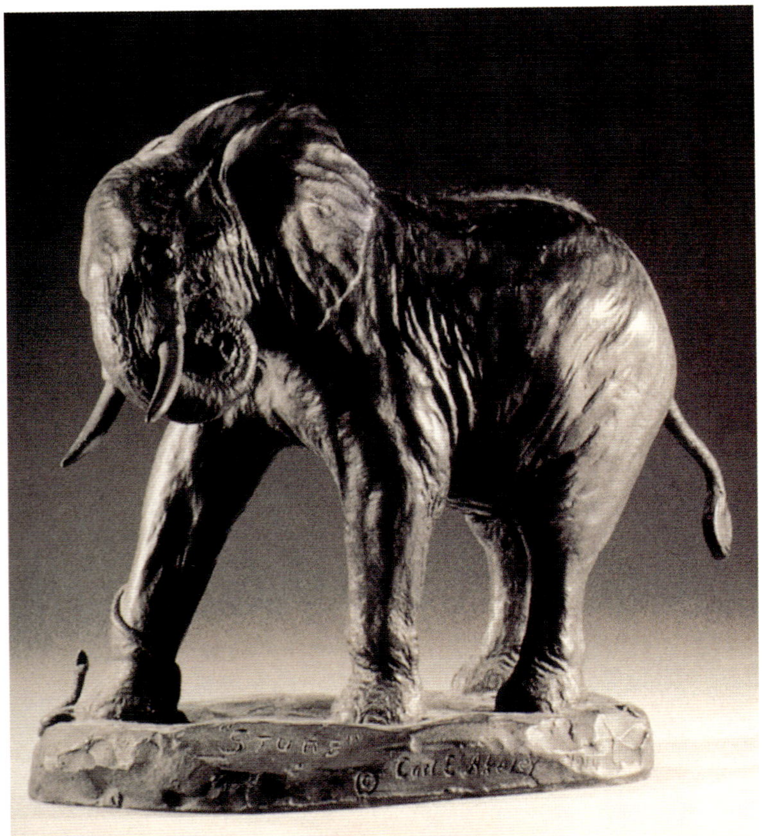

IV-155. "Stung" by Carl E. Akeley. 1914.

ANTOINE LOUIS BARYE
(1796-1875)
and
ISIDORE JULES BONHEUR
(1827-1901)

Antoine Lousi Barye and Isidore Jules Bonheur, both were fashionable nineteenth century French Salon sculptors of animals. Posthumous castings of their animal images were made by the Roman Bronze foundry. Plaster sketches of Barye's "Lion" (IV-156) and his "Trojan Horse with Arched Head" (IV-157) were acquired and recast by R.B.W., as was Bonheur's "Lion" (IV-158).

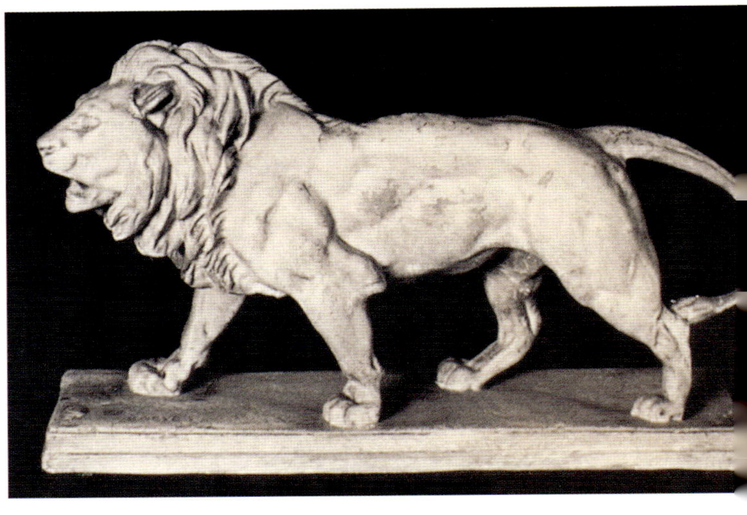

IV-156. "Lion" (plaster cast) by Antoine Louis Barye.

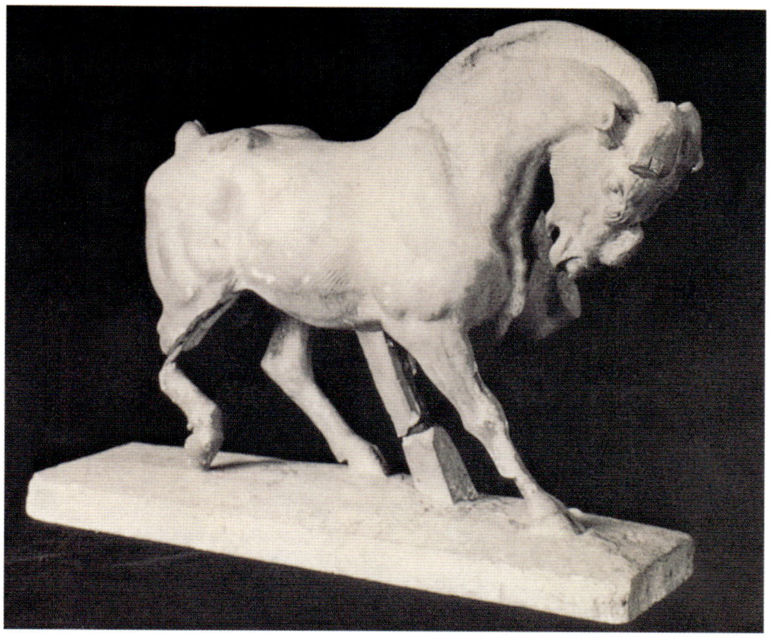

IV-157. "Trojan Horse with Arched Head" (plaster cast) by Antoine Louis Barye.

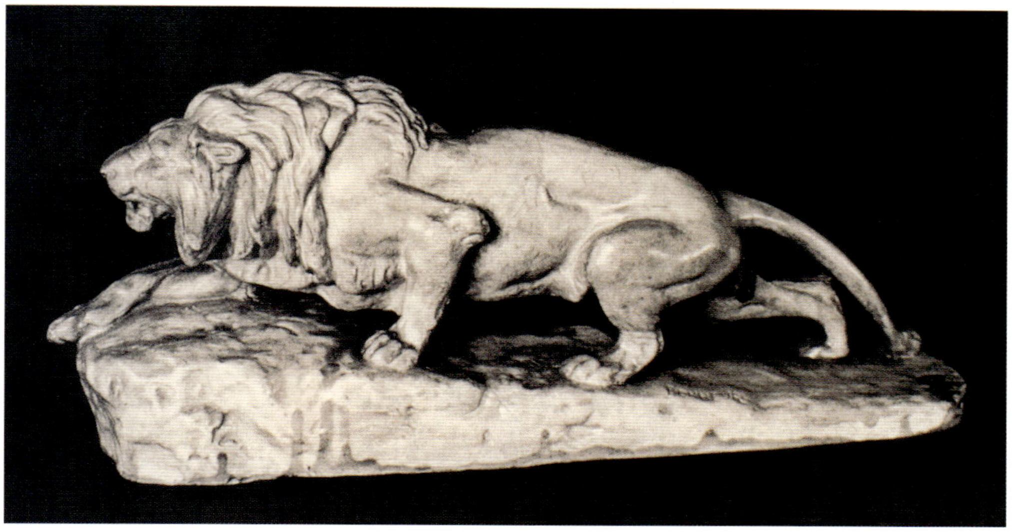

IV-158. "Lion" (plaster cast) by Isidore Jules Bonheur.

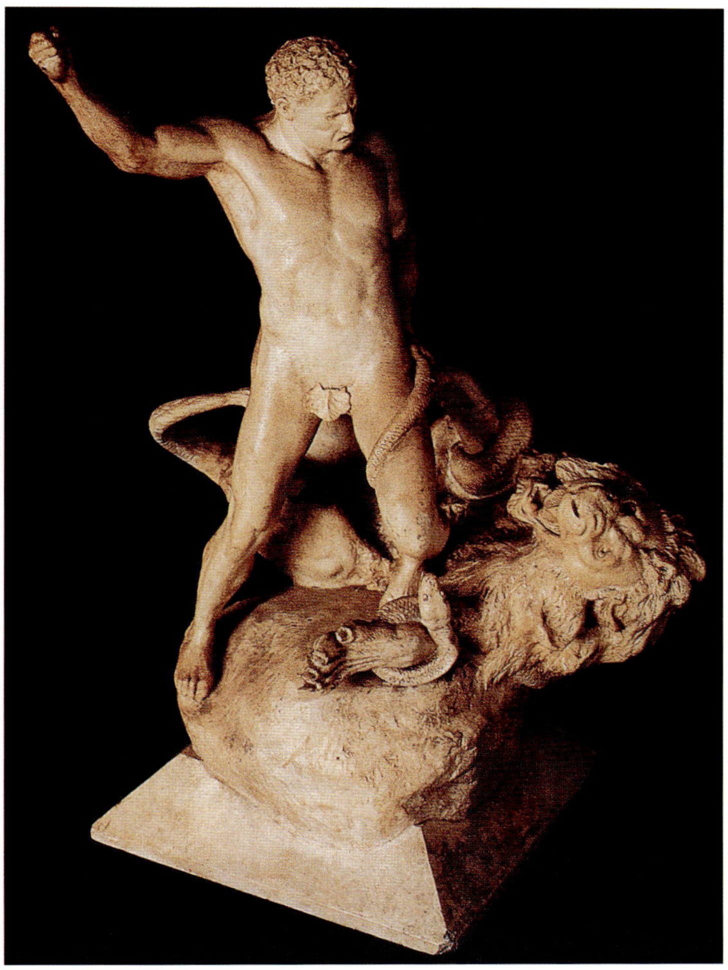

ELI HARVEY
(1860-1957)

Eli Harvey was primarily a sculptor of wildlife, including many lions; four of his lions guard the entrance to the New York Zoological Park. He spent much time sketching at the zoo and his bull elk became the symbol for the Benevolent and Protective Order of Elks of America. Among the works he cast at R.B.W. are his "Discord" (IV-159) and "Dinah" (IV-160).

IV-159. "Discord" (plaster cast) by Eli Harvey. 1911.

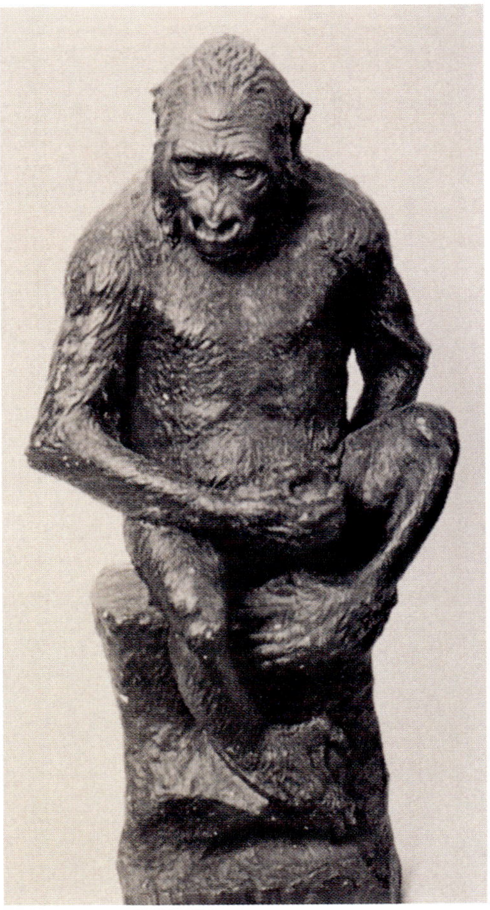

IV-160. "Dinah" (plaster cast) by Eli Harvey. 1916.

LOUIS PAUL JONAS (1894-1971)

Louis Paul Jonas, a native of Hungary, specialized in animal sculpture. He worked as a taxidermist for Carl Akeley at the American Museum of Natural History. Among his castings at R.B.W. beginning in 1915 were a zebra, a lion, a cow, a poodle, and an antelope (IV-161).

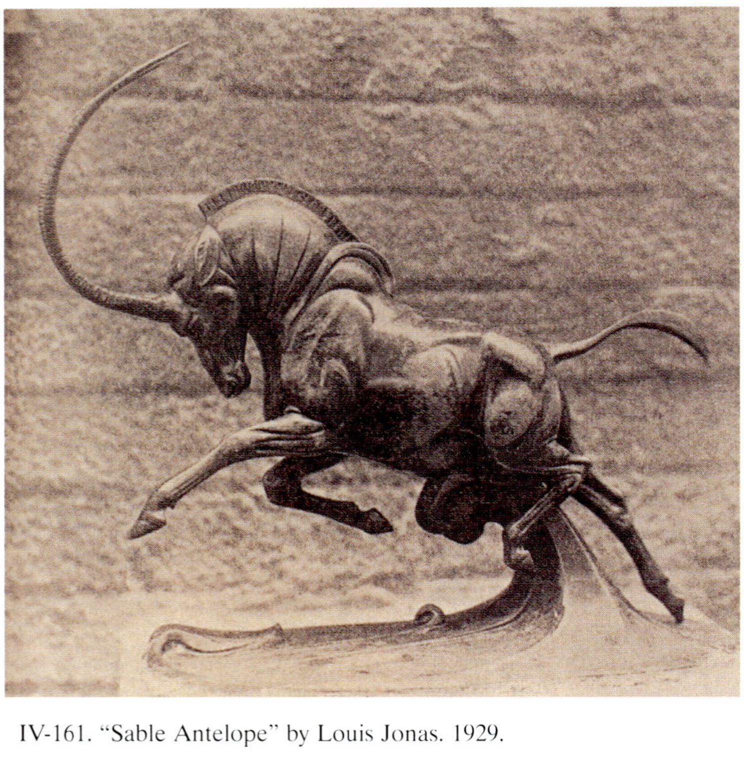

IV-161. "Sable Antelope" by Louis Jonas. 1929.

LOUIS McCLELLAN POTTER (1873-1919)

Louis McClellen Potter specialized in Indian groups, particularly Northwestern tribes. Potter spent several years in Alaska and produced a number of bronzes based on the mystical legends of Native Americans of the region, such as his "Indian Brave and Spirit" (IV-162) and "The Call of the Spirit" (IV-163). Some of these sculptures appeared as andirons, like his "Kneeling Indians" (IV-164) and "Praying Monks" (IV-165), all of which were cast by Roman Bronze Works.

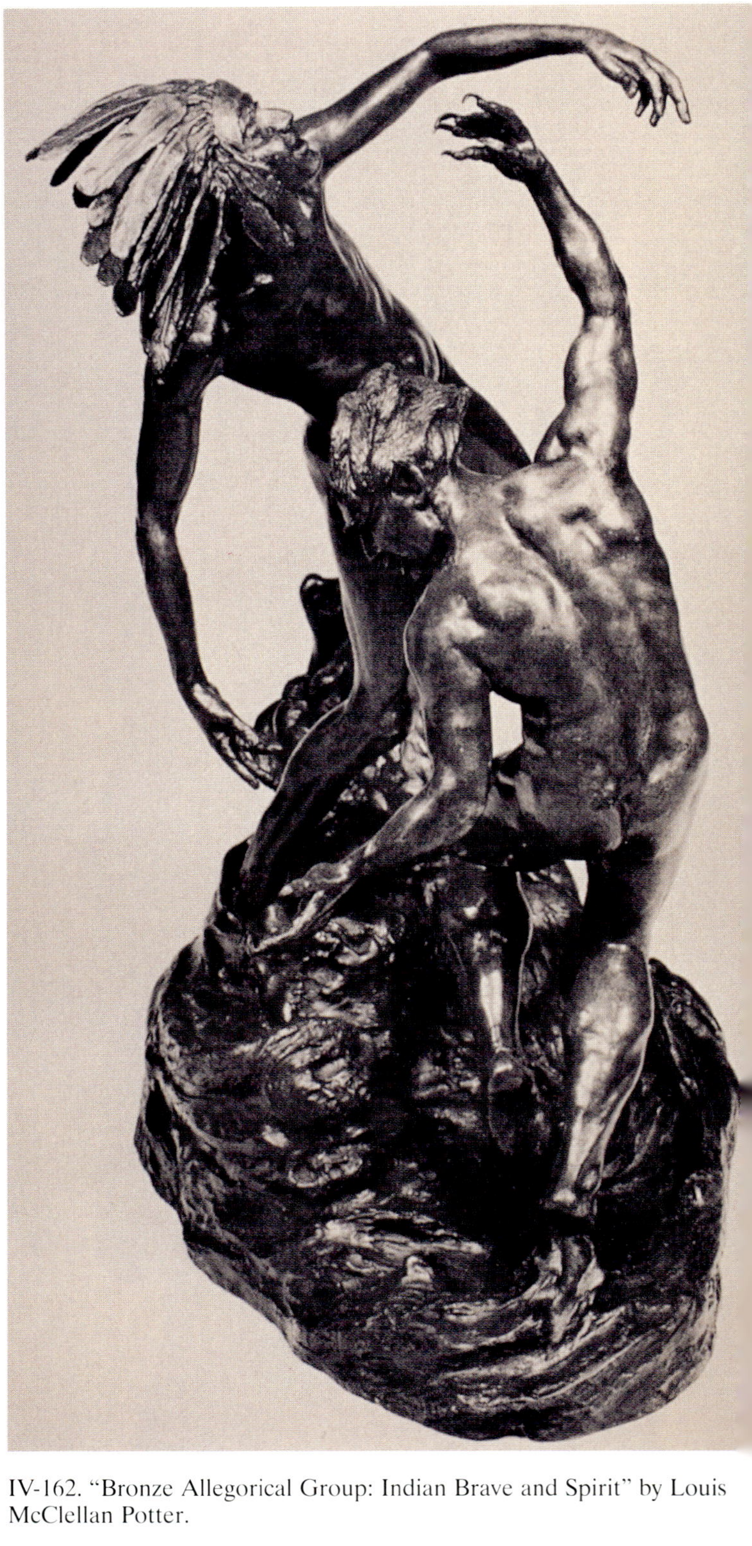

IV-162. "Bronze Allegorical Group: Indian Brave and Spirit" by Louis McClellan Potter.

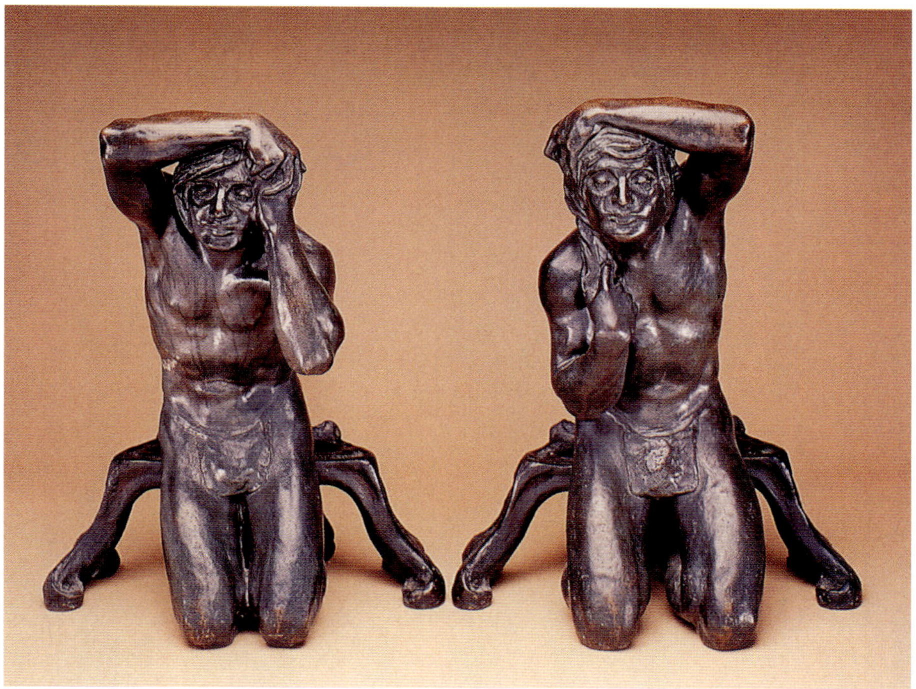

IV-164. "Pair of Bronze Kneeling Indian (Andirons)" by Louis McClellan Potter.

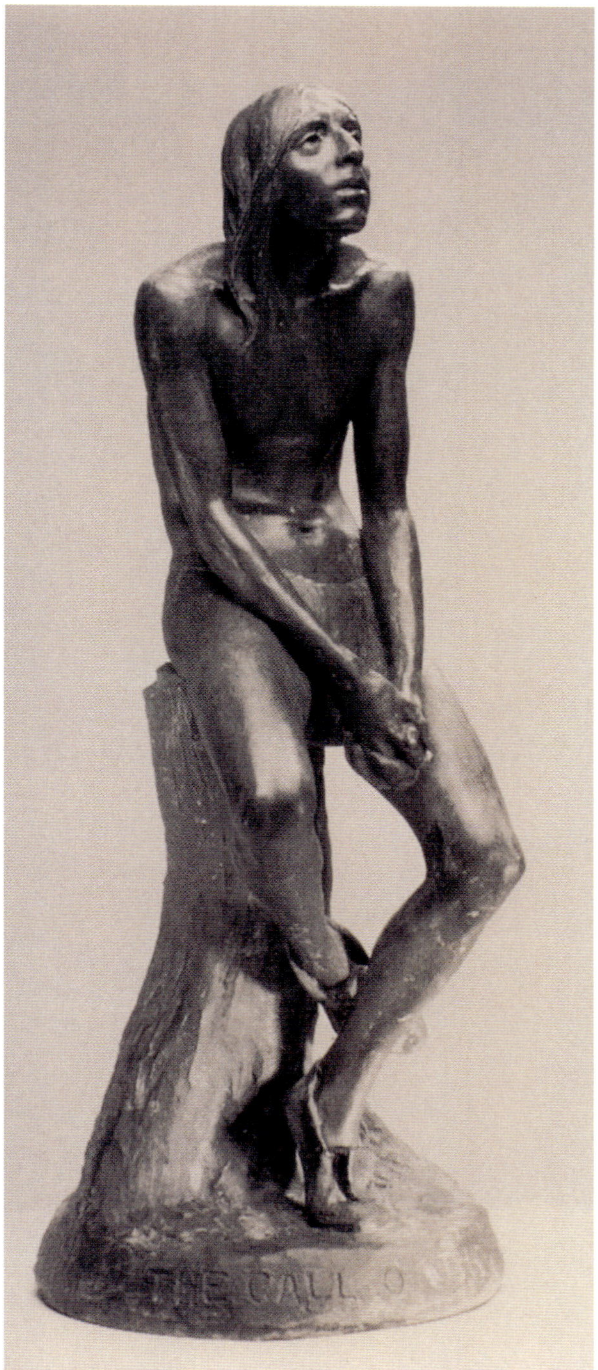

IV-163. "The Call of the Spirit" by Louis McCllelan Potter. 1907.

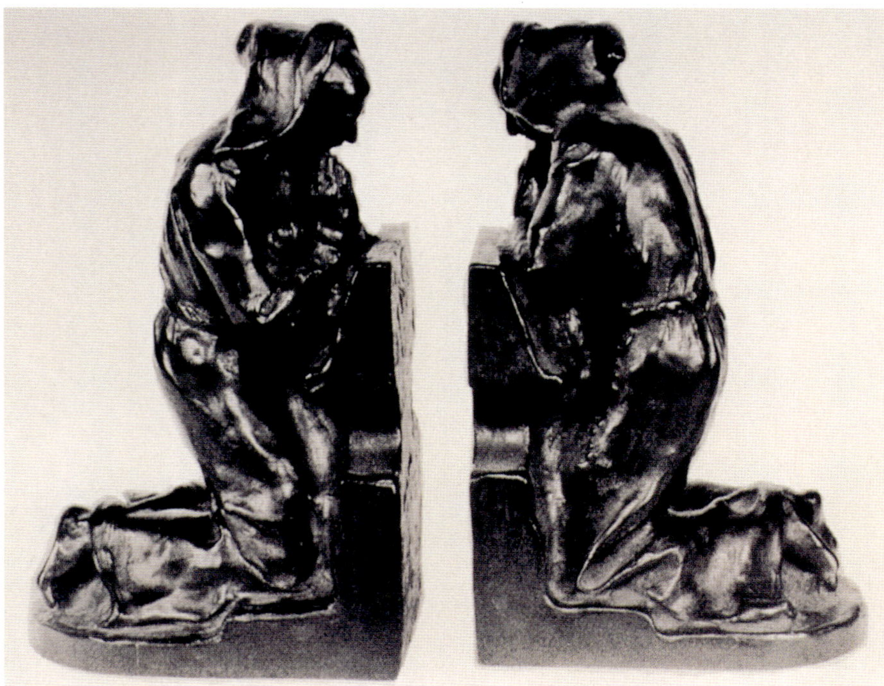

IV-165. "Praying Monks (bookends) " by Louis McClellan Potter. 1911.

CHARLES SCHREYVOGEL (1881-1912)

Charles Schreyvogel is particularly known for a Western image called "The Last Drop" (IV-166); it was modeled by the Munich-trained sculptor in 1900 and cast at R.B.W. in 1903. He made several other Western statues, including "White Eagle" (IV-167).

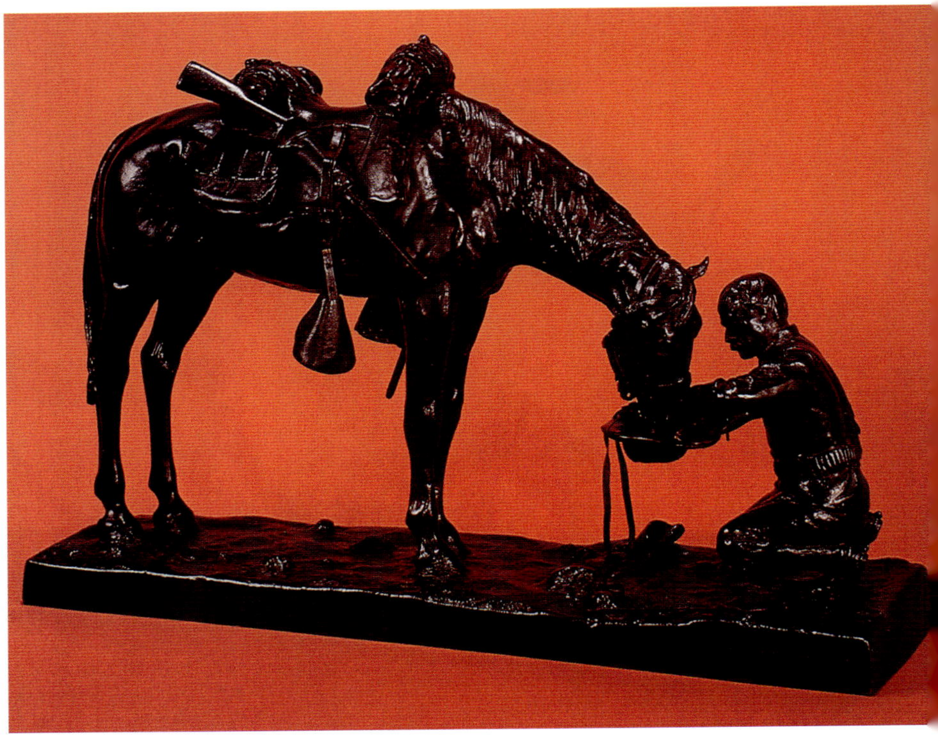

IV-166. "The Last Drop" by Charles Schreyvogel. 1903.

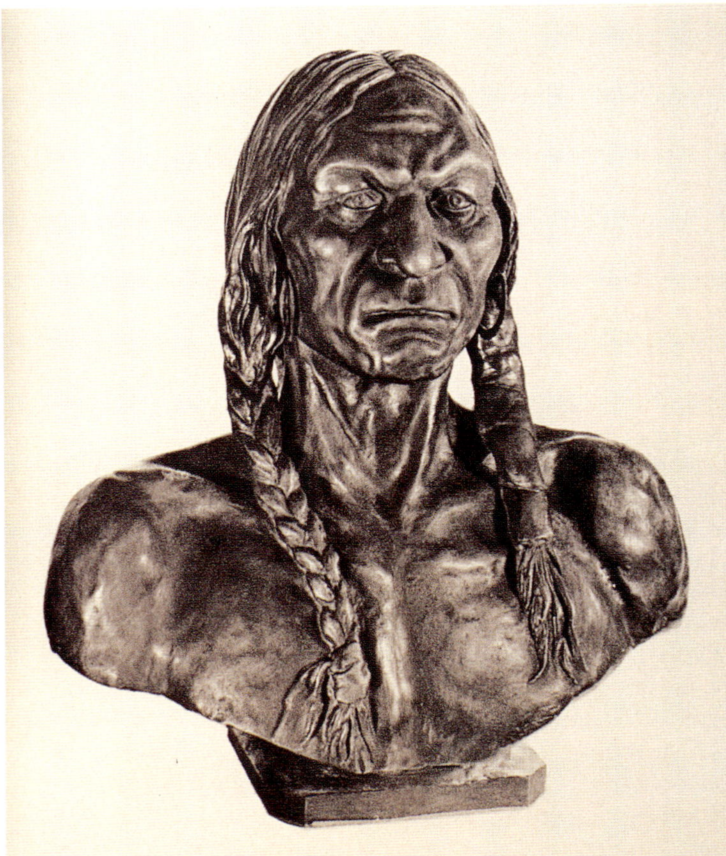

IV-167. "White Eagle" by Charles Schreyvogel. c. 1899.

CHARLES UMLAUF (1911-1982)

Charles Umlauf is represented in many Texas museums, and had thirty one-man shows. He was known for his sculpture of Western subjects, like his "Longhorn" (IV-168), which was cast at Roman Bronze Works in 1951.

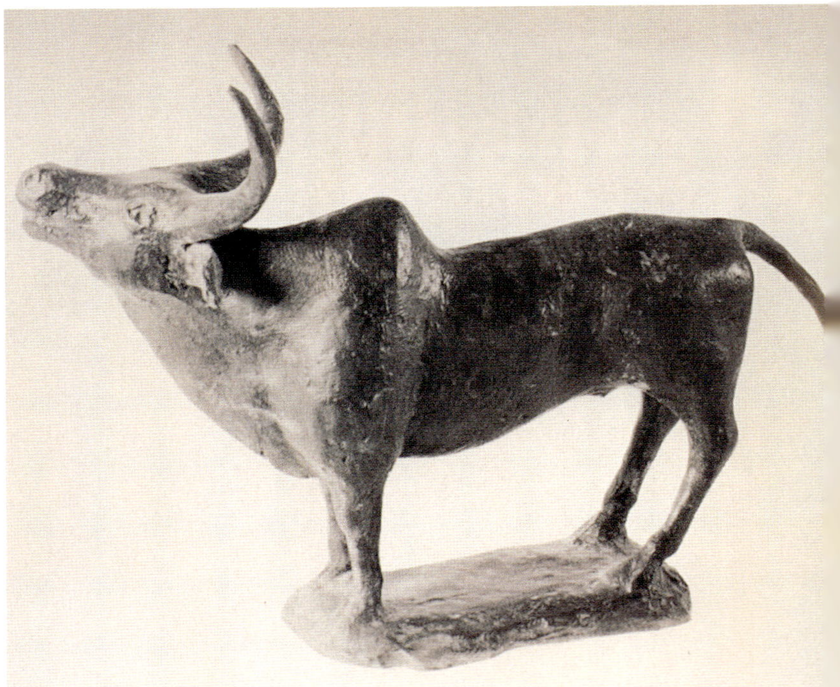

IV-168. "Longhorn" by Charles Umlauf. 1951.

V. MODERNISM IN AMERICAN SCULPTURE: NEW MODES OF EXPRESSION FROM 1913

Just as the people of America have widely differing views and values, American art of the 20th century has reflected a diverse population with diverse ideas. From strongly patriotic or socially critical realism to self-expressive abstractionism, to satirical pop art, the various styles of American art have reflected sharply differing views about what art is or should be. These debates continue in contemporary art. But at no time was the division between realism and abstraction more fraught with political and social debate and animosity than in the years following the Armory Show in New York in 1913.

It was this exhibition, featuring abstract art among its many other works, that introduced America to the kind of painting and sculpture that was already widely accepted in Europe. Before 1913 American artists interested in working in non-realistic styles had exiled themselves abroad, where they could freely explore new ideas. But with the Armory Show a new kind of art had come to stay. For the handful of "modern" artists prepared to battle for "pure" self-expression, for the right to paint or sculpt what they felt, rather than what they saw before them, a long battle with the American establishment was about to begin.

The art world in New York quickly divided itself into two basically different schools of thought—realism and abstraction—that would last throughout the century. Among the early pioneers of modern art in the United States were several who were inspired by Cubist ideas and with the basic belief that form carries its own content. Having studied abroad, they brought stylistic concepts and attitudes about the very act of making art to the U.S. that were sharply different from the traditional American views. And several leading European abstract sculptors also came to live and work in the United States; among them were sculptors Alexander Archipenko, Elie Nadelman, and Gaston Lachaise, all of whom arrived early in the century.

These modernists chose to work in new, untraditional ways, and to seek the patterns and ideas of abstraction in the world around them. They felt that the rules and traditions of realism did not reflect the modern world of relativity and change. American realists (and the art establishment), on the other hand, not only rejected abstract art for its emphasis on non-objective form and on the artist's interest in self-expression, but for its lack of moral and civic subject matter. In addition, they rejected modernism as foreign and un-American, an attack on patriotism and tradition.

Modern sculptors were in a particularly difficult position. While early modernist artists received some support from such figures as Edward Steichen, who, with his wife Georgia O'Keeffe, formed the earliest abstractionist group in New York, the art world establishment was all but united in its opposition. Although painters had a few avant garde galleries to show their modernist works and a knowledgeable few collectors to buy them, sculptors found themselves particularly ostracized. Large sculpture had never been commonly purchased for the décor of private homes, either indoors or out. Sculpture had never been independent of its audience, like painting had. There were few patrons of sculpture, other than governments and institutions. Sculpture was seen primarily as a civic art, designed to grace public spaces and to commemorate important people and events. Abstract sculpture was not considered appropriate for such commemorative commissions.

With the advent of modernist sculpture with its emphasis on pure form and self-expression, its abstract nudes and other non-civic subject matter, many commissions for public sculpture simply disappeared, often to be replaced by fountains or sundials (though some traditional statues continued to be commissioned and created). In addition, journalists and art educators heaped vitriolic criticism upon modernist sculpture. Even the National Sculpture Society voted to remain a society of realist artists, as it is to this day. Nor would art museums across the country exhibit modernist works. Without exception they were wholly opposed.

The results of this opposition to modern art in the first three decades of the 20th century were obvious. Purely abstract sculpture all but disappeared in America. With the exception of a few artists who remained committed to their ideas (and were able to support themselves without public commissions or to work abroad), the world of modern sculpture in America moved away from pure formalism and abstraction.

Several of the previously leading exponents of modernism, such as Storrs, Lachaise, and Zorach, created civic works that actually embraced social and political ideas and looked not unlike those of their realist colleagues, though with a touch of modern swagger. Just as social realism had taken hold in the world of painting, many sculptors too chose to adapt their ideas to social contexts. Others, like Paul Manship, developed a semi-abstract style that drew on archaic Greek design rather than Cubism, or like Elie Nadelman, produced works influenced by folk idioms of his adopted country. A kind of American semi-modernism developed that was less controversial, easier to understand, and acceptable to the establishment.

Another outgrowth was the reduction in size of much sculpture. Unable to get commissions for large civic works, many sculptors began creating smaller pieces which could be purchased by collectors or even decorate buildings or private gardens and homes. Many small sculptures, in several editions, were created and sold.

One of the first breaks in this wall of opposition to modernism appeared when several knowledgeable figures, including Agnes Rindge, came up with the idea of a museum dedicated completely to modernism, a place for modern art to be seen by the public. "If sculpture is not designed to be a monument, a fountain, a park ornament," she wrote in 1927, "the only place for it is a museum." She was a founder of New York's Museum of Modern Art. It opened in 1929; it took many years before other cities across the country opened their own museums of modern art. But eventually the museum became one of the only places where the public could see the new sculpture.

But during the 1930s and early 1940s modernism really took hold when a number of leading European modern artists came to live and work in America. Most of them settled in New York, which became the international capital of the art world. For it was the upheaval in Europe before and during World War II that finally brought change to the sculpture world in the United States. Dozens of leading European sculptors—many of them modernists and well-known in their own countries—sought refuge in America. They brought encouragement and important influence to their beleaguered colleagues in the United States. Suddenly, there was interest in the art establishment in non-objective sculpture and painting. With colleagues like Albers, Mondrian, Chagall, Duchamp, and Lipchitz in the American firmament, modernism finally appeared to be a bona fide artistic statement. A kind of art for which American sculptors had never found acceptance was given attention at last.

Post-war New York became the center of the new. Before long, new styles like Constructivism and abstract expressionism were developing. There was clamor for new American styles of art. A number of modernists also began exploring new media that they could make in their own studios—many of which looked increasingly like machine-shops. In addition to the emphasis on hands-on carving in wood or stone that typified much modern sculpture, the development of welding and of free-floating metal forms also changed the look of contemporary sculpture. Alexander Calder (whose early metal works were cast at R.B.W.,) was a pioneer in linear, free forms, and eventually, in moving ones. Both electricity and air currents became relevant to sculpture. The new materials—to match the new concepts of art—included plastics and aluminum, tubing, and found-objects; junk heaps and trash also had a place in new forms of modernism. To many modern sculptors, foundries and their traditional brass statues seemed as far removed and academic as the familiar national monuments they produced.

The first commissions and exhibitions of modern sculpture soon followed. By 1950, modernists were no longer the outsiders. The next half-century would see America become the world's center of modern art. Soon there were two parallel universes in the art world, with realism and abstractionism occupying separate spheres, though many artists combined elements of each. It is that very variety that typifies the 20th century. Freedom of expression—so hard won by the modern movements in the United States—has enabled artists to seek their own directions, without either government interference or the approval of an "Academy." It was a long and arduous battle for the aesthetic taste and sensibilities of the nation—a battle that even today represents a central divide in American culture.

Roman Bronze Works cast some of the earliest works of the first modernists in the United States. The foundry was hospitable to these pioneers in their field; several of the sculptors became close friends of the owner. (Paul Manship actually lived with him for a time, casting his first piece at R.B.W. in 1915.) In the late 1920s and 1930s Archipenko, Nadelman, Lachaise, and other modernists cast bronze works there. Later abstractionists like Lipchitz, Noguchi, and Calder also used the foundry. But in mid-century, as modernism turned away from brass casting to newer media, and as the size and form of sculpture changed dramatically, R.B.W. lost its preeminent position among that segment of the art world. Nonetheless, its ledgers bear the names of many of America's—and the world's— most important modernists.

ALEXANDER ARCHIPENKO (1887-1964)

Alexander Archipenko was a leading figure in the history of early modern sculpture. Born in Kiev, he arrived in Paris in 1908 where he encountered Cubism; by 1909 he had turned away from his academic training toward a style of sculpture concerned with form and mass, planes, space, the convex and concave, and the interplay of light and shadow. In pursuing the ideas of Cubist sculpture and the avant-garde, he not only abstracted the forms of sculpture, but began using a variety of mixed media and inventive conceptions. In Paris he started a teaching career which would continue throughout his life.

He sent four sculptures to the Armory Show of 1913 in New York In his studios in Nice during World War I and Berlin beginning in 1921 he continued to produce strong sculptures, mostly abstracted from the human body. In 1923 he came to America, settling first in California, and then in Chicago, and in 1939, in New York. He taught at Washington University, the Bauhaus in Chicago, and Kansas University.

As a modernist he received a scathing reception from many American critics; his abstract work was far removed from the style and beliefs of the conservative art establishment. Though much of his innovative work was produced before he arrived in the United States, he continued to create original works in America. Among them were an amalgam of painting and sculpture that he called "Archipentura" that showed painting in motion, and an invention that had plastic carvings lighted from within.

As modern art finally took hold in America, Archipenko was recognized as a pioneer modernist; he had 118 one-man shows during his lifetime, exhibited at major museums, and taught at many important institutions. His relationship with Roman Bronze Works lasted for many years;

the first listing of his name was in 1930, the last in 1955. He recast a number of his early sculptures there. RBW produced a series of abstract torsos for him including "White Torso,"(V-1), "Flat Torso"(V-2), and "Reclining Torso," (V-3). Two of these small, sleek torsos were even used in Fifth Avenue store windows; their stream-lined silhouettes symbolizing modernity (V-4). Among his other works cast at the foundry were a bronze, gold-plated "Hand" (V-5), and "King Solomon" (V-6).

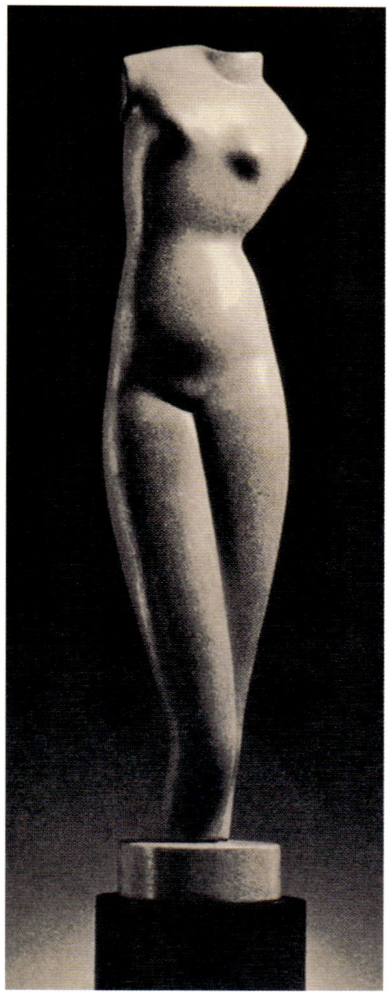

V-1. "White Torso" by Alexander Archipenko. Modeled in 1916. R.B.W. 1955.

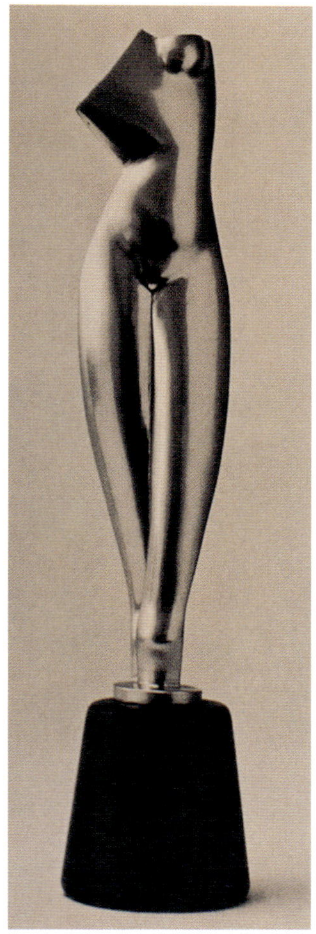

V-2. "Flat Torso" by Alexander Archipenko. Modeled in 1914. R.B.W. 1955.

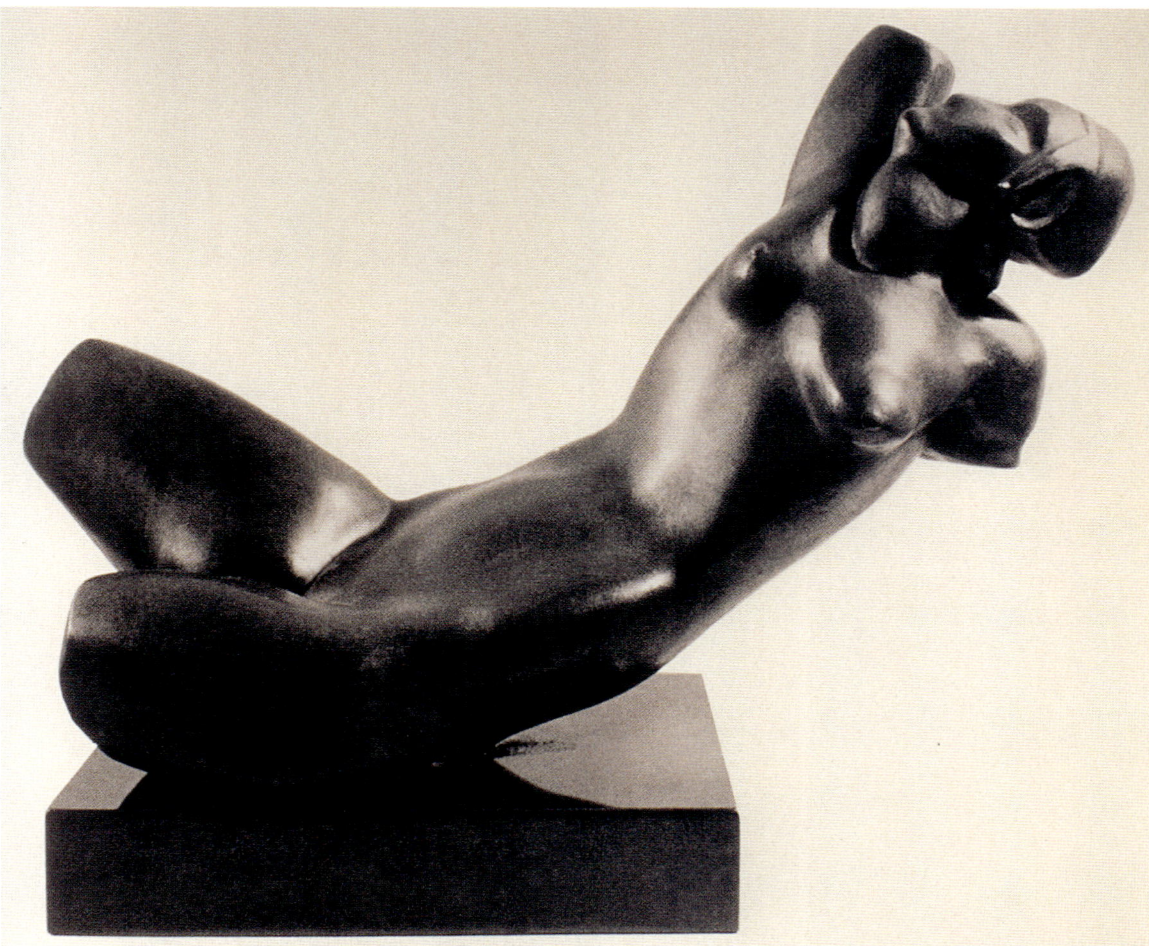

V-3. "Reclining Torso" by Alexander Archipenko. Modeled in 1922. R.B.W. 1955.

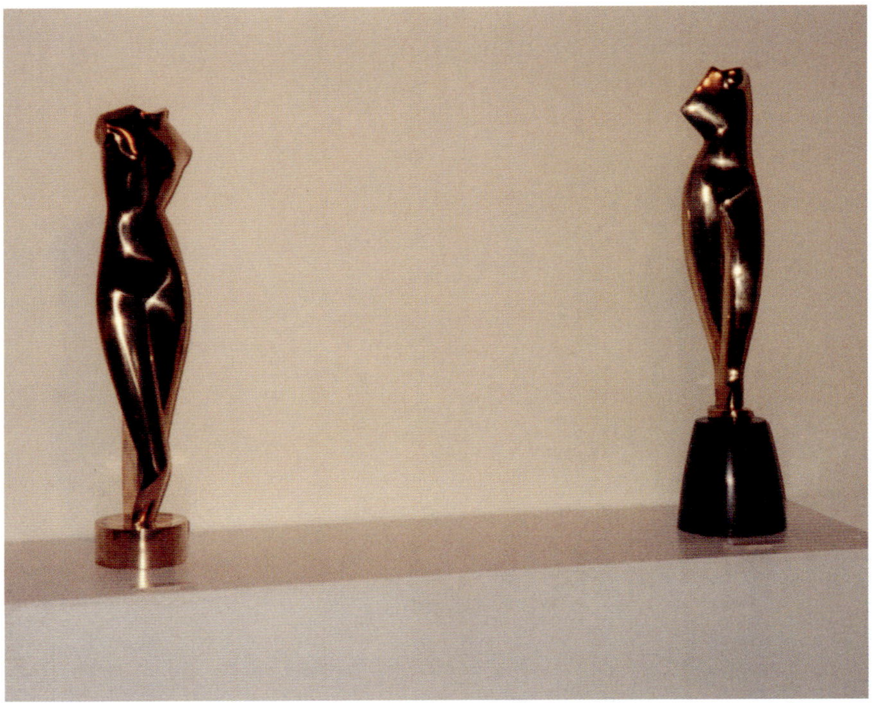

V-4. Two Archipenko statuettes used in Saks Fifth Avenue windows.

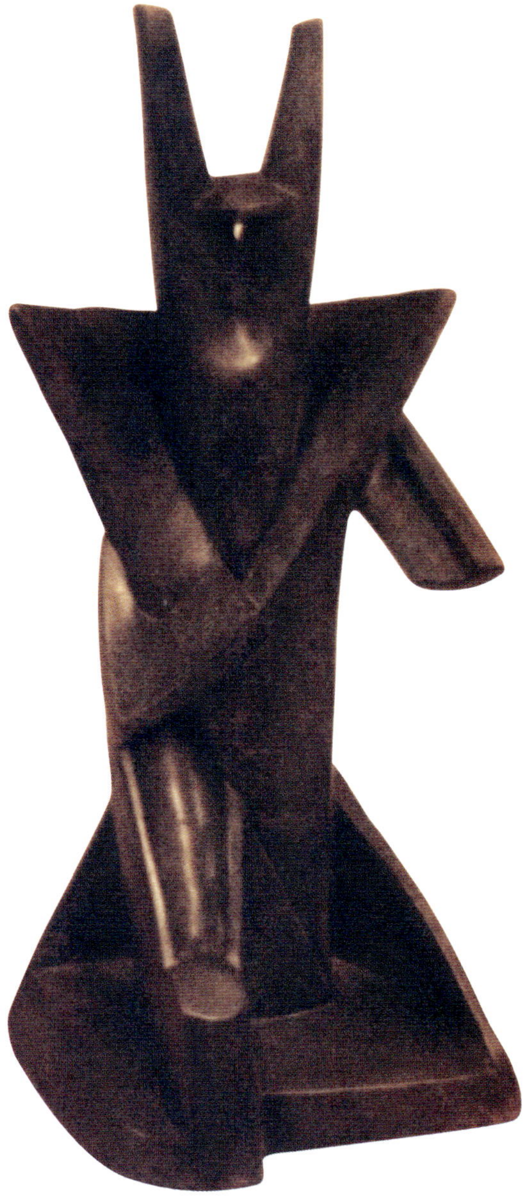

V-6. "King Solomon" by Alexander Archipenko.

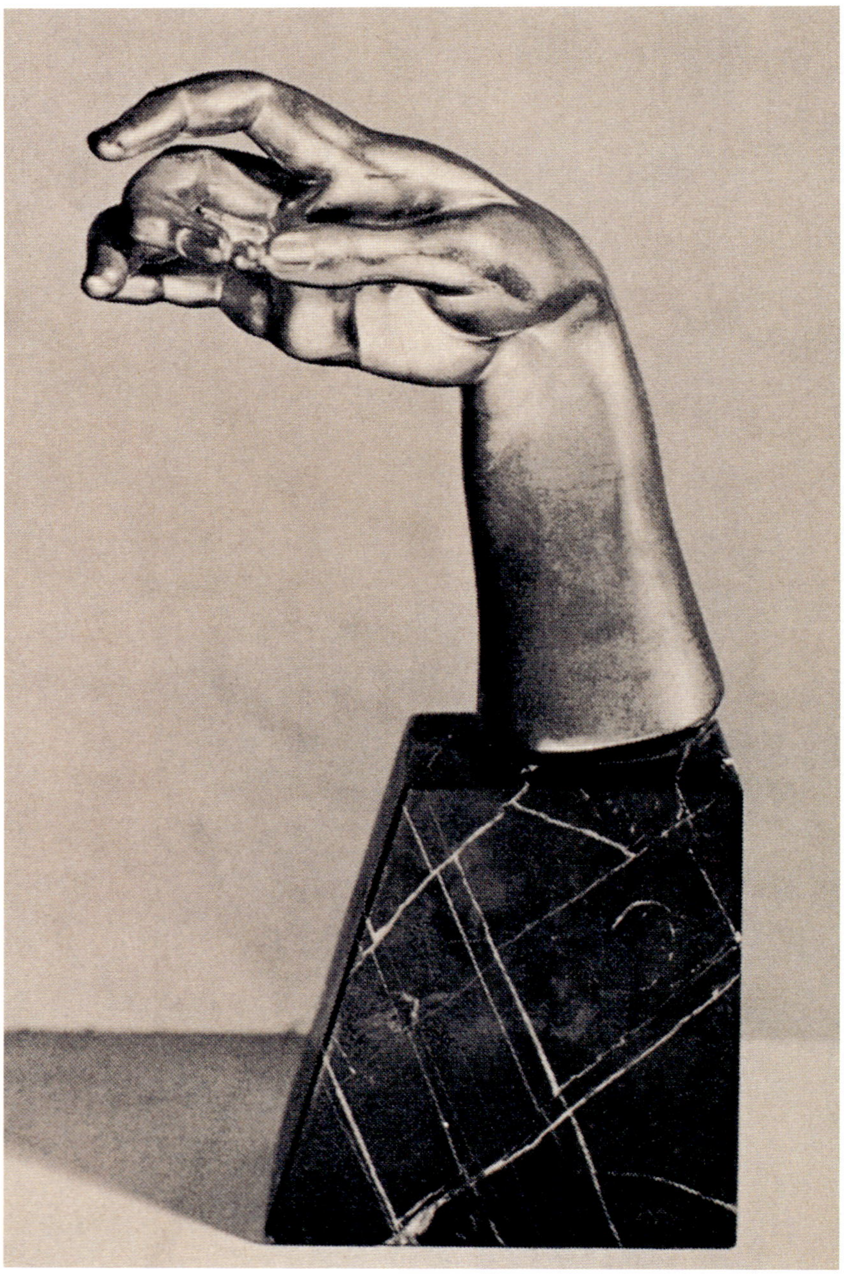

V-5. "Hand" by Alexander Archipenko.
Bronze with gold plating. c. 1928.

ELIE NADELMAN (1882-1946)

Elie Nadelman was a leading artistic force in the development of modern sculpture, and an early exponent of reorganizing natural form into simplified abstraction. By 1907 he was producing sculpture that presented the human body as a series of convex and concave shapes, abstract curves, and mass and countermass. He described his work as studies of "significant form," as "relations of form, which create a new life and have nothing to do with life in nature." His sculpture was influenced by such disparate art as classical statuary, Cubism, and Art Nouveau design. Though a major part of his career was in Europe, his work made a strong impression in the United States. He also began collecting American folk art, which he greatly admired and which eventually influenced his own work, as well.

Nadelman was born in Poland, and studied art in Warsaw, Cracow, and Munich. In Paris in 1903 he became a member of Gertrude and Leo Stein's circle, knew Picasso and Matisse, and soon became interested in avant-garde trends. He was recognized as a leading force in the modern movement. His elegant, simplified, abstracted nudes were particularly well regarded, and the stylistic elements he embraced continued to define his work for the next thirty years. In 1914, with the outbreak of World War I, Nadelman came to the United States. In New York, he cast several of his major sculptures at Roman Bronze Works, including his "Draped Standing Female Figure" (also called "Recherche des Formes") (V-7), "Horse" (V-8) made circa 1911, and "Man in a Tophat" (V-9). Helena Rubinstein also commissioned him to make sophisticated marble heads for her salons.

In the 1930s, turning away from abstraction, Nadelman began incorporating some of the ideas of American folk art into a series of small terra-cotta works. A combination of ill health, bad fortune in the stock market crash of 1929, and a devastating fire that destroyed his studio and many of his plaster and wooden pieces in 1933 brought his career to an end.

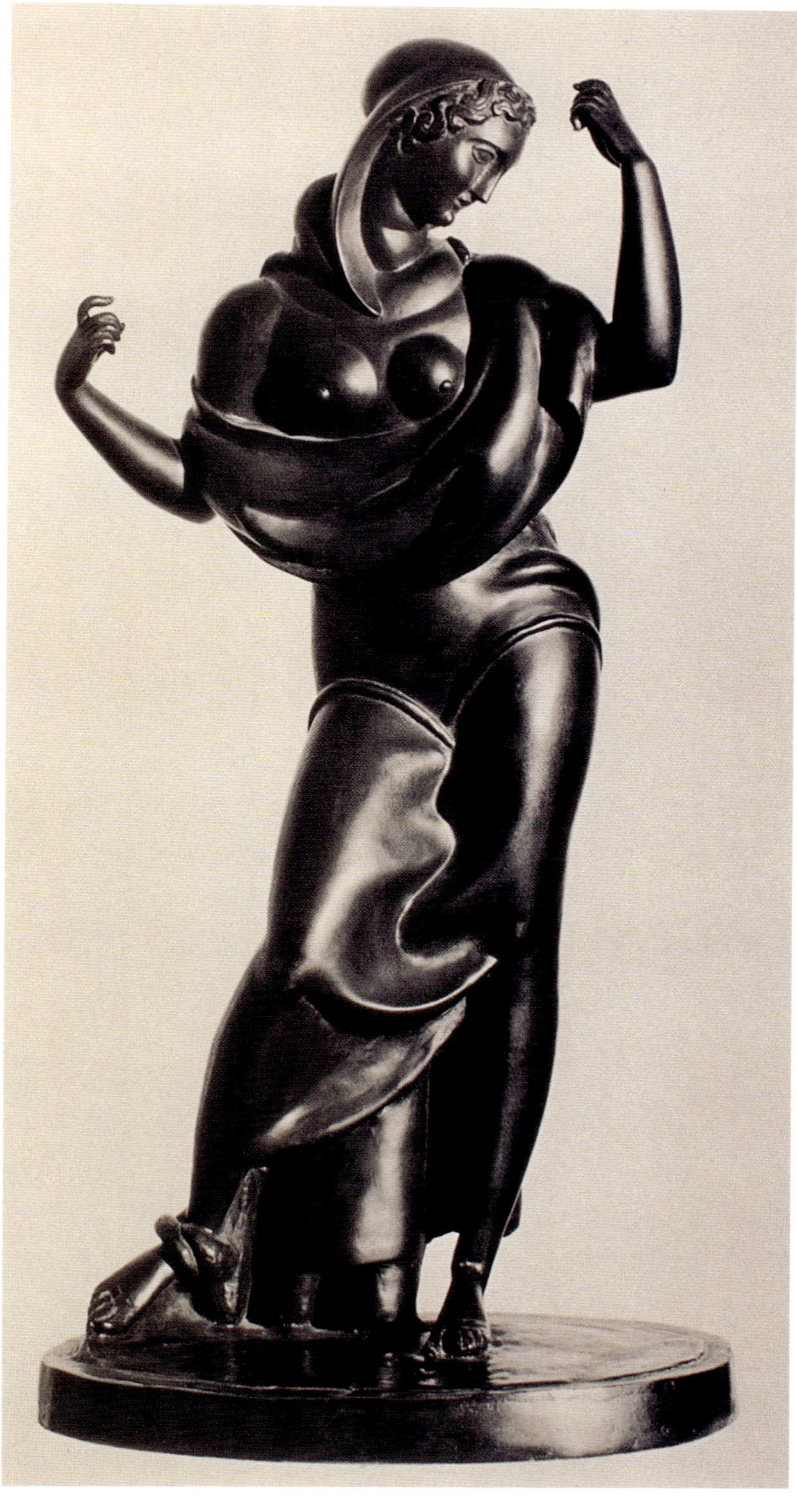

V-7. "Draped Standing Female Figure" (also called "Recherche des Formes") by Elie Nadelman. Modeled c. 1908.

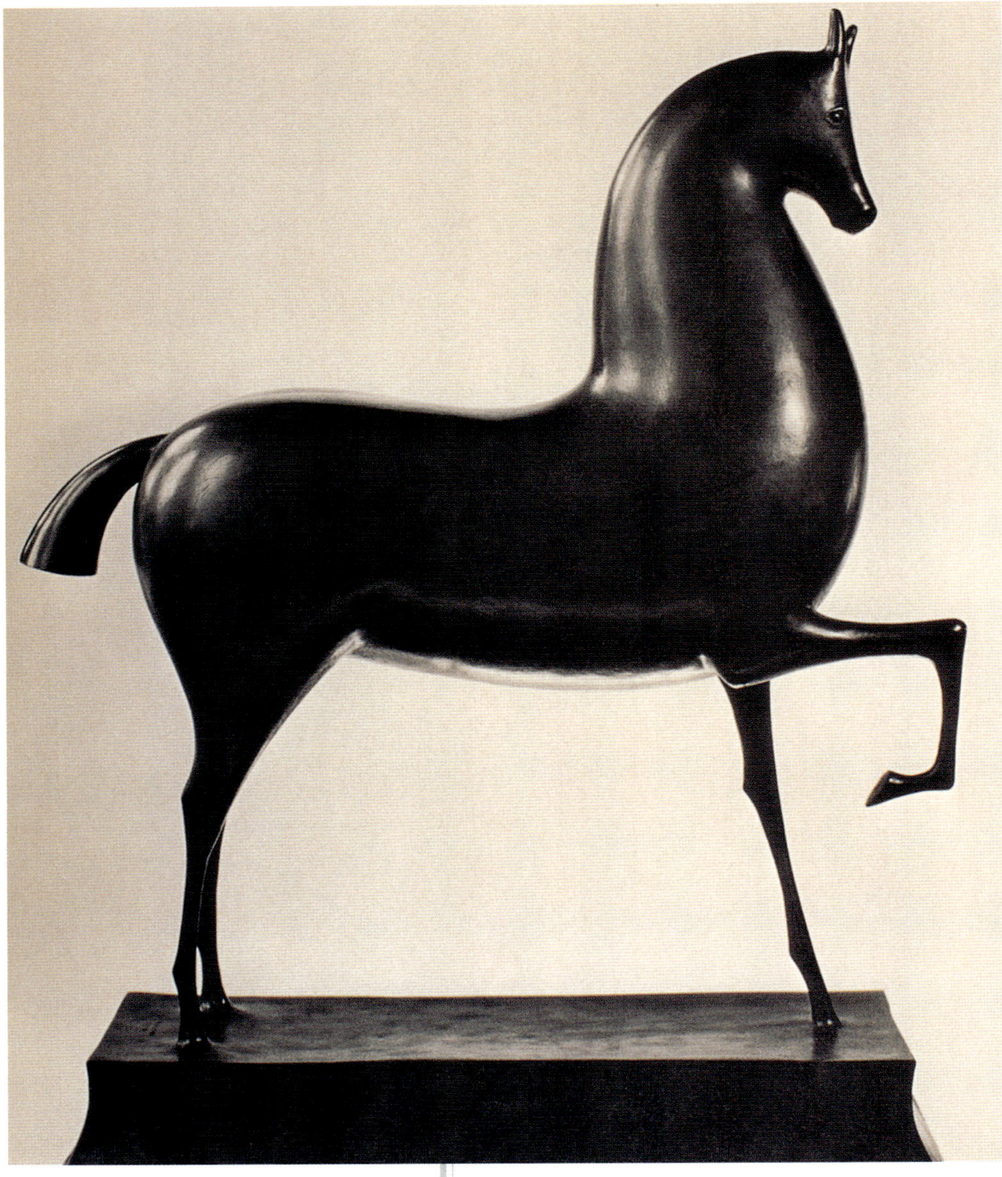

V-8. "Horse" by Elie Nadelman. c. 1911.

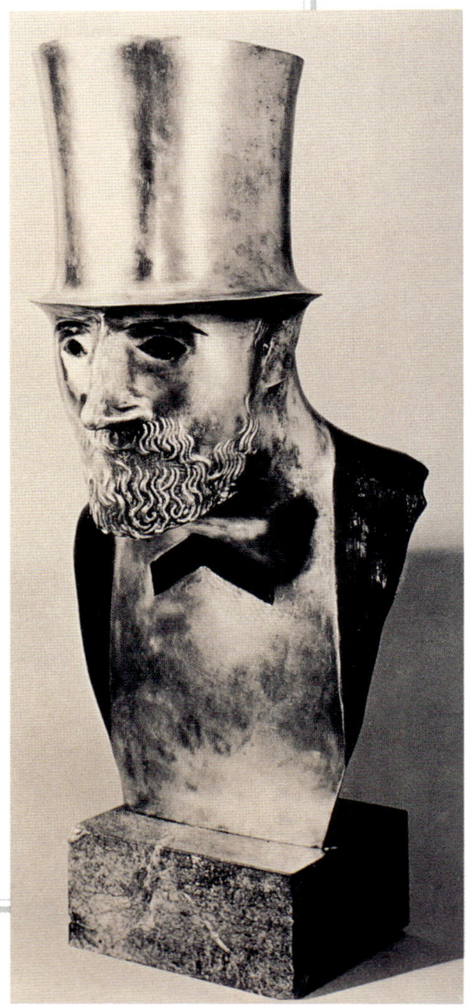

V-9. "Man in a Tophat" by Elie Nadelman. c. 1927.

GASTON LACHAISE (1882-1935)

Gaston Lachaise was a major figure in the development of modern sculpture in America. He is best known for his studies of the female body, which emphasize monumental, voluminous form. He also made portrait sculpture, animal images, and architectural sculpture. His emphasis on form and idea, rather than on realistic portrayal, made his work central to the development of modernism in American sculpture.

He arrived in 1906 in the United States from Paris where he had worked with the noted jeweler René Lalique and was trained in the Beaux-Arts and Art Nouveau traditions. In America he became an associate of Henry Kitson in Boston and then Paul Manship in New York. He showed a sculpture at the Armory Show in 1913 and went on to explore modernism through his studies of the human body. "He saw the entire universe in the form of (a) woman..." wrote fellow artist Marsden Hartley. Among the striking female figures he cast at Roman Bronze Works were numbers V-10 to V-18.

The obviously sexual aspects of his work kept Lachaise from most public commissions in America's conservative art world, and it was not until modernism was finally recognized as art—though not necessarily public art in the form of monuments—that he achieved recognition from museums. Lachaise's sculpture, however, was well-regarded by many in the more avant-garde worlds of literature and art; he exhibited at Henry Steiglitz's gallery, his work was frequently pictured in "The Dial," and he was an associate of such figures as e.e.cummings, Edgar Varese, and Georgia O'Keeffe. His first one-man show was in 1920; in 1935 the Museum of Modern Art honored him with a retrospective.

Lachaise, whose bronzes were highly polished, began casting his work at Roman Bronze Works in 1929, when Paul Manship brought him there. In addition to the female forms for which he was best known, Lachaise also cast a number of portraits—both of people he knew, such as fellow artist John Marin (V-19) and Margery Spencer (V-20)—and of idealized forms, such as "Egyptian Head" (V-21). He also created decorative works like "The Peacocks" (V-22) and a "Dolphin Fountain" (V-23), cast around 1918. Lachaise continued his association with the foundry until his death.

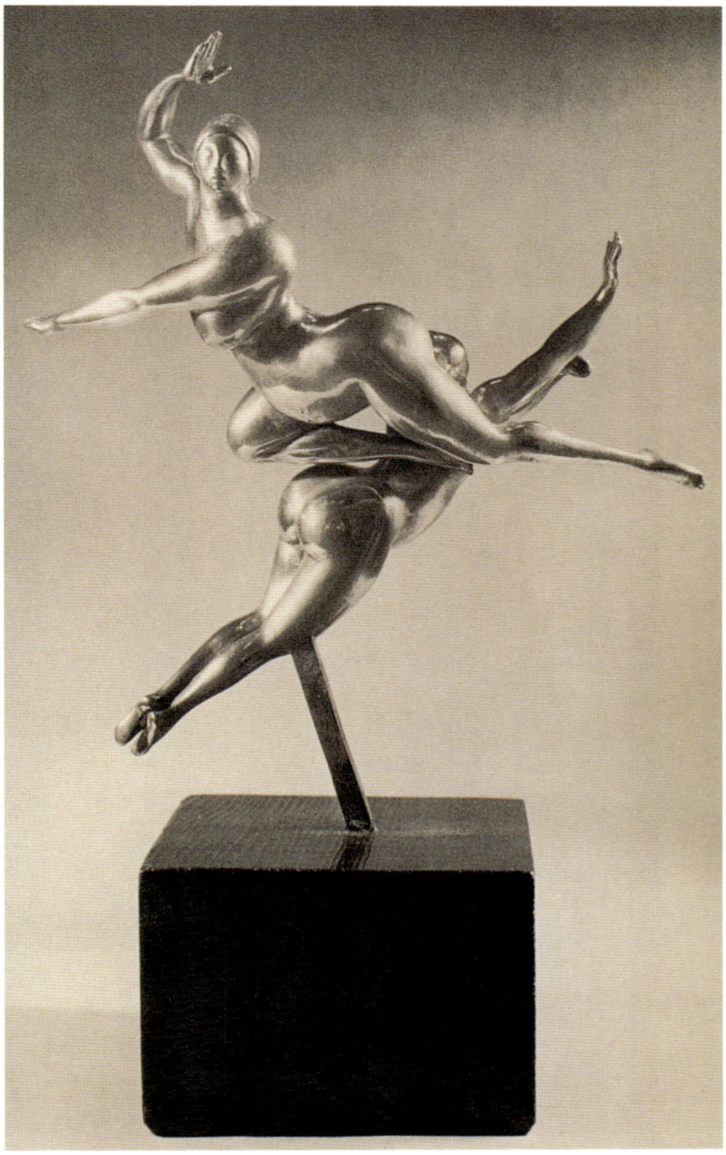

V-10. "Two Floating Acrobats" by Gaston Lachaise. 1922.

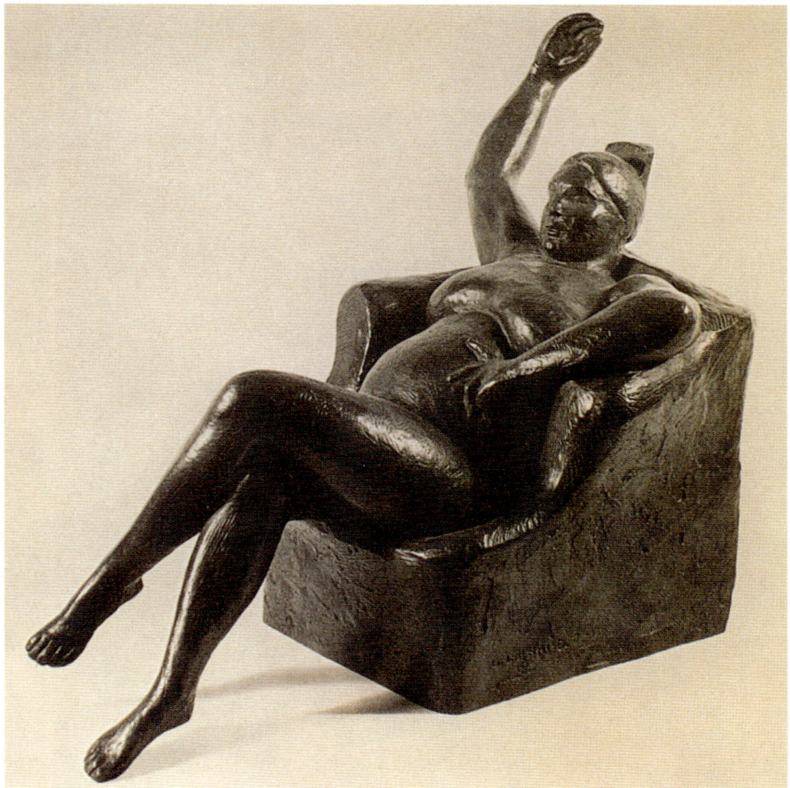

V-11. "Reclining Woman with Right Arm Raised" by Gaston Lachaise. c. 1924.

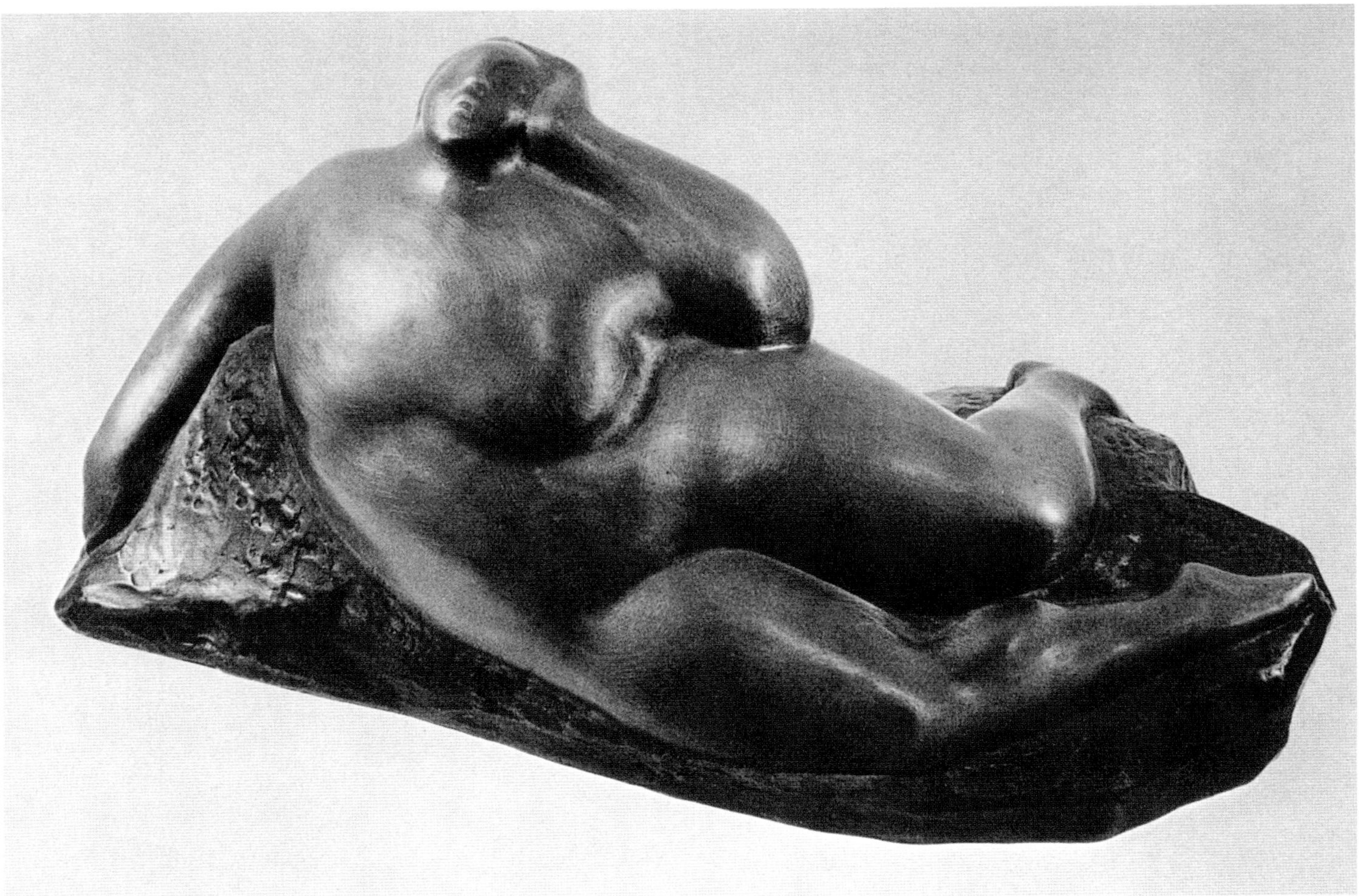

V-12. "The Mountain" by Gaston Lachaise. 1924.

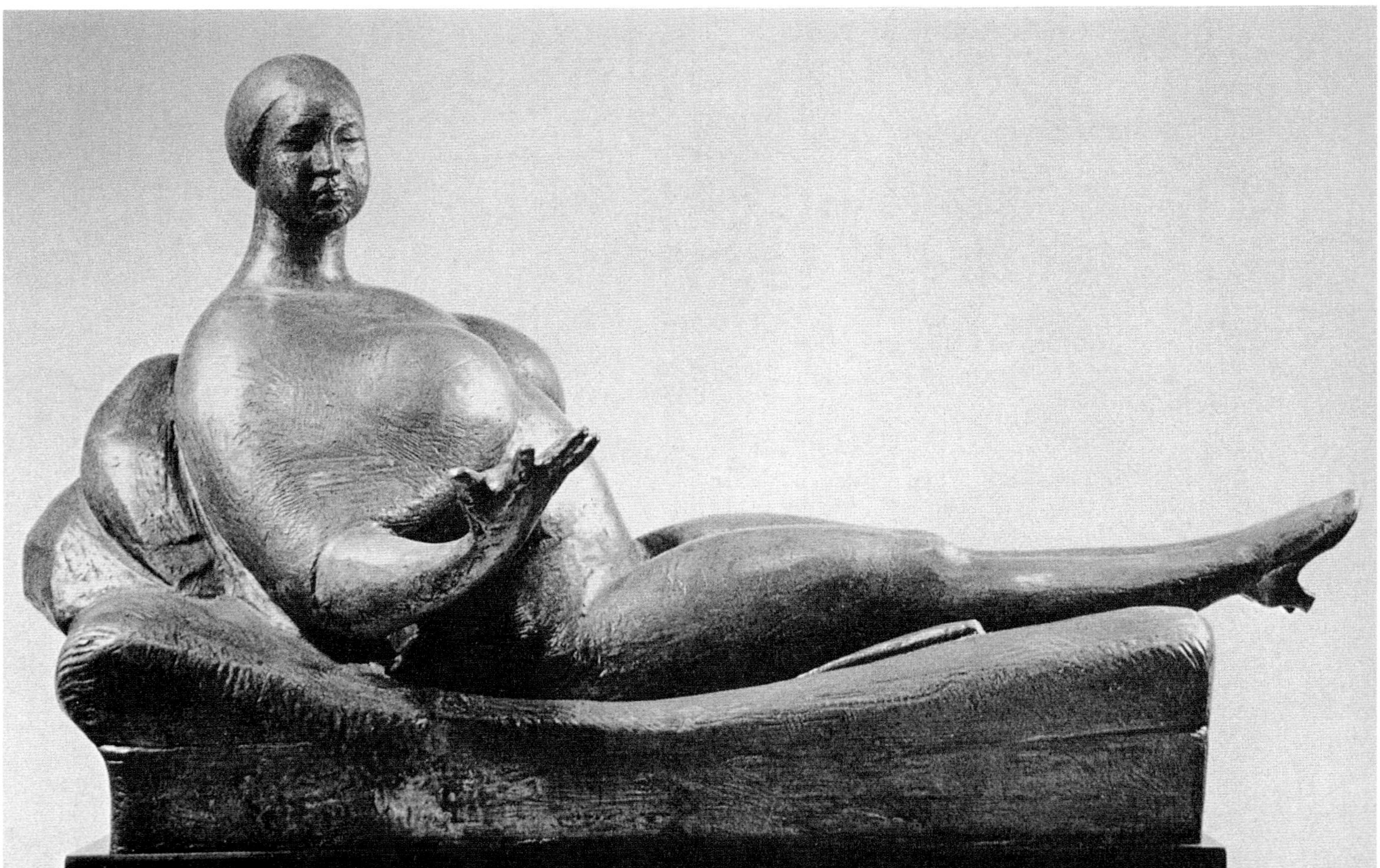

V-13. "Woman on a Couch" by Gaston Lachaise. 1928.

V-15. "Dancing Nude" (plaster bas relief) by Gaston Lachaise.

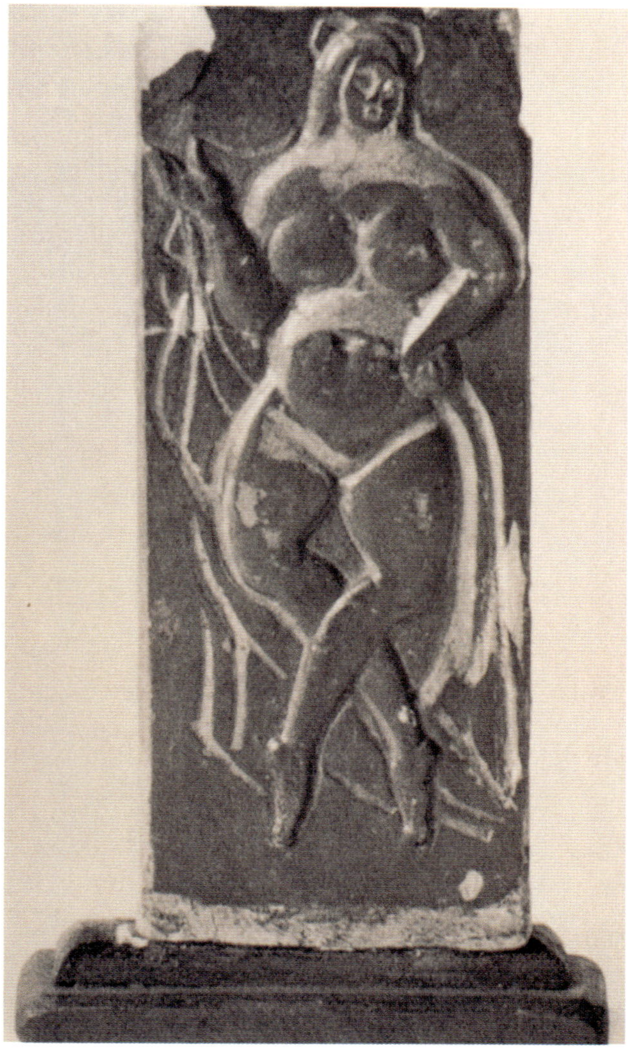

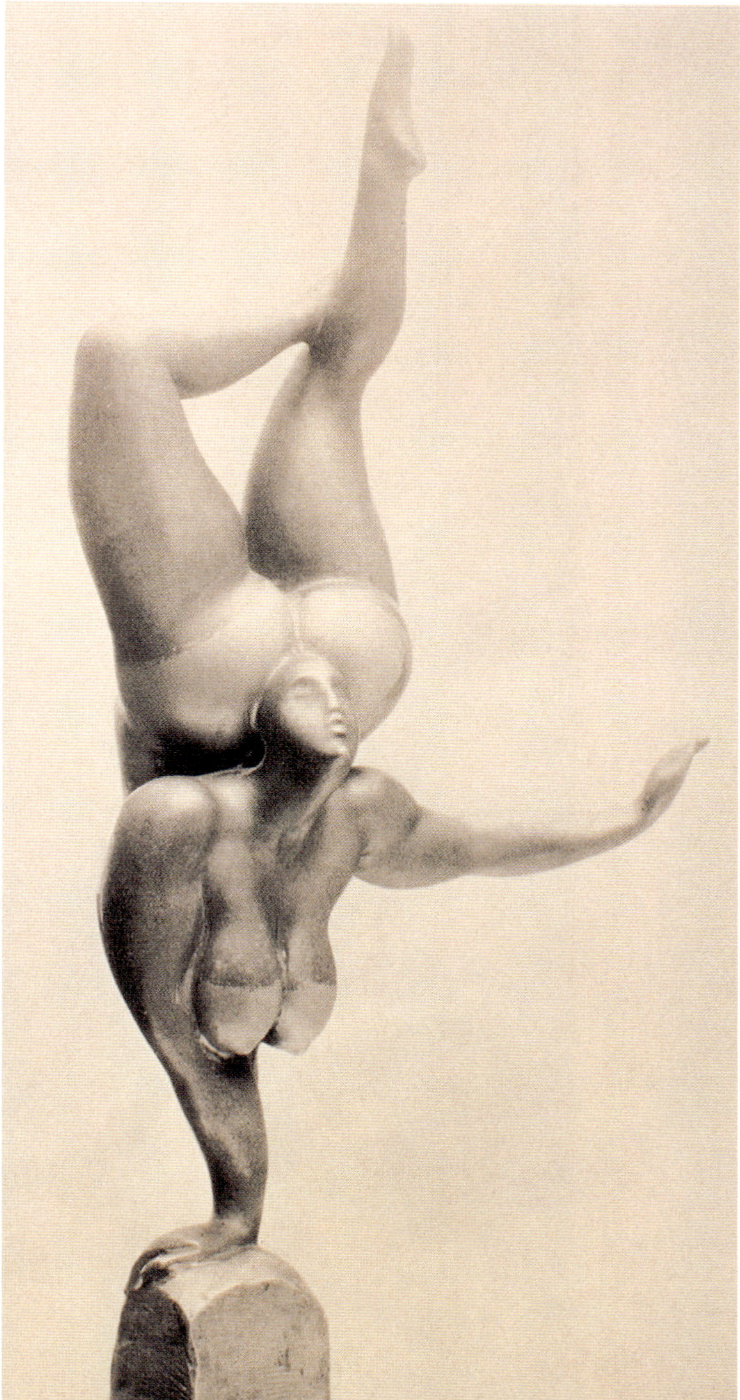

V-14. "Acrobat" by Gaston Lachaise. 1929.

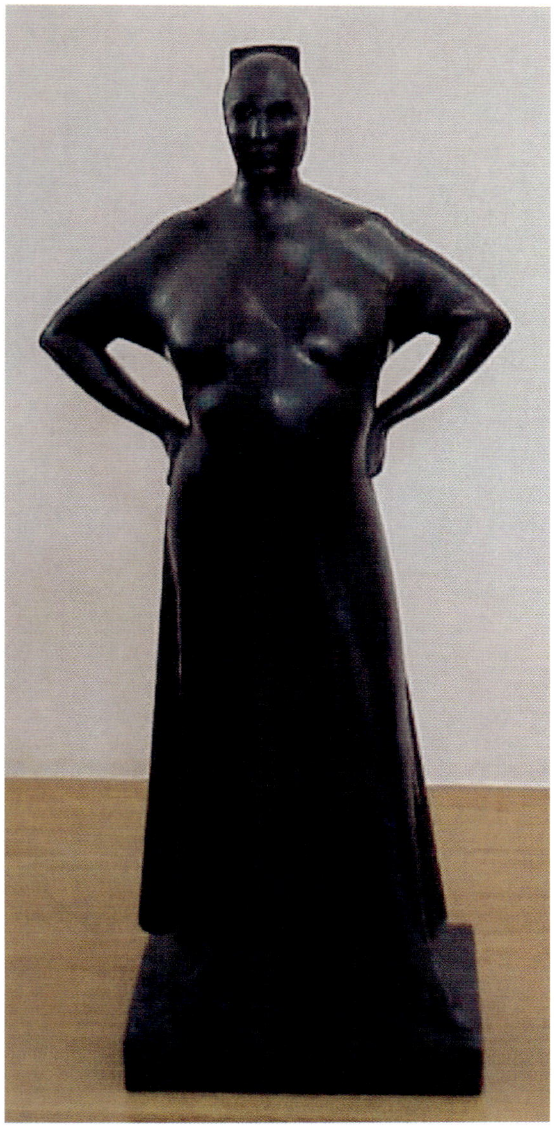

V-16. "Standing Woman" by Gaston Lachaise. c. 1927.

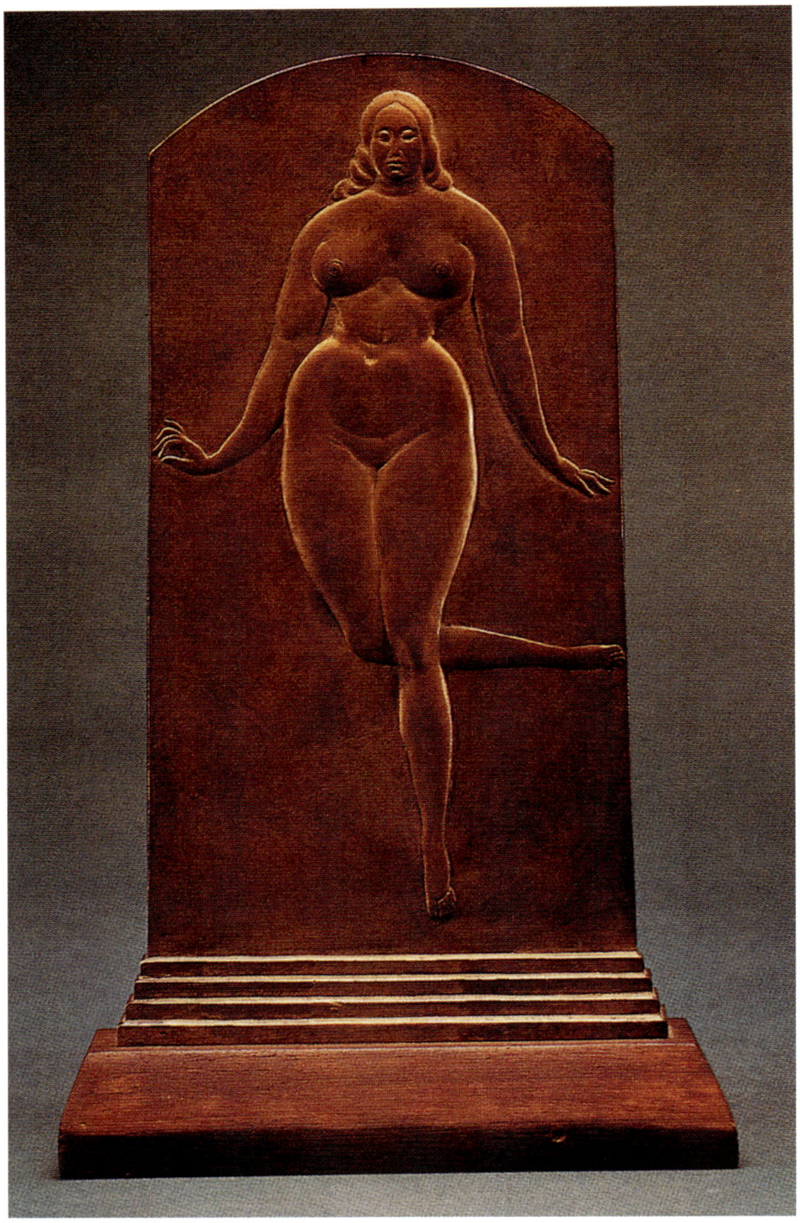

V-17. "Standing Nude" (bronze relief) by Gaston Lachaise. 1927.

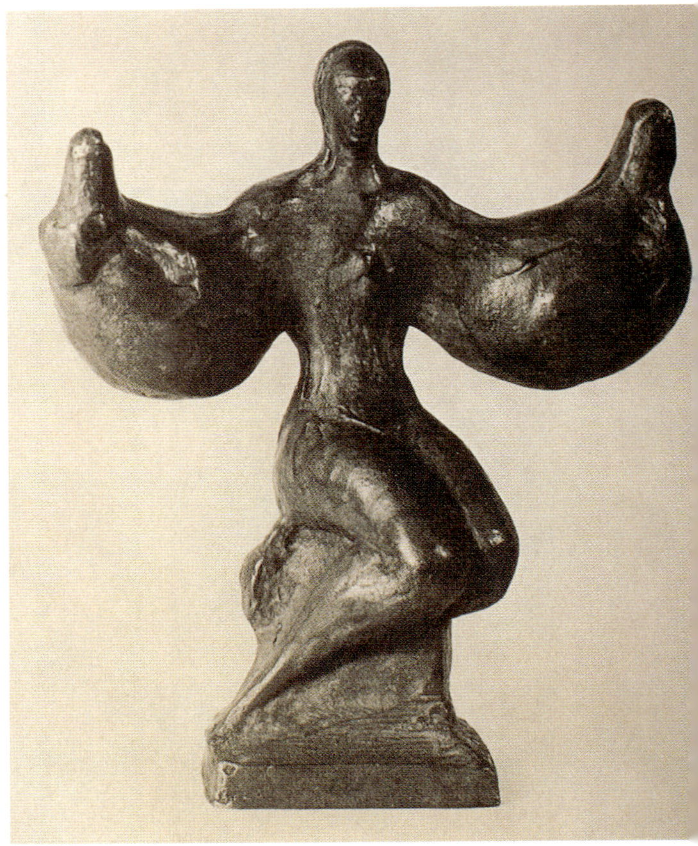

V-18. "Seated Woman Holding Breasts" by Gaston Lachaise. 1931.

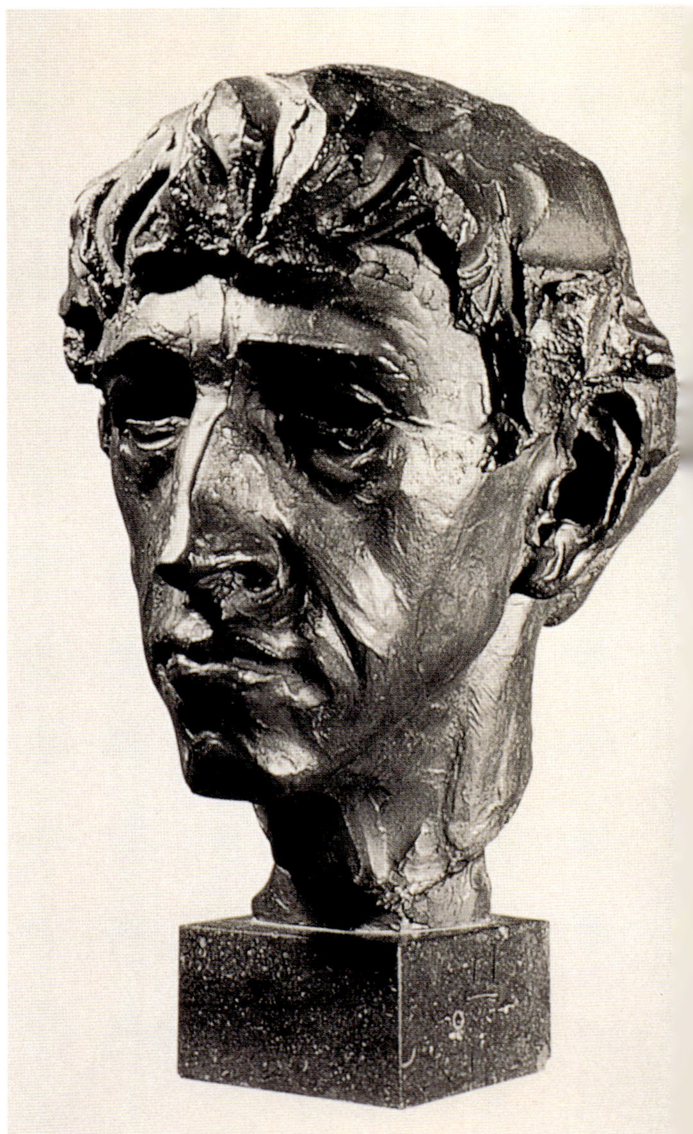

V-19. "Head of John Marin" by Gaston Lachaise. 1928.

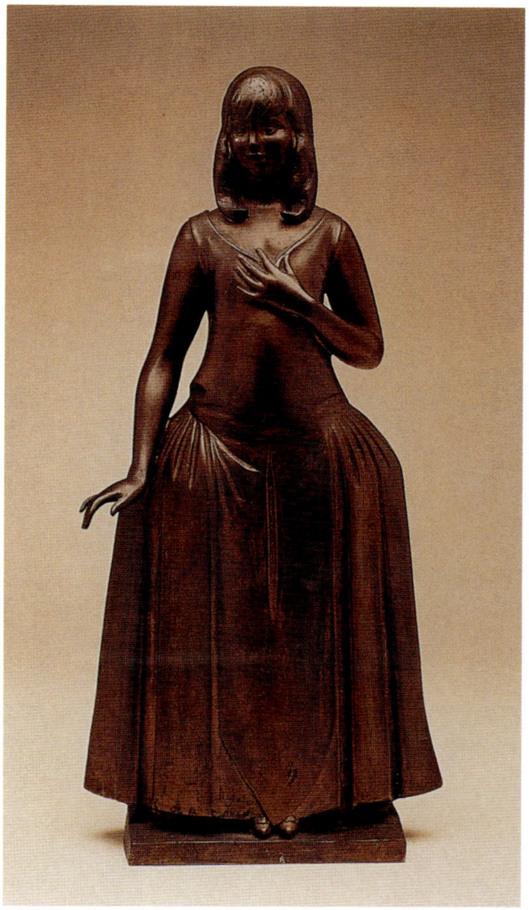

V-20. "Margery Spencer" by Gaston Lachaise. 1924.

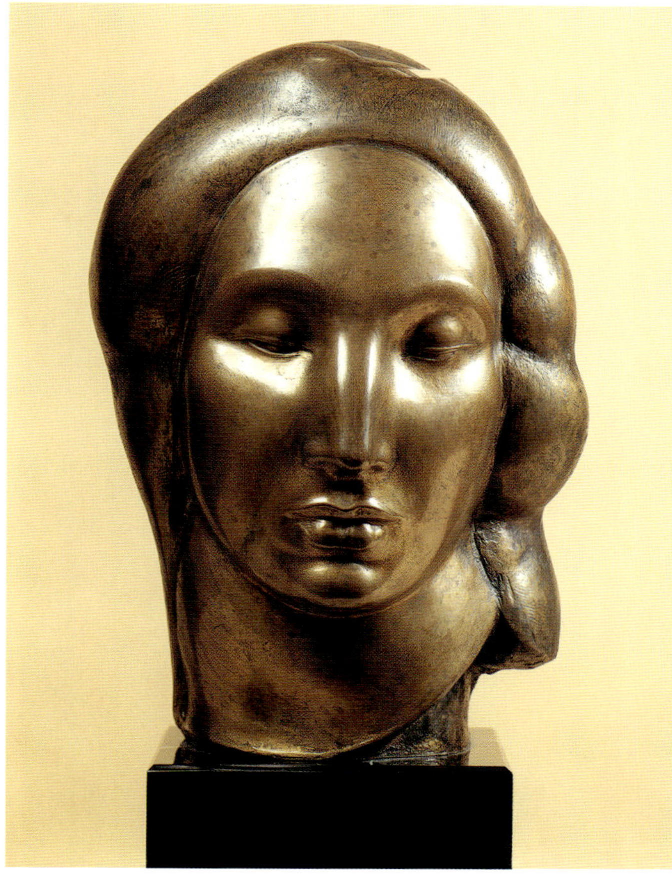

V-21. "The Egyptian Head" by Gaston Lachaise. 1923.

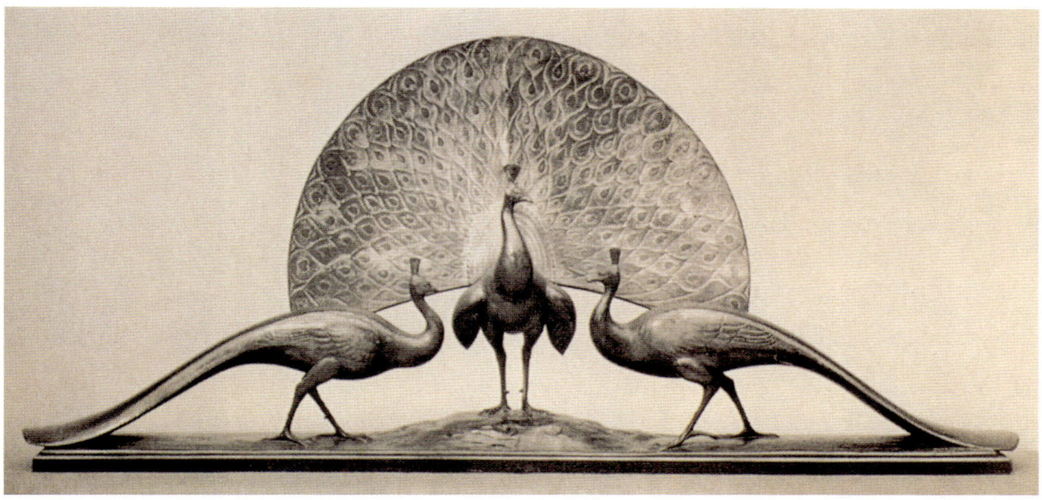

V-22. "The Peacocks" by Gaston Lachaise. c. 1918.

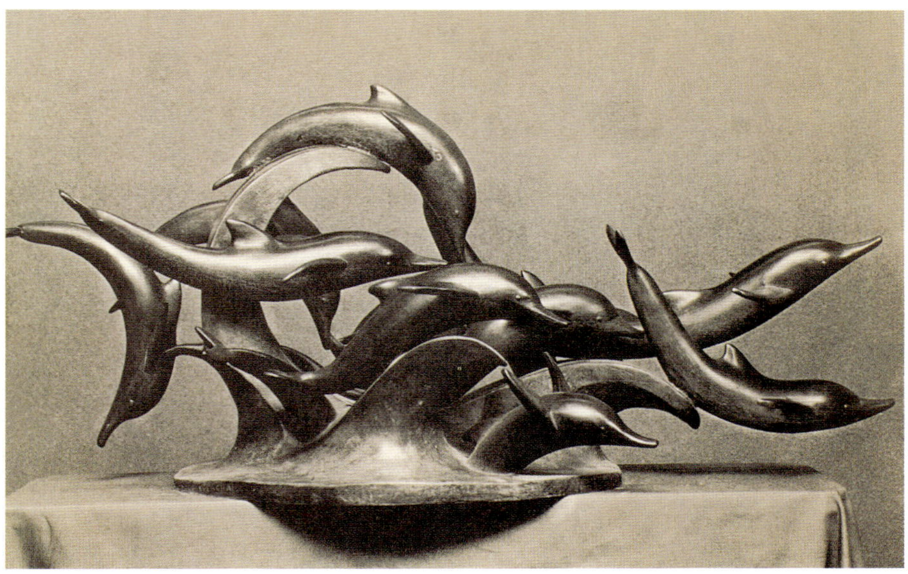

V-23. "Dolphin Fountain" by Gaston Lachaise. 1924.

JACQUES LIPCHITZ (1891-1973)

A pioneer of Cubist sculpture, Lipchitz is known both for his angular, standing Cubist figures, as well as for his later monumental, expressionistic abstractions, with their rounded and wiry shapes and forms. Lipchitz was a leading figure in the art world of avant-garde France; he continued his distinguished career in the United States after 1941.

Jacques Lipchitz was born in Lithuania and arrived in Paris in 1909. Two years later he had established his own studio, and soon met Picasso and other Cubists. By 1915 he had become a pioneer of Cubist sculpture, producing complex abstract figures with a flat, vertical presence. Many of them were cast in bronze. In the mid-nineteen twenties he began experimenting with transparent bronze sculptures, in which the open space was defined by linear contours; some had a totem-like ambiance.

By 1930 he had left Cubism and had turned to more expressive subject matter, with themes drawn from the Old Testament and classical myths. Despite the use of subject matter of personal significance, Lipchitz continued to create sculpture that was abstract and highly imaginative in form. After the fall of France to the Nazis, he emigrated to the United States. He settled in New York and began receiving commissions for work on a large scale. His monumental sculptures developed into an almost baroque style, using anthropomorphic shapes to express religious themes. During the final twenty years of his life, he received numerous awards and had many retrospective exhibitions.

Lipchitz began casting his sculpture at Roman Bronze Works shortly after his arrival in the United. States. Among his works cast there were several that he had made during his Cubist phase, including "Guitar Player in Chair (also called "Seated Man with a Guitar") (V-24), and "Harlequin with Accordion" (V-25).

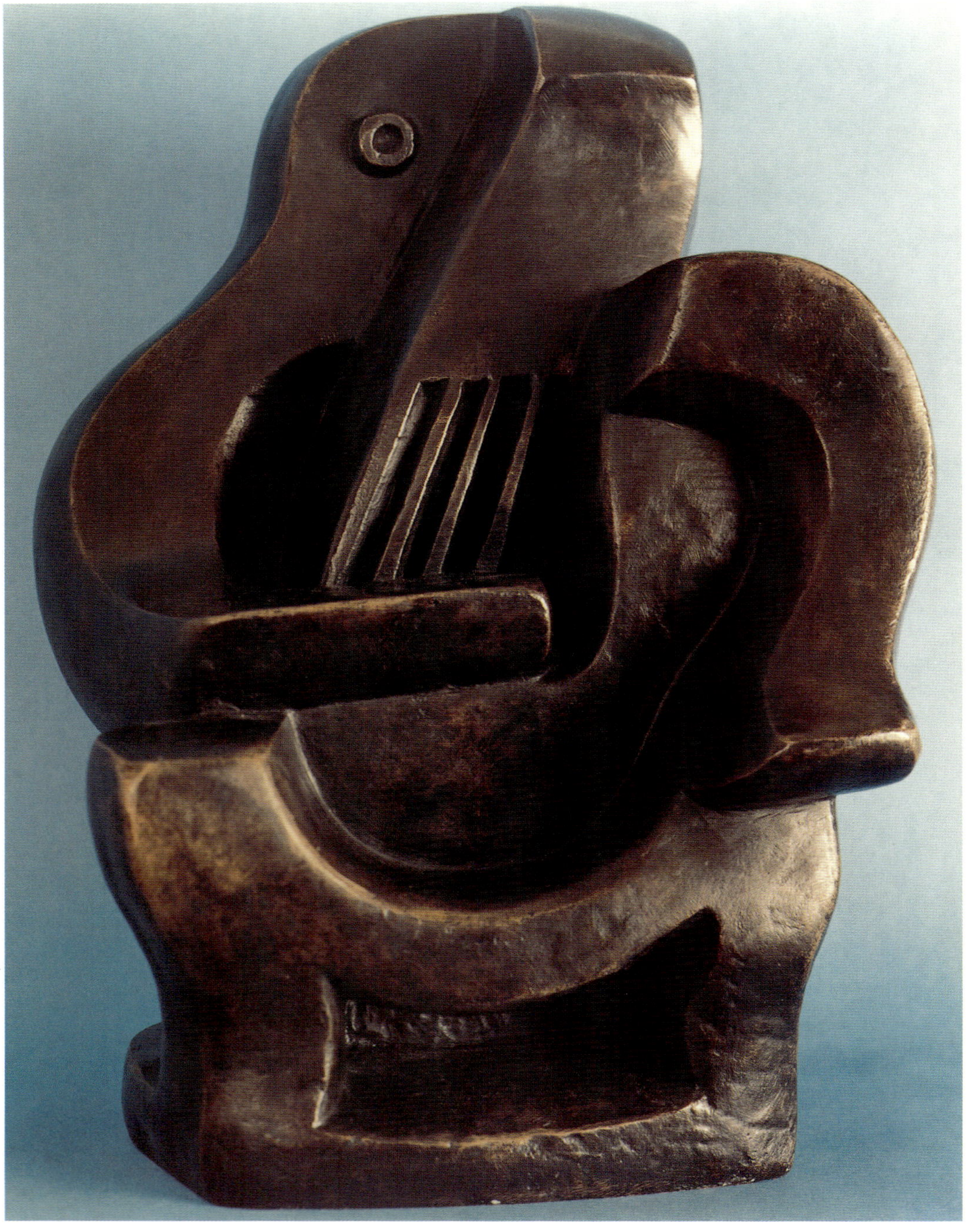

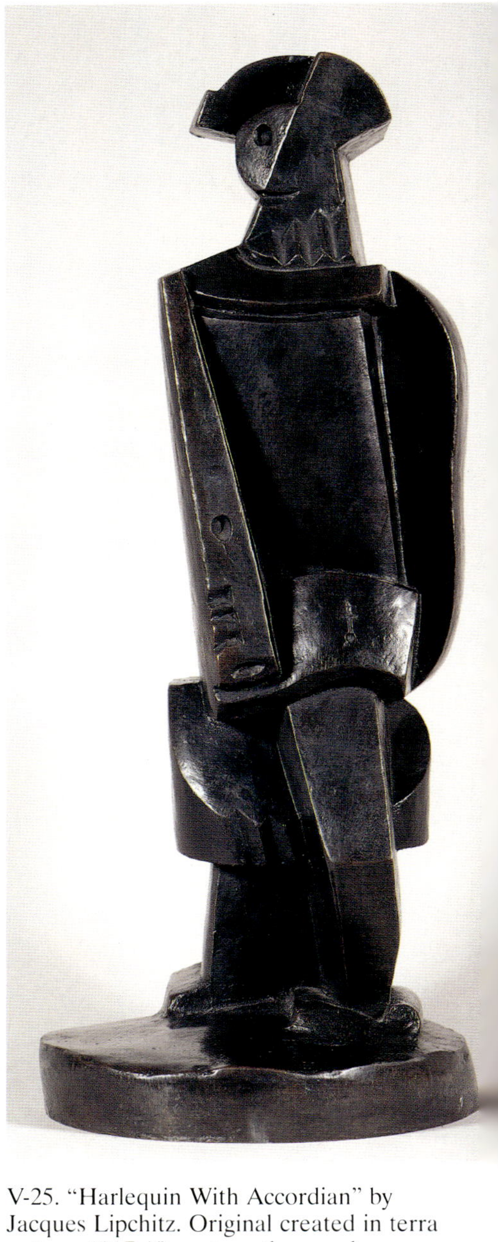

V-25. "Harlequin With Accordian" by Jacques Lipchitz. Original created in terra cotta c. 1917-18; cast posthumously.

V-24. "Guitar Player in Chair" (also called "Seated Man with Guitar") by Jacques Lipchitz. 1922.

MAX WEBER (1881-1961)

One of the first artists to translate European modernism into an American idiom, Max Weber is widely considered a major painter of the twentieth century. As a painter he explored Cubism, Futurism, and several other forms of abstraction; his work was colorful, dynamic, and considered highly avant-garde by the conservative American art world. Few realize he also made sculpture during several periods of his life. (The very sculptural qualities evident in his painting suggest his interest in three-dimensional form.) He was the first American to try non-objective sculpture that eliminated natural appearances, as well as to incorporate elements of primitivism.

Weber was born in Russia and brought to Brooklyn, New York at the age of ten. His art education began at Pratt Institute, and continued in Europe from 1905 to 1908. While in Paris he entered the world of Matisse and Cezanne, befriended the primitive master, Rousseau, and encountered Cubism and the young avant-garde. He exhibited at the major shows with the Independents in 1906 and 1907, and at the Salon D'Automne in 1907 and 1908. After returning to the United States in 1909 Weber began exhibiting his paintings; he had three one-man shows in 1915. No other artist at this time exhibited his works more widely, or was more harshly attacked by critics. Later in his career he turned to more socially conscious art; it is his stunning early abstract canvases that are now regarded as masterpieces.

Weber's first sculptures were made in 1910; they showed the influence of primitive art of Africa and Central America. Later sculptures, made in the 1940s, were polychromed plasters on a small scale; several of these works were later cast in bronze by Roman Bronze Works. Like Rousseau, Weber believed primitive art showed a direct and spiritual view of the world around us. "It is my aim, " he wrote in 1959, "to coordinate and interlock the purest plastic means of expression with spiritually informing and aesthetically inspiring content." With this objective, Weber incorporated a variety of primitive influences—from American Indian to African to Cycladic into his work.

Weber exhibited his sculpture during three particular periods of his career: in 1915, during the mid-forties, and from 1955 to 1960. His "Profile" (V-26), "Colored Bird" (V-27), and "Intersecting Profiles" (V-28) were all shown in the last period.

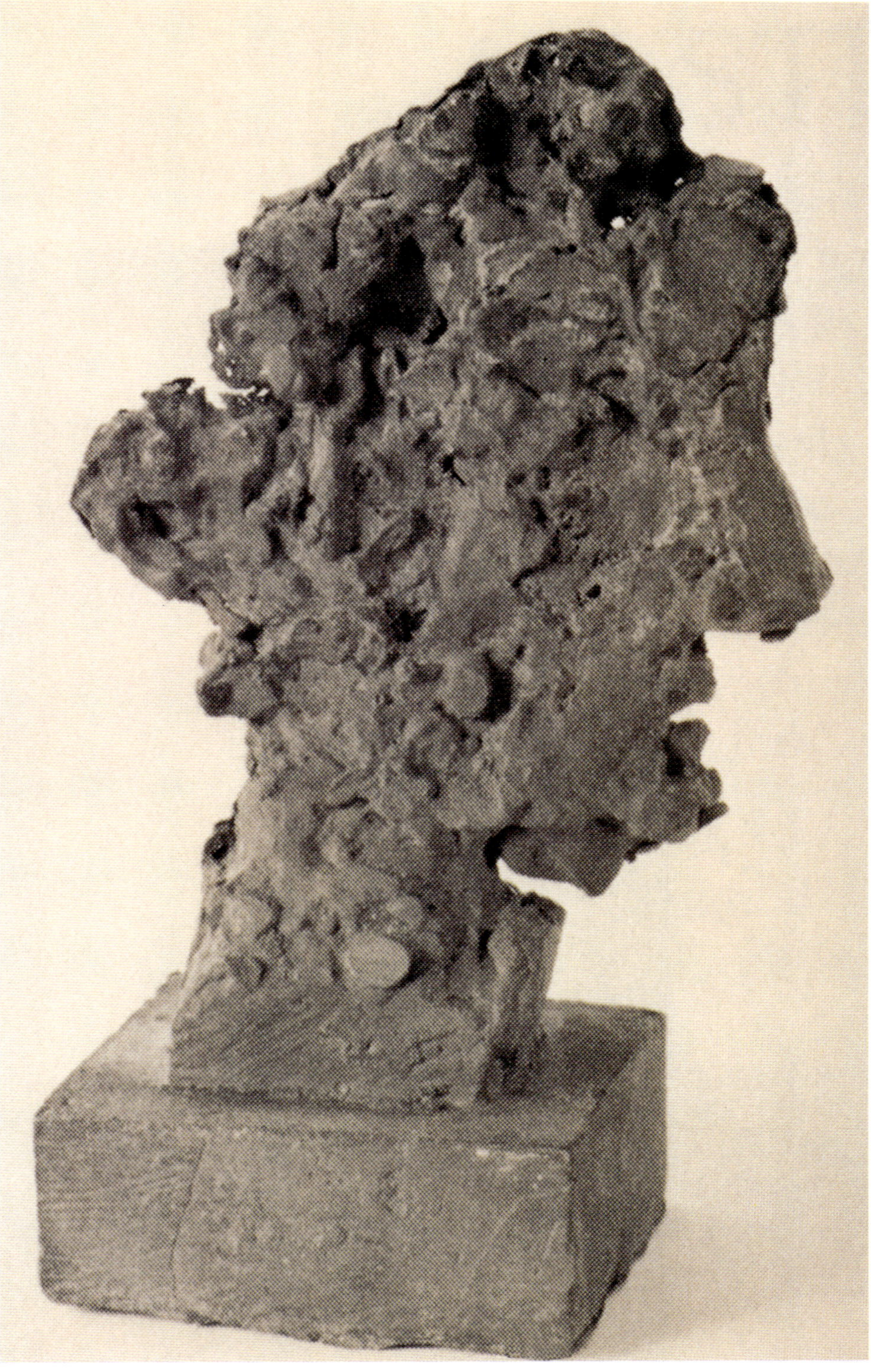

V-26. "Profile" by Max Weber. 1959.

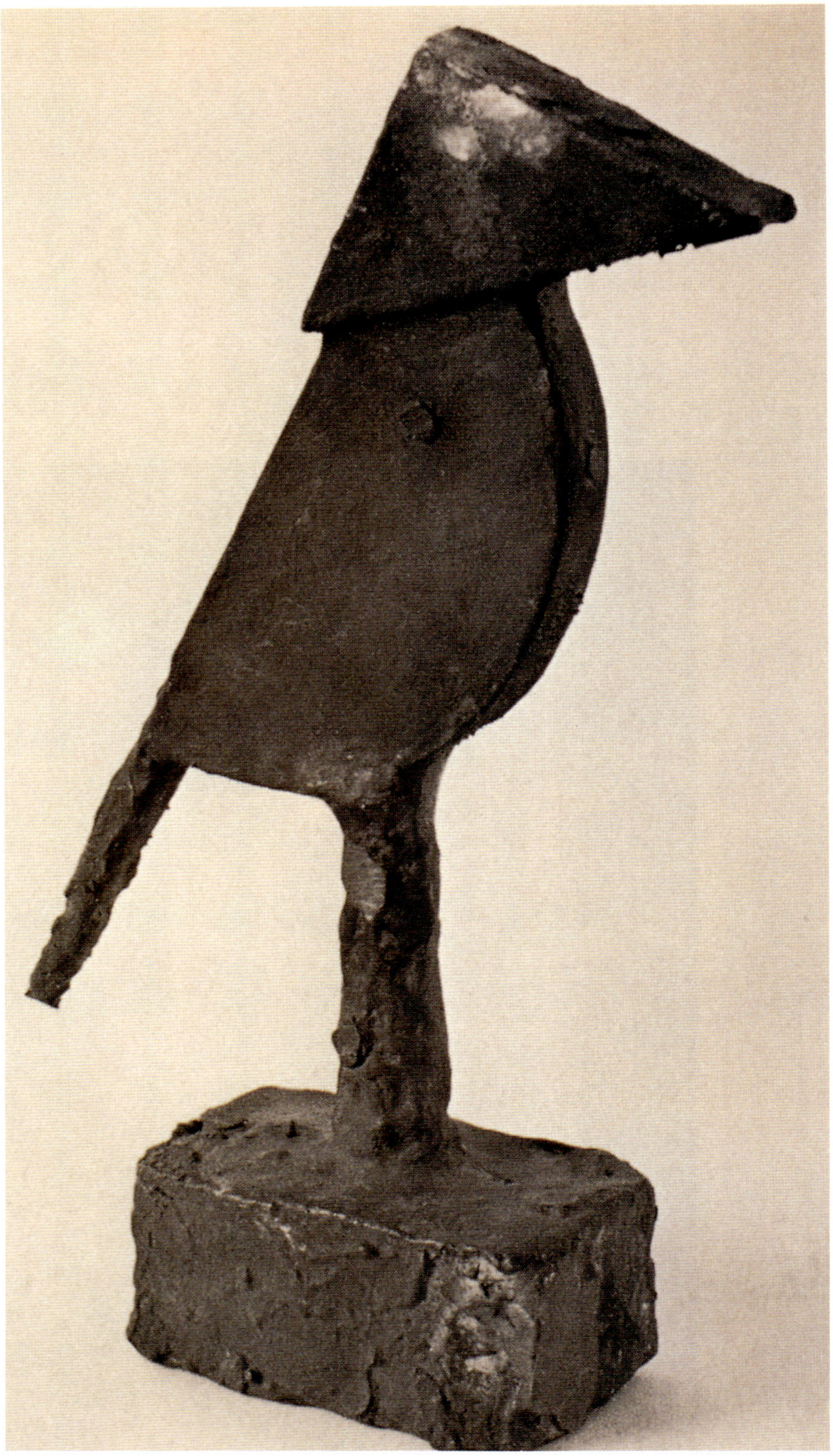

V-27. "Colored Bird" by Max Weber. 1960.

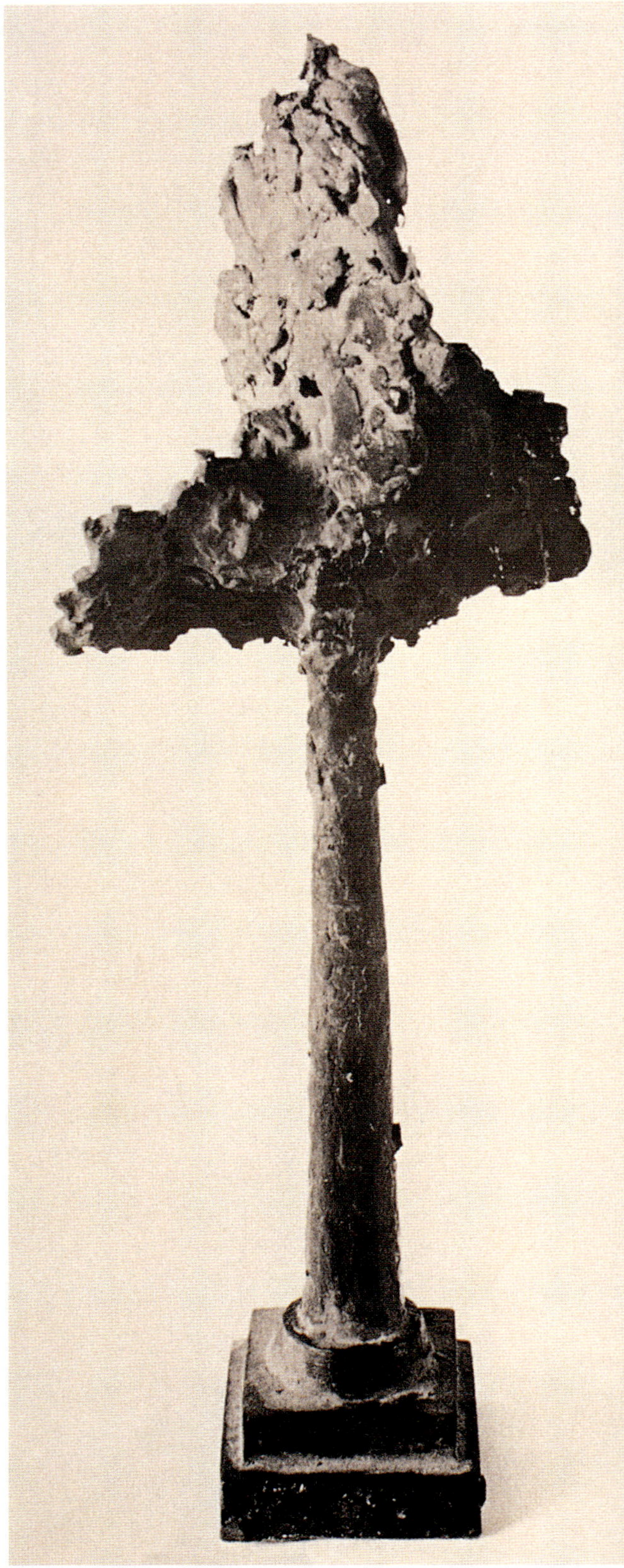

V-28. "Intersecting Profiles" by Max Weber. 1959.

ISAMU NOGUCHI
(1904-1988)

One of the twentieth century's most prodigious and independent abstract sculptors, Isamu Noguchi combined a Constructivist approach with familiar Japanese forms and a Surrealist bent. Noguchi is best known for his exceptional abstract shapes in which empty space, as well as form, plays a significant role. He saw sculpture as "creating space," and as "images of moods—moods of flowers, of the vegetative and non-vegetative aspects of nature." A master at combining sophisticated modernity with popular art forms, Noguchi was interested in assimilating art into the everyday world; in addition to his totem-like monumental sculpture, he also produced major environmental designs, gardens, furniture, lamps, playgrounds, and costumes and theatrical sets. His collaborations with Martha Graham and George Balanchine led the way to the interaction of arts so common today.

Isamu Noguchi was born in California, but spent his childhood in Japan. He was briefly apprenticed to Gutzon Borglum (who told him he would never become a sculptor, and was soon proved wrong); Noguchi's first one-man show was in 1924, only a few months after his arrival in New York. In 1927 he left for Europe where he met Alexander Calder, and became a studio assistant to Brancusi, the noted modernist sculptor. Back in New York in 1929 he began making portrait heads, but soon left for China and Japan to study Asian traditions in the arts and to explore the fusion of eastern and contemporary influences. His work has been variously described as influenced by traditional Japanese terra-cotta figures, the biomorphic forms of European Surrealists, and primitive art.

In 1935 he collaborated with Martha Graham, and began the first in a series of distinguished commissions that would include the gardens of UNESCO in Paris, New York's Chase Manhattan Bank Plaza, the Associated Press Building façade, and Hiroshima's Peace Park. He had retrospectives at the Whitney Museum in 1968 and in Japan in 1973.

At Roman Bronze Works, Noguchi cast in a variety of metals including aluminum, gold plate, and nickel, as well as bronze: a portrait head in 1933 of Count Waliefsky (V-29), "Miss Expanding Universe" (V-30) for Martha Graham in 1935, and several other works including "Death, Lynched Figure" (V-31) between 1933 and 1938.

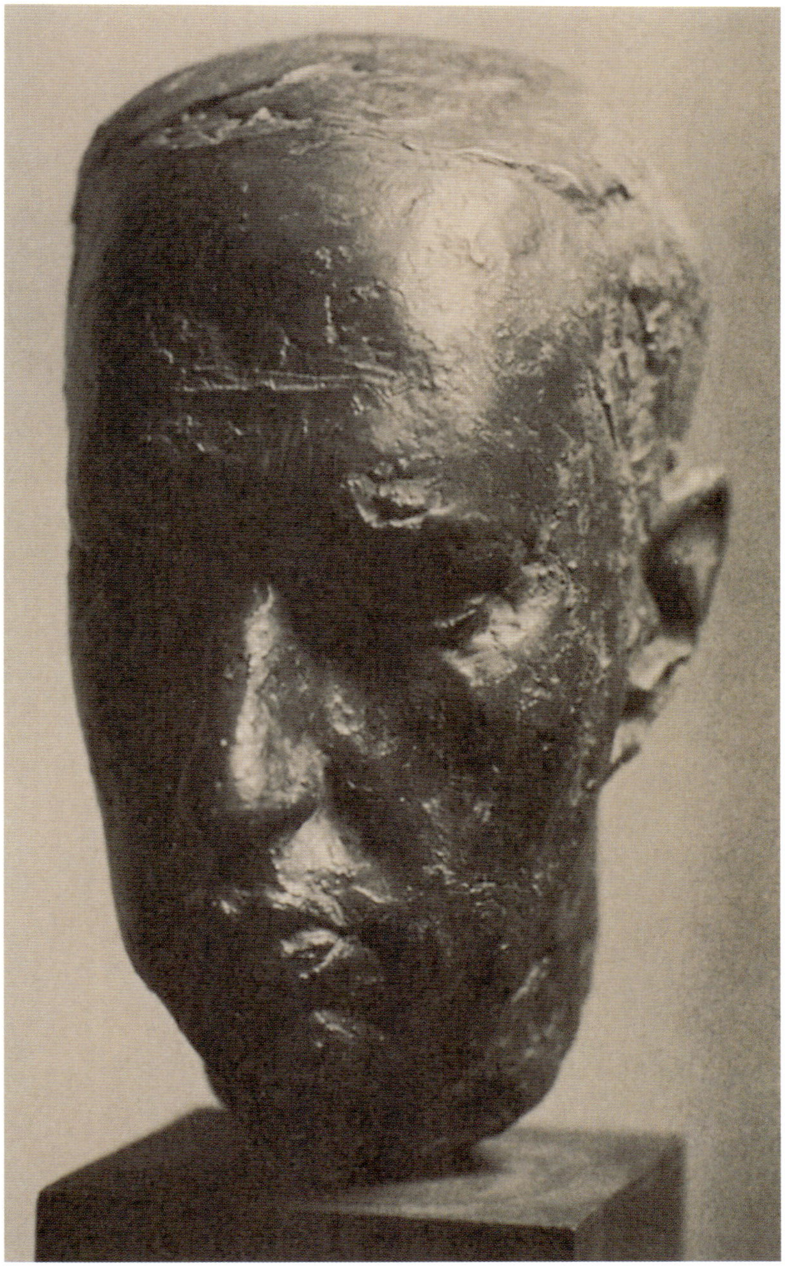

V-29. "Count Waliefsky" by Isamu Noguchi. 1929.

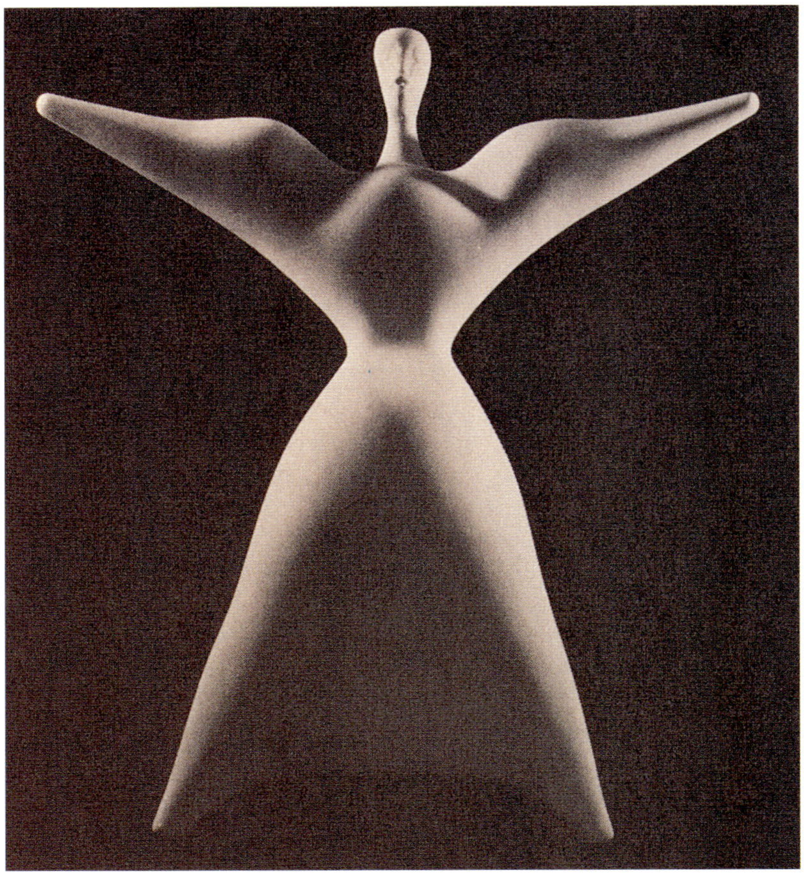

V-30. "Miss Expanding Universe" by Isamu Noguchi. 1932.

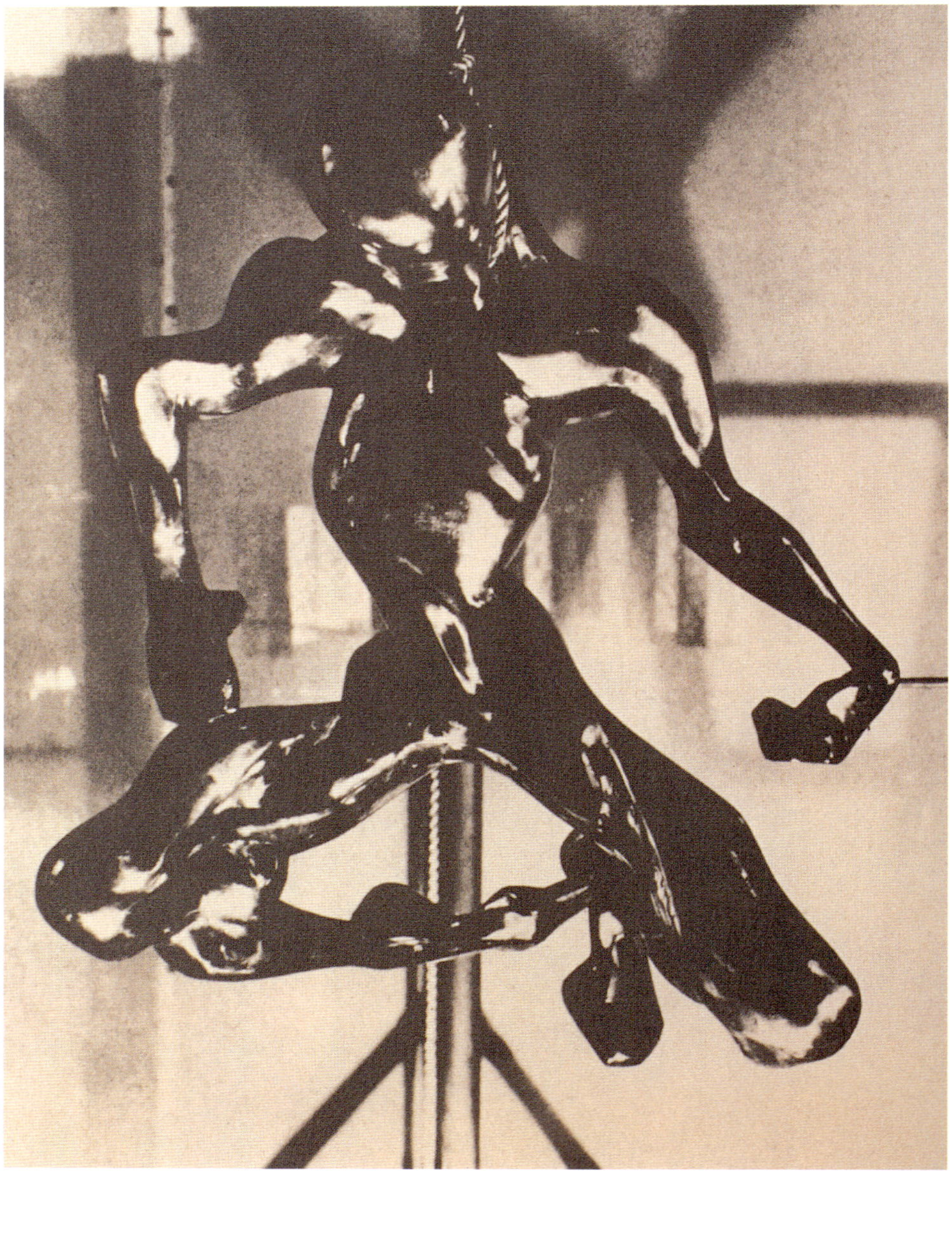

V-31. "Death (Lynched Figure)" by Isamu Noguchi. Monel metal. 1934.

ROBERT LAURENT
(1890-1970)

Robert Laurent was one of the earliest sculptors to experiment with abstraction in the United States. Born in France, he was brought as a child to America by the sculptor Hamilton Easter Field, who recognized his talent. In 1905 he returned to Paris to study, met many of the modernists, including Picasso, and first encountered Cubism and primitive art. He began carving directly in wood; he is credited with bringing the concept of direct carving to New York.

In 1915 he had his first show in New York and became a major influence on American avant-garde sculpture through his abstraction and stylization of natural form. He was influenced both by the French sculptor Maillol and by Gaston Lachaise; like them he found abstraction and monumentality in the female body's forms. Though his works were unpopular with more conservative critics and a 1932 commission for Rockefeller Center rejected (though eventually reinstated), Laurent achieved a strong reputation in the modernist art world, in particular for his large abstract nudes.

Laurent began working in bronze by the late 1920s. Among the commissions that he received were Radio City Music Hall in New York (1932), Fairmount Park in Philadelphia (1936), and the Federal Trade Commission Building in Washington (1938). Today Laurent's large nude statues (in bronze, wood, or stone) can be seen in most major sculpture collections. His work "The Awakening" (V-32) (seen here in a plaster sketch model from R.B.W.) was cast in 1935, as was his "Kneeling Figure" (V-33). He also cast an aluminum torso at the foundry and several other works between 1928 and 1938.

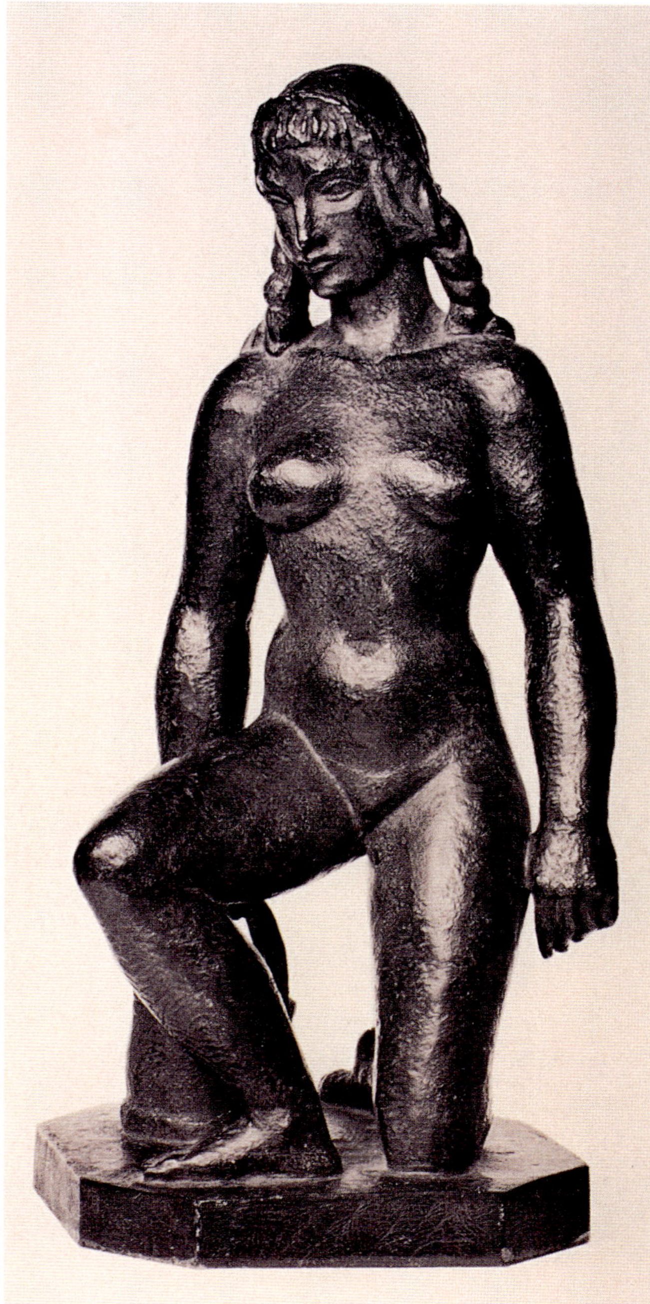

V-33. "Kneeling Figure" by Robert Laurent. 1935.

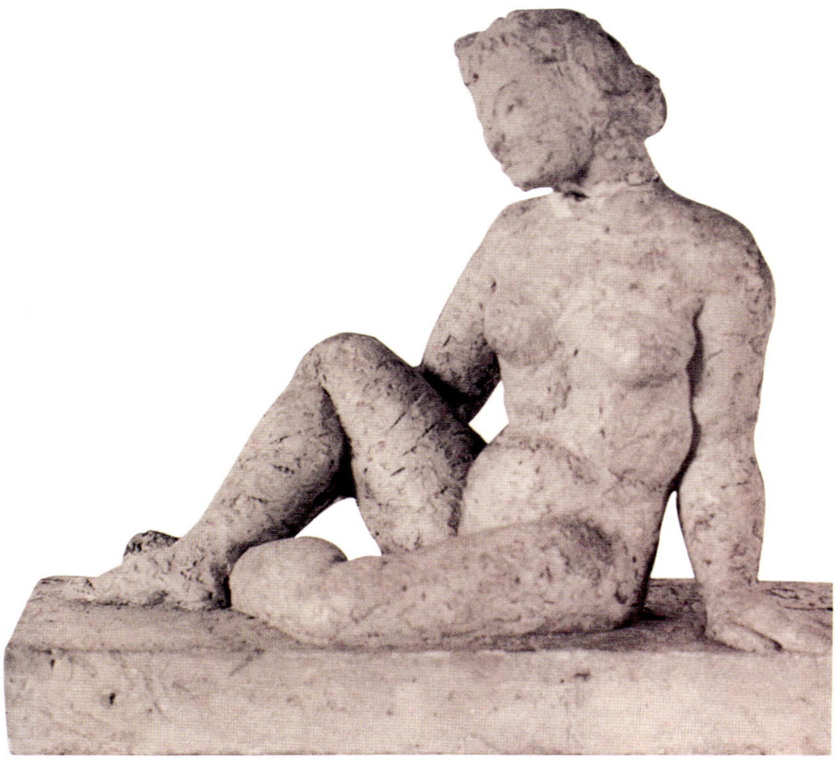

V-32. "The Awakening" (sketch model for casting) by Robert Laurent. 1935.

PAUL MANSHIP (1885-1966)

Paul Manship was the best known sculptor in America in his time. He was considered sufficiently "modern" to appeal to those who felt the old Beaux-Arts tradition was passé, but his works were grounded in traditional classicism. Though his subjects came from nature, his work was stylized and rhythmic and brought a new creative energy and elegance to familiar sculptural forms. Though he drew upon archaic Greek art rather than experimenting with pure form and space like the Cubists and Constructivists, Manship was able to create his own contemporary—but non-threatening—style that suited American tastes in the 1920s and 1930s. Its streamlined form seemed ideal for its era; it was exceptional in its craftsmanship and finish, it had a touch of primitive, as well as its classical influence, and it was ideal for public spaces. His style was quickly recognized as the "official art" of its time.

Born in St. Paul, Minnesota, Manship studied at the Art Institute there and in 1905 went to Philadelphia and then New York; he worked with several sculptors, including Gutzon Borglum, Charles Grafly, and Isidore Konti. In 1909 he won the coveted Prix de Rome, and it was in Italy and Greece that he became strongly influenced by ancient art, in particular, archaic Greek sculpture. When he returned to America in 1912, he immediately received recognition for his bronze works with their classical themes but contemporary look. In 1917 his first major patron, Albert Gallatin, admired Manship's "fire and vigor," saying his sculpture "struck a purely modern note." (Only a few years later, the same patron turned to more abstract art and artists.)

Among Manship's many classical subjects (often made with stylistic reference to archaic Greek art), was his best known work "Prometheus" (V-34) for Rockefeller Center in New York (where he also made many other smaller statues (V-35) for the complex). Other classically influenced images include a "Lyric Muse" (V-36), a "Nymph and a Satyr" (V-37), a "Centaur" (V-38), a "Hercules" (V-39), and a "Pegasus" (V-40). Even his figure called "Duck Girl" (V-41) has a classical overtones. (It was modeled after a visit to Pompeii.)

His interest in graceful design and the human body was evident in his many strikingly rhythmic works, some combining human and animal forms, such as his "Europa and the Bull (V-42), "Indian and Pronghorn" (V-43), and his well known "Dancer and Gazelles" (V-44). He frequently included flowing drapery in his studies of women, as in the rhythmic "Flight of Night" (V-45), "Salome" (V-46), and his "Briseis" (V-47). Other sculptures of the human form included "Wrestlers" (V-48), "Reclining Nude" (V-49), and "Spring Awakening" (V-50). Thematic references to motherhood and family appear in his "Little Brother" (V-51), "Mother and Child" (V-52) and his "Armillary Sphere (V-53), with its family within the great globe. More traditional works include his equestrian of General Washington (V-54) and his statuette of "David" (V-55). "The Tortoise" (V-56) is one of Manship's best loved works; it was commissioned for the Bronx Zoo.

Manship received numerous commissions, honors and awards. He had studios in New York, Paris, and Rome, where he taught at the American Academy. His fame was international; he was the only non-British artist of his time honored with an exhibition at England's Tate Gallery. For most of his life Manship was the most sought-after sculptor in America, but by the late 1940s his work was increasingly seen as academic and old fashioned, as the ideas of abstract "modern" art finally took hold.

Manship had a close association with Roman Bronze Works, beginning in 1915 and continuing to the end of his life. He actually lived with the owners of the foundry for some time, and cast a great number of pieces at R.B.W., including all of those mentioned above.

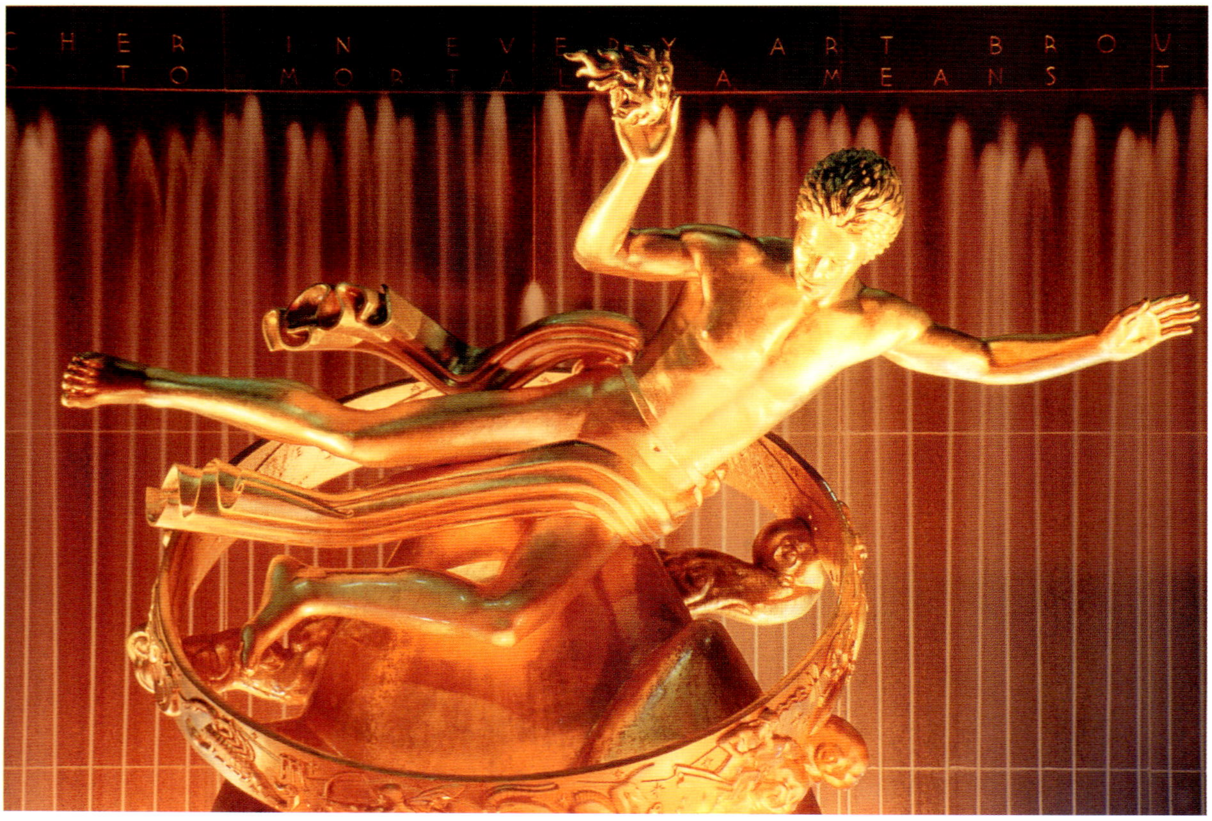

V-34. "Prometheus" by Paul Manship. 1934.

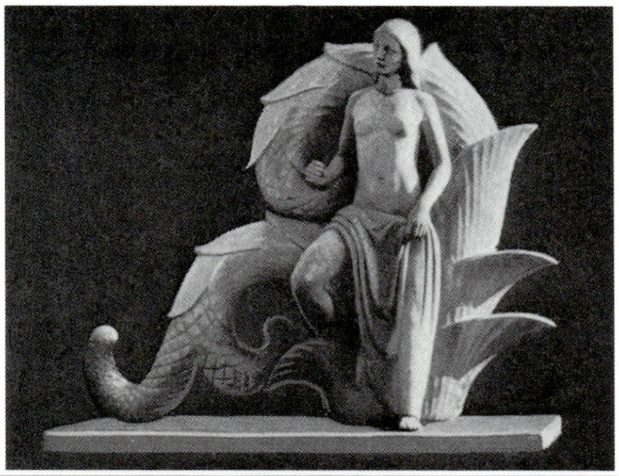

V-35. Garden sculptures for Rockefeller Center by Paul Manship, as pictured by the General Bronze Corporation. 1934.

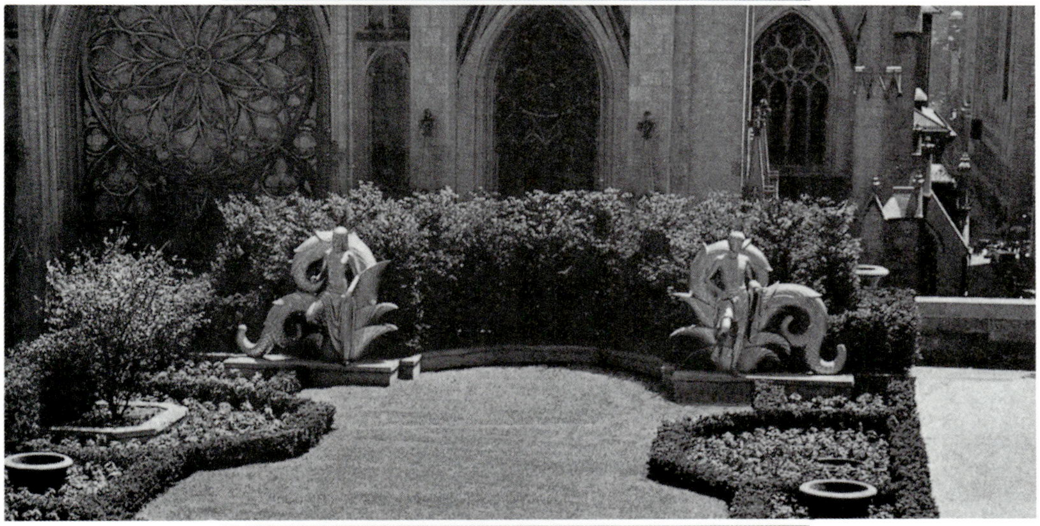

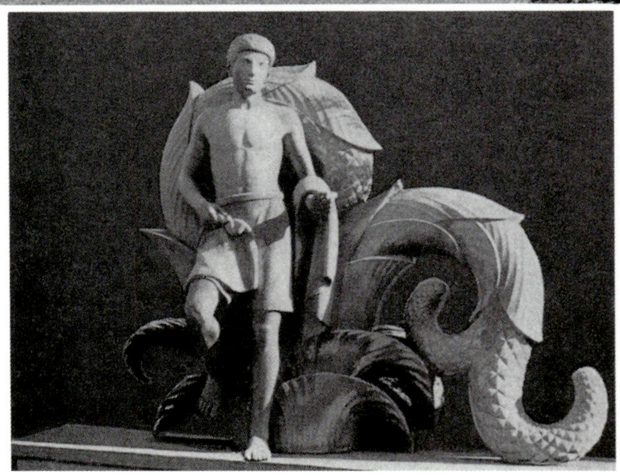

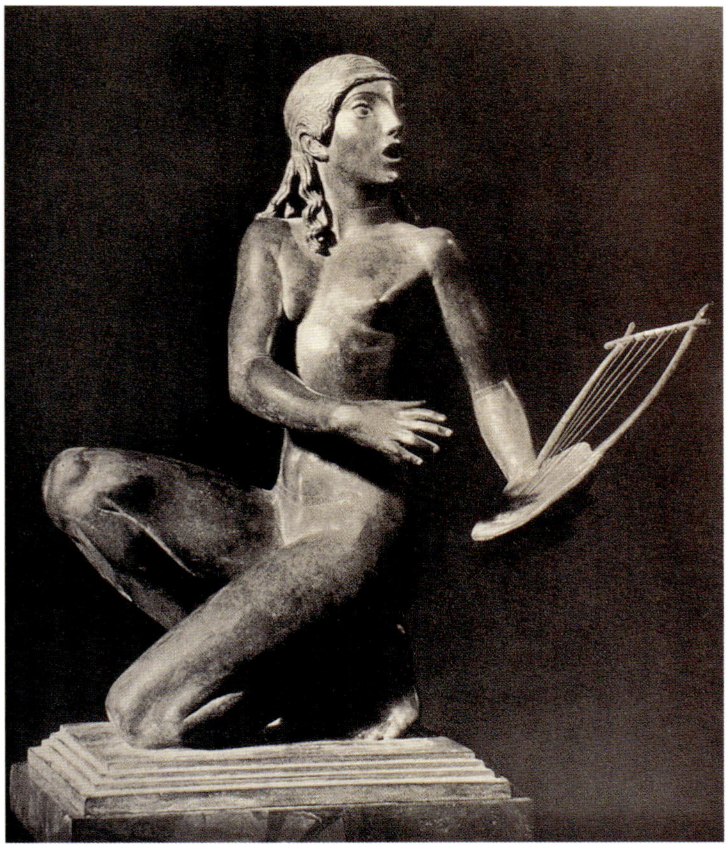

V-36. "Lyric Muse" by Paul Manship. 1912.

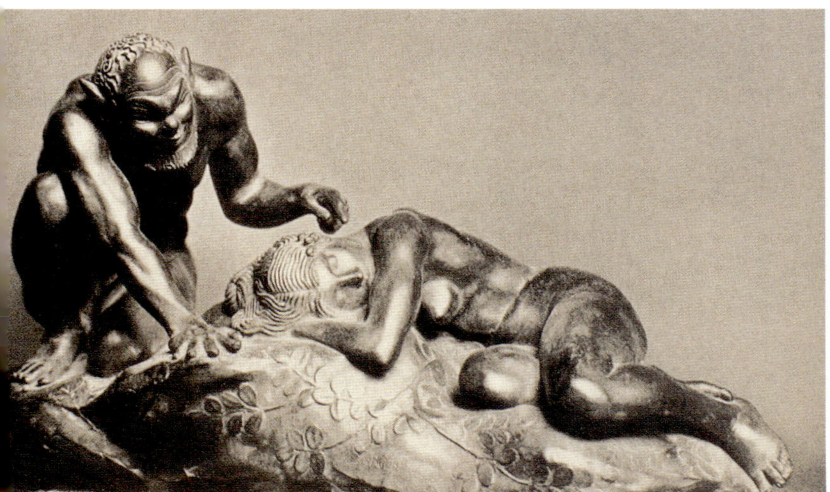

V-37. "Satyr and Sleeping Nymph" by Paul Manship. 1912. (First cast in Rome.)

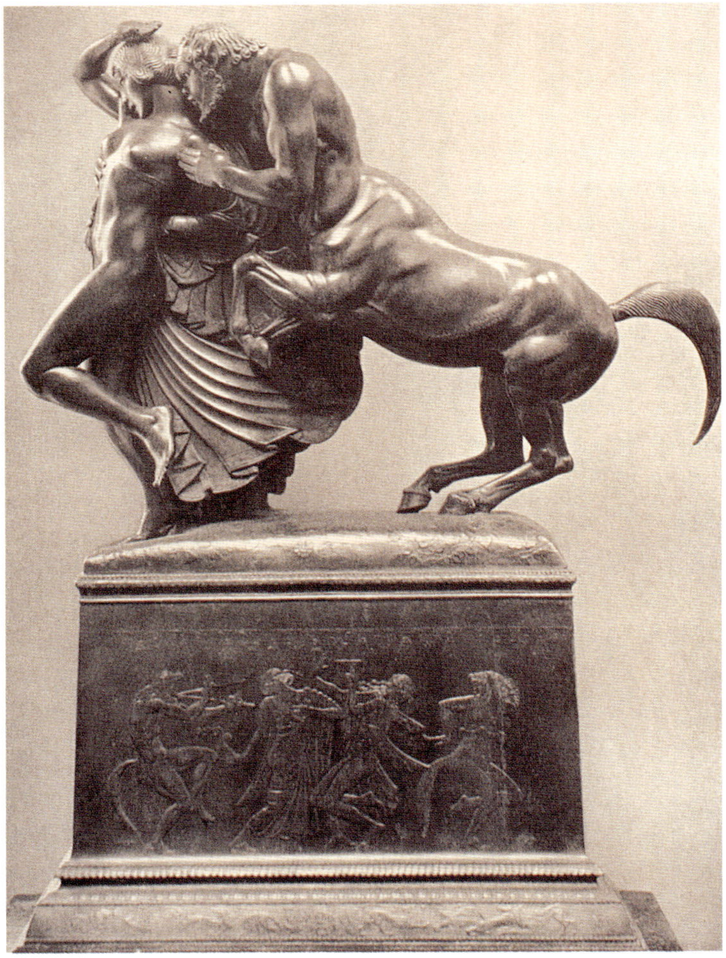

V-38. "Centaur and Dryad" by Paul Manship. 1913.

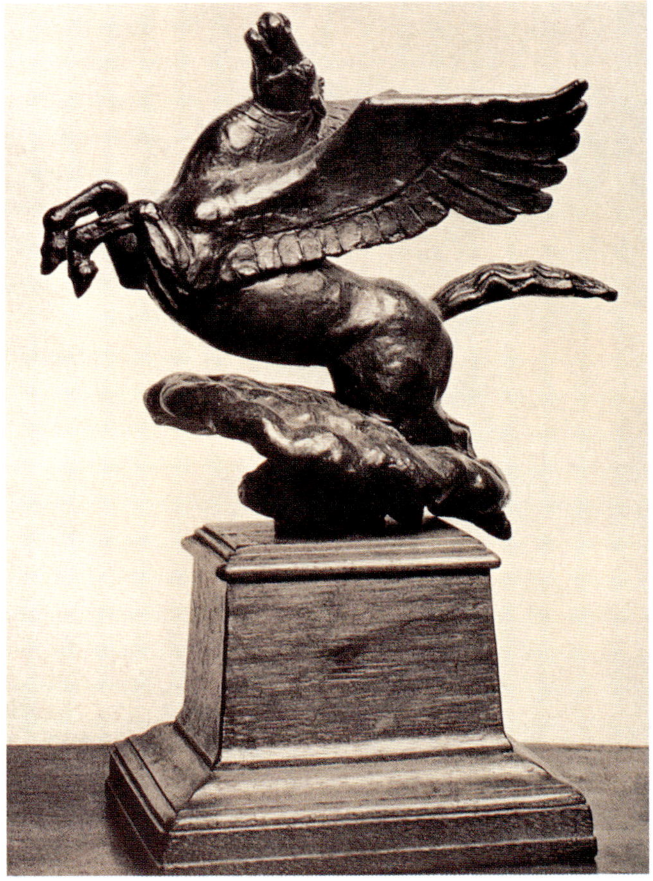

V-40. "Pegasus" by Paul Manship. 1936.

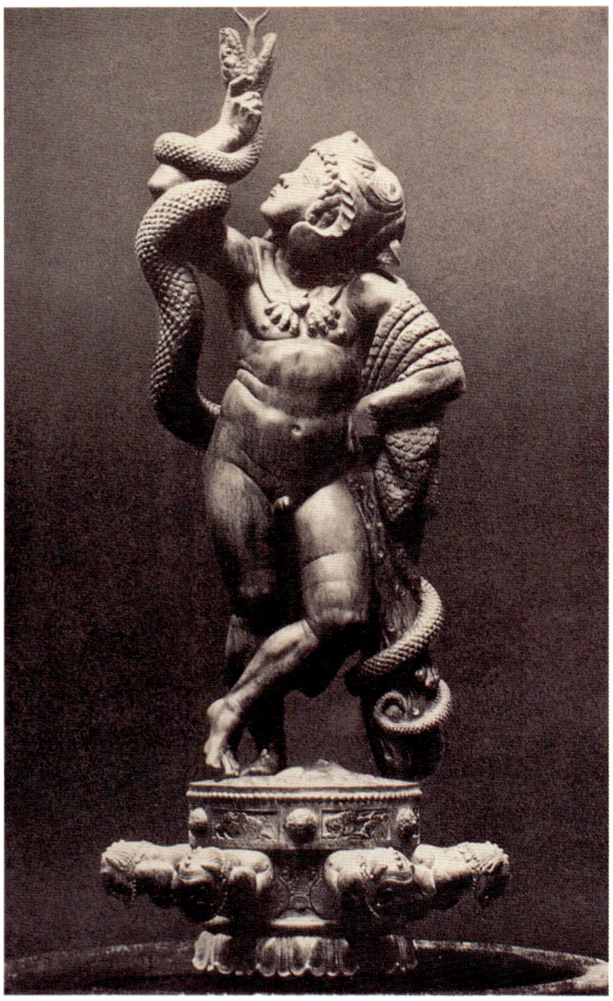

V-39. "Infant Hercules Fountain" by Paul Manship. 1914.

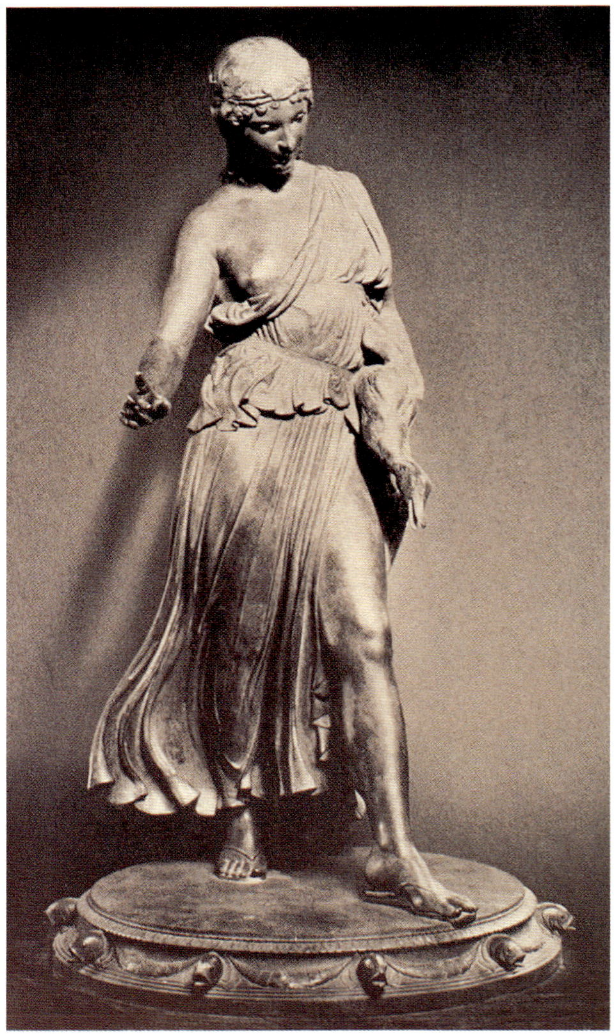

V-41. "Duck Girl" by Paul Manship. 1911. First cast in Rome.

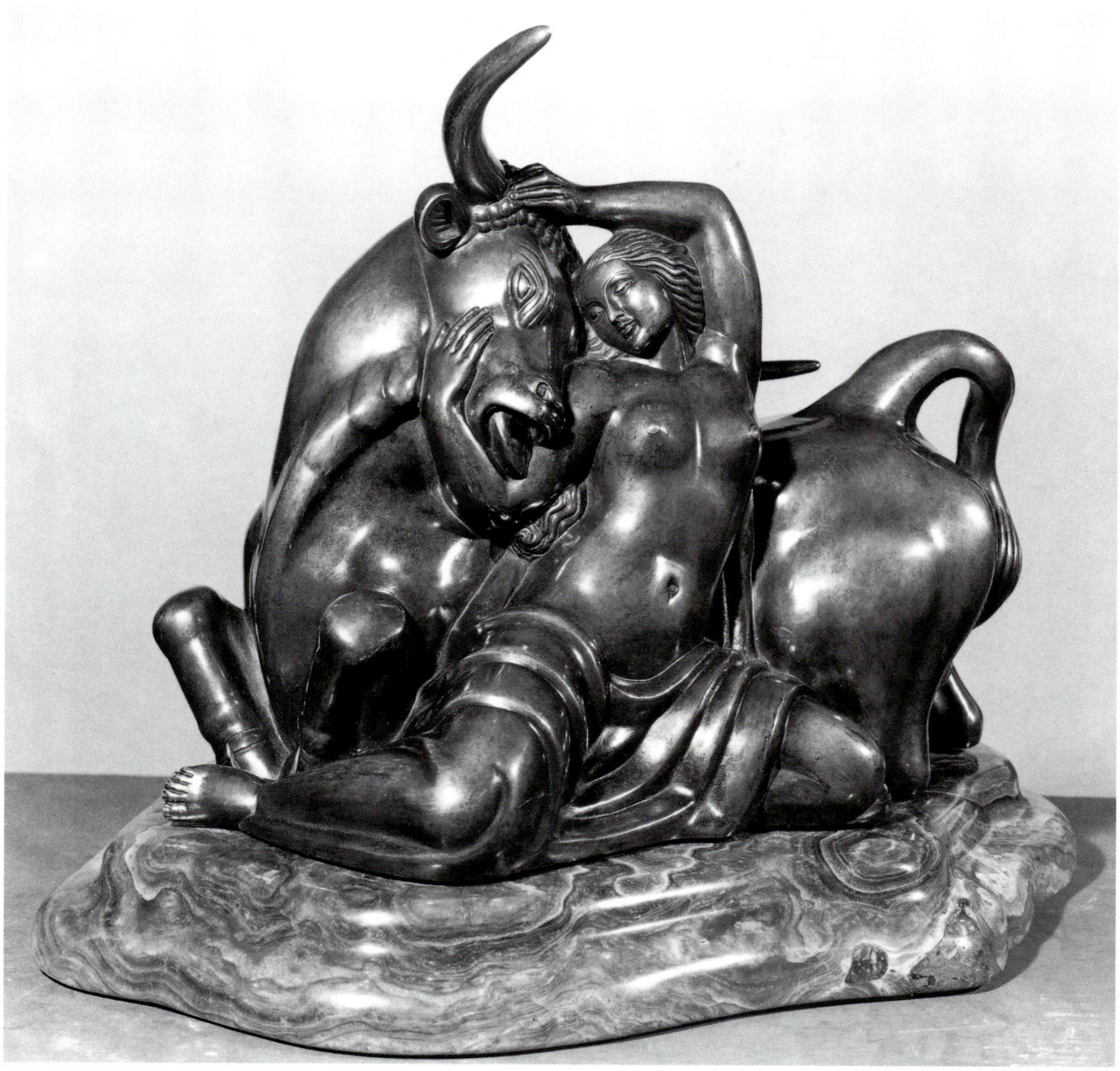

V-42. "Europa and the Bull" by Paul Manship. 1922-24.

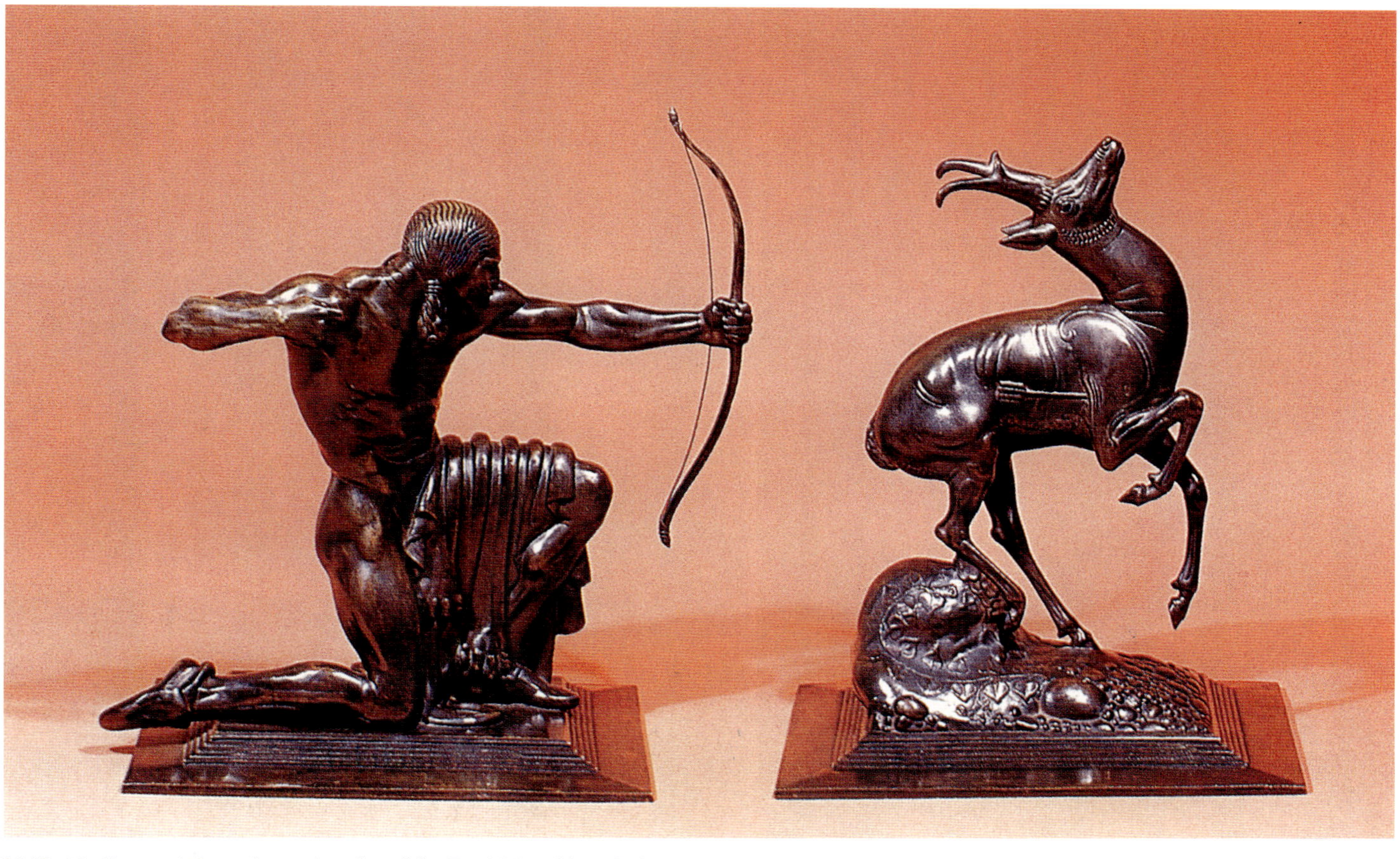

V-43. "Indian and Pronghorn Antelope" by Paul Manship. 1914.

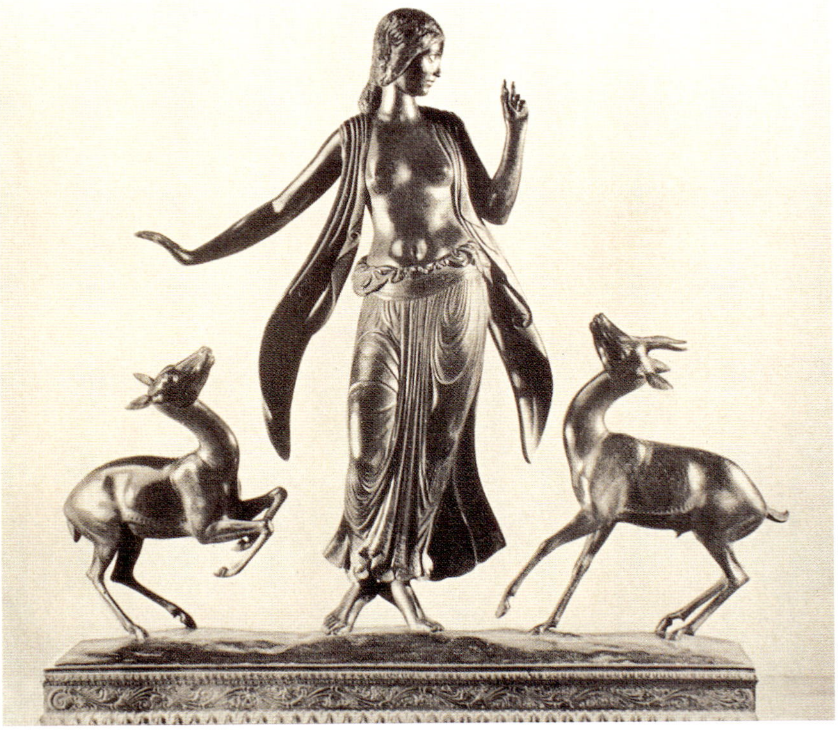

V-44. "Dancer and Gazelles" by Paul Manship. 1916.

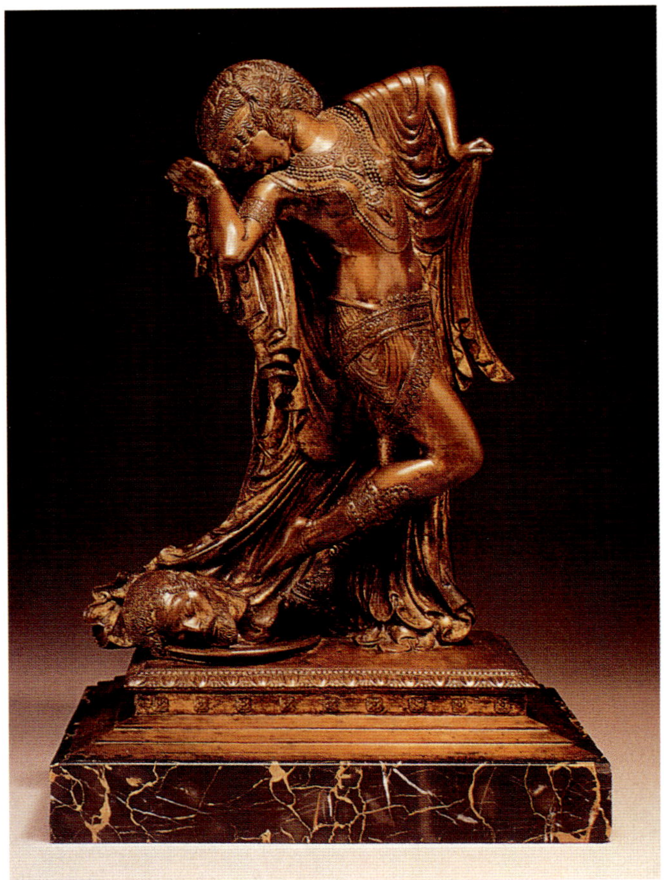

V-46. "Salome" by Paul Manship. 1915.

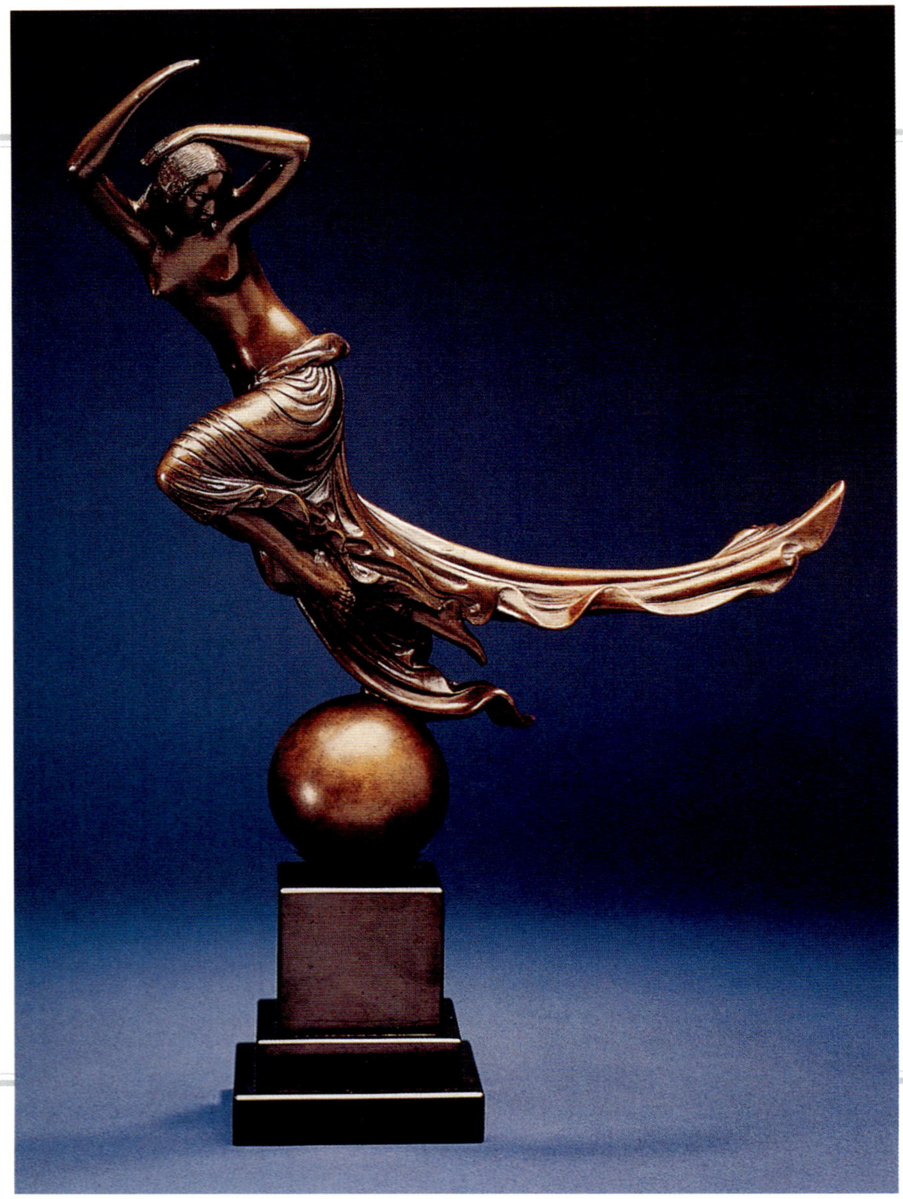

V-45. "Flight of Night" by Paul Manship. 1916.

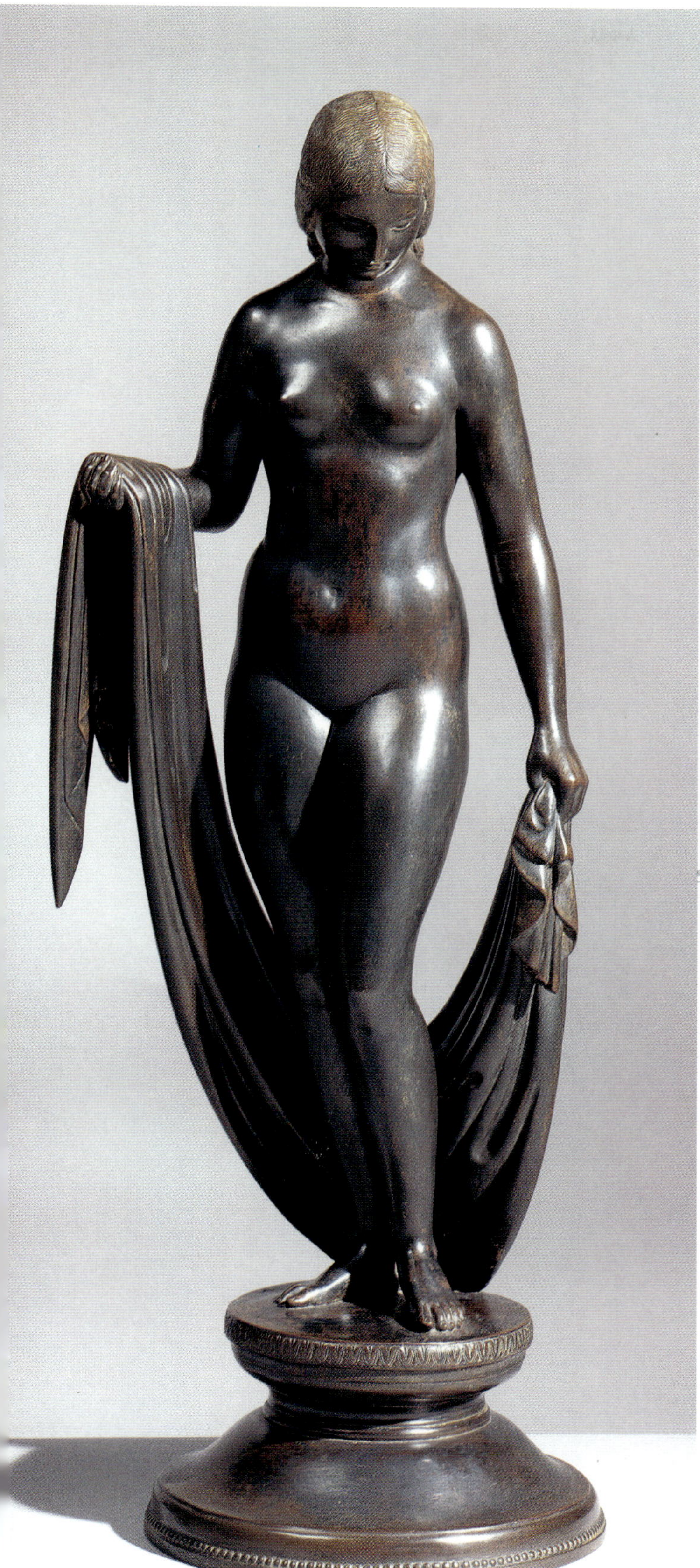

V-47. "Briseis" by Paul Manship. 1916.

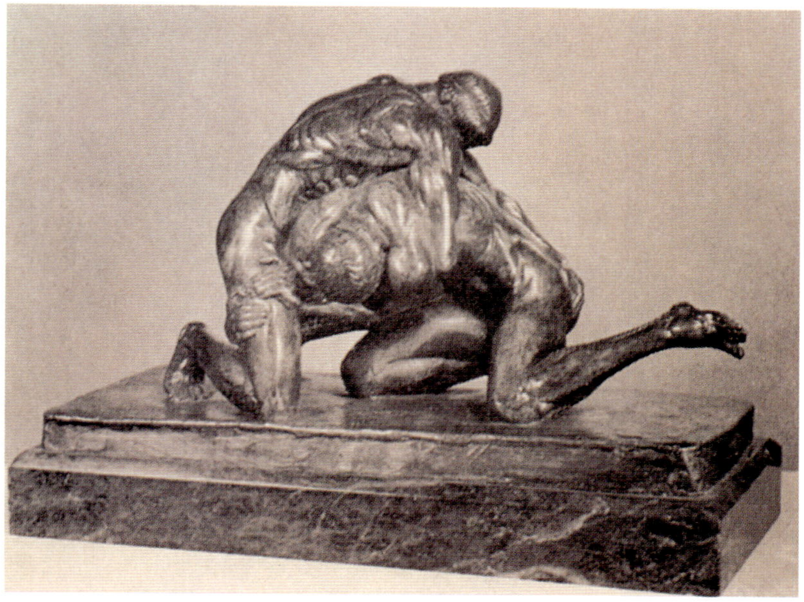

V-48. "Wrestlers" by Paul Manship. 1915.

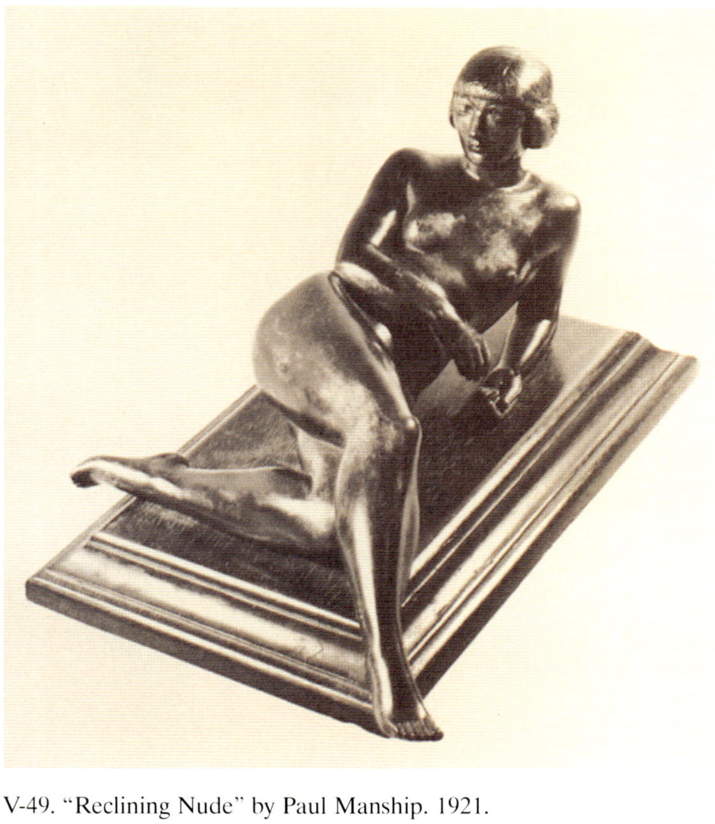

V-49. "Reclining Nude" by Paul Manship. 1921.

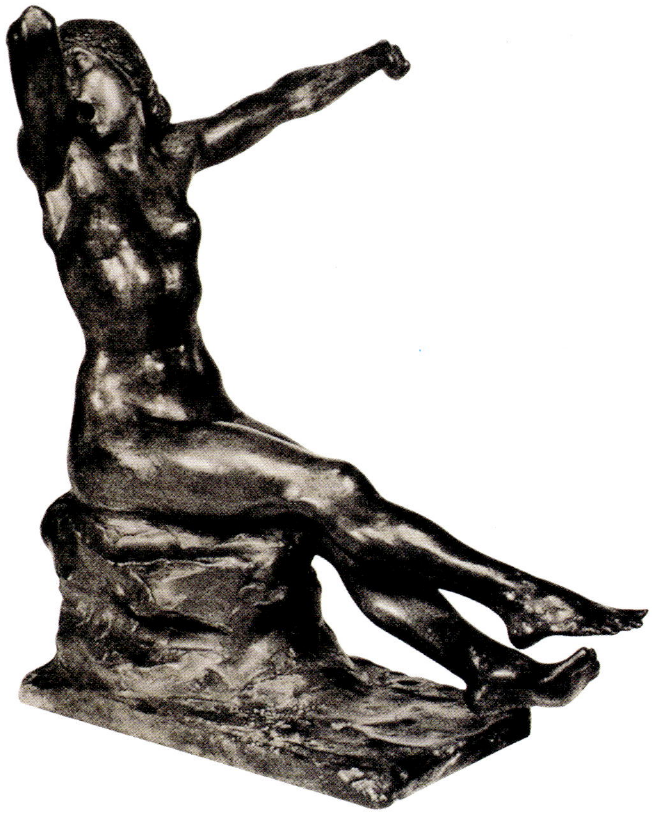

V-50. "Spring Awakening" by Paul Manship. 1912.

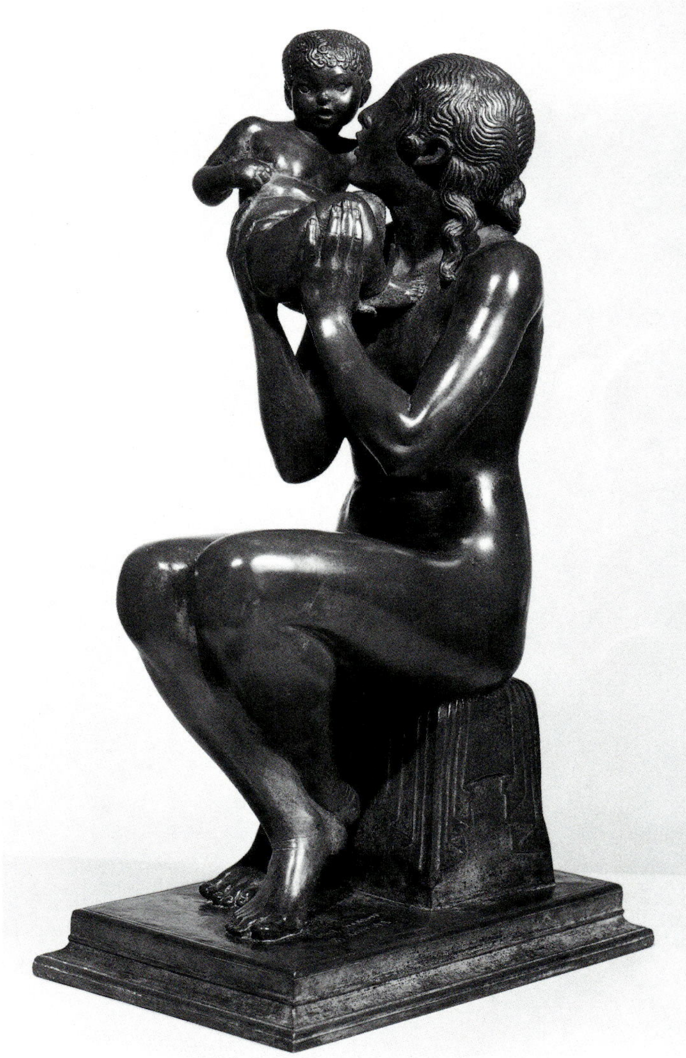

V-51. "Little Brother" by Paul Manship. 1912-1917. First cast in Rome.

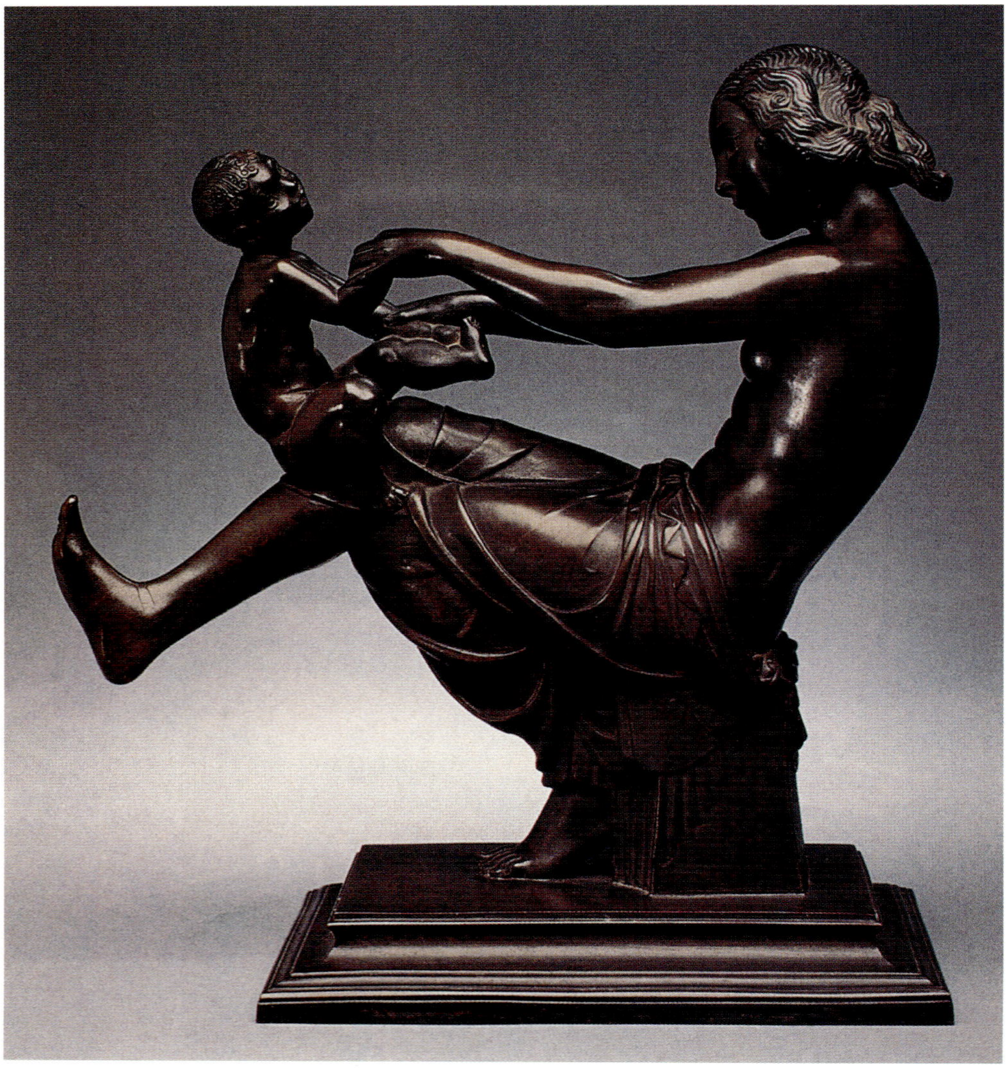

V-52. "Playfulness" by Paul Manship. 1912. First cast in Rome.

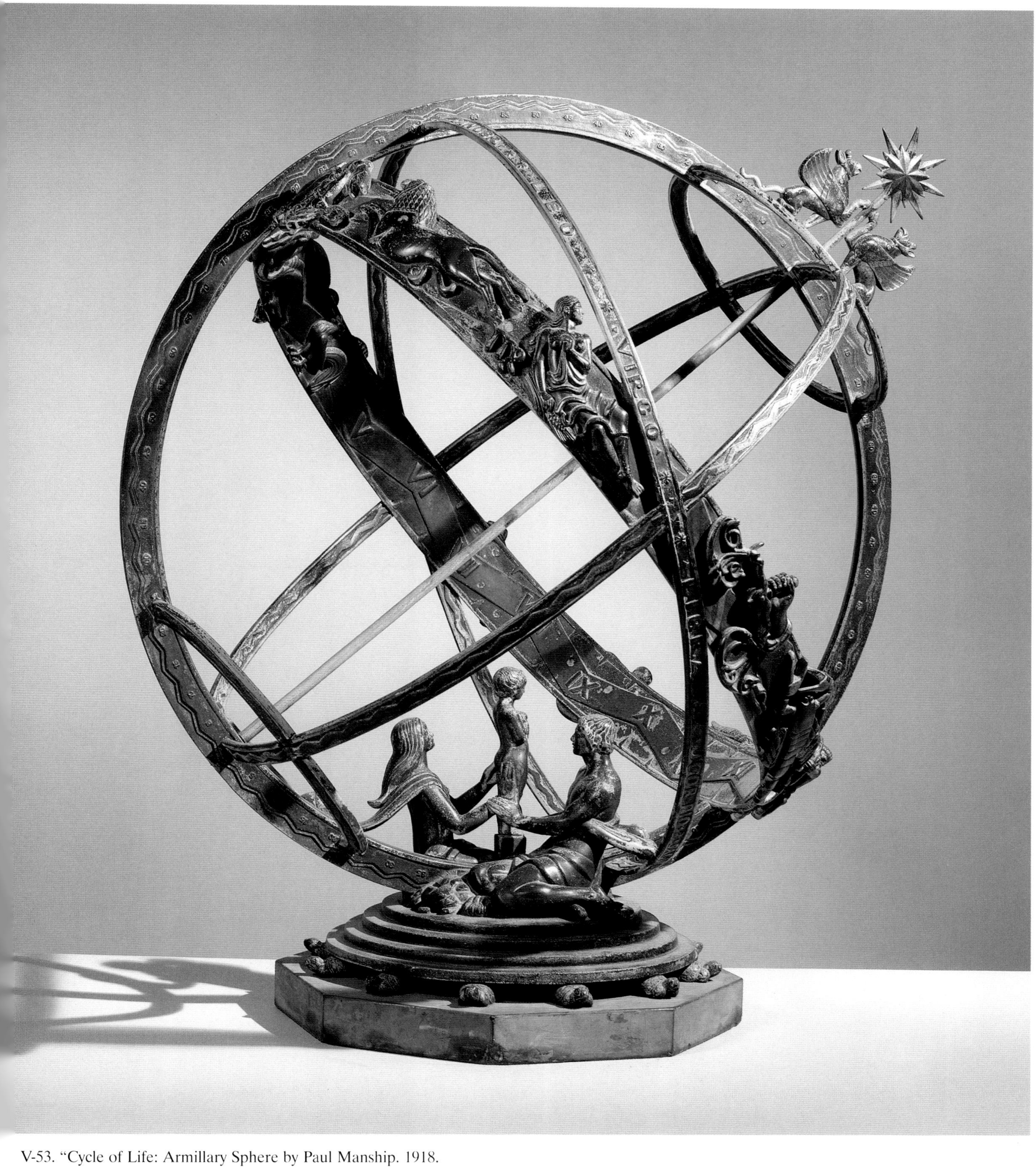

V-53. "Cycle of Life: Armillary Sphere by Paul Manship. 1918.

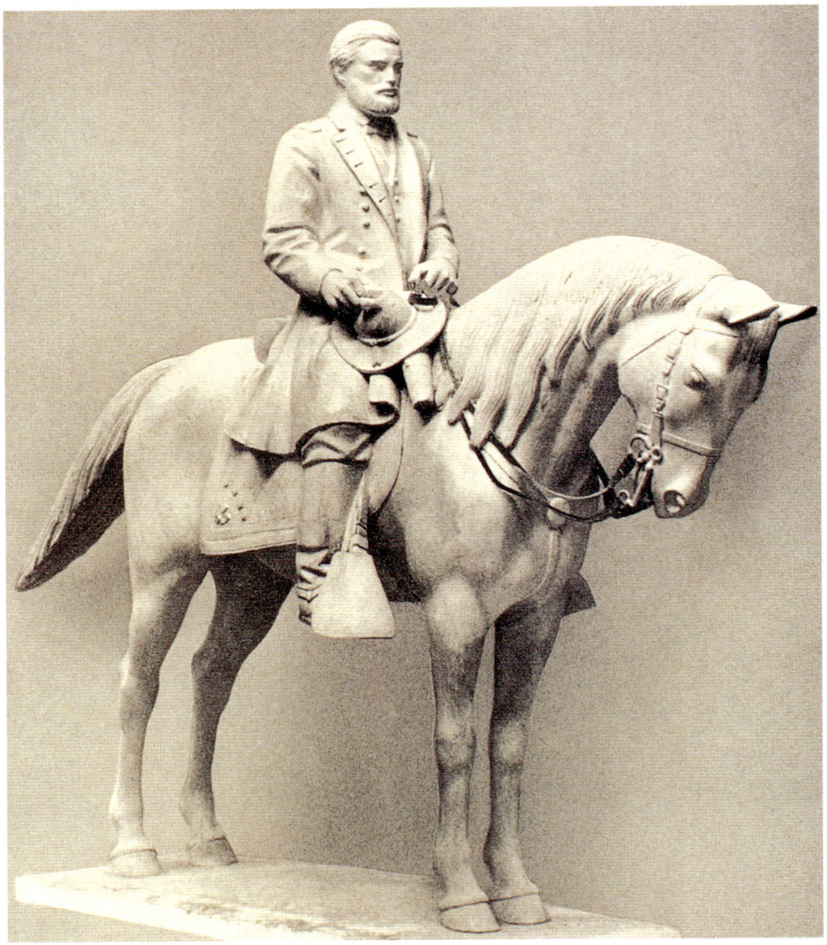

V-54. "General Grant" by Paul Manship. 1928. This was a small study for a proposed monument at Grant's Tomb, but the larger work was never made.

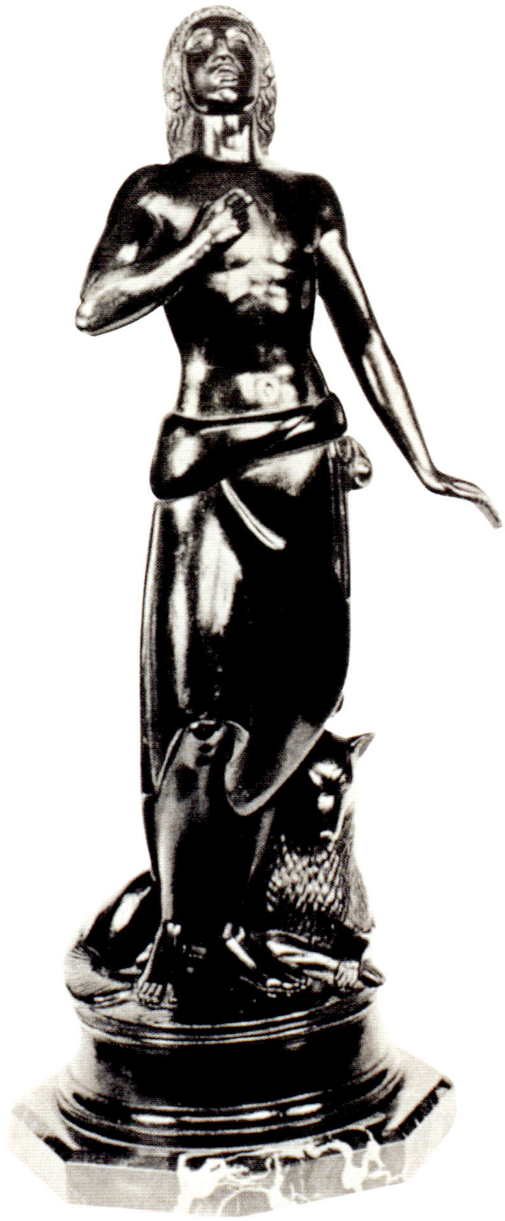

V-55. "David" by Paul Manship. 1921.

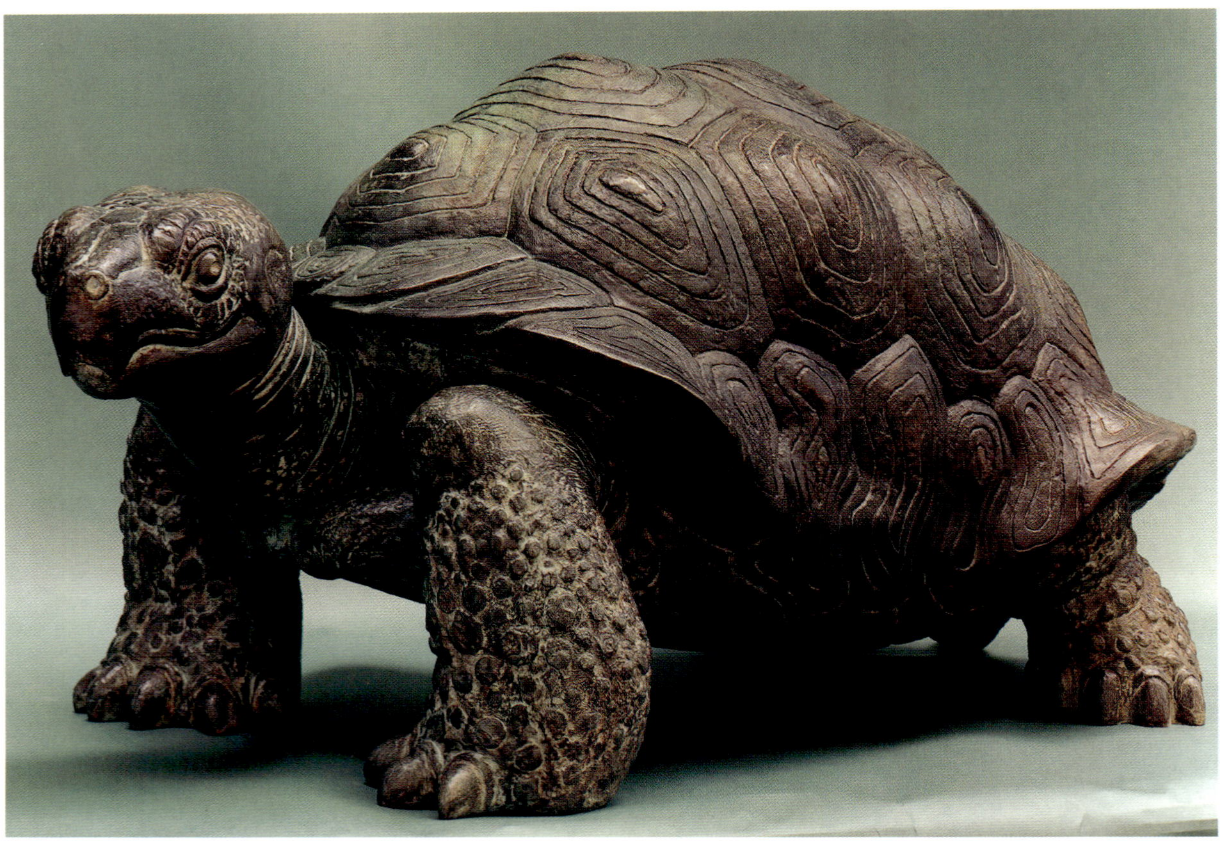

V-56. "Tortoise" by Paul Manship. Original plaster modeled in 1916.

LEE LAWRIE
(1877-1963)

Lee Lawrie is best known for his major New York landmark: the "Atlas" at Rockefeller Center, and for the semi-abstract architectural decorations he made for buildings around the country. Lawrie was considered a deft stylist who combined a somewhat Gothic style with an interest in abstract ideas and forms.

Lawrie was born in Germany and brought to the United States as an infant. He was raised in Chicago. From 1894 to 1906 he worked in the studio of Augustus Saint-Gaudens, as well as for several other sculptors. He received a degree from Yale University in 1910 and was a teacher there from 1908 to 1918.

His career as a maker of architectural decorations and sculpture for buildings was furthered by his friendship with architect Richard Goodhue; one such commission was "The Sower" (V-57) for the Nebraska State Capital. He was an innovator of the flying buttress in decorative architecture. In New York he made the reredos for St. Thomas' Church, and was a sculpture consultant for the World's Fair of 1939.

His sculpture was grand in conception. His "Four Horsemen of the Apocalypse" (V-58) of 1923 is one of his most interesting works; in it he attempted to abstract the apocalyptic idea as the riders and their horses degenerate from warriors to imminent destruction.

But Lawrie's most recognizable contribution is his heroic bronze "Atlas;" (V-59 and V-60 in process); it was one of many pieces of sculpture cast at Roman Bronze Works for Rockefeller Center. His sketches for the Center (V-61) show his involvement in planning the Center's sculptural aspects. Lawrie cast bronze works at R.B.W. between 1908 and 1935.

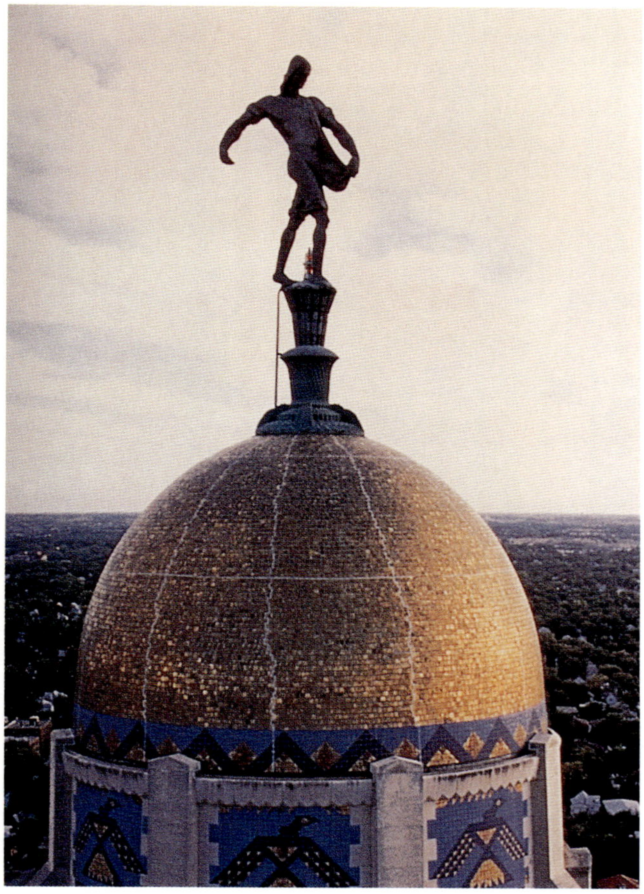

V-57. "The Sower" at the Nebraska State Capitol, by Lee Lawrie. 1933.

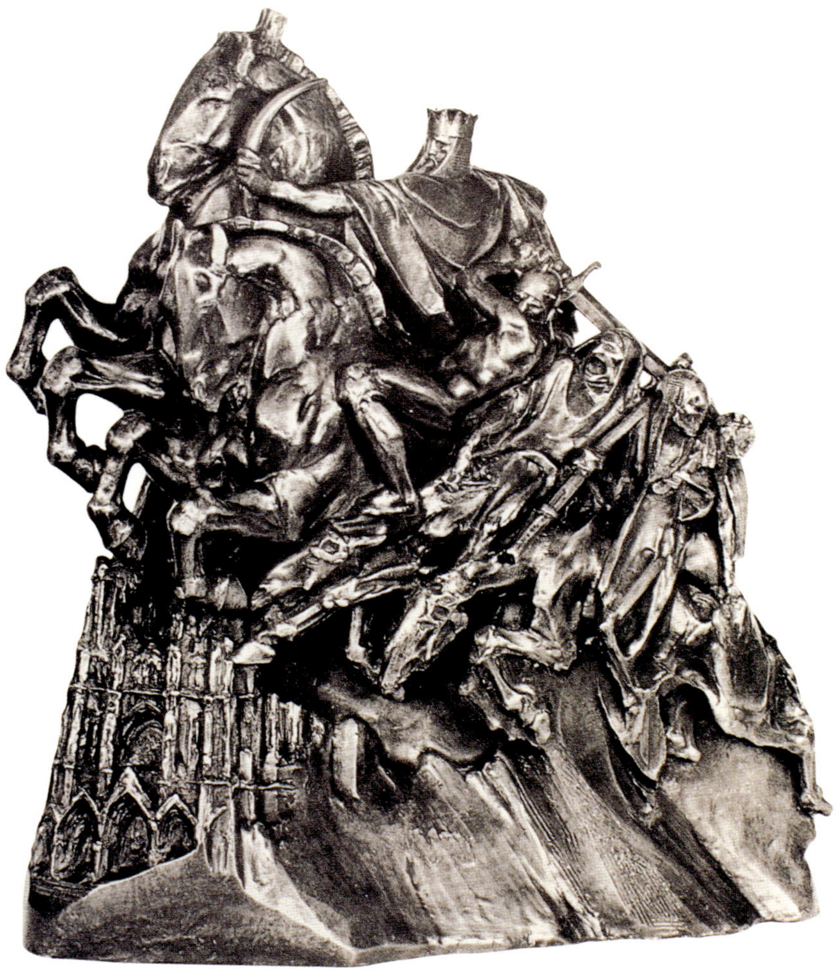

V-58. "The Four Horsemen" by Lee Lawrie. 1923.

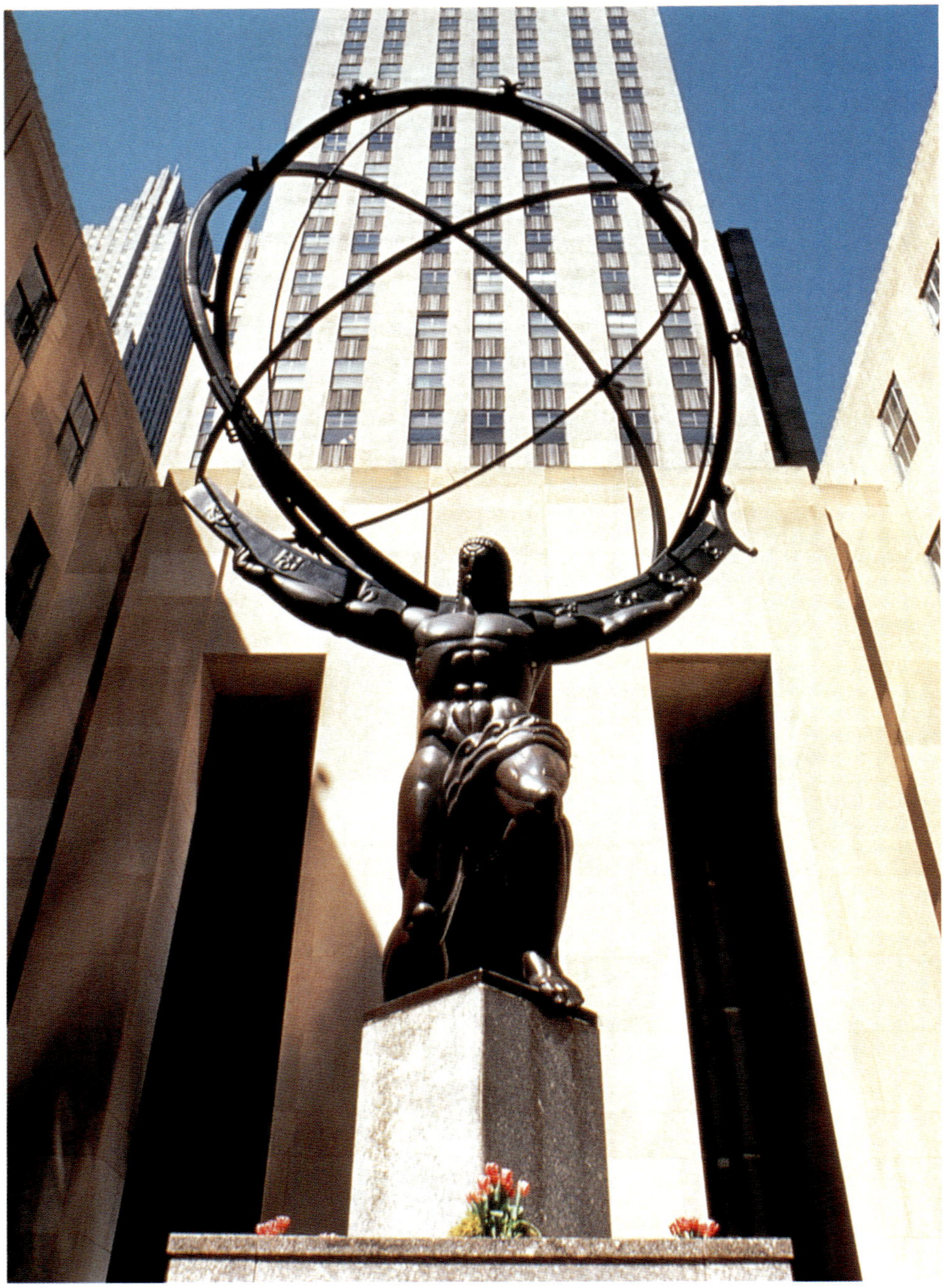

V-59. "Atlas" by Lee Lawrie. Erected 1937.

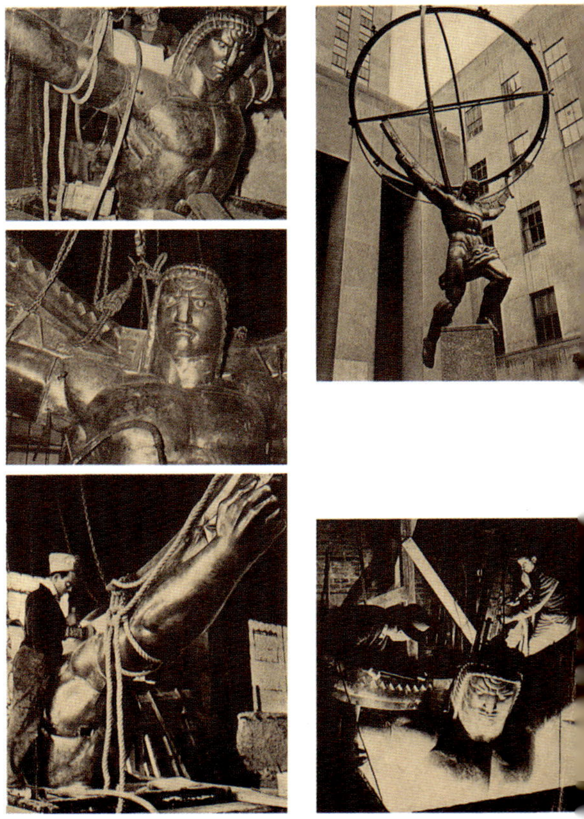

V-60. The process of placing "Atlas" by Lee Lawrie at Rockefeller Center, 1937.

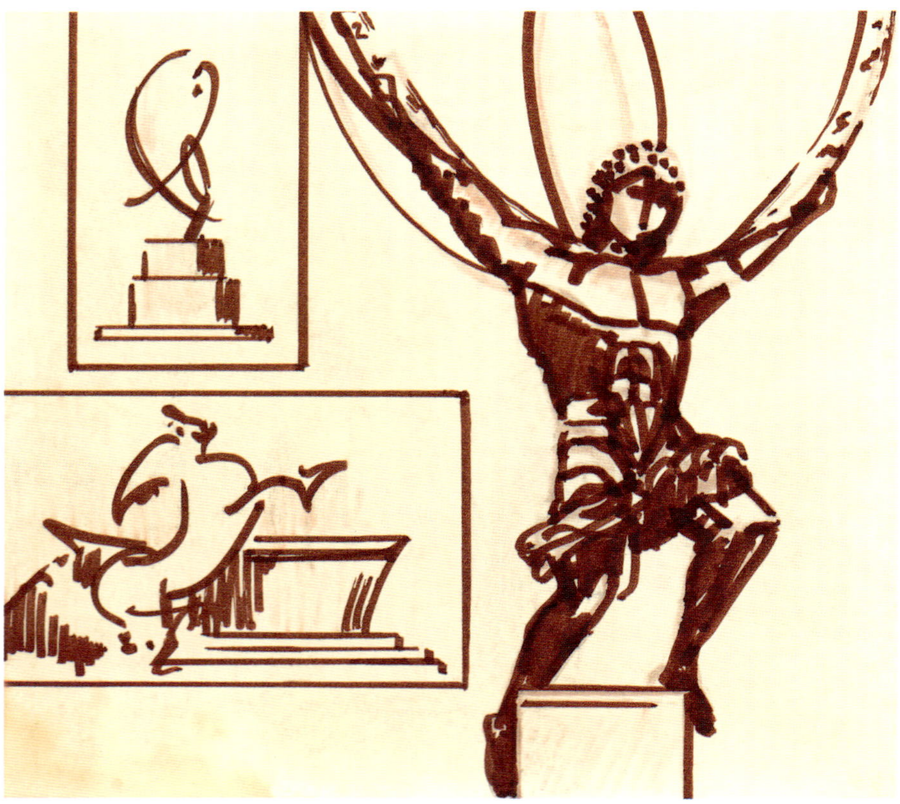

V-61. Drawings by Lee Lawrie for Rockefeller Center's sculpture.

WILLIAM ZORACH (1887-1966)

William Zorach's sculpture is noted for a restrained, dignified monumentality. While he began his career influenced by Cubism, his mature style was more traditional, though he continued to abstract and simplify natural forms. Zorach's emphasis was on figure sculpture, for the most part carved directly in stone, (suggesting ancient Egyptian and archaic Greek sculpture). A number of his works were also cast in bronze.

Zorach was born in Lithuania, arriving in the United States in 1891. He was a student at the Cleveland Art Institute before attending the National Academy of Design in New York. His first trip to Paris in 1910 introduced him to avant-garde art and he began his career as a painter (he exhibited his painting at the Armory Show of 1913). In 1917 he began carving directly in wood and by 1924 was carving in stone.

During the 1930s Zorach was a leader of the modernist movement in American art, but his work was more traditional both in theme (the human body, motherhood, children and animals) and in concept than that of many of his contemporaries. He began teaching at the Art Student's League in 1929 and continued for thirty years. His monumental works brought him a number of public commissions, including the New York World's Fair of 1929; he was awarded a show at the Whitney Museum in 1959, and was elected to the National Academy of Arts and Letters.

In 1932 his now well known heroic nude figure "Spirit of the Dance " (V-62) was commissioned and rejected by Radio City Music Hall; after a public outcry it was accepted. His "Mother and Child" (V-63) was a recurrent theme in Zorach's work. Other sculptures, including a head of St. John the Baptist, a baseball player, and a "Seated Figure" (V-64) were among the sculptures cast at Roman Bronze Works beginning in 1928. In 1934 he cast "A Kneeling Girl with a Cat" in aluminum there.

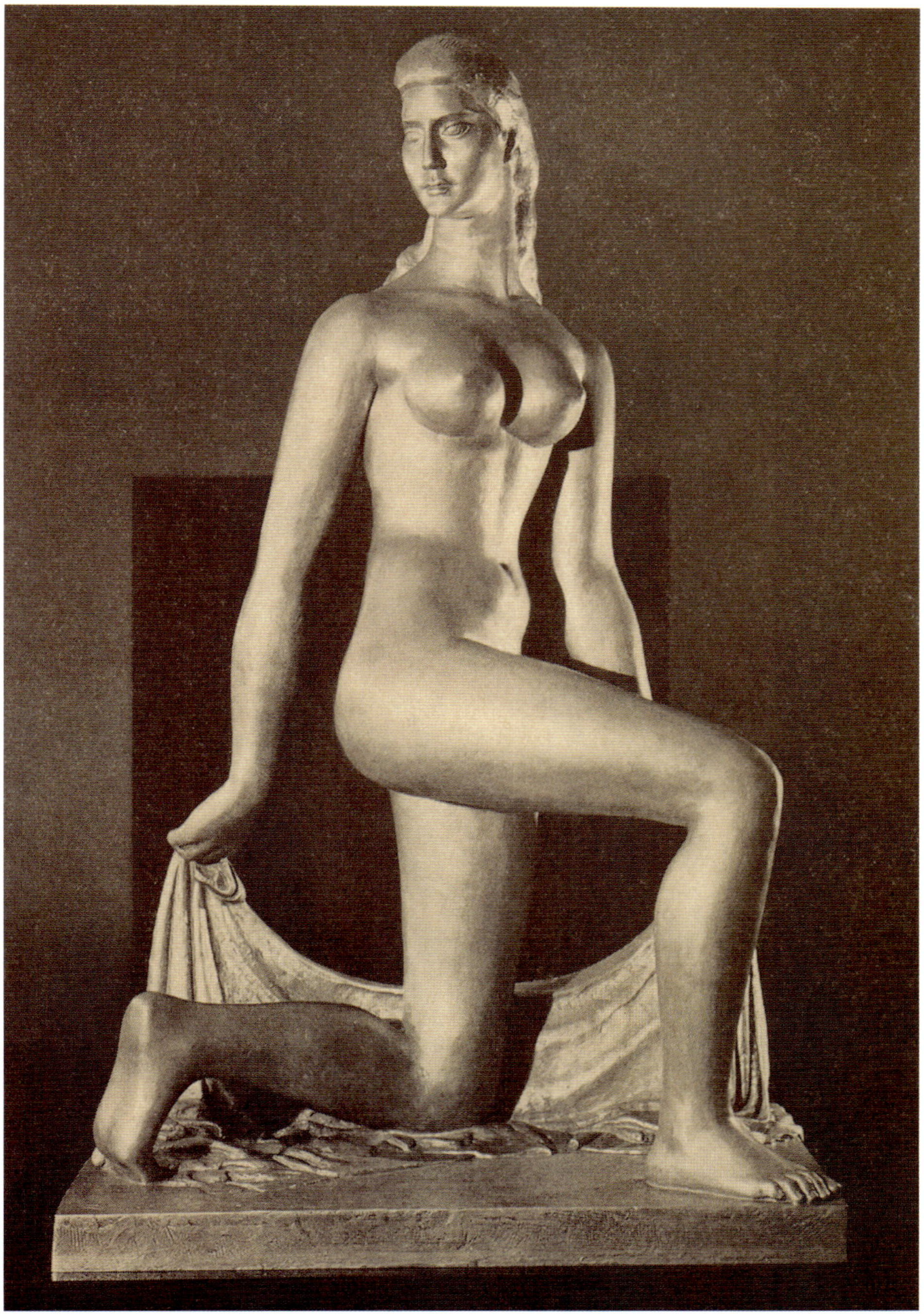

V-62. "Spirit of the Dance" (also called "Dancing Girl") by William Zorach. 1932-33.

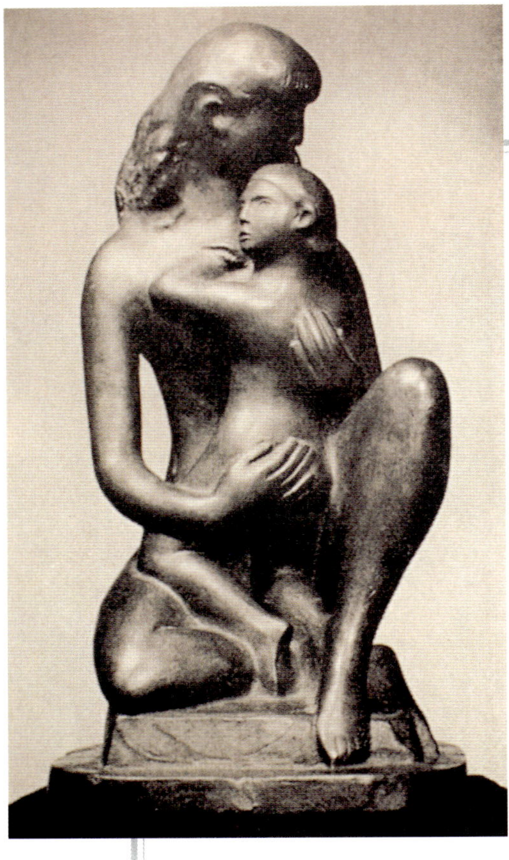

V-63. "Mother and Child" (clay model) by William Zorach. 1927-30.

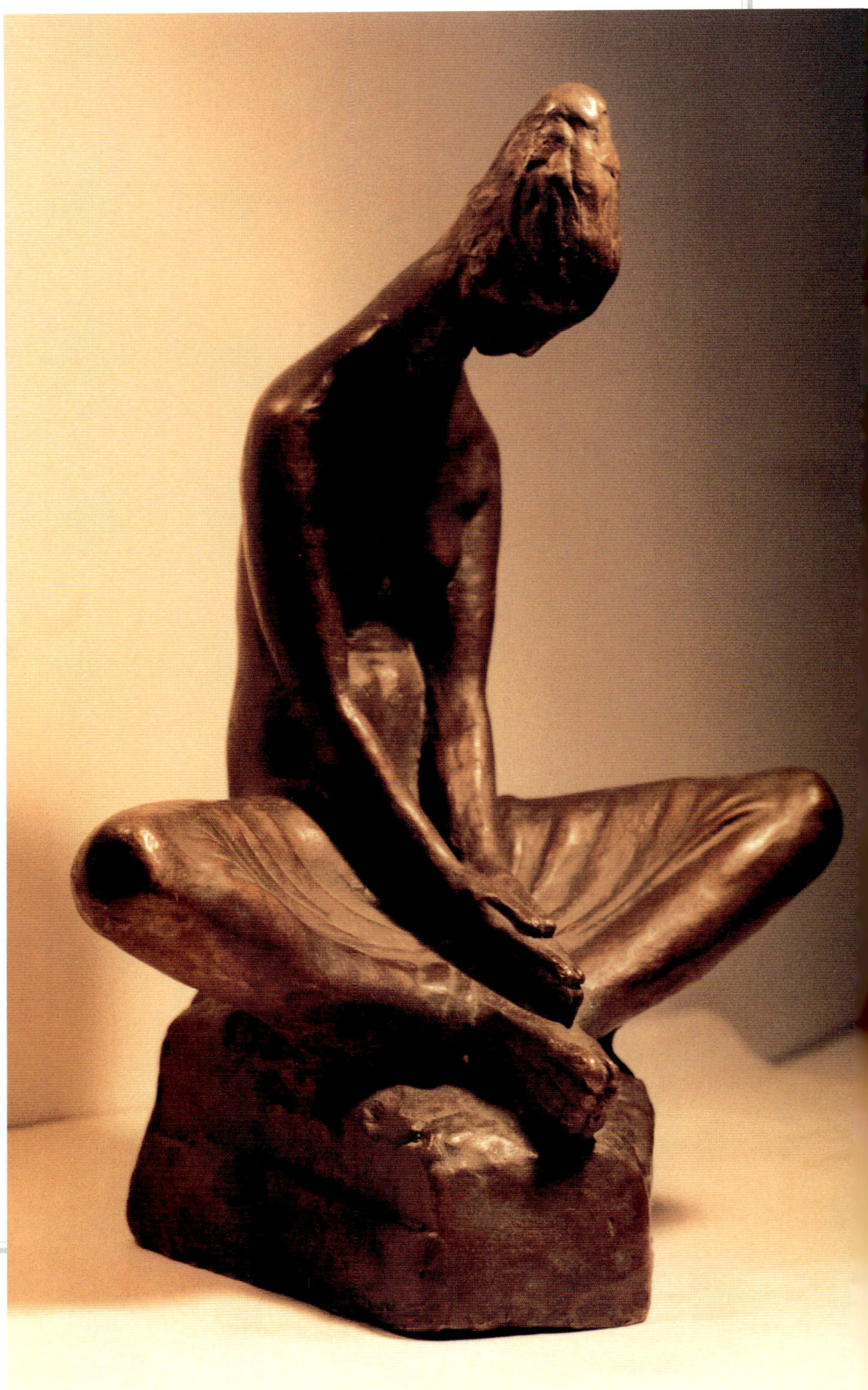

V-64. "Seated Figure" by William Zorach.

HUGO ROBUS (1885-1964)

Hugo Robus devoted most of his sculptural career to small-scale sculptures depicting the human body. His delicate pieces often suggest finely crafted jewelry, with forms that are simplified and abstracted.

Robus was born in Cleveland and at an early age worked for a jewelry maker. He studied both at the Cleveland Institute of Art and in New York at the National Academy of Design, before leaving for Paris to study in1912. In France Robus was a student of Antoine Bourdelle at the Grande Chaumière. He returned to New York in 1915 as a painter. It was not until 1920 that he devoted himself to sculpture. As a sculptor he was unknown for many years, working as a craftsman in precious metals. In the 1930s he began exhibiting his small pieces, many of them female torsos.

Over the next twenty-five years he cast a number of pieces at R.B.W. Almost all of them abstracted the female body like "Girl Washing her Hair, Soap in her Eyes" (V-65), "Woman Without a Head" (V-66), and "Standing Nude" (V-67). Other semi-abstract evocations of women include "Reflections" (V-68) and "Dreamer" (V-69).

Despite their formalist style, several of Robus' works were clearly expressive of emotions, including his "Supplication " (V-70), "Quiet Form #2" (V-71) and his "Laughing Boy" (V-72),

By the 1950s he had received many awards, and had developed a reputation for expressing "the nostalgic mysteries of our inner emotions and dreams," as his colleague William Zorach commented. Robus had major exhibitions at the Corcoran Gallery in Washington (1958), and the Whitney Museum (1960).

At Roman Bronze Works, a number of Robus bronzes were cast, beginning in 1952 (a bust for the Schubert Theater), and continuing until long after his death.

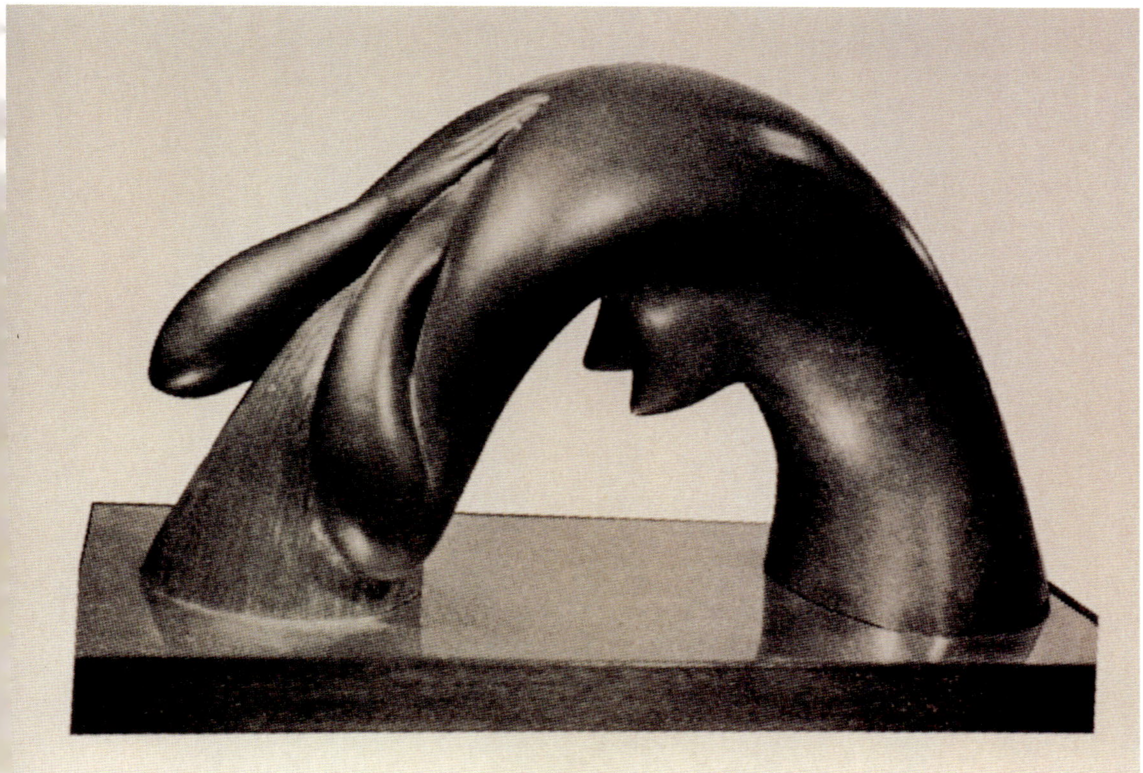

V-65. "Girl Washing Her Hair, Soap In Her Eyes" by Hugo Robus. 1933.

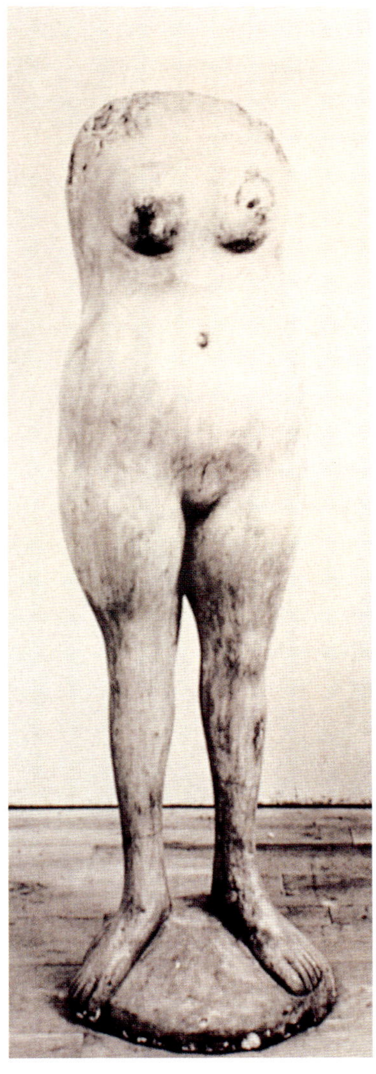

V-66. "A Woman Without a Head" (plaster cast) by Hugo Robus.

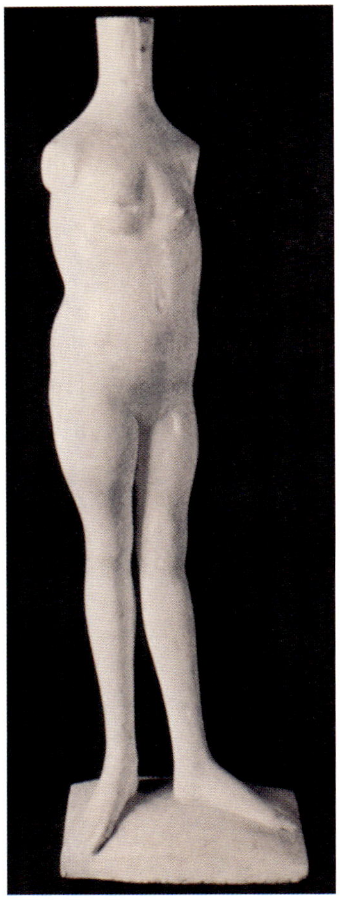

V-67. "Nude Standing Torso" (plaster cast) by Hugo Robus.

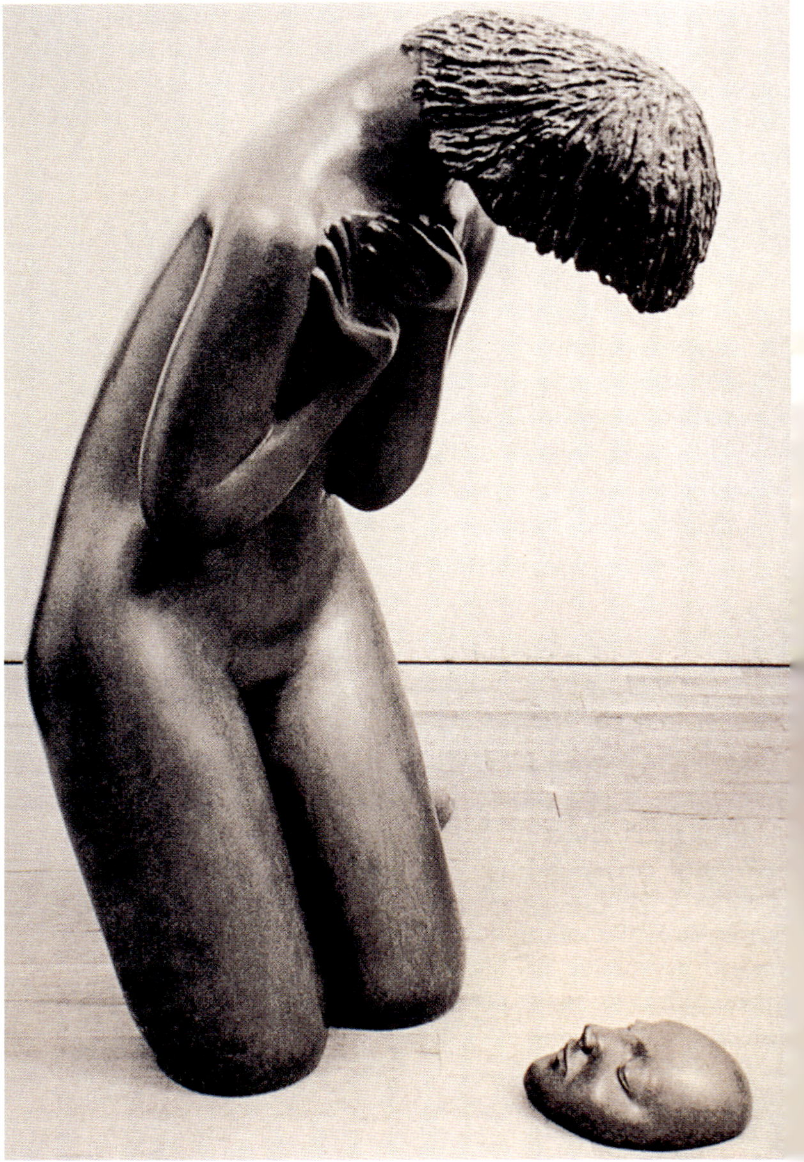

V-68. "Vanity" (also called "Reflections" by Hugo Robus. 1963.

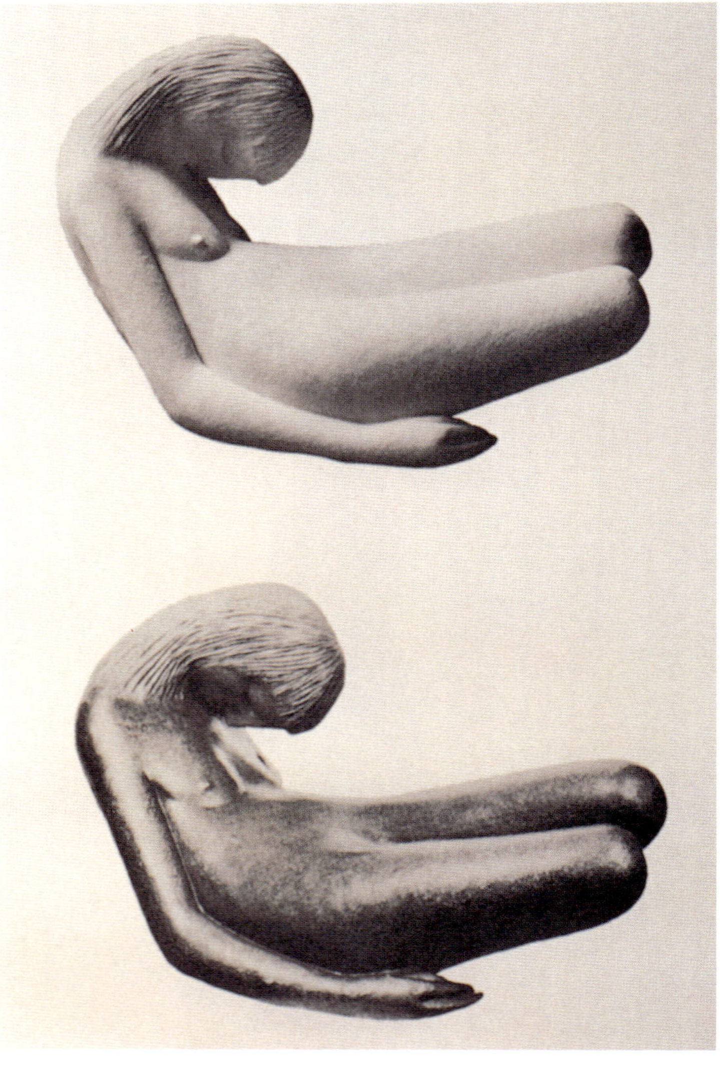

V-69. "Dreamer" by Hugo Robus. 1963.

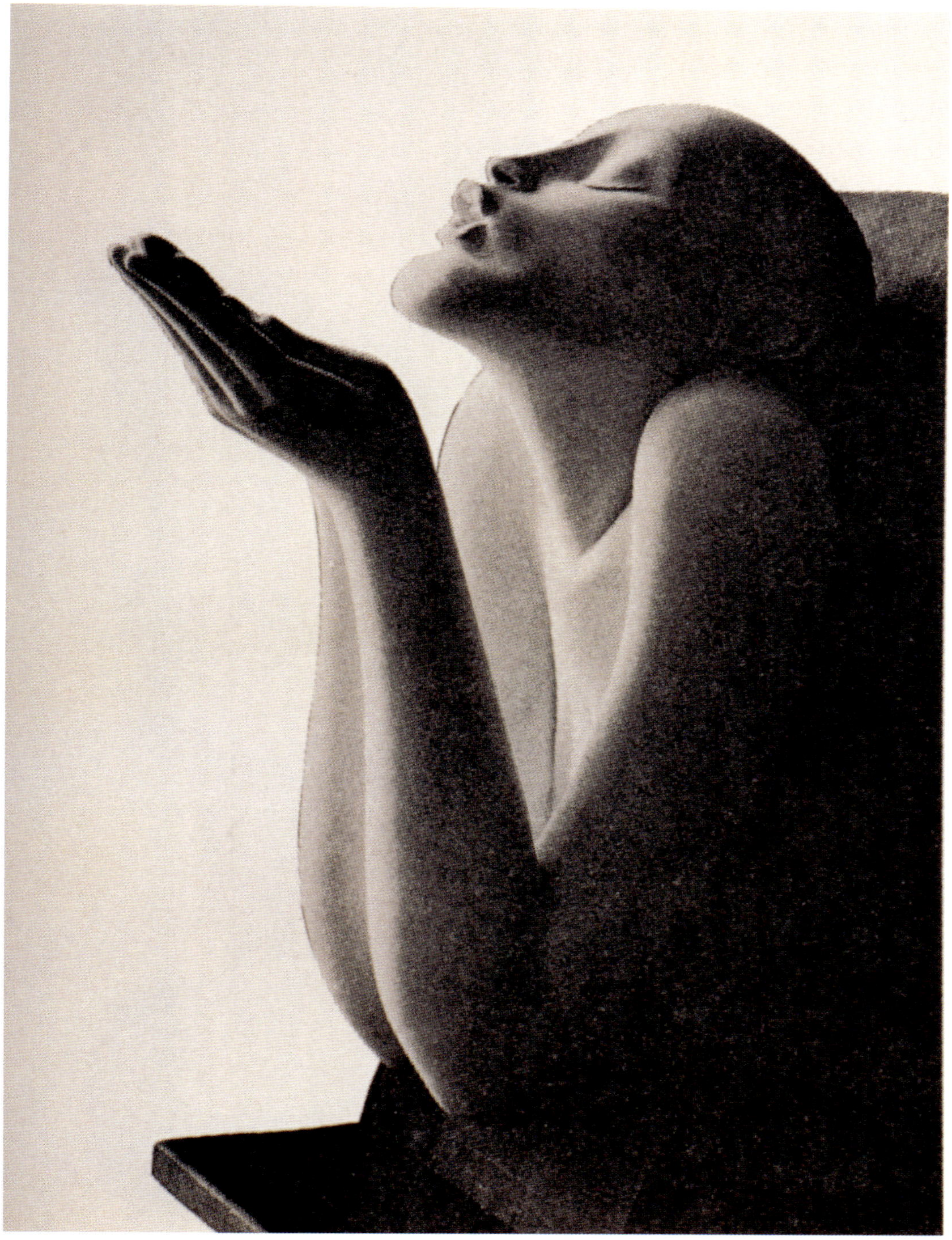

V-70. "Supplication" by Hugo Robus. Originally cast 1941; second casting at R.B.W. in 1954.

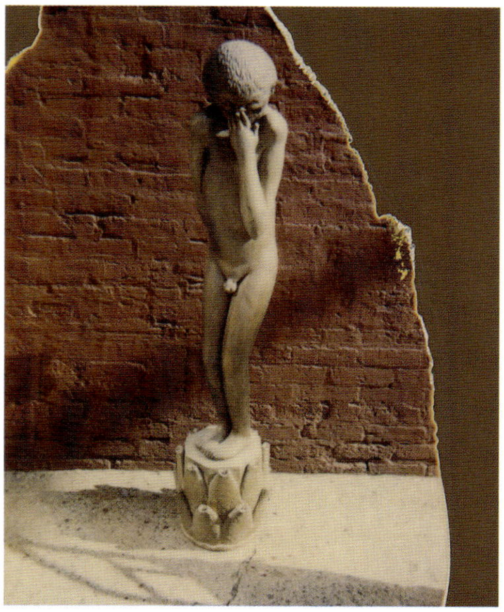

V-72. "Laughing Boy" (also called "Spirit of Youth") by Hugo Robus. 1930.

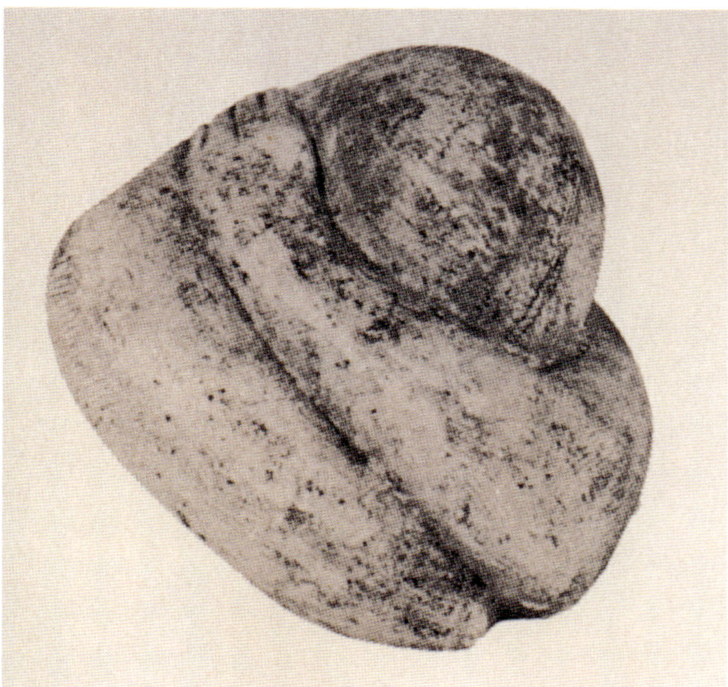

V-71. "Quiet Form # 2" by Hugo Robus. 1963.

ALEXANDER CALDER (1898-1976)

Known throughout the world as the sculptor who made sculpture move, Calder was the inventor of mobiles, abstract three-dimensional works of art that were delicately balanced to move in air and wind. In addition to his mobiles, he made many abstract works he called stabiles, mostly of metal.

Calder (the son of sculptor Alexander Stirling Calder) was born in Philadelphia and educated as a mechanical engineer, but he also painted and drew. Between 1923 and 1926 he studied at the Art Student's League and then left for Paris. His miniature animated circus made of wires and small forms and his wire portraits brought him to the attention of the art world. His first one-man show was in New York in 1928 but he returned often to France; he had homes in Connecticut and France for most of his life.

By the 1930s he had begun making wire and metal works that were motorized; later he turned to wind- and air-operated mobiles. Both his stabiles and mobiles showed his interest in abstract form and line, fragility and monumentality; his originality in concept and design made his work arguably the most recognizable sculpture of his time. Today Calder's work is in most major museums, sometimes indoors and sometimes as modern monuments in outdoor environments.

In 1944 he "made things (in plaster) which were mobile objects and had them cast in bronze—acrobats, animals, snakes, dancers, a starfish, and tightrope performers," according to his autobiography. Twenty four years later Roman Bronze Works (with the help of Sculpture Services of New York and the sculptor himself) cast a number of these smaller stabile sculptures that Calder had made in the 1940s (numbers V-73 to V-90). These whimsical early works (some of which have separate balancing parts) share his sense of delicate balance and movement with many of his monumental mobiles and stabiles.

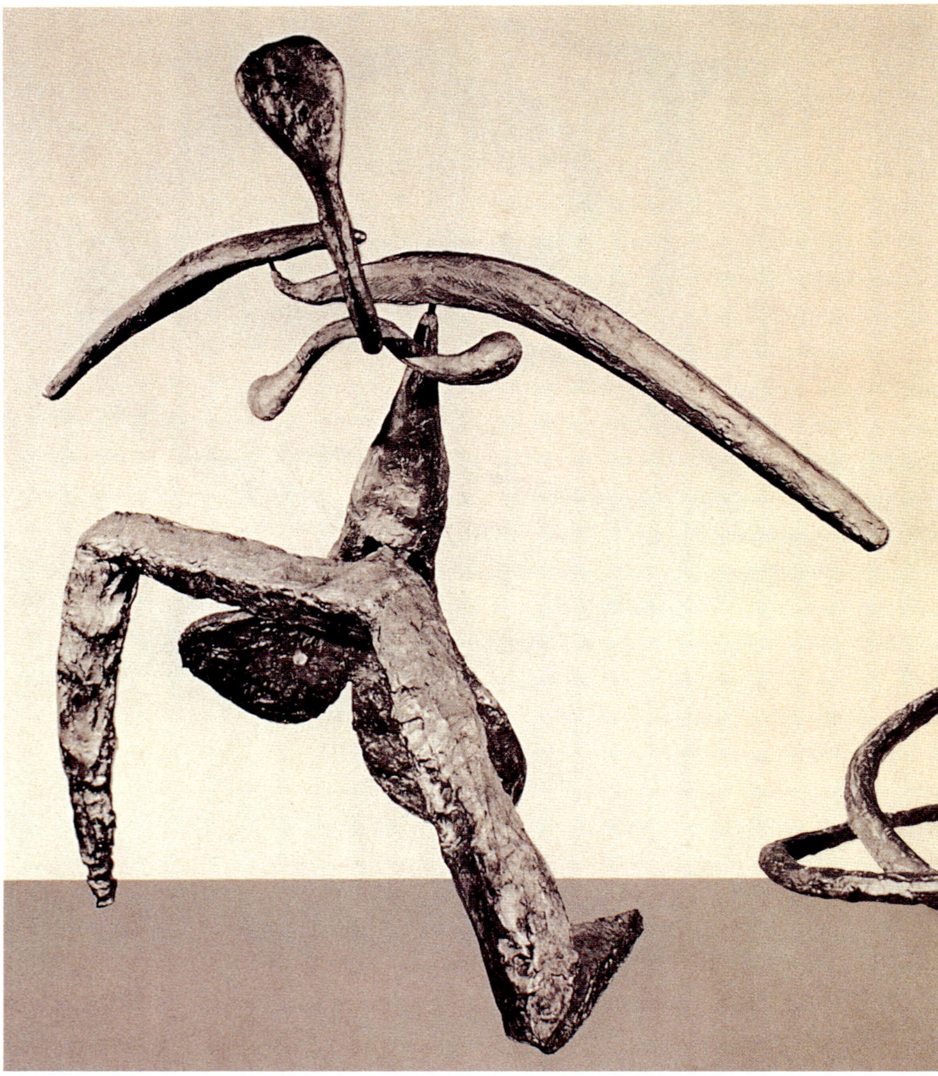

V-73. "On One Knee"...in six parts, by Alexander Calder. 1944.

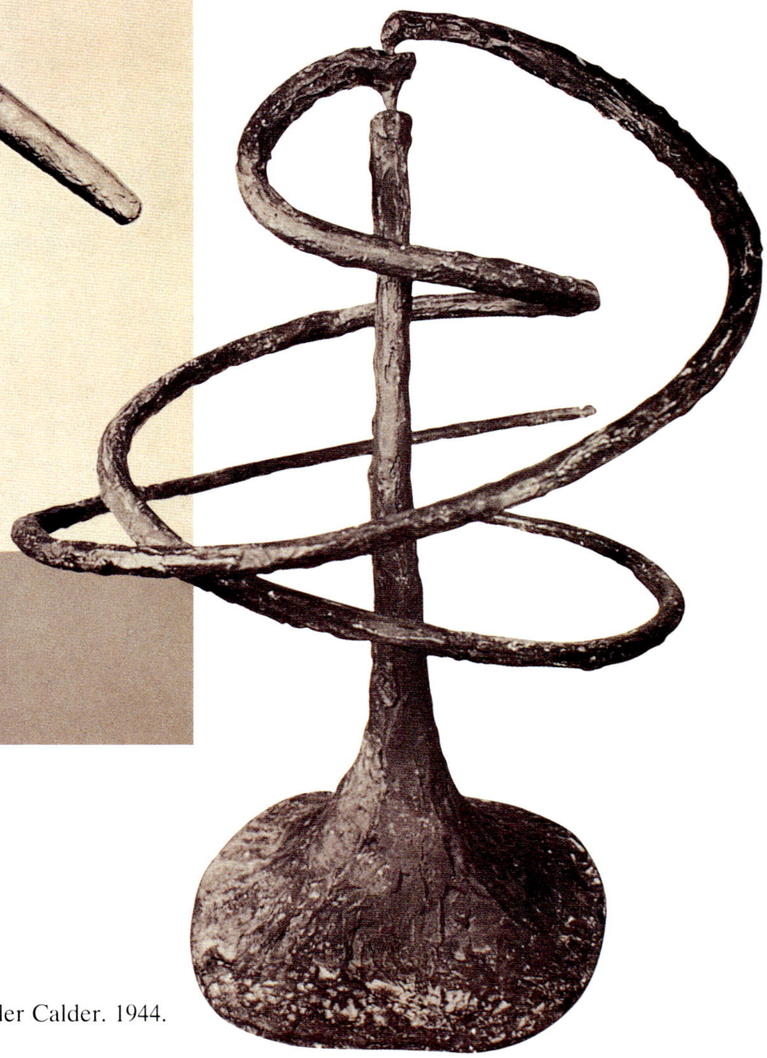

V-74. "Double Helix"...in three parts, by Alexander Calder. 1944.

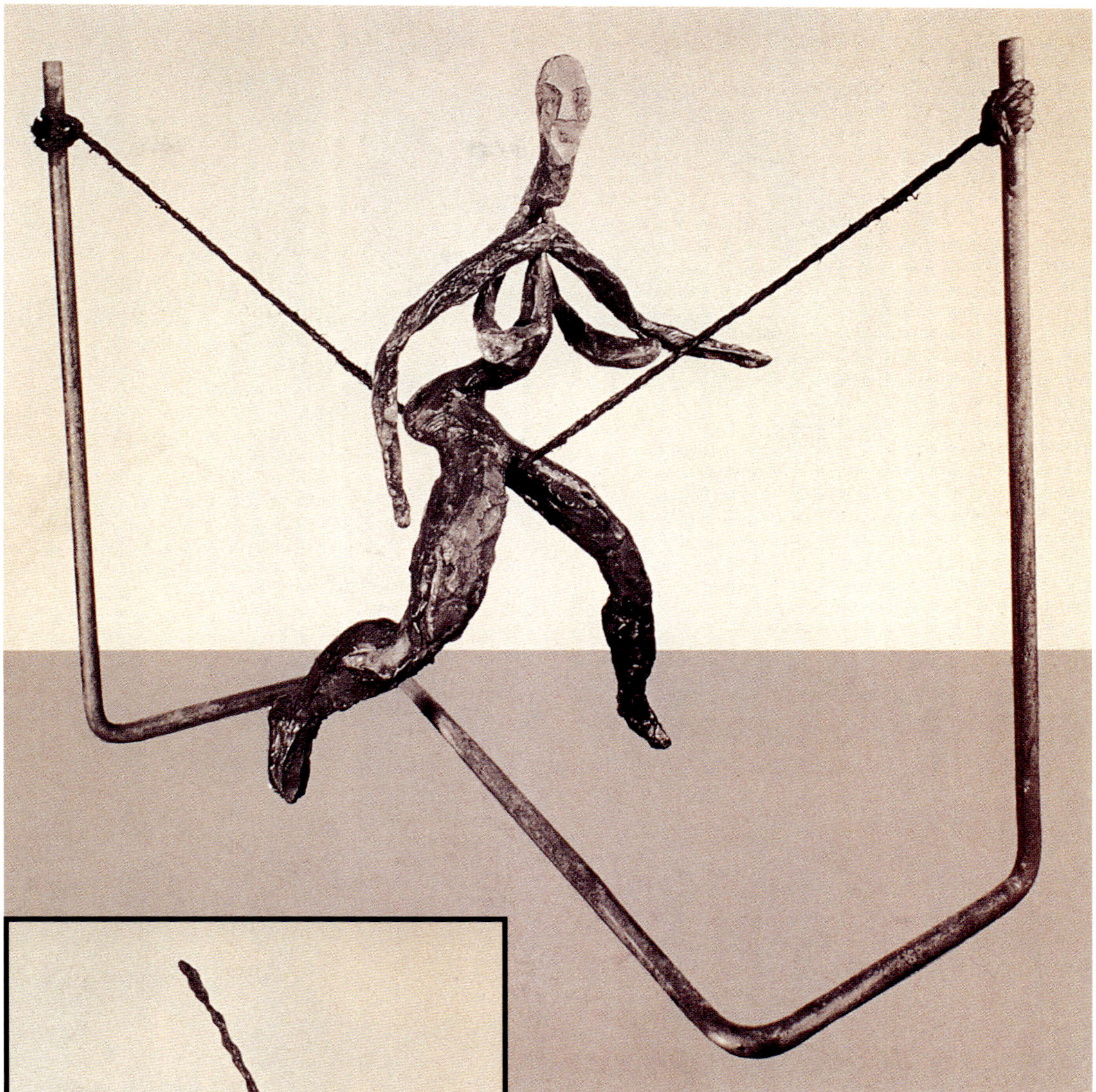

V-76. "Woman on Cord" …in three parts plus base, by Alexander Calder. 1944.

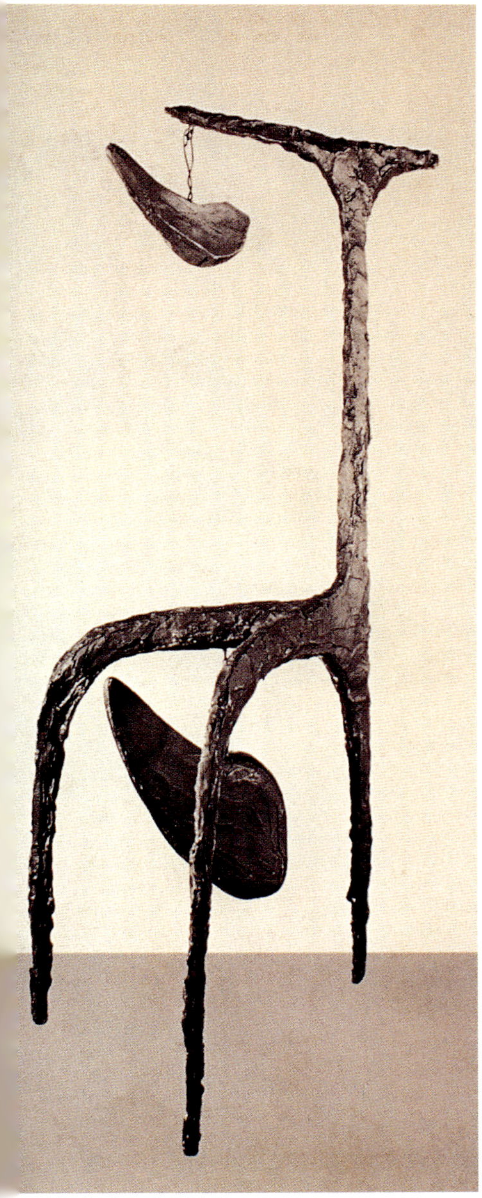

V-75. "The 'T'"…in three parts, by Alexander Calder. 1944.

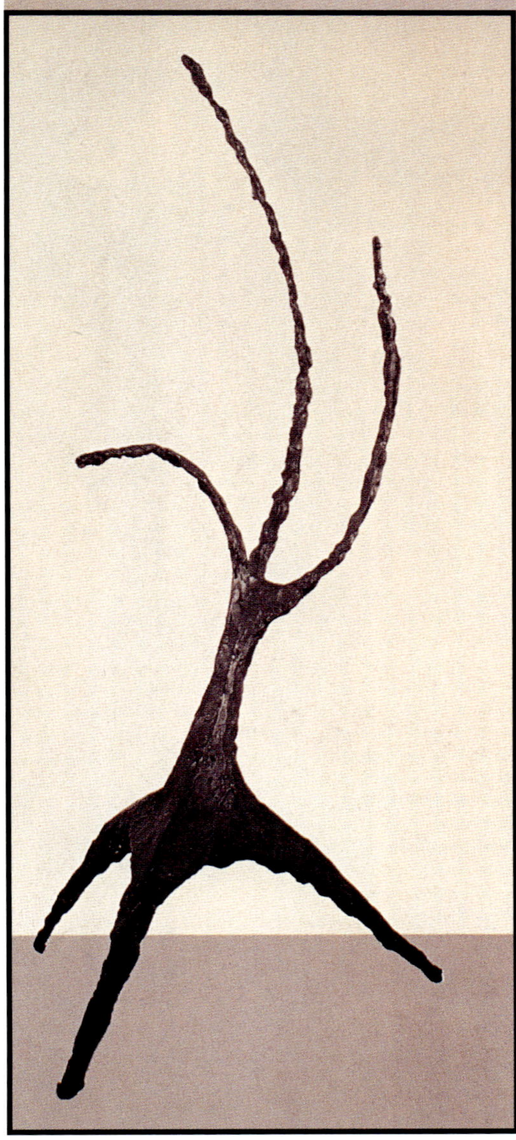

V-77. "Three Fingers" by Alexander Calder. 1944.

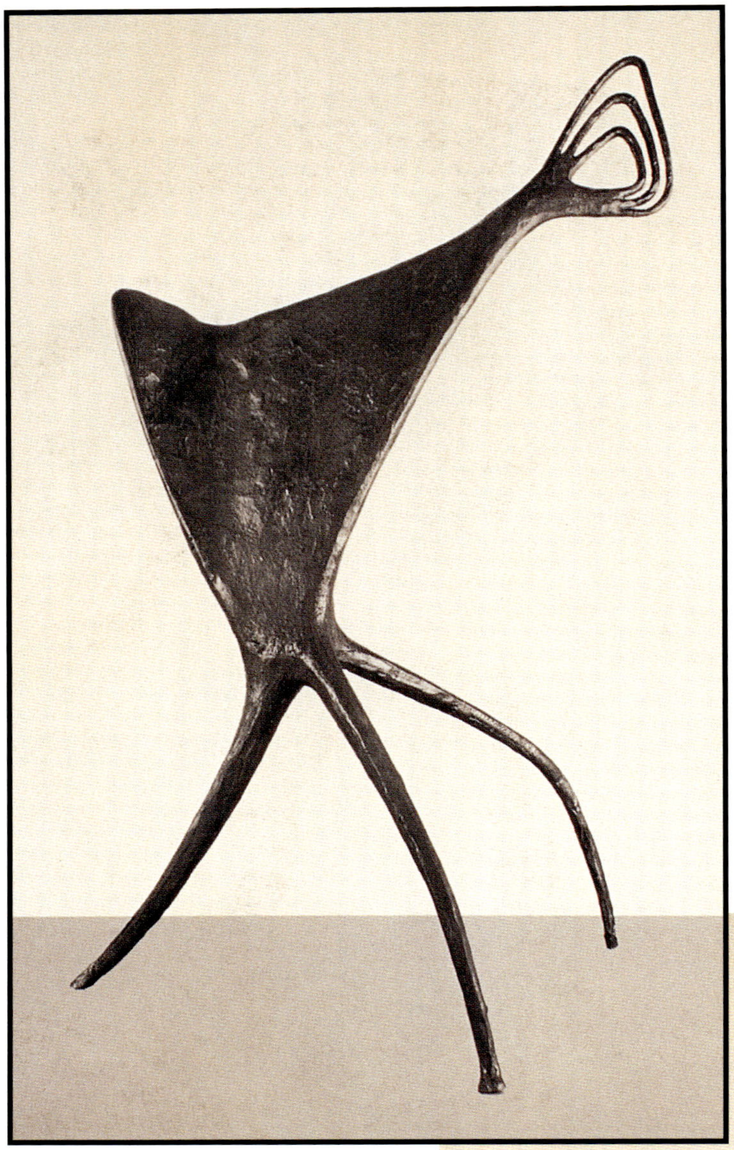

V-78. "The Fawn" by Alexander Calder. 1944.

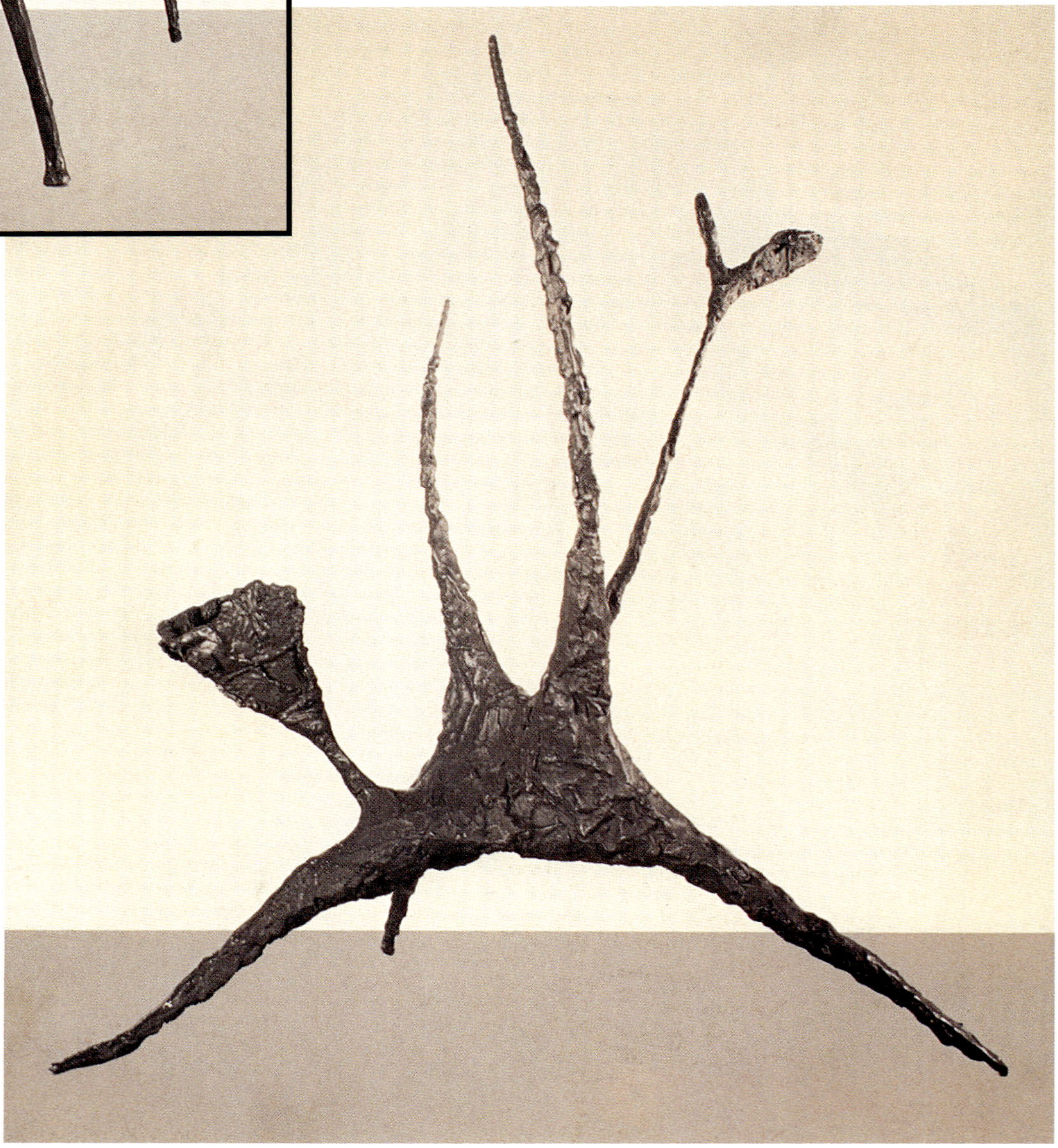

V-79. "The Flower" by Alexander Calder. 1944.

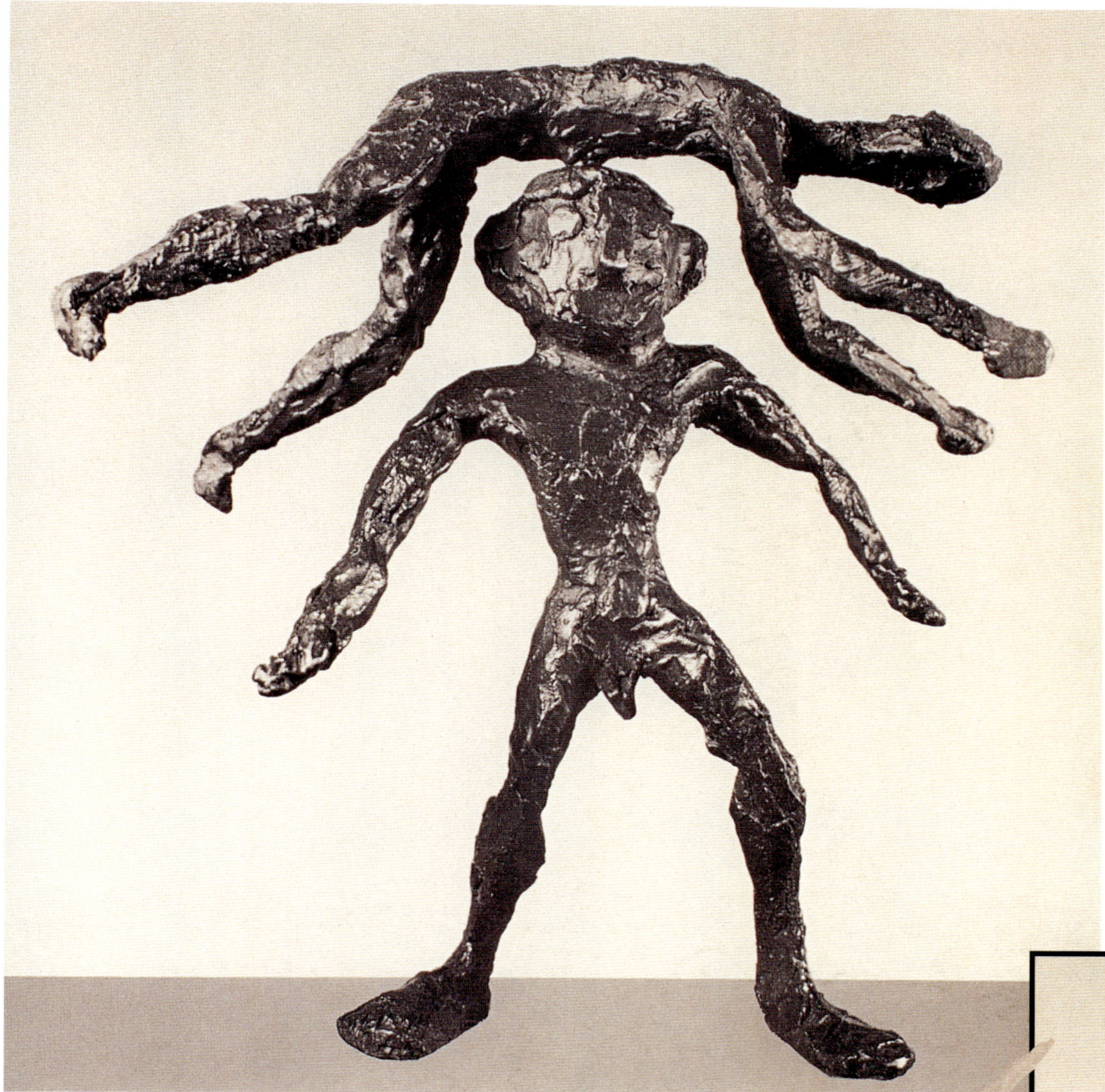

V-80. "The Acrobats"...in two parts, by Alexander Calder. 1944.

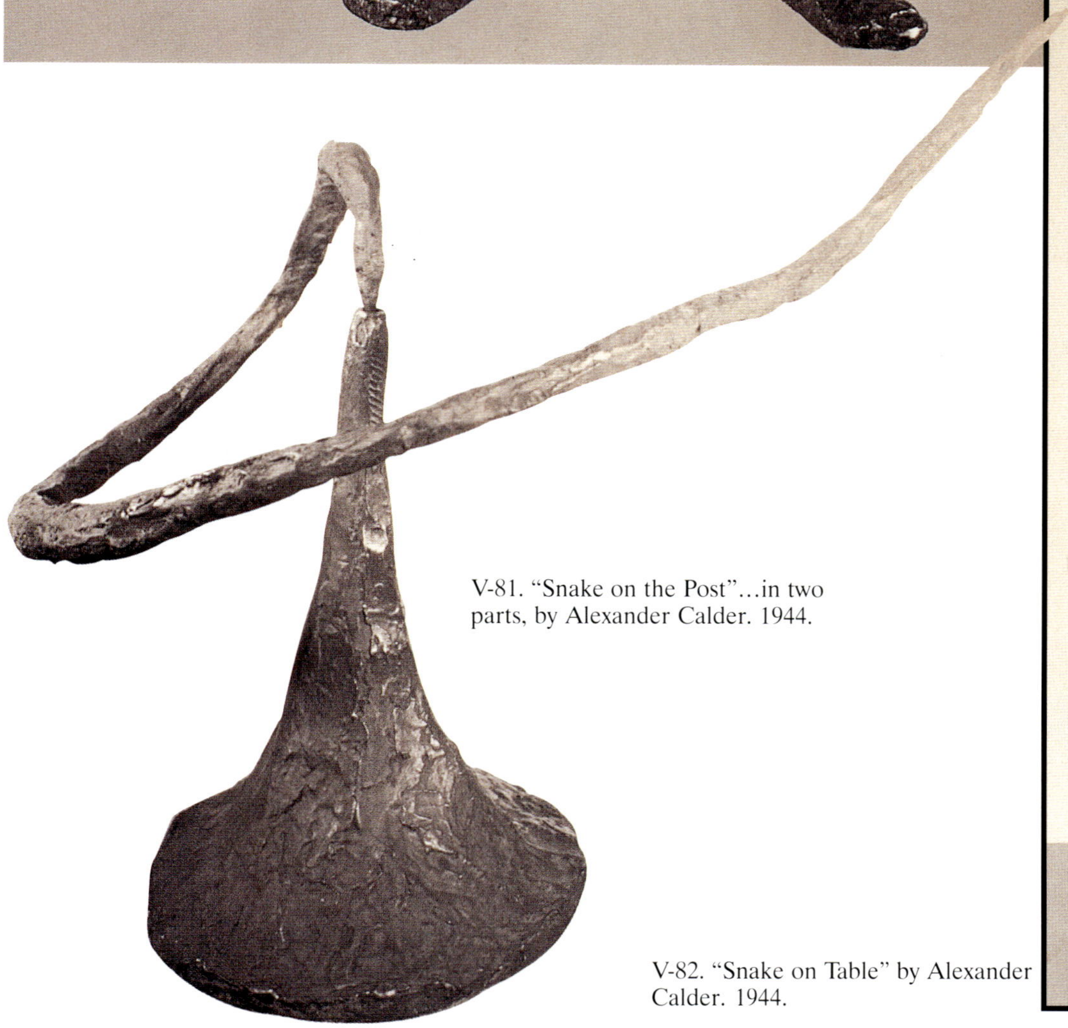

V-81. "Snake on the Post"...in two parts, by Alexander Calder. 1944.

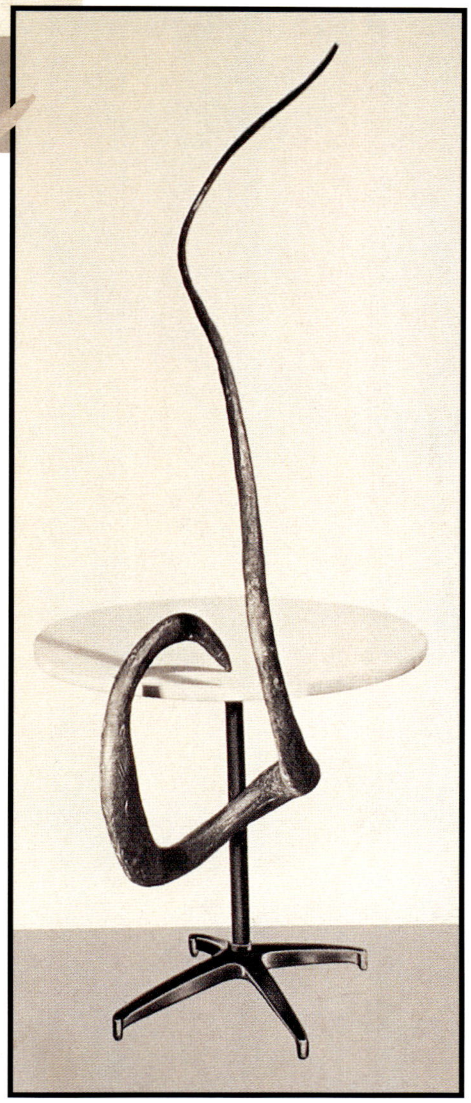

V-82. "Snake on Table" by Alexander Calder. 1944.

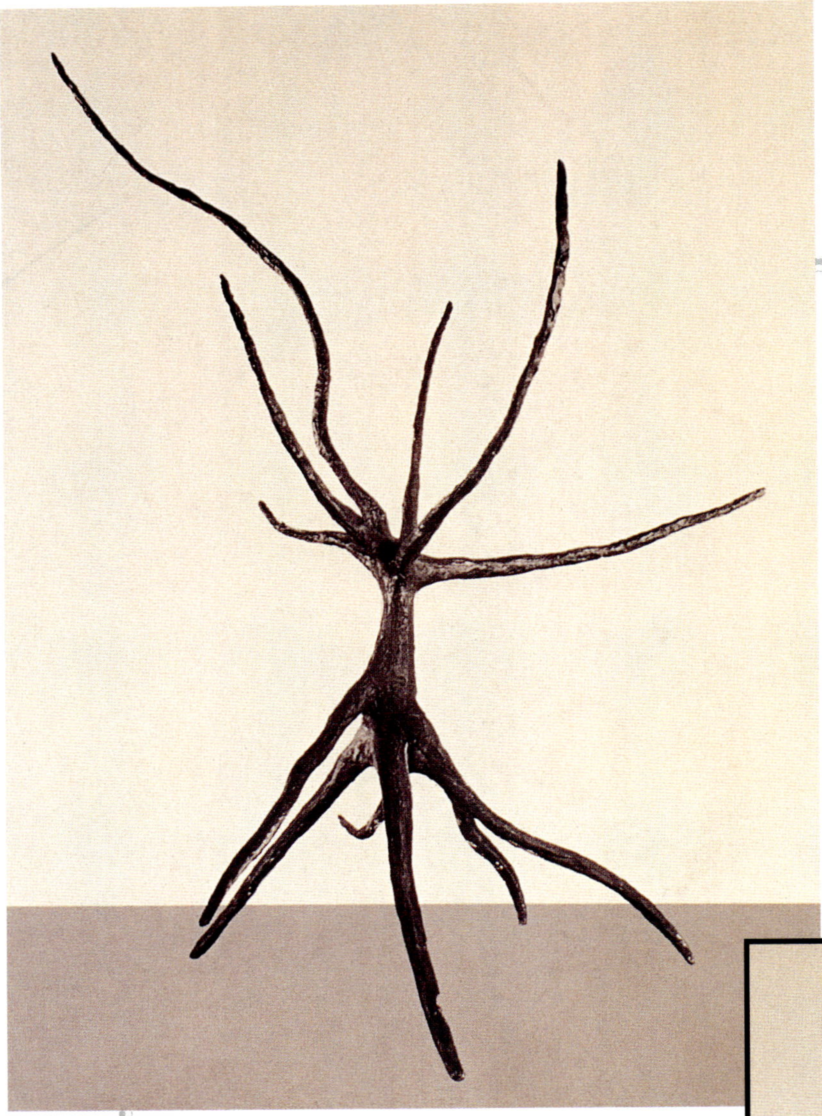

V-83. "The Snag" by Alexander Calder. 1944.

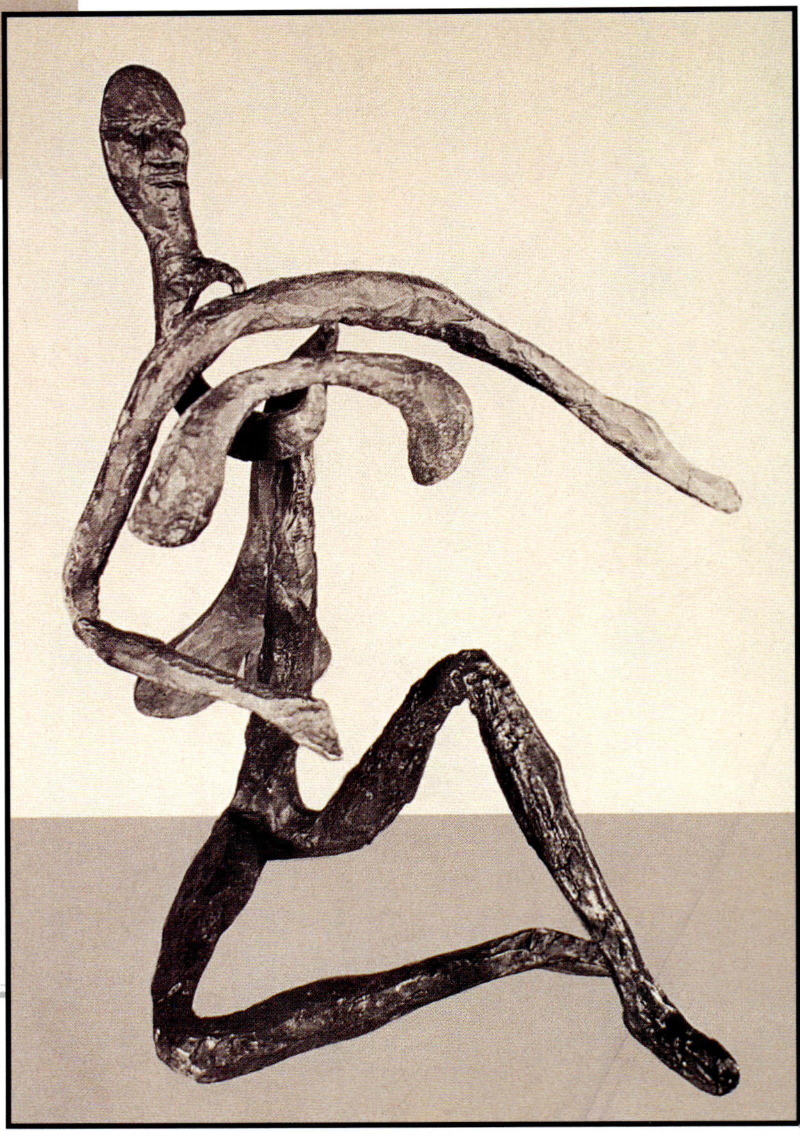

V-84. "Seated Woman"...in five parts, by Alexander Calder. 1944.

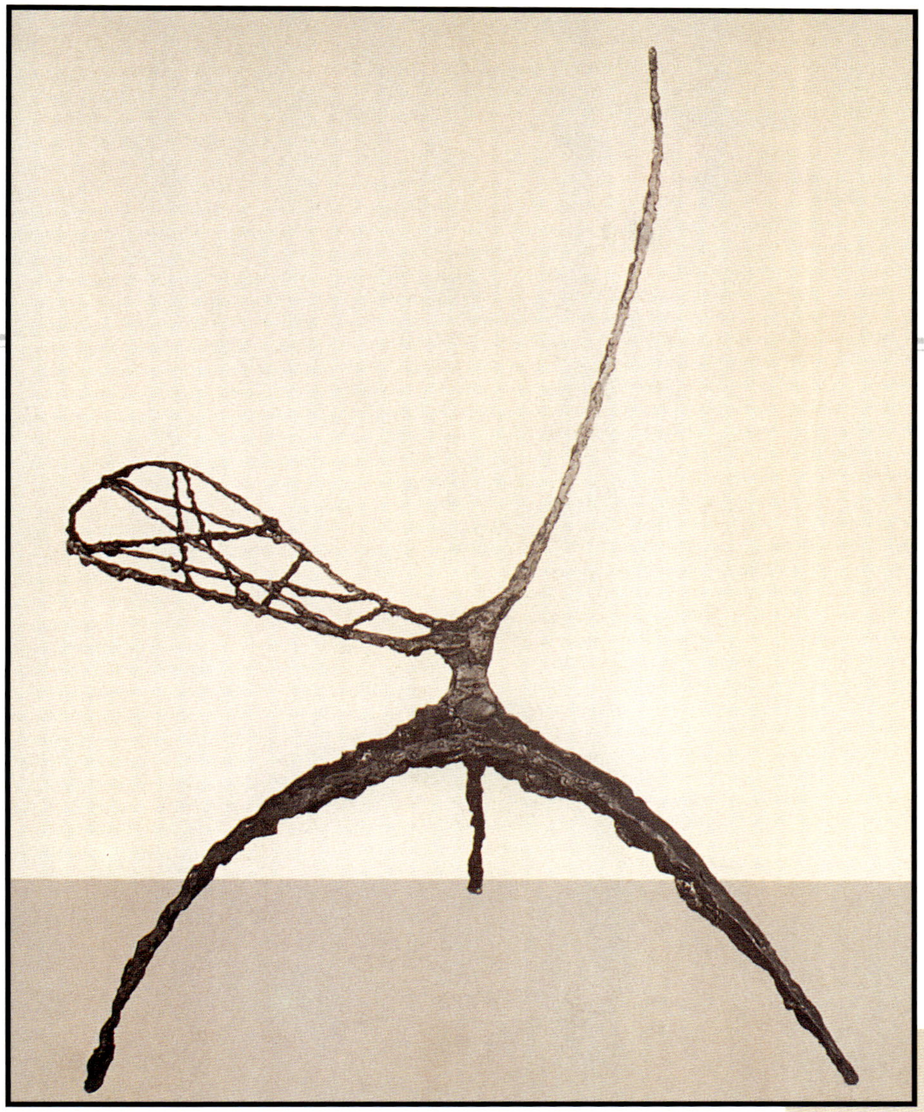

V-85. "The Racket" by Alexander Calder. 1944.

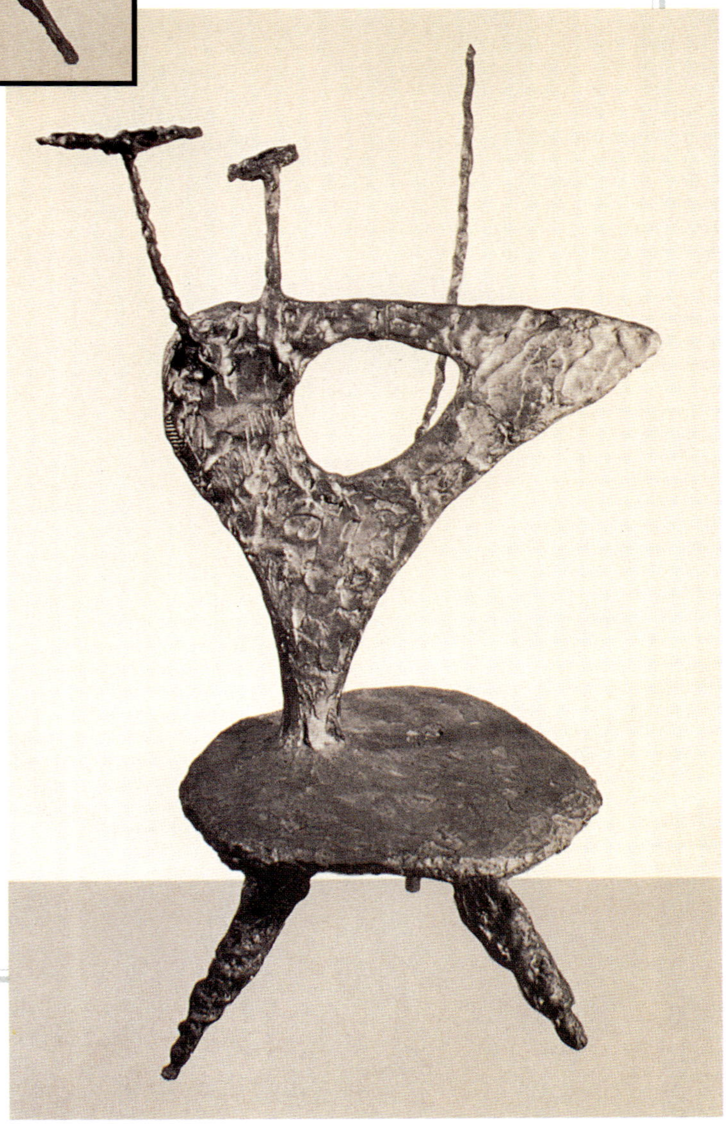

V-86. "The Chicken" by Alexander Calder. 1944.

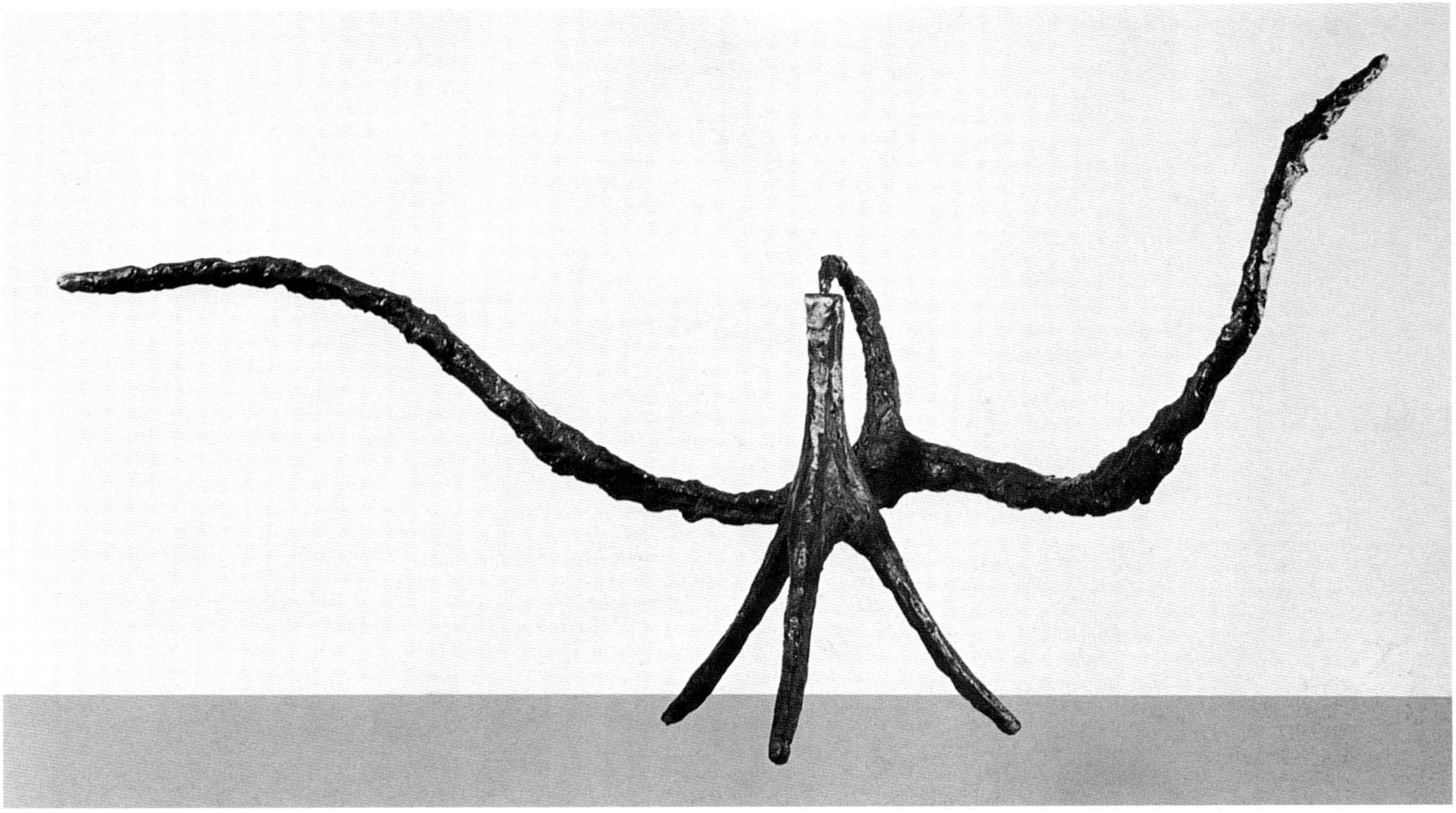

V-87. "The Antlers" by Alexander Calder. 1944.

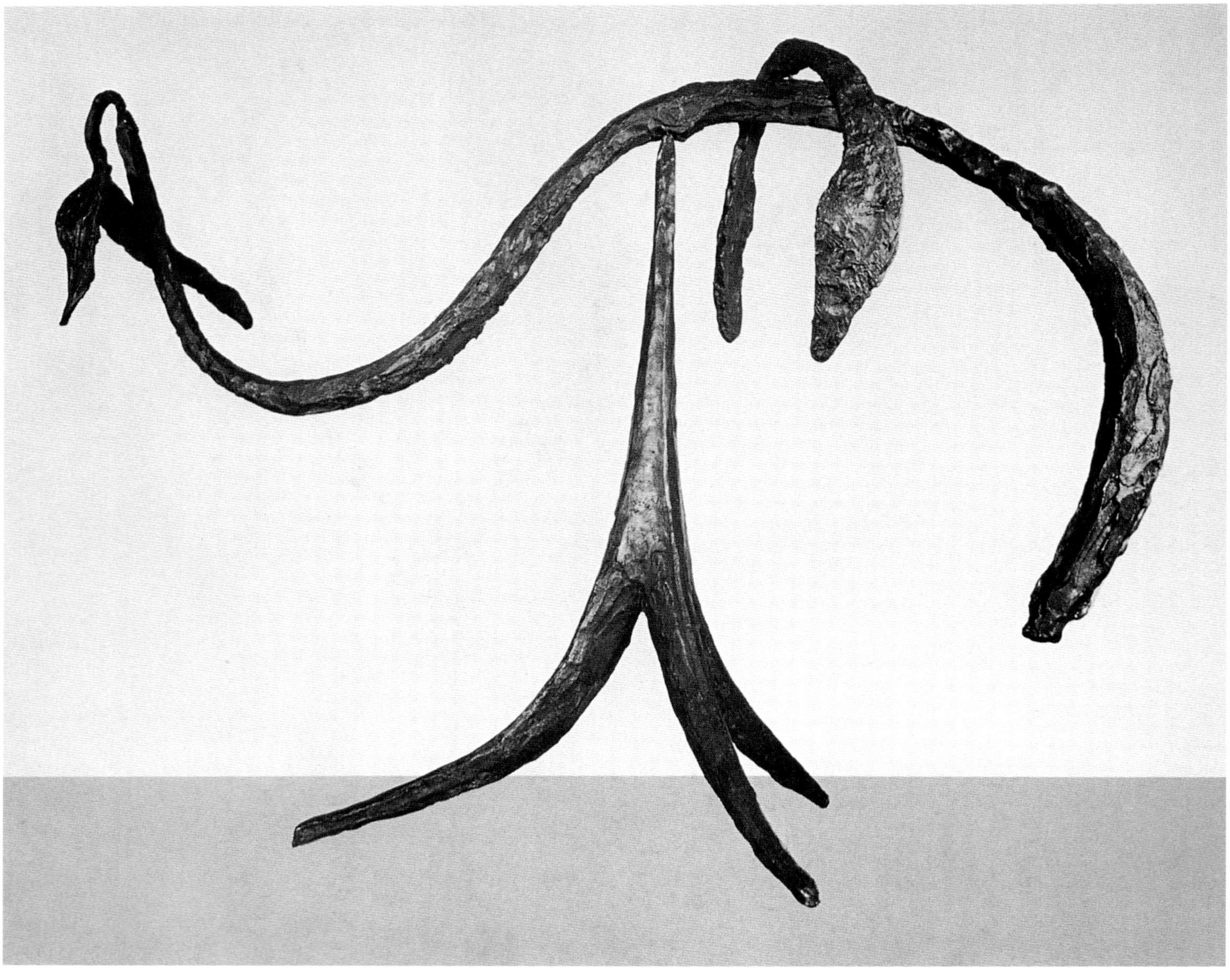

V-88. "The Vine"...in four parts, by Alexander Calder. 1944.

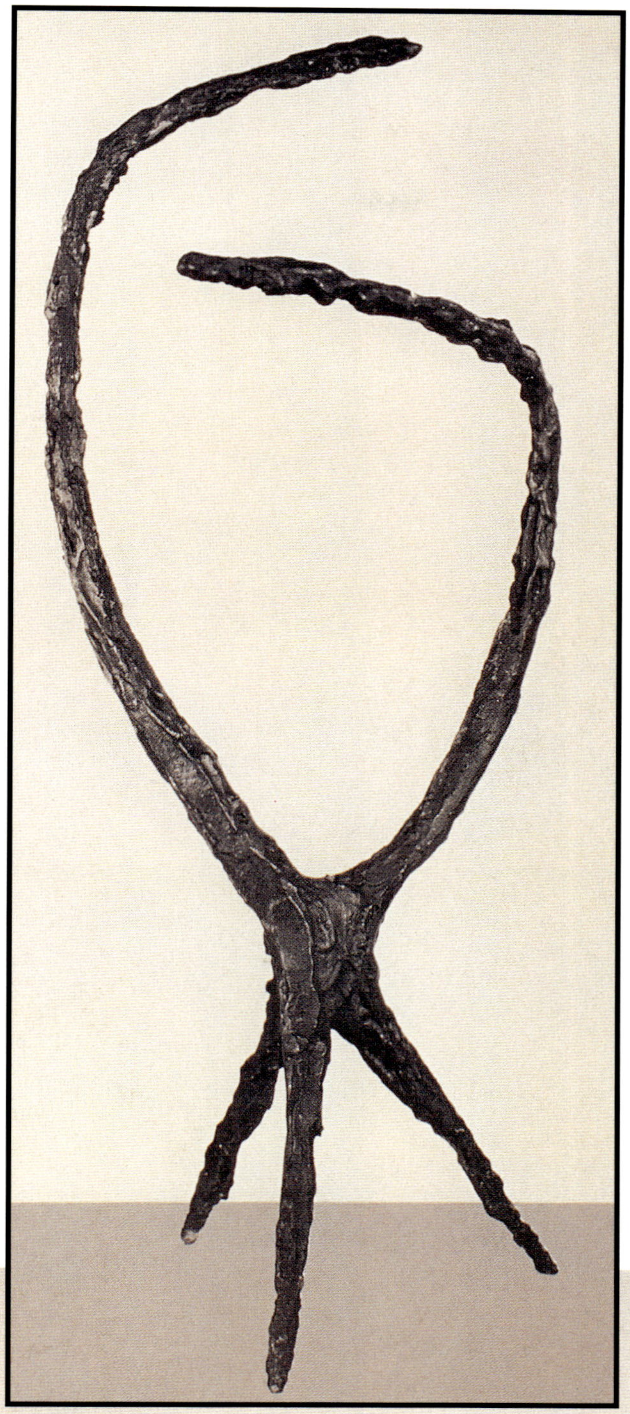

V-89. "Upraised Arms" by Alexander Calder. 1944.

V-90. "The Handstand"...in two parts, by Alexander Calder. 1944.

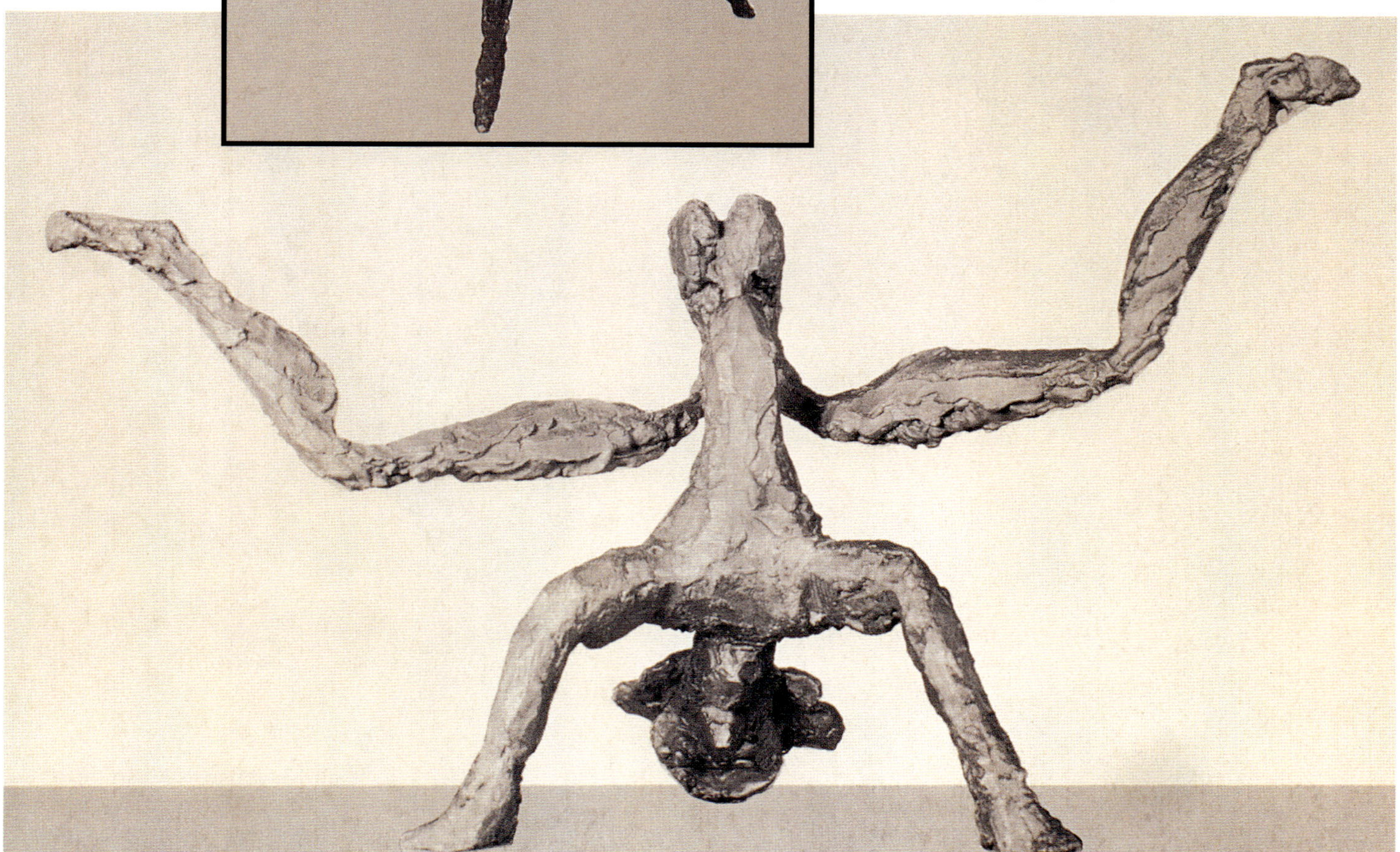

REUBEN NAKIAN (1897-1986)

Reuben Nakian was a sculptor whose career mirrored the changing American art world. He was born on Long Island, New York, and spent most of his career in New York City. He was an apprentice to Paul Manship and was a close friend of Gaston Lachaise; the three sculptors all worked at Roman Bronze Works beginning in 1917, and Nakian and Lachaise shared a studio in the 1920s.

His early works, made in the 1920s and 1930s, were bronze and marble portrait heads and full-length figures. His early sculpture had recognizable subject matter and highly finished surfaces. After meeting a number of more avant-garde artists, including Arshille Gorky and Willem deKooning, however, Nakian turned to a more modern idiom; his works lost their smooth surfaces and became more improvisatory. By the 1940s his mature style was a sculptural equivalent to action painting, emphasizing spontaneity, and what he described as "drawing in stone." He also produced monumental abstractions with rough surfaces, as well as terra cotta reliefs, sometimes suggesting mythological subjects. His choice of media ranged from marble, clay, and plaster to metal, terra cotta, paper, and even styrofoam.

Nakian's works were widely accepted; he exhibited at the Whitney Museum and the Museum of Modern Art in New York, and the Hirshhorn Museum and Corcoran Gallery of Art in Washington, among many other American museums. He showed some forty of his works at the Sao Paulo Bienal in 1961, and in 1968 represented the United States at the Venice Biennale with ten monumental sculptures. His sculpture is in the collections of many major museums and he was widely honored. He was a member of the National Academy of Arts and Letters.

Among the sculptures he cast at Roman Bronze Works were his portrait head of Marcel Duchamps (V-91), "Hecuba" (V-92), "Voyage to Crete" (V-93), "Garden of the Gods I" (V-94), and "Nymph and Dolphins" (V-95) and "Satyricon I" (V-96).

V-91. "Portrait Head of Marcel Duchamp" by Reuben Nakian. 1943.

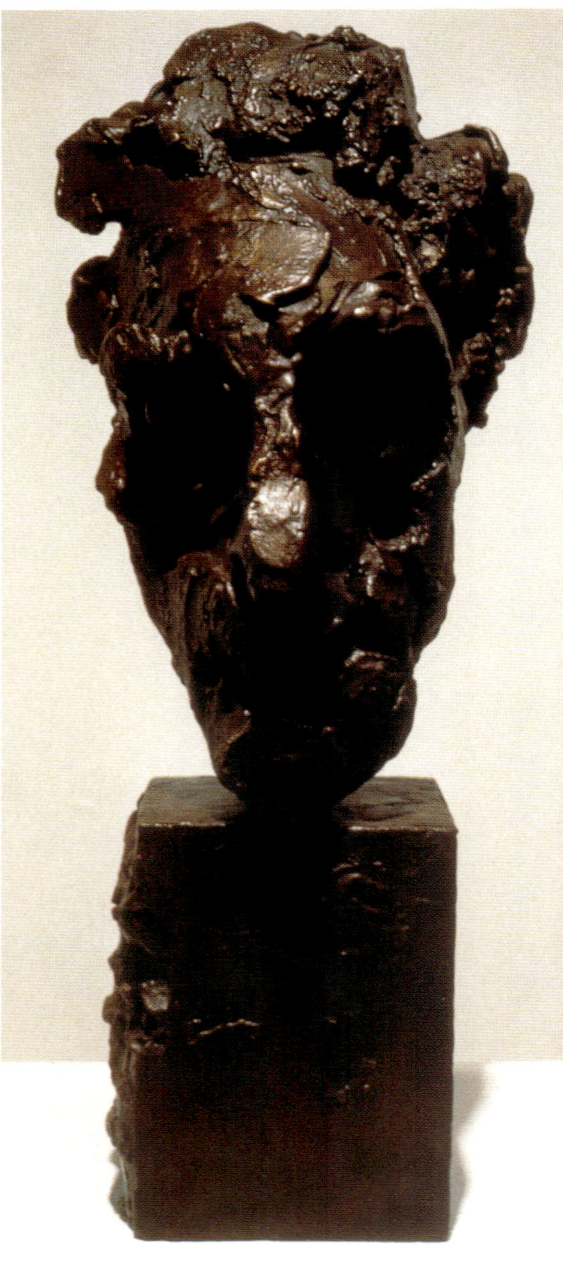

V-92. "Hecuba" by Reuben Nakian. 1960-62.

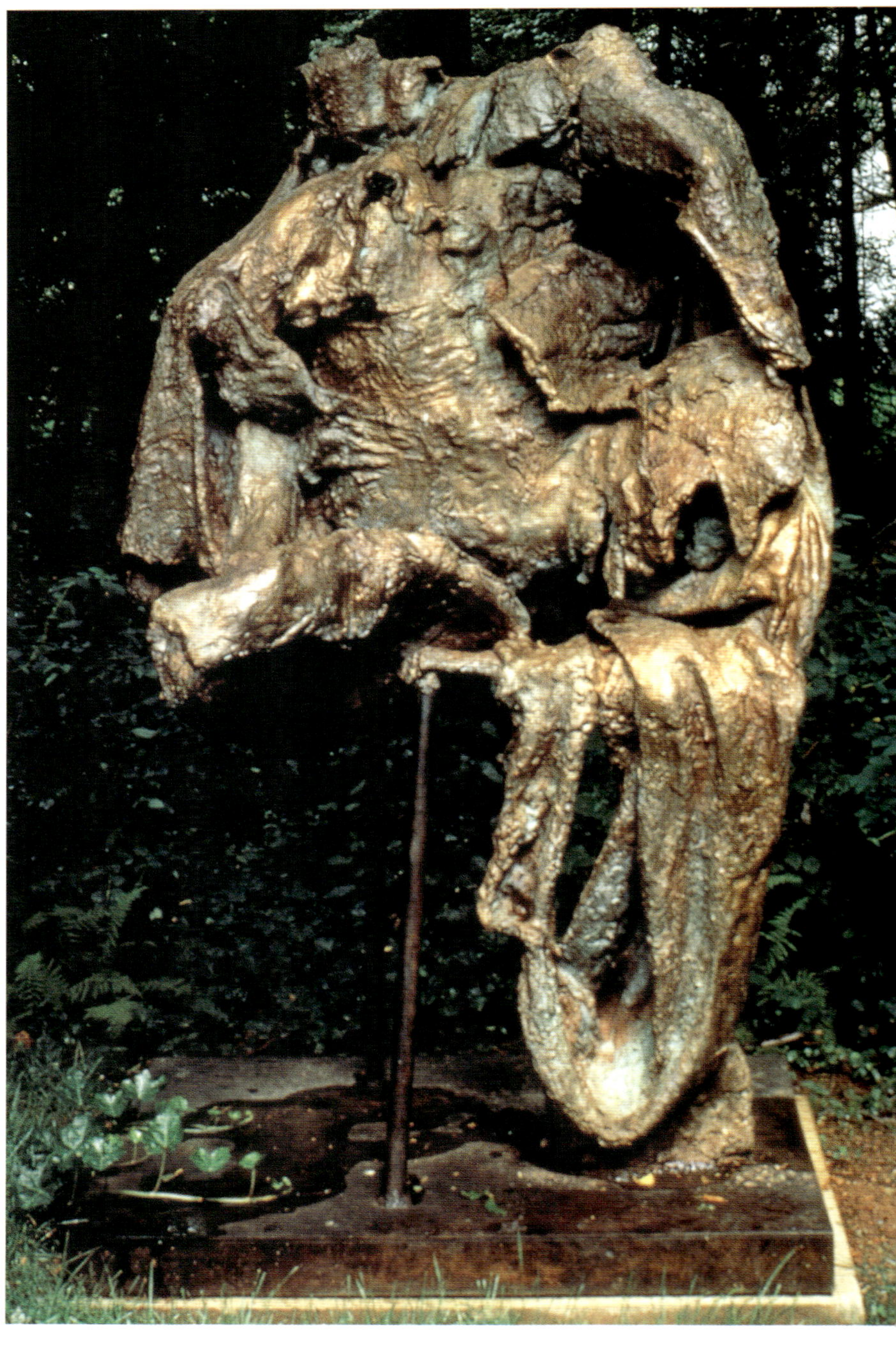

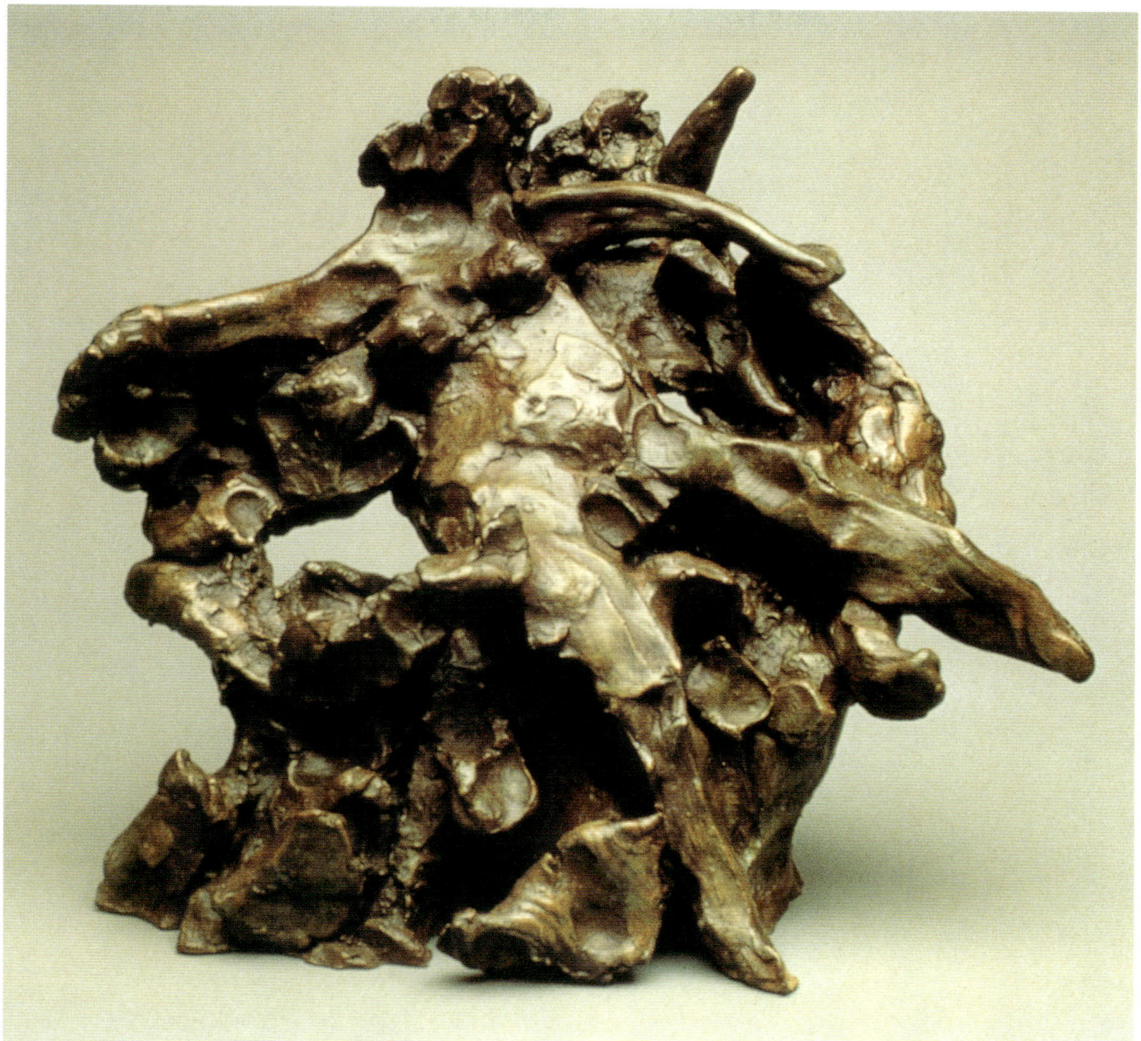

V-93. "Voyage to Crete" by Reuben Nakian. 1962-65.

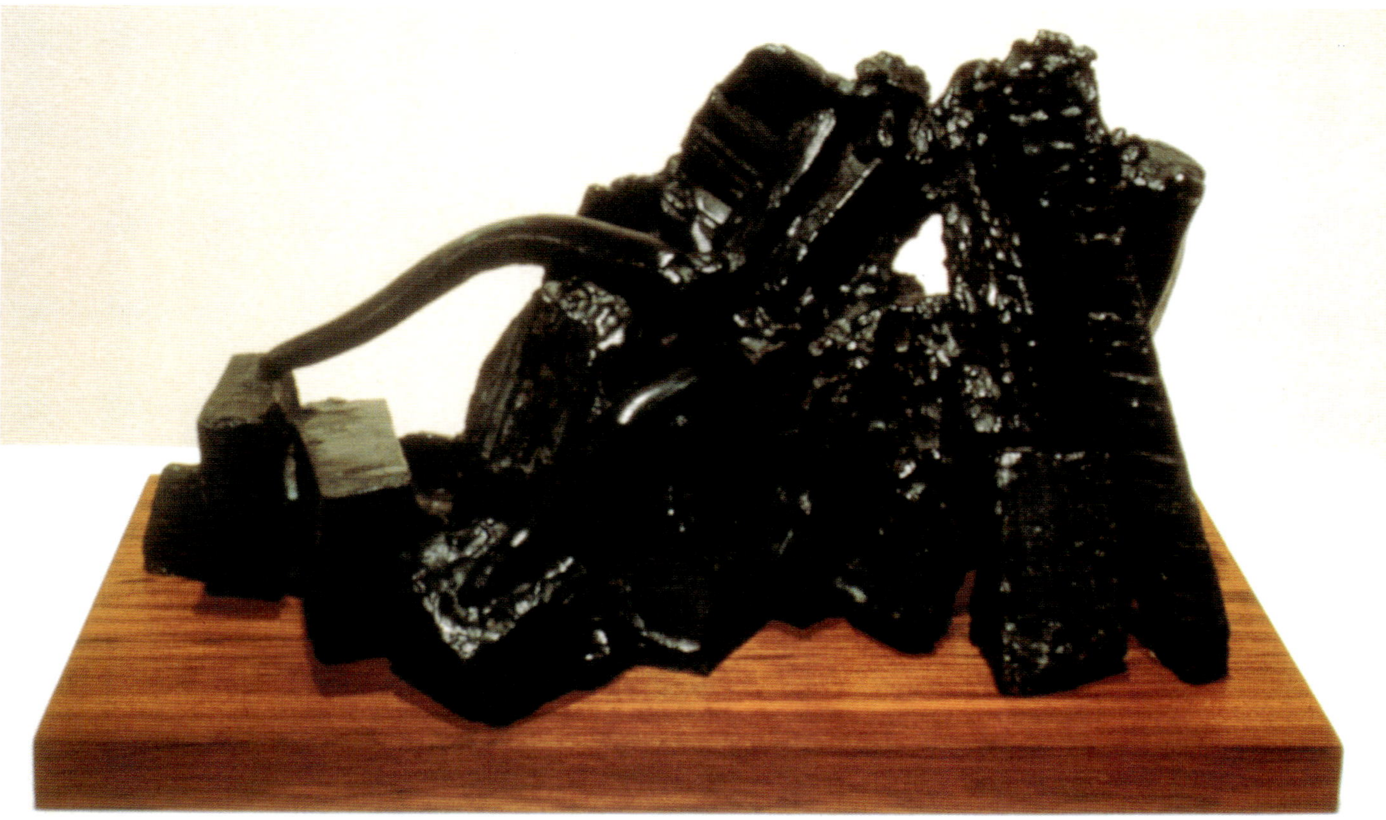

V-94. "Garden of the Gods I" by Reuben Nakian. 1979.

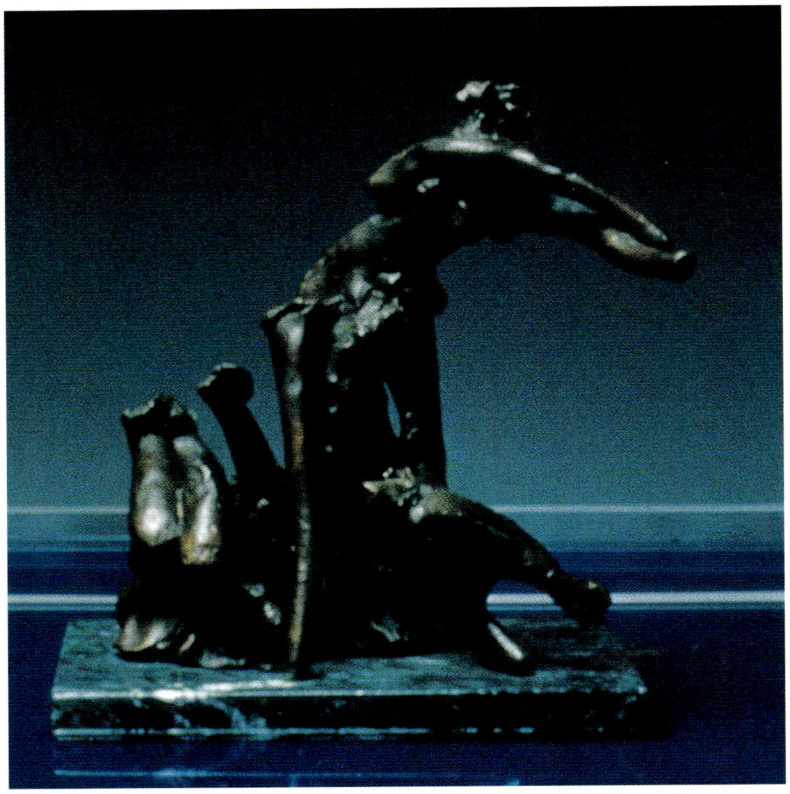

V-95. "Nymph and Dolphins" by Reuben Nakian. 1980.

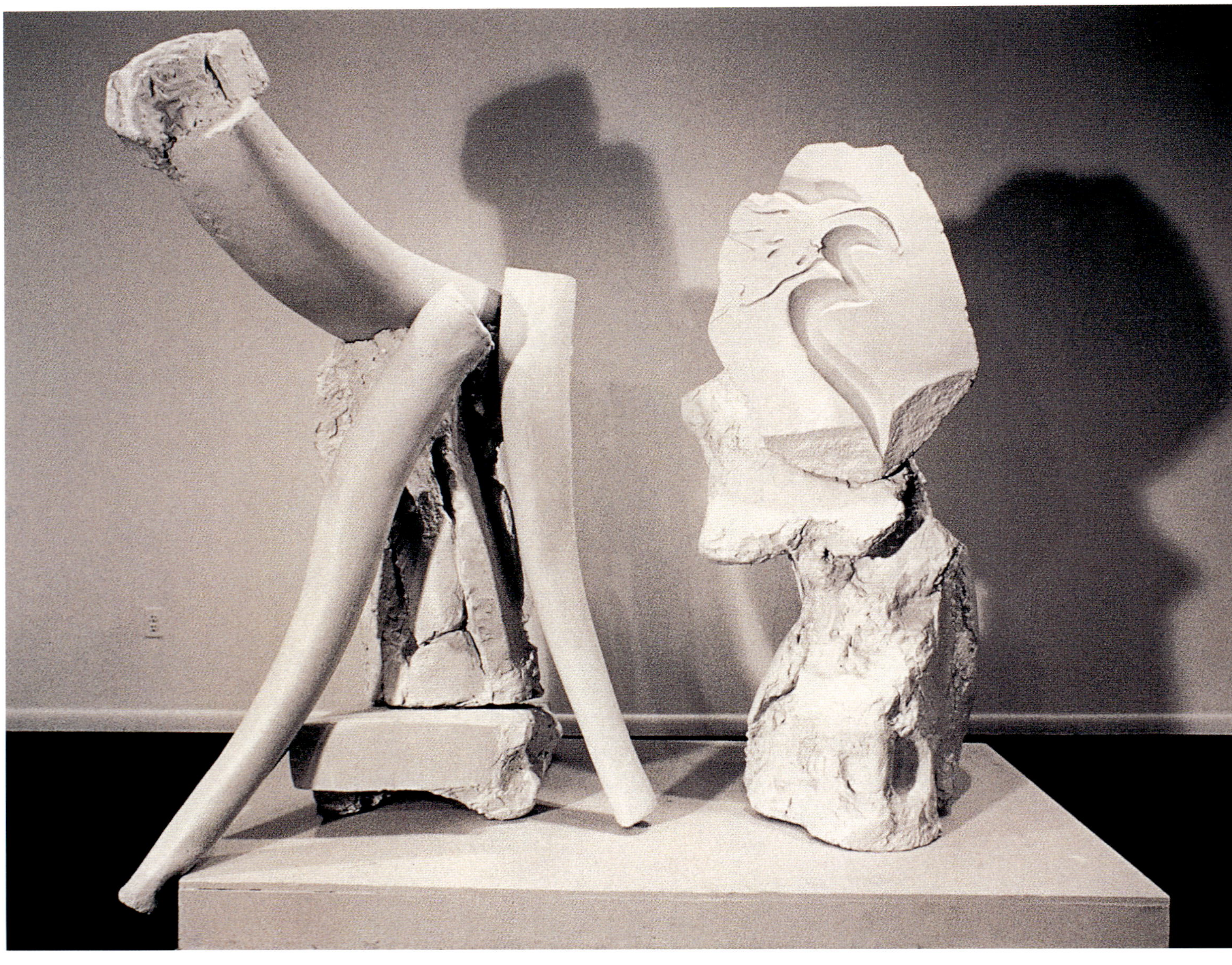

V-96. "Satyricon I" by Reuben Nakian. 1981.

OTHER MODERNIST SCULPTORS:

JOHN STORRS, MAX BECKMANN, CHARLES ALSTON, MARY CALLERY, CARL MILLES, JOSÉ DE CREEFT, and MAURICE STERNE

Sculptors of abstract and semi-abstract works (before 1960) who cast only a few of their pieces at Roman Bronze Works include:

JOHN STORRS (1885-1956) was an American sculptor who lived most of his life in France. A favorite pupil of Rodin, he turned toward Cubism early in his career, and began experimenting with abstract form and the organization of planes and contours. Though never well received in America, today Storrs' Cubist sculptures, like his "Standing Gendarme" (V-97) (which was cast by Roman Bronze Works) are considered seminal examples of modernism. During World War II, Storrs was imprisoned in a Nazi concentration camp; he died in France.

MAX BECKMANN (1884-1950), the great German Expressionist painter, also made a number of sculptures. One series was cast at Roman Bronze Works; this example is "Adam" (V-98).

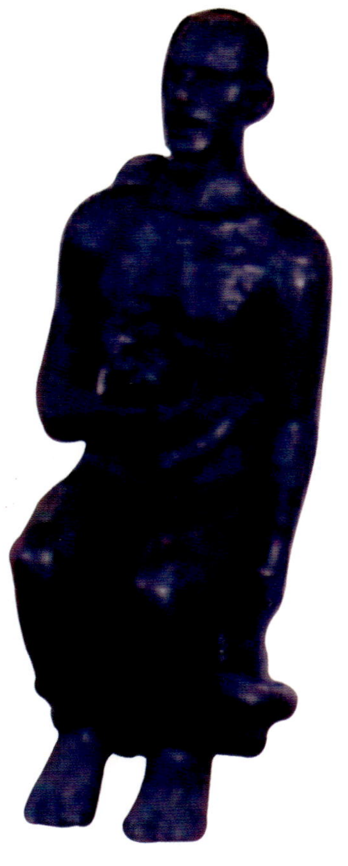

V-98. "Adam" by Max Beckmann.

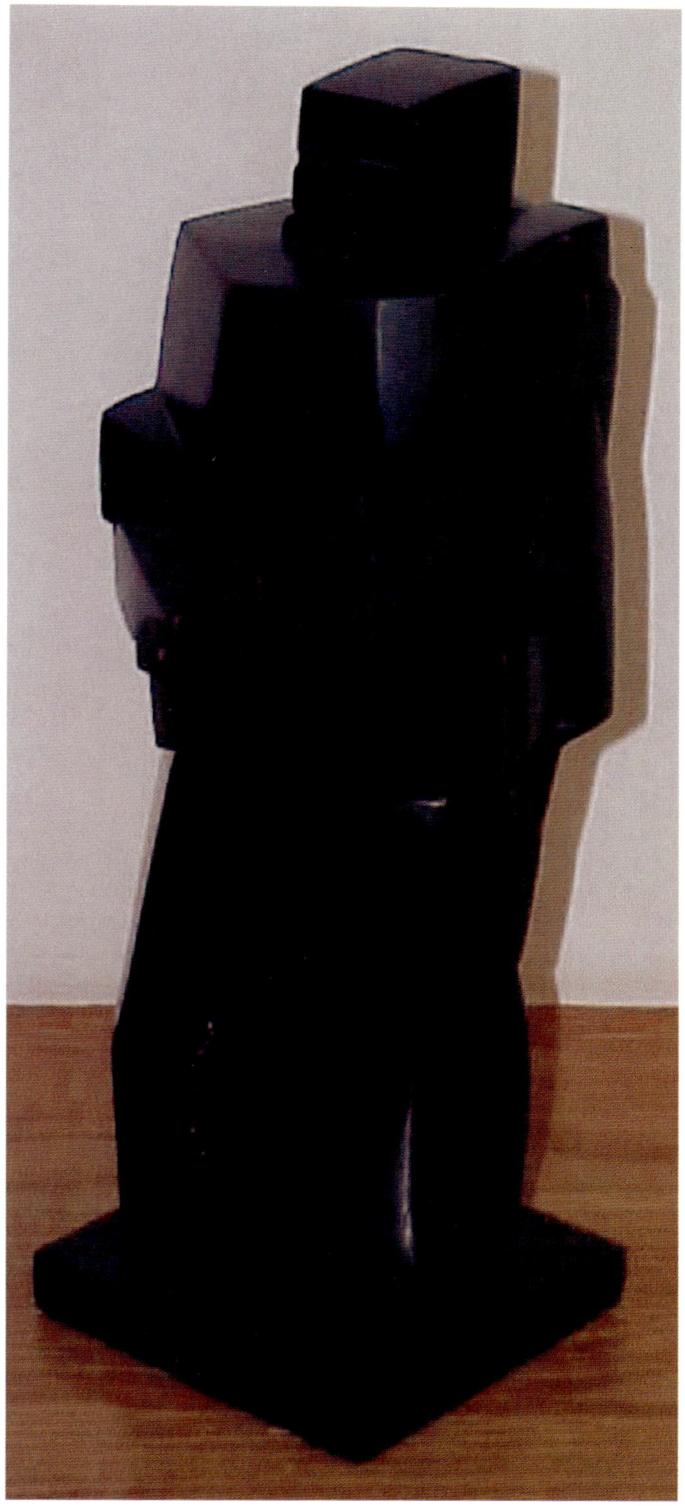

V-97. "Standing Gendarme" by John Storrs. c. 1925.

CHARLES ALSTON (1907-1977) was an African-American painter and sculptor whose works included murals for Harlem Hospital, paintings in abstract and semi-abstract style, as well as sculpture such as a "Head" cast at R.B.W. in 1957.

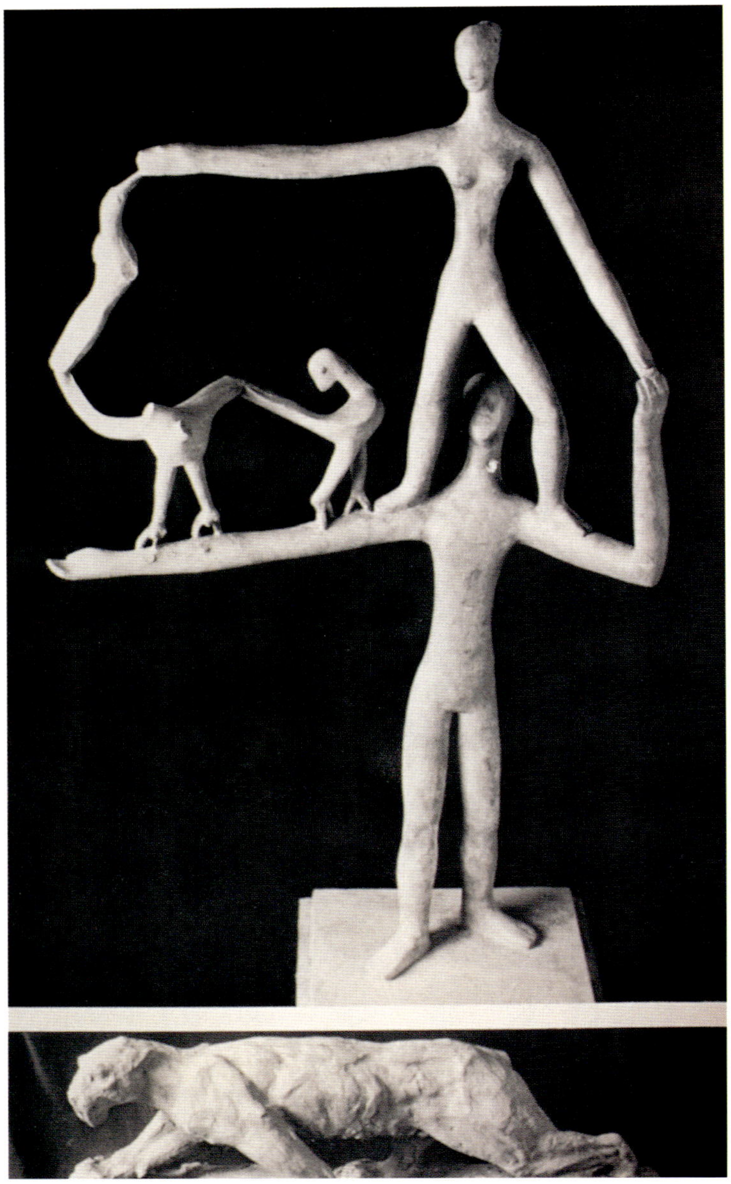

V-99. "Balancing Act" (plaster group of figures and birds) by Mary Callery. 1957.

MARY CALLERY (1903-1977) was a non-objective sculptor who worked in a constructivist, linear style. Her works were sometimes described as "drawings in space." Her first one-person show was in 1944, and she contributed pieces to numerous museum and exposition shows in the United States and Europe, including the Museum of Modern Art, and the Brussels World's Fair. Her "Balancing Act" (V-99) and "Fables of La Fontaine" (V-100) were cast at Roman Bronze Works in the 1950s, as were a star, a frog, and a mother and child.

CARL MILLES (1875-1953) was a native of Sweden who was, from 1931, a professor at the Cranbrook Academy in Michigan. Though he retained a classically oriented style, he employed elements of modernism with boldly simplified forms. Among works he cast at R.B.W. were fountain figures, an astronomer, and a unicorn, in addition to the Baird Memorial Fountain for Ann Arbor.

JOSÉ DE CREEFT (1900-1982) was primarily a carver of stone. He was an influential figure in modernist sculptural forms, which were nonetheless based on nature. He cast a few works at R.B.W. including a figure of a dog in 1930.

MAURICE STERNE (1878-1957) though primarily known as a painter, and draftsman, Sterne received many major commissions as a sculptor. He was granted the first one-man exhibition accorded an American at the Museum of Modern Art in 1933. His "Welcoming the People" for Fairmount Park, Philadelphia was cast at R.B.W. in 1939.

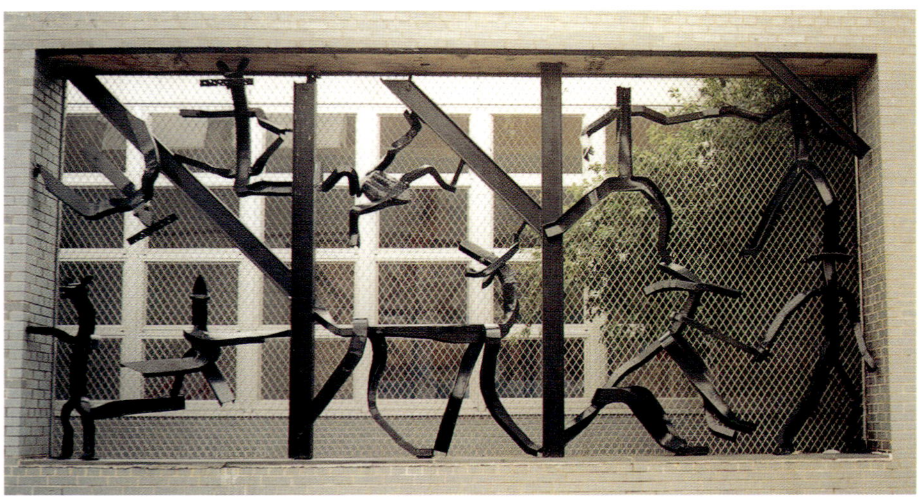

V-100. "The Fables of La Fontaine" by Mary Callery. 1954.

VI. WITHIN THE REALIST FRAMEWORK: GENRE, CLASSICISM, ALLEGORY, AND THE EXOTIC

While the modernists fought for acceptance and commissions, the vast majority of American sculptors remained firmly rooted within the realist framework. They were backed by the art world establishment, including the National Sculpture Society, the commission givers, the critics, and the art schools. Until mid-century the major commissions for monuments, museums, gallery exhibitions, sculpture gardens, and private collections went to sculptors working in realistic modes. Art schools furthered the traditional; their teachers were frequently the very sculptors who had based their careers in academic styles. With such art education, combined with the usual two or three years spent in academic training in Europe, most young American sculptors were firmly grounded in realistic styles. Their patrons, local and federal governments, as well as the great majority of critics, encouraged realism; they preferred American sculpture to be representational, recognizable, and representative of national ideals and tastes.

For, despite the introduction of European modernism and increasing exposure to different types of art in the first half of the century, most Americans remained unchanged in their preference for recognizable painting and sculpture. The reasons given for this sense of national taste were many: strong dislike for foreign influences, respect for the taste of the ordinary citizen (as opposed to the aesthete's), desire for art that was moral and uplifting or nostalgic in theme, and above all, the choice of art that seemed homegrown and indicative of America itself. The battle between realism and modernism, as we have seen, was a vociferous one that drew critics and artists alike into angry aesthetic, political, and social debate.

Within the realistic framework there was increasing interest in the collecting of art for both public and private venues. In fact, sculpture experienced something of a renaissance in America during the first half of the century. Many individual sculptors, no longer dependent solely on creating the nation's monuments, were able to make careers doing smaller works cast in bronze for private homes and gardens, and public buildings and parks. Individual collectors, as well as public organizations, began to buy sculpture. The sculptor was increasingly seen as a professional, a person whose work represented national taste and sophistication.

Realist sculptors took many different stylistic directions. Realism reflected diverse views of American society, with sculptors using their work to convey a variety of ideas: patriotism, optimism, sympathy for the poor and downtrodden, or nostalgia for the past. Some worked in genre or in a narrative style, a thoroughly American tradition, updating their work to reflect daily life in the twentieth century. Some remained firmly in the Neo-Cclassical or Beaux-Arts traditions of the previous century. Many worked in symbolic or allegorical styles, using ancient themes to express timeless beliefs and legends. Some stretched their subject matter to include foreign and exotic civilizations, which they portrayed as carefully as their counterparts in the animalier field. Continuing in traditions that were clearly articulated in the first decades of the twentieth century, the work of these sculptors ranged from graceful Neo-Classical silhouettes symbolizing civic virtues to cutely whimsical "child gods;" from glamorous portrayals of the socially prominent to statues of the working man; from the naturalistic "Westerns" and animals we have already discussed to exotic dancers and natives of foreign lands.

The tradition of genre art was well-established in the United States, both in painting and sculpture. Genre artists told stories using recognizable characters—the old farmer, the naughty child, the family doctor. Nostalgia for a simpler time and a humorous outlook made genre art very popular. R.B.W. produced many copies of John Rogers' genre works of the previous century. Some twentieth century sculptors continued this tradition; among them was Bessie Potter Vonnoh, whose sculptures of devoted mothers and small children were tremendously popular. But as society changed and social realism depicting a more industrialized nation developed, genre art reflected a more diverse world, with sculptors like Mahonri Young and Max Kalish portraying laborers, and Abstenia St. Leger Eberle picturing immigrants on New York's Lower East Side. There was occasional story-telling art or art with a political message (as in James Earle Fraser's in "The End of the Trail," or Hermon MacNeil's "The Slave Block,")" but social realism and political art never became a dominant movement in sculpture as in the world of painting.

Neo-Classicism continued to prevail as a preferred style, not only for monuments, but for smaller works as well. Using time-honored traditions of Greek-robed goddesses and nude maidens to represent the civic virtues and the legends of the past, numerous American sculptors continued to work in academic styles learned abroad and influenced by sojourns in Italy, Greece, and France. Typical of these neoclassical sculptors were Frederick MacMonnies, Edward McCartan, and Paul Jennewein who depicted Dianas and Bacchantes, Greek dancers, nymphs, and fauns (as did many of their colleagues). The ideally proportioned bodies of ancient athletes had their counterparts in the runners and wrestlers of Robert Tait McKenzie. Allegorical themes gave the many beautifully proportioned nudes a thematic raison-d'etre. Even the small "putti" of Italian art were recreated for fountains and public gardens.

Exotic and foreign themes also appealed to sculptors in the first half of the century. The mysterious lands of Asia and Africa were visited by several sculptors, such as Malvina Hoffman and Alan Clark, who meticulously documented the sights they saw. (Hoffman's sculptures of native dancers and warriors in exotic costumes were so accurate as to be used in museum displays.) Some sculptors absorbed stylistic devices of exotic art from other parts of the world, incorporating them into their own work.

The following pages provide a sampling of the many smaller pieces created by America's best -known realist sculptors and cast at Roman Bronze Works. Along with the century's great sculptors of monuments, these men and women were the makers of most of the notable three-dimensional art in realistic styles to be found in the nation's collections. (Genre artists, are followed by neo-classical, allegorical, and exotica sculptors, in chronological order by date of birth.)

JOHN ROGERS (1829-1904)

John Rogers was the premier genre sculptor in the nineteenth century; between 1860 and 1893 the American public bought some eighty thousand of his putty-colored statuettes. At prices ranging from five to fourteen dollars, buyers were able to have small sculptures depicting everyday events with careful realism and a certain humorous charm or pathos. "The Rogers (sculpture) was indeed almost as much of a necessity to a well-appointed Victorian parlor as the antimacassar and the chromo of George Washington," according to one commentator. His naturalistic groups of figures were captured in life-like action in scenes from daily life, literature, or sports.

Rogers was born in Salem, Massachusetts, became a machinist and draftsman, and a master railroad mechanic in Missouri. He began modeling small clay figures and genre groups for his own amusement. When he lost his railroad job he studied sculpture briefly in Paris and Rome in 1858, but found his own taste for narrative genre art completely at odds with the neo-classical training in Europe. He returned home, set up a studio in New York and with his first Civil War subjects suddenly found himself successful. He began making plaster casts and patented his designs; he was the first American sculptor to mass-produce his work. Rogers established his own factory, producing as many as 12,000 copies of a group sculpture from molds. The sculptures were sold at general stores or from a catalog. He also made several monuments, including a life-size Abraham Lincoln, but he was best known for his genre sculpture. In 1893 Rogers was a founder of the National Sculpture Society.

The New York Historical Society owns 37 of the original bronzes, and dozens of his plasters. Roman Bronze Works acquired some of his plasters in the early 1900s and reproduced them in bronze. Among them are "Uncle Ned's School" (VI-1), and "Mail Day" (VI-2).

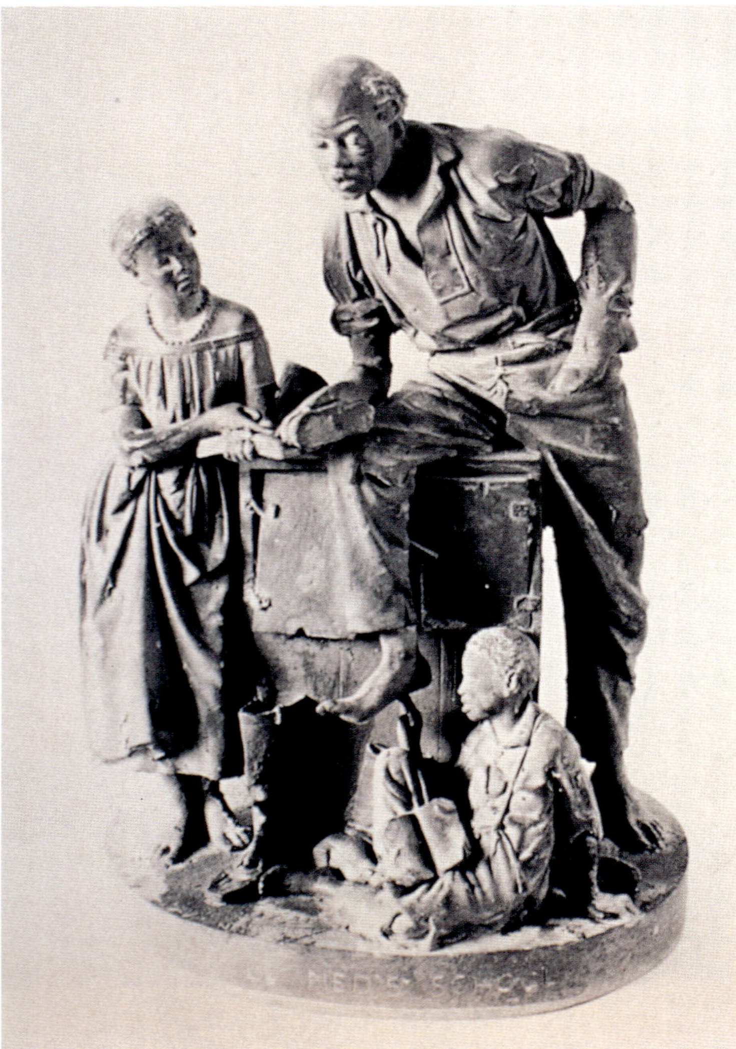

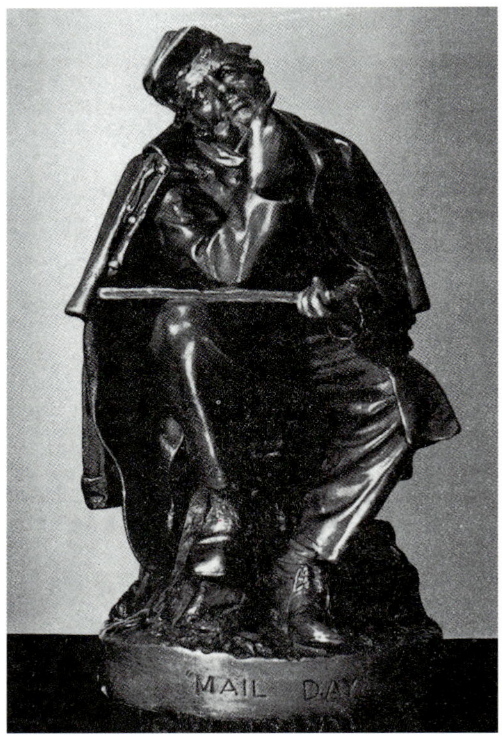

VI-1. "Mail Day" by John Rogers.

VI-2. "Uncle Ned's School" by John Rogers. Cast acquired by R.B.W. in the early 1900s and reissued in the 1970s.

THOMAS EAKINS (1844-1916)

America's foremost realist painter in the late 19th and early 20th century, Thomas Eakins is rarely mentioned as a sculptor. But his emphasis on sculptural form and three-dimensional pictorial space in his painting suggests an interest in the plastic arts as well as painting and photography. In fact, Eakins made a number of small sculptural works, and, like his painting, they are detailed and realistic, with anatomy or scenes of daily life as their subjects.

Eakins was born in Philadelphia, where he lived and taught for most of his life. His early education was in drawing at the Pennsylvania Academy of Fine Arts, and the study of anatomy at the Jefferson Medical College. He had three years of training in Paris in academic painting and sculpture, before returning to Philadelphia in 1870. Eakins' first works were portraits of friends and family, and observant studies of light, form, and structure. His interests were in an almost scientific rendering of reality, and to that end he made numerous studies of exact perspective and anatomy; he also used photography to enhance the accuracy of his perceptions. His portraits were uncompromising in their careful analysis and deep psychological insight. His emphasis on naturalistic portrayals of the human body led to protests at the Pennsylvania Academy of Fine Arts, where he had become a teacher, and his most striking works were rejected by a prudish late-Victorian society. But Eakins' contribution to American realist painting was unmatched.

As a sculptor, Eakins produced both genre pieces, "Knitting" (VI-3) and "Spinning" (VI-4), and anatomical studies, "Cast of a Hand" (VI-5), and three horse studies: "Skeleton of a Horse" (VI-6), "Clinker" (VI-7), and "Relief of a Horse" (VI-8). He also cast a relief called "Arcadia" (VI-9) and "Youth Playing Pipes" (VI-10). All of these works (originally made between 1876 and 1892) were posthumously cast by Roman Bronze Works at the request of the artist's wife.

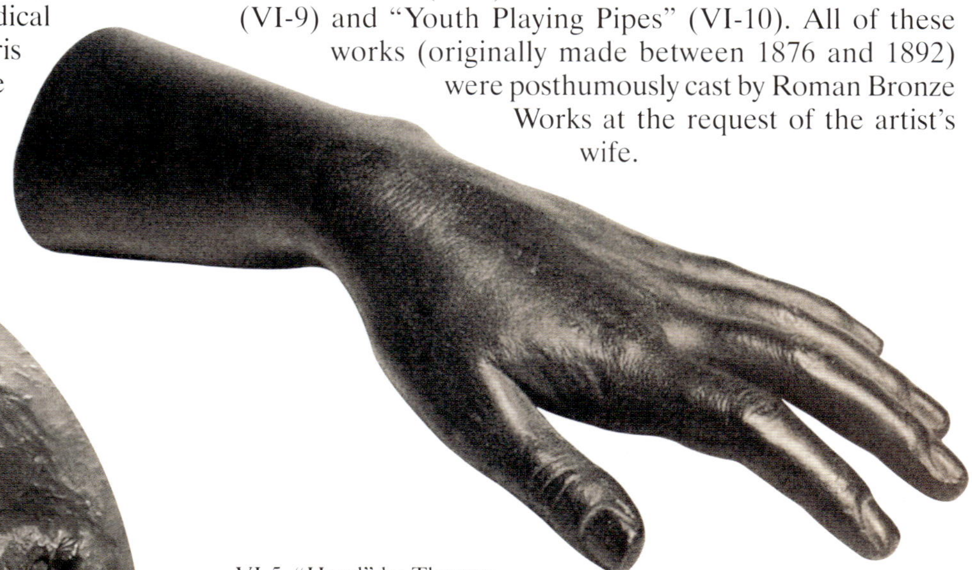

VI-5. "Hand" by Thomas Eakins. Originally cast c. 1878. Cast by R.B.W. 1928-30.c

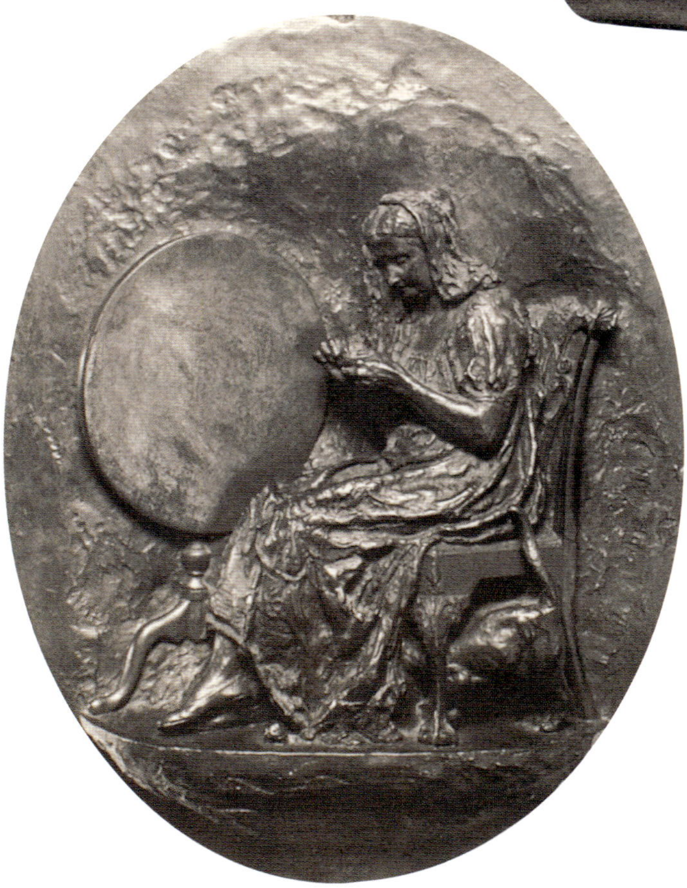

VI-3. "Knitting" by Thomas Eakins. Originally cast in 1882-3; recast by R.B.W. in 1930.

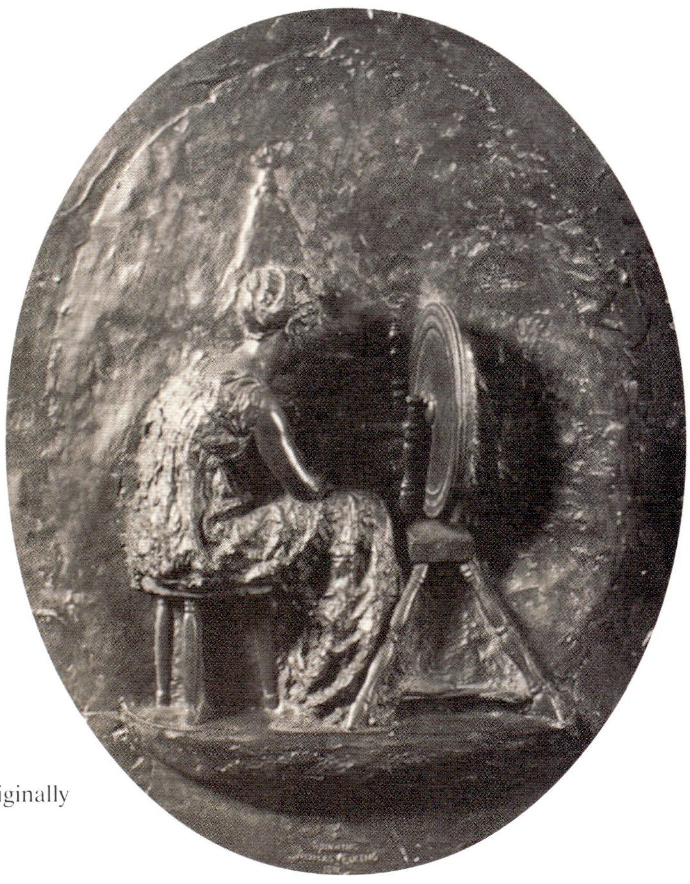

VI-4. "Spinning" by Thomas Eakins. Originally cast in 1882-3; recast by R.B.W. in 1930.

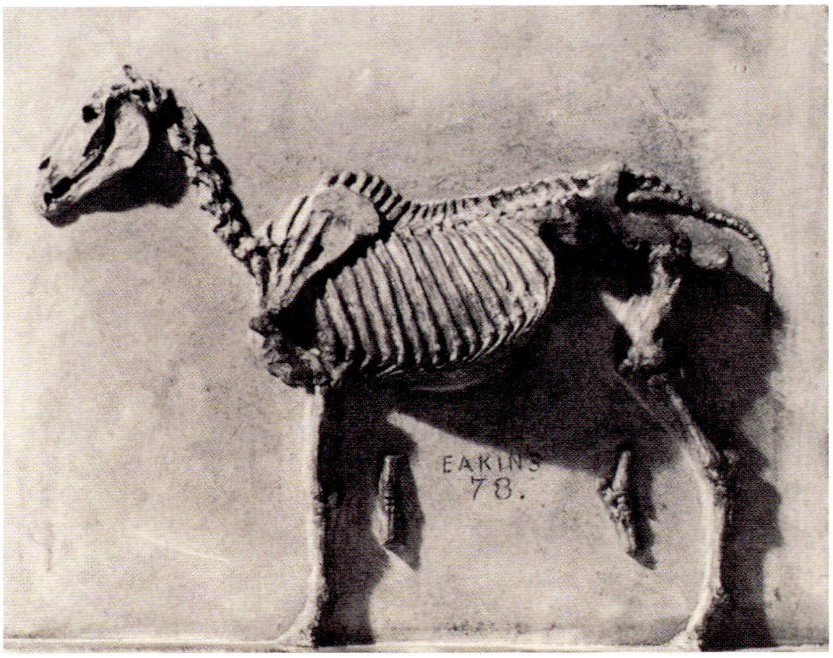

VI-6. "Skeleton of a Horse" by Thomas Eakins. Original 1878; recast by R.B.W. in 1928-30.

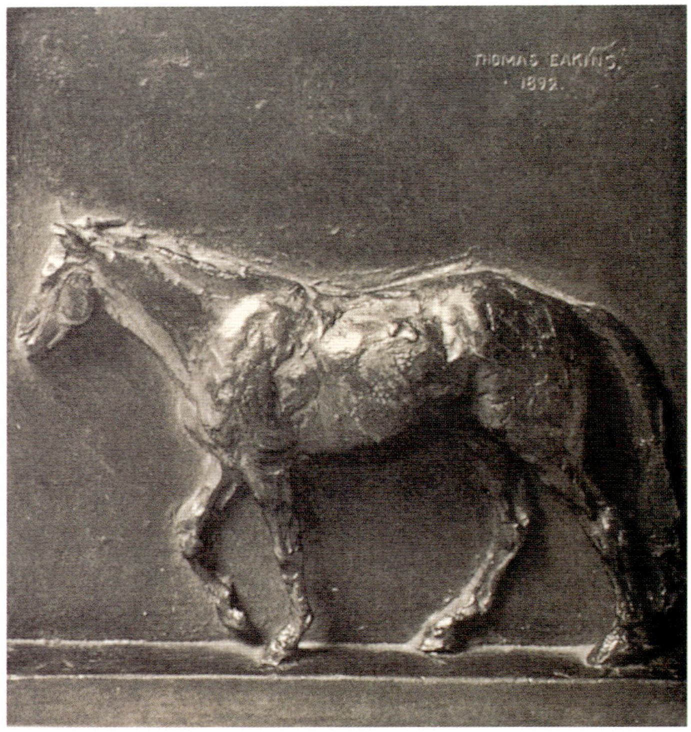

VI-7. "Clinker" by Thomas Eakins. Original c. 1892; cast by R.B.W. in 1930.

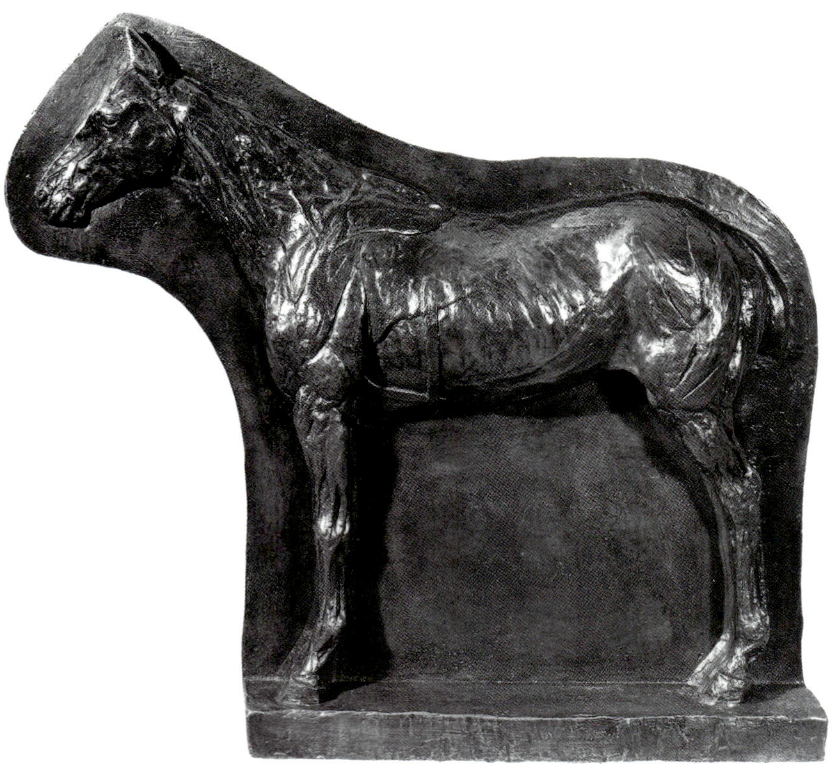

VI-8. "Relief of a Horse" by Thomas Eakins. Original 1882; cast by c. 1928-30.

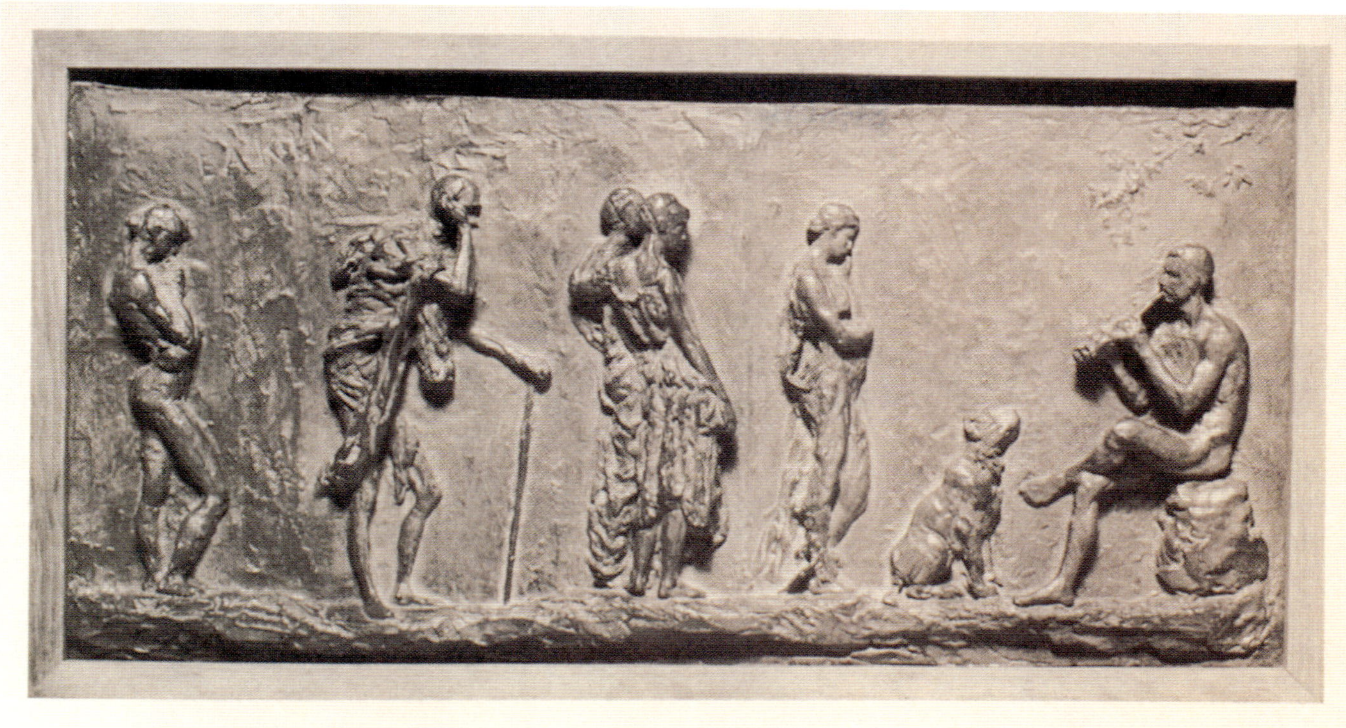

VI-9. "Arcadia" by Thomas Eakins. Original 1883; recast by R.B.W. 1928-30.

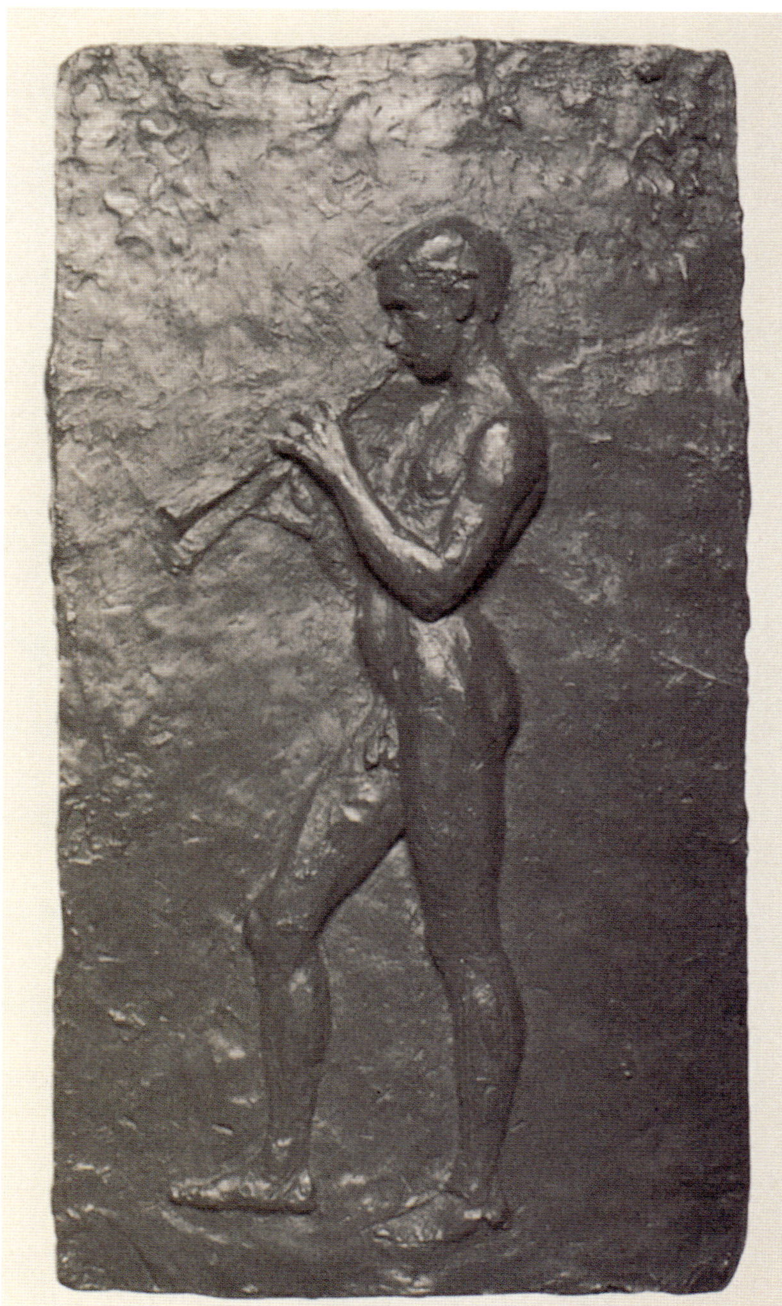

VI-10. "Youth Playing Pipes" by Thomas Eakins. Original 1883; recast by R.B.W. in 1930.

BESSIE POTTER VONNOH (1872-1955)

Bessie Potter Vonnoh was the most prolific and popular of genre sculptors in the first three decades of the 20th century. Her subjects—primarily mothers and small children—were drawn from domestic life, and her style was warm, engaging, and intimate, with a slight Impressionistic influence. In the tradition of genre art, the scenes she portrayed were loving, recognizable, somewhat nostalgic, and the public loved them. She also made a few portrait busts and several full-sized fountain figures.

Born in St. Louis, Vonnoh received her first training at the Art Institute of Chicago, where she studied sculpture with Lorado Taft and became his assistant. She opened her own studio in 1894 and soon realized that her forte was not for heroic works but for smaller statues portraying "a sort of delicate domesticity." Her genre pieces were produced in multiples. In Paris in 1895 (when she met Rodin) she exhibited her sculpture "The Young Mother" (VI-11); it won her a bronze medal at the Paris Exposition. For some thirty years following she produced dozens of genre pieces. She received numerous medals and awards, and was elected to the National Institute of Arts and Letters and was an academician of the National Academy of Design.

Vonnoh began her relationship with Roman Bronze Works in 1907, and continued to cast her works there until 1939. Among the dozens of her small sculptures cast by R.B.W. were many images of fashionable women (numbers VI-12 to VI-19); mothers and small children (numbers VI-20 to VI-24); young girls (numbers VI-25 to VI-28); fountain figures (numbers VI-29 and VI-30); and medallions (VI-31).

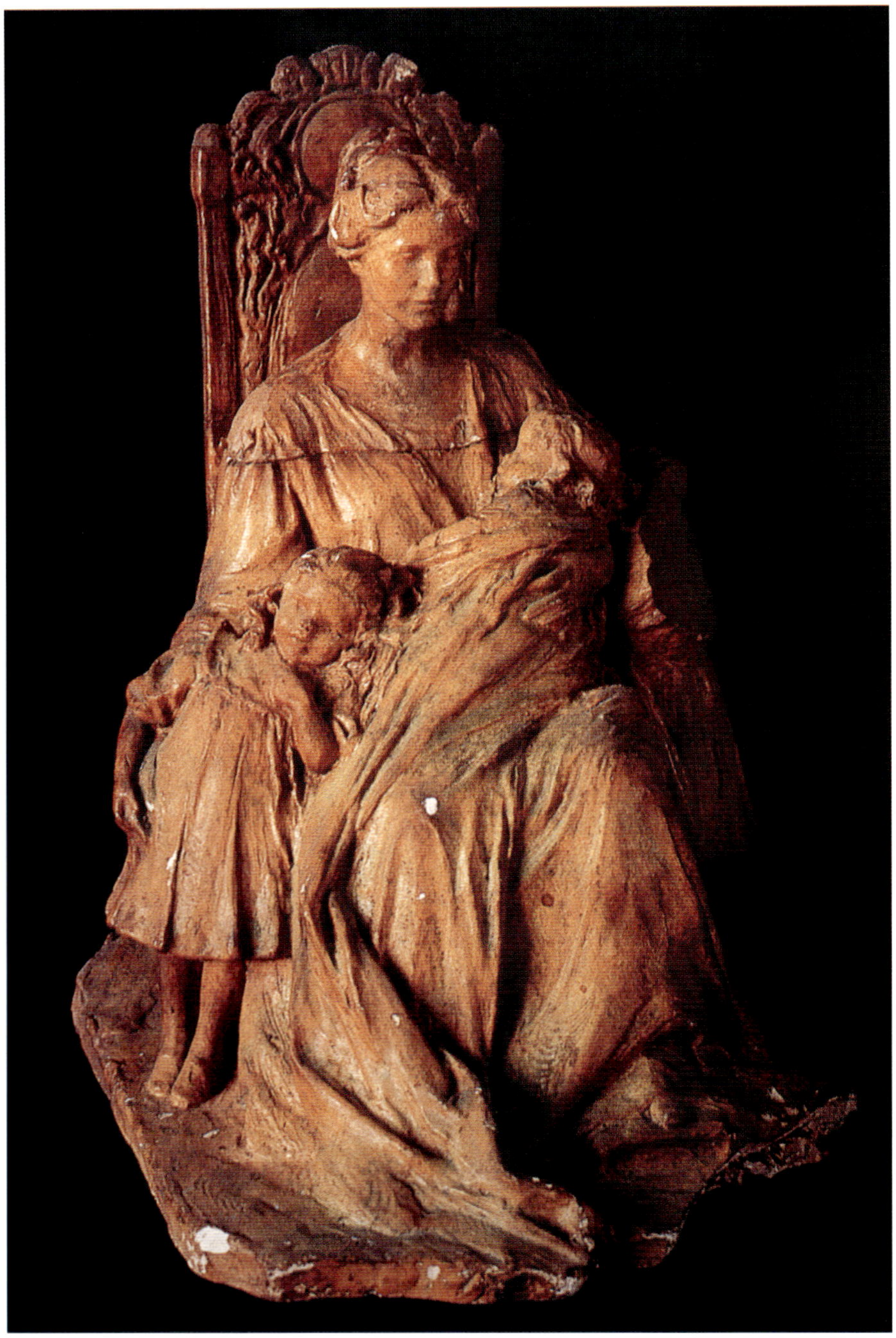

VI-11. "The Young Mother" by Bessie Potter Vonnoh. Originally cast 1896; R.B.W. subsequently made thirty casts.

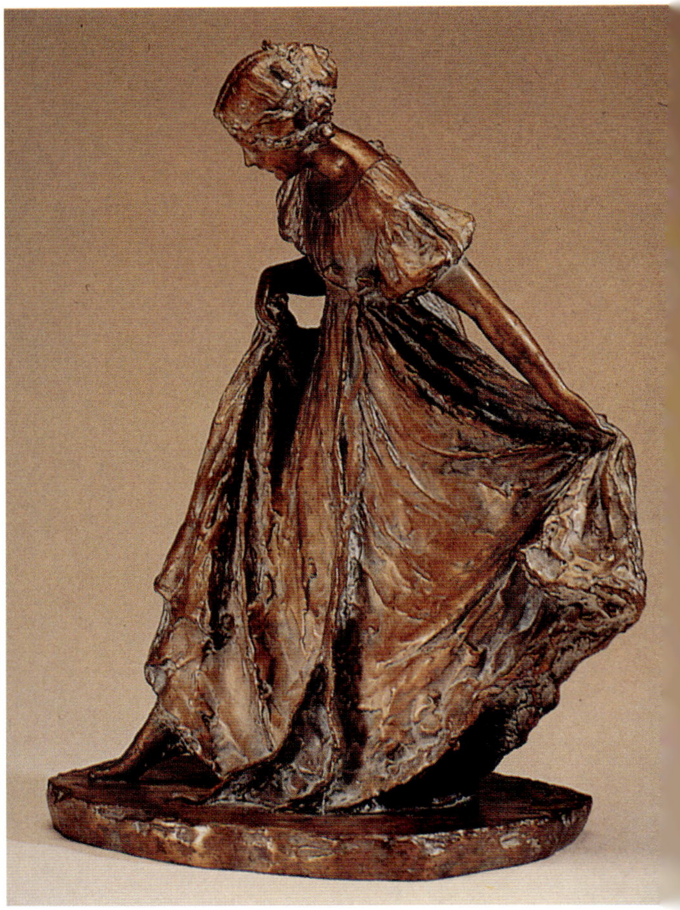

VI-12. "Girl Dancing" by Bessie Potter Vonnoh. Modeled in 1898, this was one of Vonnoh's first successful works.

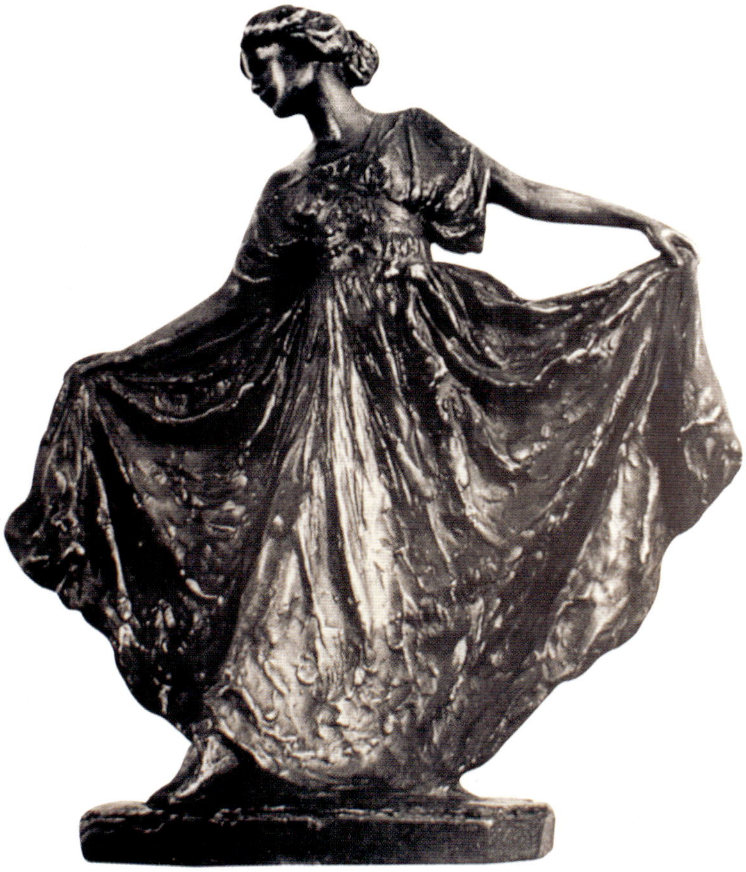

VI-13. "Girl Dancing" by Bessie Potter Vonnoh. 1900; this model posthumously cast.

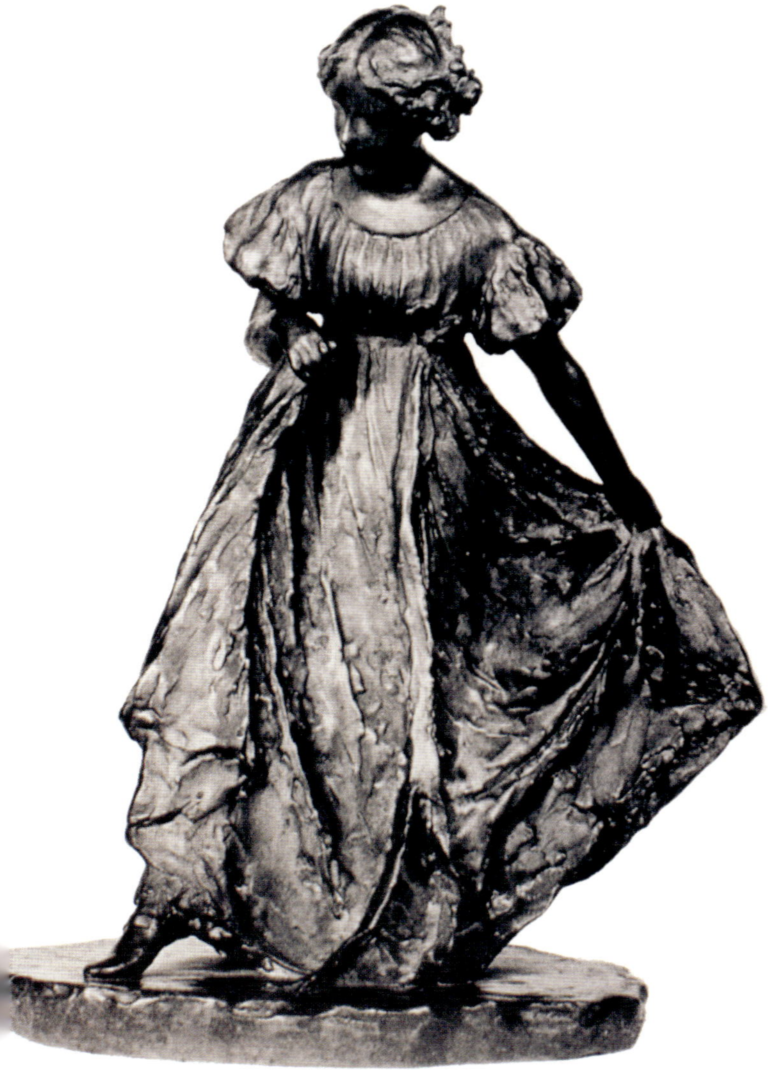

VI-14. "Party Dress" by Bessie Potter Vonnoh. c. 1919.

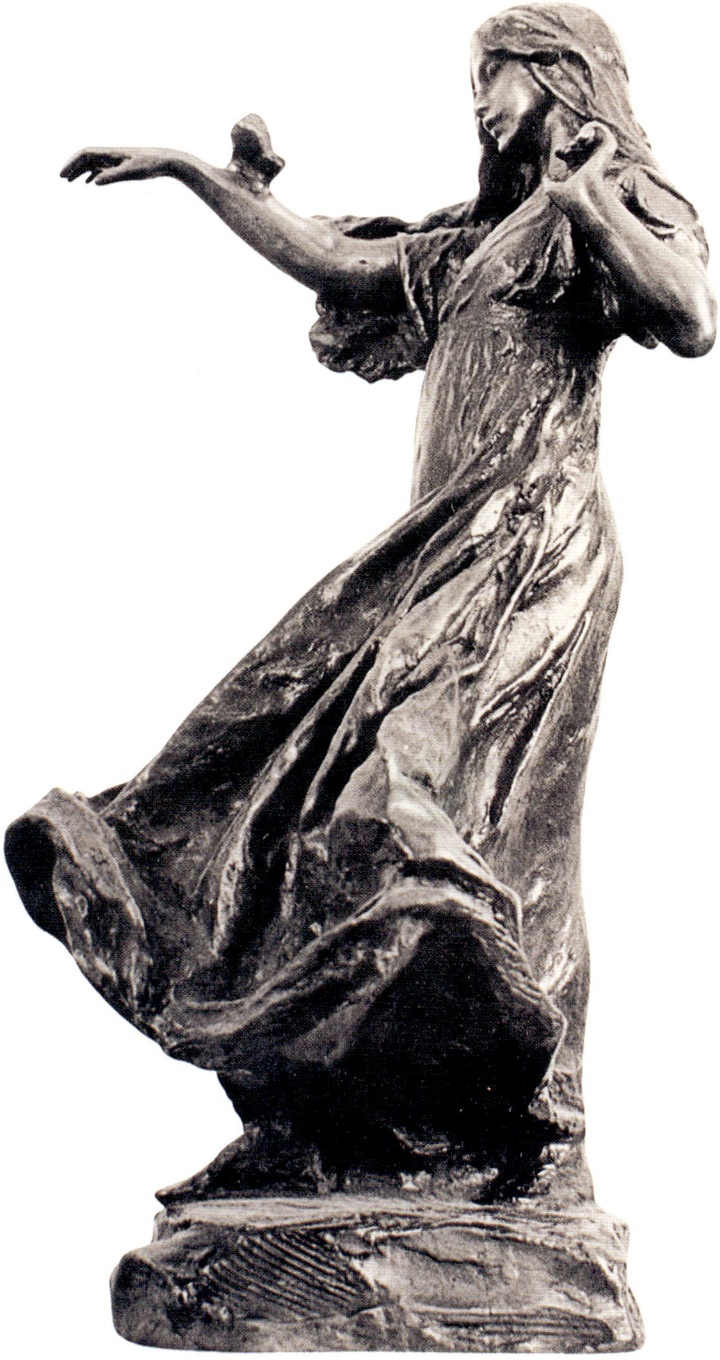

VI-15. "Butterflies" by Bessie Potter Vonnoh. Modeled c. 1900; posthumously cast.

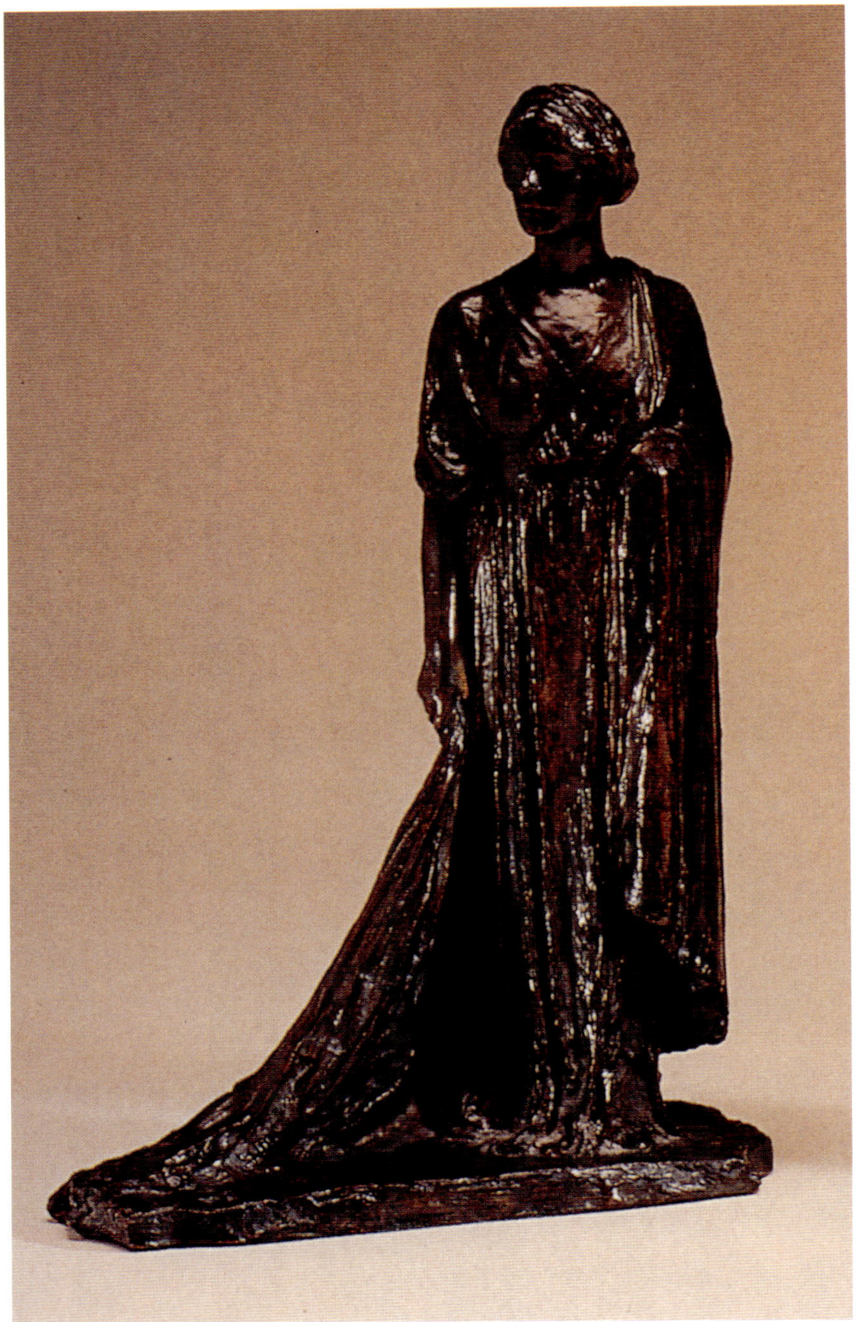

VI-16. "A Fashionable Woman" by Bessie Potter Vonnoh. 1911.

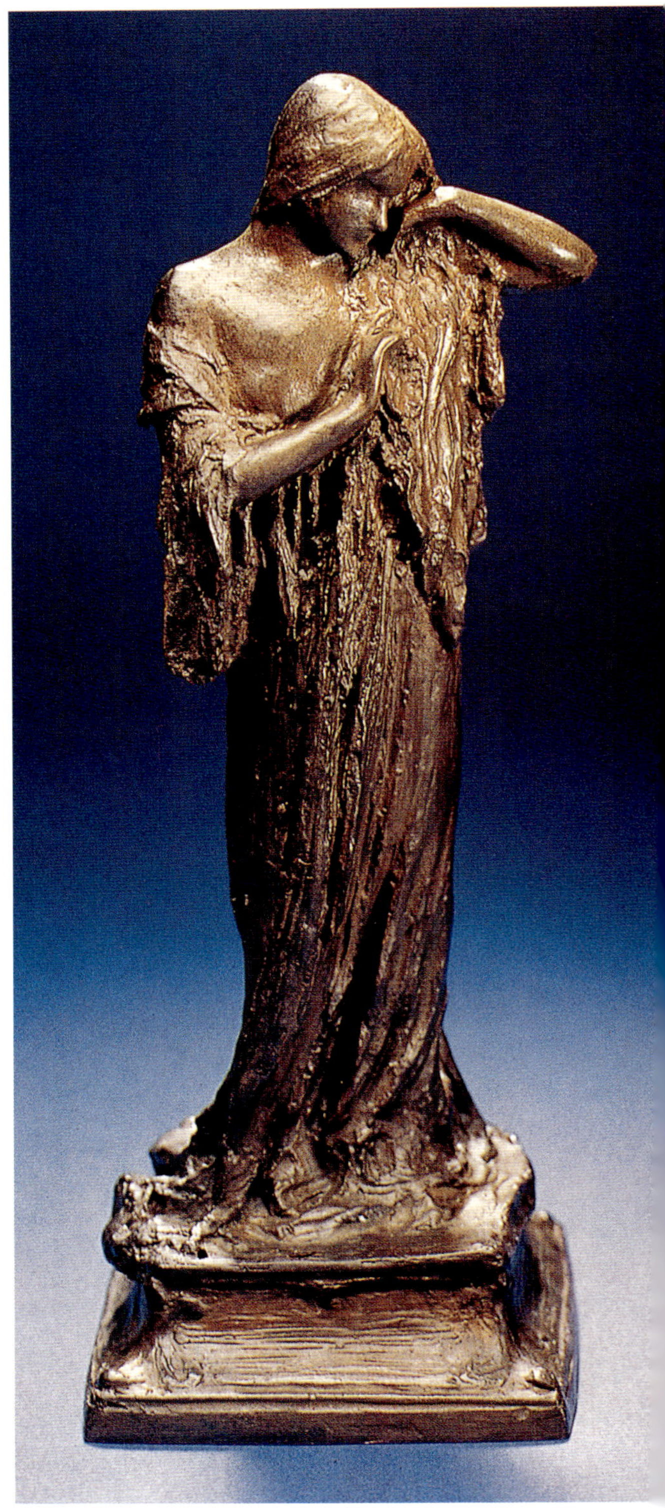

VI-17. "Alicia" by Bessie Potter Vonnoh.

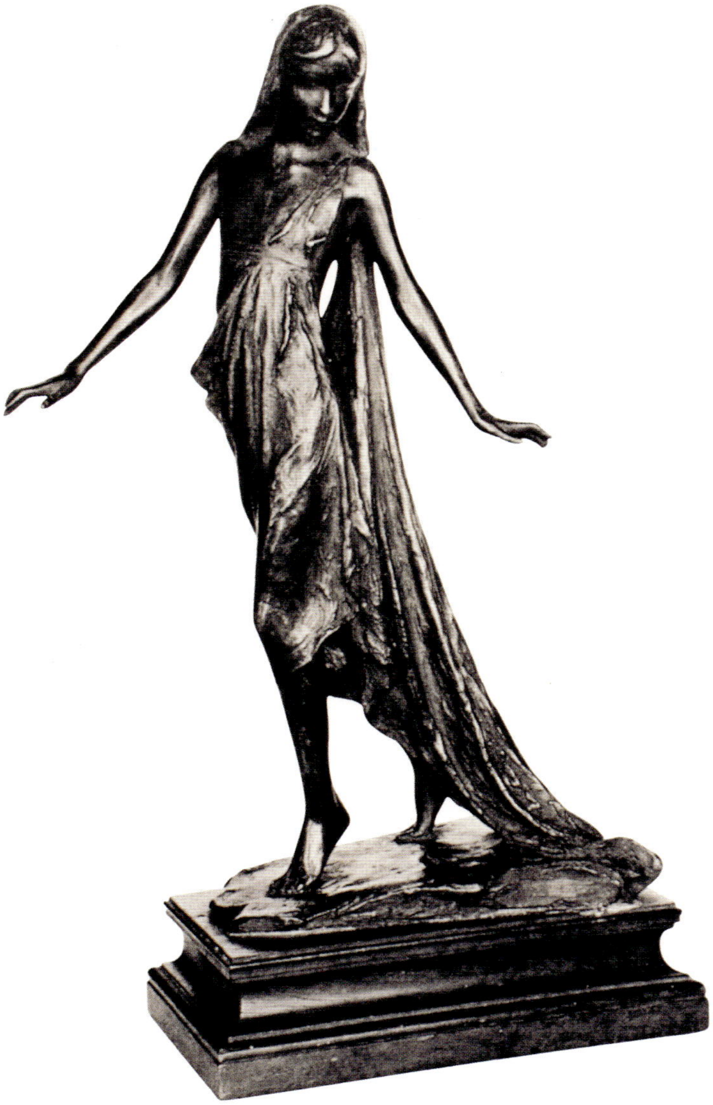

VI-18. "Dainty Steps" by Bessie Potter Vonnoh. c. 1925.

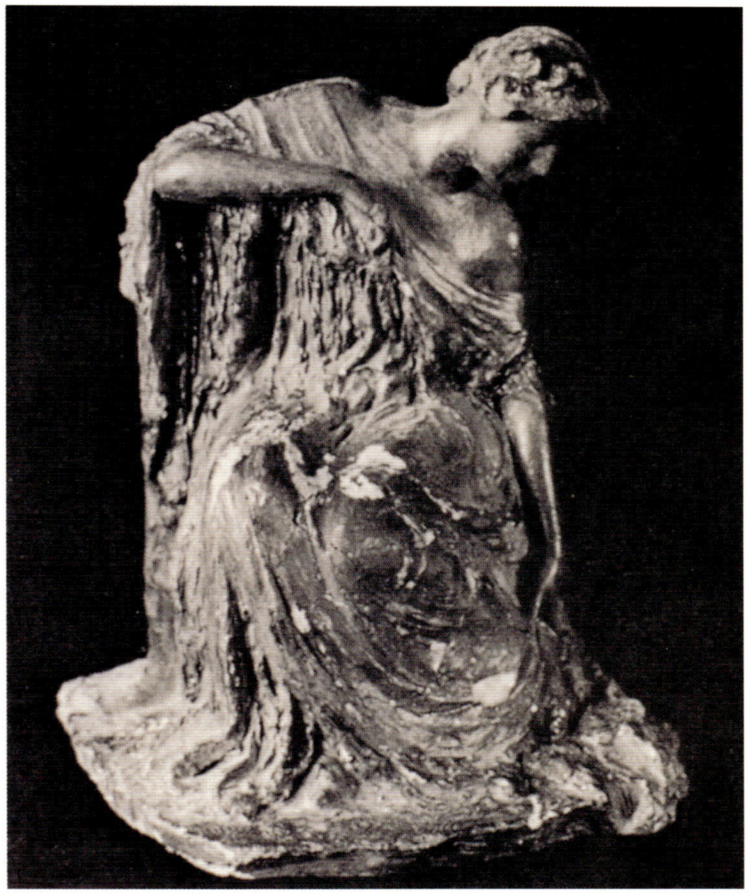

VI-19. "Draped Seated Woman Leaning Over" (plaster cast) by Bessie Potter Vonnoh.

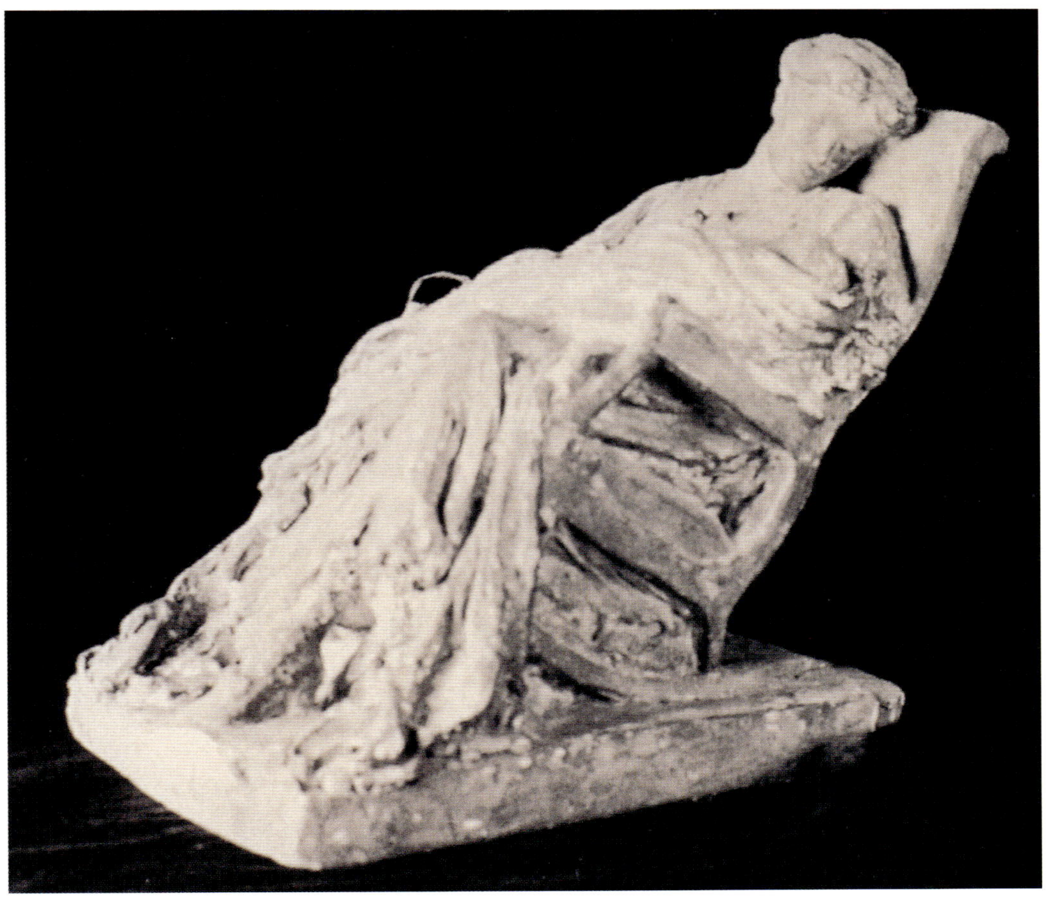

VI-20. "Mother and Child in Rocking Chair" (plaster cast) by Bessie Potter Vonnoh.

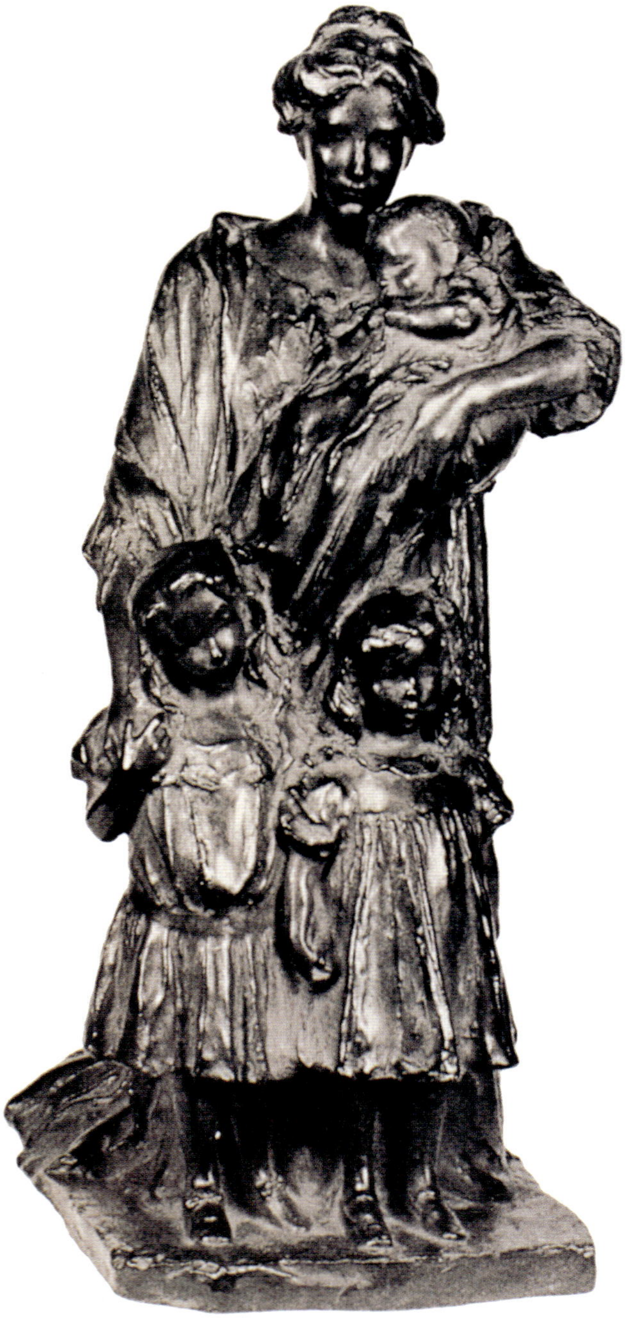

VI-21. "Mother of Three" by Bessie Potter Vonnoh. c. 1900.

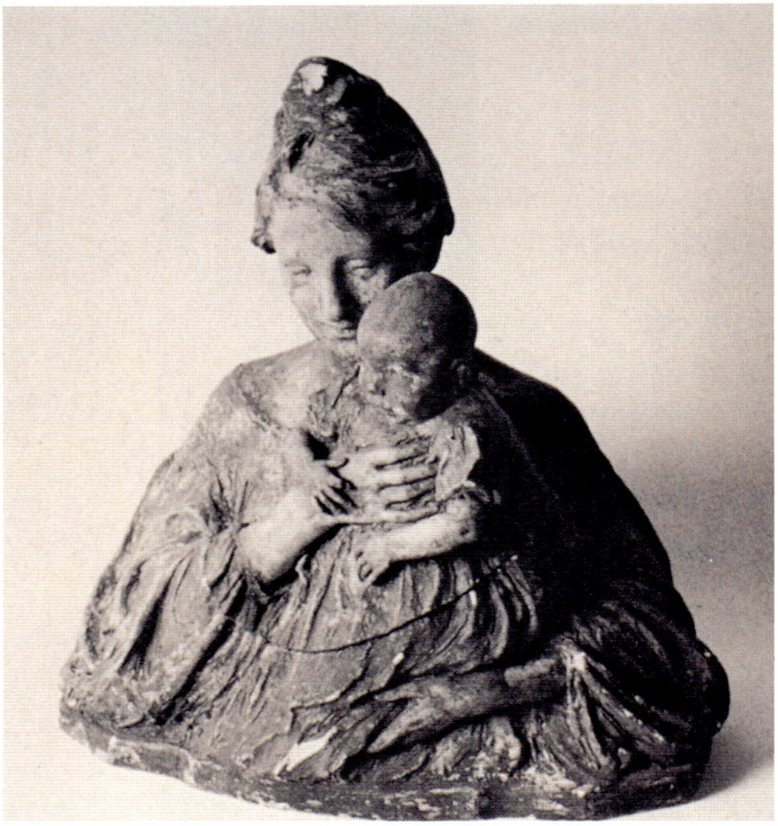

VI-22. "Bust of Mother Holding Baby" (plaster cast) by Bessie Potter Vonnoh.

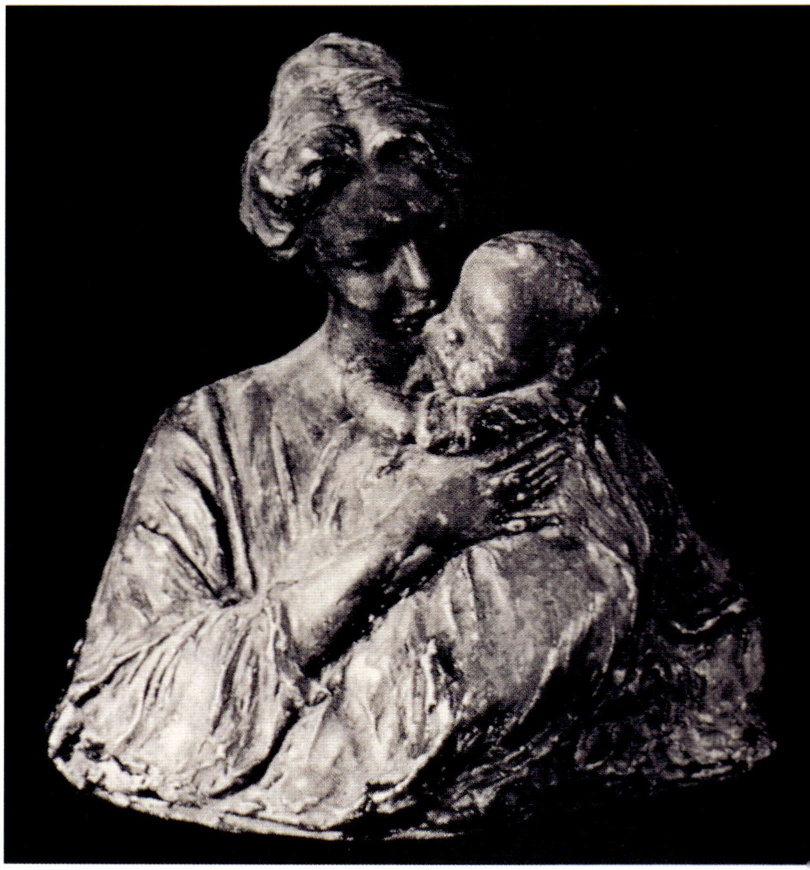

VI-23. "Mother and Child" (plaster cast) by Bessie Potter Vonnoh.

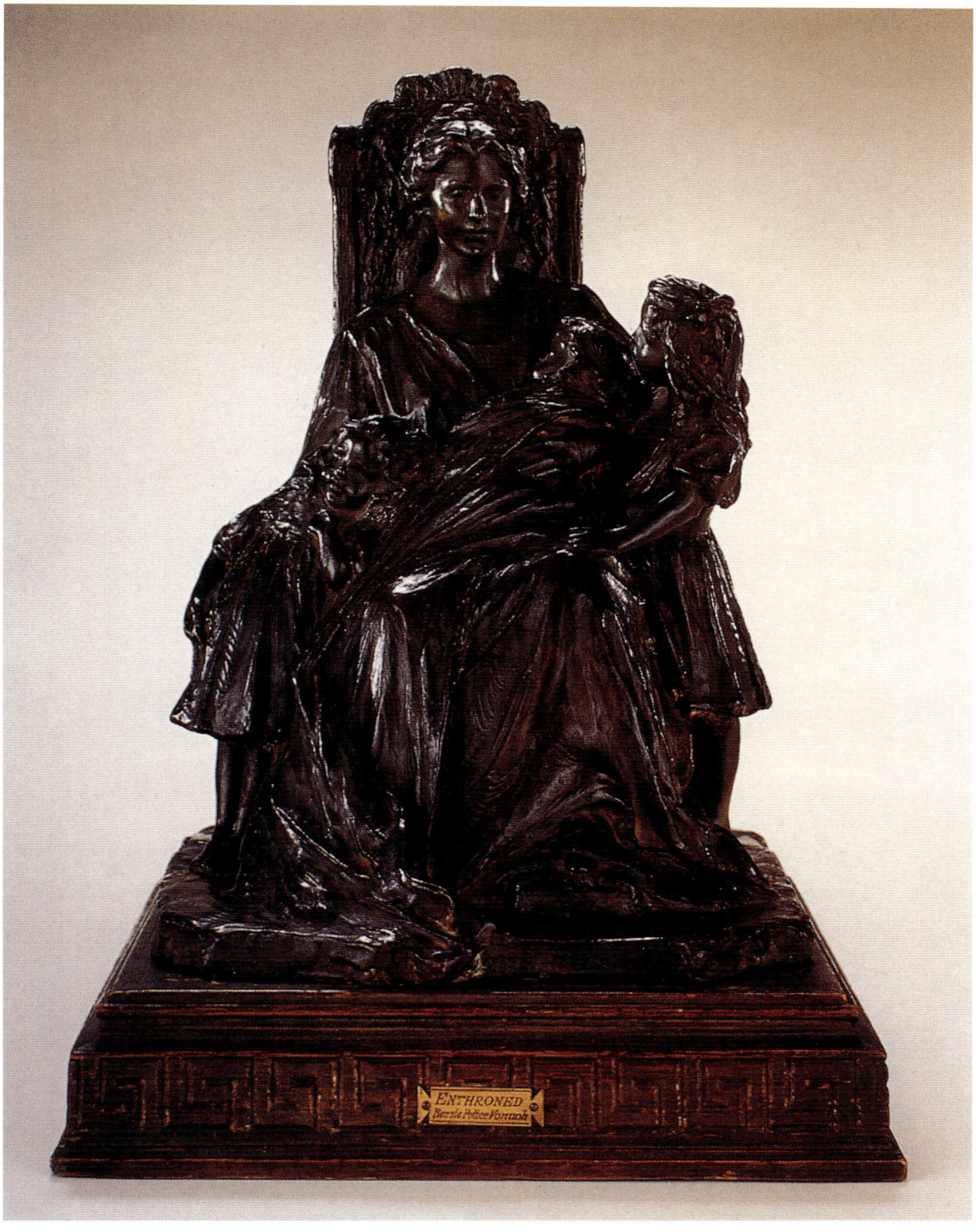

VI-24. "Enthroned" by Bessie Potter Vonnoh. Seventeen casts dated 1903-1916; two additional casts in 1929.

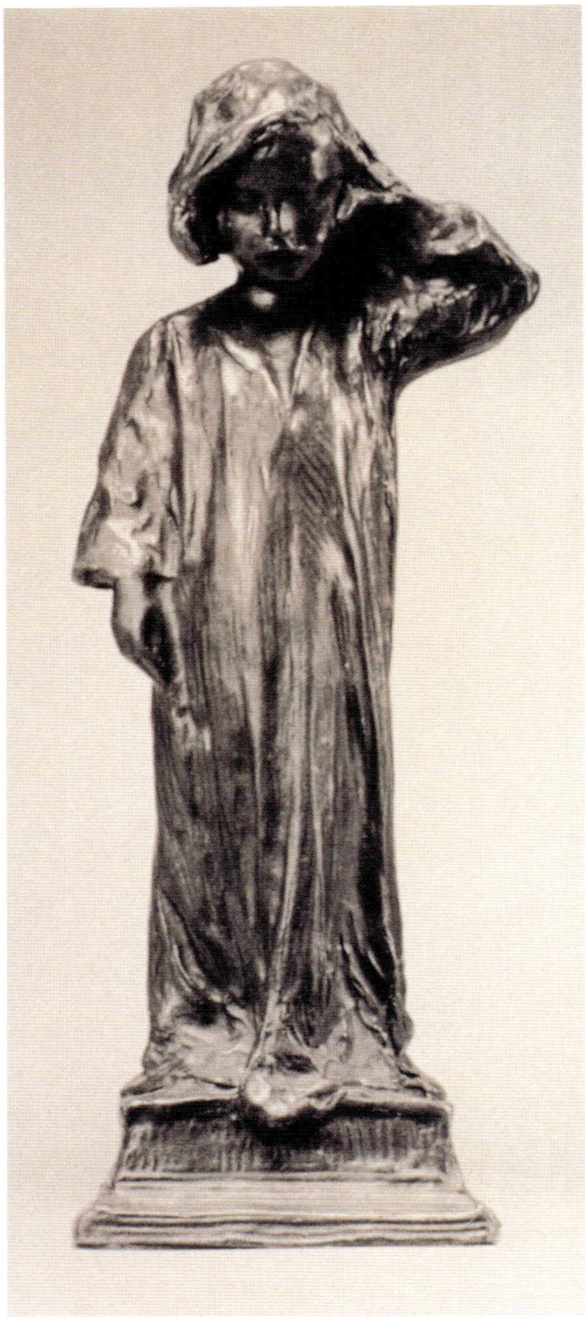

VI-25. "Good Night" by Bessie Potter Vonnoh. Twenty-nine cast before 1916.

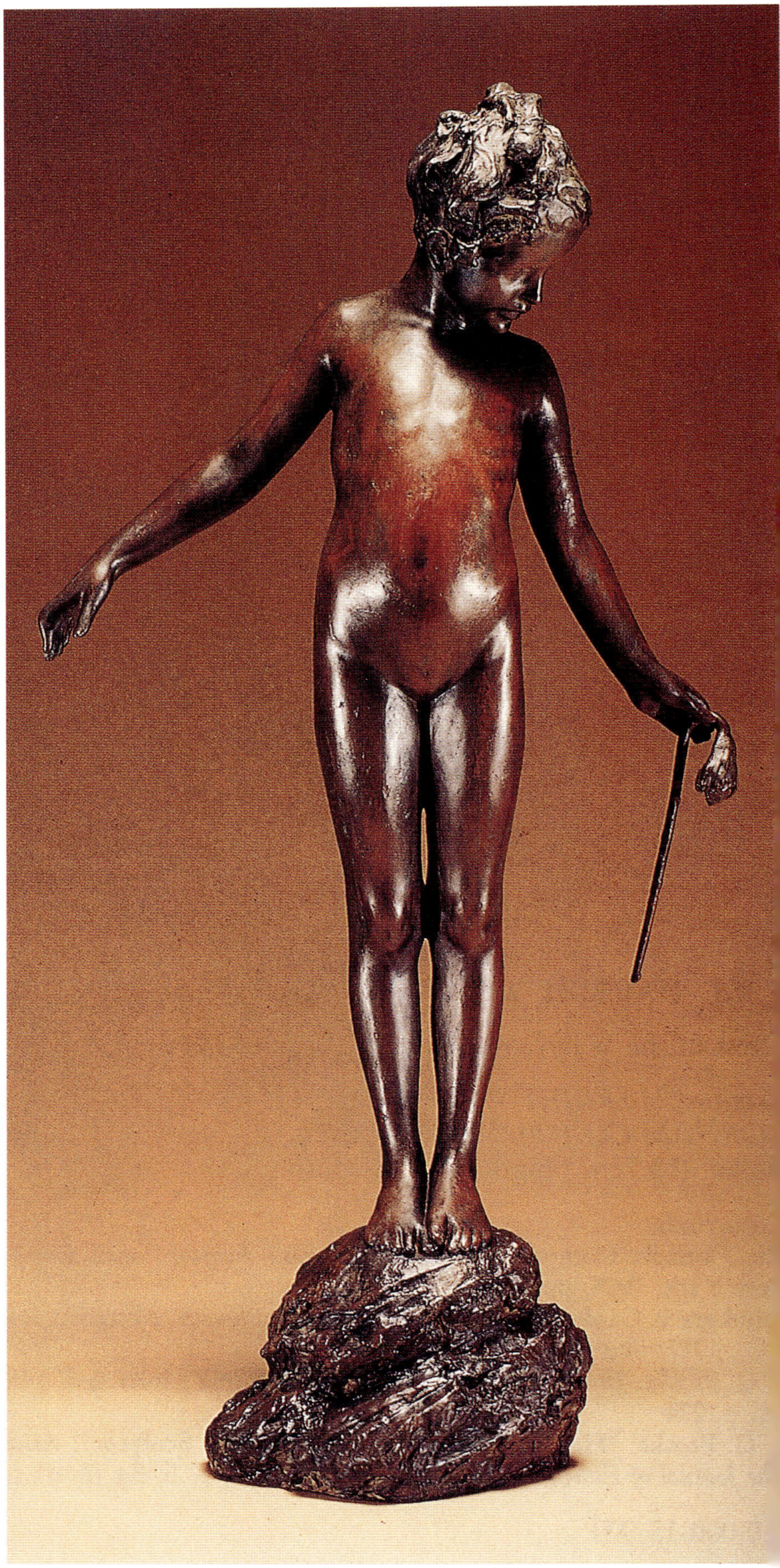

VI-27. "Water Lilies" (a bronze fountain figure) by Bessie Potter Vonnoh. First exhibited in 1913.

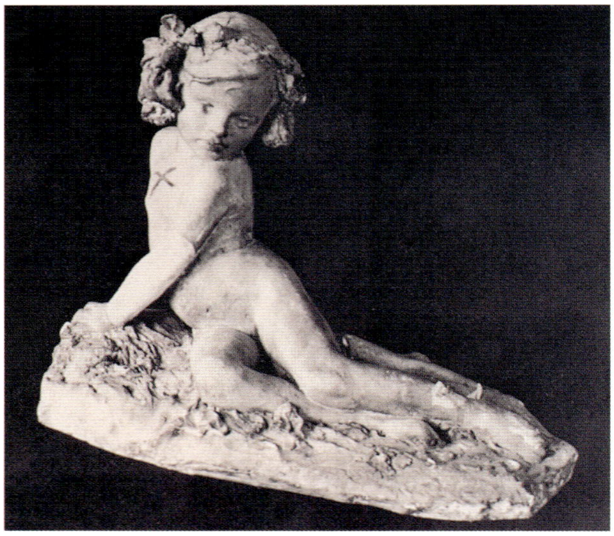

VI-26. "A Chance Acquaintance" (plaster cast) by Bessie Potter Vonnoh.

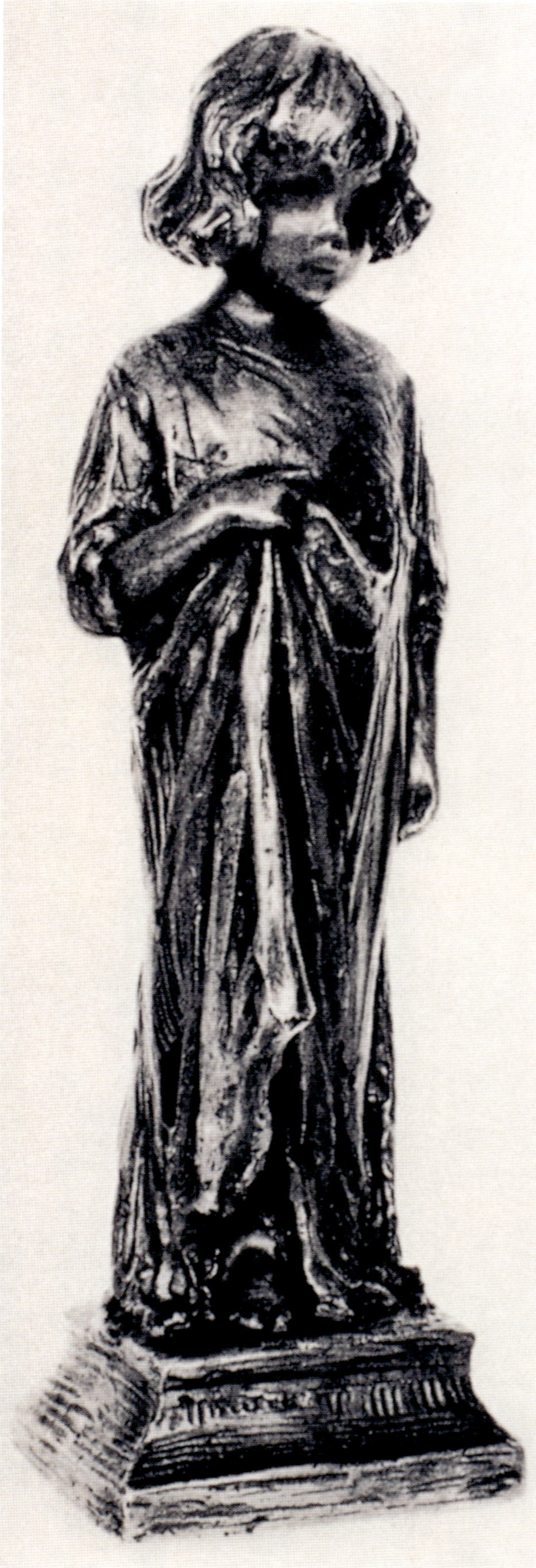

VI-29. "Sea Sprite" (a fountain group) by Bessie Potter Vonnoh.

VI-28. "Good Night Girl" by Bessie Potter Vonnoh. c.1930.

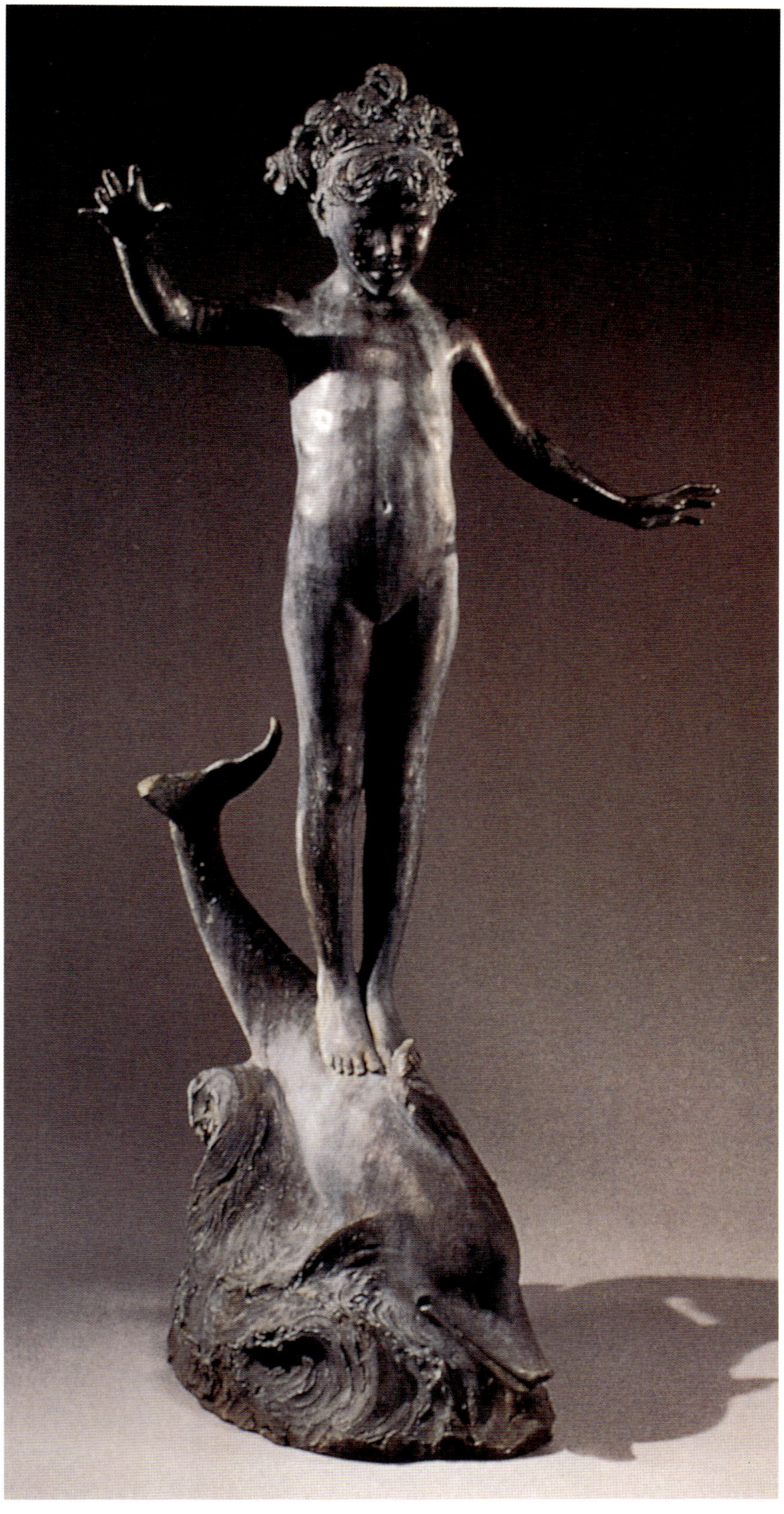

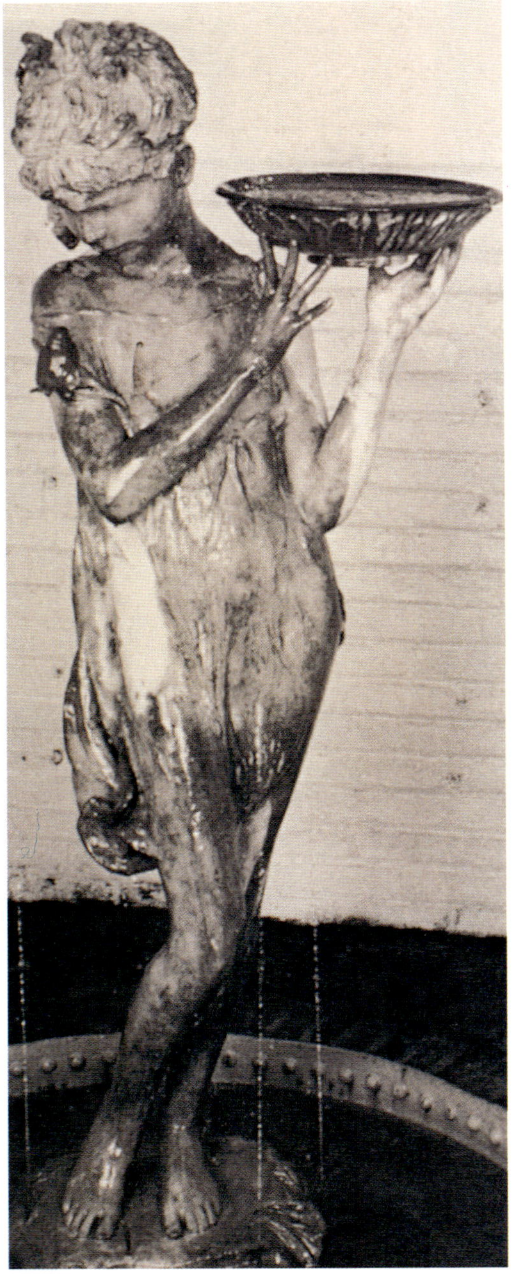

VI-30. "Frances Hodgson Burnett Memorial Fountain," Central Park, New York City, by Bessie Potter Vonnoh. 1926-36. Up to ten copies were cast.

VI-31. "Profile of a Woman" (medallion) by Bessie Potter Vonnoh.

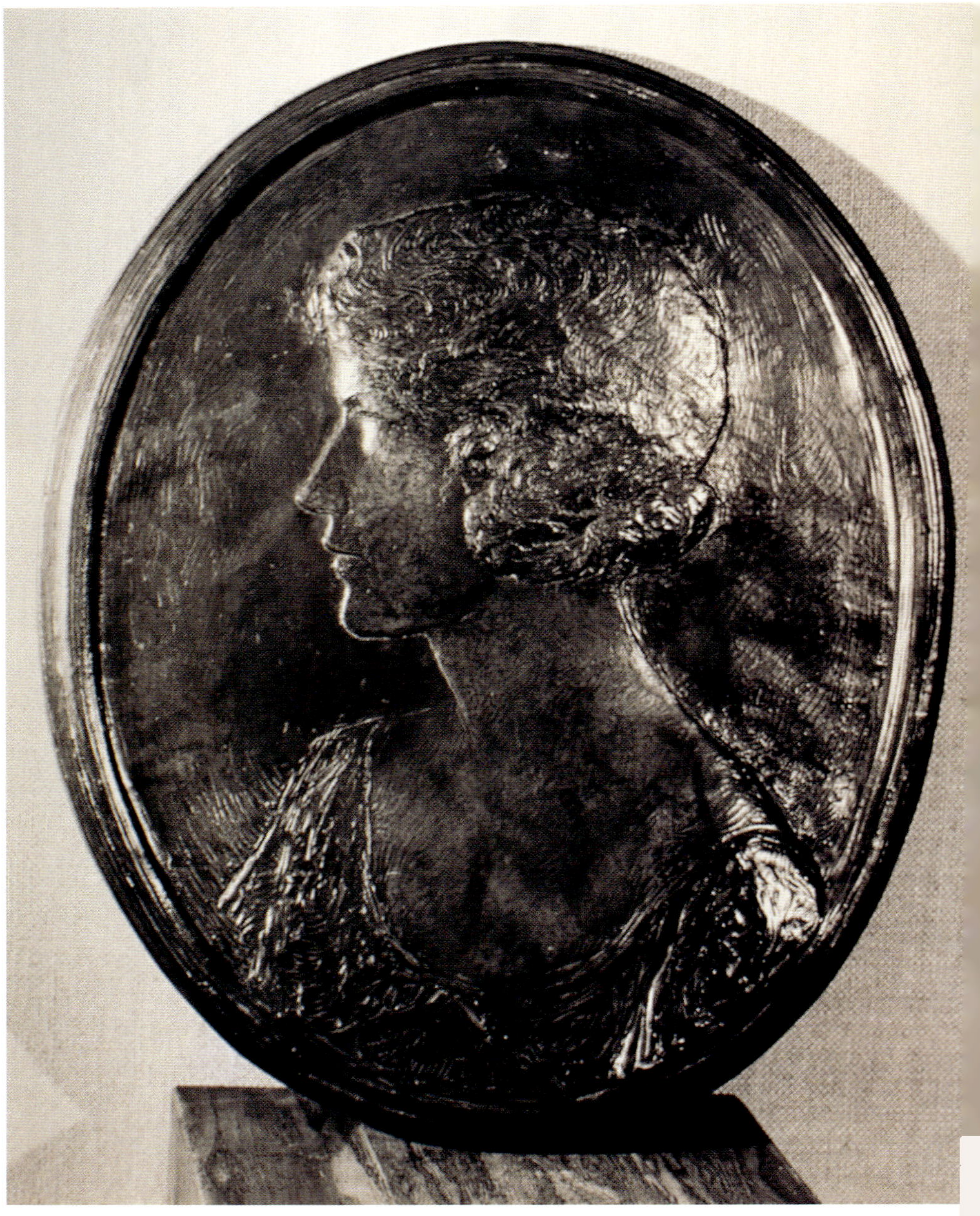

MAHONRI YOUNG (1877-1957)

Mahonri Young (see page 107) brought the images of genre art away from the domestic scene to the working world. His stevedores and road laborers were drawn from real life. He also made a number of Western images of cowboys and pioneers, and several sculptures of gritty prizefighters.

Young's style of genre art was naturalistic, based on observation. "In everyday language," he said, "I am eye-minded." His best known works, aside from his Mormon monuments, are his sculptures of working men. While later genre artists may have stressed the sociological implications of depicting the working classes, Young was primarily interested in the sculptural quality of the human body as it labored. Balance, effort, and muscularity interested him, as did the inherent dignity of his subjects.

Among the striking images that Young produced were his early "Bovet Arthur –A Laborer" (VI-32) which won him a major prize, and his working figures: "A Stevedore" (VI-33), "Man with a Pick" (VI-34), "Man with a Wheelbarrow" (VI-35), "The Rigger" (VI-36), and "The Driller" (VI-37), all cast by Roman Bronze Works. He also cast a figure of Pan, an organ grinder, and a violinist at R.B.W. These and a number of other works (including his cowboy sculptures, and several boxers, "Right to the Jaw," "The Knockdown," etc.) were cast at R.B.W. between 1910 and the 1950s.

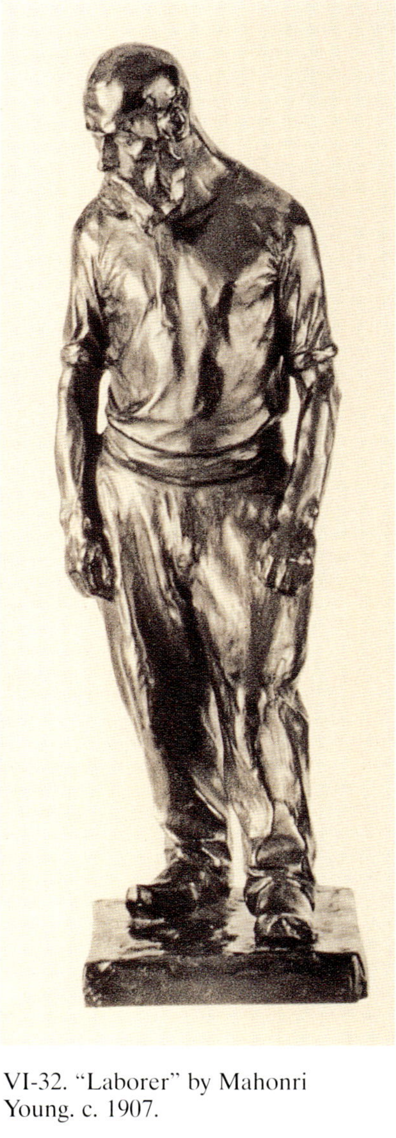

VI-32. "Laborer" by Mahonri Young. c. 1907.

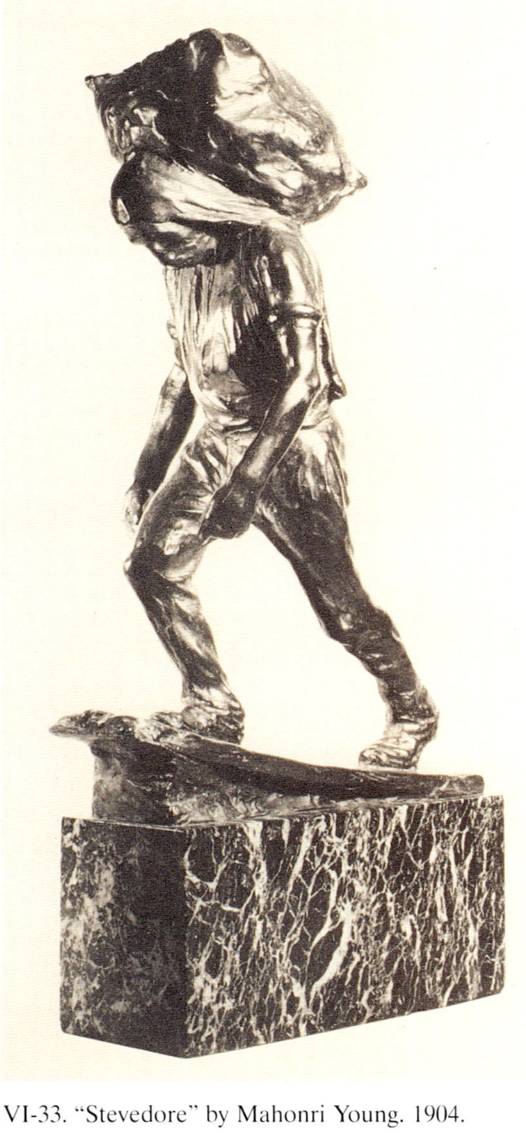

VI-33. "Stevedore" by Mahonri Young. 1904.

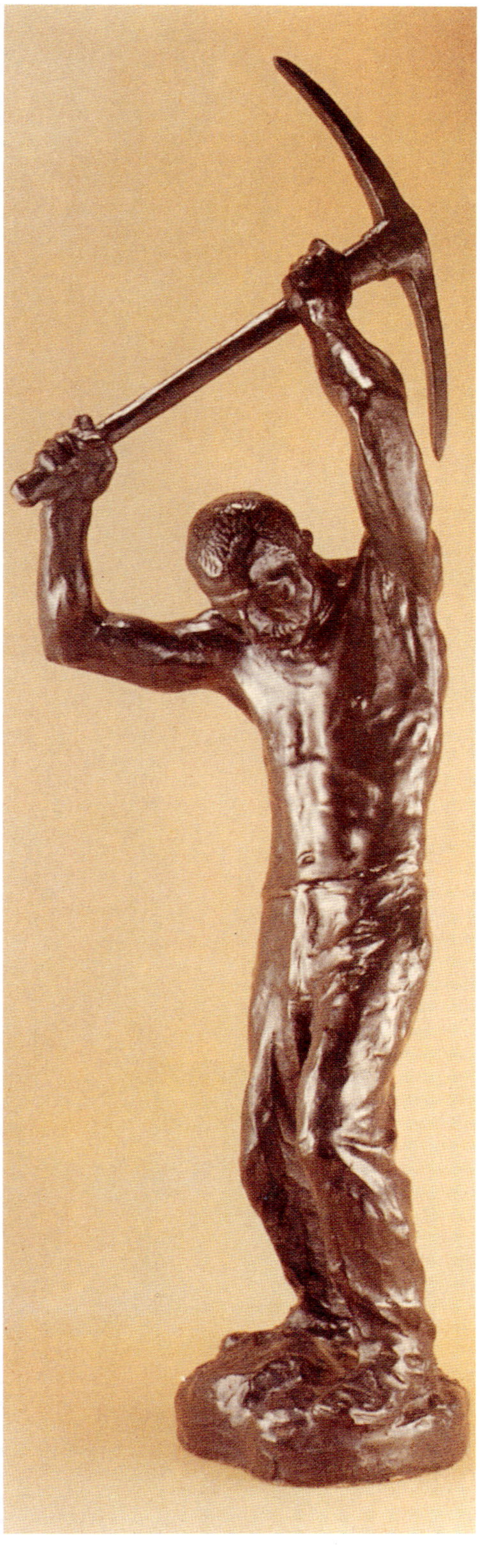

VI-34. "Man With a Pick" by Mahonri Young. 1915.

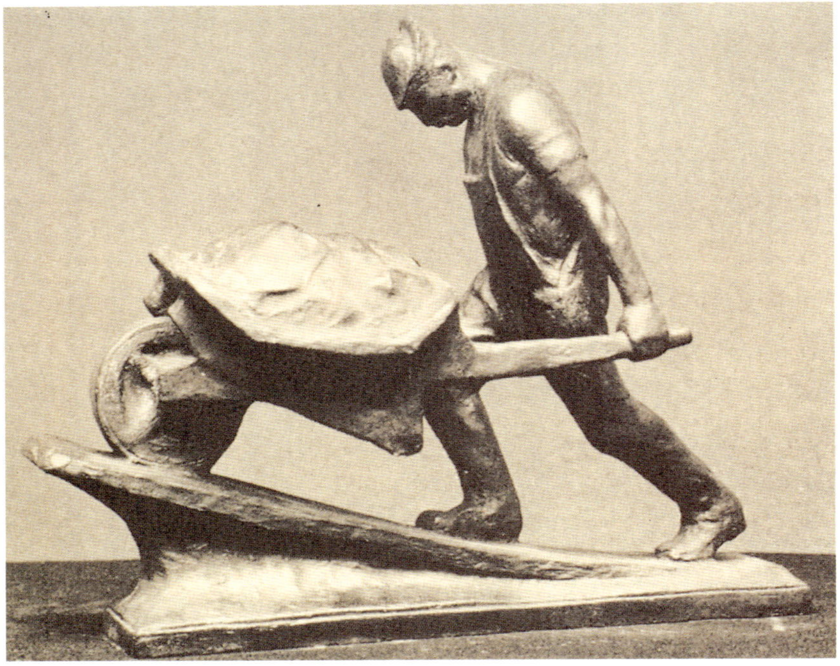

VI-35. "Man With a Wheelbarrow" by Mahonri Young. 1915.

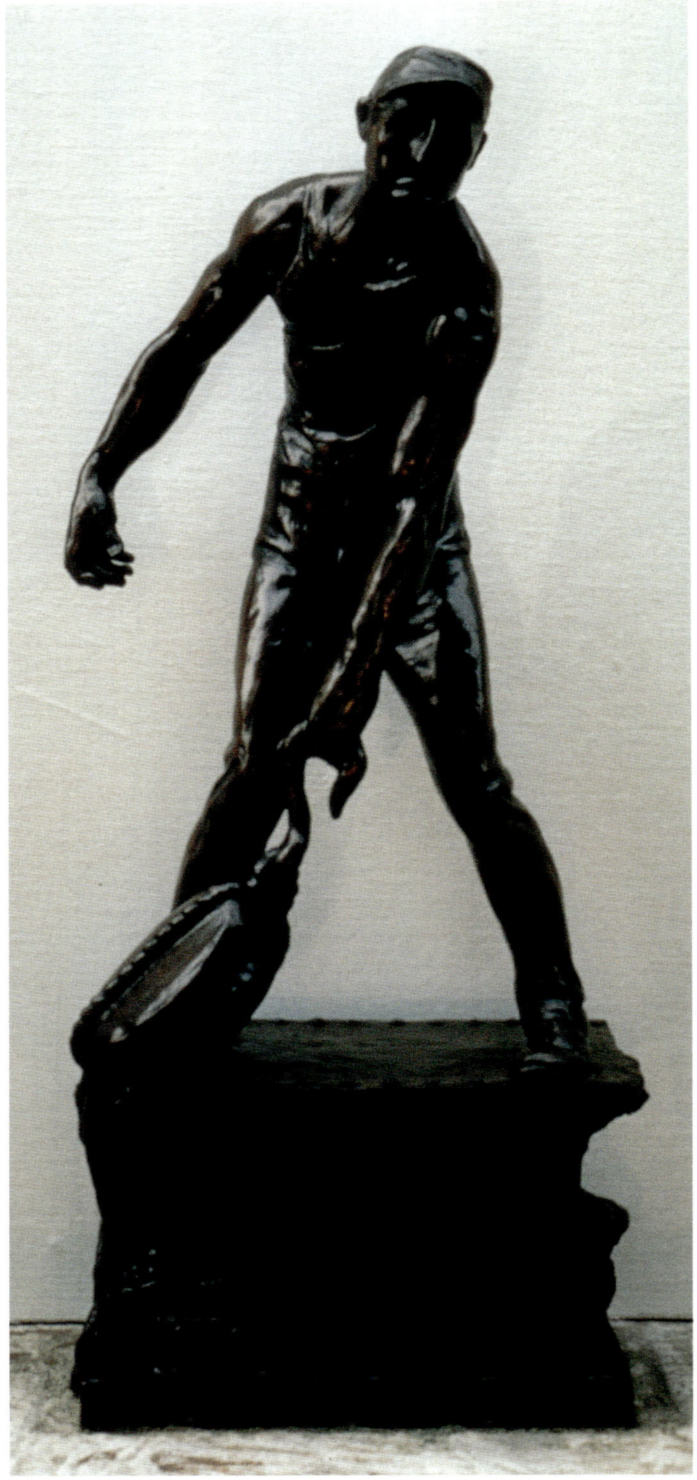

VI-36. "The Rigger" by Mahonri Young. 1917.

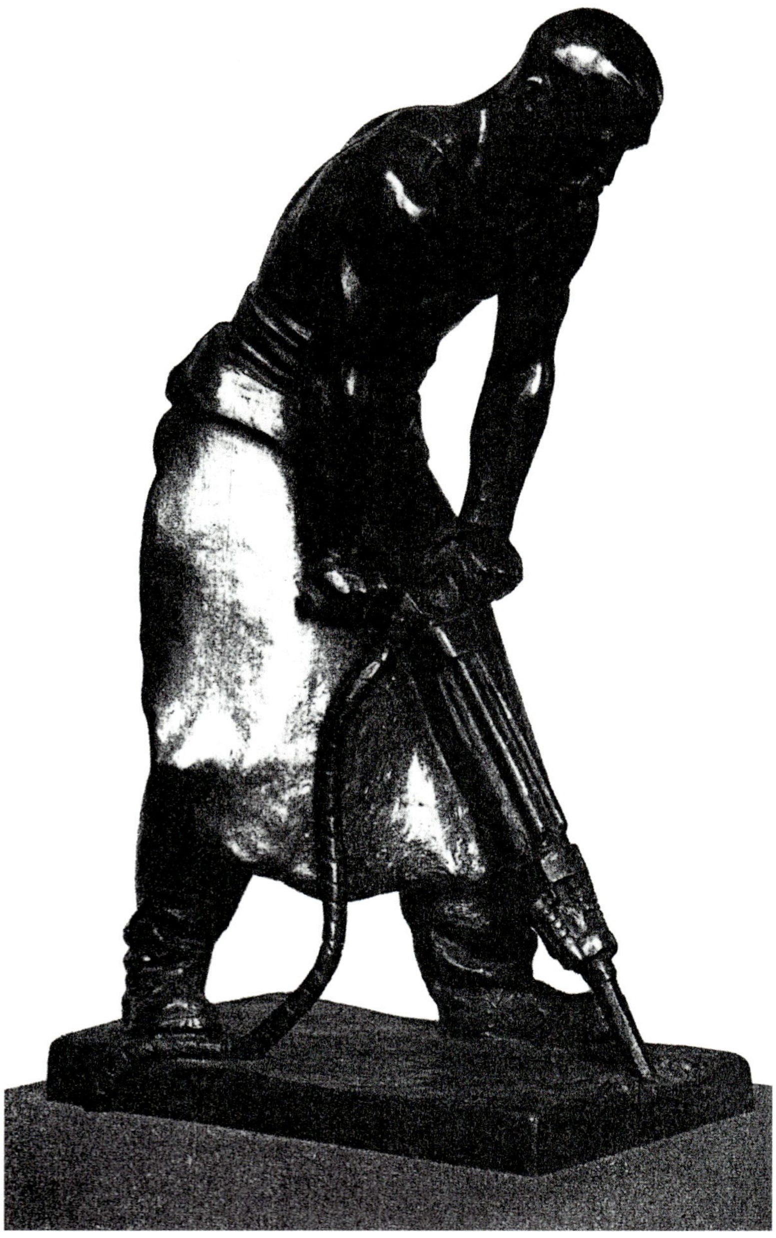

VI-37. "The Driller" by Mahonri Young. c. 1928-30.

ISIDORE KONTI
(1862-1938)

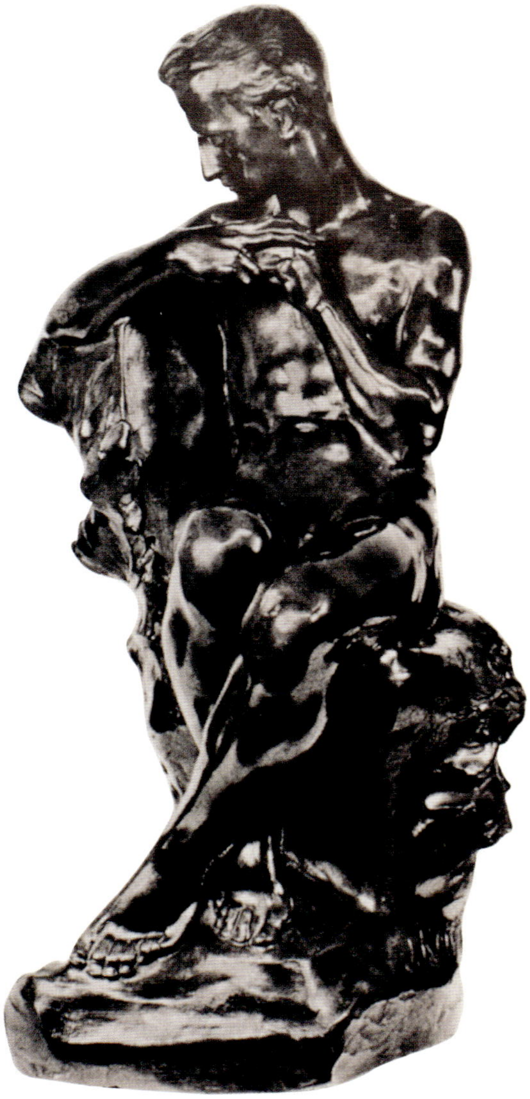

VI-38. "Genius of Immortality" by Isidore Konti. c. 1912.

Isidore Konti, like many European-trained sculptors of his time, worked in a decorative style that was often allegorical or fanciful. He made both architectural reliefs and monuments, and his fountains and statuary were popular in the early decades of the 20th century. Konti's work was delicate, almost Baroque in its ornamental curves and modulated surfaces. In the New York area numerous public buildings and private mansions (and their gardens) were ornamented by Konti's work.

Born in Vienna of Hungarian parents, Konti spent two years studying in Rome, where he was greatly influenced by Renaissance art. In 1891 he arrived in the United States and became an assistant to Karl Bitter, working with him on the sculpture for the Columbian Exposition and then in his New York studio. He went on to create groups of figures and fountains for several other major expositions; he won a gold medal at the St. Louis Exposition. Working in both marble and bronze, Konti designed interior architectural sculpture for several notable buildings, as well as allegorical figures, portrait busts of prominent Americans, war memorials, and many smaller bronzes with symbolic themes.

Konti cast more than twenty different sculptures at Roman Bronze Works between 1905 and 1928; among them were (numbers VI-38 to VI-44).

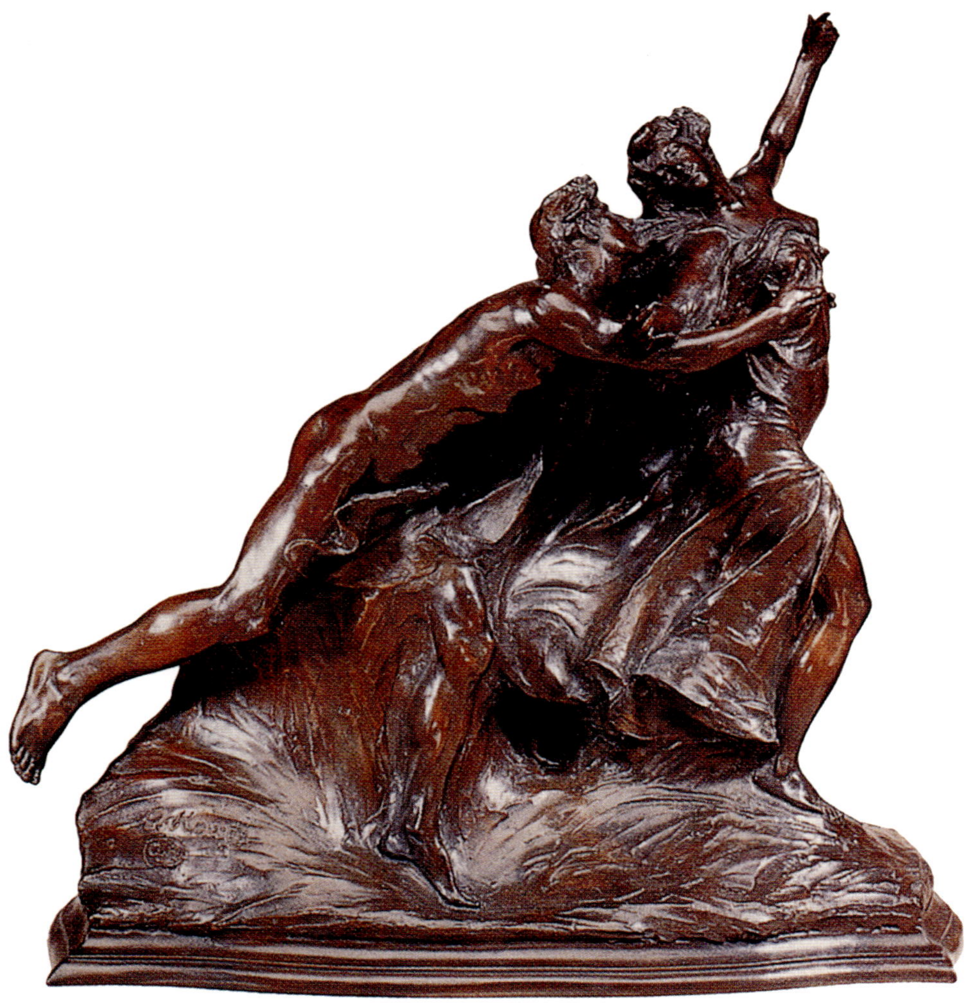

VI-39. "Illusion: Pursuit of Happiness" by Isidore Konti. 1912.

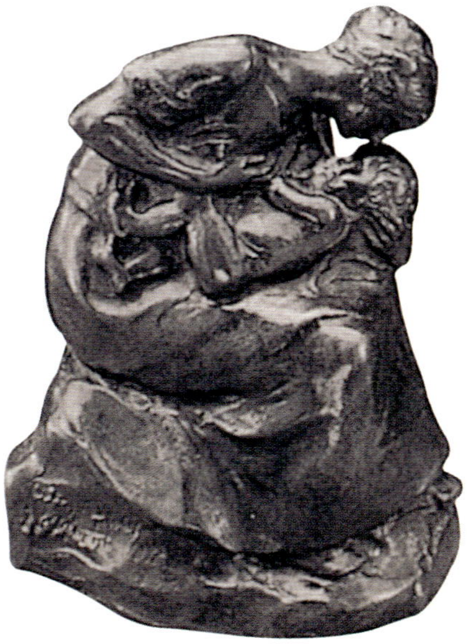

VI-40. "Mother and Baby" by Isidore Konti. 1911.

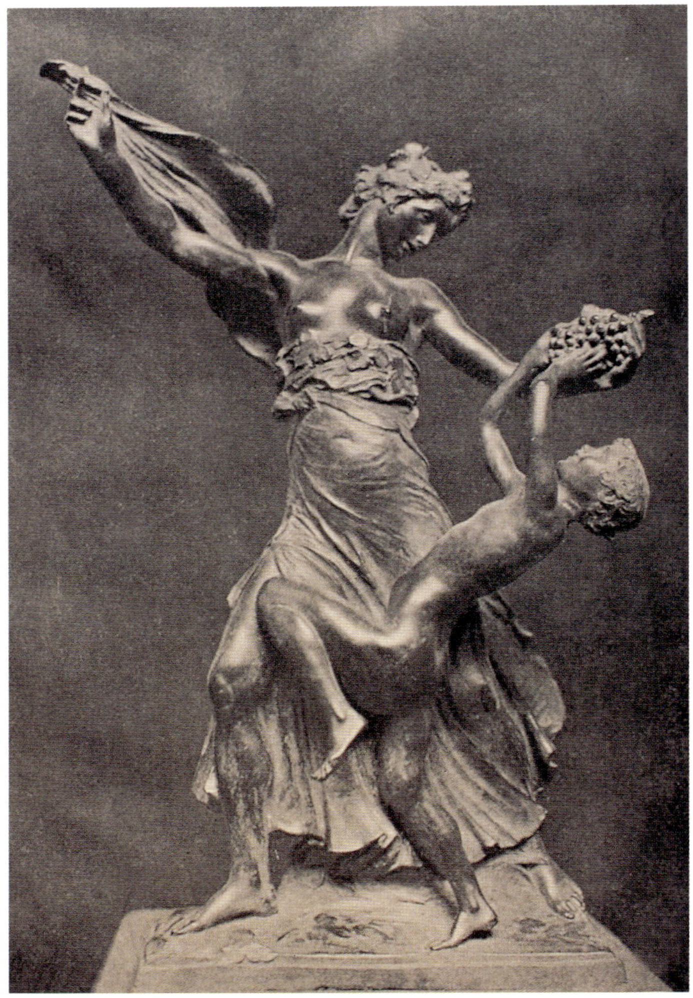

VI-42. "Harvest" by Isidore Konti.

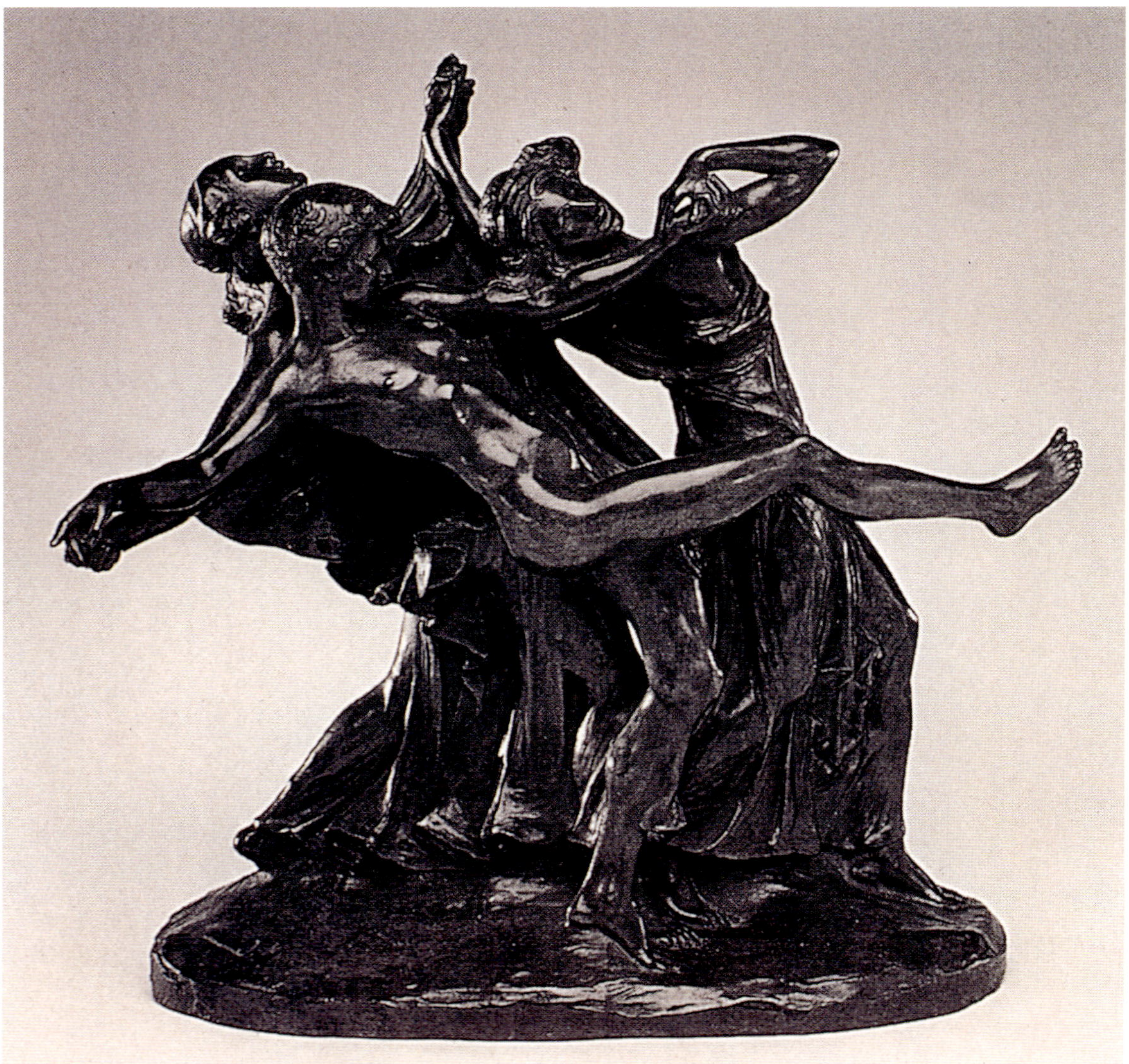

VI-41. "Allegro" by Isidore Konti. 1914.

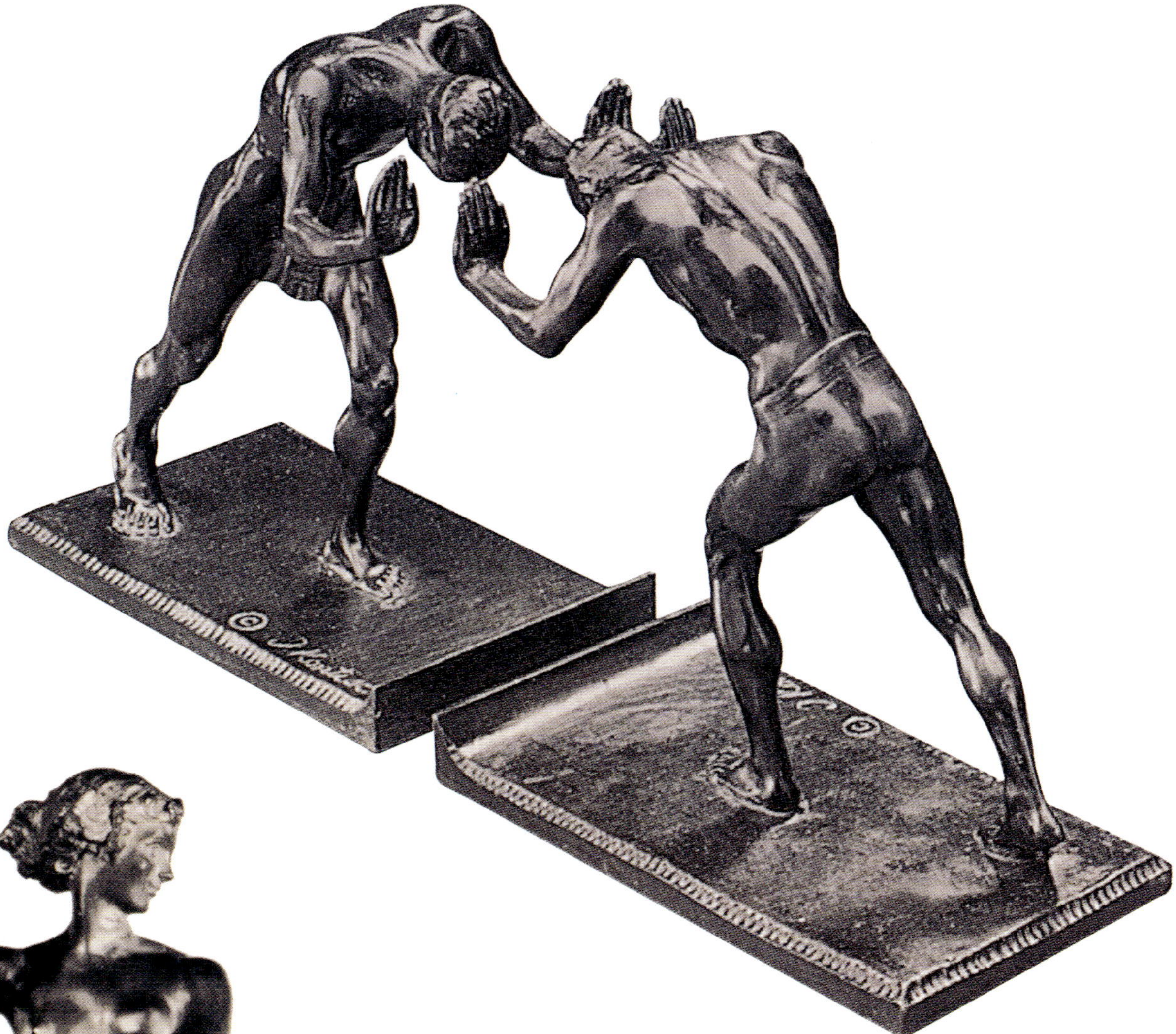

VI-43. "Men Pushing" by Isidore Konti. 1913.

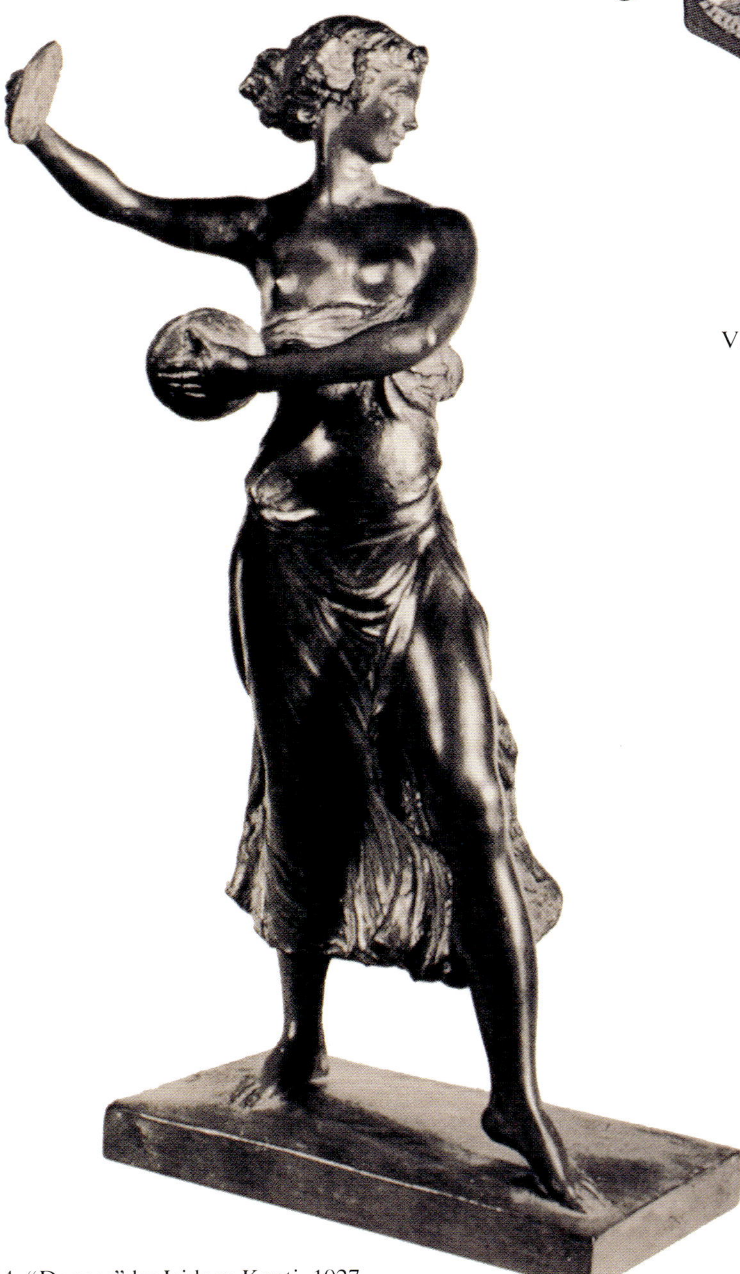

VI-44. "Dancer" by Isidore Konti. 1927.

FREDERICK WILLIAM MacMONNIES (1863-1937)

Frederick MacMonnies was an American-born sculptor who spent many years in France and reinforced the supremacy of the Beaux-Arts style of sculpture in the United States. He had a very successful career producing statuary in a decorative, often allegorical, style, as well as monuments and memorials. Though many of his works had subjects drawn from Greek myth, he never embraced the Neo-Classicism of earlier generations of sculptors; he preferred the ornate naturalism and roughly modeled contours of Beaux-Arts sculpture.

MacMonnies was born in Brooklyn, New York, and became a studio assistant to Augustus Saint-Gaudens in 1880. After studying in New York he left for Europe, where he was trained in Munich and at the Ecole des Beaux-Arts in Paris. He exhibited in Europe, winning coveted awards at Antwerp and Paris, and becoming a Chevalier of the Legion of Honor. He won recognition in the United States for his massive Columbian Fountain at the Exposition in Chicago in 1893; it contained an ornate barge filled with allegorical ladies. But he remained in Paris, opening a studio there where he received an enormous number of American commissions. These included several major monuments: for New York City, "Nathan Hale" (VI-45); for Boston, "Bacchante and Infant Faun" (VI-46); for Washington, "Shakespeare" (VI-47); and for Denver. He returned to the United States at the outbreak of World War I, and set up his studio in New York, where he made paintings as well as sculpture. One of his major post-war pieces was a monument to the Battle of the Marne (1926). He was a member of the National Institute of Arts and Letters.

MacMonnies faced controversy with several of his works; objections were made to their nudity, the appearance of drunkenness, and perceived insults to womanhood in a few of his allegorical statues. MacMonnies produced many sculptures with classical themes. Among the bronzes cast at Roman Bronze Works were his "Diana" (VI-48), "Pan of Rohallion" (VI-49), "Young Faun and Heron" (VI-50), "Infant Faun" (VI-51) and "Boy and Duck" (VI-52).

Many of MacMonnies' smaller works were cast at Roman Bronze Works between 1902 (when R.B.W. cast four copies of "Boy and Duck", two "Shakespeares" and nine "Bacchantes") and 1916.

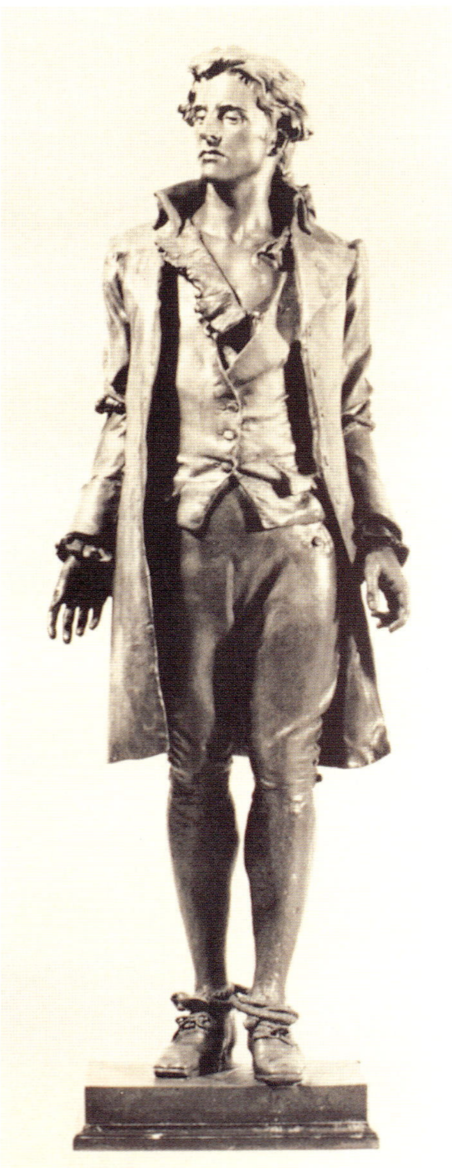

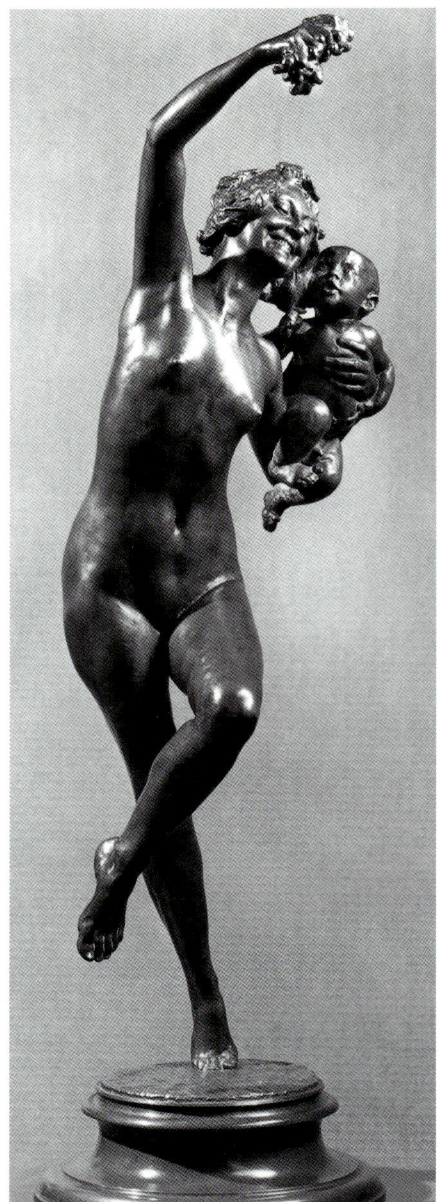

Left:
VI-45. "Nathan Hale" by Frederick MacMonnies. Modeled c. 1890; cast by R.B.W. 1915-16.

Right:
VI-46. "Bacchante with Infant Faun" by Frederick MacMonnies. Modeled 1894; cast by R.B.W. 1902. Originally designed for the Boston Public Library, it was later cast in smaller form in both America and France.

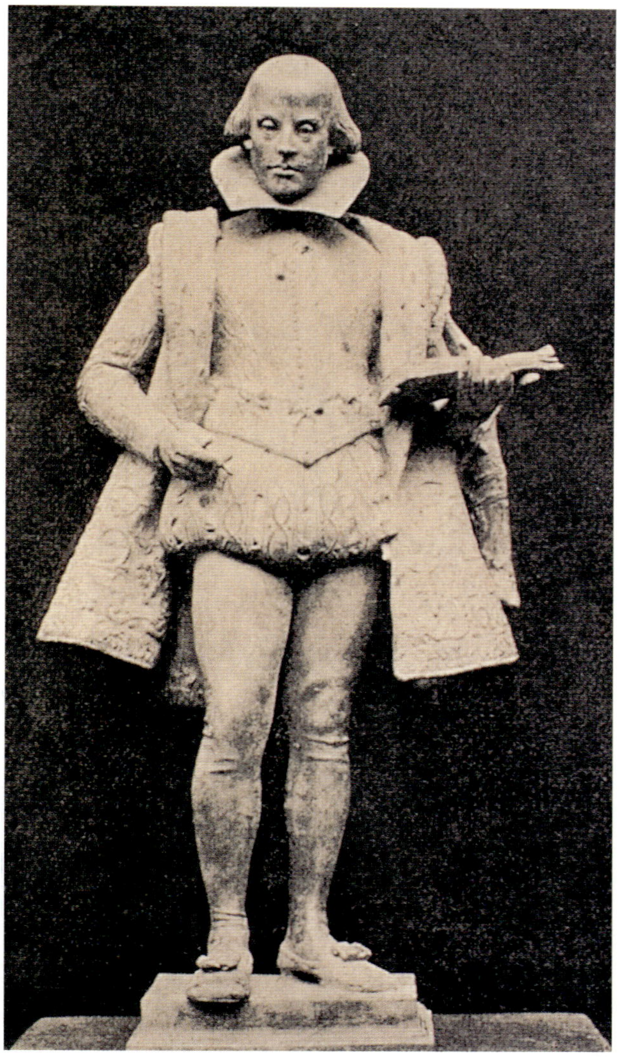

VI-47. "Statue of Shakespeare," at the Congressional Library, Washington, by Frederick MacMonnies. 1902-1904.

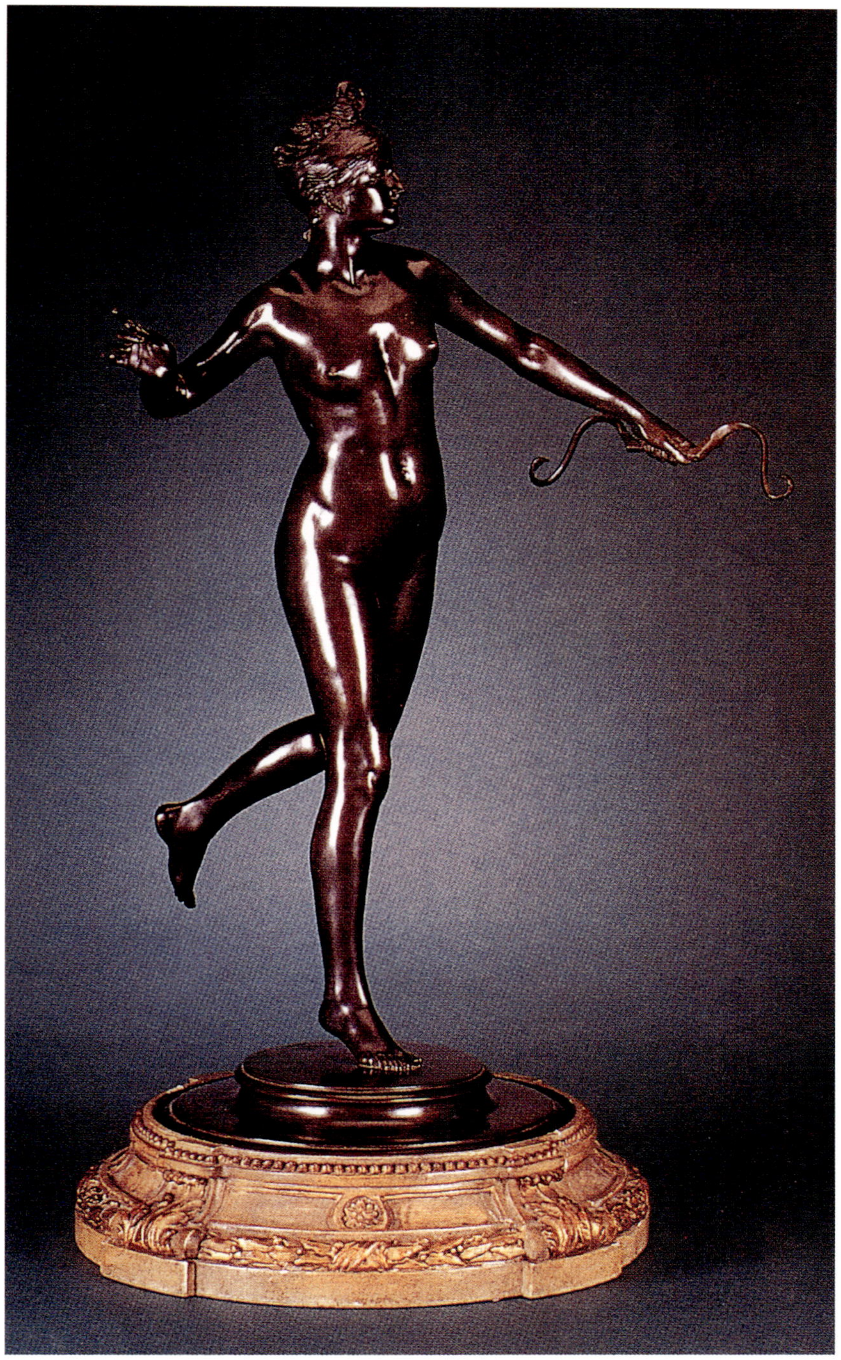

VI-48. "Diana" by Frederick MacMonnies. Modeled 1894; cast by R.B.W. 1902-04.

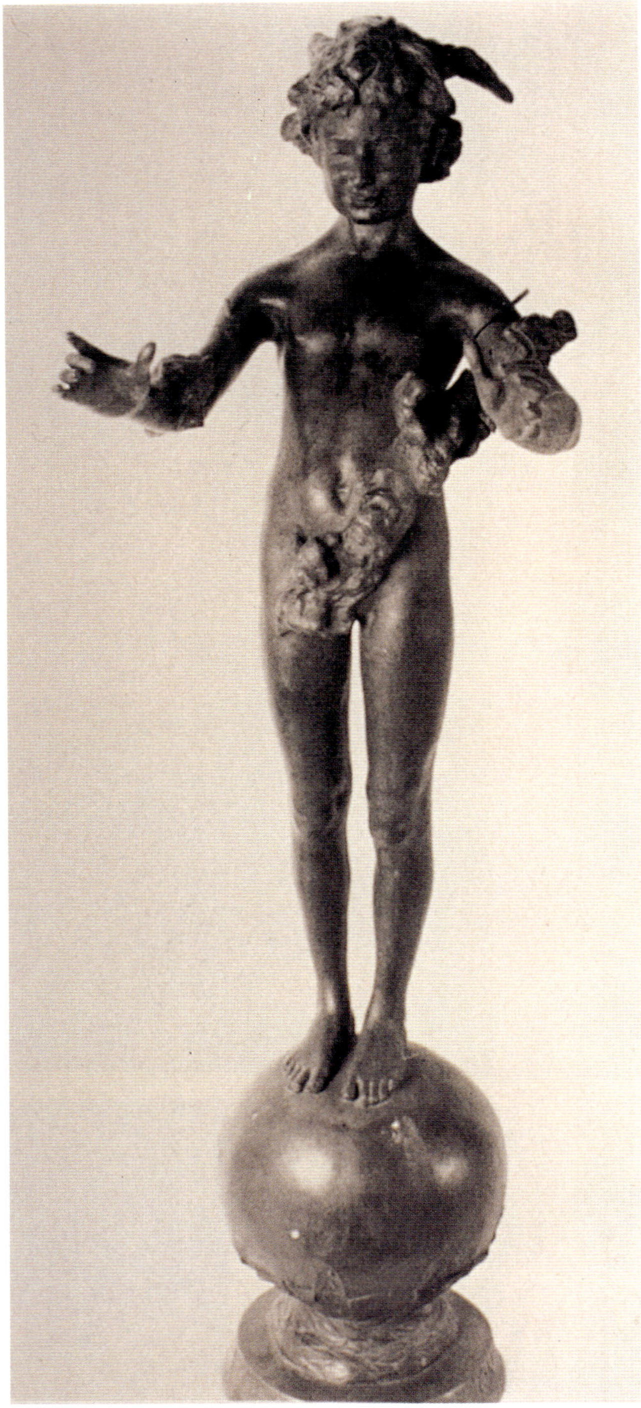

VI-49. "Pan of Rohallion" by Frederick MacMonnies. Originally modeled 1890-94; cast by R.B.W. 1915-16. There are several versions.

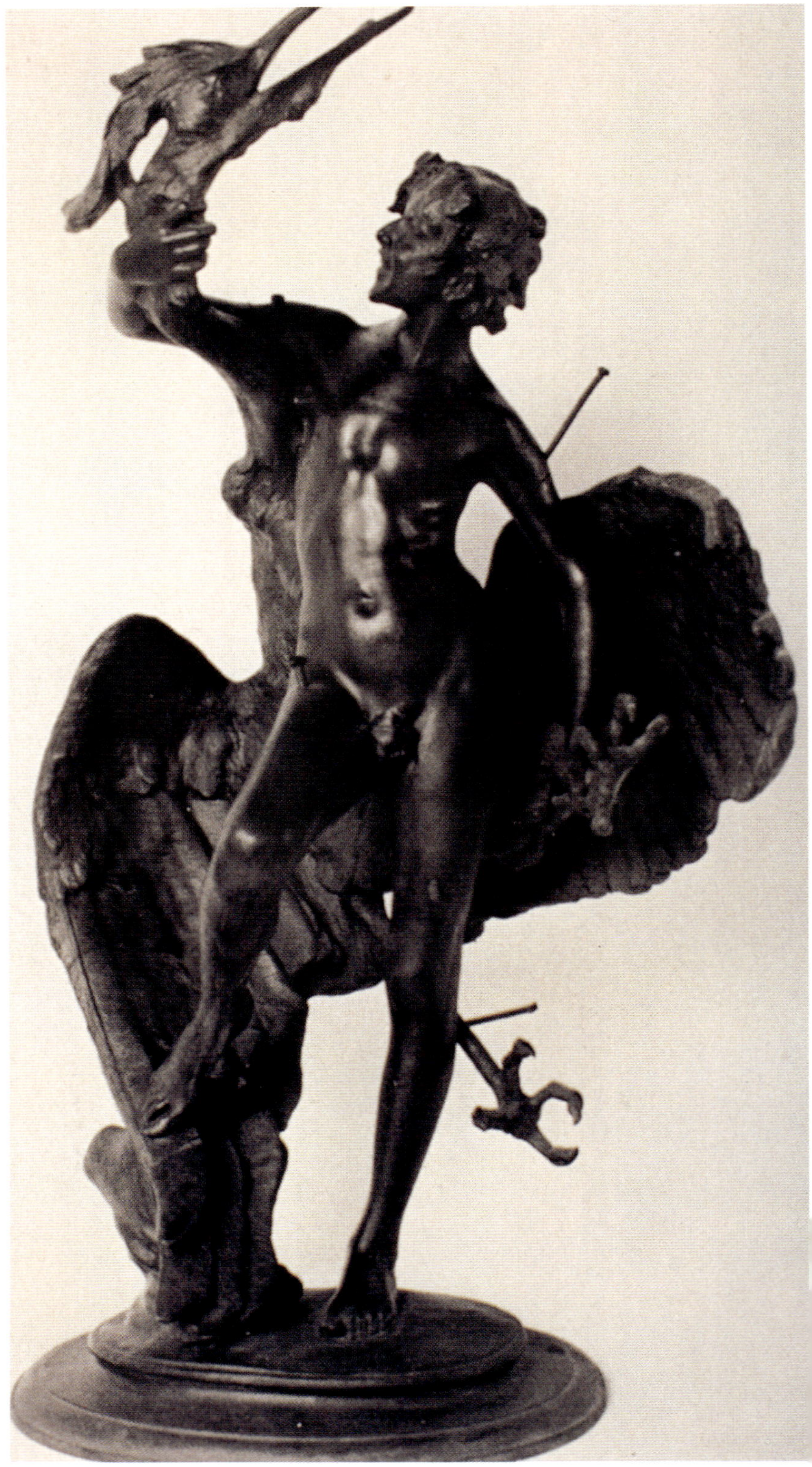

VI-50. "Young Faun and Heron" by Frederick MacMonnies. Original modeled in Paris 1890-94; given to R.B.W. to use as a pattern to cast "Boy and Heron." Four copies cast between 1904 and 1917.

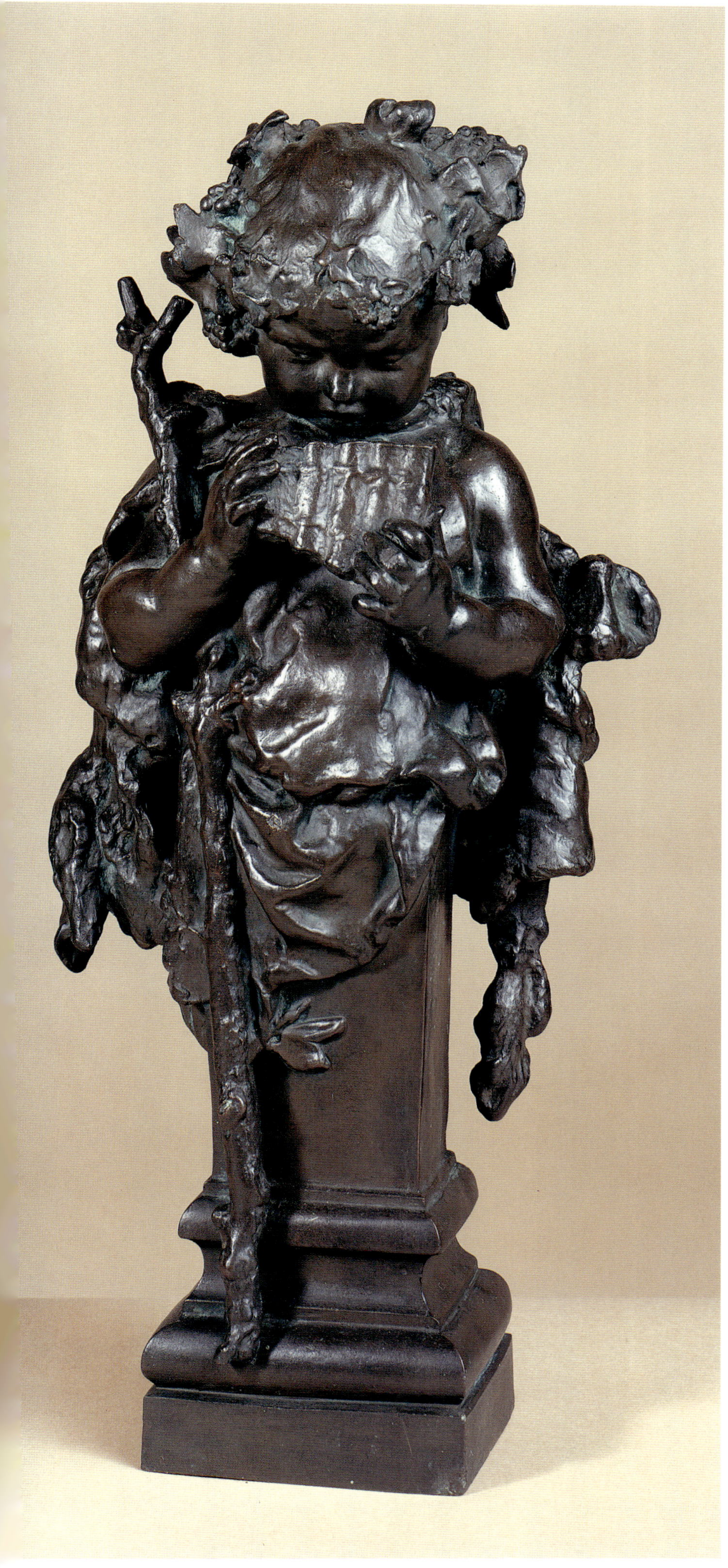

VI-51. "Infant Faun" by Frederick MacMonnies. c. 1915.

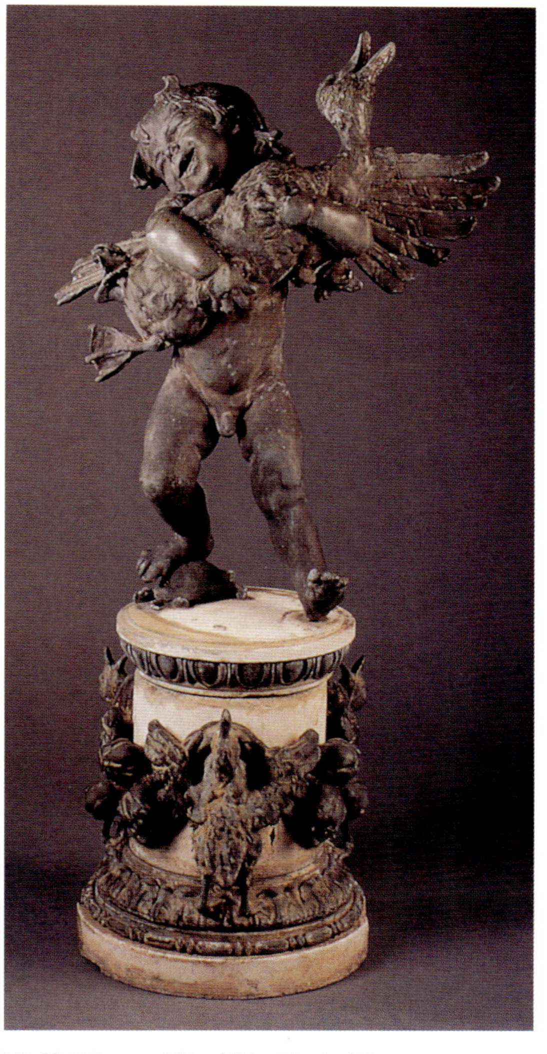

VI-52. "Boy and Duck" by Frederick MacMonnies. 1907.

PAUL TROUBETZKOY (1866-1938)

Paul Troubetzkoy, primarily a sculptor of small bronzes, is best known for his statuettes of fashionable women, children and animals, and for his portrait busts of notable artists and writers. He also made a number of Western images and sculptures of dancers; the portrayal of movement and action interested him and was well-suited to his Impressionistic technique. Influenced by Rodin, he emphasized the play of light on the rough surfaces of his small-scale sculptures.

Troubetzkoy, the son of a Russian prince and an American woman was born in Italy, and was mostly self-taught, modeling animals and portrait busts at an early age. As a young man he had his own studio in Milan where he produced spontaneous, impressionistic likenesses and statuettes. His style, resembling snapshots rather than finished portraits, was fresh and characterized by a surface flourish; it emphasized the personality of his sitter. His sculpture "The Indian Scout" (VI-53), made long before he ever visited the United States, won him a gold medal in Rome in 1894. Several years later Troubetzkoy went to Moscow where he opened a studio, befriended Tolstoy, and became a favorite portraitist of the Russian nobility. His equestrian statue of Czar Alexander was dedicated in St. Petersburg in 1909; it was one of his only monumental works.

He moved his studio to Paris in 1904 after winning a grand prize at the Paris Exposition. It was in France that his ability as a portraitist fully developed; among his sitters were Rodin, George Bernard Shaw, and Anatole France. He was able to capture the individuality of character in his impressionistic way, and he had many commissions in Paris and other European cities.

In 1914 he came to the United States, spending long periods in California and seeing his first Western sights; a "Bucking Broncho" is listed, as is his "Standing Indian" (VI-54), which suggest some of the images that he saw.) His fascination with movement was evident in many statuettes of dancers including "Mademoiselle Sversky" (VI-55). But he was best known for his portrait statuettes of fashionable women: among them were "Mrs. George Waterman" (VI-56); he sculpted many other elegant clothed female subjects, such as "Standing Girl" (VI-57), and a "Fashionable Lady" (VI-58). He also made several animal sculptures; R.B.W.'s ledgers include listings for dog and elephant sculptures (VI-59).

Troubetzkoy lived and worked in the United States from 1914 to 1920, before returning to Europe. He used the Roman Bronze Works foundry for his casting between 1915 and 1917.

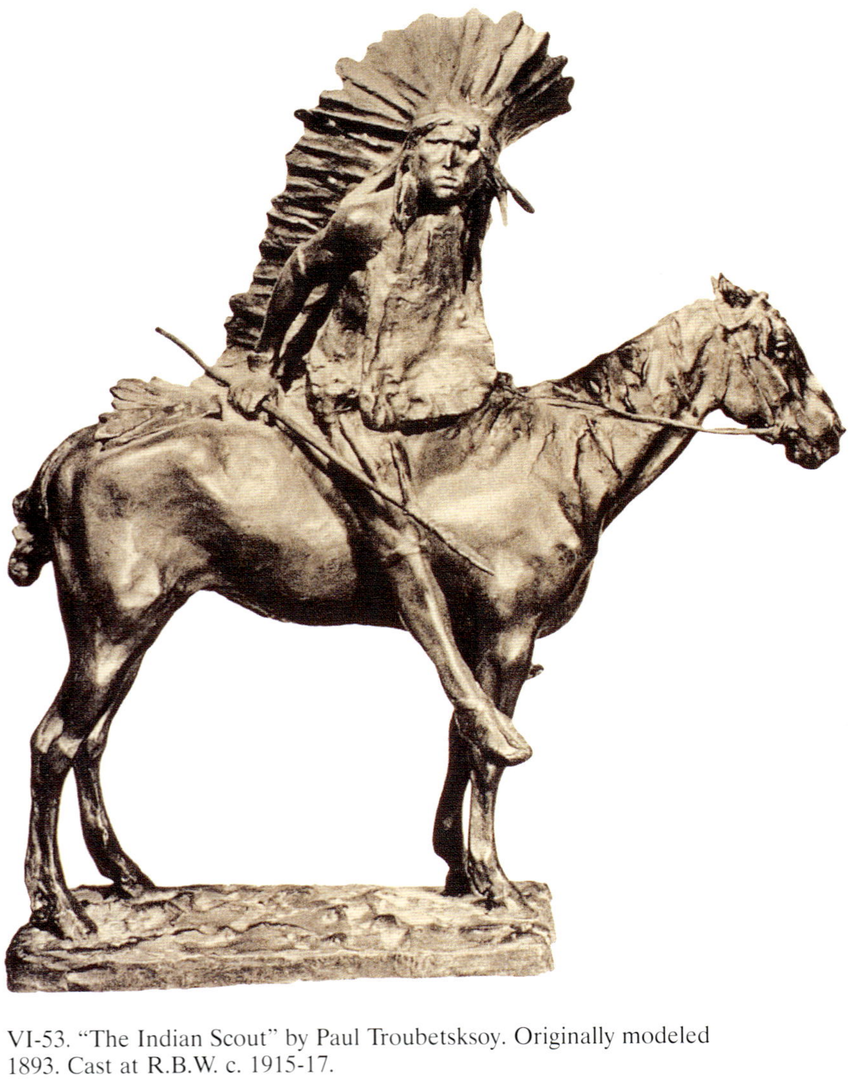

VI-53. "The Indian Scout" by Paul Troubetsksoy. Originally modeled 1893. Cast at R.B.W. c. 1915-17.

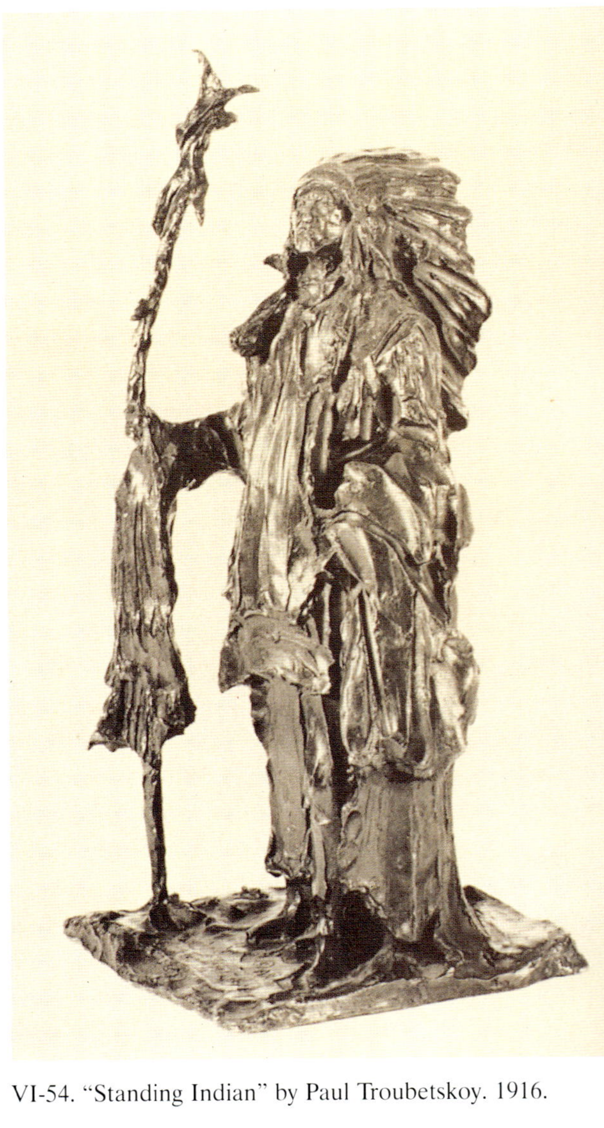

VI-54. "Standing Indian" by Paul Troubetskoy. 1916.

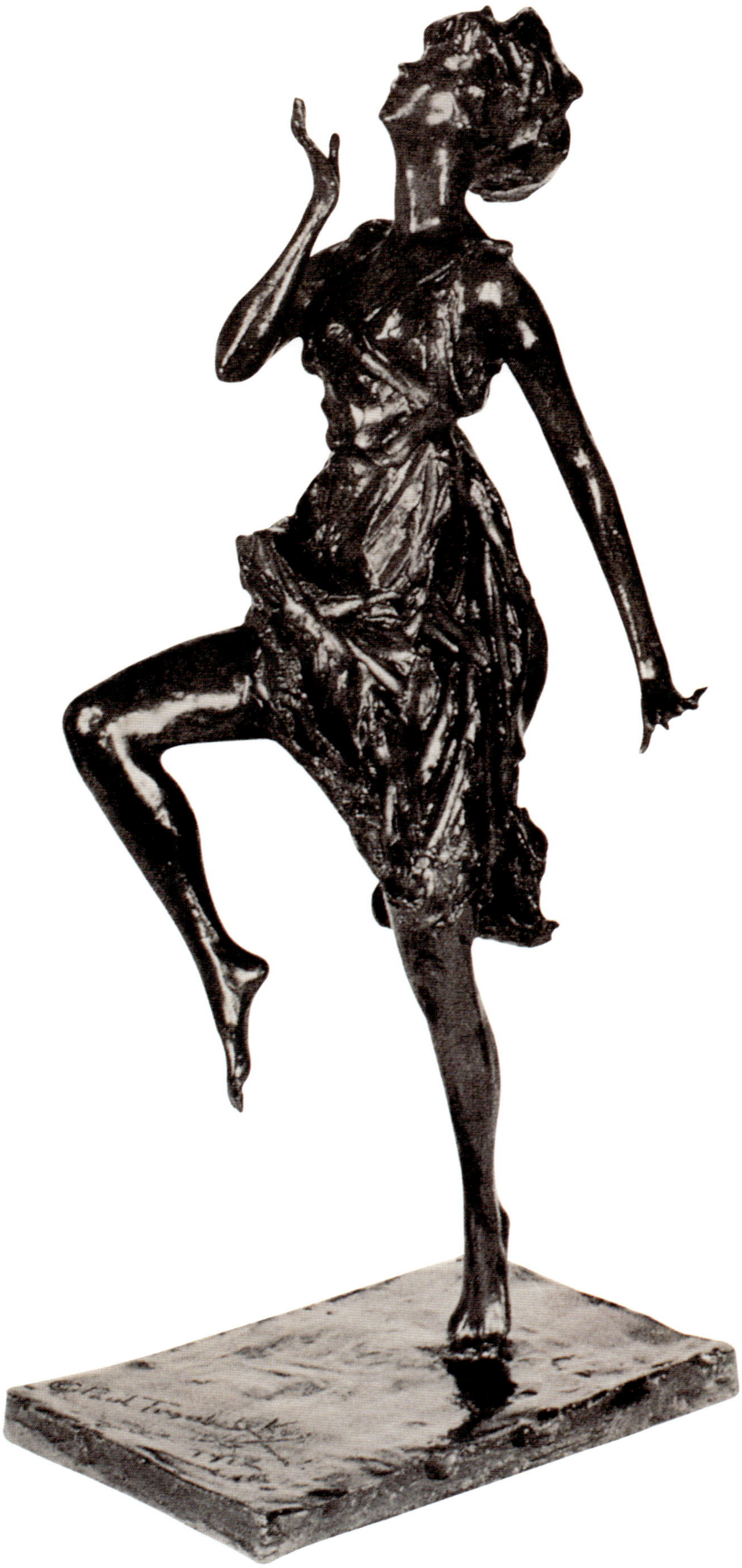

VI-55. "Mademoiselle Sverski" by Paul Troubetskoy. 1912.

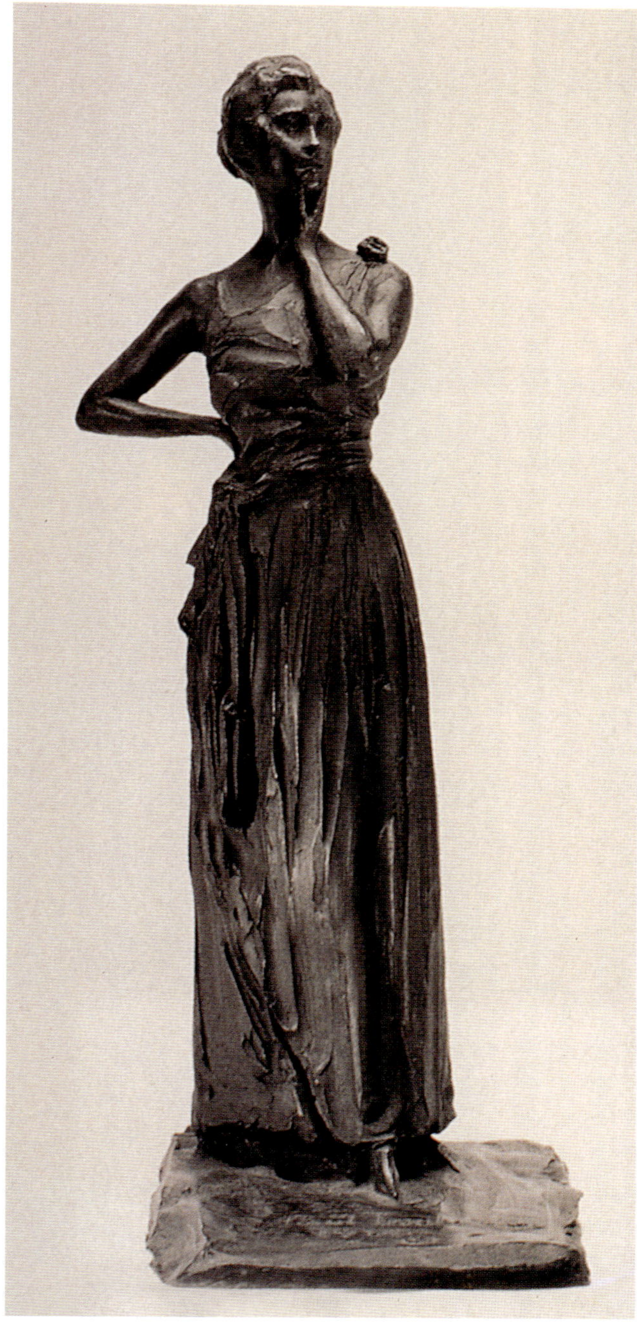

VI-56. "Portrait of Mrs. George (Daisy) Waterman" by Paul Troubetskoy. 1914.

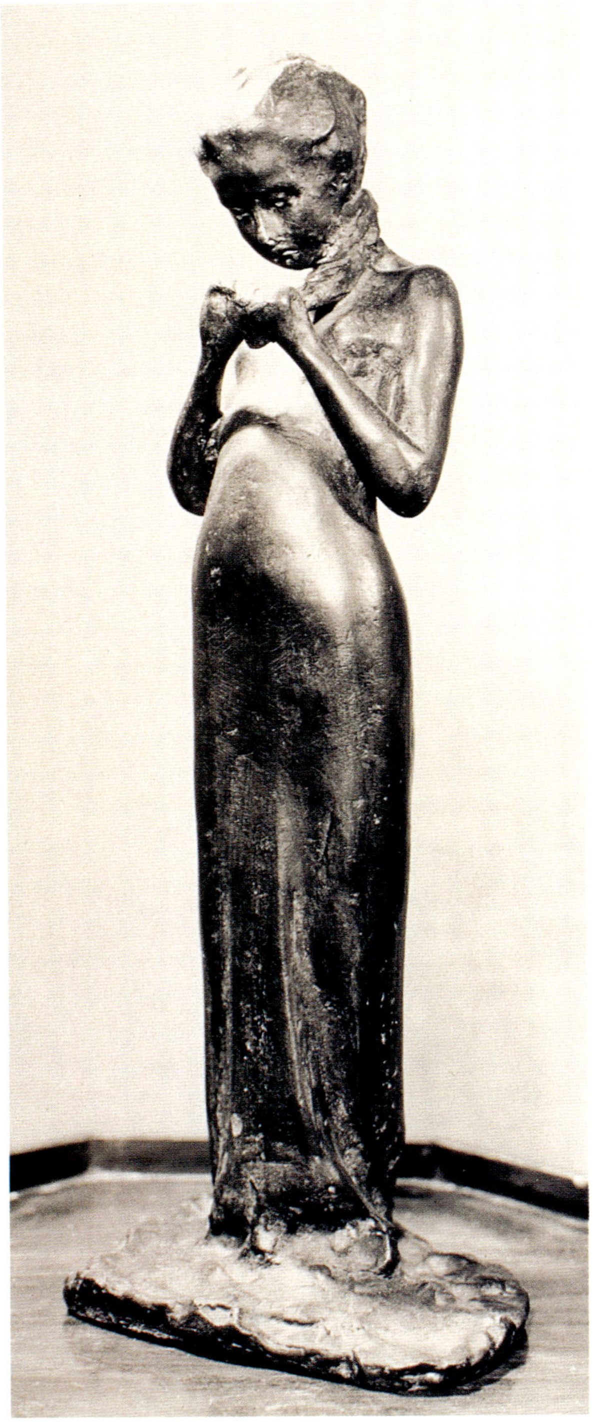

VI-57. "Standing Girl" (also called "Grief") by Paul Troubetskoy. c.1915-17.

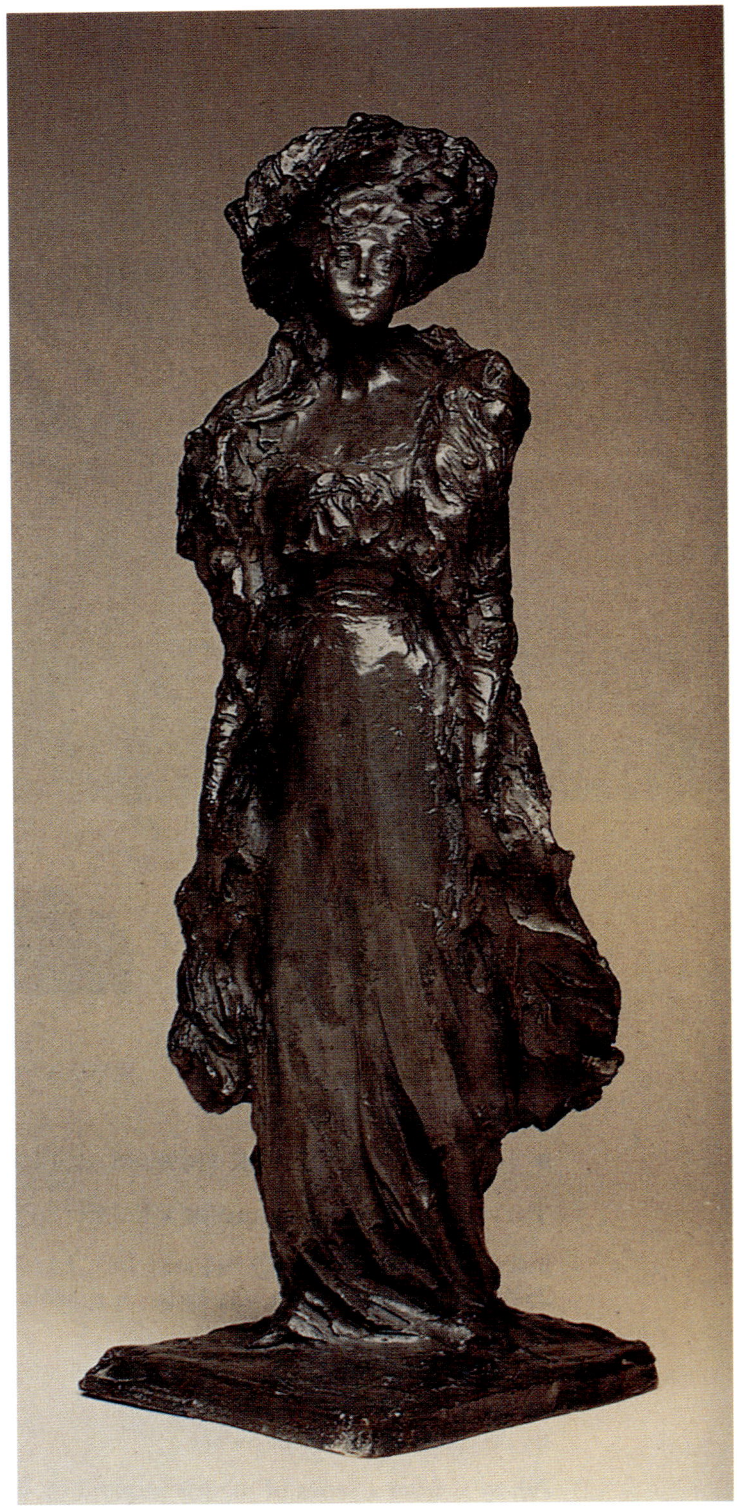

VI-58. "A Fashionable Lady" by Paul Troubetskoy. 1920.

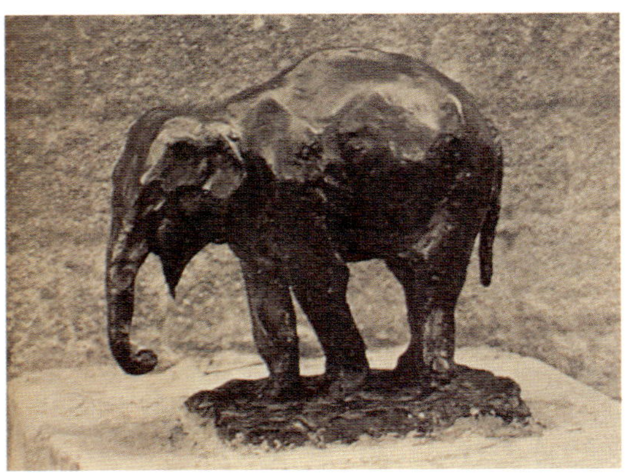

VI-59. "Elephant" by Paul Troubetskoy. Originally modeled 1889; recast by R.B.W. 1916.

ROBERT TAIT McKENZIE (1867-1938)

Robert Tait McKenzie was a doctor, a physical educator, and an anatomist, as well as a sculptor. In his statues—many of which portray athletes—he combined his many interests and talents. His knowledge of anatomy, his understanding of the human figure in motion, and the revival of the Greek Olympic Games in 1896 led McKenzie to his favorite subject. Favoring a sleek, Neo-Classical style, McKenzie sculpted idealized human figures suspended in glorious moments of athletic grace. He became America's leading sculptor of sports subjects (numbers VI-60 to VI-67). He also made several full-sized portrait sculptures, including a well-known image of Benjamin Franklin (numbers VI-68 and VI-69), portrait busts like that of Sarah Lippincott Biddle (VI-70), and a much loved World War I memorial called "Blighty" (VI-71).

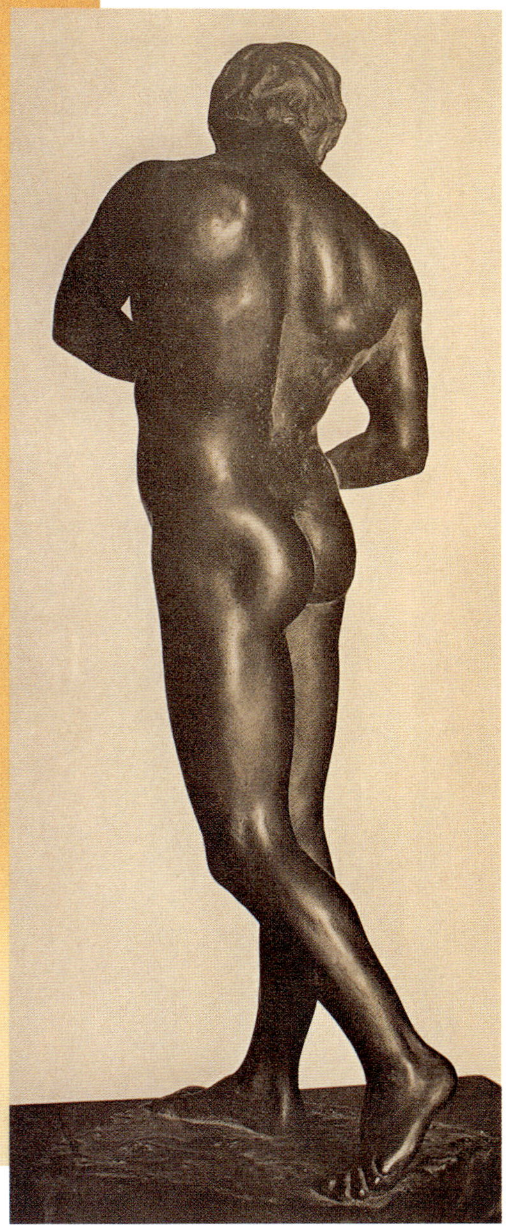

VI-61. "The Athlete" by Robert Tait McKenzie. 1913.

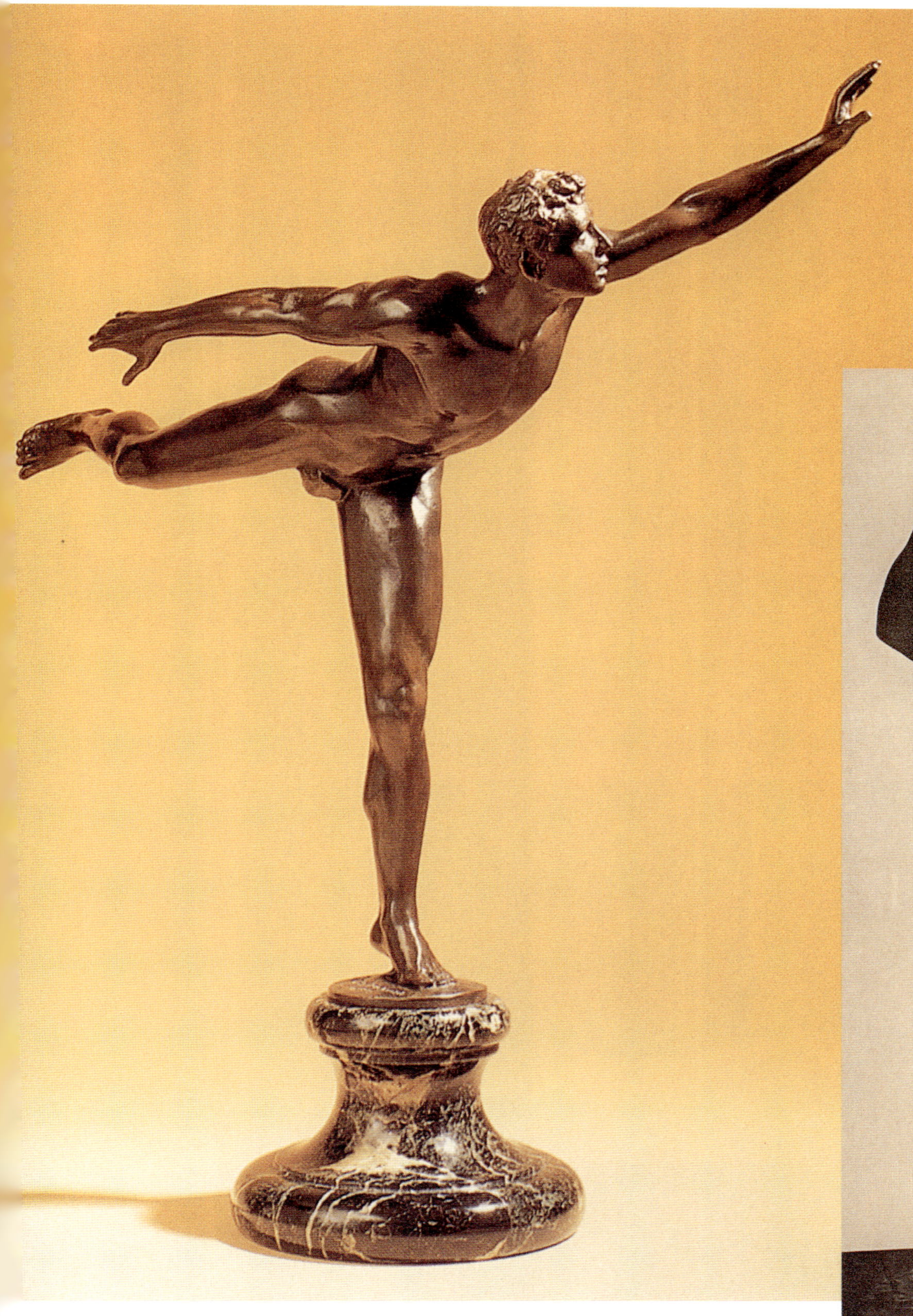

VI-60. "Flying Sphere" by Robert Tait McKenzie.1919.

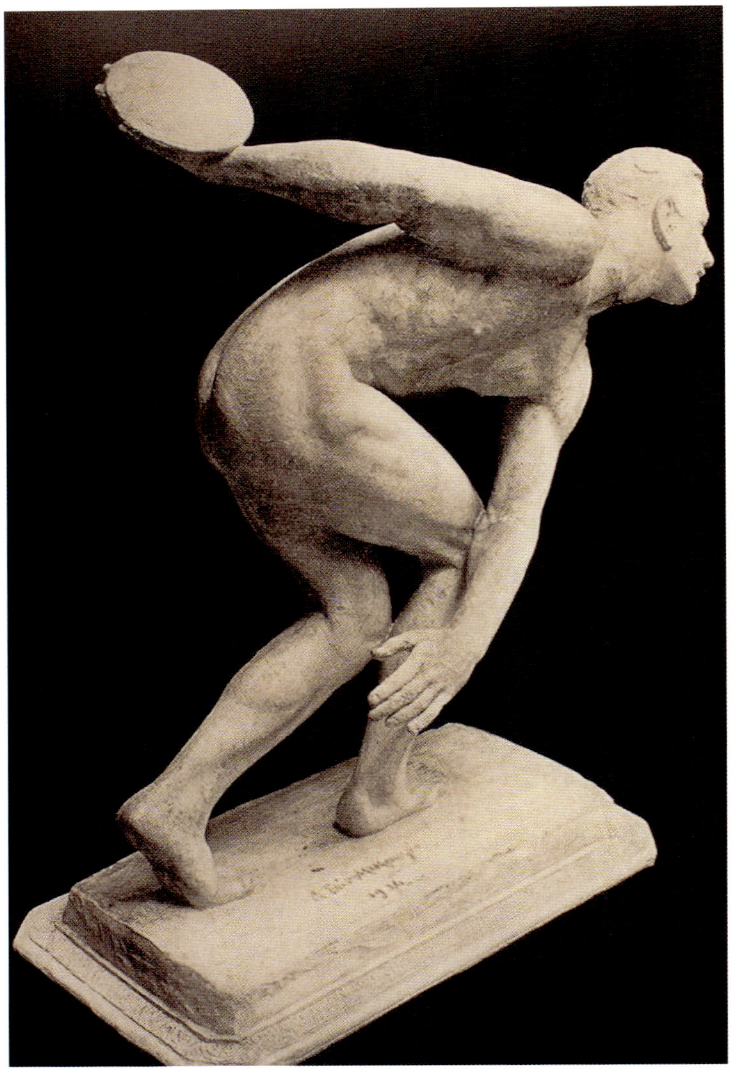

VI-62. "The Modern Discus Thrower" by Robert Tait McKenzie. c. 1909-17.

McKenzie, a native of Ontario, Canada, graduated as a doctor of medicine from McGill University, where he became a demonstrator in anatomy and a medical director of physical education. In 1904 he became director of physical education at the University of Pennsylvania, remaining there until 1930. After World War I, using both his sculptural and medical knowledge, he began experimenting with plastic surgery on soldiers disfigured in the war. Throughout his career in medicine and physical education he completed many monuments and smaller sculptures; his first work in bronze was cast at Roman Bronze Works in 1907 (a juggler).

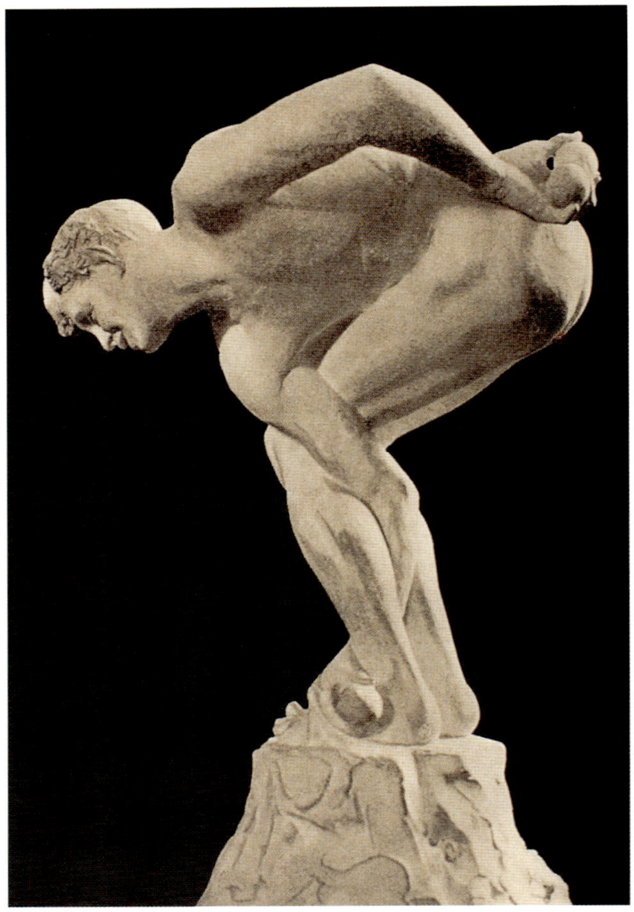

VI-63. "The Supple Juggler" by Robert Tait McKenzie. 1907.

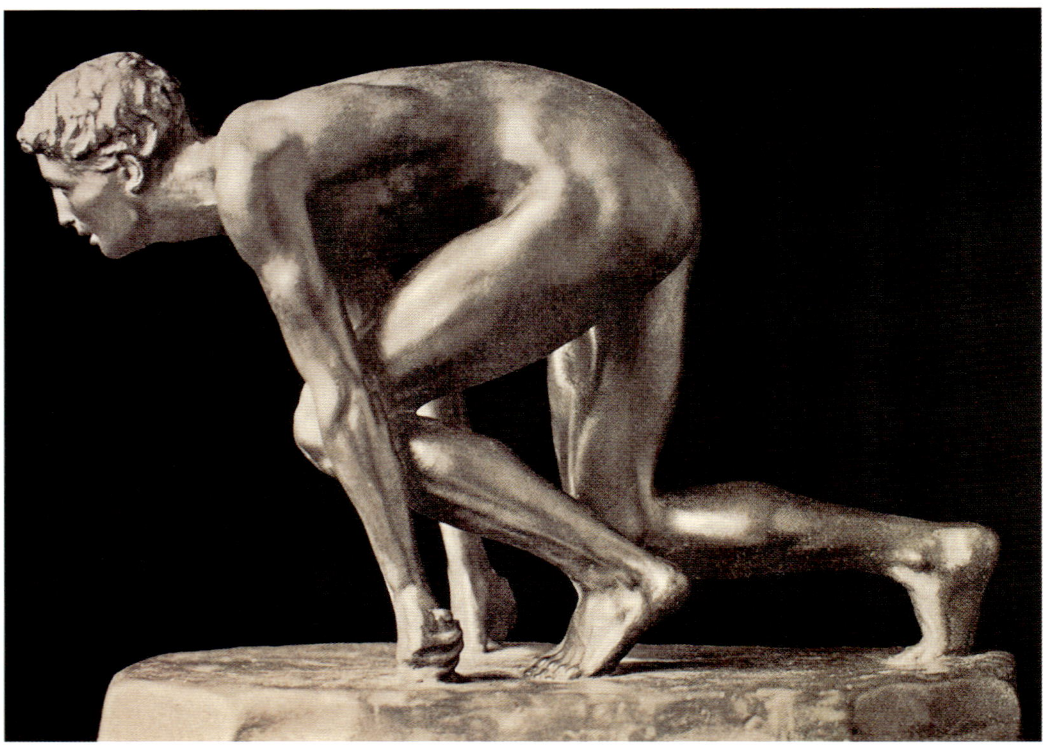

VI-64. "The Sprinter" by Robert Tait McKenzie. 1909.

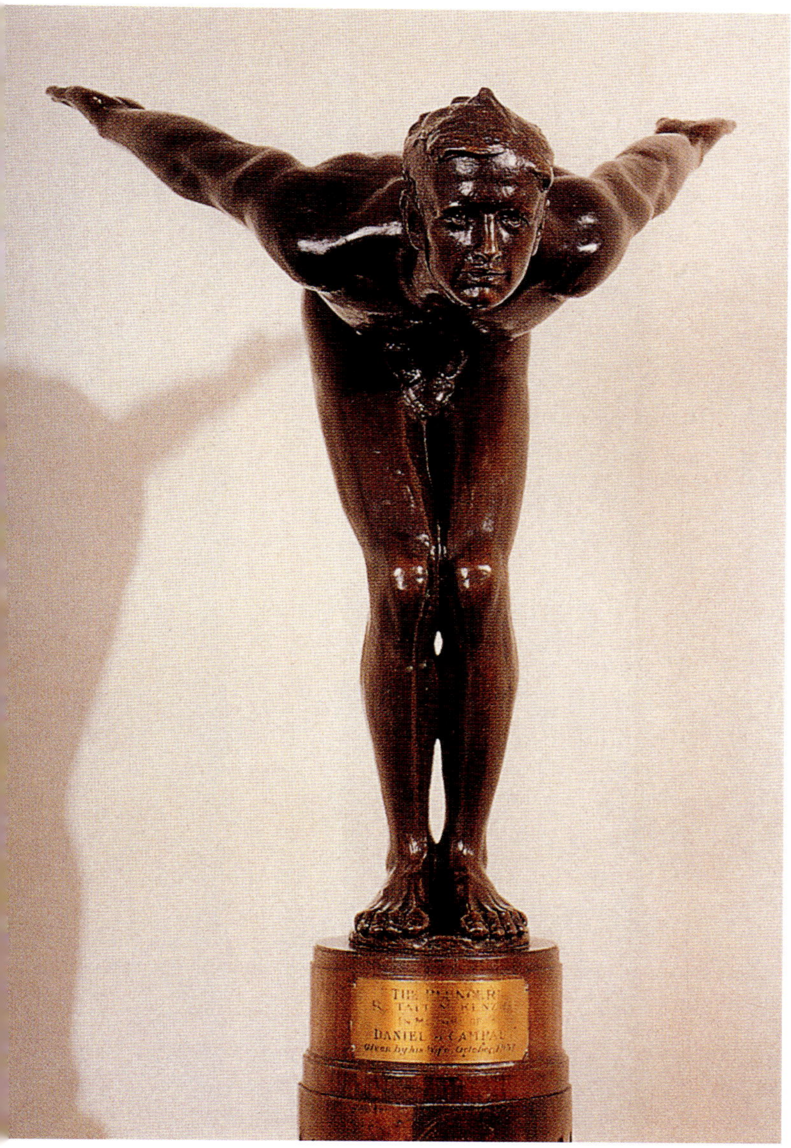

VI-65. "The Diver (also called "The Plunger") by Robert Tait McKenzie. 1925.

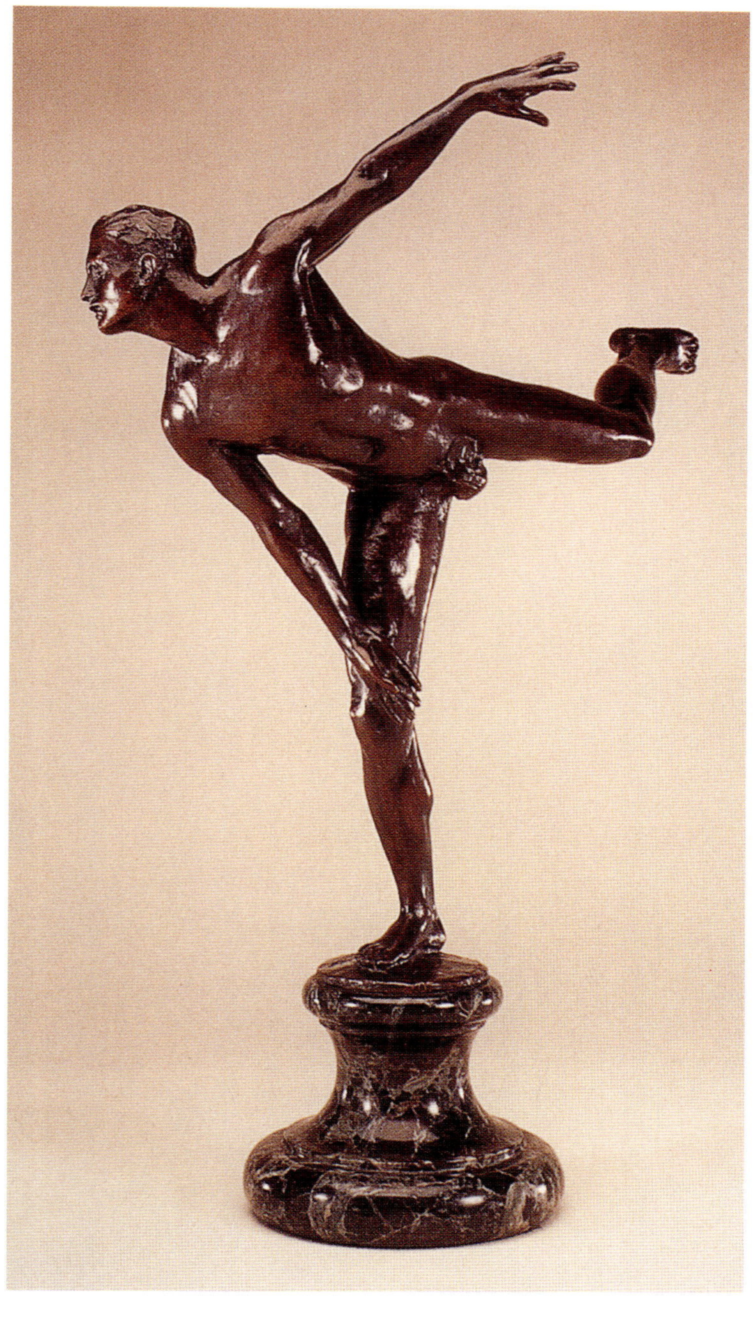

VI-66. "Javelin Cast" by Robert Tait McKenzie. 1923.

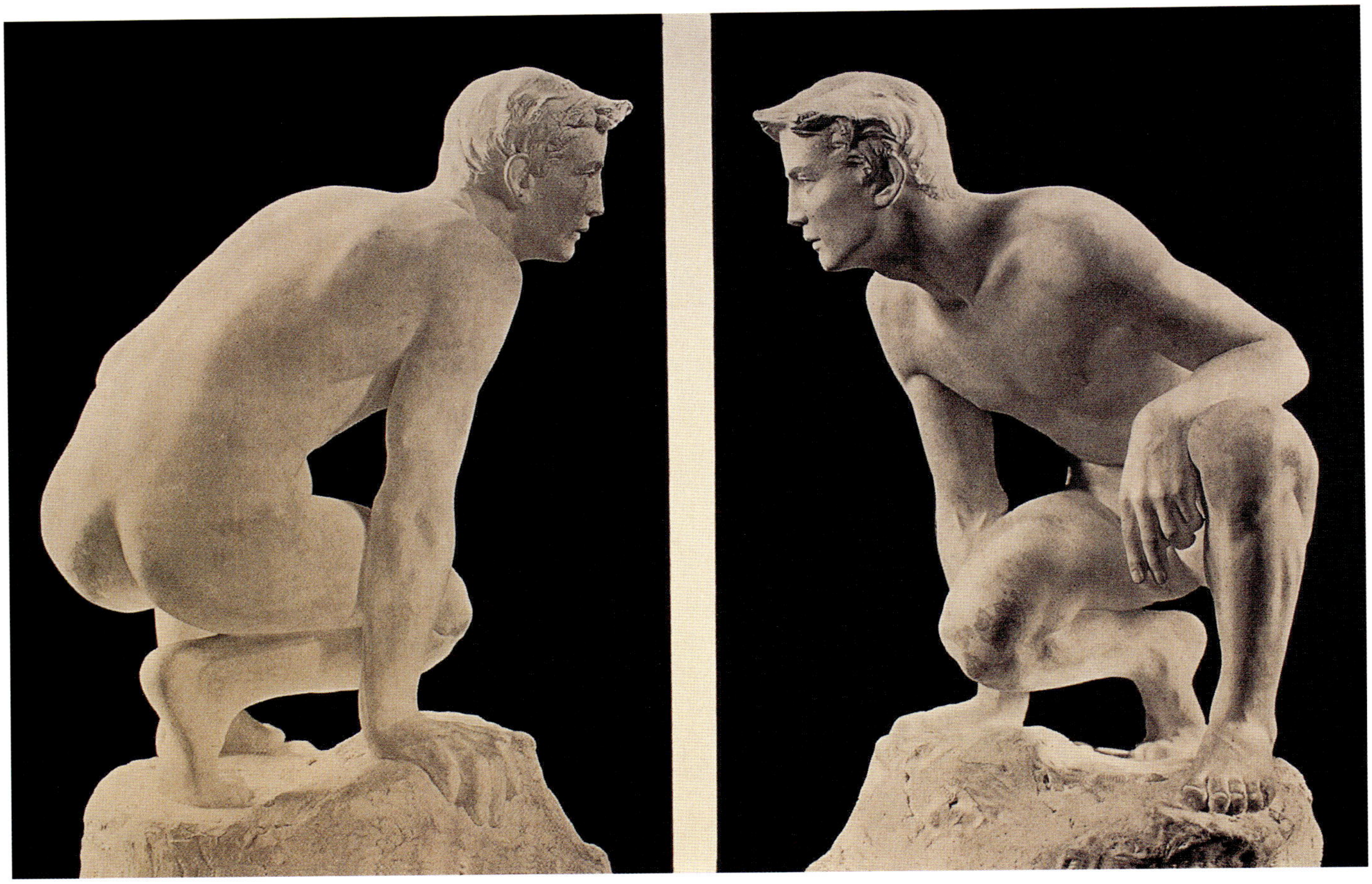

VI-67. "The Relay" by Robert Tait McKenzie. 1909.

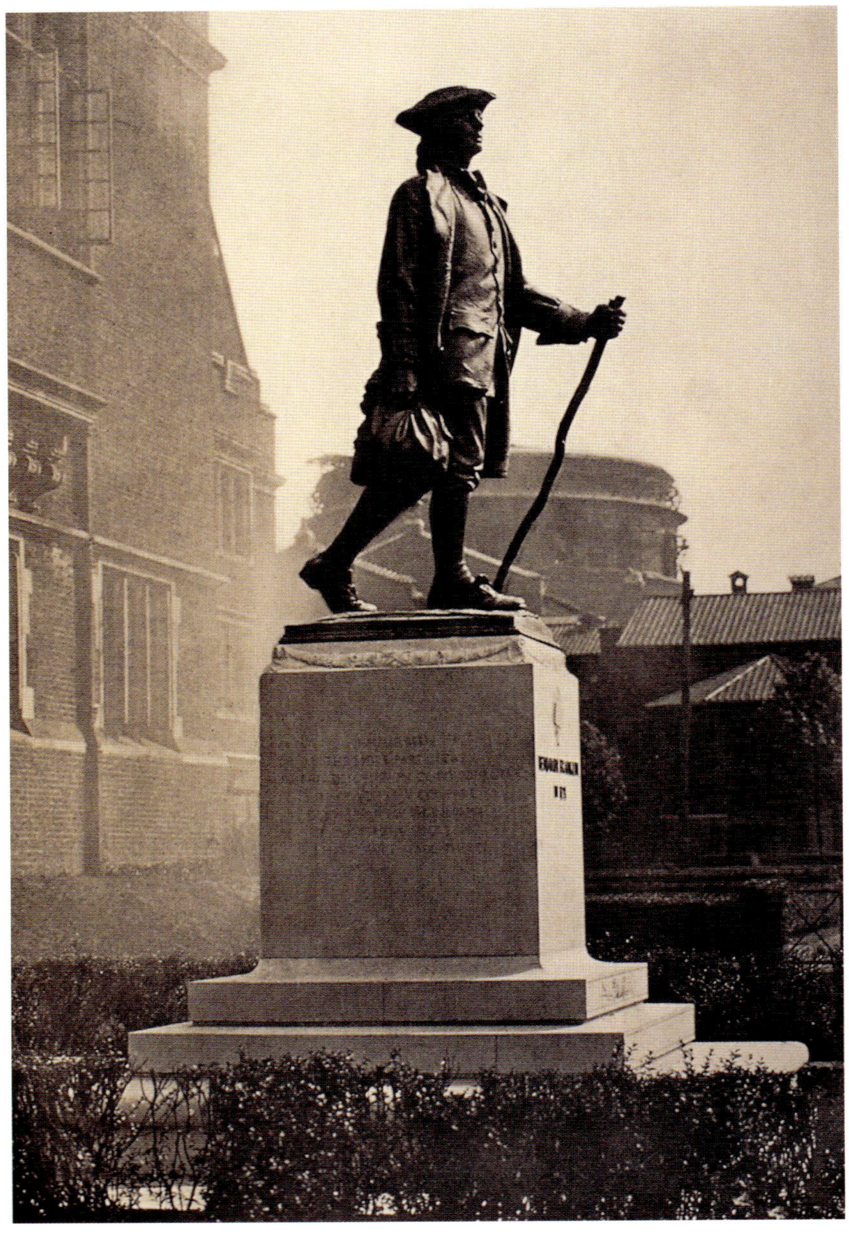

VI-68. "The Youthful Franklin" by Robert Tait McKenzie. 1913.

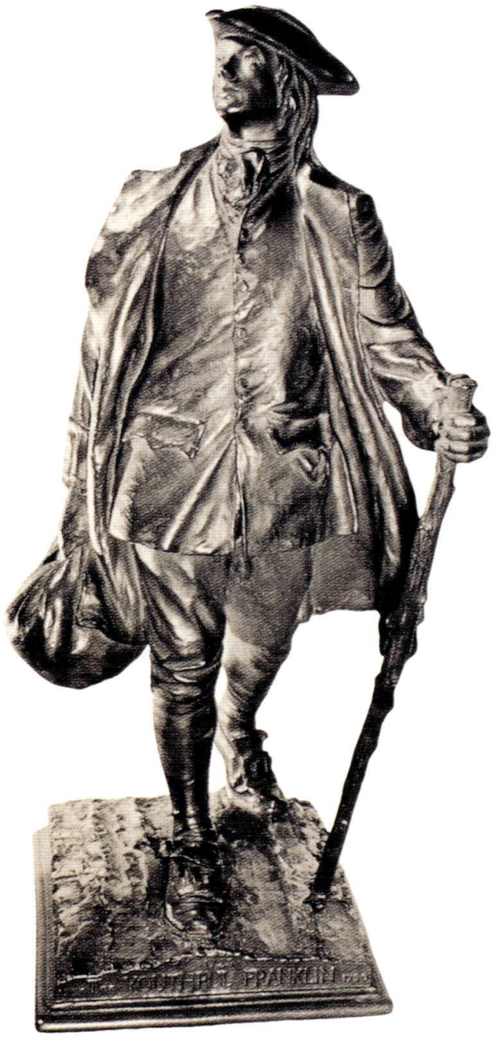

VI-69. "The Youthful Franklin" (bronze statuette) by Robert Tait McKenzie. 1914.

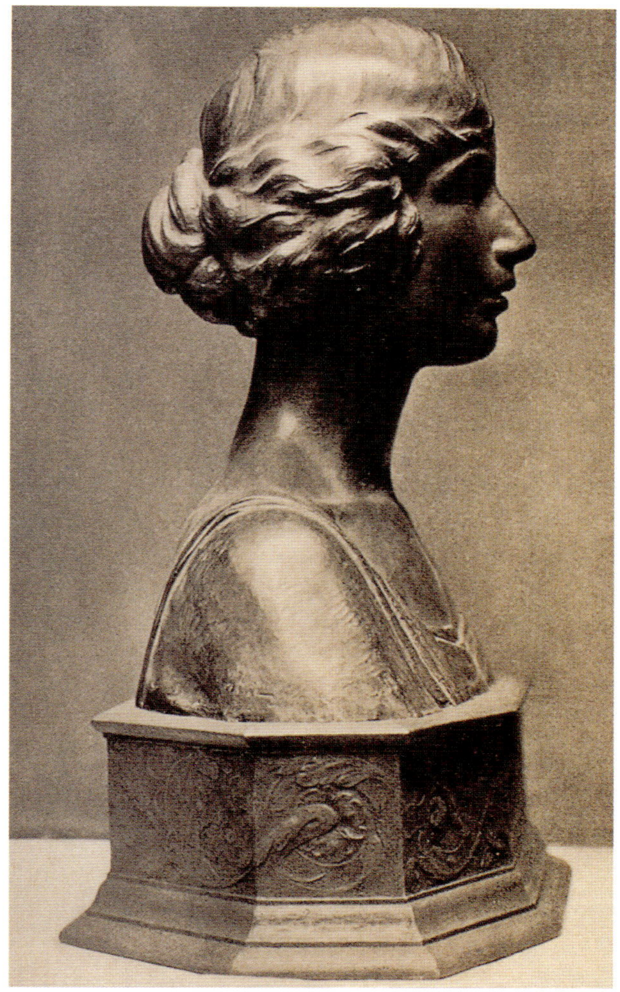

Right:
VI-70. "Sarah Lippincott Biddle" by Robert Tait McKenzie. 1909.

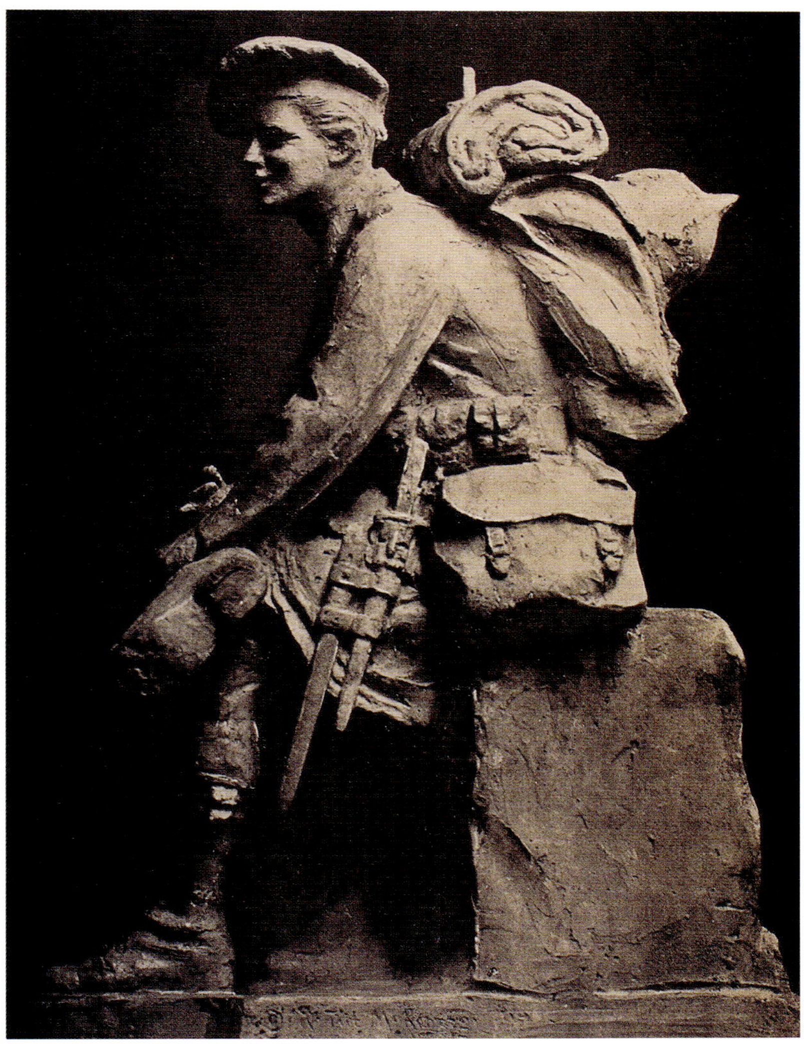

VI-71. "Blighty" by Robert Tait McKenzie. 1917-18.

GERTRUDE VANDERBILT WHITNEY (1875-1942)

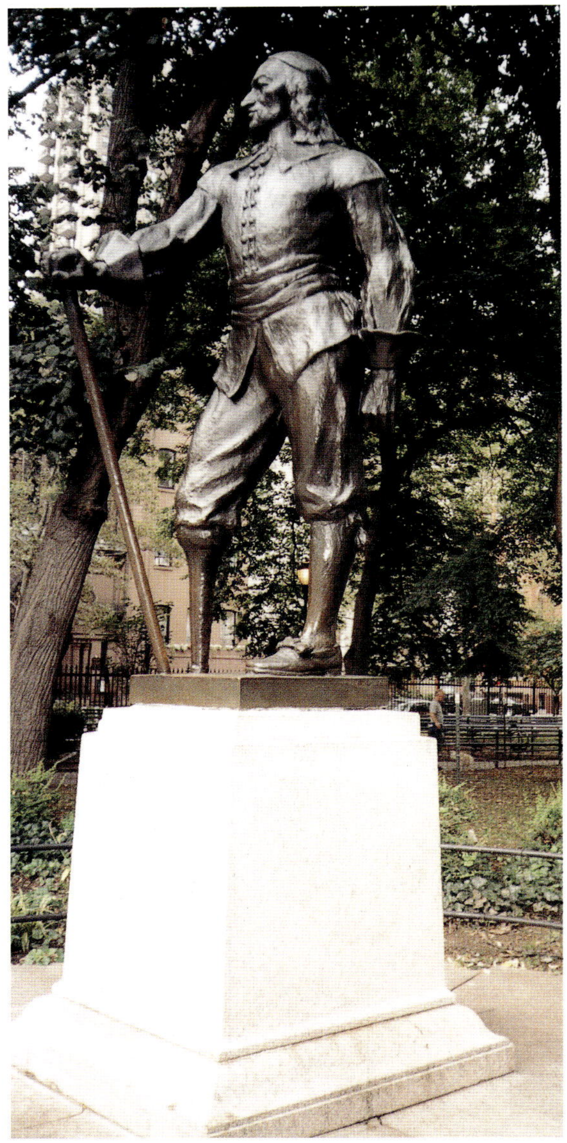

VI-73. "Pieter Stuyvesant" by Gertrude Vanderbilt Whitney. 1936.

Gertrude Vanderbilt Whitney was a sculptor of monuments, architectural and fountain sculptures, and portrait memorials. Her work was characterized by its simplified, heroic style, and its Rodin-like rough surfaces. Though her early work was in a naturalistic style, she later veered away from ideal figures to a more realistic genre style. Born in New York into a wealthy and social family, Whitney became both a sculptor and an extraordinary figure in the worlds of art and philanthropy.

She was trained as a sculptor at the Art Student's League and with James Earle Fraser in New York; she continued her education in Paris, where she worked with Andrew O'Connor and Rodin. During World War I she set up a field hospital in France. The models she sketched at the hospital eventually became parts of her first large war monuments. In New York's Greenwich Village, where she maintained a studio from 1907 until her death, she produced a series of prize-winning monuments and commissions. Among her best-known works are her "Titanic Memorial" in Washington, a statue of Buffalo Bill Cody (VI-72), and a statue of Pieter Stuyvesant (VI-73). Whitney was noted for her contributions to the art world as well; she was a sponsor of numerous young modern-thinking artists, and in 1914 set up a gallery for their works called the Whitney Studio Club. It would become, in 1930, the Whitney Museum of Art, today one of the world's leading museums of contemporary art.

Whitney cast several sculptures at Roman Bronze Works beginning with her sketch for "Titanic Head" (VI-74) in1912. These included a "Daphne" in 1936, and her last piece cast there in 1940, called "The Spirit of Life."

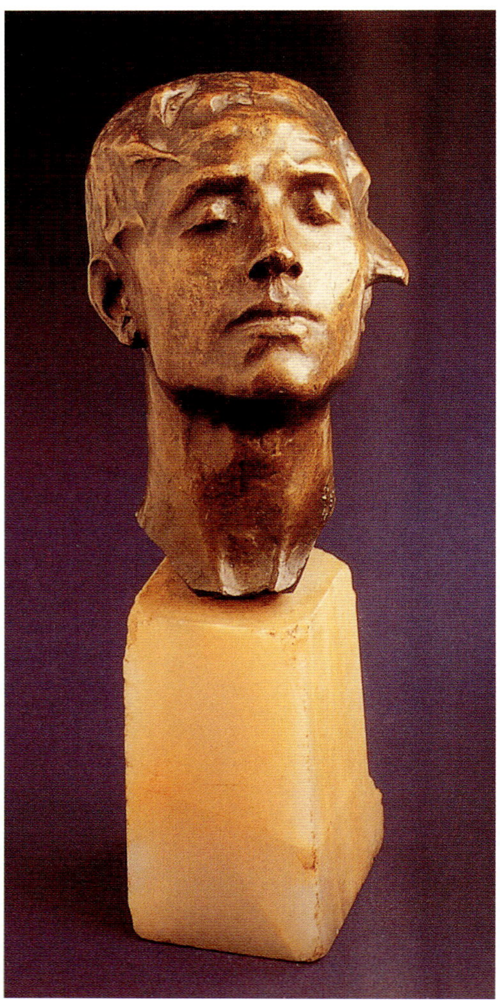

Left:
VI-74. "Titanic Head" by Gertrude Vanderbilt Whitney. 1912.

Right:
VI-72. "Buffalo Bill Cody—The Scout" by Gertrude Vanderbilt Whitney. 1923.

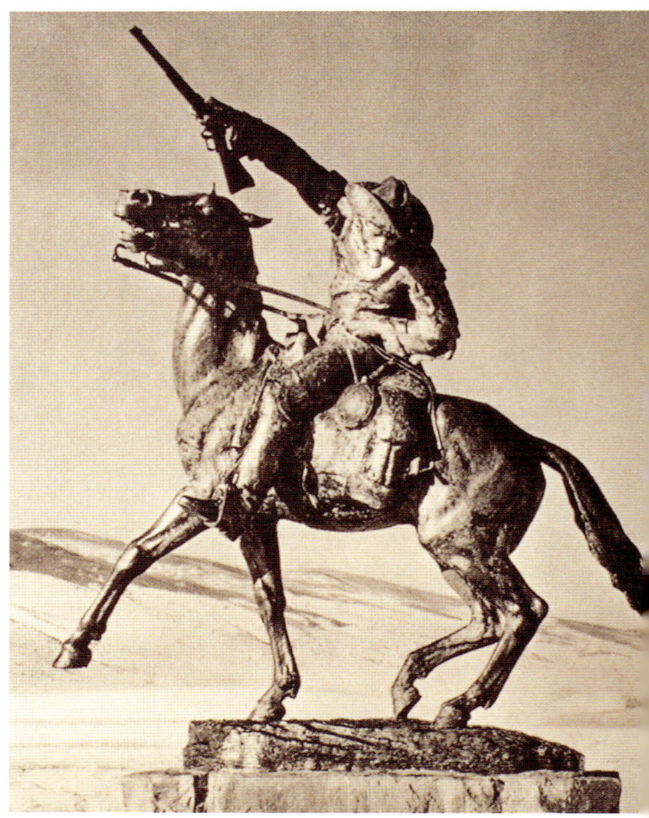

JAMES EARLE FRASER (1876-1953)

James Earle Fraser was a noted maker of portrait busts and commemorative statues, as well as the buffalo and Indian head nickel, but he is perhaps most famous for his sculpture (in two versions), "The End of the Trail" (VI-75 and VI-76). A small version was first exhibited in Paris in 1898; the heroic monument was exhibited in at the Pan-Pacific Exposition in 1915. This heartfelt work, created when the artist was seventeen, became one of the best known sculptures in America. Depicting the tragic fate of the American Indian, it captured awards and national attention, and was cast and recast in five different sizes throughout the century. It is in the collection of museums around the country. Fraser's early work, though distinctly realistic in style, was noted for its grasp of character and colorful technique; later his style became more precise and formal.

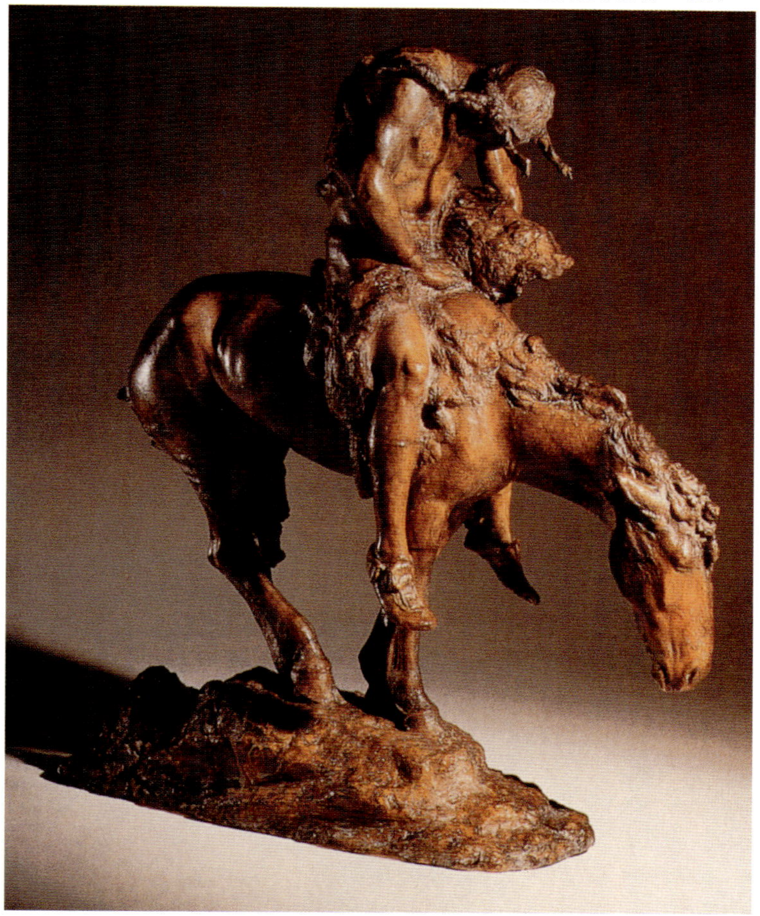

VI-75. "End of the Trail" (plaster cast) by James Earle Fraser. 1918.

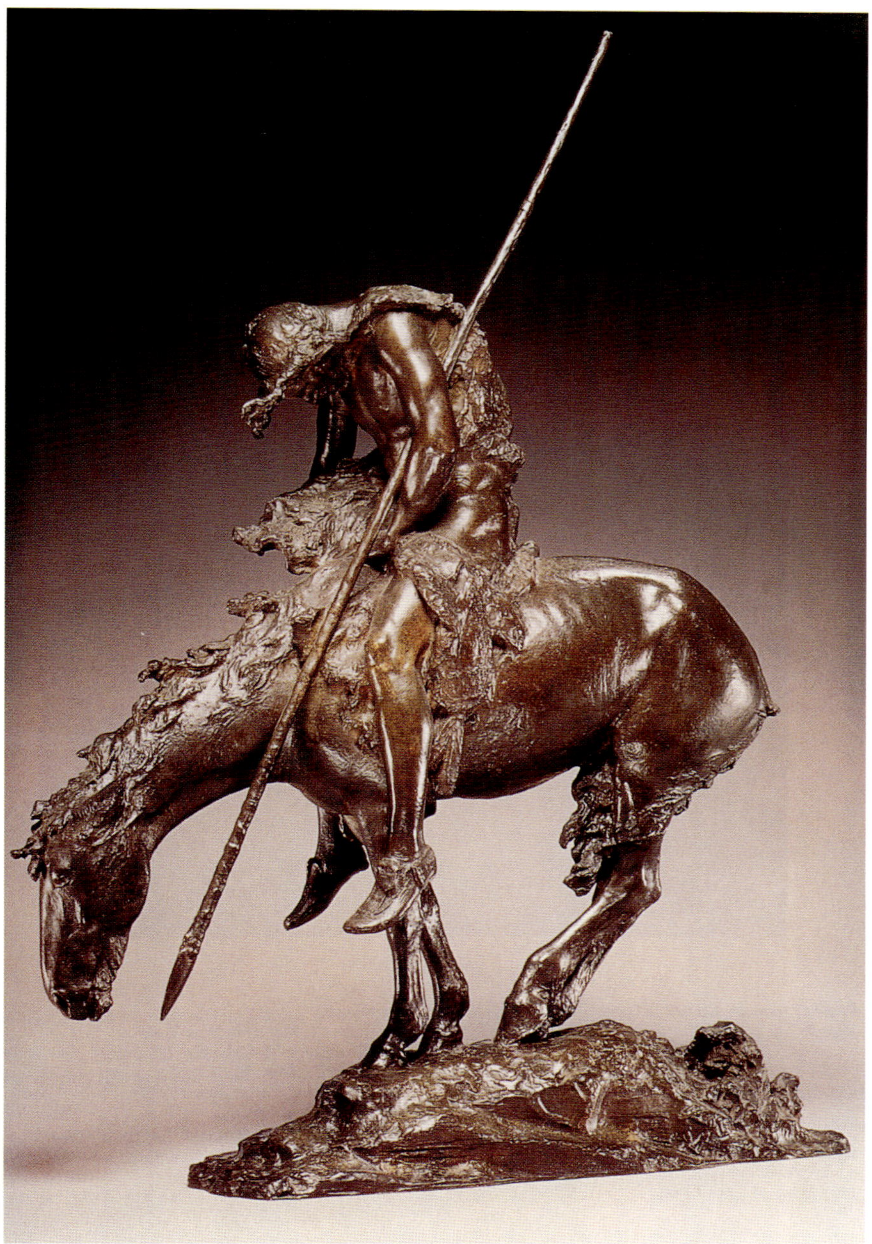

VI-76. "End of the Trail" (bronze) by James Earle Fraser. 1918.

Fraser was born in Minnesota, spending several years of his youth in the frontier territory of Dakota, where he developed a life-long fascination with Western American life. He studied art in Chicago, and left for the École des Beaux-Arts in Paris in 1895. "The End of the Trail" not only garnered him an award; it also brought him to the attention of Augustus Saint-Gaudens, for whom he became an assistant both in Paris and, in 1900, in New Hampshire. In 1902 Fraser opened his own studio in New York. He received numerous commissions. His works ranged widely, from portrait commemoratives, including Theodore Roosevelt (VI-77), Alexander Hamilton, Thomas Jefferson, Albert Gallatin (VI-78), Thomas Edison, to monuments to the pioneers (VI-79), to the Theodore Roosevelt equestrian group at the American Museum of Natural History, to medal and coin designs for the United States Mint. He also made a number of ideal nude figures, such as "The Water Girl" (VI-80). Fraser received many awards, and was president of the National Sculpture Society.

At Roman Bronze Works, Fraser cast only a few works beginning in 1905, but several were cast many times through 1980.

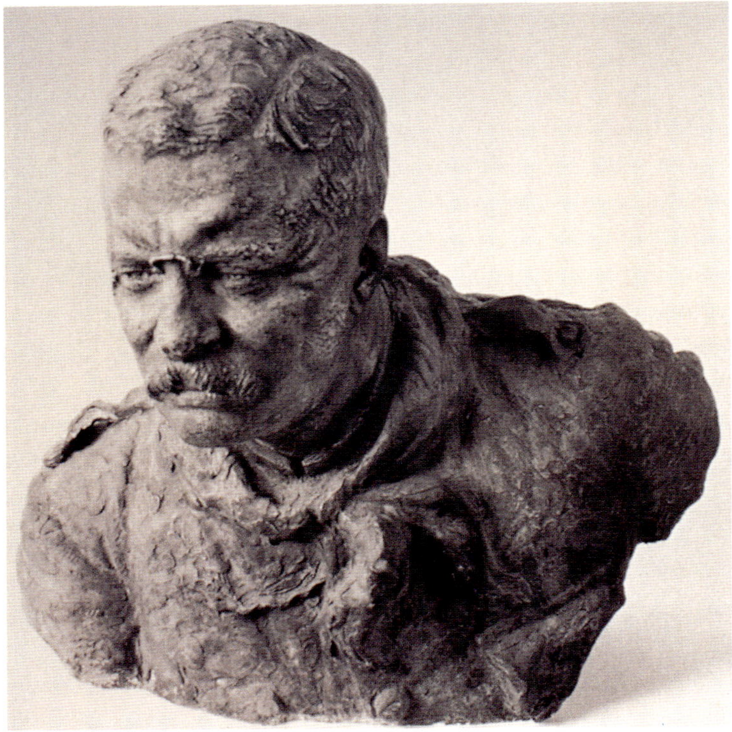

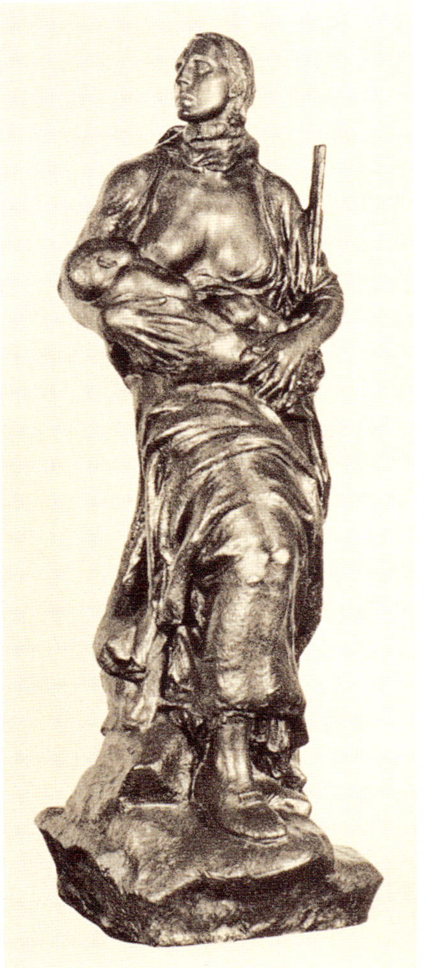

Left:
VI-77. "Theodore Roosevelt as a Rough Rider" (plaster cast) by James Earle Fraser. Cast in bronze at R.B.W. in 1920.

Below left:
VI-78. "Albert Gallatin," Treasury Building, Washington, D.C., by James Earle Fraser. 1947.

Right:
VI-79. "Pioneer Woman" by James Earle Fraser. 1927.

Below right:
VI-80. "The Water Girl" by James Earle Fraser. c. 1980.

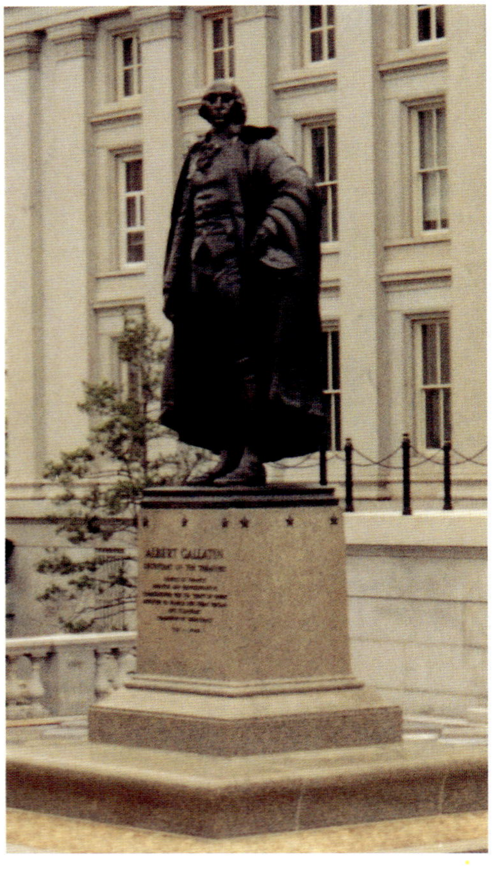

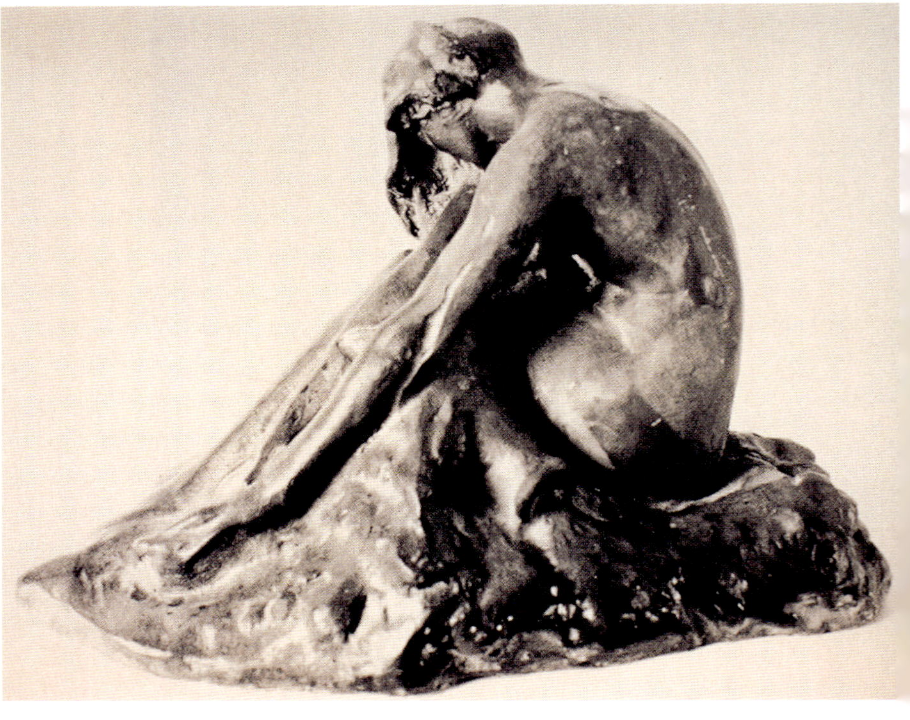

EDWARD McCARTAN (1879-1947)

Best known for his ornamental architectural sculpture, Edward McCartan was a designer of fountains and pediments, and neo-classical and allegorical bronzes. His work was decorative; it often featured nude figures and ancient Greek characterizations, among them several variants of the Diana theme. His style was reminiscent of 18th century France, with its emphasis on Neo-Classical decorative design and linear elegance.

McCartan was born in Albany, New York, studied sculpture with George Grey Barnard and Hermon MacNeil before leaving for Paris in 1907. After three years at the Ecole des Beaux-Arts, McCartan returned to New York and became associated with the Society of Beaux-Arts Architects, training craftsmen in ornamental modeling and carving. He received several commissions for architectural decorations, including the U.S. Department of Labor and Interstate Commerce Building. He was a member of many artistic organizations, including the National Institute of Arts and Letters.

As a sculptor of free-standing statuary, McCartan was particularly drawn to classical figures; among them several different images of Diana with animals (numbers VI-81, VI-82, and VI-83), "A Nymph and a Frog" (VI-84), and a "Pan" (VI-85), as well as studies of the female form: "Girl Drinking from a Shell" (VI-86) and "Female Figure En Pointe" (VI-87). McCartan cast a number of works, including fountain figures, portrait commemoratives, and animals at Roman Bronze Works. His first work cast there was a portrait of fellow artist, Thomas Eakins in 1907; his last was in the 1940s, but copies were made by R.B.W. as recently as 1979.

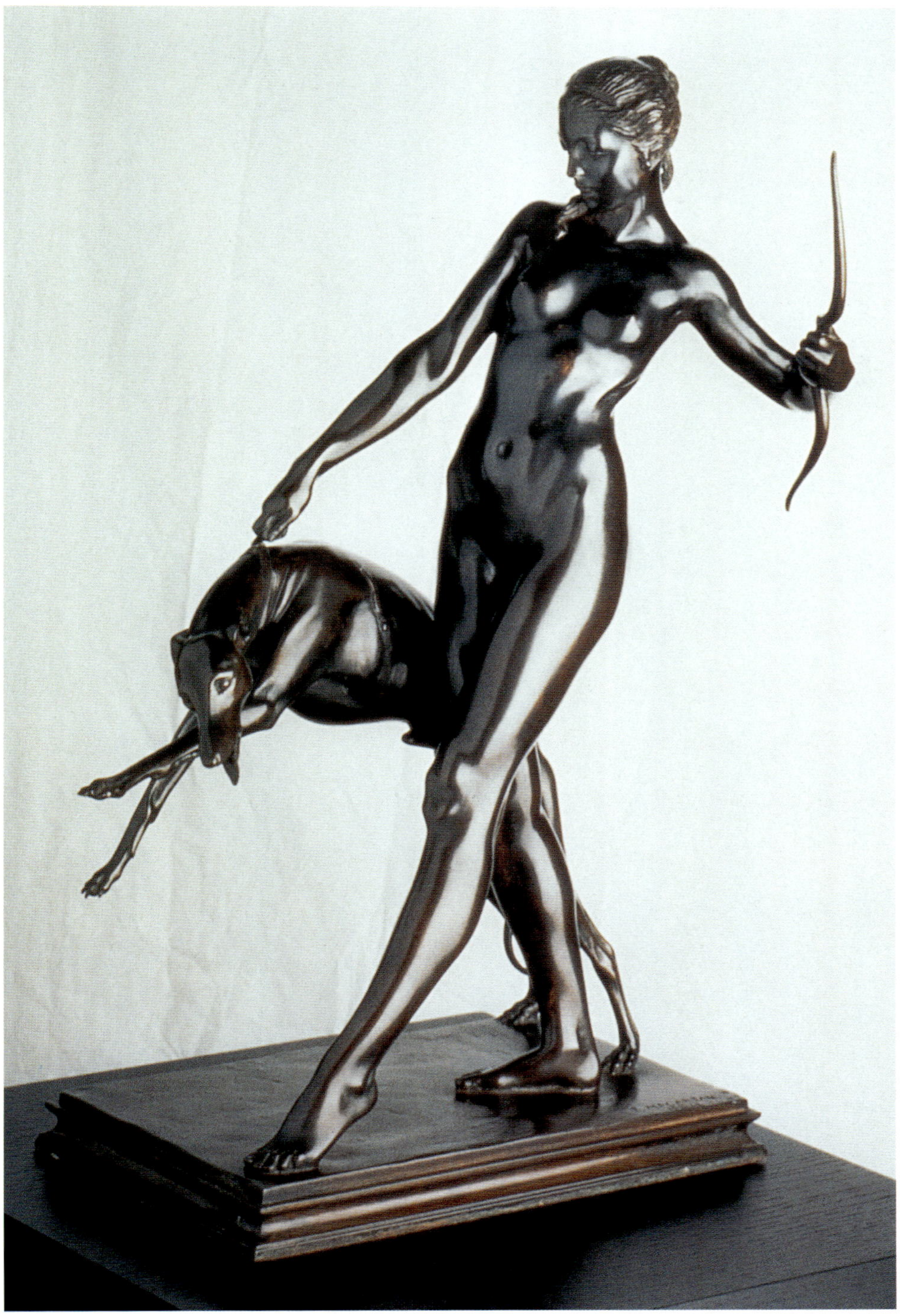

VI-81. "Diana with Hound" by Edward McCartan. 1920.

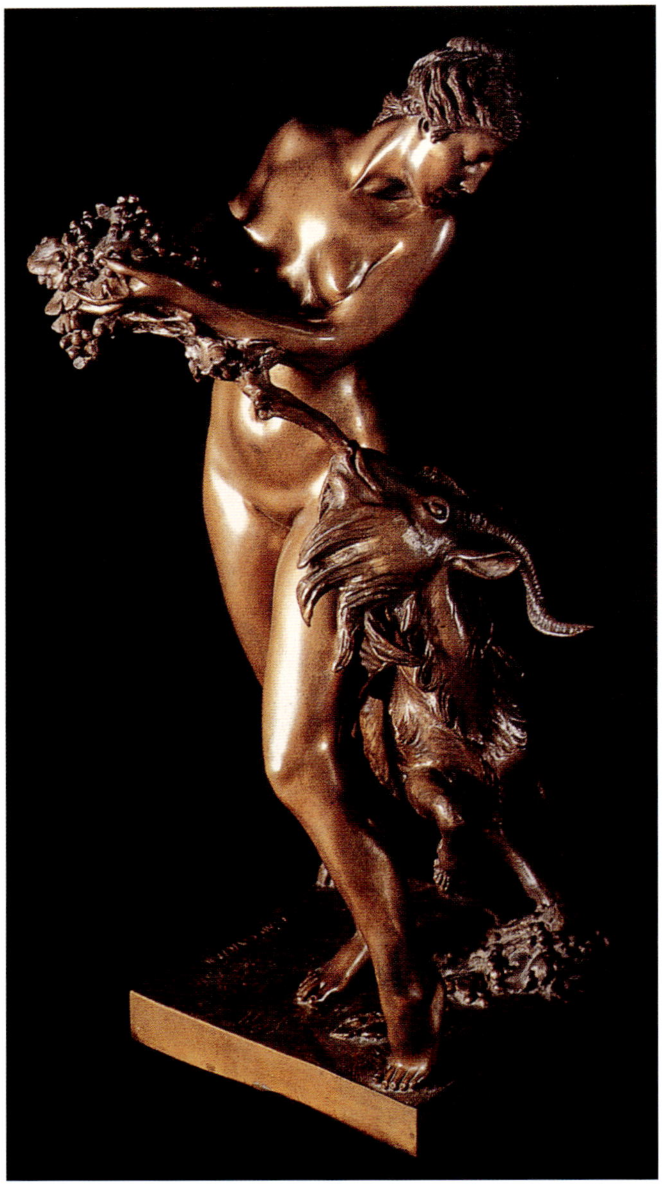

VI-82. "Diana and the Goat" by Edward McCartan. Cast 1978-79.

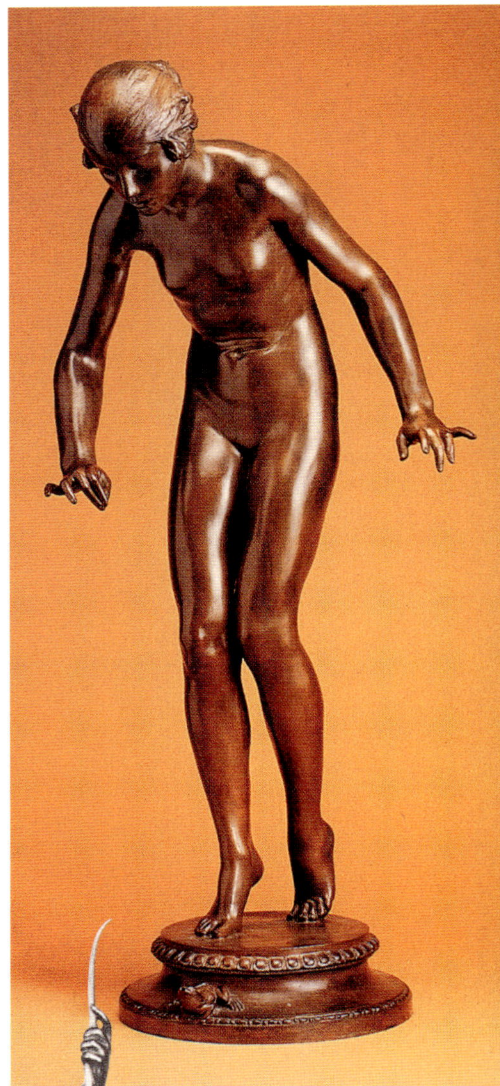

VI-84. "A Nymph and a Frog" by Edward McCartan. C. 1936-38.

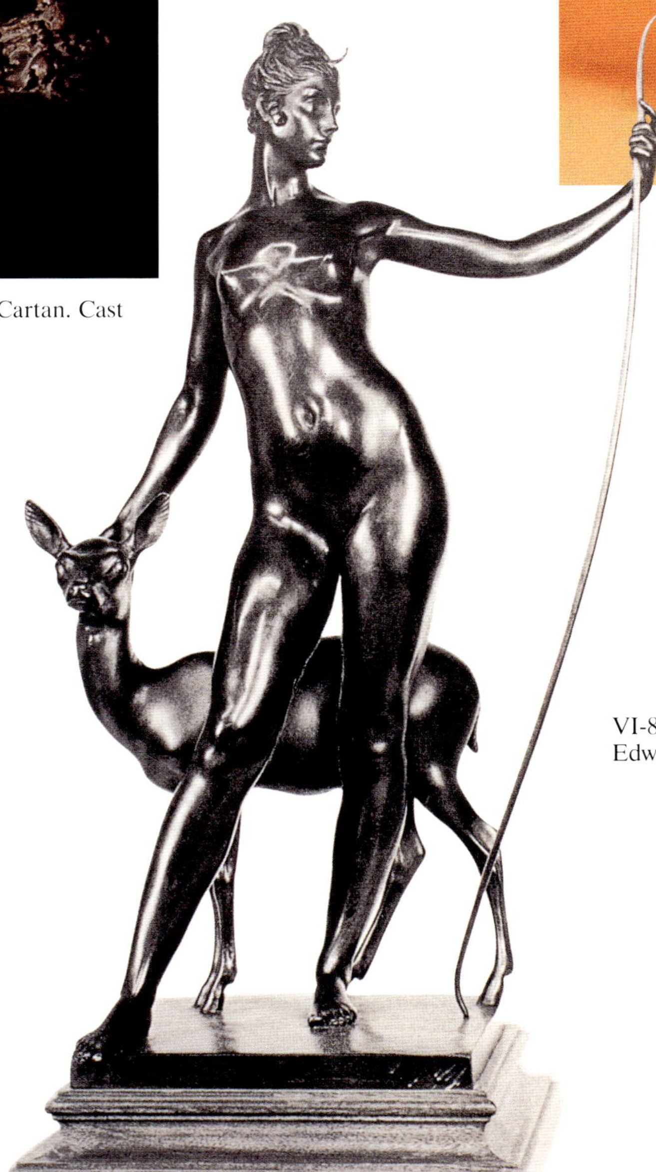

VI-83. "Diana with Doe" by Edward McCartan. 1924.

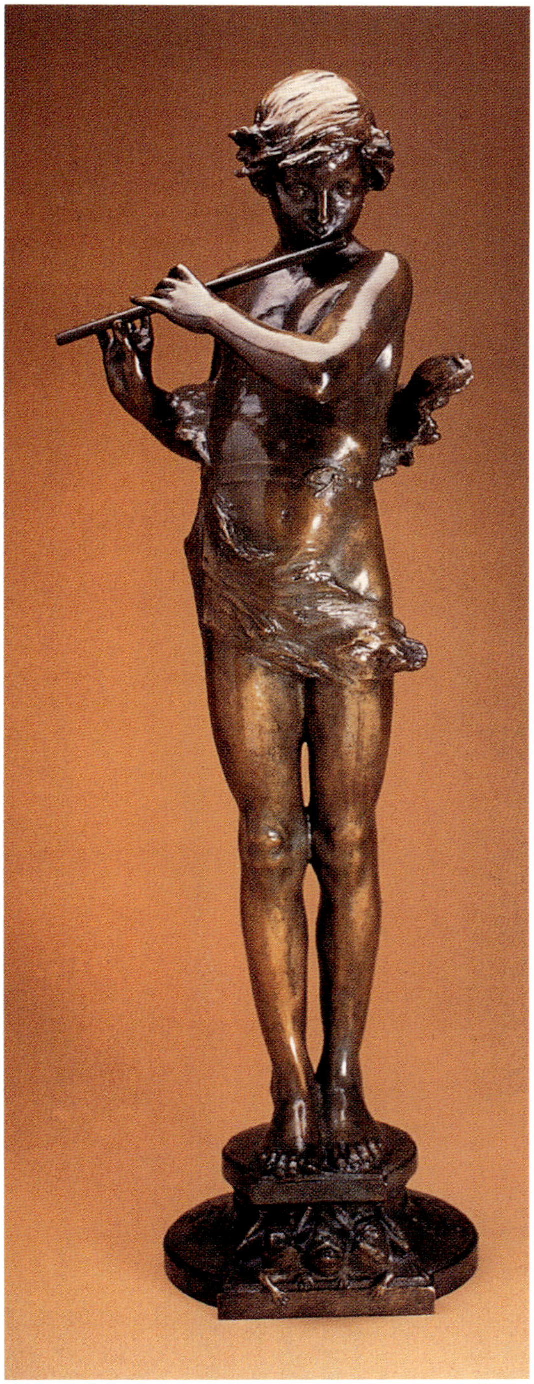

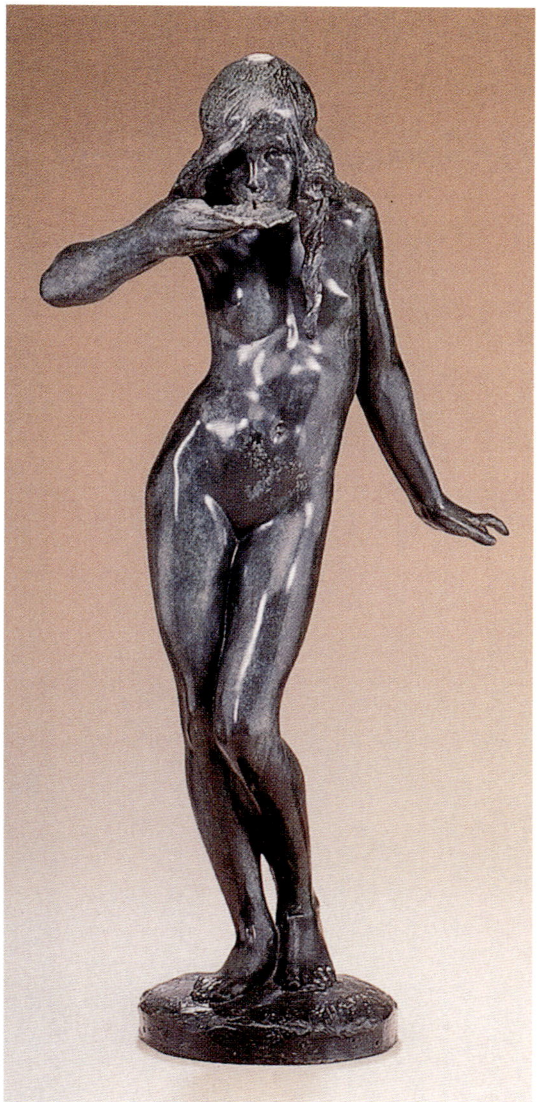

VI-86. "Girl Drinking from a Shell" by Edward McCartan. 1916.

VI-85. "Pan" (fountain figure) by Edward McCartan. 1915.

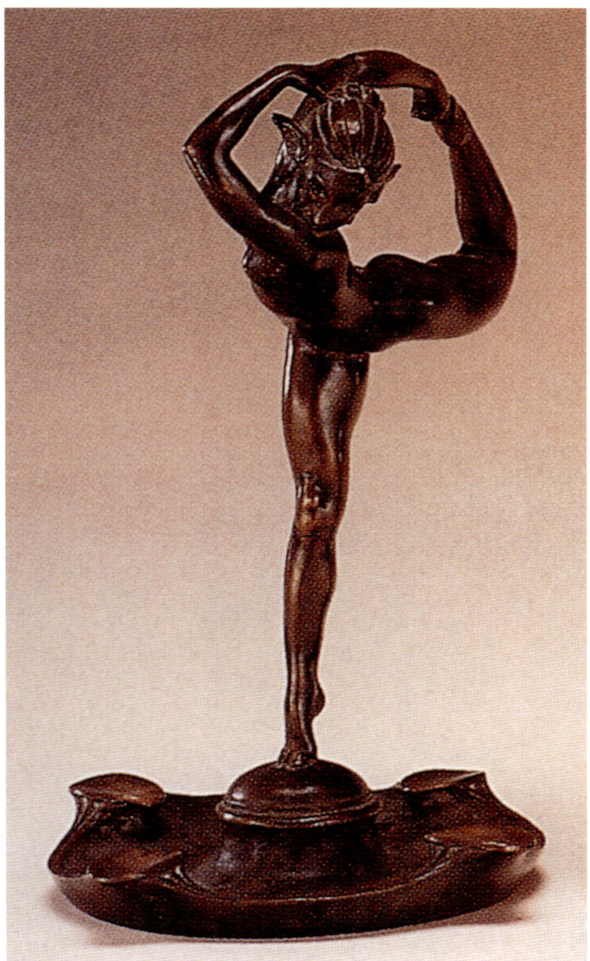

VI-87. "Female Figure en Pointe in a Vide Poche" by Edward McCartan.

ALBIN POLASEK (1879-1965)

Albin Polasek was a native of Moravia; his work included several religious statues, commemorative portrait statues, and allegorical nudes. He worked both in marble and bronze, and some of his work reflects Slavonic Christian images of his youth. His nudes, several of which were cast at Roman Bronze Works, show a stylistic interest in the Neo-Classical and symbolic.

Polasek began his career in the United States in 1901 working as a wood carver for an altar factory. He then studied with Charles Grafly at the Pennsylvania Academy of Fine Arts, and won a scholarship for two years at the American Academy in Rome. In 1916 he began his longtime position as head of the Department of Sculpture at the Chicago Institute of Art. Polasek made a number of portrait statues and medallions of well-known Americans (Daniel Boone, Abraham Lincoln, and Zachary Taylor, for example) and made memorial statues both for the United States and Czechoslovakia. Polasek had a career filled with awards, prizes, and decorations. He was a member of the National Academy of Design, the American Academy in Rome, and numerous other artistic organizations.

Polasek's first sculpture at Roman Bronze Works was cast in 1914 (a Neptune Fountain); he continued to use the foundry through 1957. Among his sculptures cast at R.B.W. were his symbolic piece "Man Carving his Own Destiny"(VI-88), and nudes such as "Unfettered" (VI-89), "Bubble" (VI-90), "Forest Idyll" (VI-91), and "Maiden of the Roman Compagne" (VI-92).

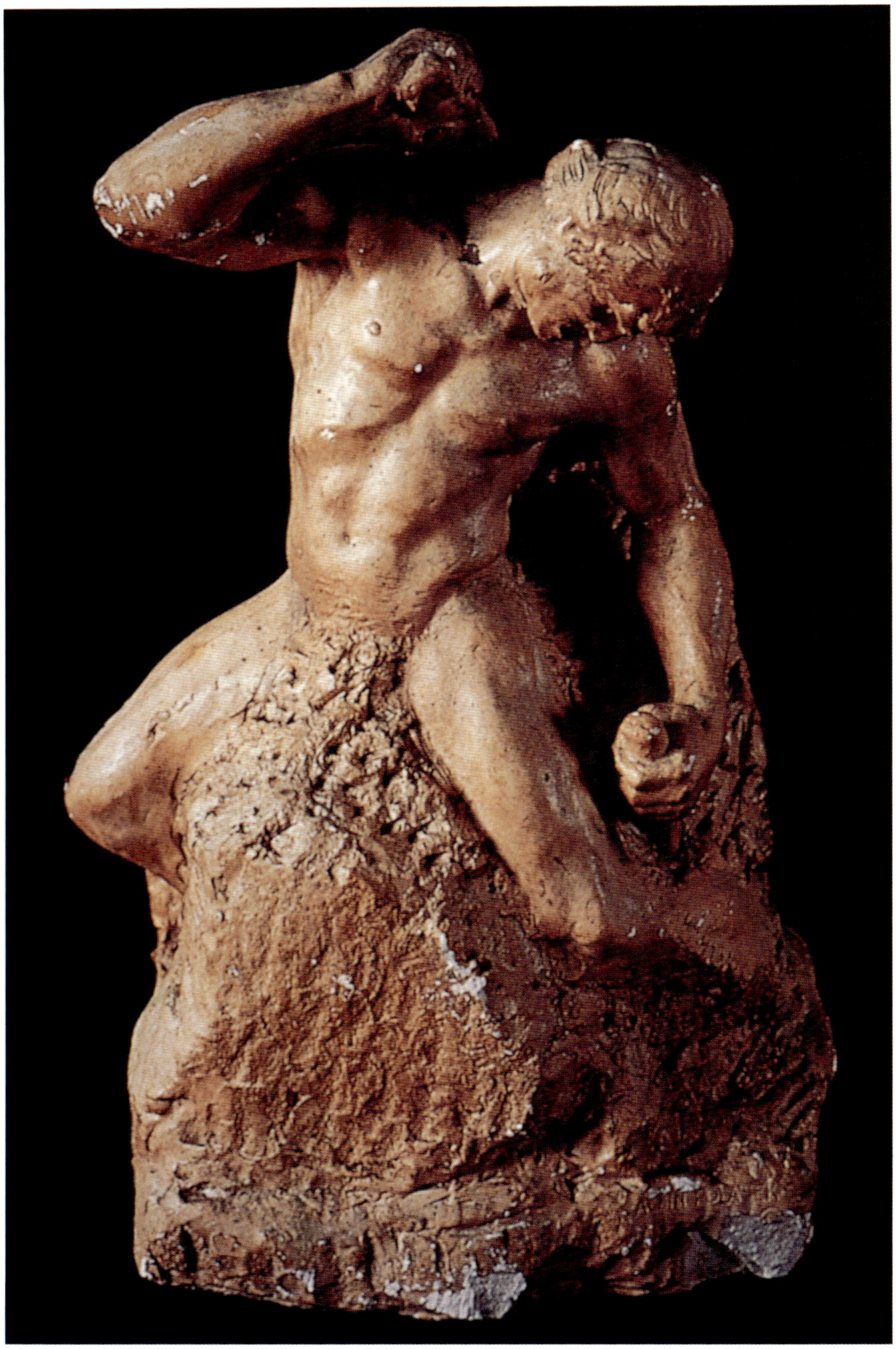

VI-88. "Man Carving his own Destiny" by Albin Polasek. 1917.

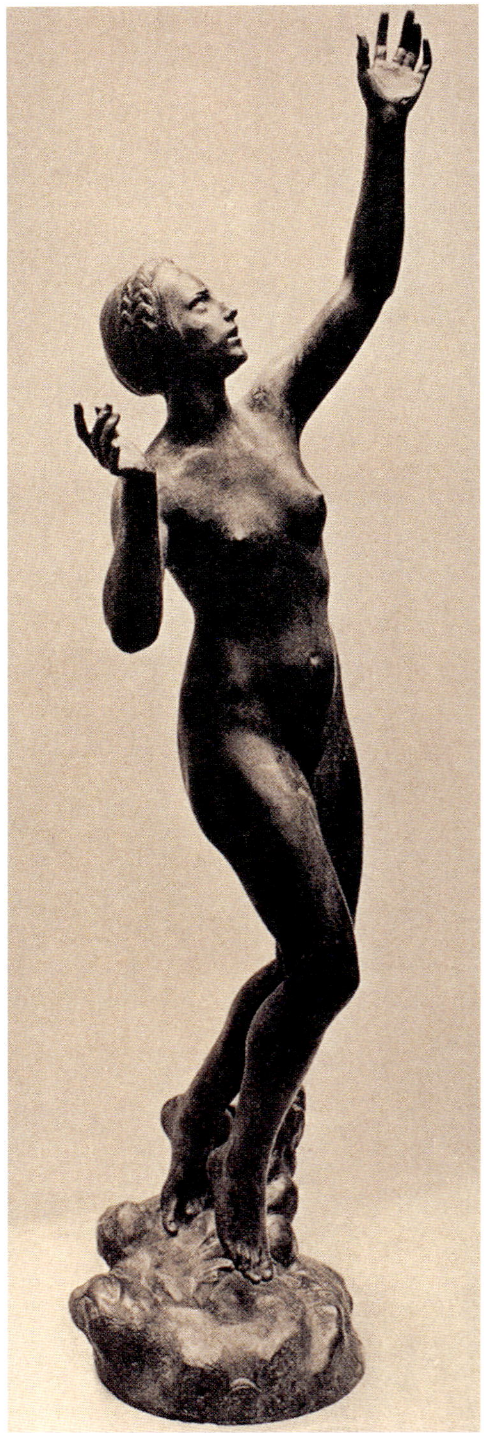

VI-89. "Unfettered" by Albin Polasek. 1932.

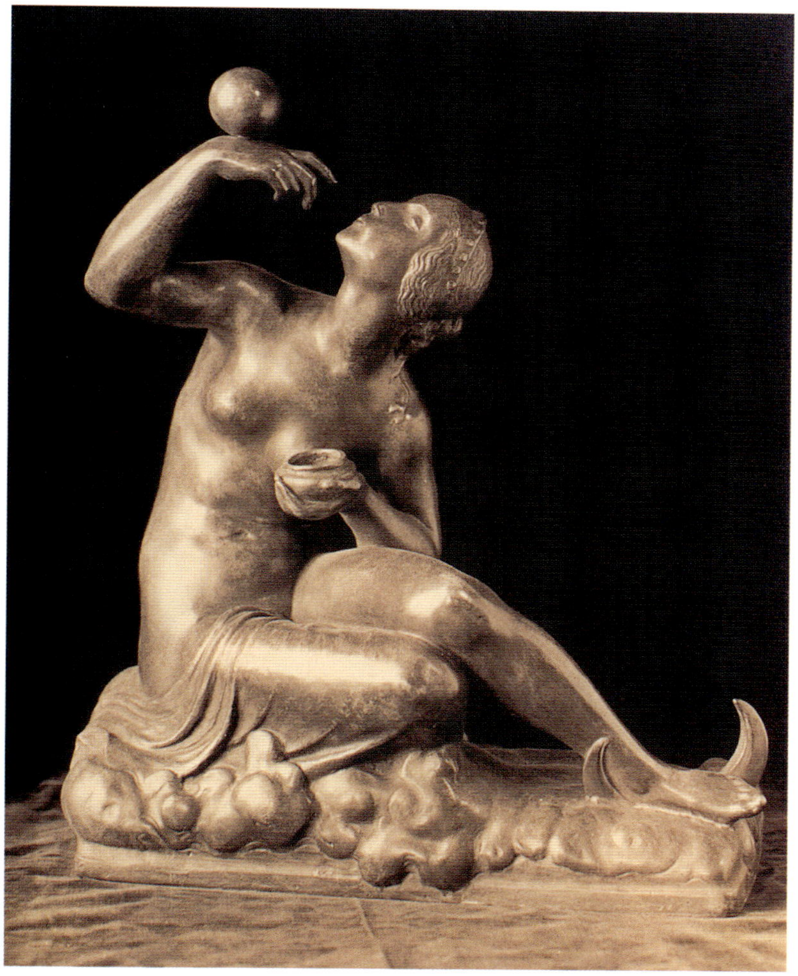

VI-90. "The Bubble" by Albin Polasek. 1917.

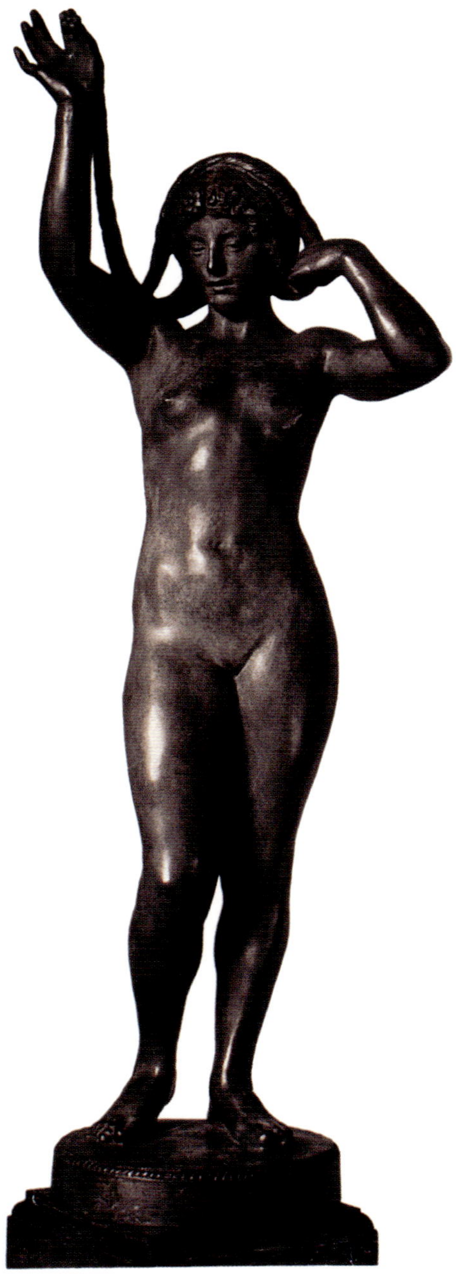

VI-92. "Maiden of the Roman Compagna" by Albin Polasek. 1916.

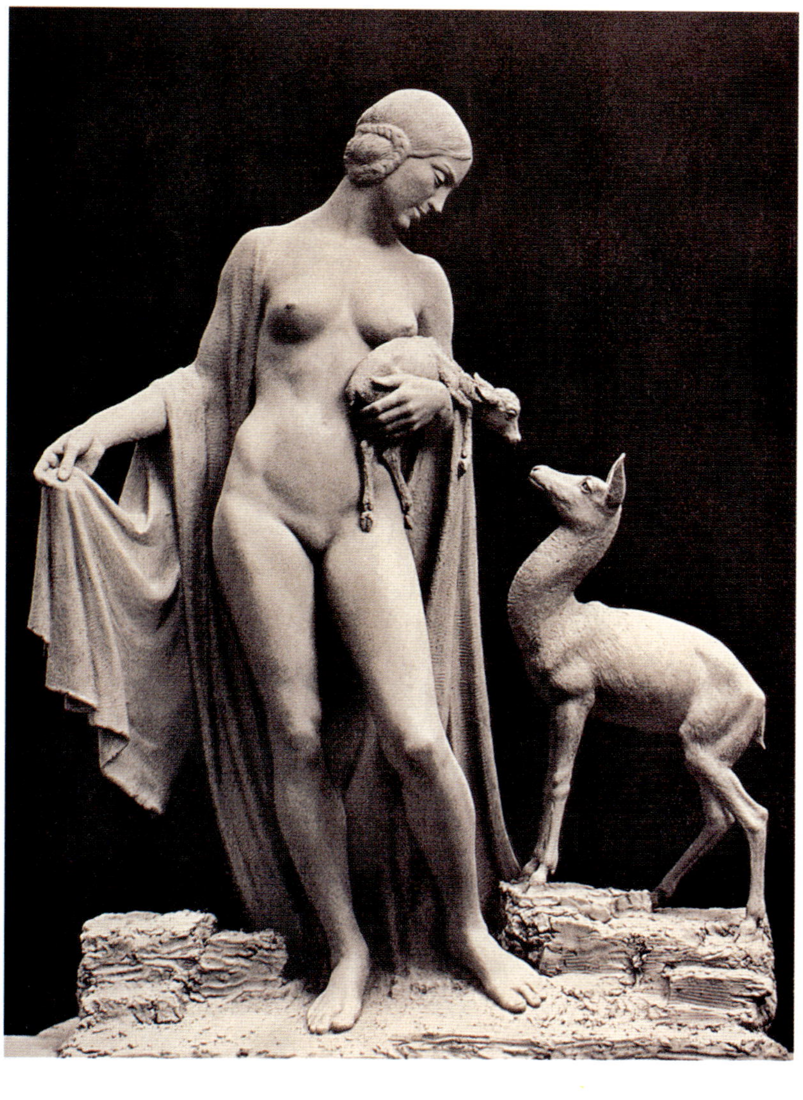

VI-91. "Forest Idyll" by Albin Polasek. 1930.

JOHN GREGORY
(1879-1958)

A sculptor who devoted himself primarily to architectural sculpture and garden figures, John Gregory worked in a classical mode. Many of his sculptures were based on ancient myth. His style was a form of stylized classicism, occasionally featuring formalized groups of figures.

Gregory was born in England, came to America at the age of fourteen, and by 1900 was a student of John Massey Rhind and later George Grey Barnard and Hermon MacNeil at the Art Student's League. After further study in London and Paris, Gregory returned to New York and became an assistant to several important sculptors: MacNeil, Gutzon Borglum, and Herbert Adams. He spent three years as a Fellow at the American Academy in Rome, served as a camouflage designer in World War I, and, in 1921, received his first major architectural commission. His career brought him several honors, the presidency of the National Sculpture Society, and membership in the National Academy of Design and the National Institute of Arts and Letters.

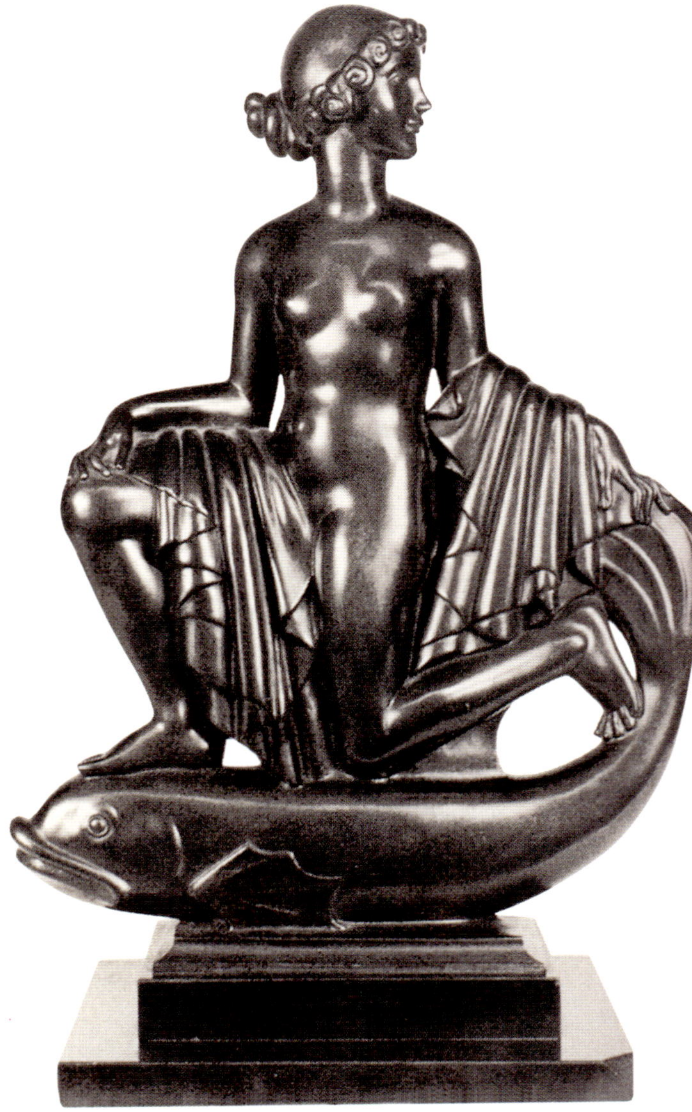

VI-94. "Toy Venus" by John Gregory. 1923.

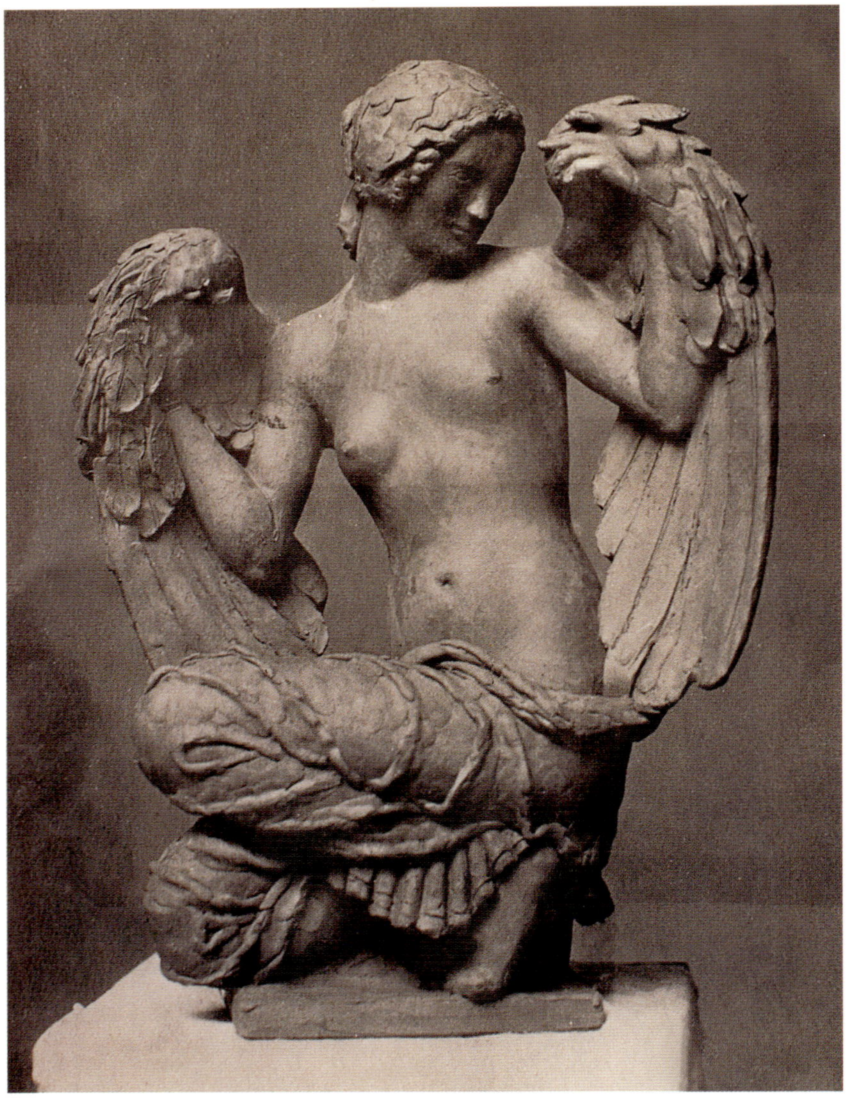

VI-93. "Philomena" by John Gregory. 1919-20.

As an architectural and commemorative sculptor, Gregory created classical figures and panels, representatives of the Four Seasons, and group scenes from Shakespeare's plays for various public and private buildings. He also sculpted several ancient Greek mythological figures such as "Philomena" (VI-93), "Toy Venus" (VI-94), and his group piece "Orpheus" (VI-95). Studies of the female form include "Summer" (VI-96) and "Seated Nude" (VI-97). Other works included figures of Confucius and of Christ, cast at the foundry in the late 1920s.

Gregory began casting in at R.B.W. in1908 and continued through 1930.

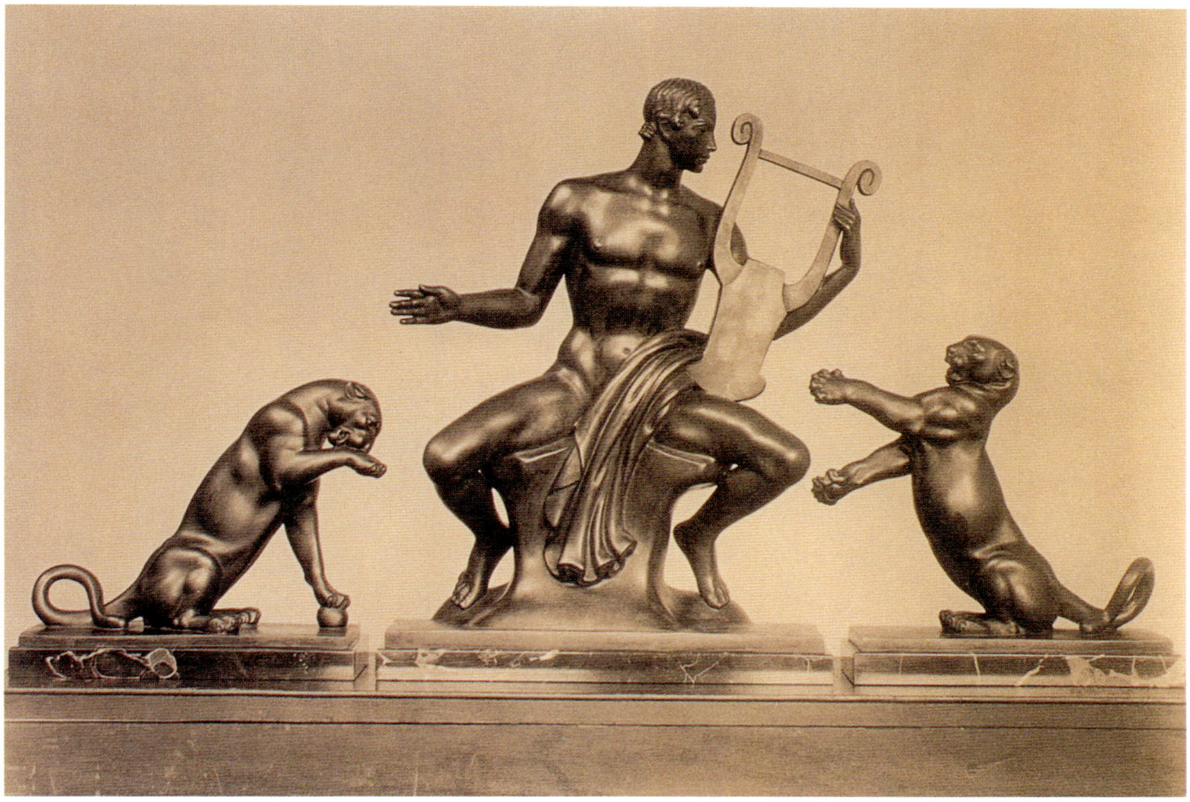

VI-95. "Orpheus" by John Gregory. First cast 1918.

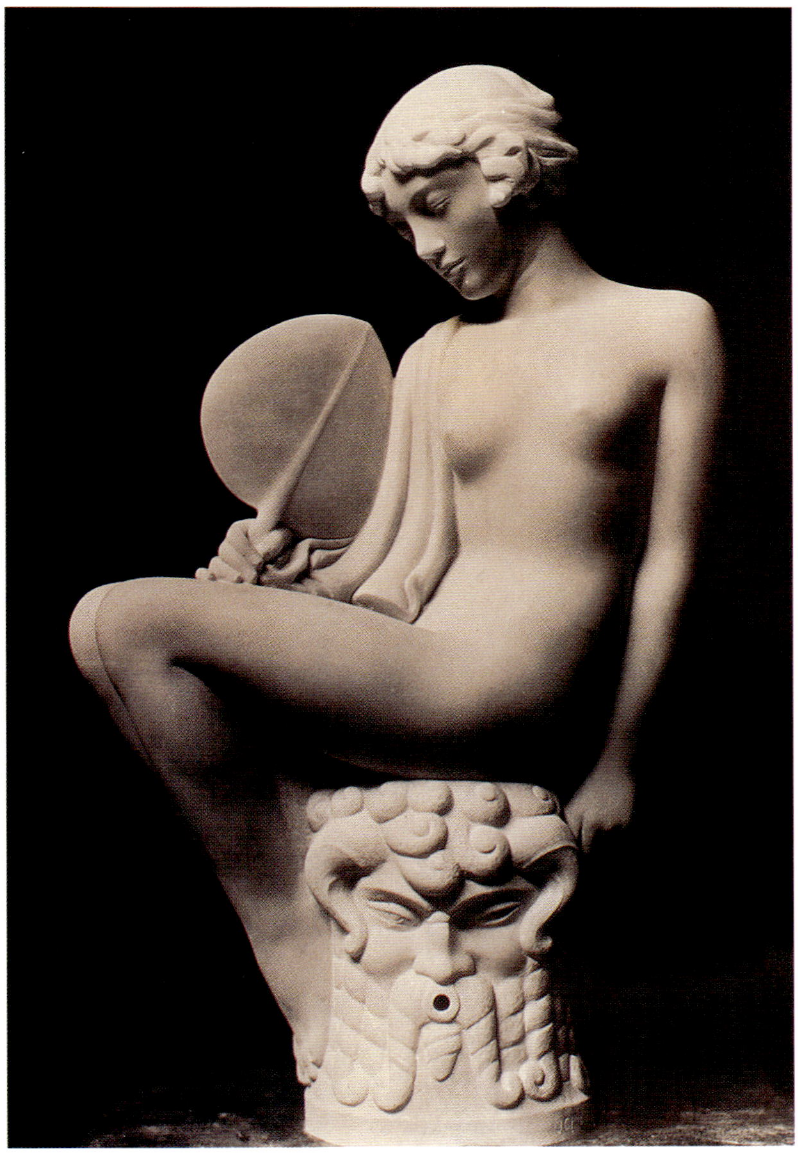

VI-96. "Summer" by John Gregory. 1926.

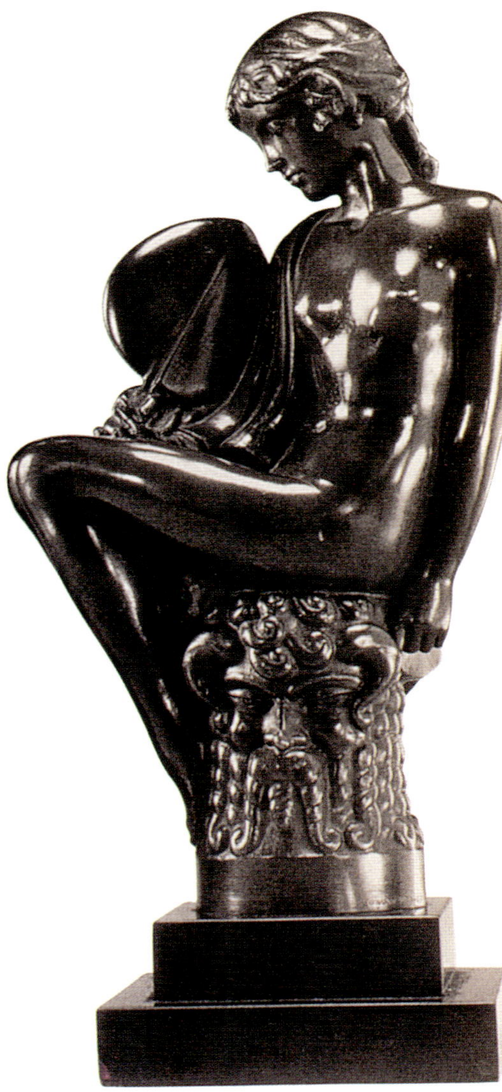

VI-97. "Seated Nude" by John Gregory. 1927.

HARRIET FRISHMUTH (1880-1980)

Noted for her genre pieces, and some portrait busts (including one of President Wilson), Harriet Frishmuth was one of the few American women to have a successful career as a sculptor in America early in the twentieth century. She worked in a realistic, lyrical style with flickering surfaces, often picturing nude young women as dancers and nymphs cavorting with nature.

Born in Philadelphia, she moved to New York to study at the Art Student's League with Hermon MacNeil and Gutzon Borglum, and then to Paris to work with Rodin, among other mentors. She first exhibited at the Paris Salon in 1903. In New York she became a member of the National Sculpture Society and the National Academy of Design. She was the recipient of several awards.

In New York she sculpted portrait heads, fountain figures, medals, and various images of a popular dancer called "Desha" (VI-98), who also posed for "The Vine" (VI-99). Other examples of her delicate female figures in moments of motion include "Allegra" (VI-100), which became a companion piece to her popular "Desha," "The Fisherman's Daughter" (VI-101), "Playdays" (VI-102), "Crest of the Wave" (VI-103) and "Bubble Dance" (VI-104). Frishmuth cast these and several additional bronze pieces at Roman Bronze Works between 1928 and the 1950s.

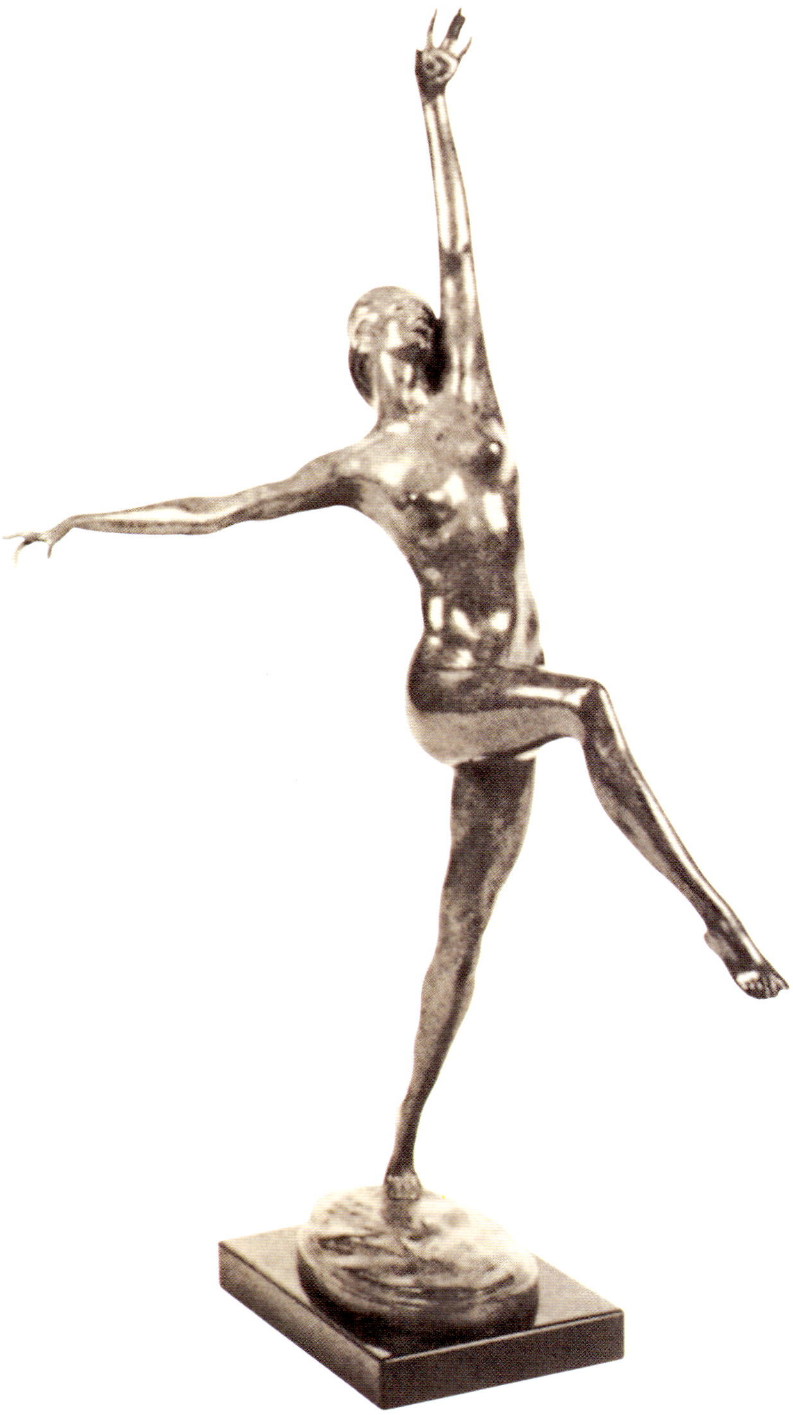

VI-98. "Desha" by Harriet Frishmuth. 1927.

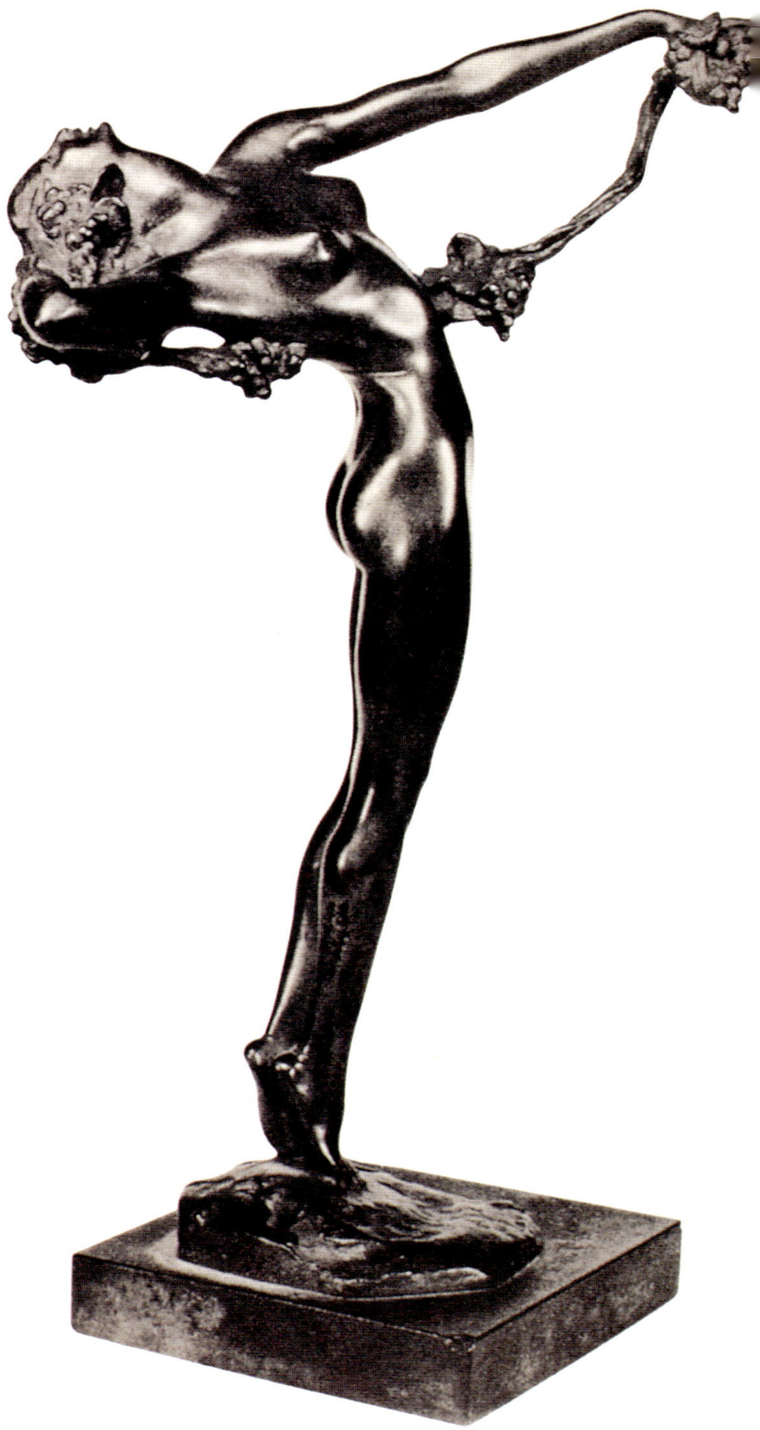

VI-99. "The Vine" by Harriet Frishmuth. c. 1920.

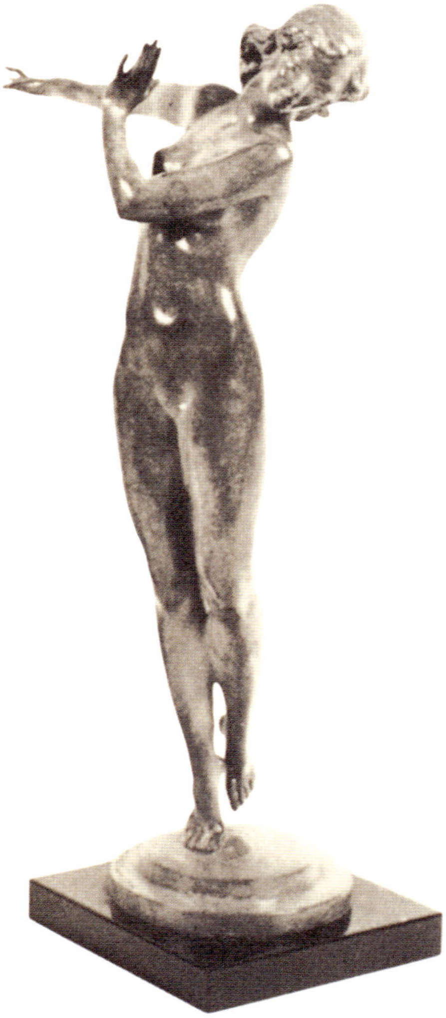

VI-100. "Allegra" by Harriet Frishmuth. 1929.

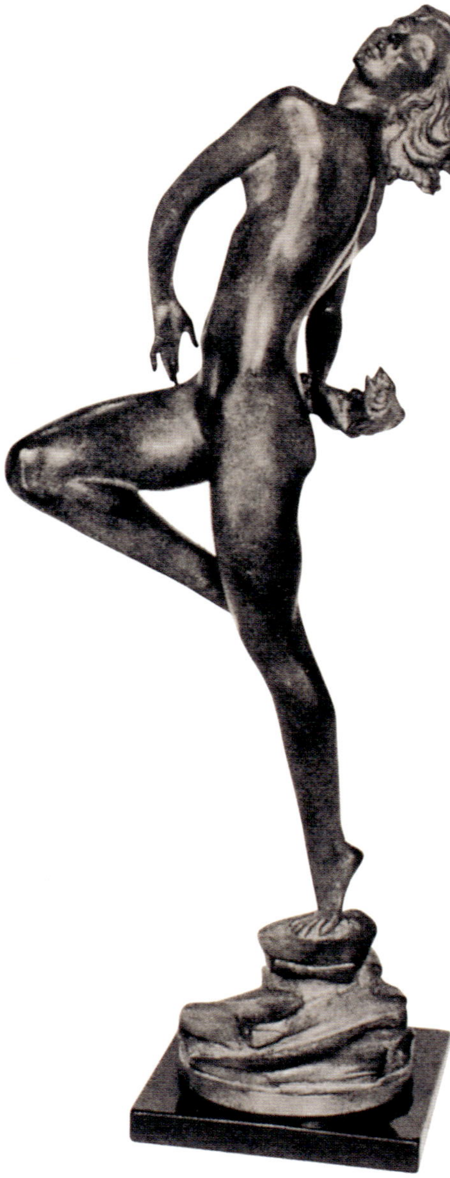

VI-101. "The Fisherman's Daughter" by Harriet Frishmuth. 1929.

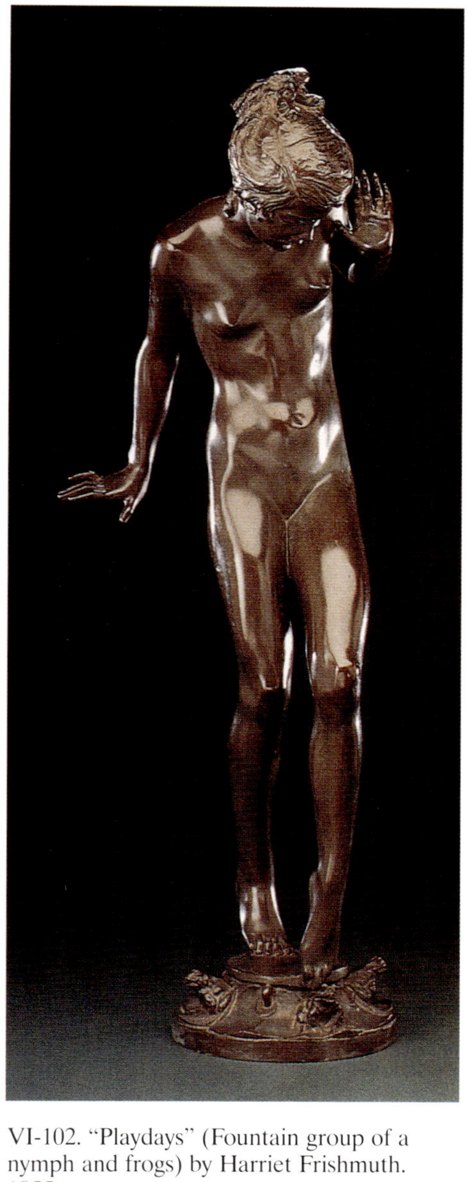

VI-102. "Playdays" (Fountain group of a nymph and frogs) by Harriet Frishmuth. 1955.

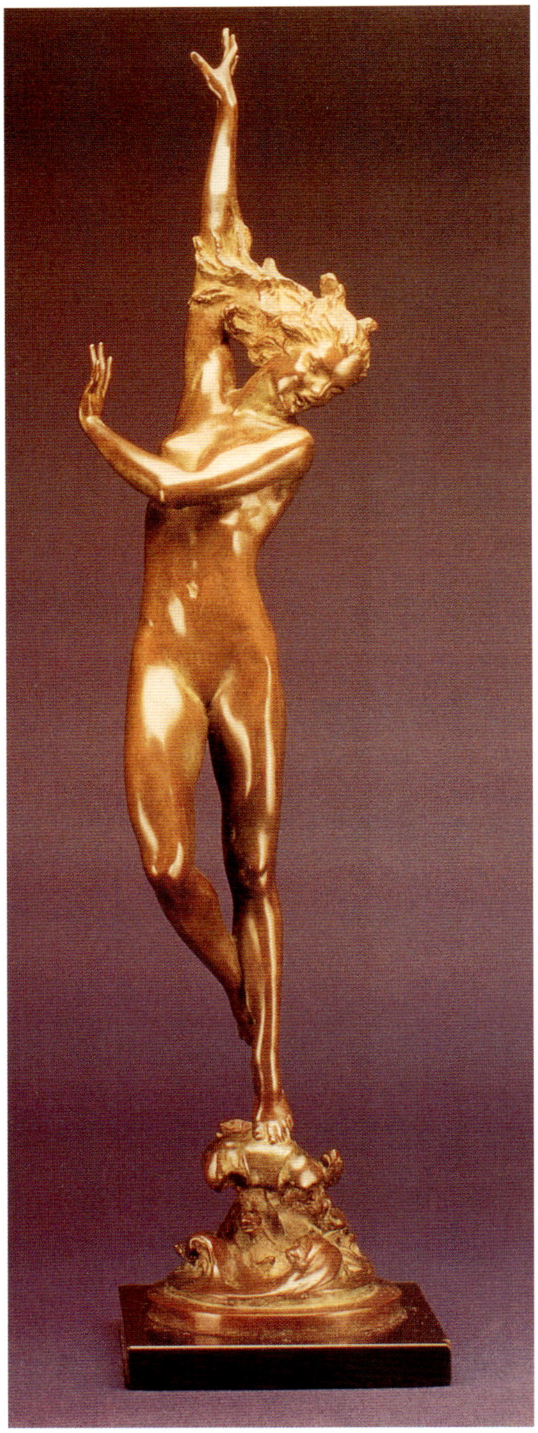

VI-103. "Crest of the Wave" by Harriet Frishmuth. 1925.

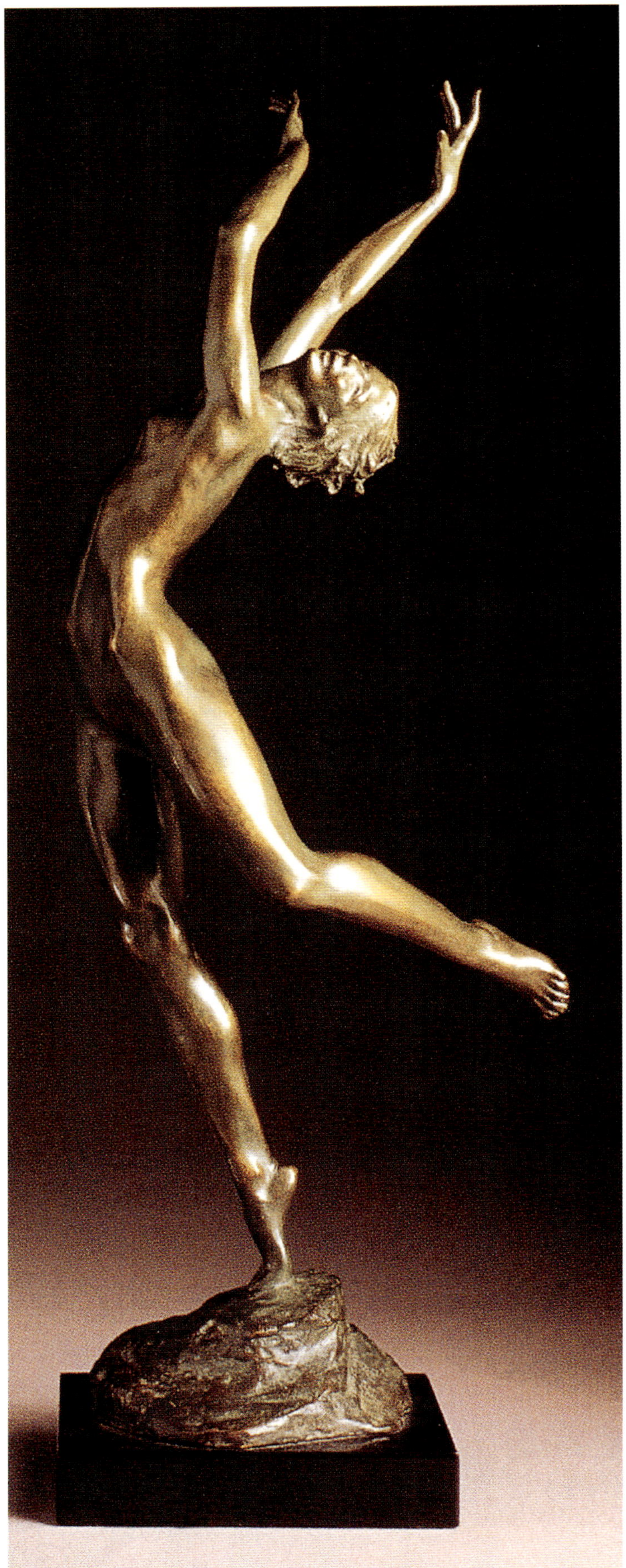

VI-104. "Bubble Dance" by Harriet Frishmuth. 1921.

JOSEF MARIO KORBEL (1882-1954)

Josef Mario Korbel specialized in allegorical statues. His work had a pure neo-classical look, emphasizing the ideal in a serene, soft-edged, and often symbolic manner. He produced a number of allegorical figures, some garden statuary, commemoratives, and a soldier's monument, but is best known for his female nudes.

Korbel was born in Bohemia, Czechloslovakia. He arrived in the United States at eighteen but returned to Europe for study in Germany and France. He settled in New York in 1913. His first major commission, in 1922, was for the garden of a large estate in Michigan, where he began his career sculpting allegorical figures representing Music, Dawn, Morning, and Evening. He went on to create a number of symbolic sculptures of female nudes with musical connotations, such as "Nocturne" (VI-105) and "Andante" (VI-106), and various other studies of the human body (numbers VI-107 to VI-111). He also made commemorative statues, such as his "Pioneer Woman" (VI-112). "The Kiss" (VI-113) was a venture into a more impassioned style, but most of his work retained a cool classical ambience.

Korbel, who was a member of the National Academy of Design and the National Sculpture Society, cast these and many other statues at Roman Bronze Works beginning in 1915 and continuing until his death. Some of his statues were cast there posthumously through 1978, such as his "Standing Woman" (VI-114) and "Standing Nude" (VI-115).

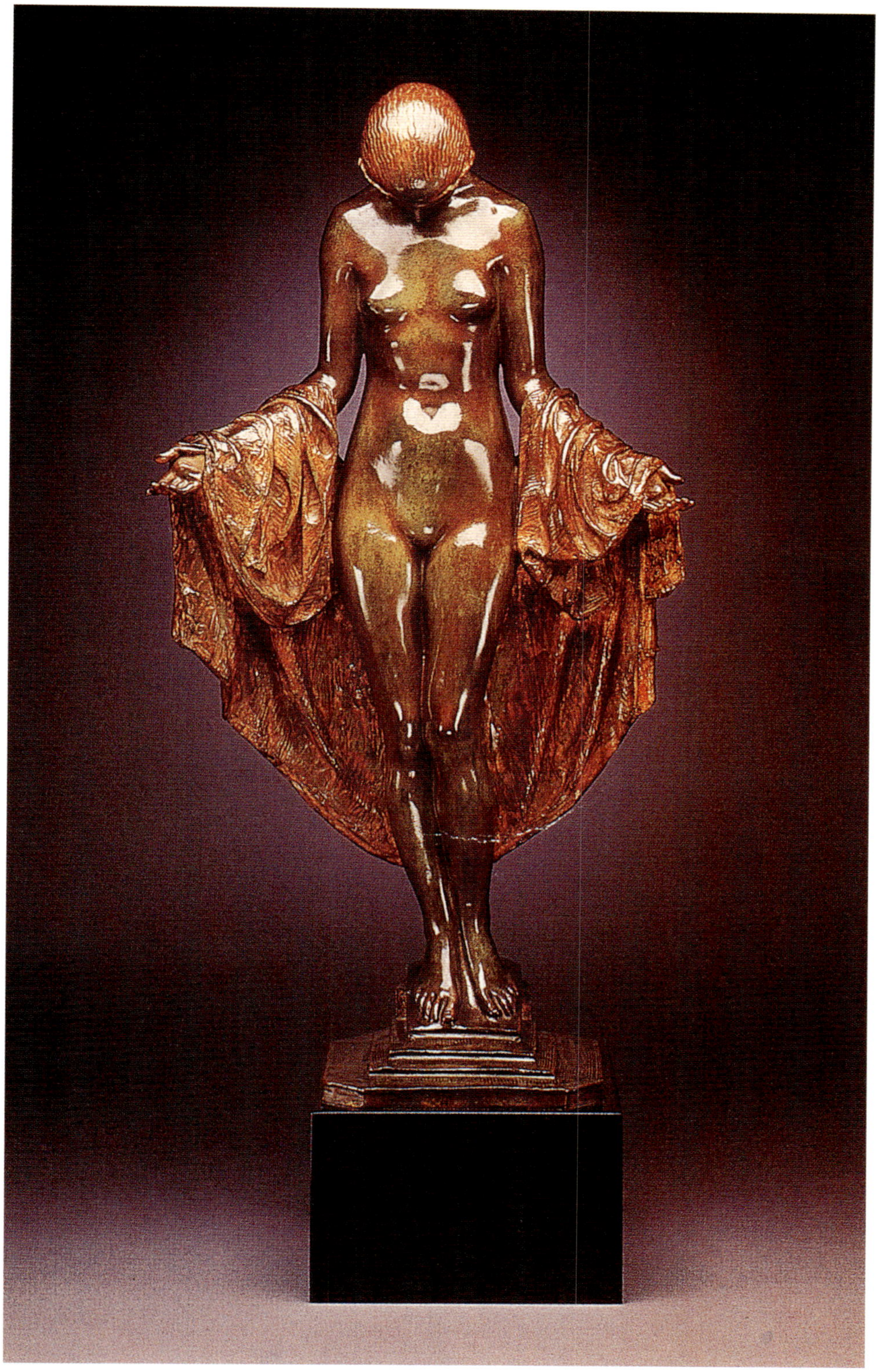

VI-105. "Nocturne" by Josef Mario Korbel. 1920.

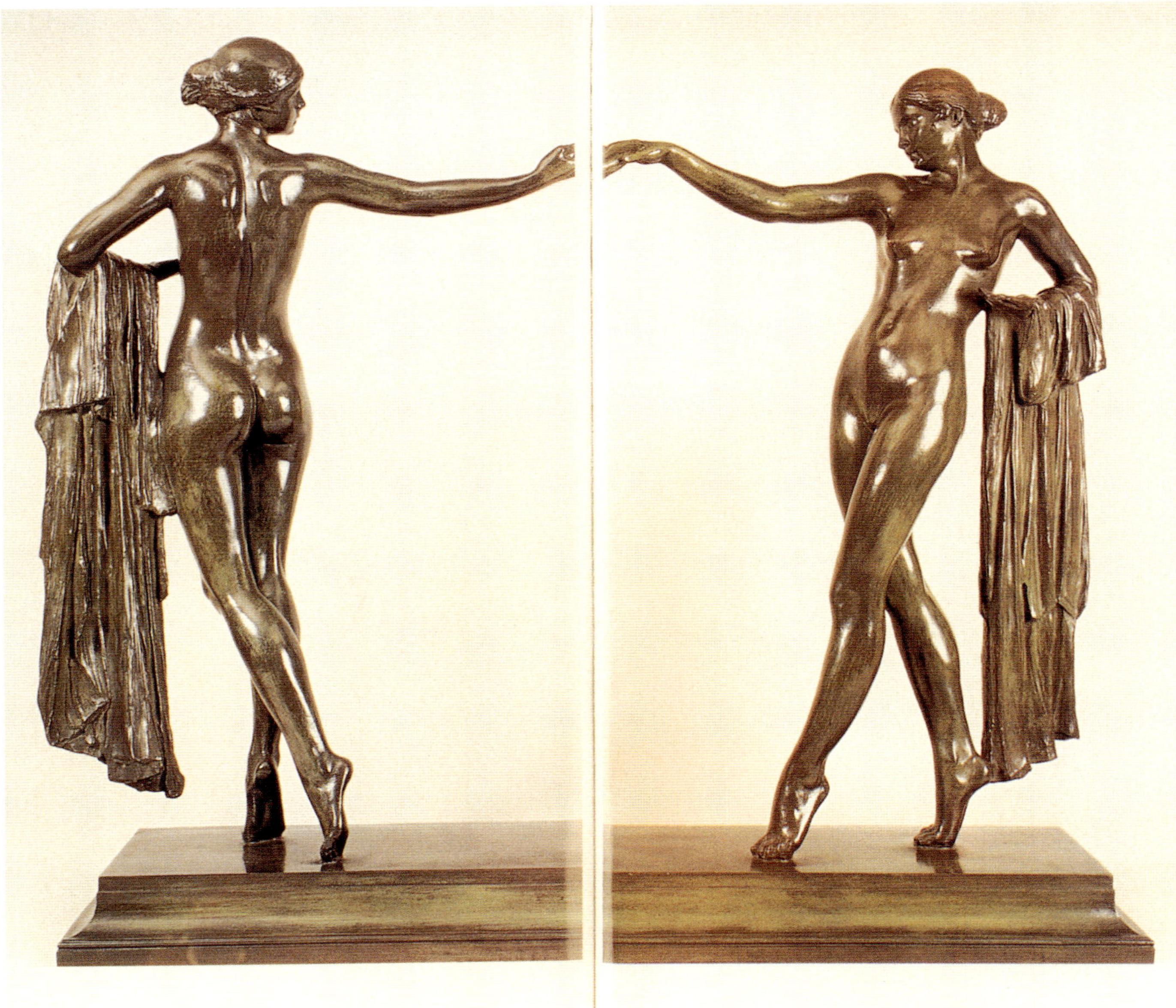

VI-106. "Andante" by Josef Mario Korbel. 1917.

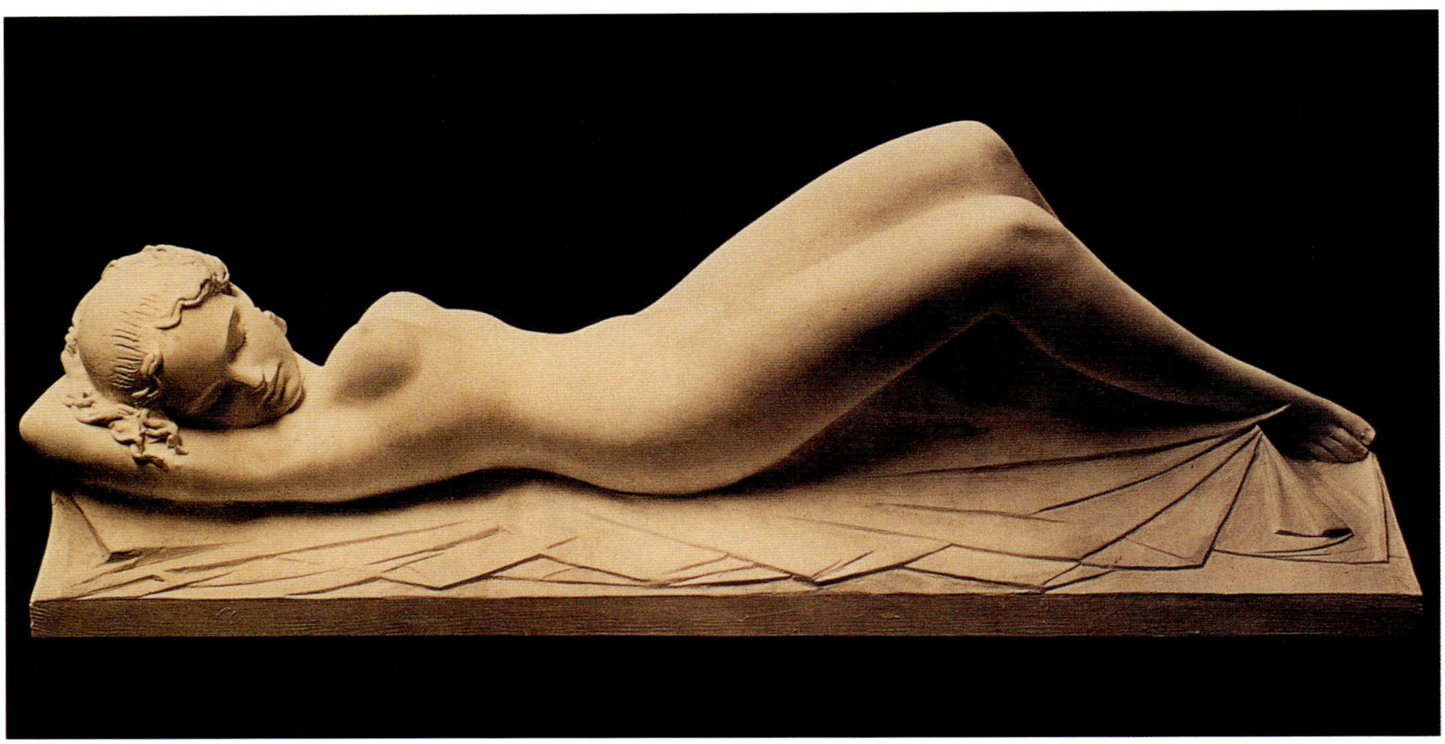

VI-107. "Siesta" by Josef Mario Korbel. 1929.

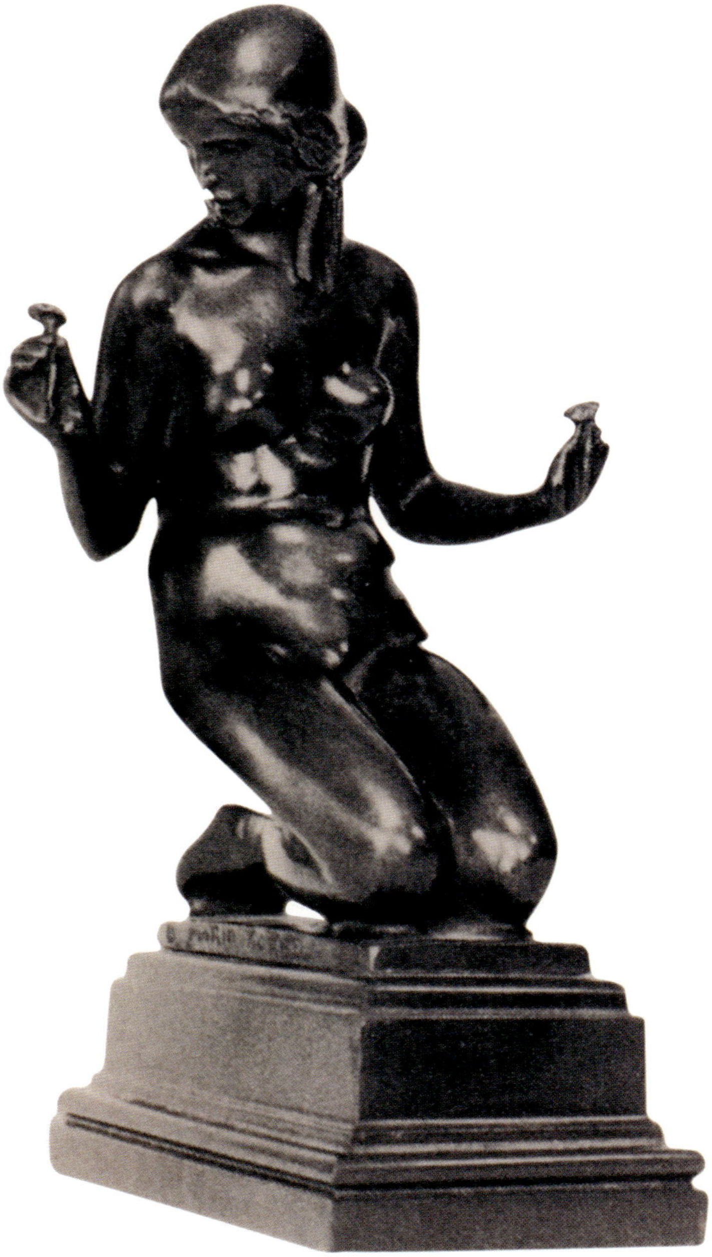

VI-108. "Flowers" by Josef Mario Korbel. c. 1915.

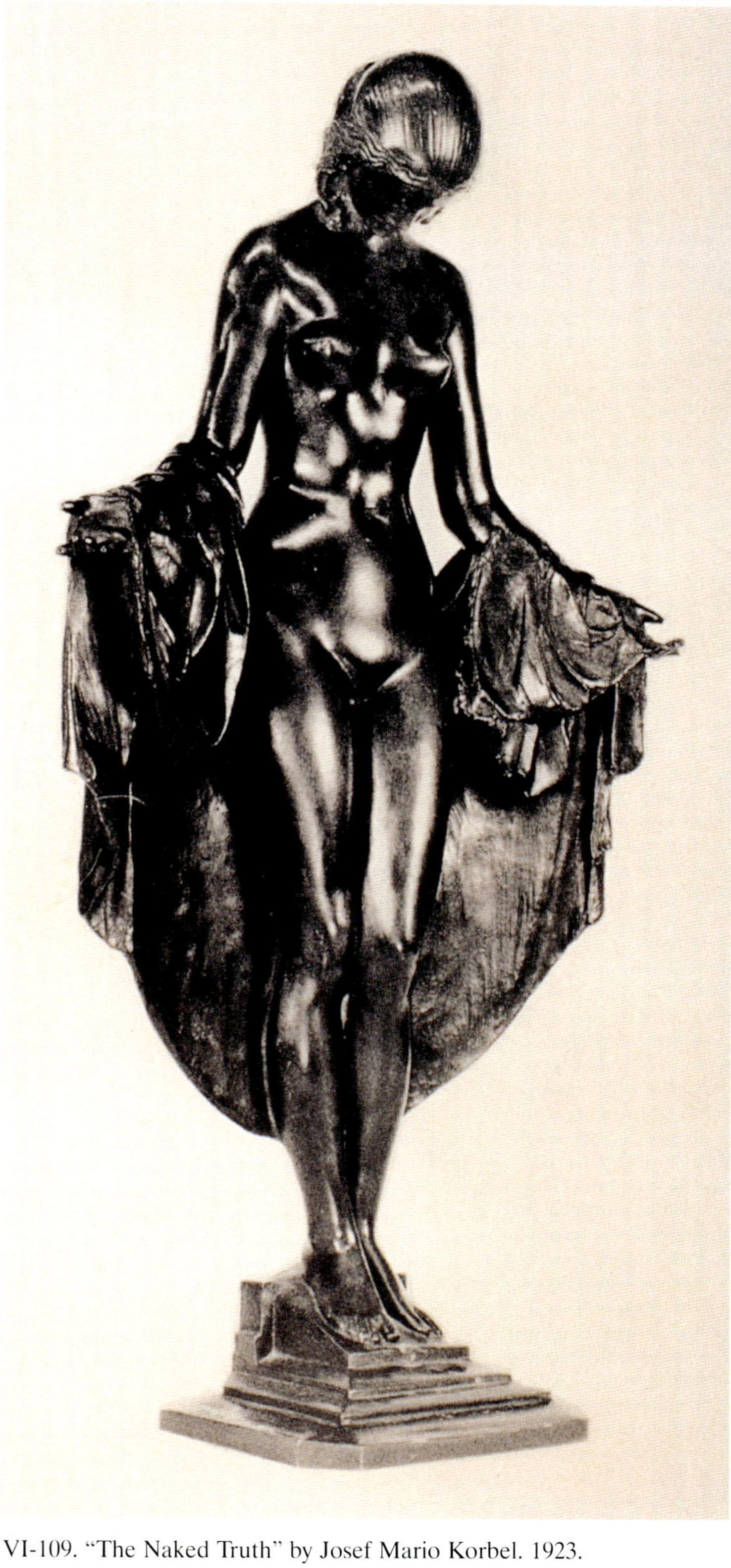

VI-109. "The Naked Truth" by Josef Mario Korbel. 1923.

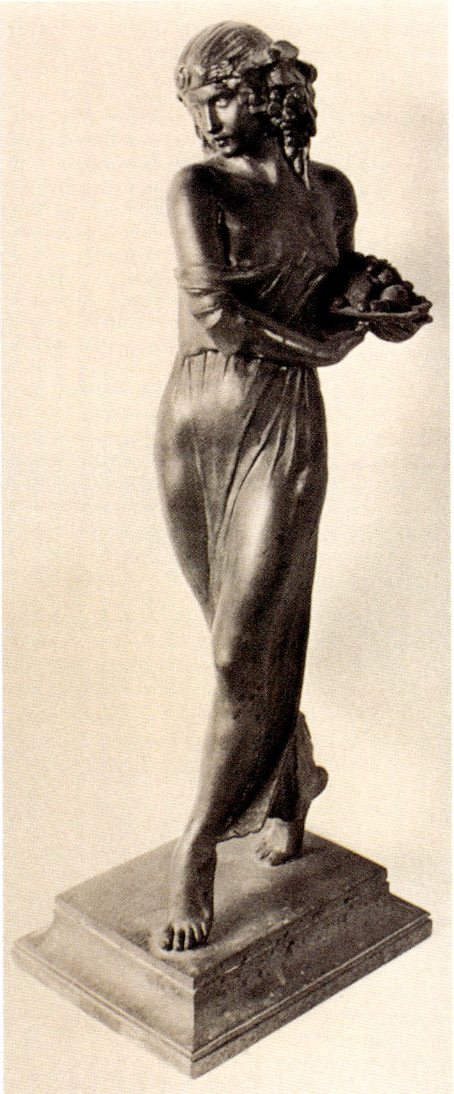

VI-110. "Woman with Grapes" by Josef Mario Korbel.

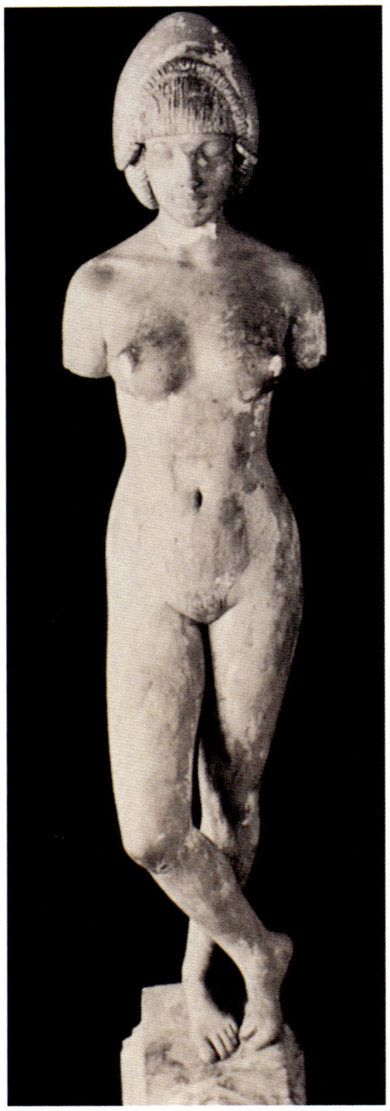

VI-111. "Art Deco Female" (plaster sketch for a large bronze) by Josef Mario Korbel.

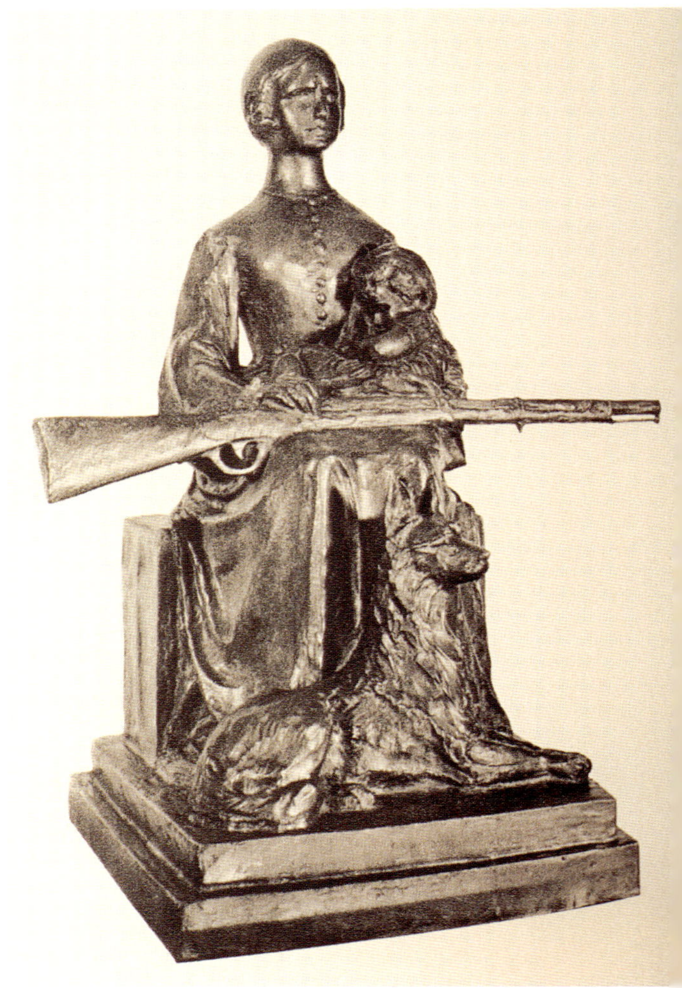

VI-112. "Pioneer Woman" by Josef Mario Korbel. 1927.

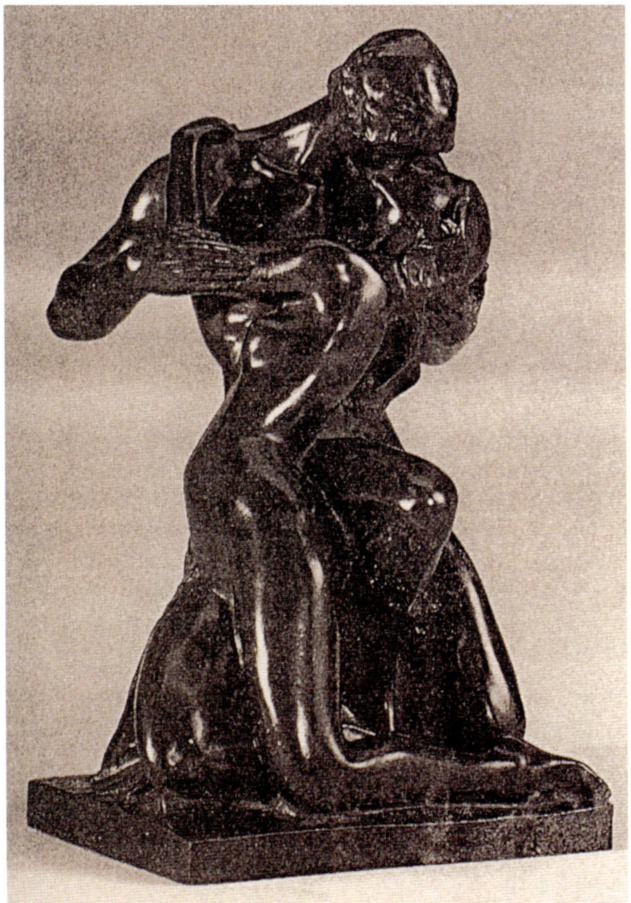

VI-113. "The Kiss" by Josef Mario Korbel. c. 1929.

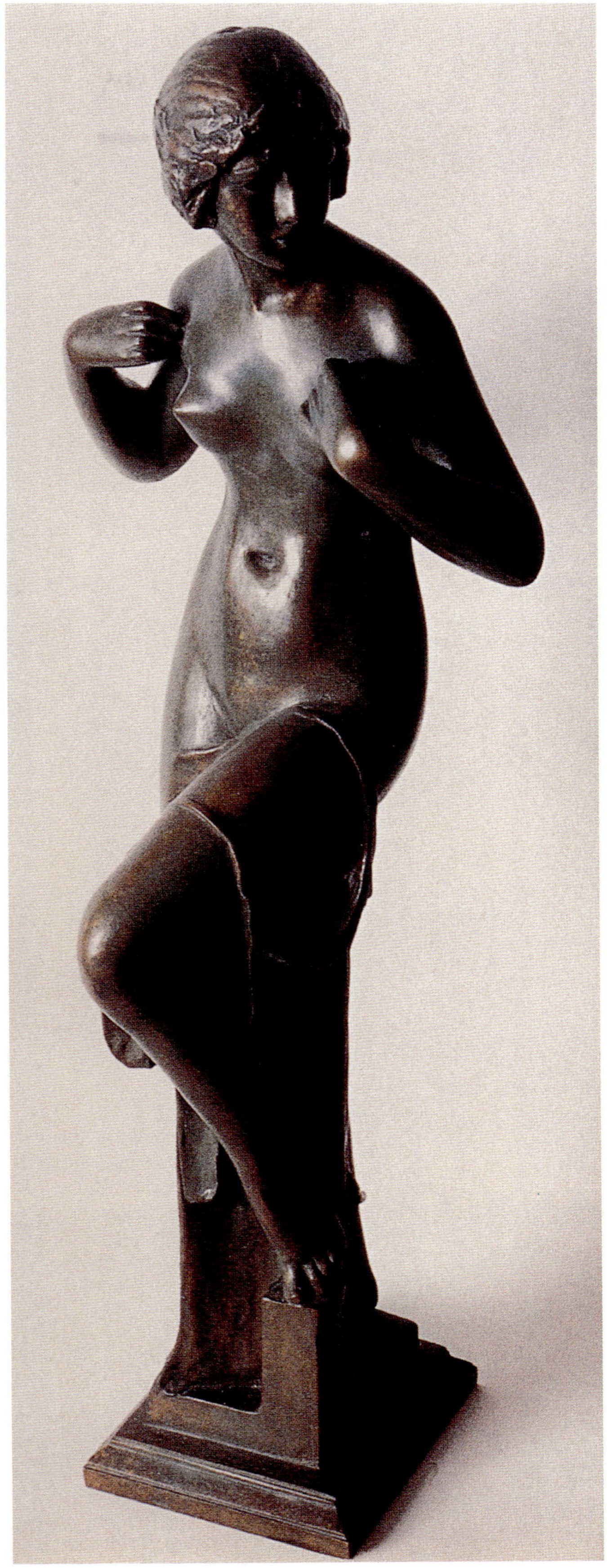

VI-114. "Standing Nude" by Josef Mario Korbel. Cast posthumously in the 1970s.

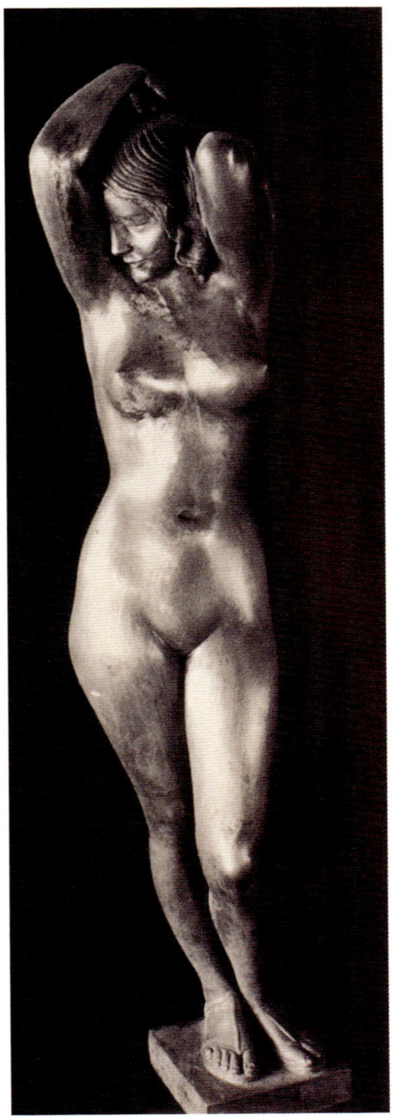

VI-115. "Standing Woman" by Josef Mario Korbel. Cast posthumously in 1978.

MALVINA HOFFMAN (1887-1966)

Malvina Hoffman has a unique place in the history of American sculpture: she became the expert sculptor who realistically pictured men and women of different races and nationalities from around the world. Her accurate portrayals of native figures from Asia, Africa, Oceania, North America, and Europe were cast in bronze and placed in the Field Museum in Chicago. Hoffman also made a series of dramatic dance sculptures; several were based on the ballets of Anna Pavlova and Nijinsky. Her style was spirited and bold, and she was able to project personality and expression within the naturalism of her work.

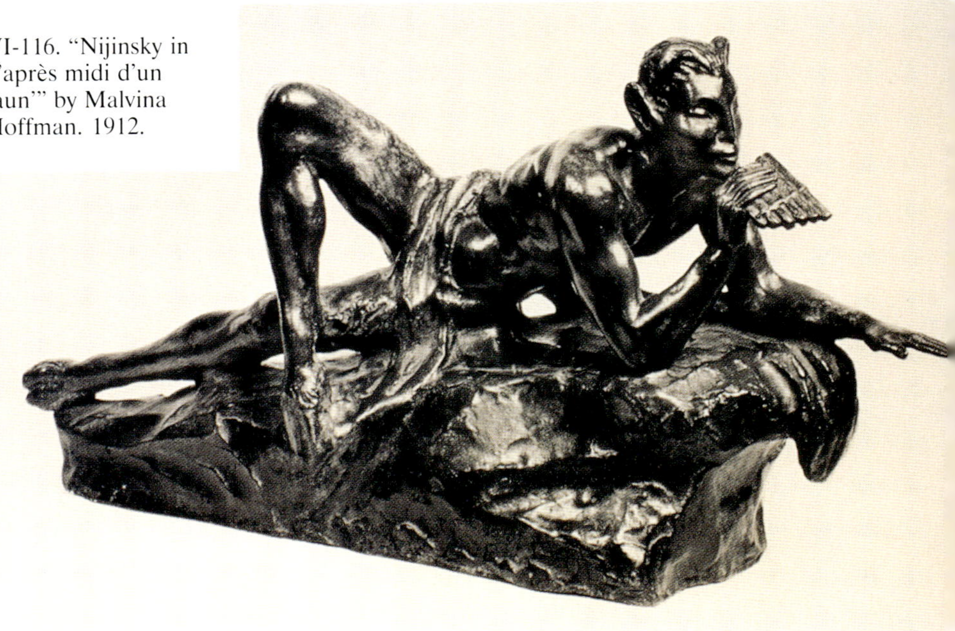

VI-116. "Nijinsky in 'l'après midi d'un faun'" by Malvina Hoffman. 1912.

Hoffman was born in New York, and studied there with Gutzon Borglum, Herbert Adams, and in Paris with Rodin. He advised her to study dissection at a medical school, and she early learned anatomy and the accuracy that was to define her style of quasi-scientific sculpture. She also learned the technical side of casting at a foundry.

In Paris before World War I, she became acquainted with the great dancers of the age and first began using them as subjects. Her sculptures "Nijinsky in l'après midi d'un faun" (VI-116), "Mask of Pavlova" (VI-117), "Pavlova Gavotte" (VI-118), "Russian Dancers" (VI-119), and "Bacchanale Russe" (VI-120) were based on her associations with the dancers. A brief foray into classical subjects produced works like "Boy and Panther Cub" (VI-121).

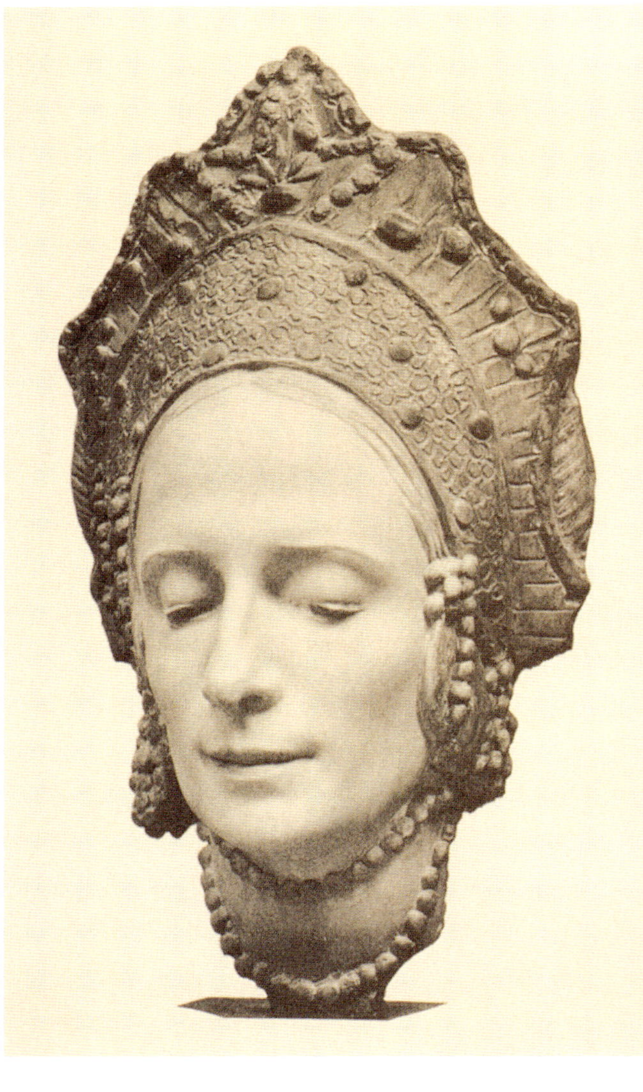

VI-117. "Mask of Pavolova" by Malvina Hoffman. 1924.

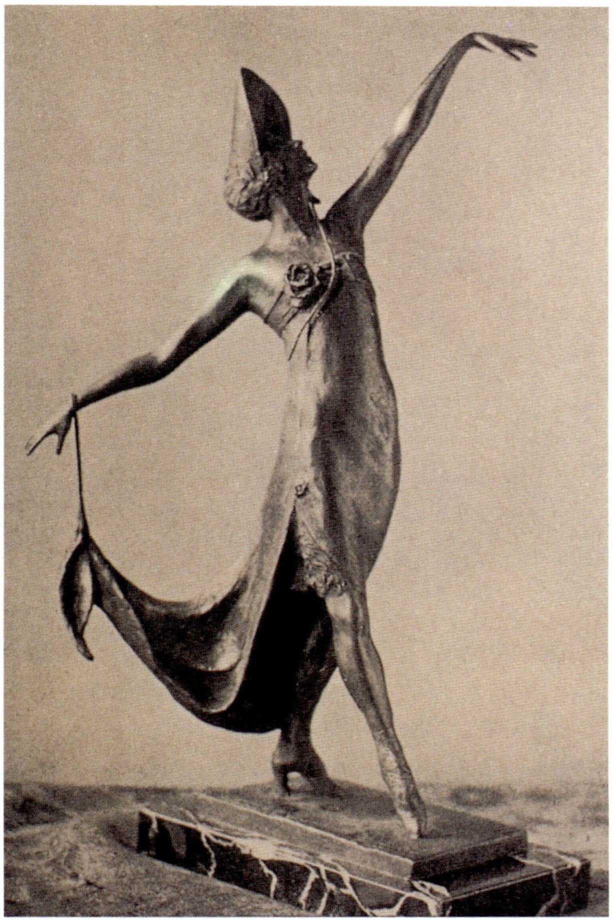

VI-118. "La Gavotte" by Malvina Hoffman. Modeled 1915; cast 1918.

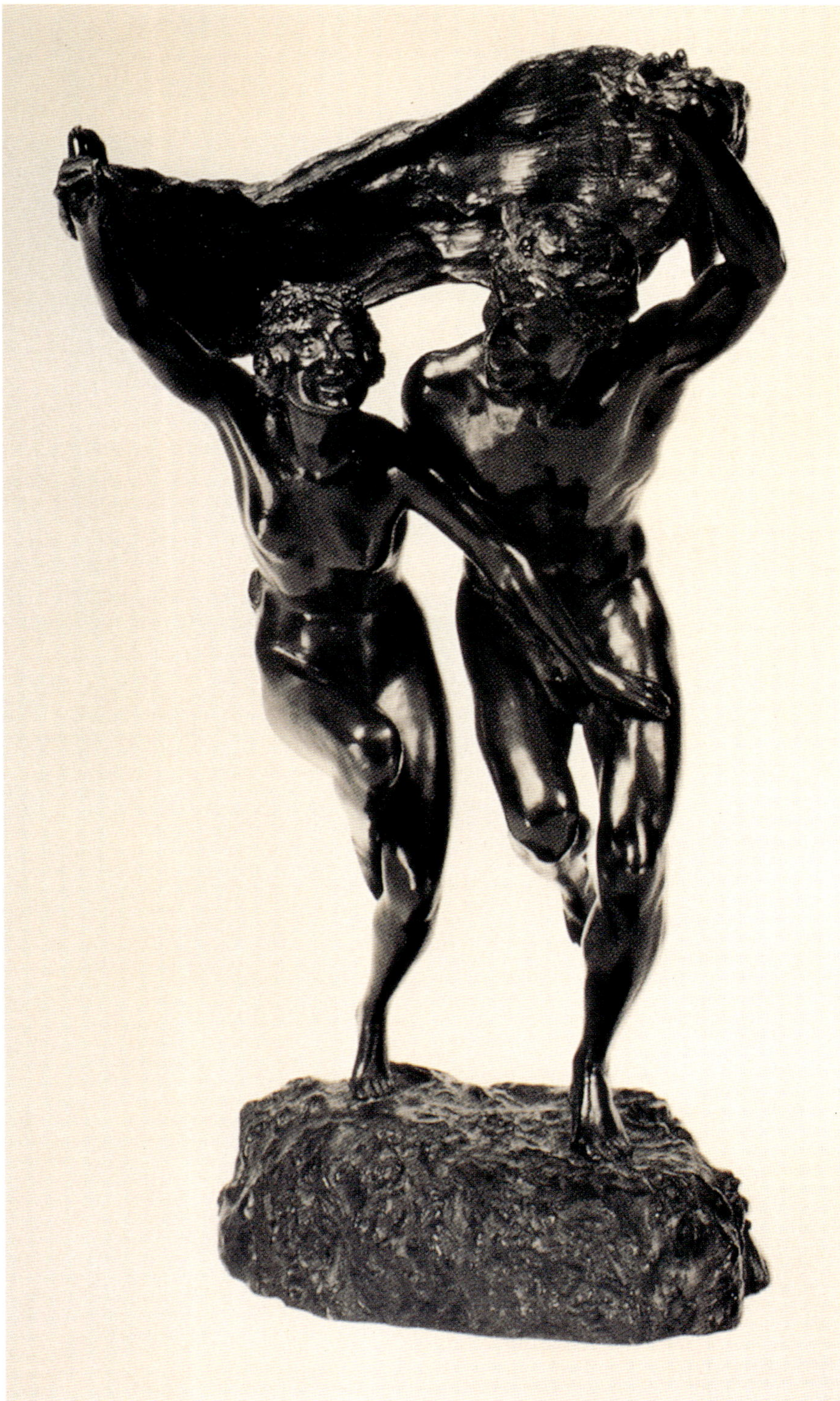

Right:
VI-119. "Bacchanale Russe" by Malvina Hoffman. 1912. This sculpture portrays Pavlova and Michael Mordkin.

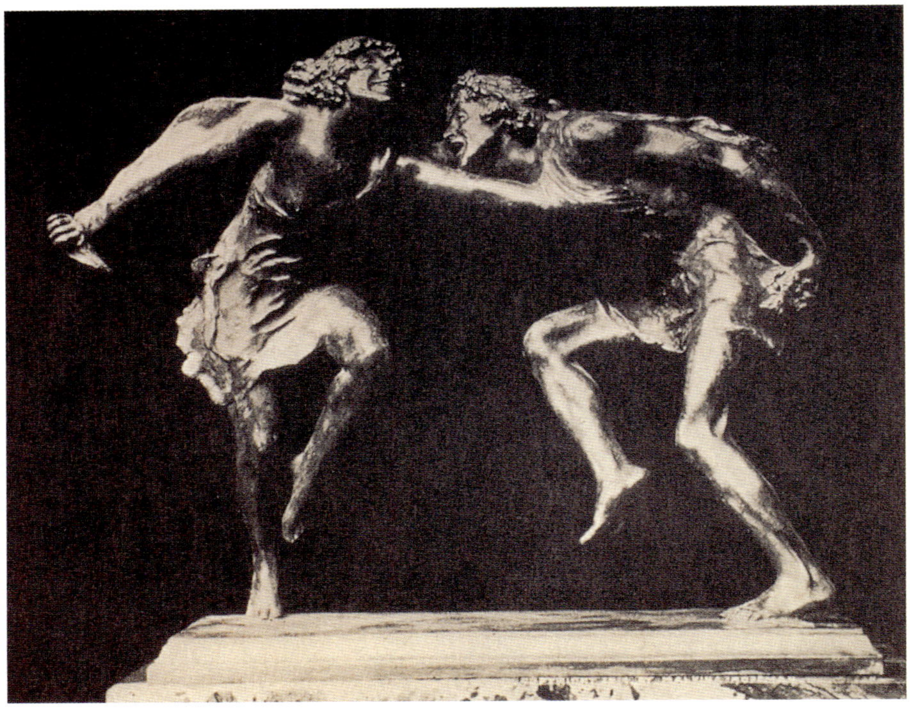

VI-120. "Russian Dancers" by Malvina Hoffman. c. 1917. This was one of a series of works she did representing Pavlova and Mordkin in different ballets.

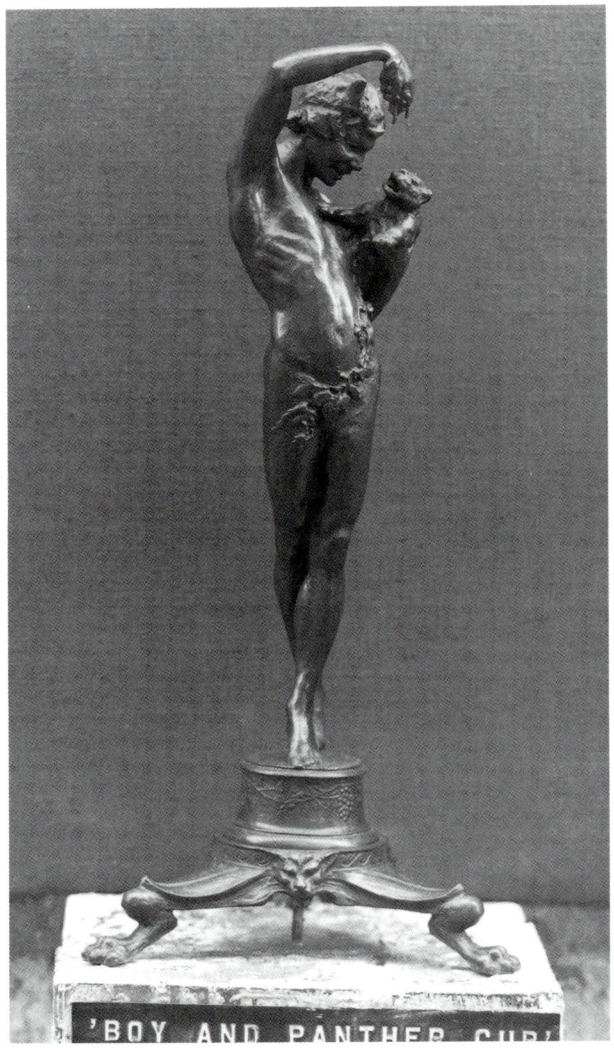

VI-121. "Boy and Panther Cub" by Malvina Hoffman. 1915.

Hoffman returned to New York at the outbreak of war, and began work for the Red Cross, which sent her to Yugoslavia in 1919. There she met and sculpted "Paderewski The Artist" (VI-122), one of her best known works. She later returned there and made several portraits of her friend and teacher, the sculptor Ivan Mestrovic (VI-123).

Hoffman's ability to create highly naturalistic images –such as the "Hand" (VI-124)—can be seen in her studies of peoples of the world, for which she became best known. In 1926 during a visit to Africa she began making portrait heads of native peoples; these led to her commission to sculpt the Races of Man for permanent exhibition at the Field Museum in Chicago. From 1930 she traveled round the world sketching and sculpting and producing more than one hundred figures of people in their native environments (numbers VI-125 to VI-132), ranging from a bullfighter in Spain to a Balinese dancer. She also wrote several books about her experiences.

Hoffman first cast "a group of dancers" at Roman Bronze Works in 1912. Her association with the foundry continued through the 1950s, with dozens of sculptures cast there.

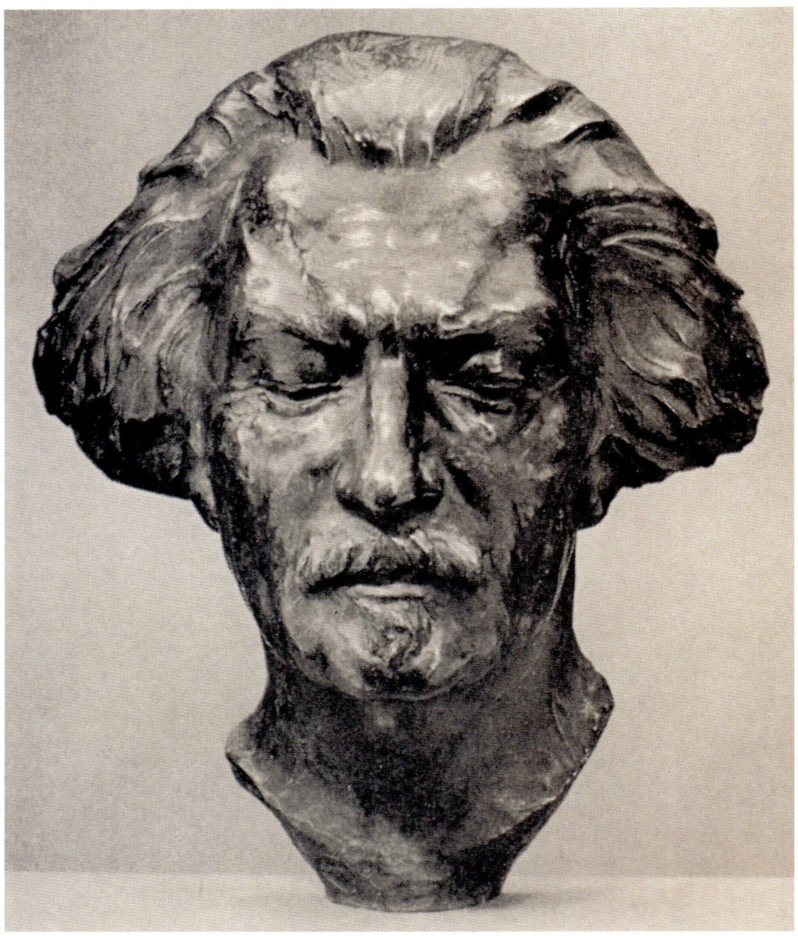

VI-122. "Paderewski the Artist" by Malvina Hoffman. 1923.

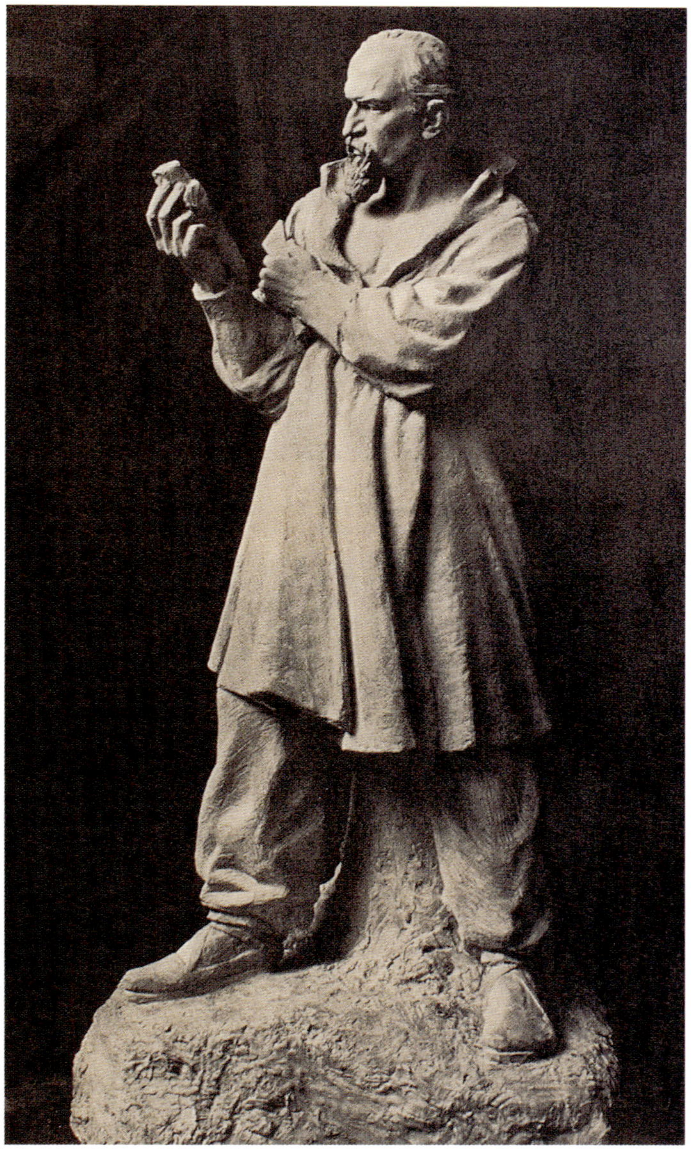

VI-123. "Ivan Mestrovic" by Malvina Hoffman. One of several studies Hoffman did of the Yugoslav sculptor. c. 1925.

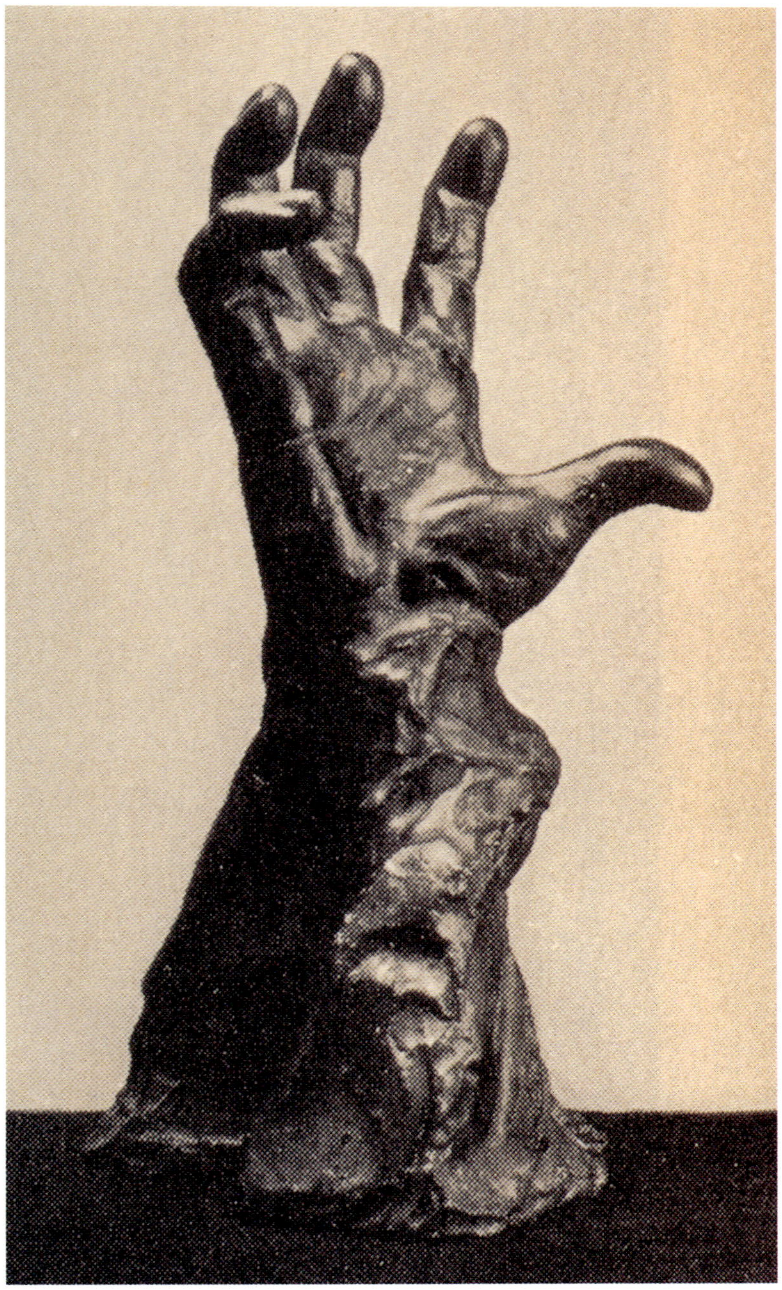

VI-124. "Right Hand" by Malvina Hoffman. This is an over-life sized bronze of the right hand of Jean Julien Lemordant. 1934.

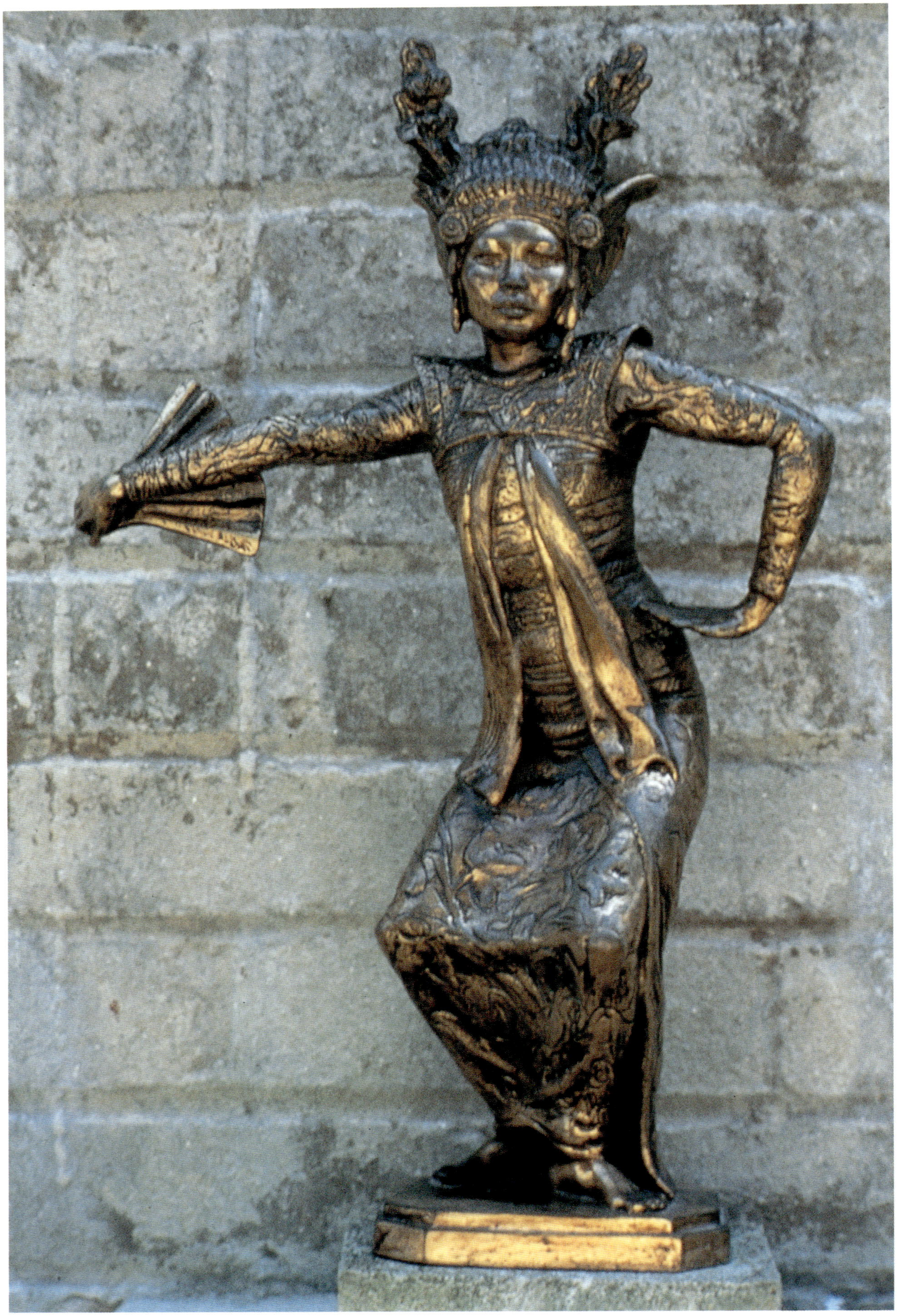

VI-125. "Bali Dancer" by Malvina Hoffman. c. 1936.

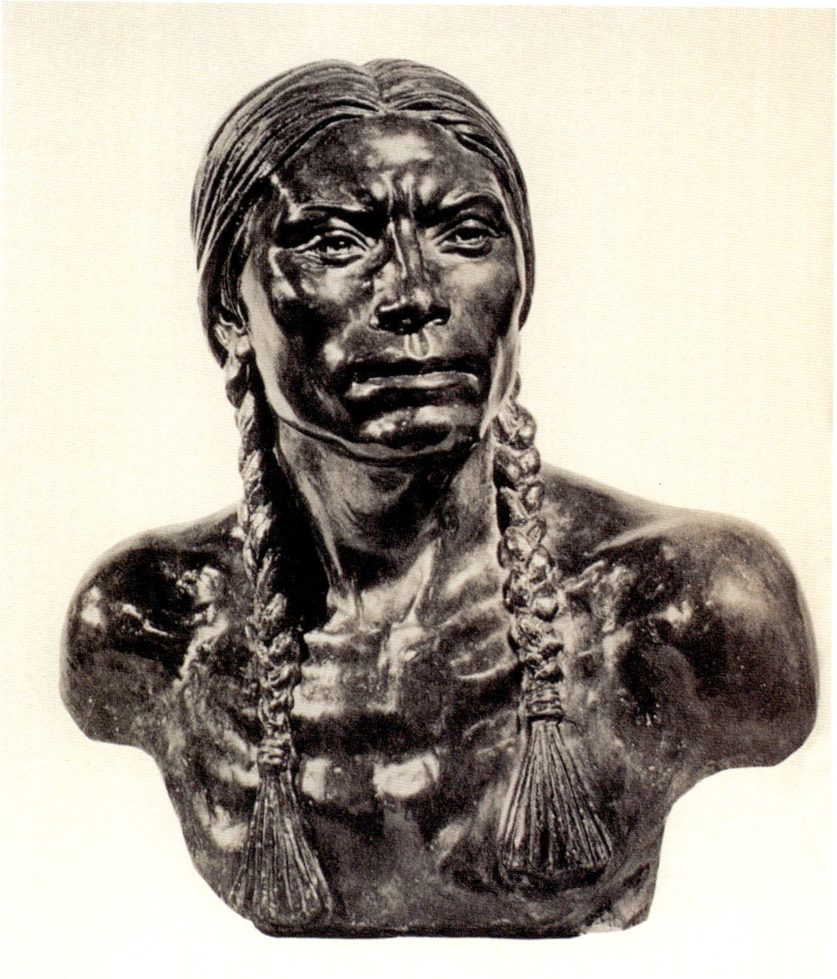

VI-127. "Apache: Jicarilla" by Malvina Hoffman. 1934.

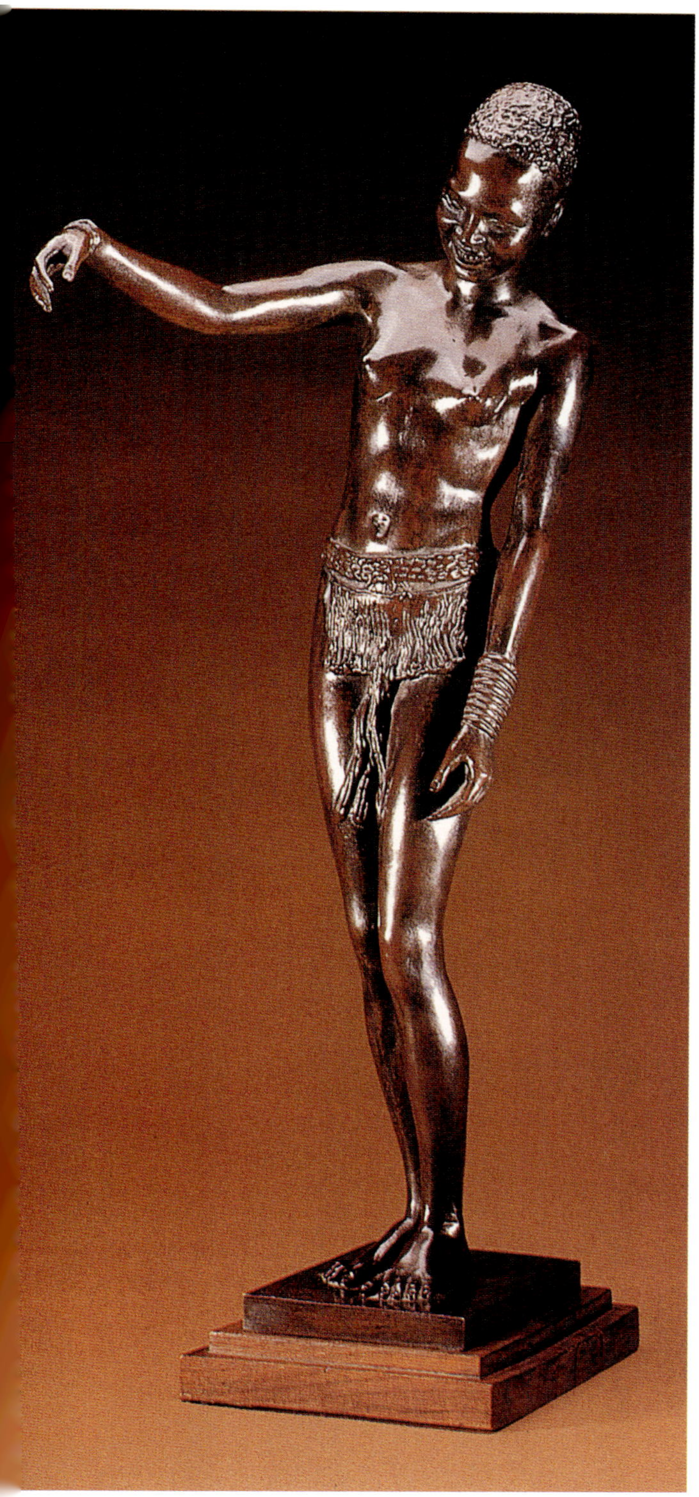

VI-126. "Daboa" by Malvina Hoffman. 1931.

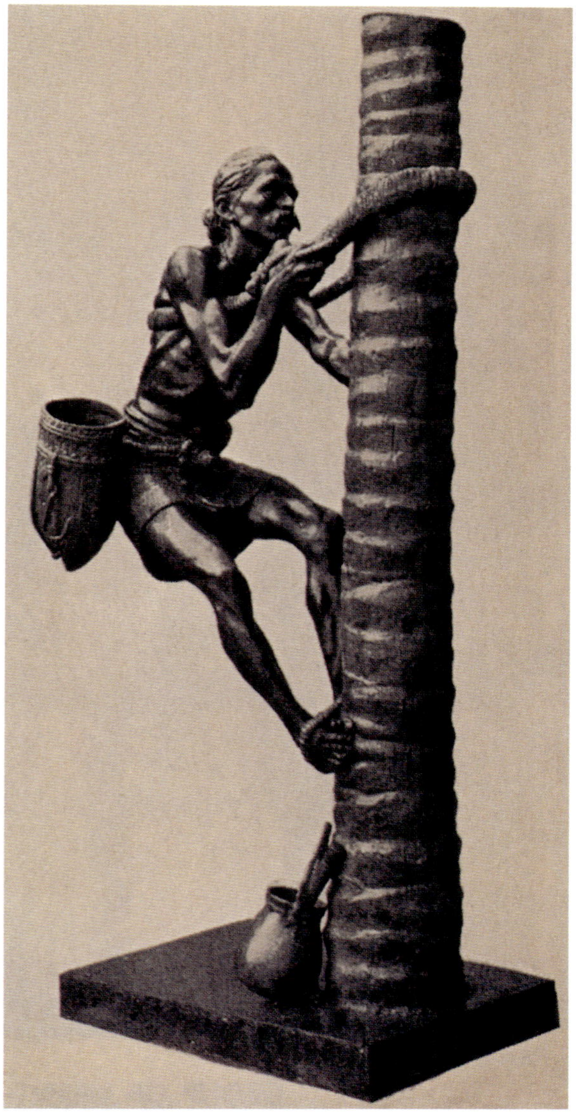

VI-128. "Tamil Climbing a Palm Tree to Collect Toddy for Wine" by Malvina Hoffman.

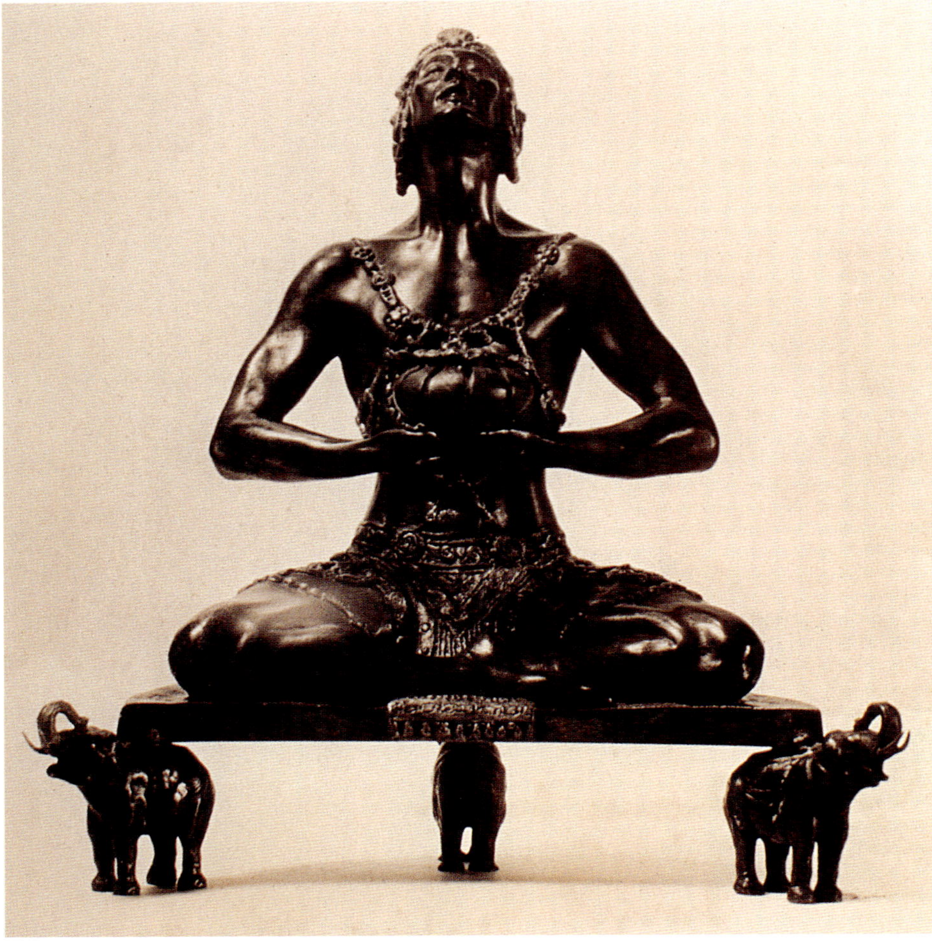

VI-129. "Hindu Incense Burner" by Malvina Hoffman. c. 1920.

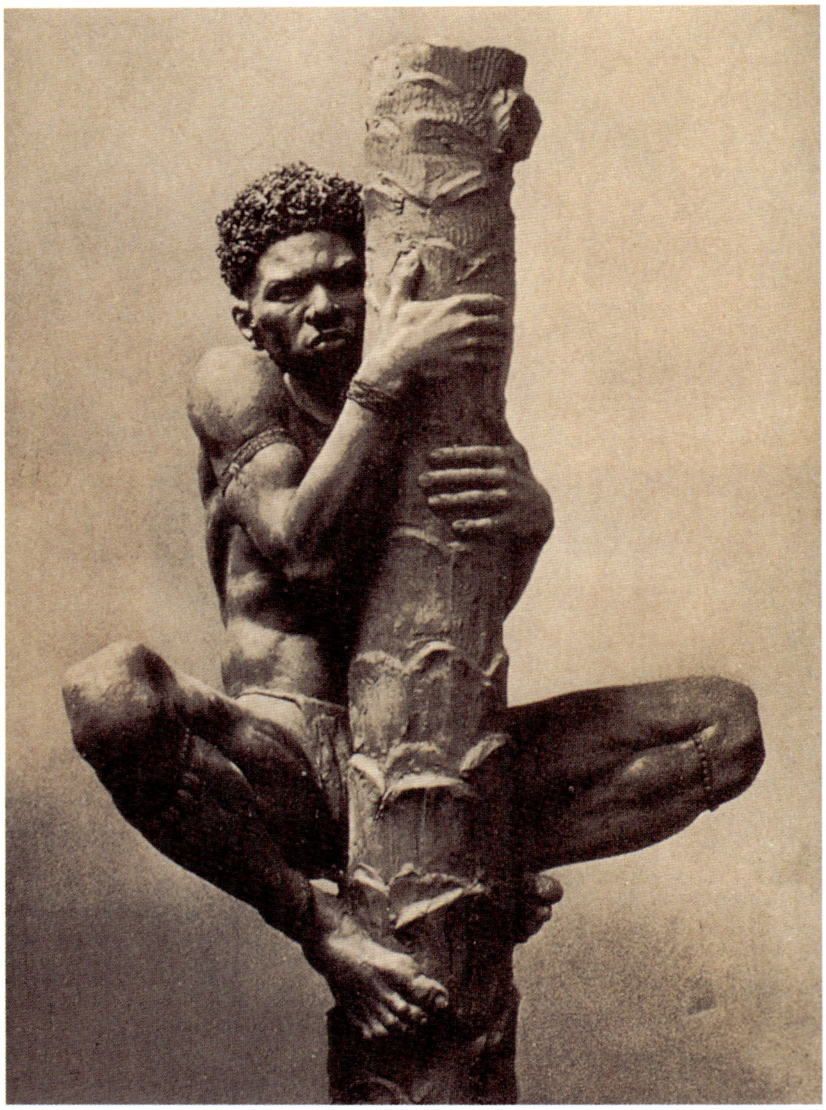

VI-130. "Solomon Islander Climbing a Palm Tree, A Ring in his Nose, a Plug in his Ear" by Malvina Hoffman. c. 1930.

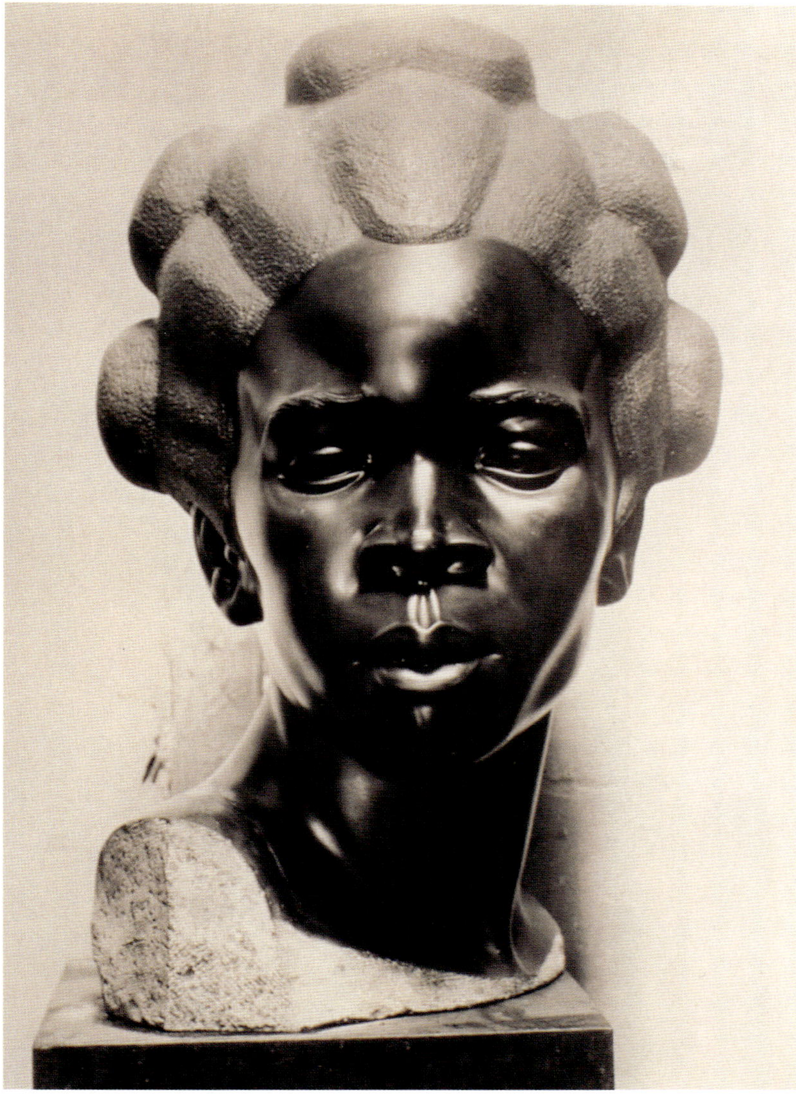

VI-131. "Martinique Dancer" by Malvina Hoffman. c. 1929.

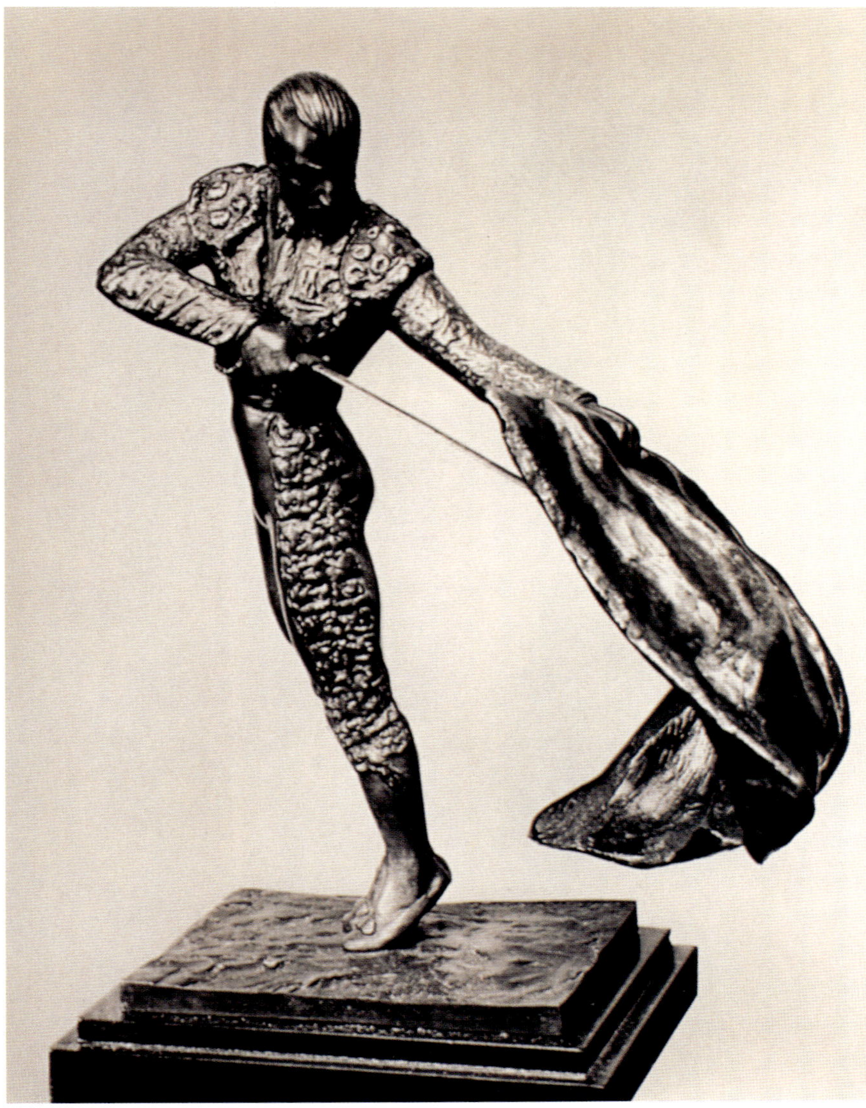

VI-132. "The Bullfighter" by Malvina Hoffman. 1955.

ANTHONY DE FRANCISI (1887-1964)

Though particularly well known as a medalist, Anthony de Francisi also had a career as a maker of memorial and garden statues. His work was usually characterized by smoothly stylized detail and decorative design (occasionally with an Eastern flavor). De Francisi's work ranged widely from his 1926 Murphy Flagpole in Union Square in New York (VI-133), to portraits of Joan of Arc (VI-134), and "Abraham Lincoln" (VI-135), to ideal nudes like "Cup of Basil" (VI-136) to garden pieces.

But de Francisi also reached into the vocabulary of modernism in a few of his works, with his rough surfaced, less-defined "Lovers on the Rocks" (VI-137) and his formalistic "Crouching Nude" (VI-138). His "Birds in Flight" (VI-139), which was cast in gold-leafed aluminum, with mirror and onyx, was a striking departure from his more realistic work.

As a medalist, de Francisi received numerous commissions, including the 1921 "peace" dollar for the United States Mint, and the World War II Discharge Button given to sixteen million men and women of the Armed Forces. He was the recipient of numerous awards for his medals.

The sculptor was born in Italy, emigrated to the United States in 1903, studied with a number of leading New York sculptors, and became an assistant to Philip Martiny, Hermon MacNeil, Charles Niehaus, and Adolph A.Weinman. He was an instructor at Columbia University for many years. At Roman Bronze Works he began casting in 1917 with "La Bayadere" (VI-140), and continued through the 1950s.

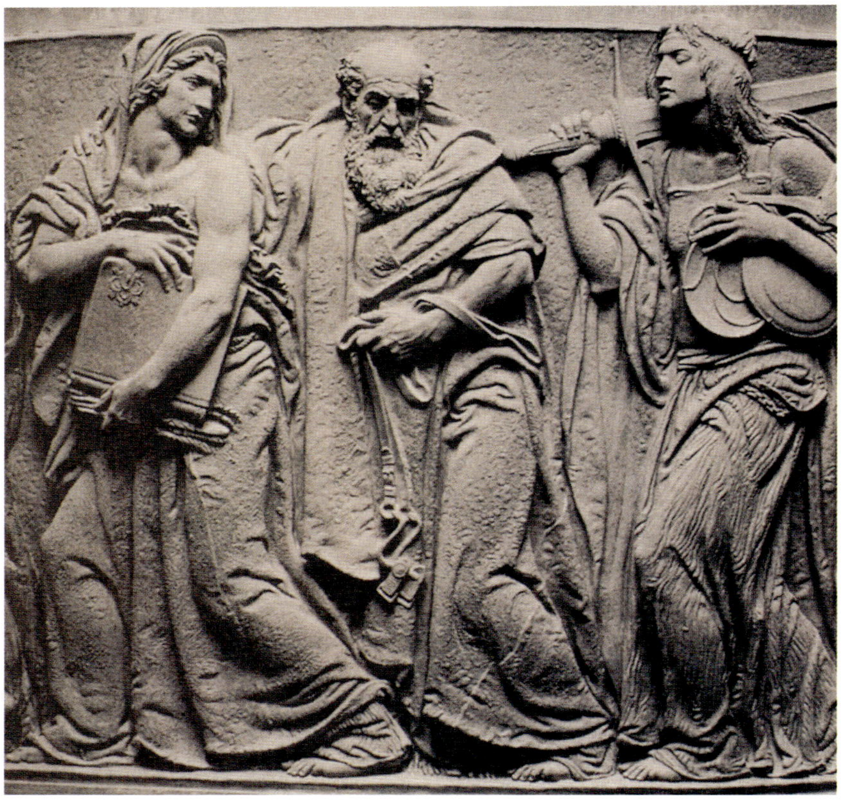

VI-133. "Charles F. Murphy Memorial Flagpole," New York City, by Anthony De Francisi. 1926.

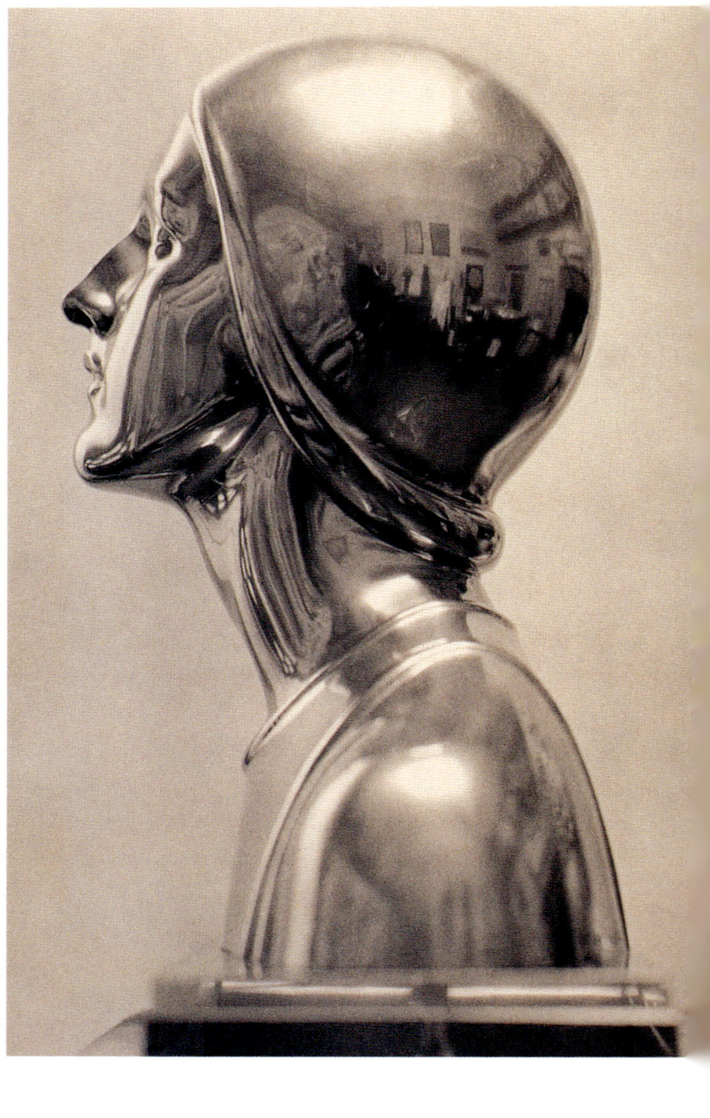

VI-134. "Joan of Arc" by Anthony De Francisi. 1937.

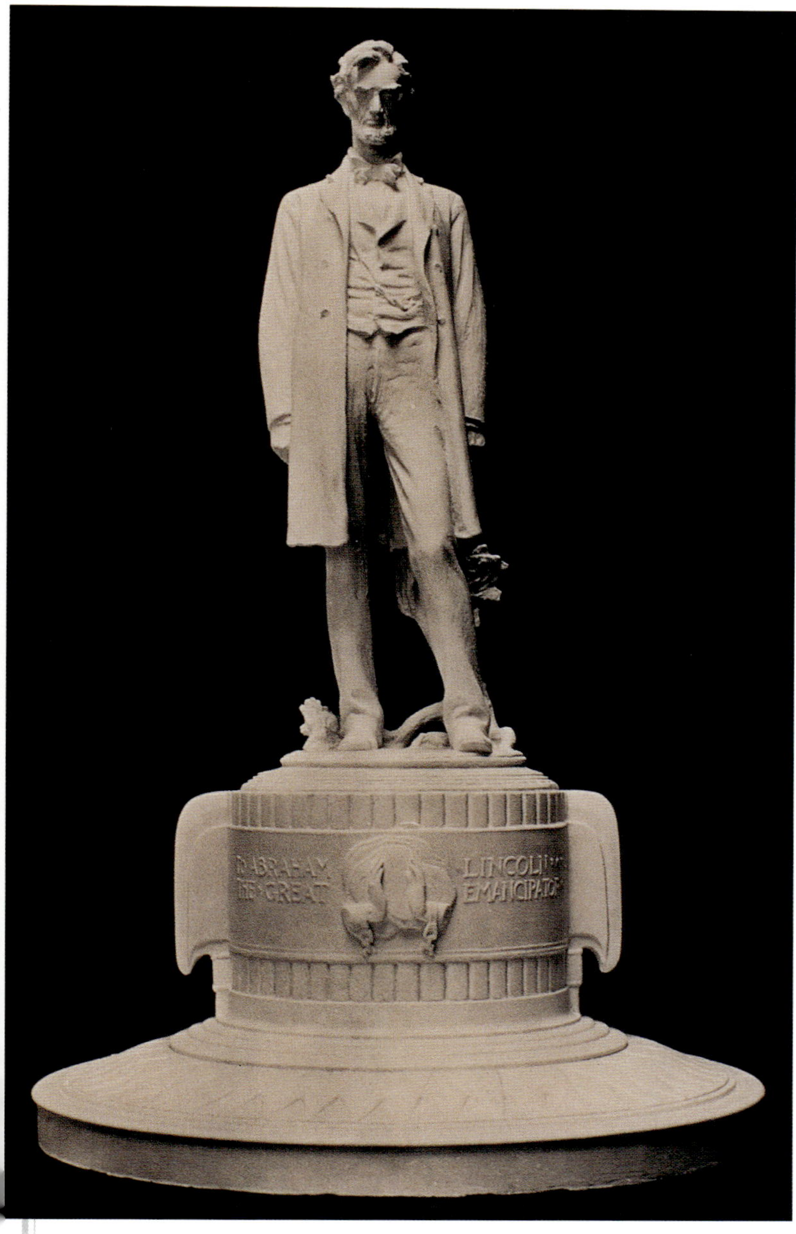

VI-135. "Abraham Lincoln" by Anthony De Francisi. 1956.

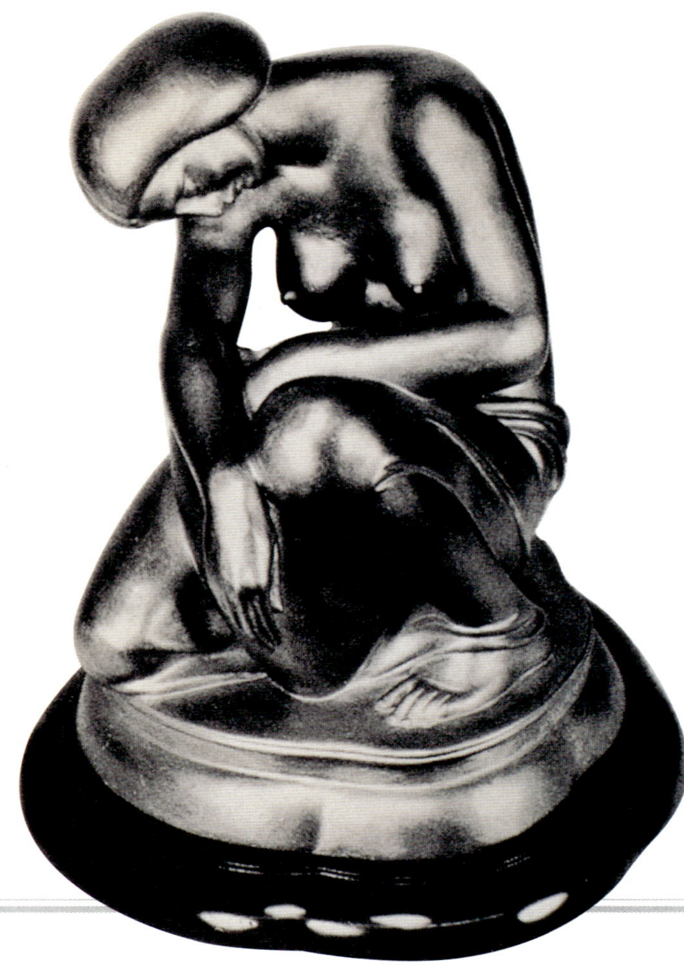

VI-136. "Cup of Basil" by Anthony De Francisi. 1940.

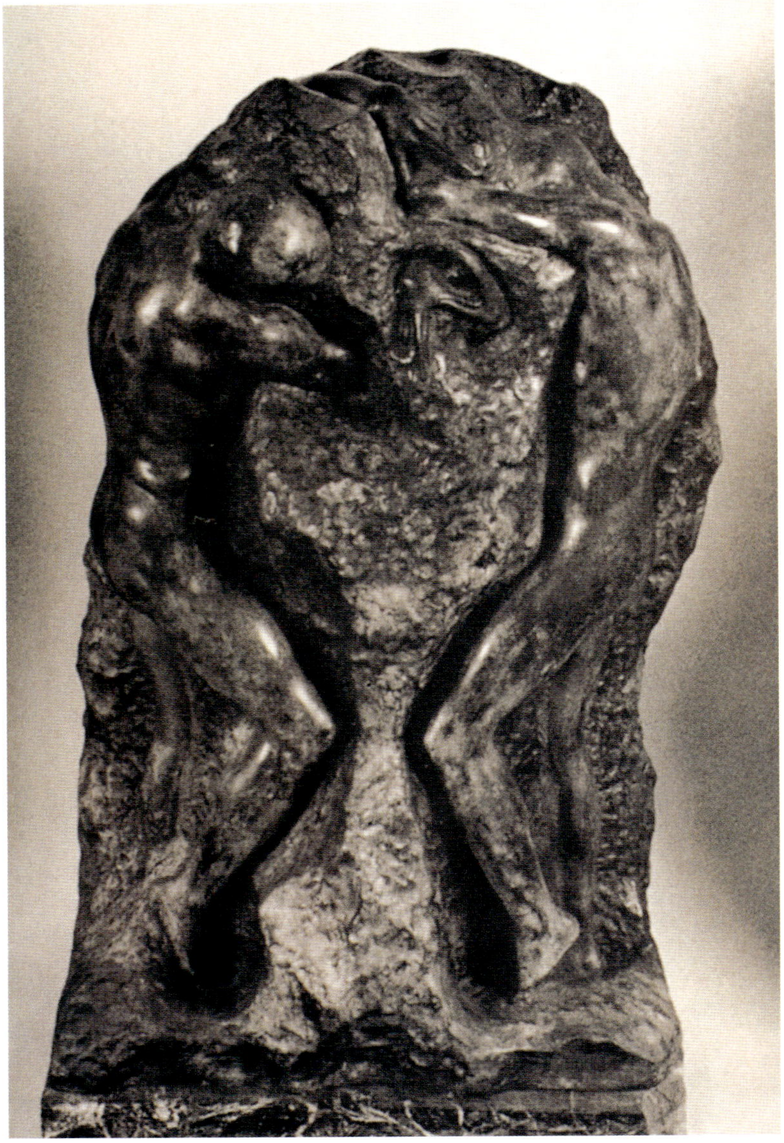

VI-137. "Lovers on the Rocks" by Anthony De Francisi.

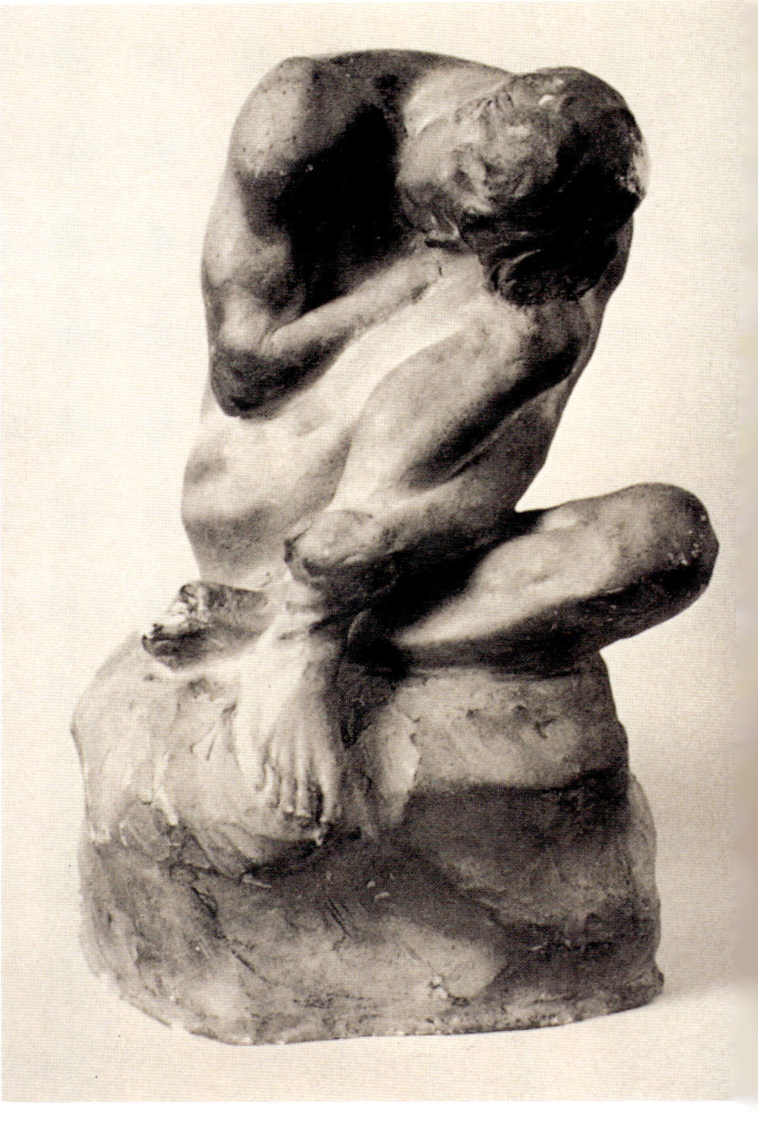

VI-138. "Crouching Nude" (plaster cast) by Anthony De Francisi.

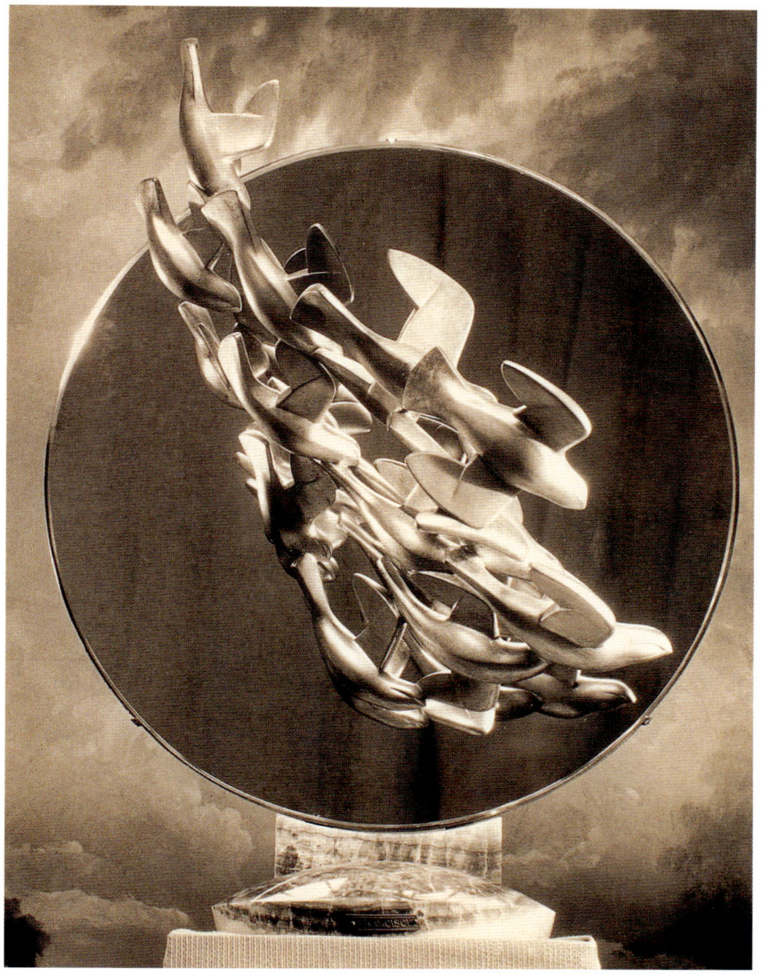

VI-139. "Birds in Flight" (mirror, onyx base, and gold-leafed aluminium) by Anthony De Francisi. 1936.

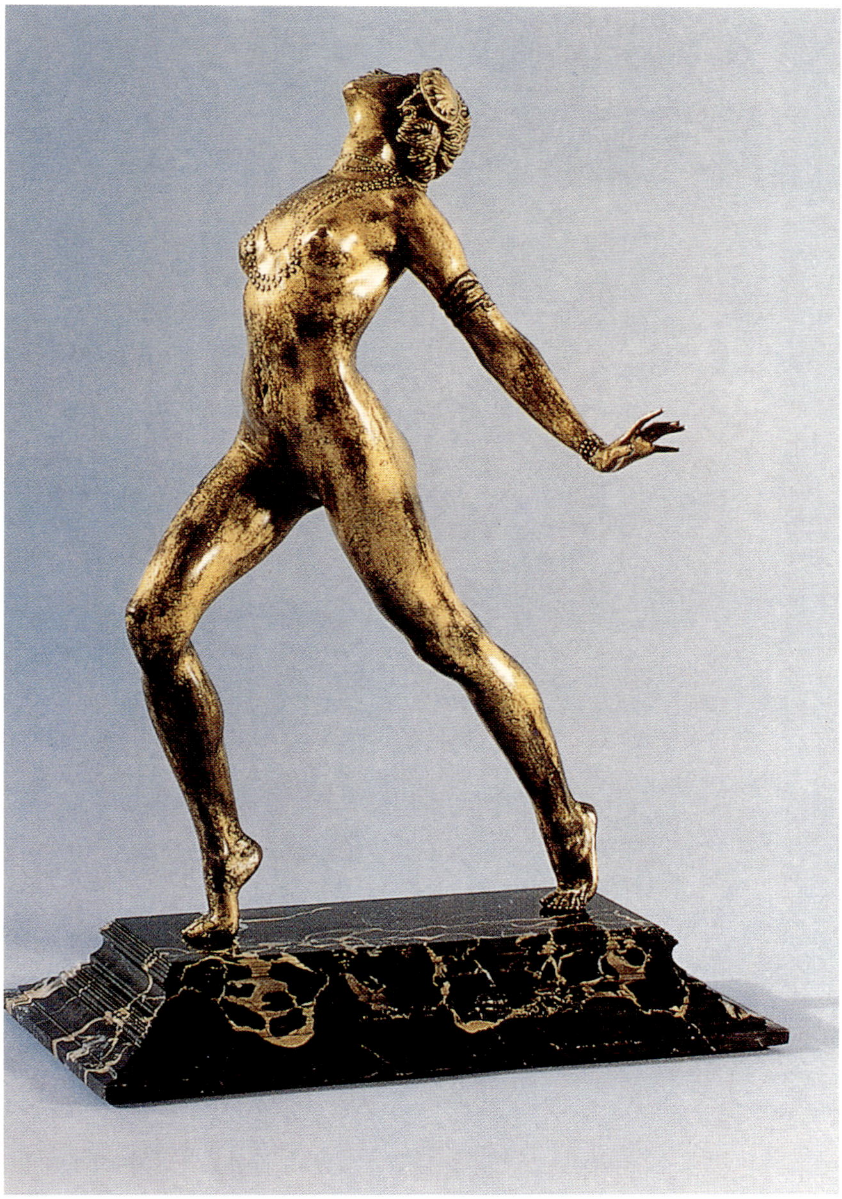

VI-140. "La Bayadere" by Anthony De Francisi. 1917.

GLEB DERUJINSKY (1888-1975)

Gleb Derujinsky combined a career as a portraitist and a maker of Neo-Classical statuary of nudes, animals, classical subjects and religious figures. He worked in a strongly linear and rhythmic style, with smooth surfaces. "Modeling," he wrote, "is like melody that must pass from one perfect note to another." He used a variety of media, working in alabaster, bronze, wood, and terra cotta.

Born in Russia, Derujinsky studied in Paris and Leningrad. He arrived in the United States in 1919, and had his first New York show two years later. He became a specialist in portrait busts and heroic commemorative statuary; among his illustrious subjects were Presidents Theodore Roosevelt (VI-141) and Franklin Roosevelt, and composers Serge Rachmaninoff, and Sergei Prokofiev. Around 1920 he portrayed a number of dancers and subjects from the Ballet Russes, such as his "Ruth Page as Columbine" (VI-142).

Derujinsky also made a group of whimsical and lyrical statues with titles like "Ecstasy" and "Odalisque." Among his Neo-Classical subjects were his "Three Graces" and his rhythmically balanced "Diana" (VI-143 and 144), as well as ideal nudes like "Innocence" (VI-145). His works can also be seen at a number of religious institutions. Other listings include a "Dragon and a Female," a "Cupid and Dolphins," "Spiders," and a "Prometheus."

During his career he won many awards, including a Gold Medal at the Paris World's Fair in 1937. Derujinsky cast a large number of pieces at Roman Bronze Works (beginning with a statue of a dog in 1929) and ending in the 1940s.

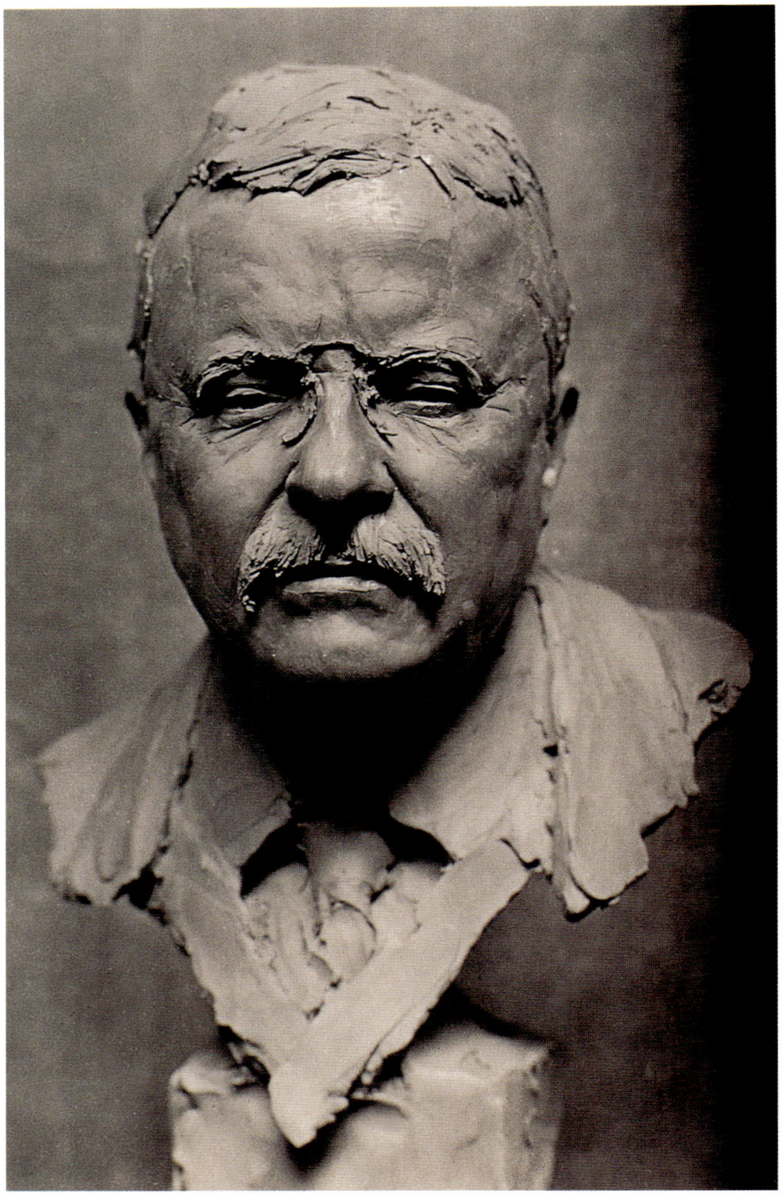

VI-141. "Theodore Roosevelt" by Gleb Derujinsky.

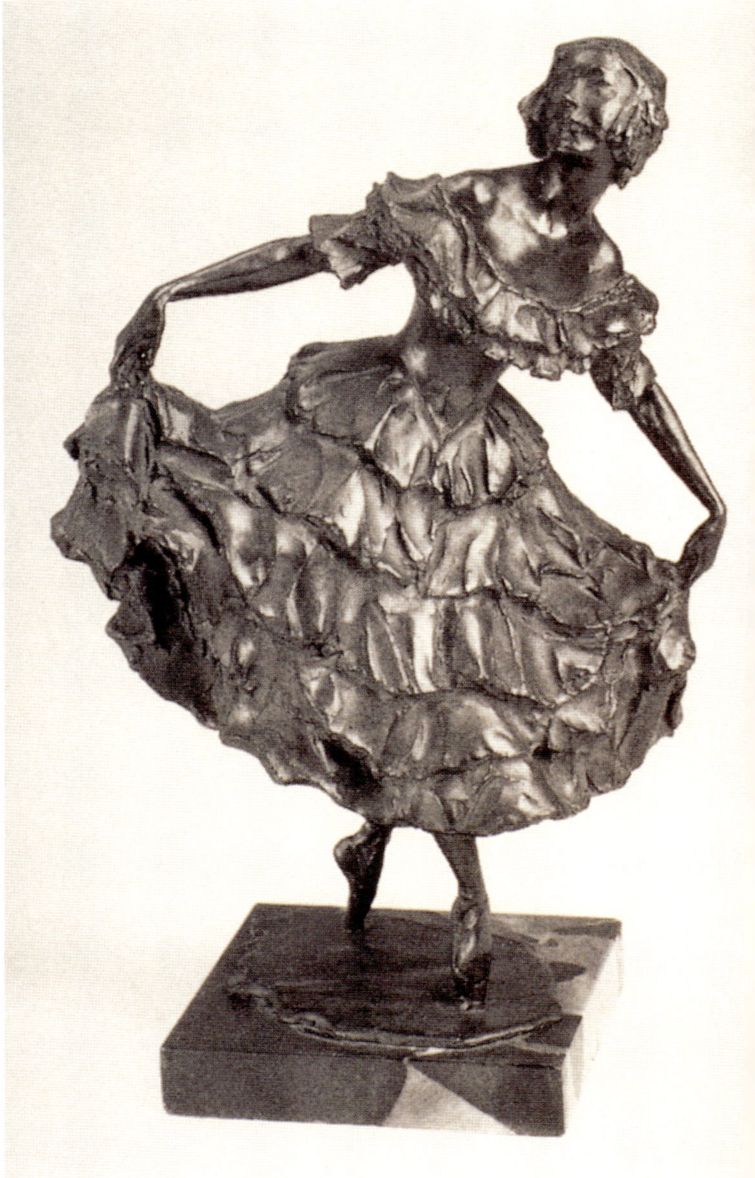

VI-142. "Ruth Page as Columbine" by Gleb Derujinsky. c. 1920.

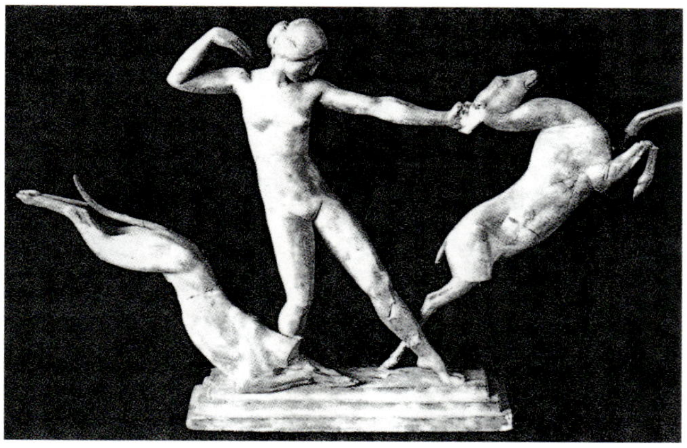

VI-143. "Diana" (plaster study for bronze "Diana") by Gleb Derujinsky. c. 1925-27.

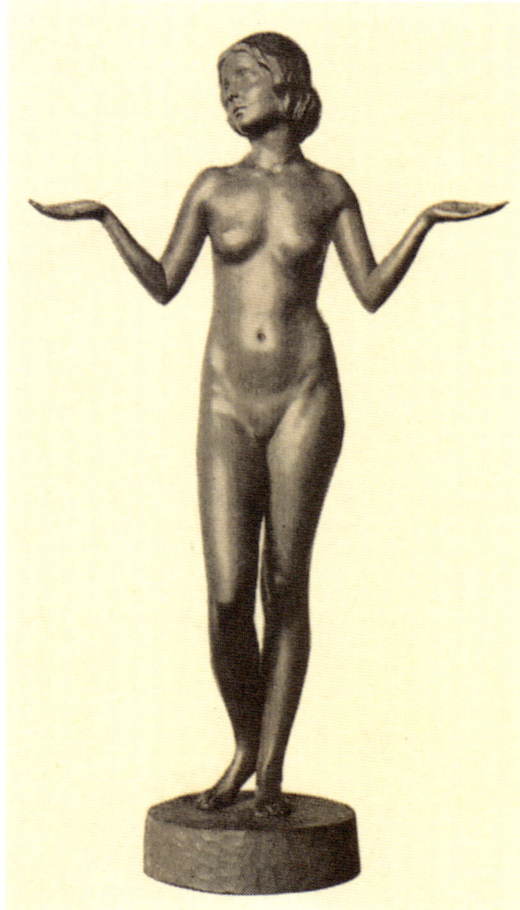

VI-145. "Innocence" by Gleb Derujinsky. c. 1928.

VI-144. "Diana" (bronze) by Gleb Derujinsky. c. 1925-27.

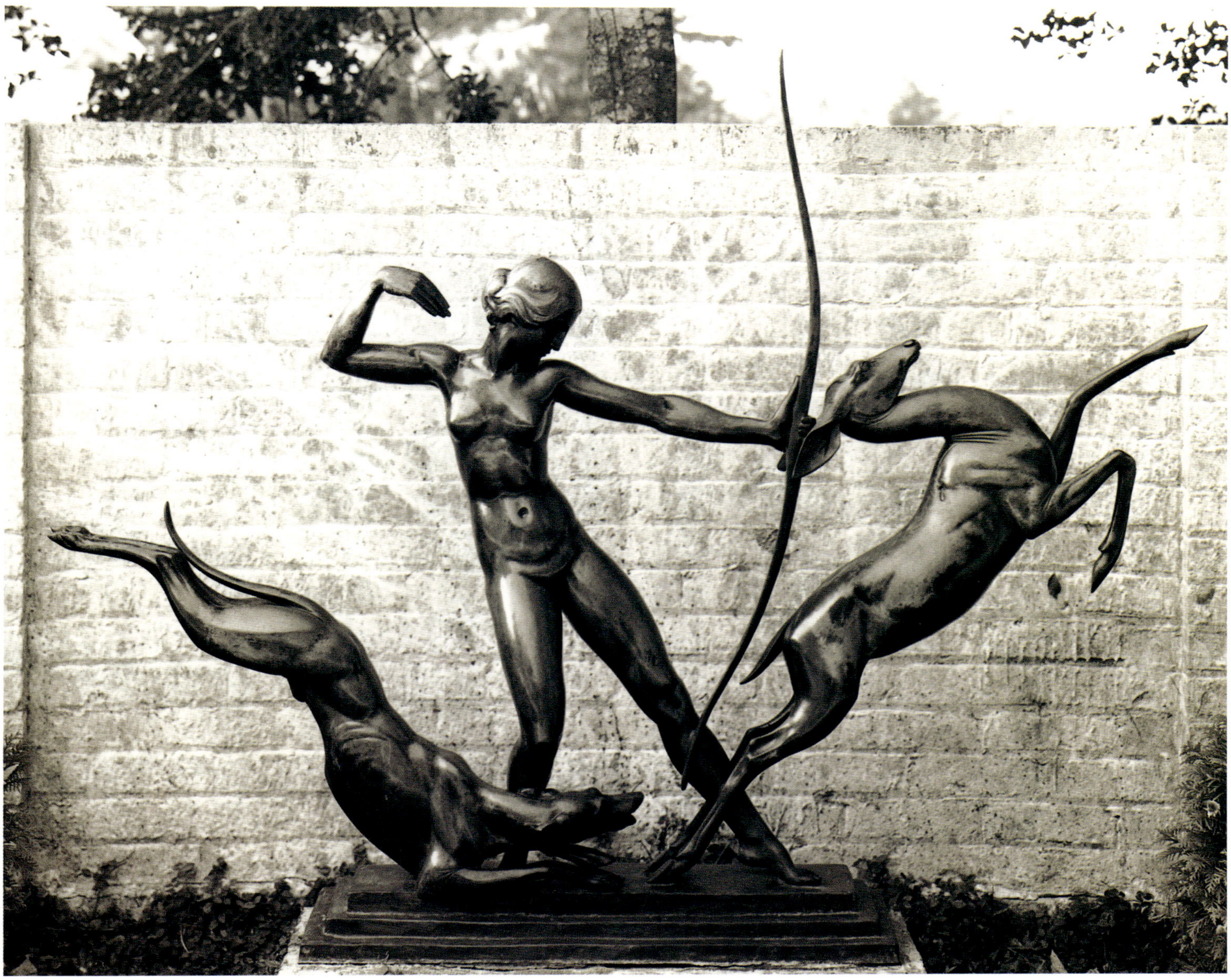

PAUL JENNEWEIN (1890-1978)

Paul Jennewein was a sculptor whose work had an ornamental quality, an emphasis on linear rhythm, and frequently allegorical themes. He produced both architectural and monumental sculpture, as well as smaller bronzes, often in a symbolic or Neo-Classical style. Like Paul Manship, he worked within the realist framework in a highly decorative style; during his lifetime he was recognized as a leader of academic sculpture in the United States.

Jennewein was born in Germany and came to the United States in 1907. He studied at the Art Student's League and the National Academy of Design in New York, and at the American Academy in Rome for four years (1916-1920). The Neo-Classical style and the decorative motifs of Roman art remained a source of inspiration to him.

He opened a studio in New York where for many years he worked on a series of commissions for public monuments; among them were a war memorial for Barre, Vermont, and mythological terra cotta figures for the pediment of the Philadelphia Museum of Art. Other notable commissions included the bronze doors for the British Empire Building at Rockefeller Center and the architectural eagle for the Rayburn Office Building in Washington. Jennewein received numerous honors and was a member of the National Academy of Design, the National Institute of Arts and Letters.

He also began making garden and fountain statuary with whimsical design, such as his most popular work, "Cupid and Gazelle" (VI-146). Many of his pieces had classical subjects. His "Ackroteria" (VI-147), "Faun" (VI-148), and "Greek Dancer" (VI-149) typify his taste for carefully balanced, ornamental works with classical overtones.

Among his best known sculptures, however, is one based on his interest in American Indians: his "Indian and Eagle" (VI-150) has been cast numerous times. The first listed casting by Jennewein at Roman Bronze Works was his "Noyes Armillary Sphere" (VI-151) in 1931. He used the foundry through the 1940s.

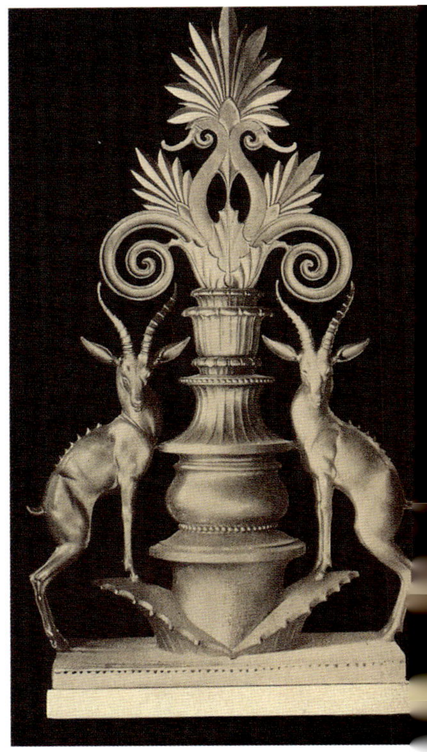

VI-147. "Ackroteria" by Paul Jennewein.

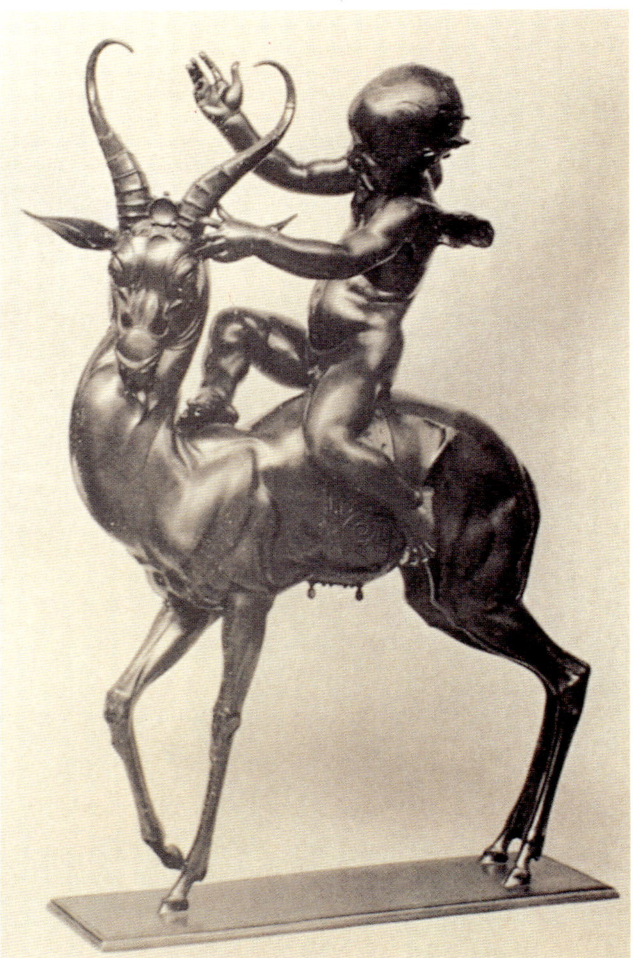

VI-146. "Cupid and Gazelle" by Paul Jennewein. 1929-30.

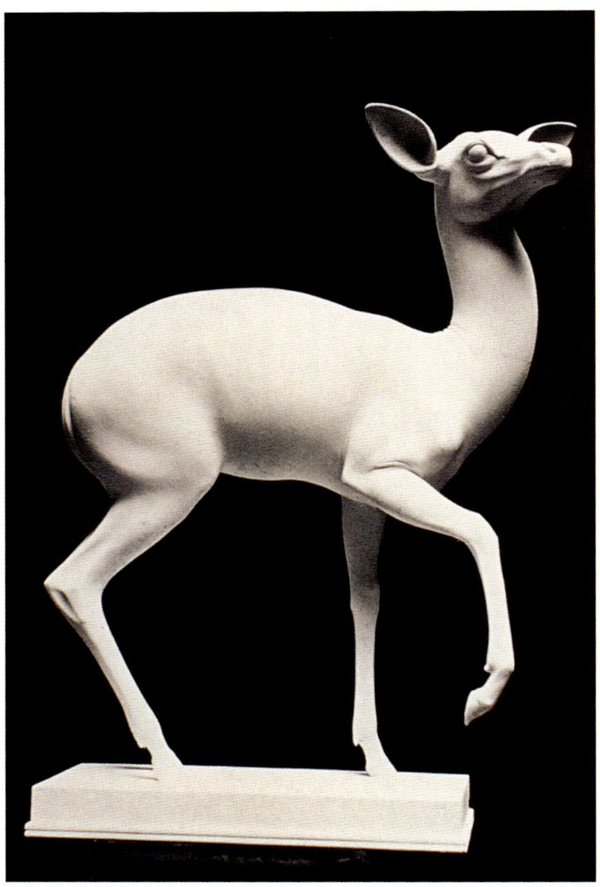

VI-148. "Fawn" by Paul Jennewein. 1937.

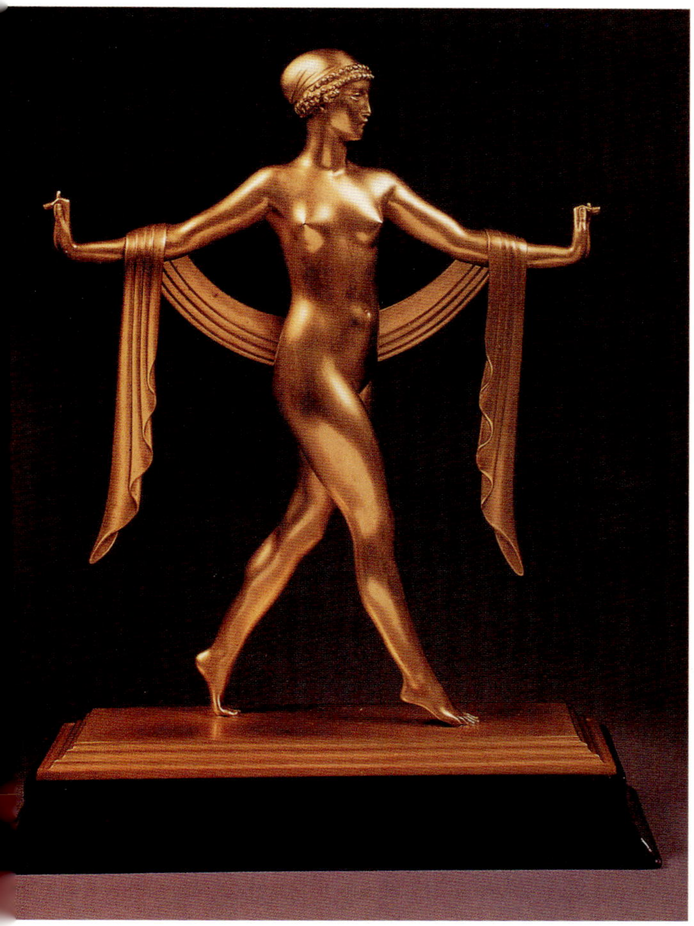

VI-149. "Greek Dancer" (also called "Golden Diana") by Paul Jennewein. c.1925-1926.

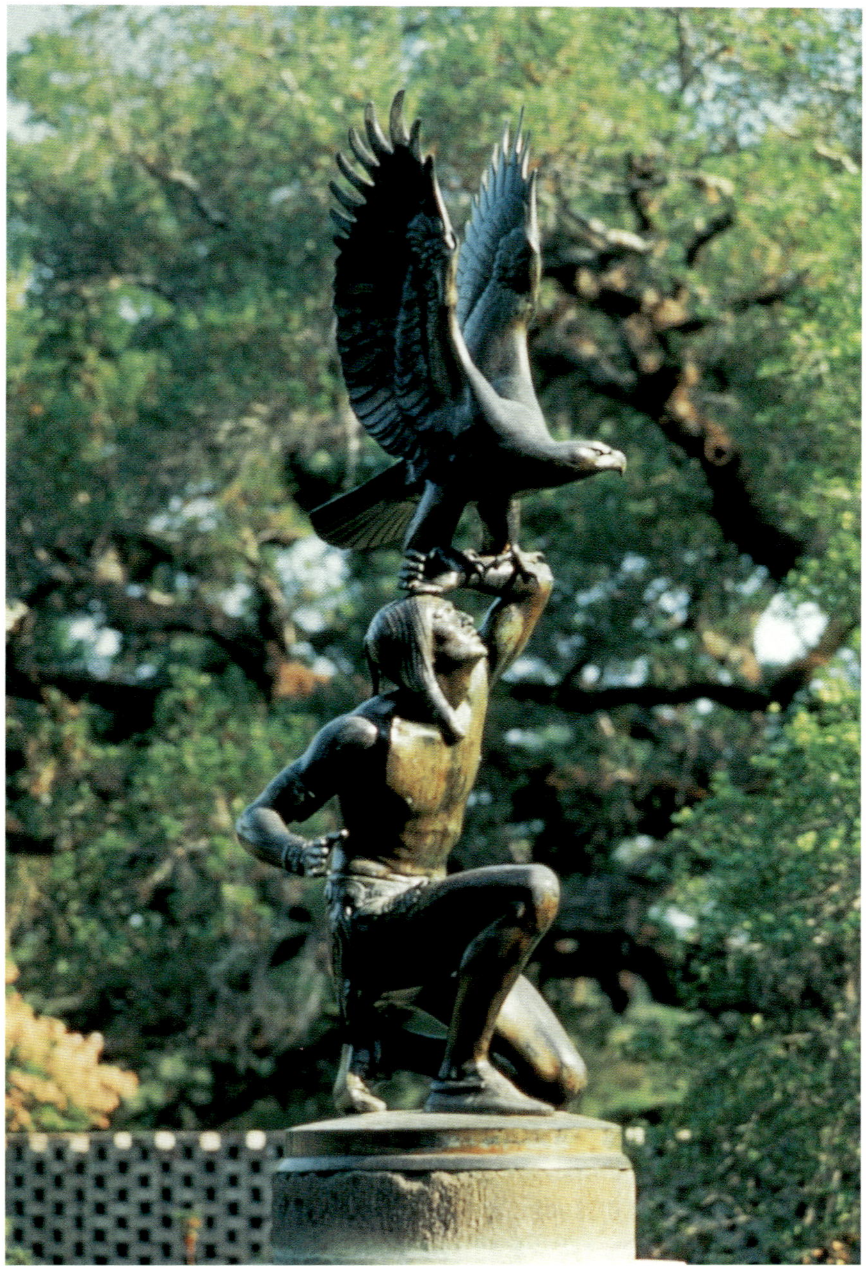

VI-150. "Indian and Eagle" by Paul Jennewein. 1929.

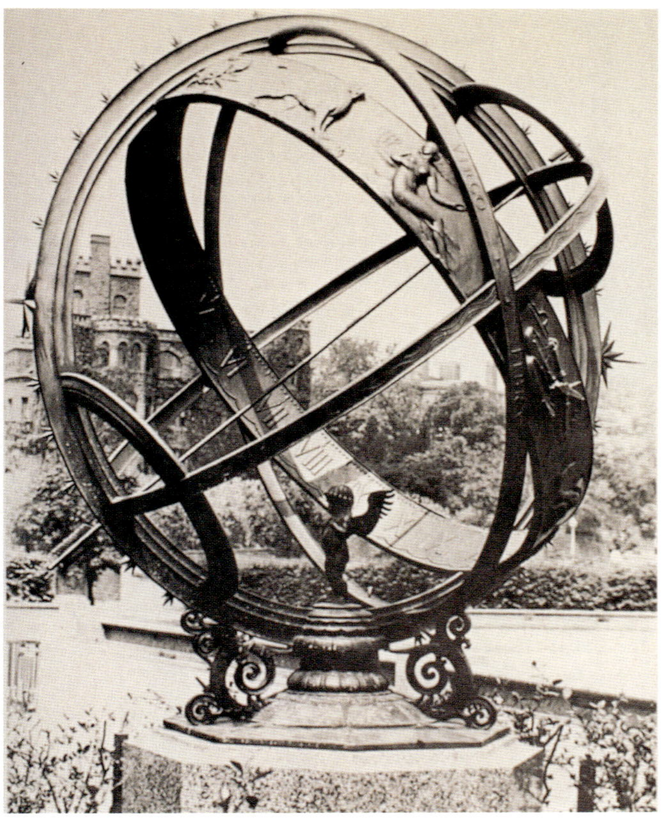

VI-151. "Noyes Armillary Sphere" by Paul Jennewein. 1931.

RENÉ PAUL CHAMBELLAN (1893-1955)

Although he was best known in America for his statuary at New York City's Rockefeller Center, René Chambellan also produced several other well-known pieces both in the United States and France. His style was rhythmic and fanciful, with highly decorative surfaces.

Chambellan was educated at the Académie Julian in Paris, and was also a student of Solon Borglum, with whom he collaborated on a dedication panel for Pershing Stadium in Vincennes, France. Chambellan designed medals, memorials, groups, figures, and plaques for such sites as Princeton University, Catholic Cemeteries in Chicago, and the Russell Sage Foundation Building in New York.

His contributions to the art at Rockefeller Center included carved limestone insets for the facades of several of the buildings. But it was his series of bronze fountain figures for Rockefeller Center's shallow pools for which he is best remembered. These graceful tritons and nereids riding on the backs of whimsical sea creatures (numbers VI-152, 153, and 154) surround the central plaza. The sequence of the stylized fountain figures symbolize Leadership, Will, Thought, Imagination, Energy, and Alertness. They were all cast at Roman Bronze Works.

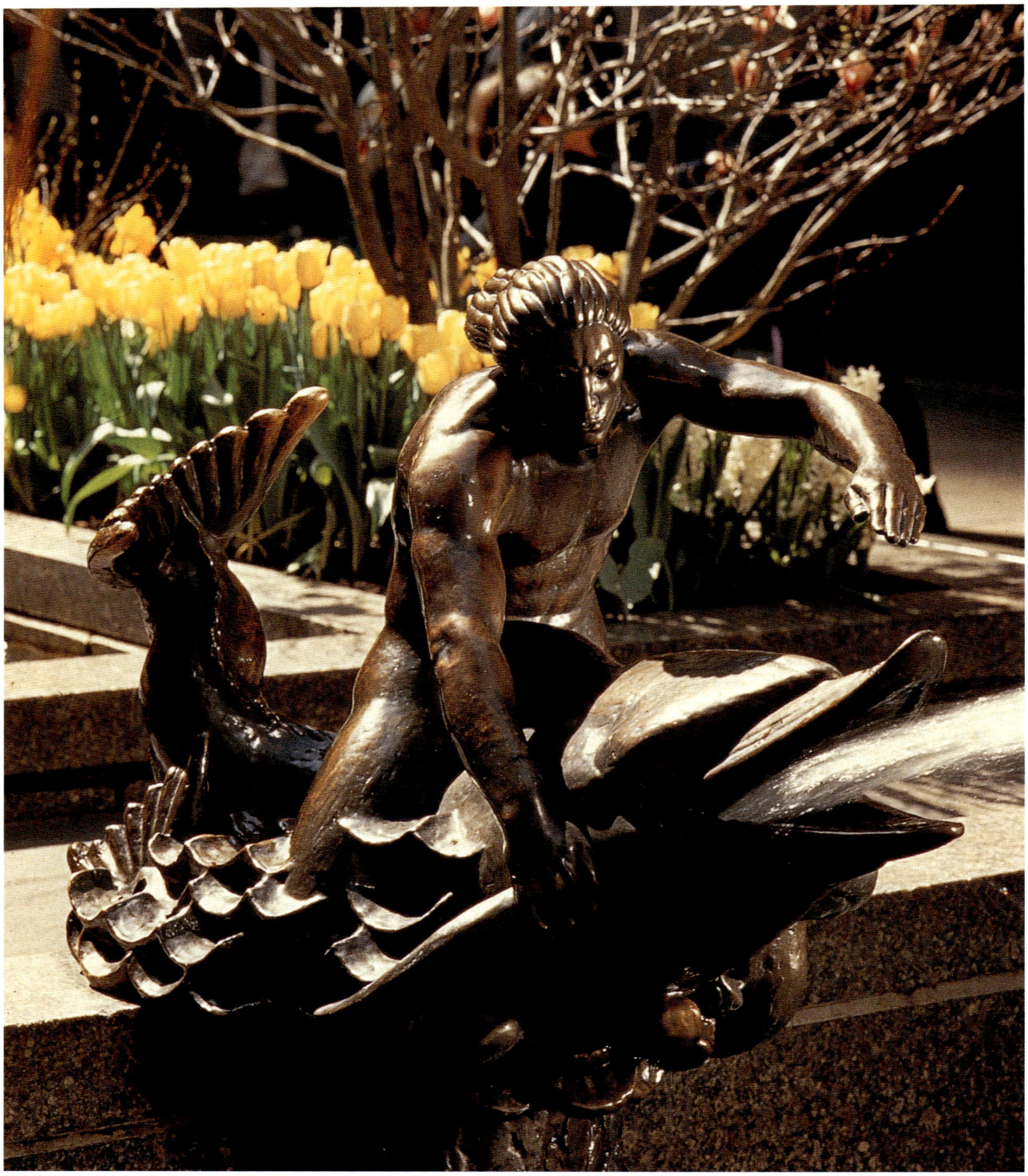

VI-152. "Triton and Dolphin" (Channel Garden figure, Rockefeller Center, New York City) by René Paul Chambellan. c. 1935.

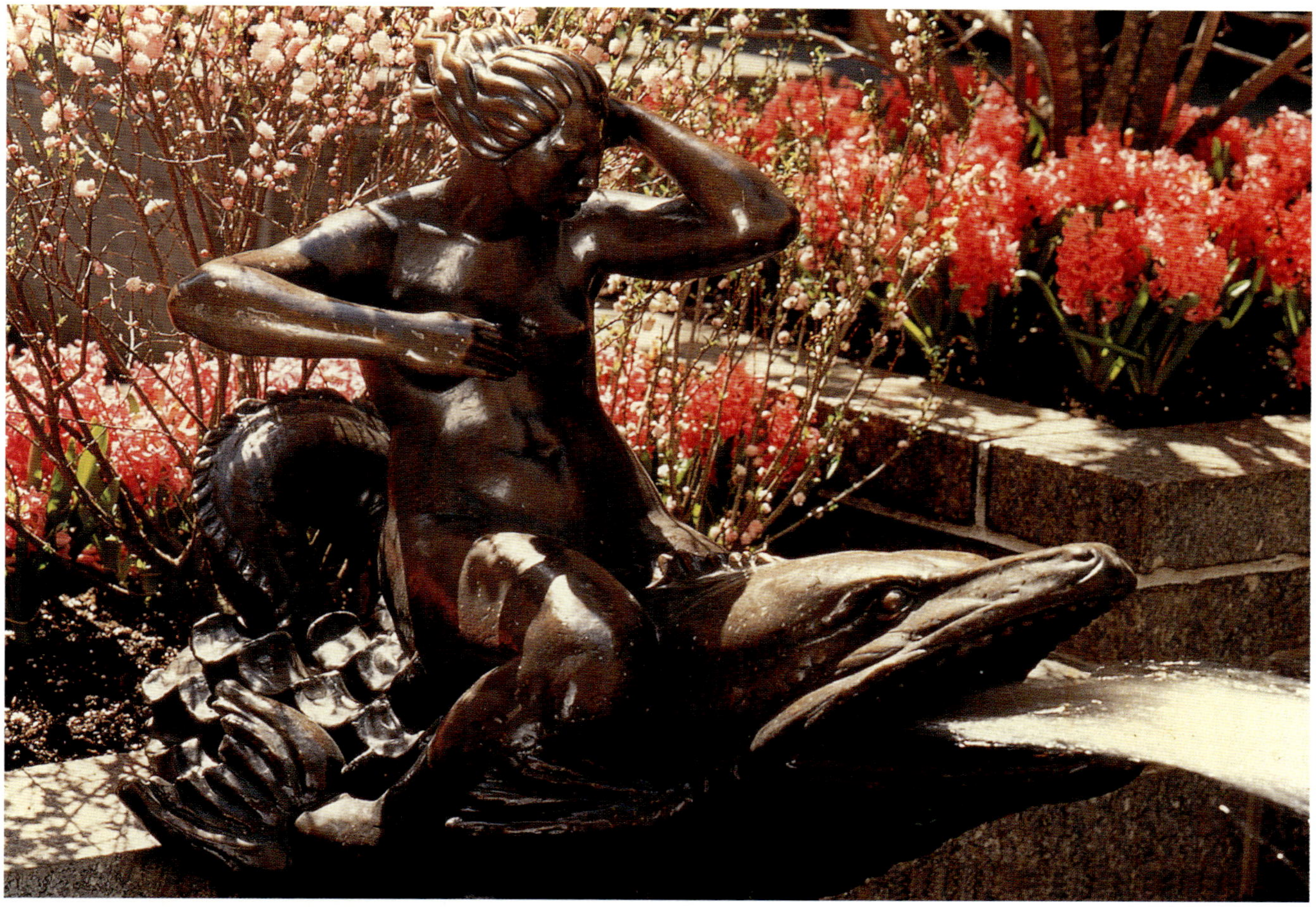

VI-153. "Nereid and Dolphin" (Channel Garden figure, Rockefeller Center, New York City) by René Paul Chambellan. c. 1935.

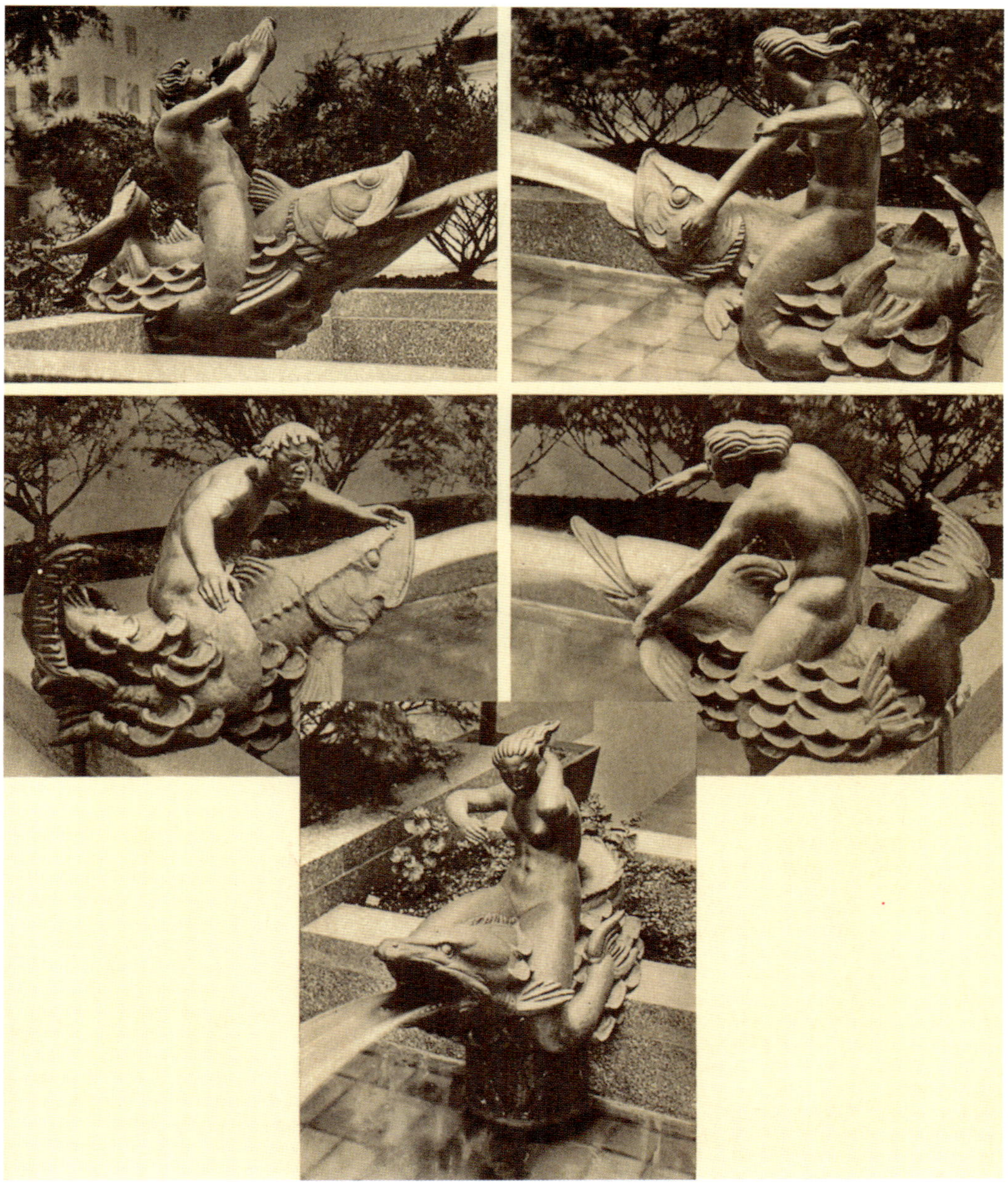

VI-154. Five of the series of "Tritons and Nereids with Dolphins" by René Paul Chambellan, c. 1935, as presented in a flyer by the General Bronze Corporation.

ALAN CLARK (1896-1950)

Alan Clark was born in Montana. He was a student of Albin Polasek at the Art Institute of Chicago and Robert Aitken at the National Academy of Design. Though thoroughly grounded in academic realism, he had an interest in Asian art and studied both Japanese and Chinese sculpture. His works were primarily small, free-standing bronze sculptures, many with a definite appreciation for the exotic.

Clark's interest in Asian art was increased by a three-year visit to China, Korea, and Japan, as well as Southeast Asia between 1924 and 1927. While in the Orient he studied Buddhist art and wood carving, and made several sculptures of well-known actors and dancers, including one of the famous Chinese beauty "Yang Kwei Fei" (VI-155). After his return, he opened a studio in Santa Fe, New Mexico, and applied the techniques of Asian woodcarving to his portraits of American Indians; he also produced a number of bronzes of native Americans like his "Male Indian" (VI-156).

Clarks's bronze work ranged from traditional realist nudes, such as "Standing Woman" (VI-157) and "Nude Woman Holding a Falcon" (VI-158), to the elegant "Seated Deco Woman" (VI-159). His best known bronzes were his exotic "Parvati: A Balinese Dancer" (VI-160), and perhaps his most famous work "Kongo Voodoo" (VI-161). His smaller works included a pair of bookends depicting a "Satyr and Nymph" (VI-162) All of these sculptures were cast at Roman Bronze Works between 1927 and the 1940s.

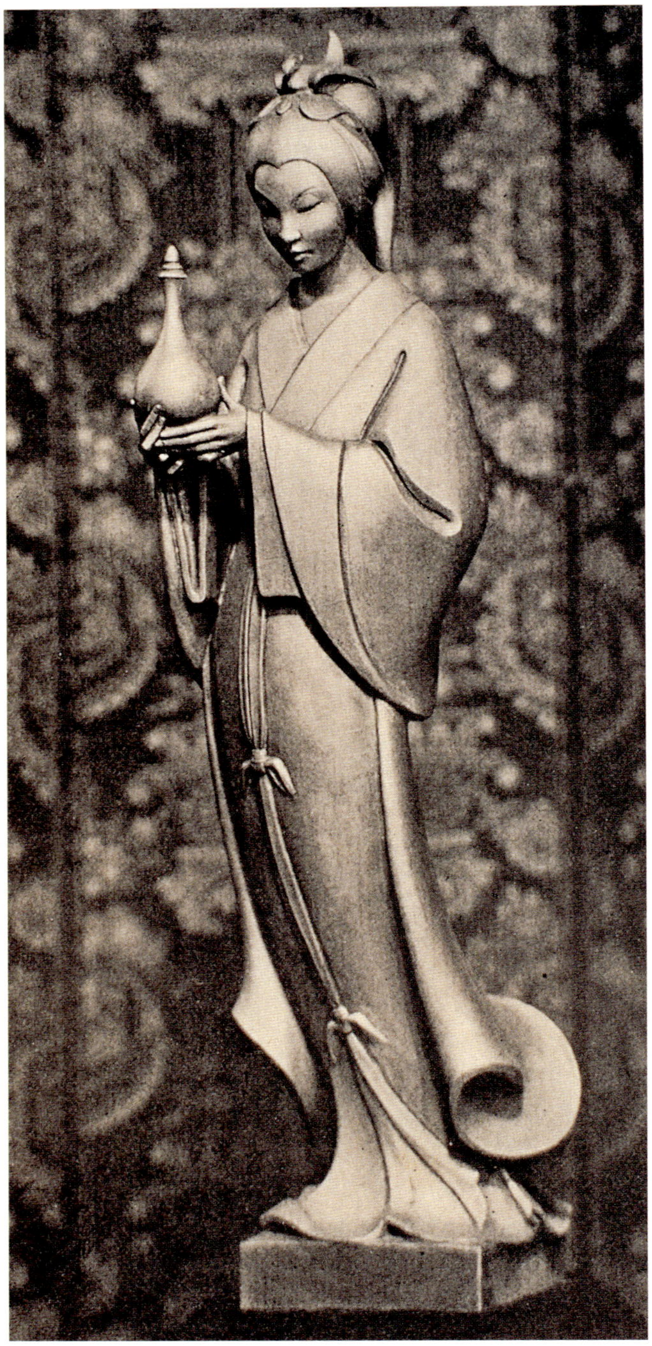

VI-155. "Yang Kwei Fei" by Alan Clark. 1928.

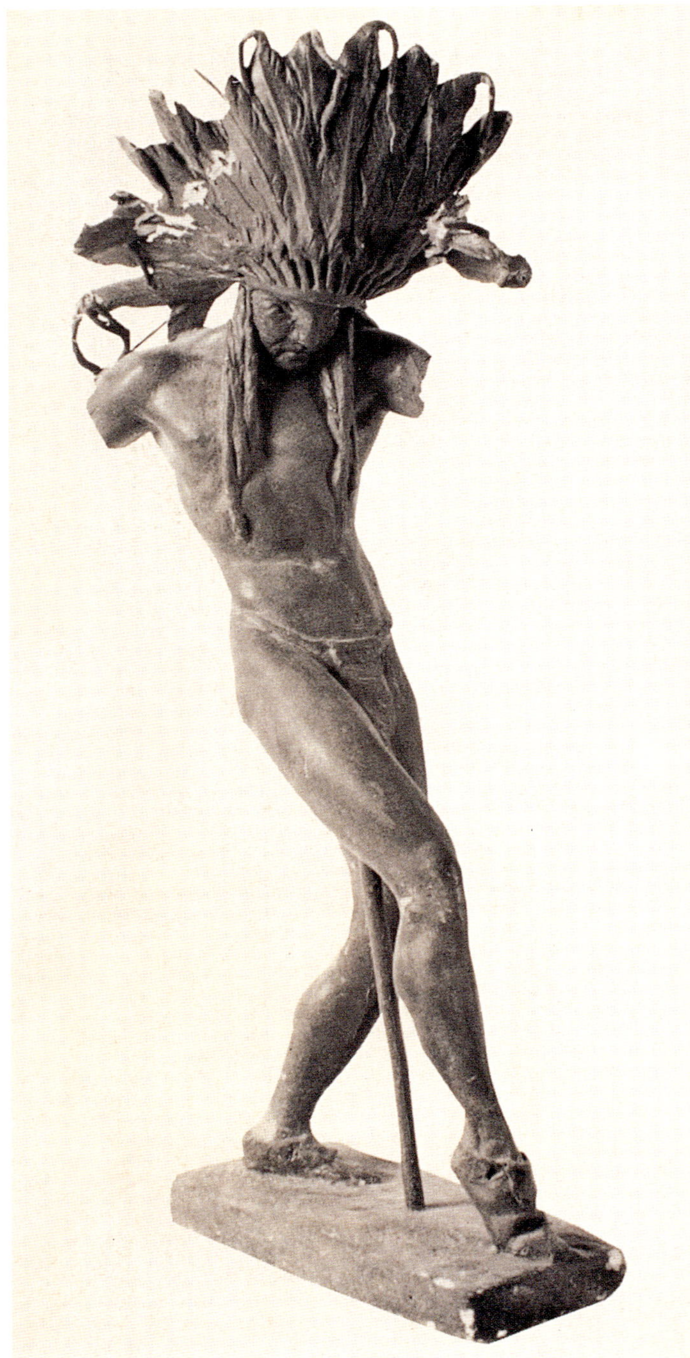

VI-156. "Male Indian" (plaster cast) by Alan Clark. c. 1928.

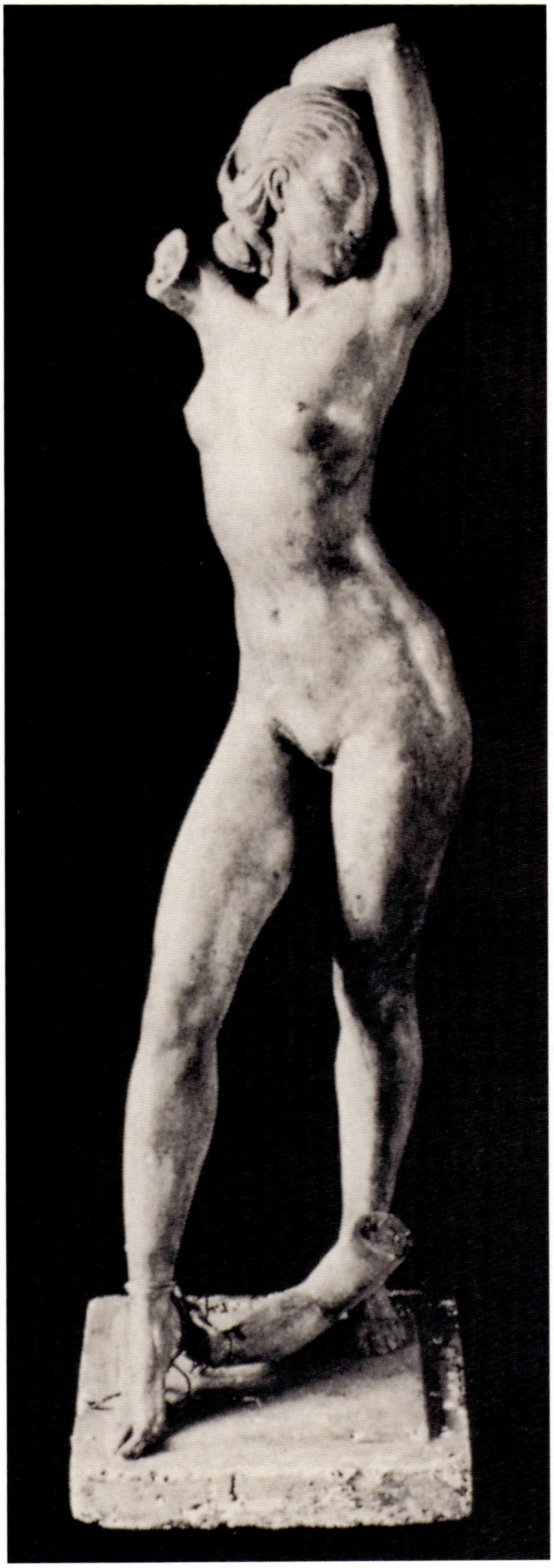

VI-157. "Standing Woman" (plaster cast) by Alan Clark. 1930s.

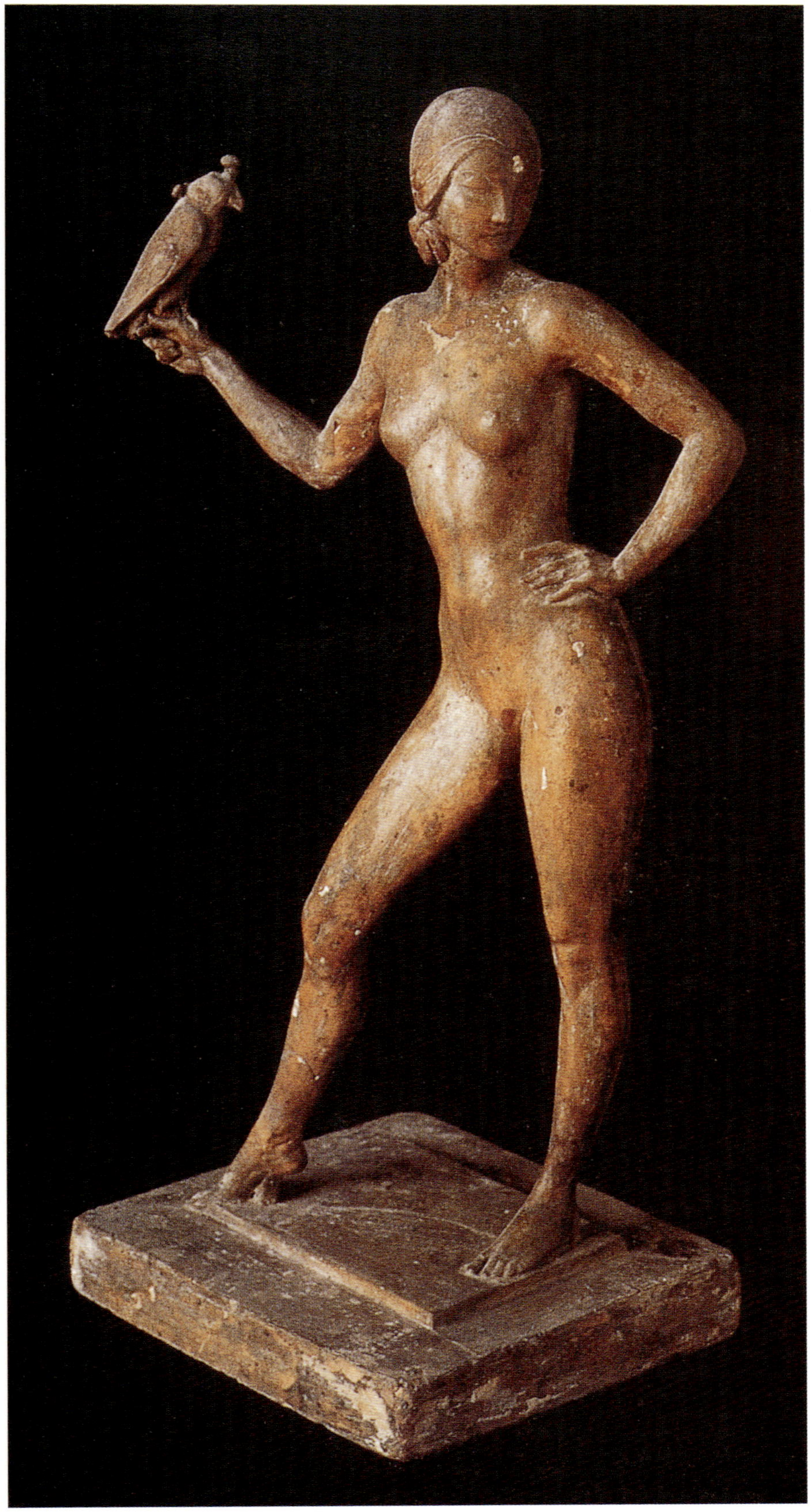

VI-158. "Nude Woman Holding a Falcon" (plaster cast) by Alan Clark.

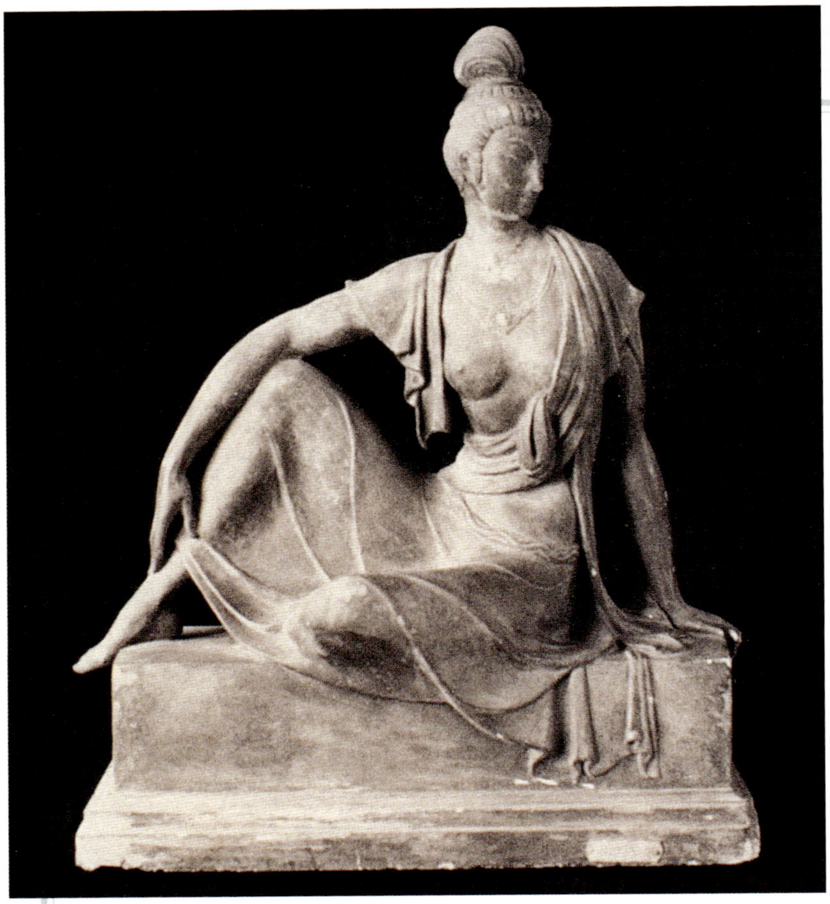

VI-159. "Seated Deco Woman" (Study for a Garden Pool) by Alan Clark. The sculpture depicts Bodhivisahva Avalokitesvara. Modeled in China in 1925; R.B.W. 1928.

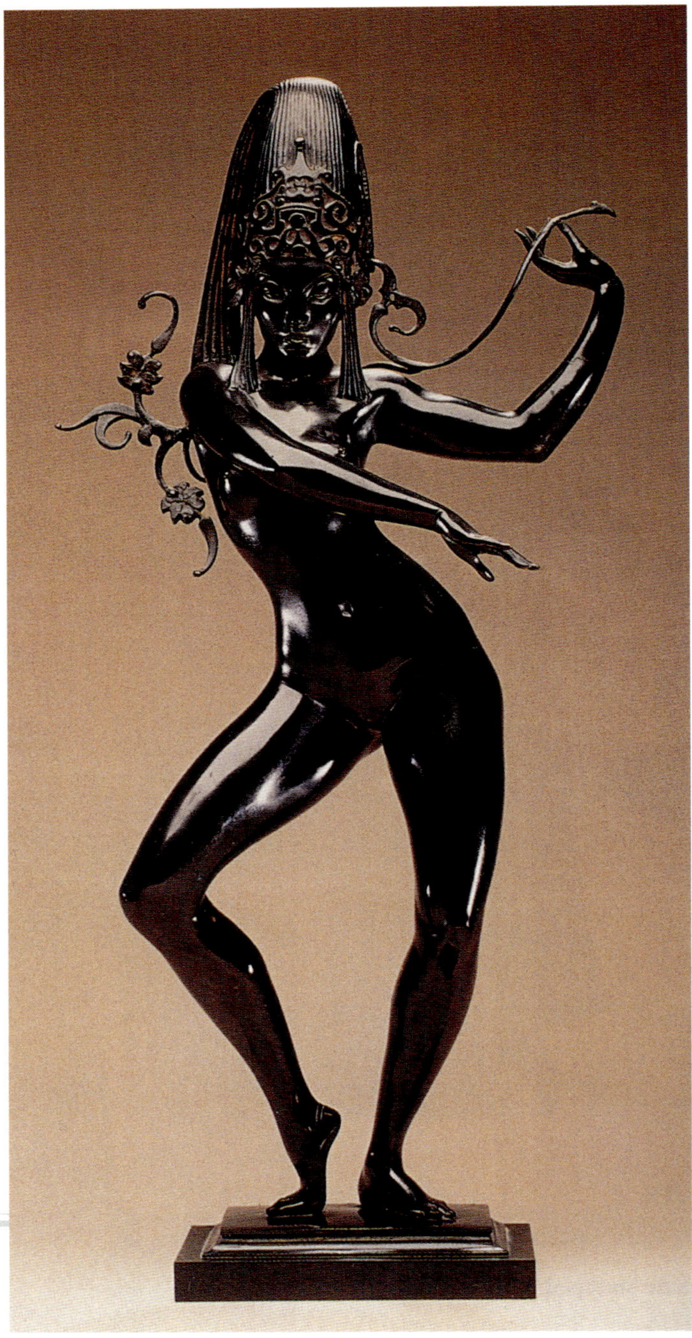

VI-160. "Balinese Dancer" (also called "Parvati") by Alan Clark. 1927.

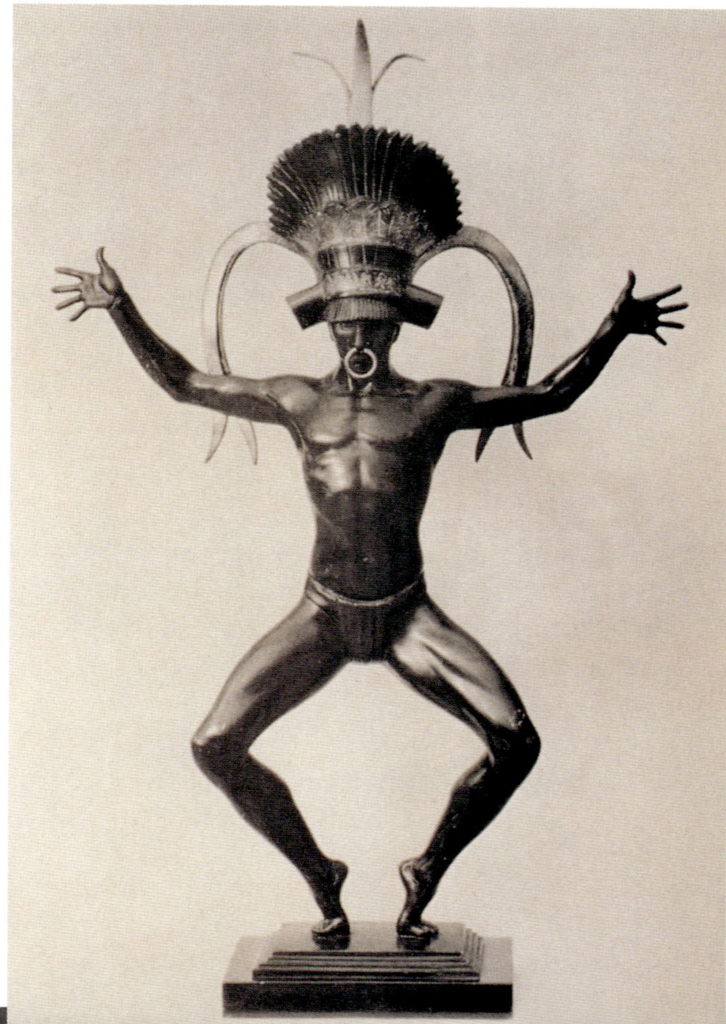

VI-161. "Kongo Voodoo" by Alan Clark. 1927.

VI-162. "Satyr and Nymph" by Alan Clark. c. 1925. R.B.W. c. 1927.

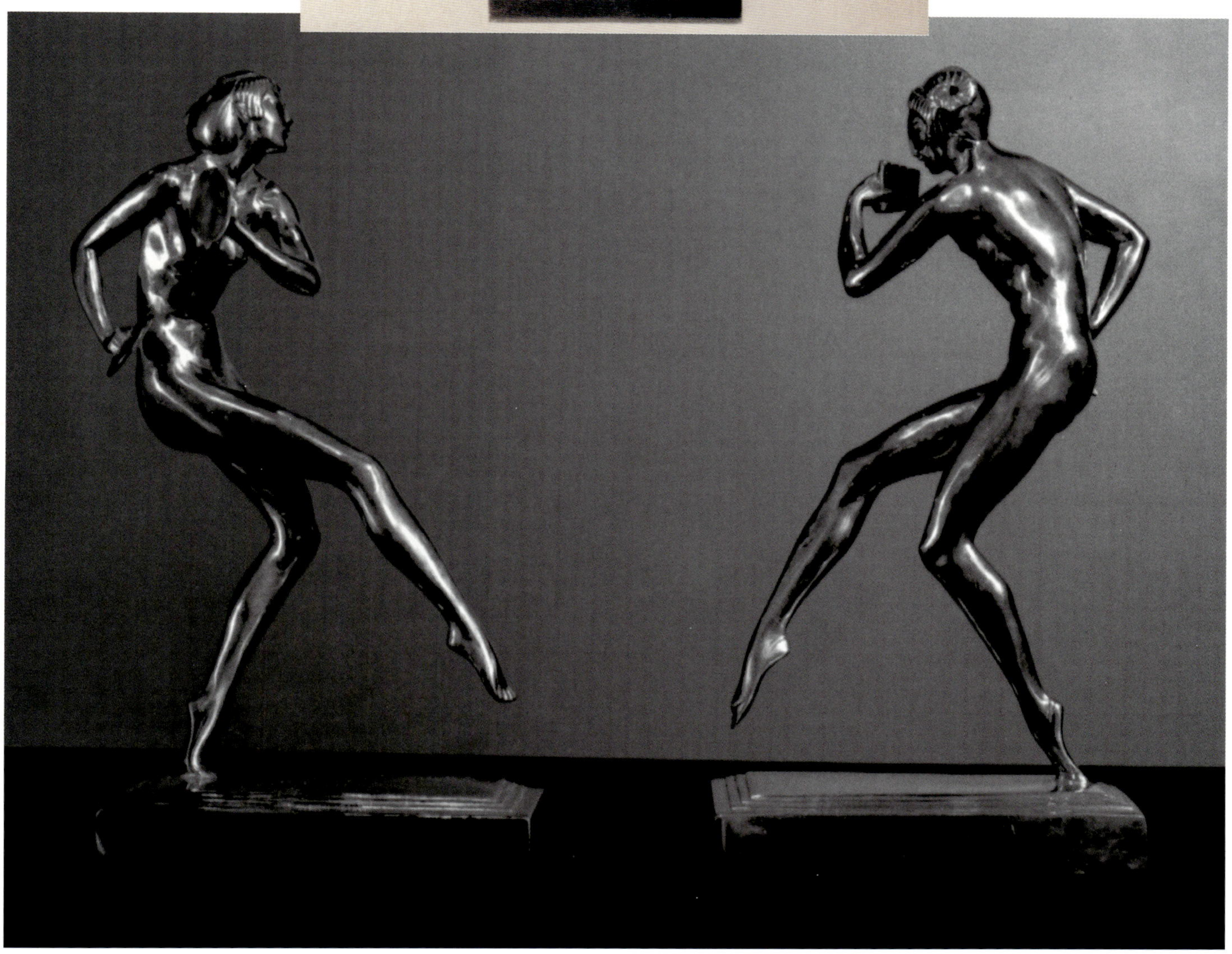

AVARD FAIRBANKS (1897- 1987)

A native of Utah, Avard Fairbanks was noted for making many public monuments, including several relating to pioneers and American history. In addition to monuments, he made medals, portrait busts, fountain pieces, and a tabernacle door. His style was realistic and often strongly narrative, with an emphasis on character and drama.

Fairbanks, who was born in Provo, was first trained by his father, a painter, and later at the Art Student's League with James Earle Fraser and the National Academy of Design. In Paris he studied at the École des Beaux-Arts, returning home when World War I began. He graduated from Yale University and also studied in Florence. In New York he met, and learned from several leading sculptors, including Alexander Proctor, the Borglum brothers, Hermon MacNeil, and Cyrus Dallin. Early in his career he made a number of animal sculptures (several were shown at the Panama-Pacific Exposition), and a series of works on Hawaiian subjects. He also began a long career as a teacher of sculpture at several universities; in 1929 he became a professor at the University of Michigan, where he remained for many years.

His interest turned to memorials to pioneers like his own ancestors; among them were his "Pioneer Group" (VI-163) and his "Winter Quarters" (VI-164) and his portrait of the American missionary and pioneer, Marcus Whitman (VI-165). These three monuments, though designed for Western sites, were cast at Roman Bronze Works. His head of Abraham Lincoln (VI-166) was typical of his strong and realistic style. He also made several impressive war monuments. In a more lyrical mode is his "Rain" (VI-167). The R.B.W. ledgers list Fairbanks between 1936 and the 1950s.

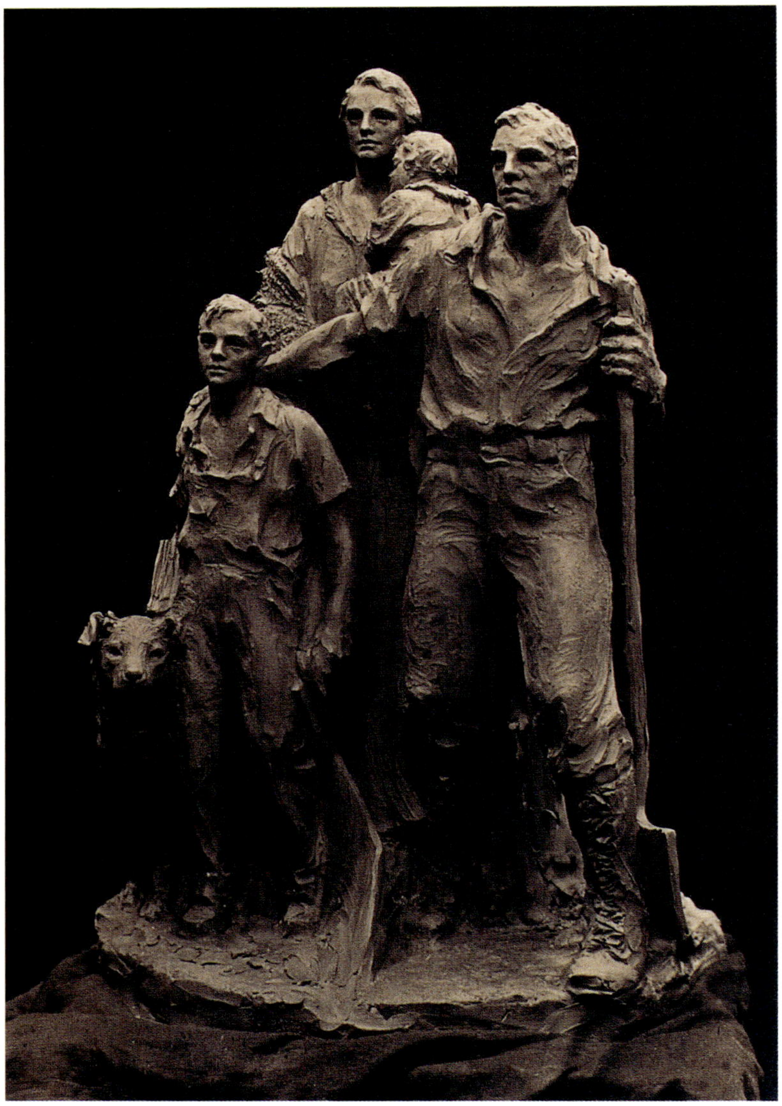

VI-163. "Pioneer Family" (Bismarck, North Dakota) by Avard Fairbanks. 1936.

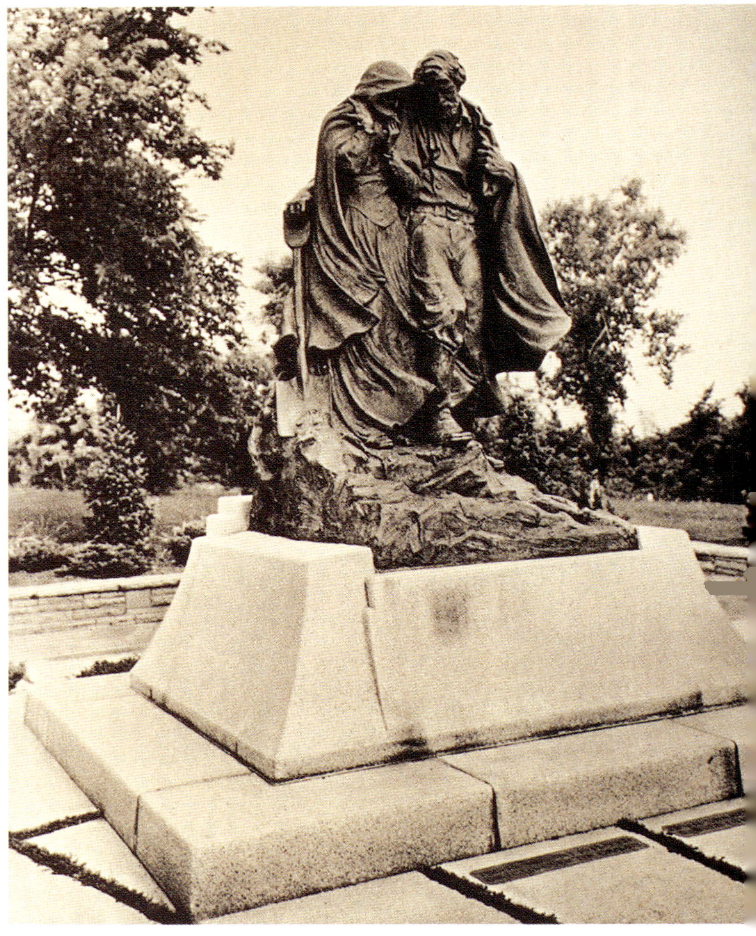

VI-164. "Winter Quarters" (Florence, Nebraska and Salt Lake City, Utah) by Avard Fairbanks. 1935.

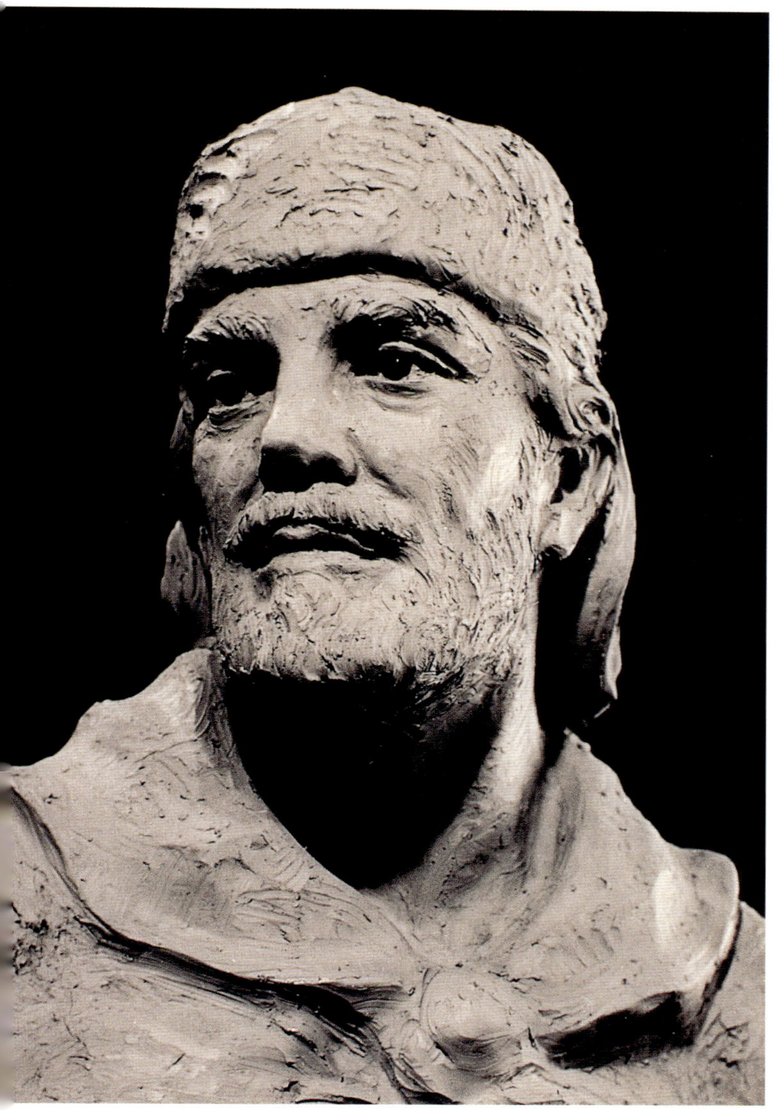

VI-165. "Marcus Whitman" by Avard Fairbanks. 1950.

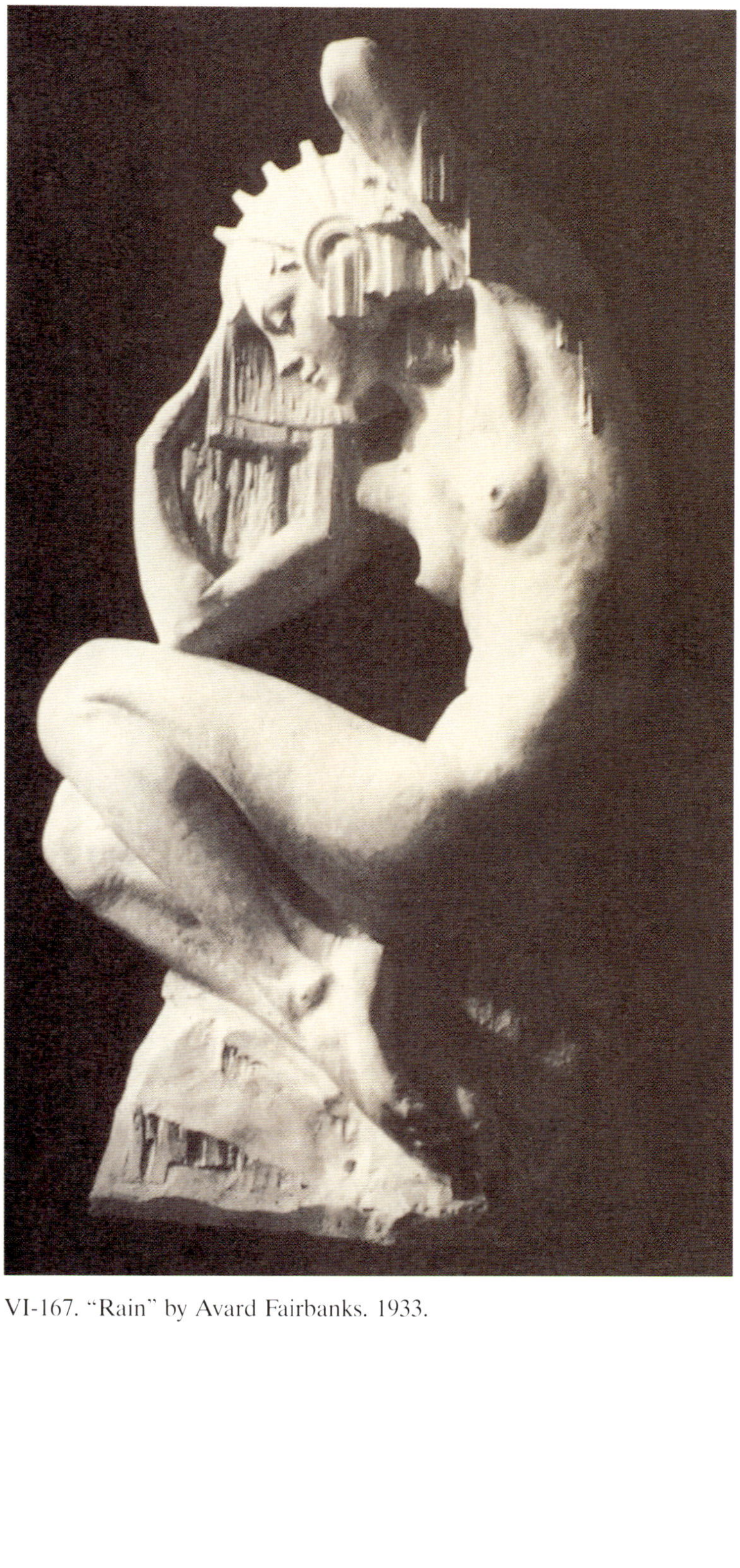

VI-167. "Rain" by Avard Fairbanks. 1933.

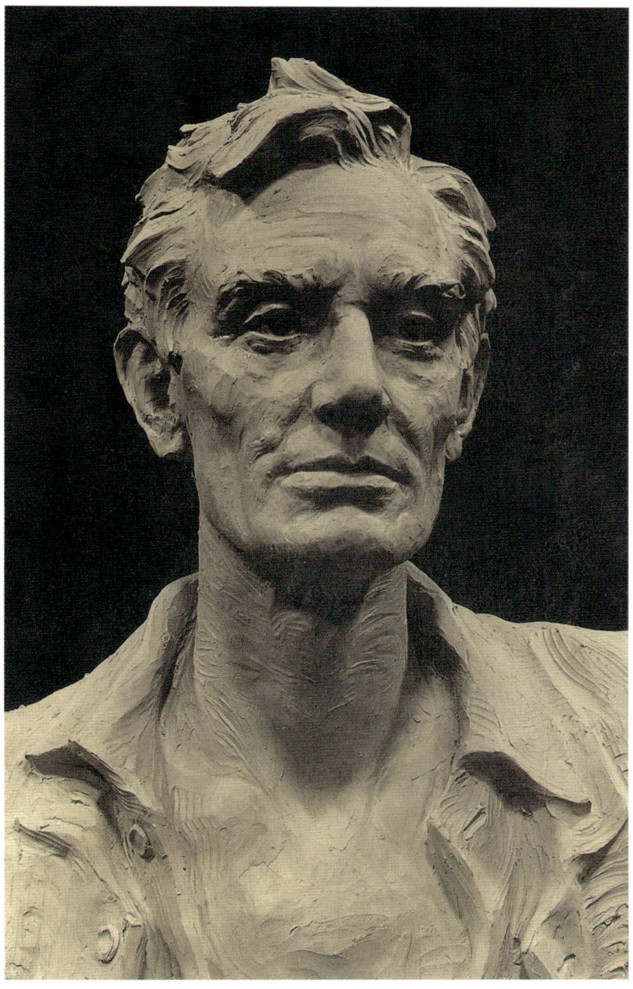

VI-166. "Abraham Lincoln" by Avard Fairbanks. 1941.

WHEELER WILLIAMS (1897-1972)

Wheeler Williams was a sculptor, painter, and lithographer whose own style of conservative realism and strong antipathy to modern art made him a somewhat controversial figure. His own taste was for strongly naturalistic monuments and "baby deities." These whimsical little pieces, which he called "The Childhood of the Gods," (VI-168 and 169) represented ancient gods as small children, and they were very popular. He also created a variety of architectural and animal statuary, and several fountain pieces.

Williams, who was active in many sculptural organizations, testified strongly against modern art in the United States Congress in the early 1950s, when abstract American painting and sculpture had begun to be widely shown and was seen by conservatives as "un-American."

Williams was born in Chicago, educated at Yale University and the École des Beaux-Arts in Paris. Among his many commissions were the entrance to the Holland Tunnel, the pediment for the Interstate Commerce Building in Washington, and stylized statues of the Four Seasons. Animal sculptures included his "Sea Lion Fountain (VI-170) and his "Black Panther" (VI-171). He also made a number of portraits, including his study of the boxer Knute Hansen, "The Bronze Boxer" (VI-172) and "Bronze Bust of a Woman" (VI-173). In addition to these sculptures, Williams cast many of his animal pieces and several nude figures, including a "Reclining Pan" and a "Narcissus" at Roman Bronze Works between 1929 and 1940.

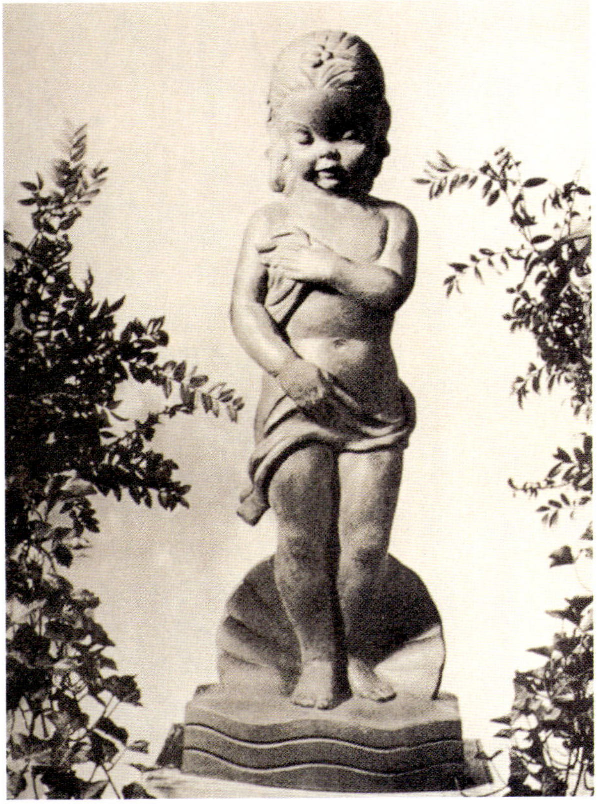

VI-168. "Venus" (One of eight sculptures in "Childhood of the Gods" series) by Wheeler Williams. 1939.

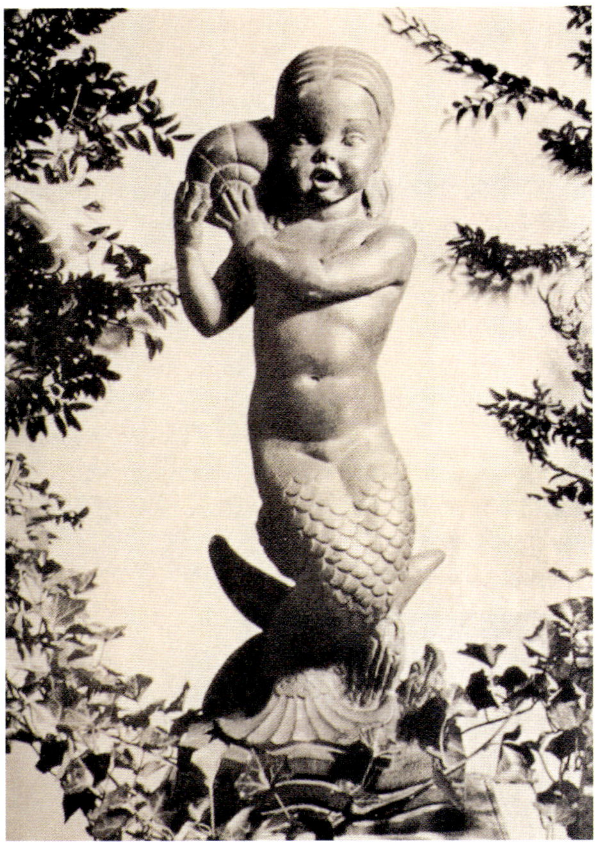

VI-169. "Amphitrite" (One of eight sculptures in "Childhood of the Gods series) by Wheeler Williams. 1939.

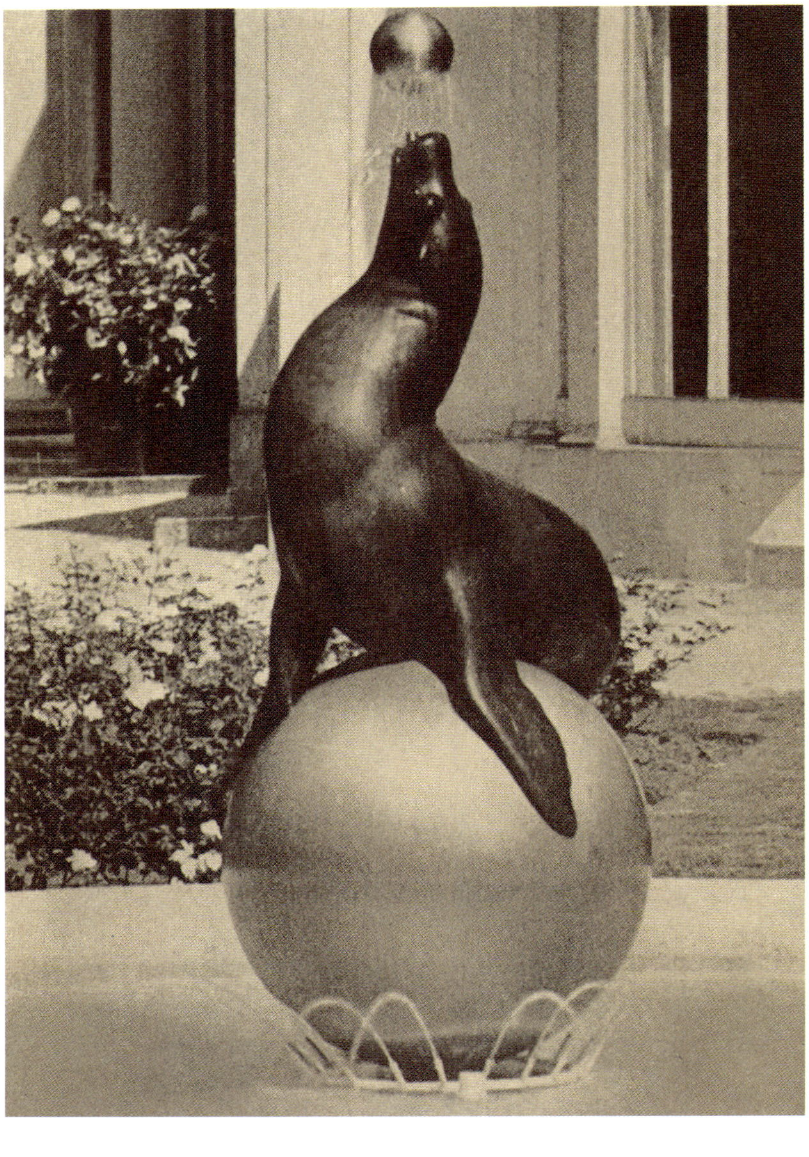

VI-170. "Sea Lion Fountain" (bronze figure on aluminium ball) by Wheeler Williams.

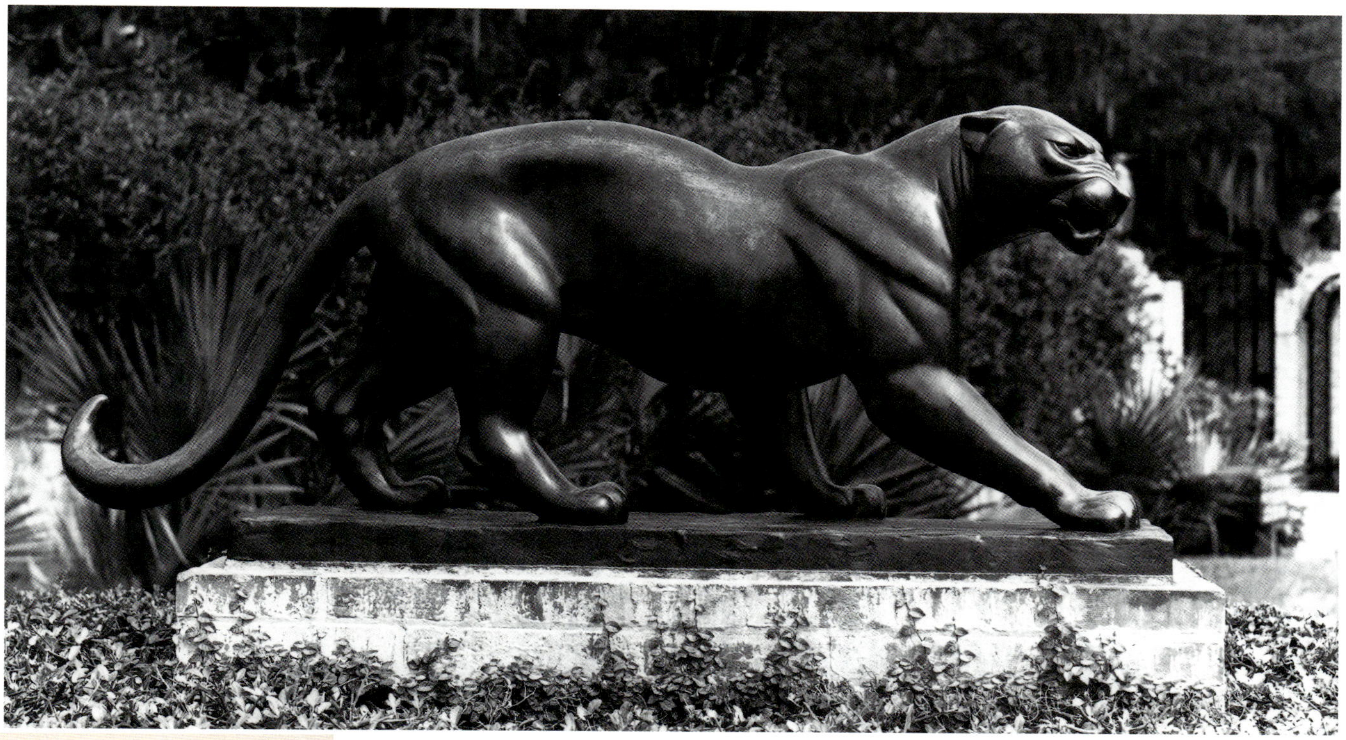

VI-171. "Black Panther" by Wheeler Williams. 1933.

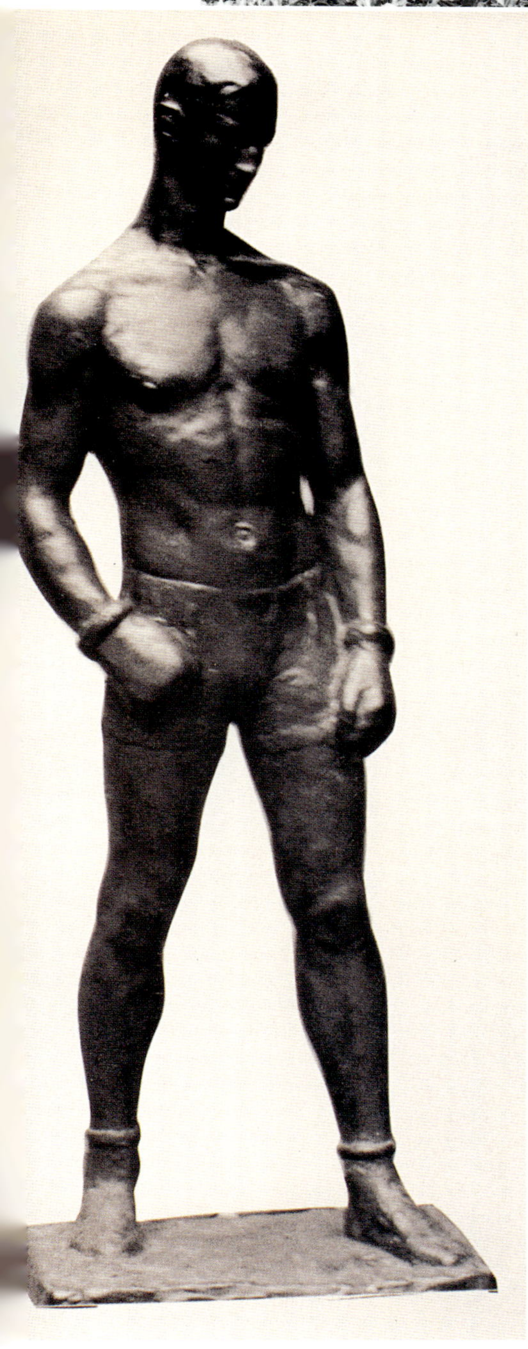

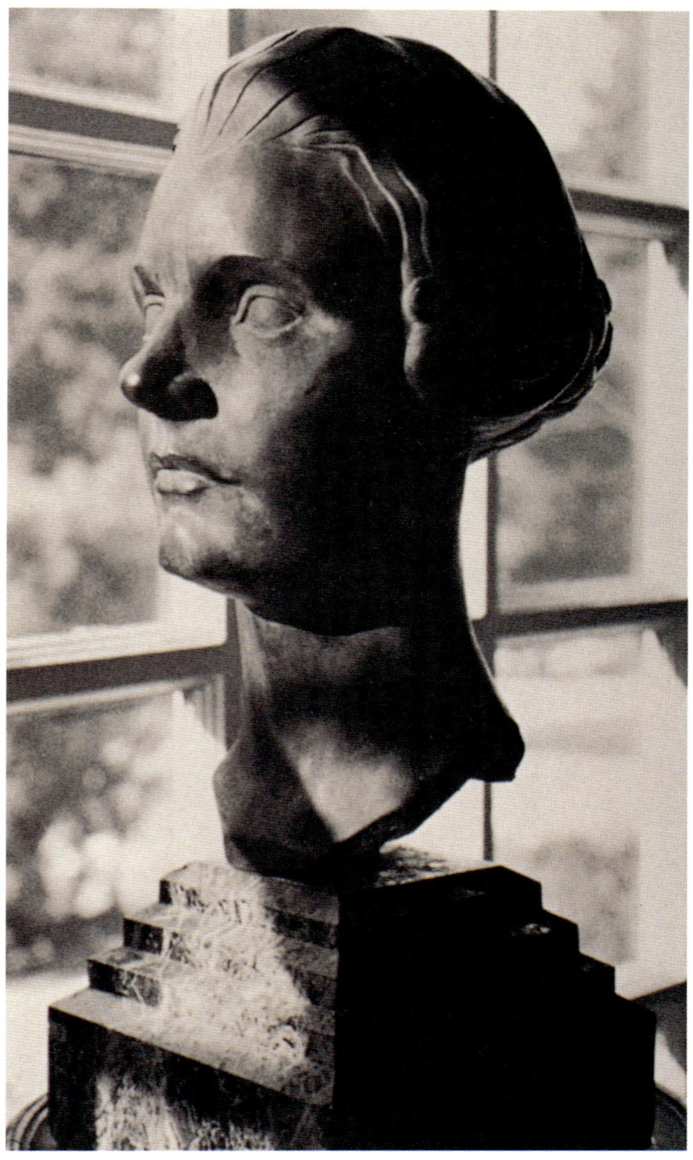

Left:
VI-172. "The Bronze Boxer" by Wheeler Williams. Modeled 1928.

Right:
VI-173. "Bronze Bust of a Woman" by Wheeler Williams. Modeled 1923.

JOSEPH KISELEWSKI (1901-1988)

Joseph Kiselewski was a traditionalist, working in a carefully crafted, realist style. He made a number of large architectural sculptures, fountain pieces, allegorical nudes, medals, ecclesiastical works, and portrait busts. He cast some of his smaller sculptures at Roman Bronze Works.

Born in Minnesota, Kiselewski was educated at the Minneapolis School of Art, the National Academy of Design, in Paris, and as a recipient of the Prix de Rome at the American Academy there. For four years he was an assistant to Lee Lawrie.

His commissions were varied: he modeled statues of Christ, and a "Madonna and Child" (VI-174) for religious institutions, a statue of General Pulaski for the city of Milwaukee, and numerous architectural sculptures. In 1932 his massive monument "Victory and Peace" was unveiled at the Bronx Courthouse. He made the fisheries pediment for the Commerce Building in Washington (after a drawing by James Earle Fraser) and sculptures for the House Chamber of the Capitol Building.

He also worked in low relief, creating architectural medallions and a sundial and several medals. Among his best-known sculptures were his Sea Horse Fountains (VI-175) made for the pools at the Radio Corporation of America's World's Fair display in 1939. Kiselewski's many portrait busts included likenesses of Sinclair Lewis, Oliver Wendell Holmes, and Lee Lawrie. Other small pieces, all cast at Roman Bronze Works, include his "Water Baby Fountain" (also called "Girl on a Dolphin") (VI-176), and two versions of the ideal female called "Dawn" (VI-177 and 178).

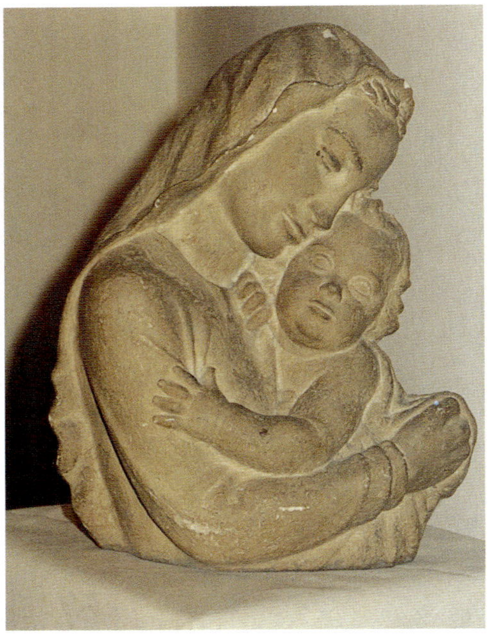

VI-174. "Madonna and Child" by Joseph Kiselewski.

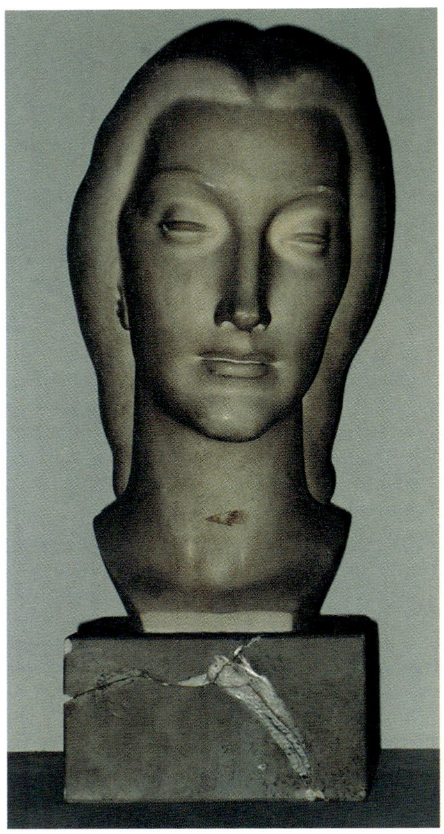

VI-177. "Dawn" by Joseph Kiselewski. c. 1933.

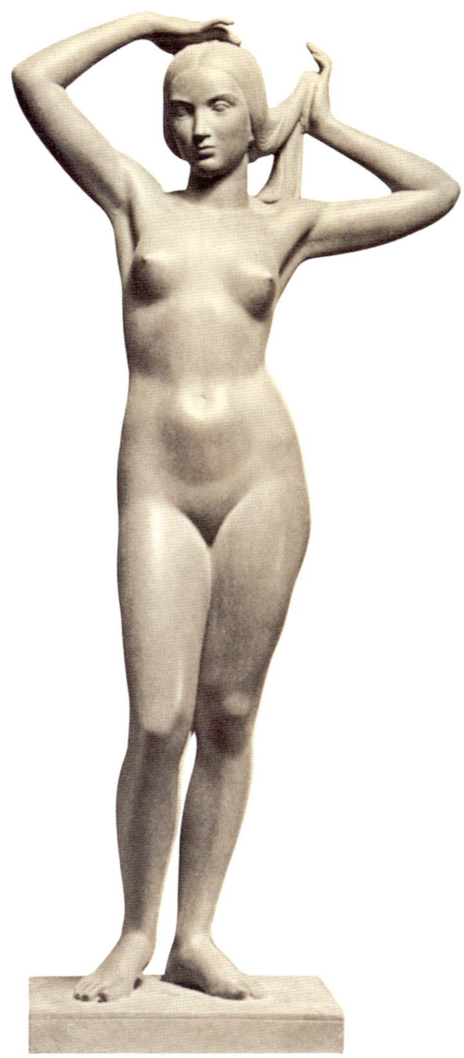

VI-178. "Dawn" by Joseph Kiselewski. 1933.

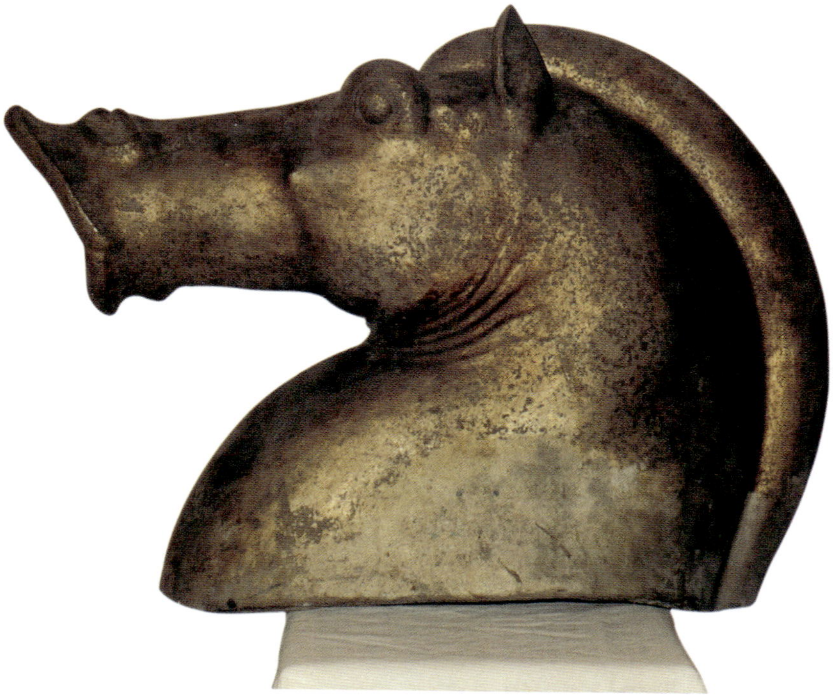

VI-175. "Sea Horse" by Joseph Kiselewski. c. 1937.

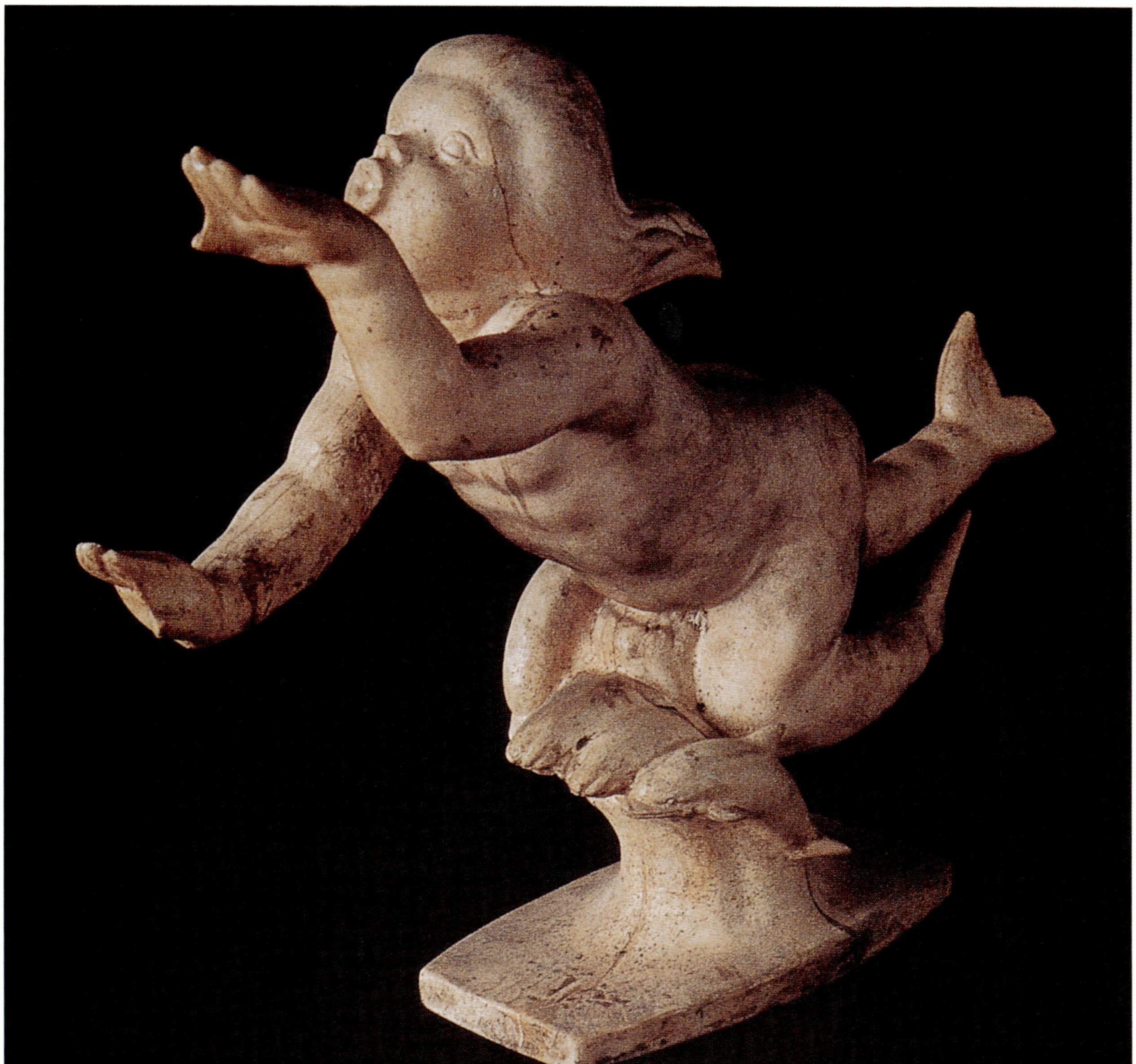

VI-176. "Water Baby Fountain" by Joseph Kiselewski.

ARCHIMEDES GIACOMANTONIO (JOCK MANTON) (1906-1962)

Archimedes Giacomantonio was a colorful figure in the world of sculpture; he produced portrait images of many of the most illustrious figures of his time, including Harry Truman, Dwight D. Eisenhower, and Martin Luther King. He also made monuments and memorial sculptures, including a standing figure of Christopher Columbus (VI-179), and ecclesiastical and Biblical sculptures, like his "Lot's Wife" (VI-180), as well as occasional nature studies (VI-181). But he was best known for his associations with, and portraits of famous people. He had a separate career in show business for much of his life. His portraits were uncannily lifelike and realistic, capturing the personality as well as the appearance of sitters like Pat McGlade (VI-182).

Giacomantonio, who adopted the name Jock Manton, was born in New Jersey after his father, a composer and conductor, had accompanied Giacomo Puccini from Italy to the United States and had decided to settle in America. By eighteen, the young sculptor had befriended the opera star Enrico Caruso who offered to sponsor him, had made a colossal head of Beethoven, and had received a scholarship to study in Rome. In Italy he worked with Vincenzo Gemito, of whom he made a portrait (VI-183).

On shipboard, returning to America, he met the movie actress Janet Gaynor, and a lifelong friendship brought about his introduction to other stars, and a number of commissions. During World War II, he produced anatomically correct statues for the training of medics and worked with a team of surgeons who reconstructed the faces of disfigured soldiers. His portrait of General Wainwright for West Point led to commissions to make likenesses of presidents Theodore Roosevelt (VI-184), Woodrow Wilson, Franklin Roosevelt, Dwight D.Eisenhower, and Harry Truman (VI-185). Giacomantonio had a long and close association with Roman Bronze Works.

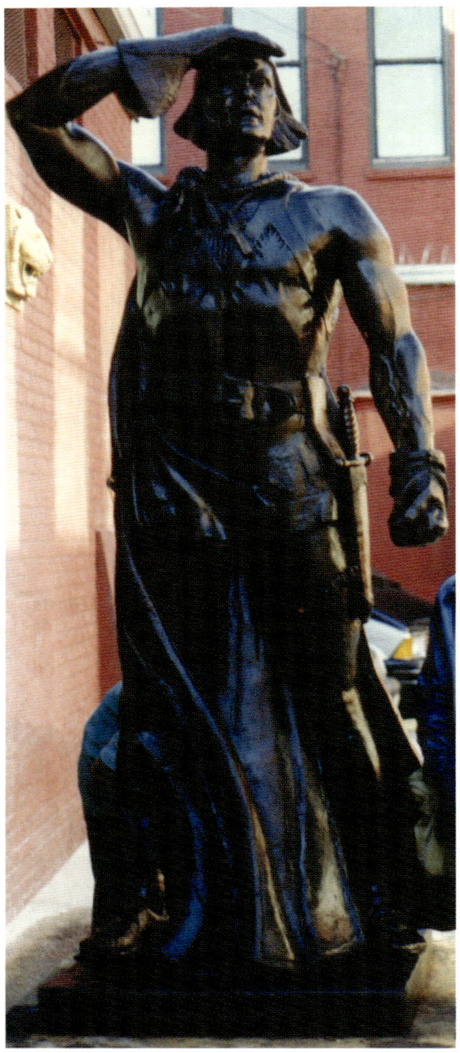

VI-179. "Christopher Columbus" by Archimedes Giacomantonio. 1952.

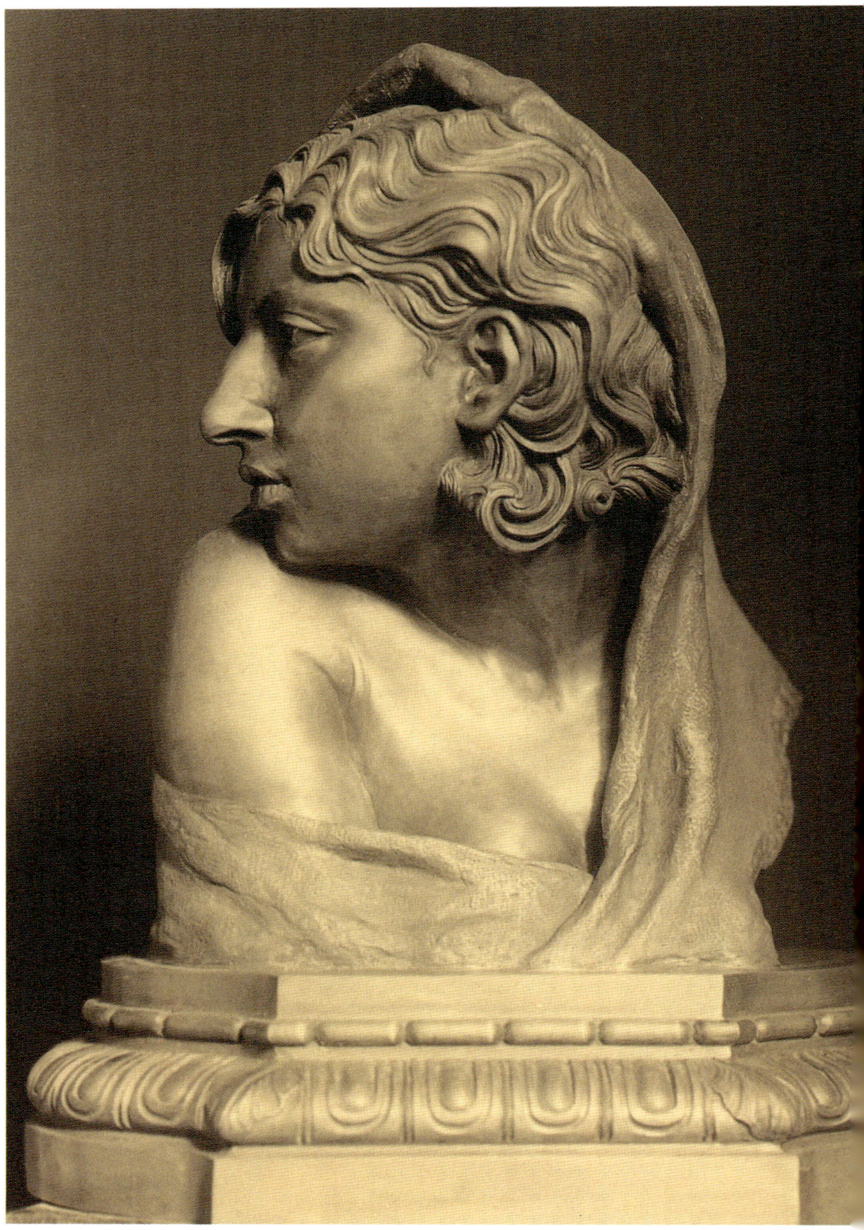

VI-180. "Lot's Wife" by Archimedes Giacomantonio.

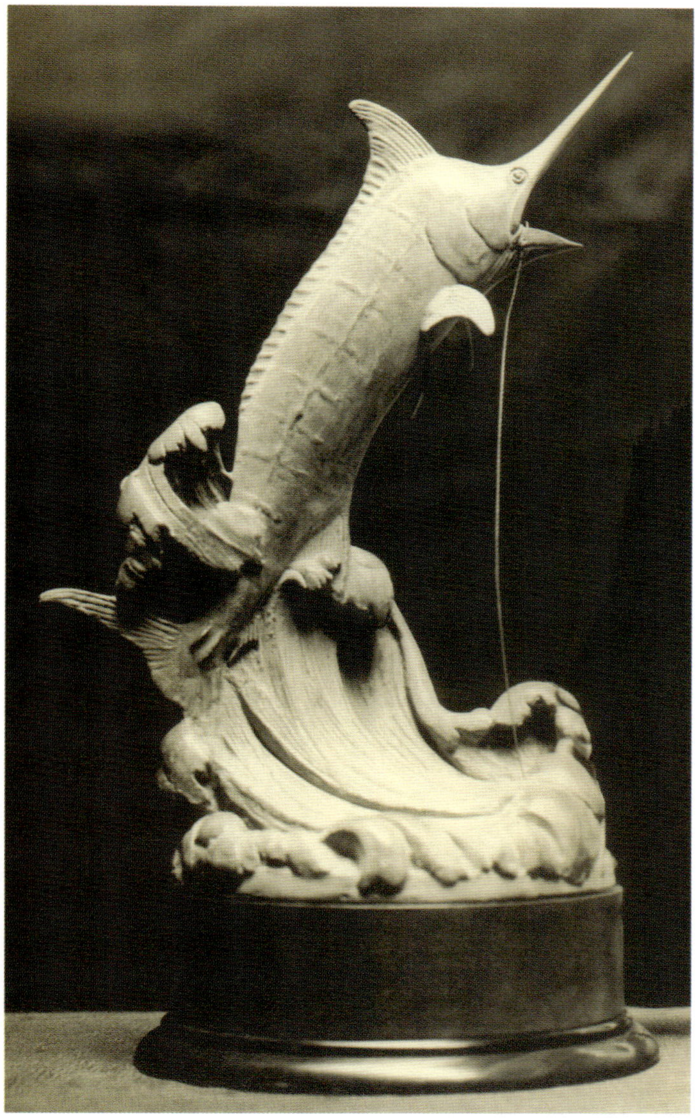

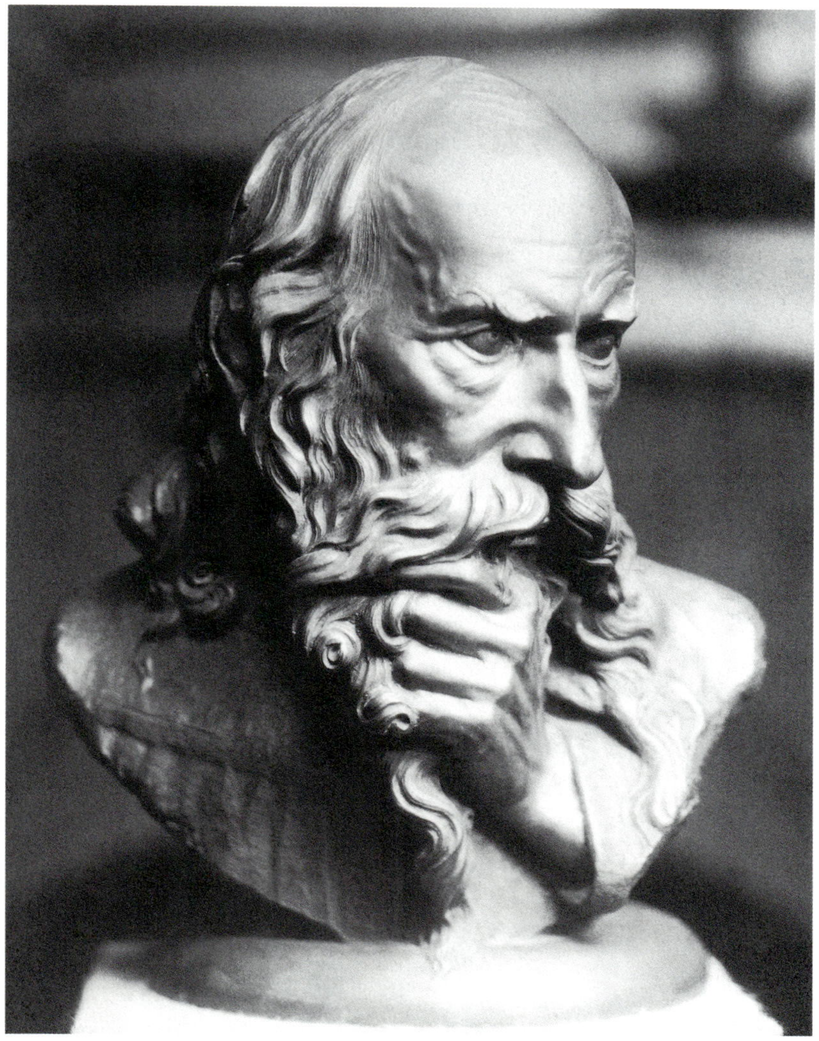

VI-183. "Bust of Vincenzo Gemito" by Archimedes Giacomantonio. 1949.

VI-181. "Fish Fountain" by Archimedes Giacomantonio.

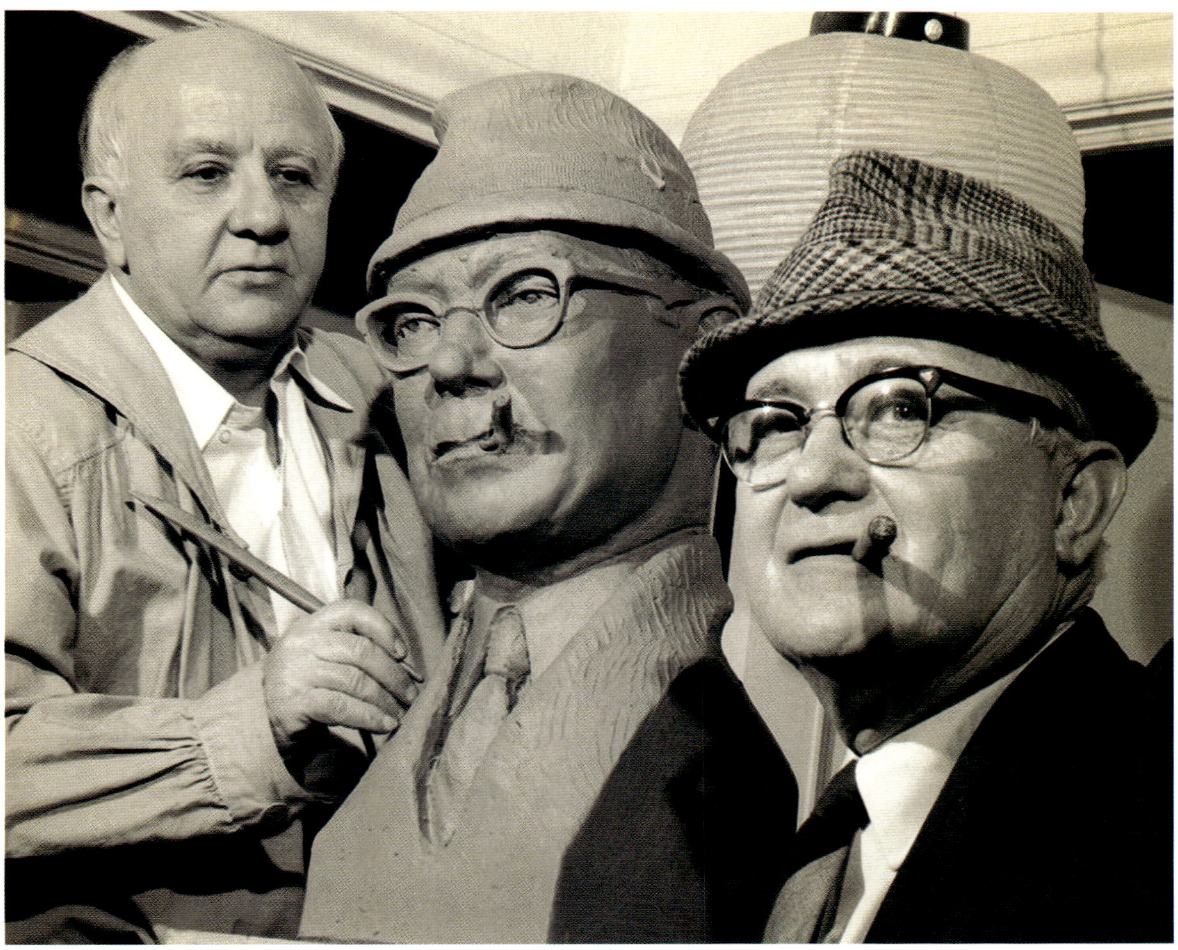

VI-182. "Pat McGlade" (with the sculptor) by Archimedes Giacomantonio.

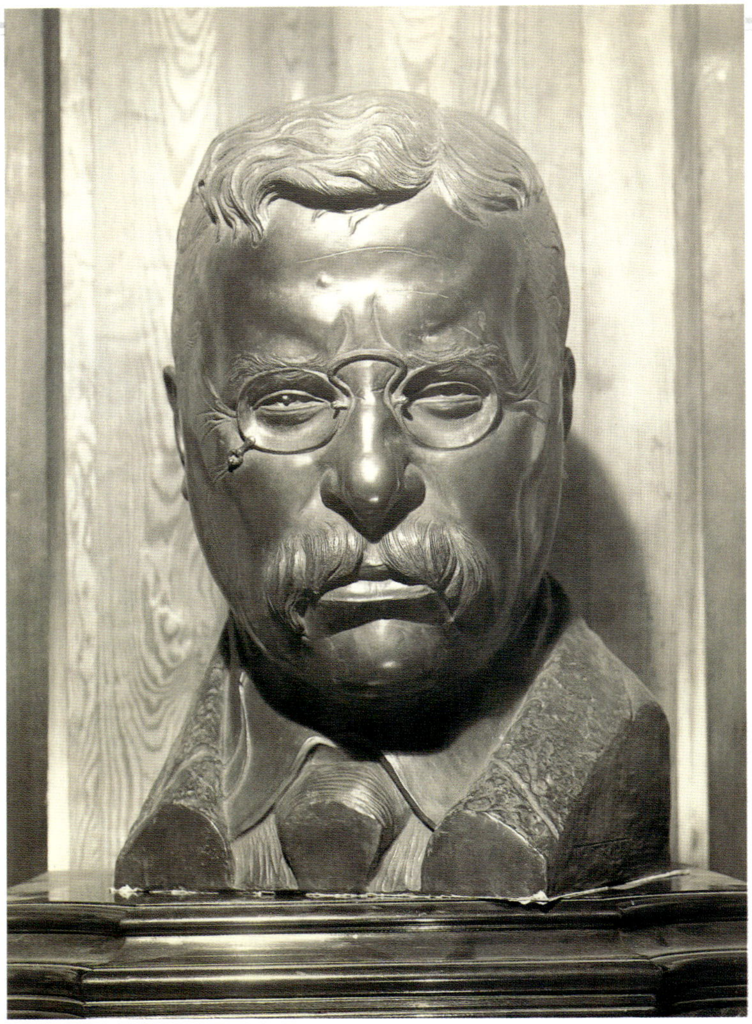

VI-184. "Theodore Roosevelt" by Archimedes Giacomantonio.

VI-185. "Harry S. Truman" (with the sculptor) by Archimedes Giacomantonio. 1957.

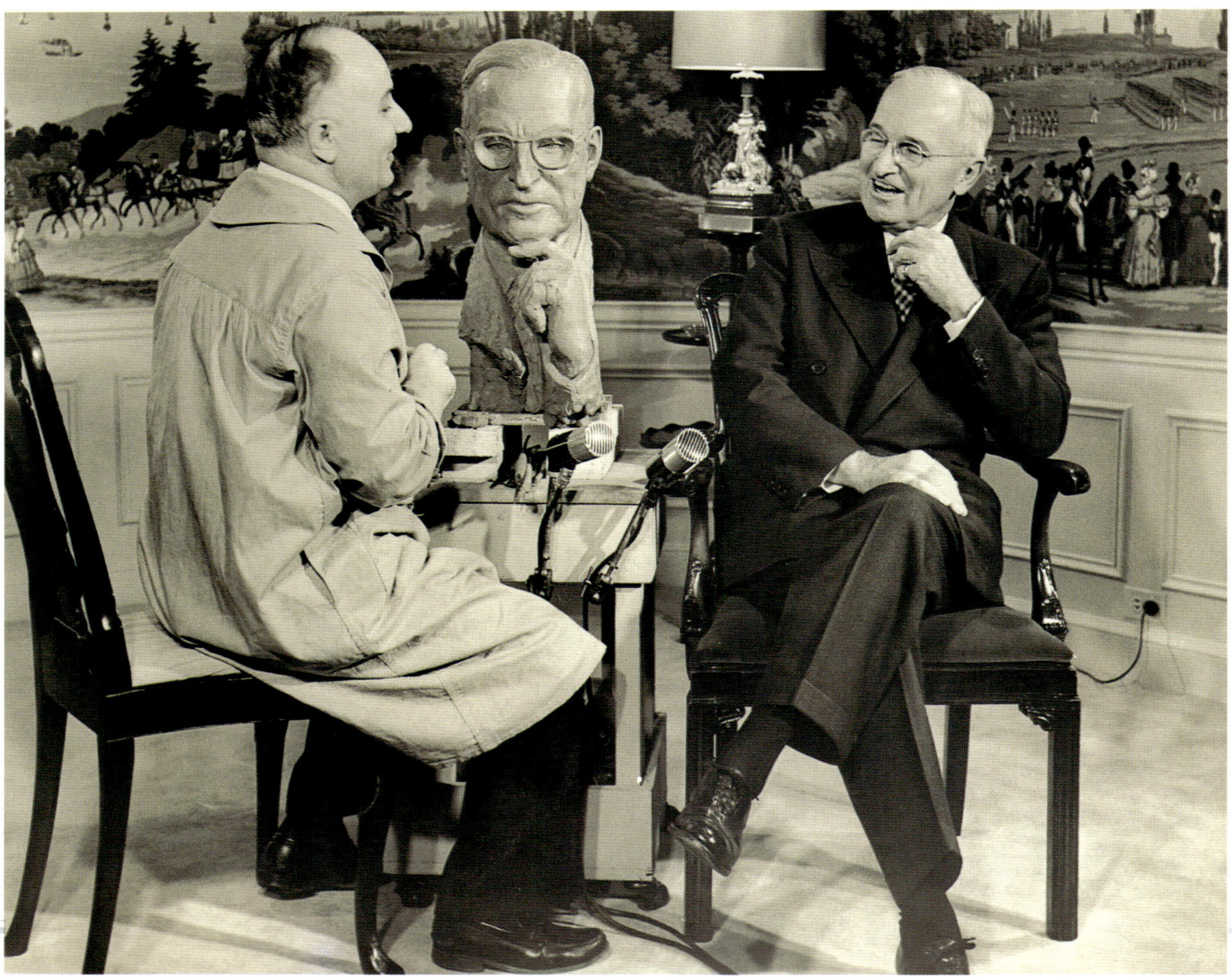

OTHER REALIST SCULPTORS:

CHESTER BEACH, EDWARD BERGE, GUSTAV BOHLAND, JOSEPH ARTHUR COLETTI, POMPEO COPPINI, HUNT DIEDERICH, ABSTENIA ST. LEGER EBERLE, MAX KALISH, NATHANIEL KAZ, LEO LENTELLI, HENRY LINDER, EVELYN BEATRICE LONGMAN, BORIS LOVETT-LORSKI, ELEANOR MELLON, ATTILIO PICCIRILLI, JOSEPH POLLIA, BRENDA PUTNAM

Among the sculptors who worked in various styles of realism but cast only a few of their pieces at Roman Bronze Works were:

CHESTER BEACH (1881-1956)

Chester Beach was a native of San Francisco who sculpted in several different types of media; his work included ivory statuettes, bronze monuments, medallic art, and marble commemoratives. His style was equally varied, ranging from symbolic nudes to realistic portrait busts and statues of working men, to impressionistic images. He was a president of the National Sculpture Society, and a member of the National Academy of Design. He used Roman Bronze Works as his foundry between 1909, when he cast "Long Hair" (VI-186), and 1931.

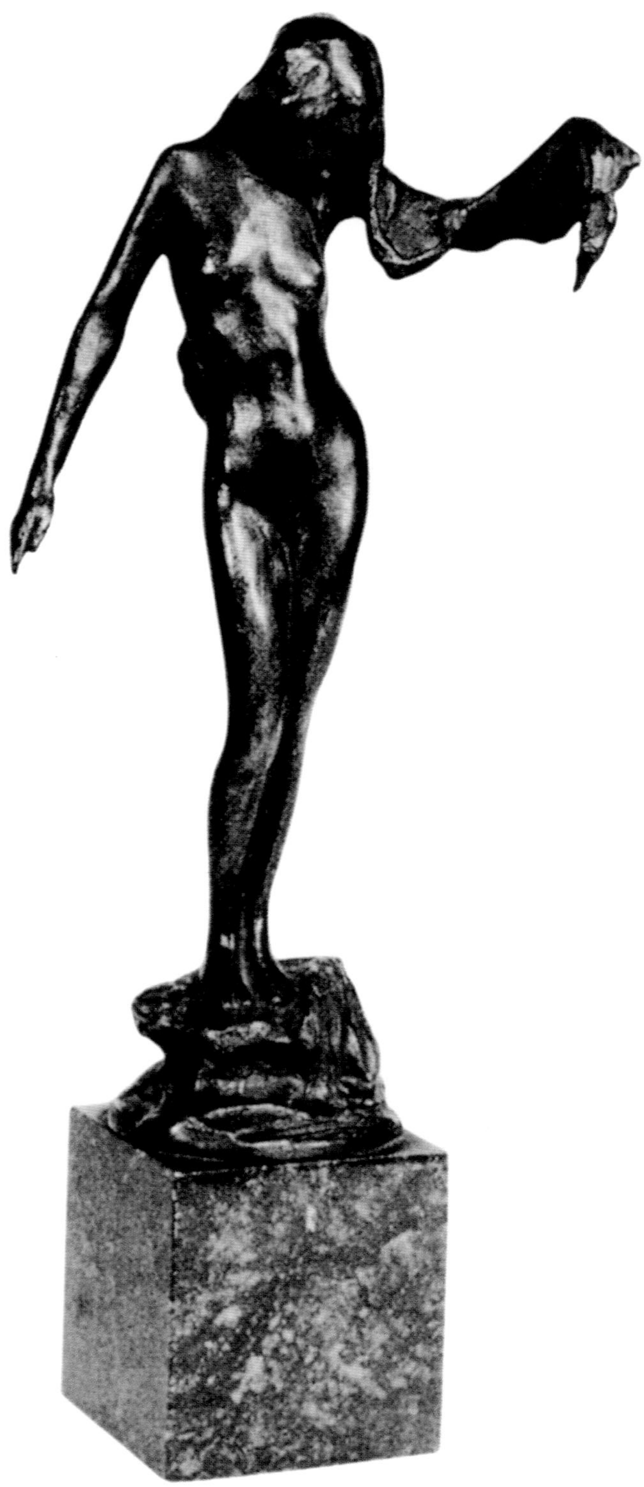

VI-186. "Long Hair" by Chester Beach. 1909.

EDWARD BERGE (1876-1924)

Edward Berge was born in Baltimore and studied at the Maryland Institute of Art before leaving for Paris. There he worked with Verlet and Rodin. He specialized in medallions and plaques, as well as figure sculpture and garden statuary. Among the works he cast at R.B.W. was a neo-classical nude figure (VI-187). But Berge also favored a certain whimsicality, as evidenced in his fountain figures and pieces with small, cute children: "Duck Mother" (VI-188), "Wildflower" (VI-189), and "Fishing Boy" (VI-190). R.B.W. files include Berge commissions beginning in 1914; several works were cast posthumously up until 1978.

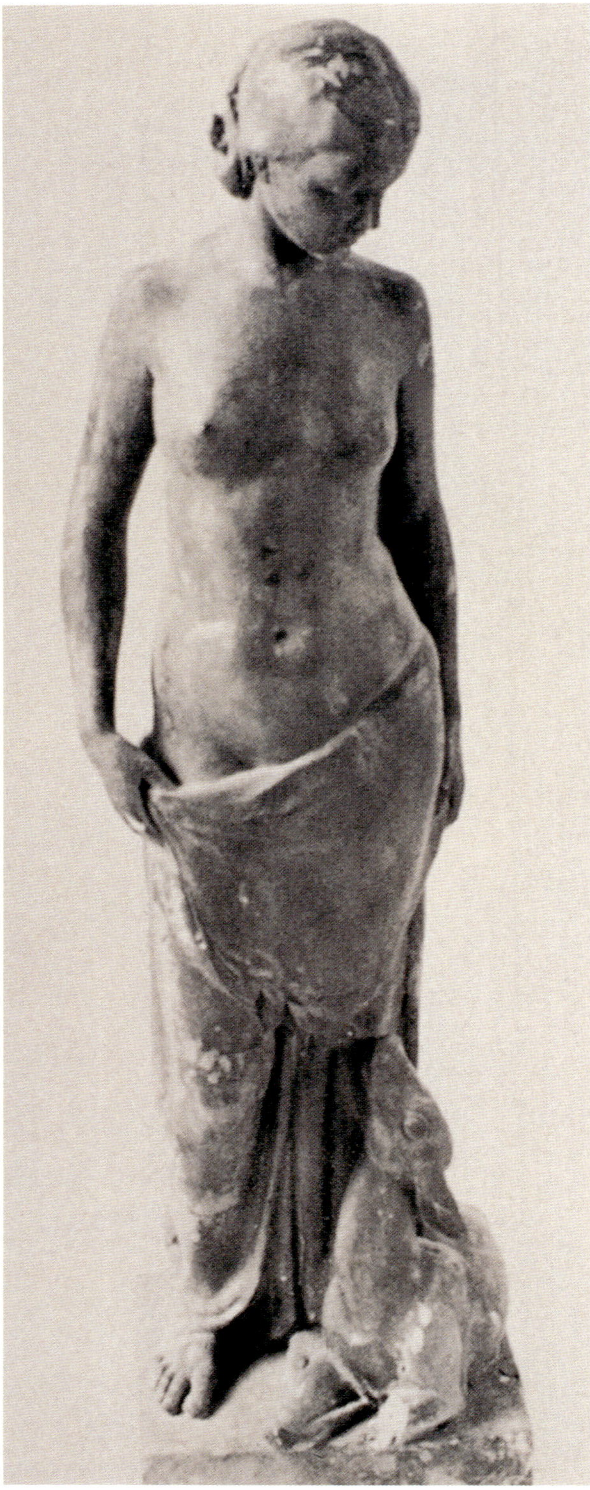

VI-187. "Serenity" by Edward Berge. 1955-56.

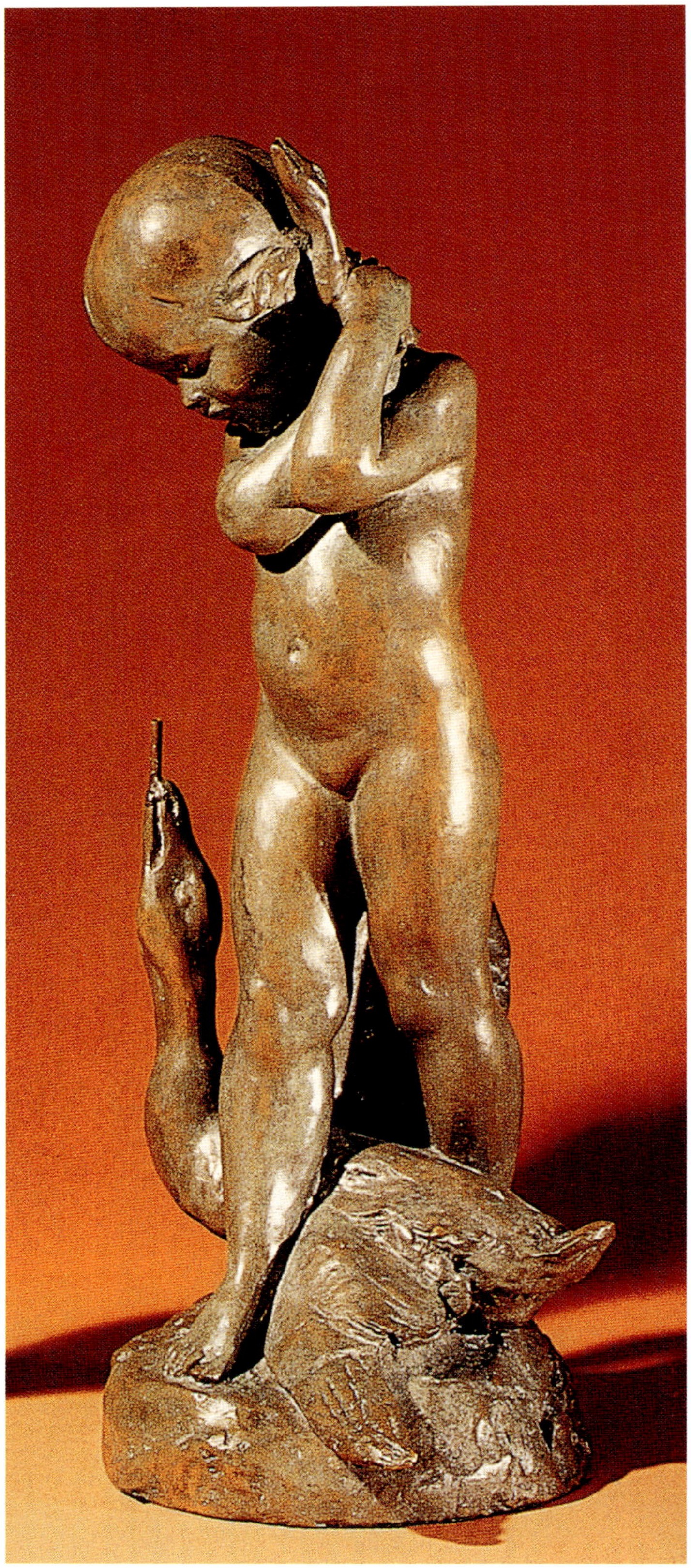

VI-188. "Duck Mother" (fountain figure, plaster cast) by Edward Berge. 1929.

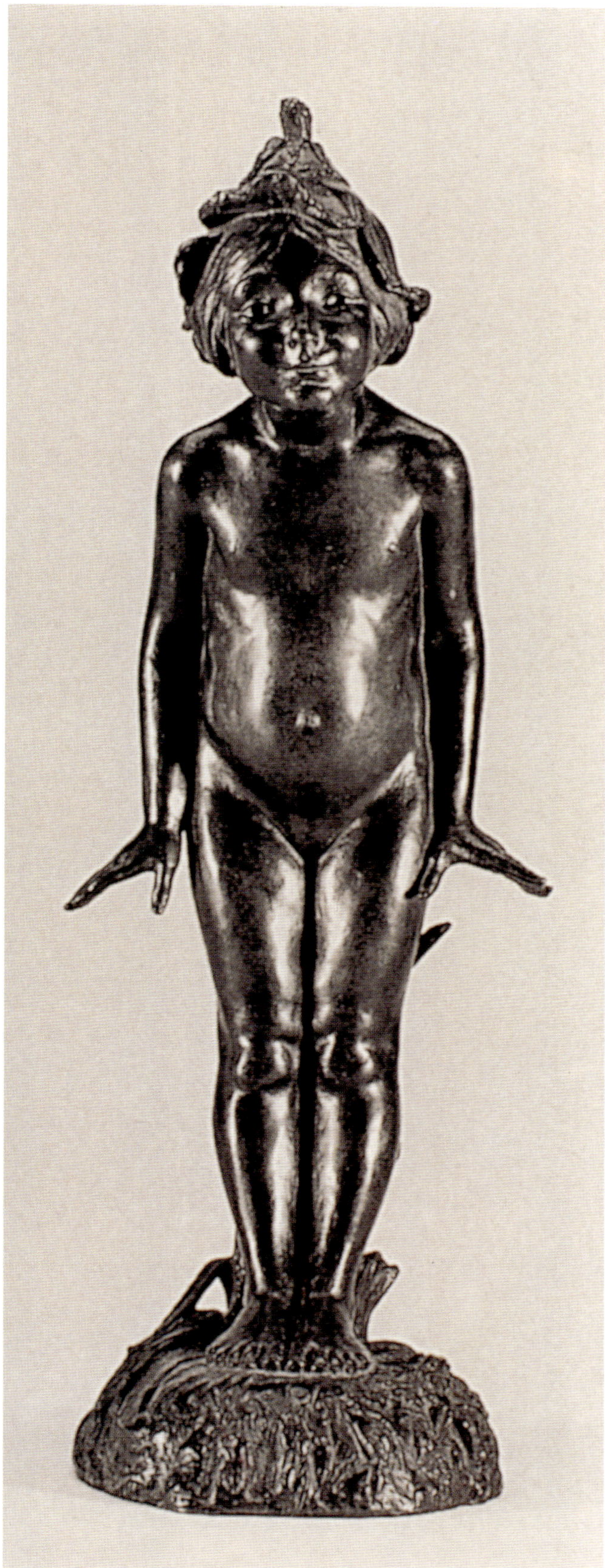

VI-189. "Wildflower" by Edward Berge. 1915.

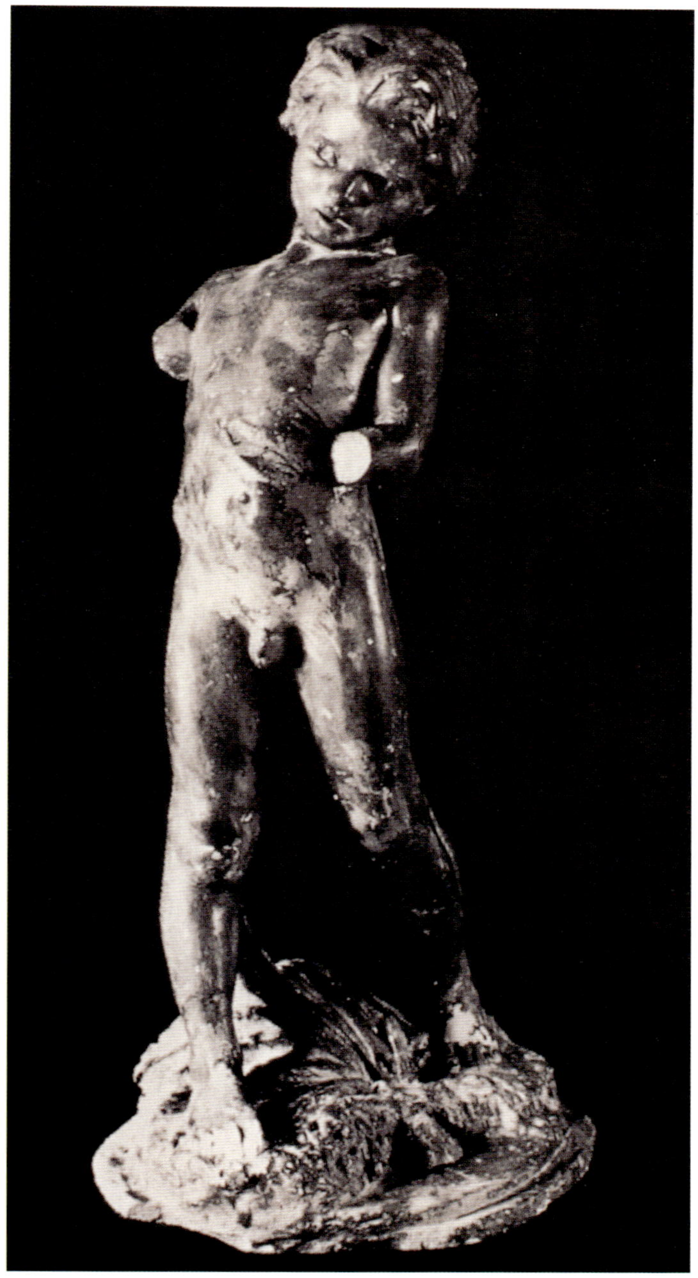

VI-190. "Fishing Boy" (plaster cast) by Edward Berge.

GUSTAV BOHLAND (1897-1959)

Gustav Bohland, a native of Austria, made a number of public monuments, in addition to smaller bronzes that suggested a taste for the symbolic and exotic. Among the latter, cast at Roman Bronze Works between the 1930s and 1950s, were his "Dawn of an Era" (VI-191) and his "Mask Dancer" (VI-192).

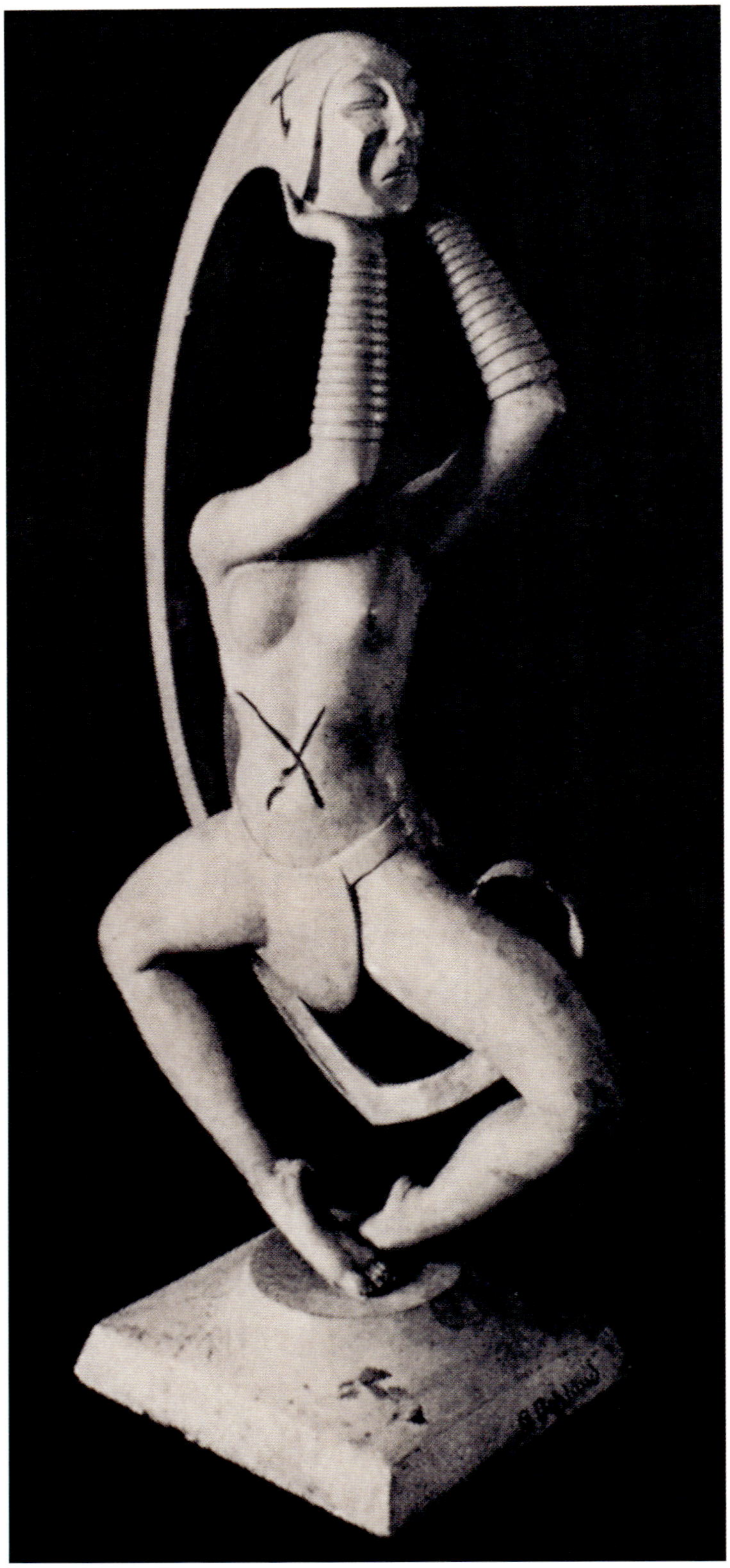

VI-191. "Dawn of an Era" by Gustav Bohland.

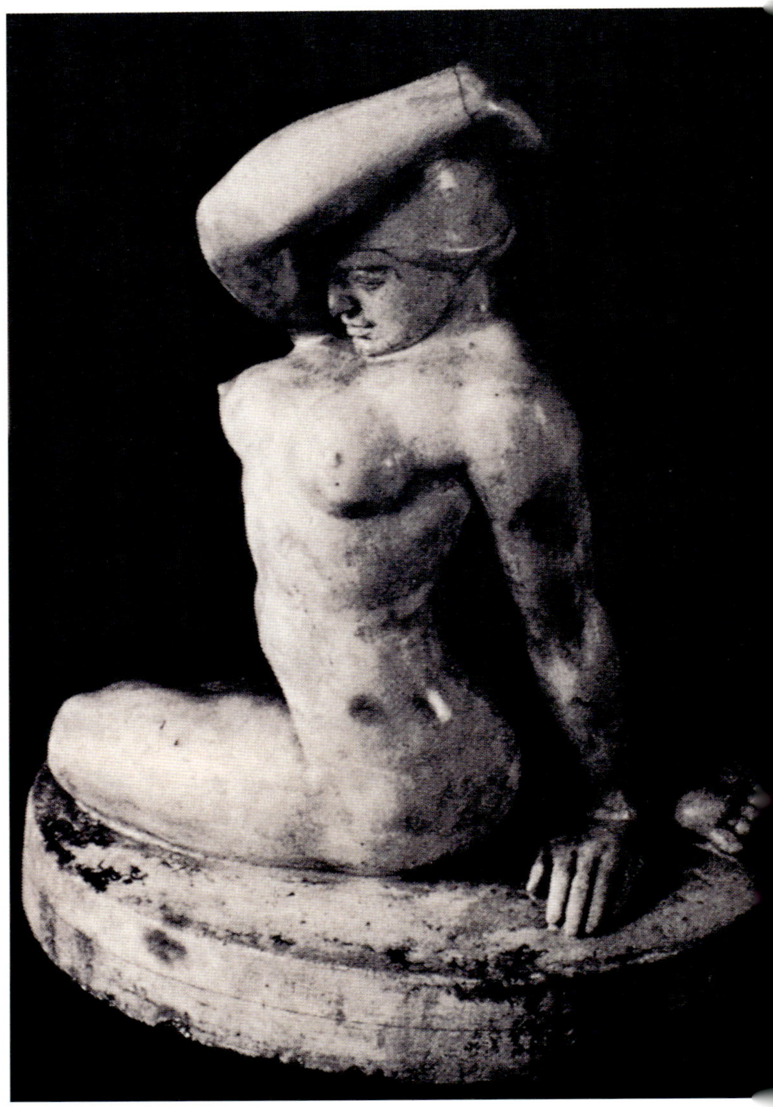

VI-192. "Mask Dancer" by Gustav Bohland. 1933.

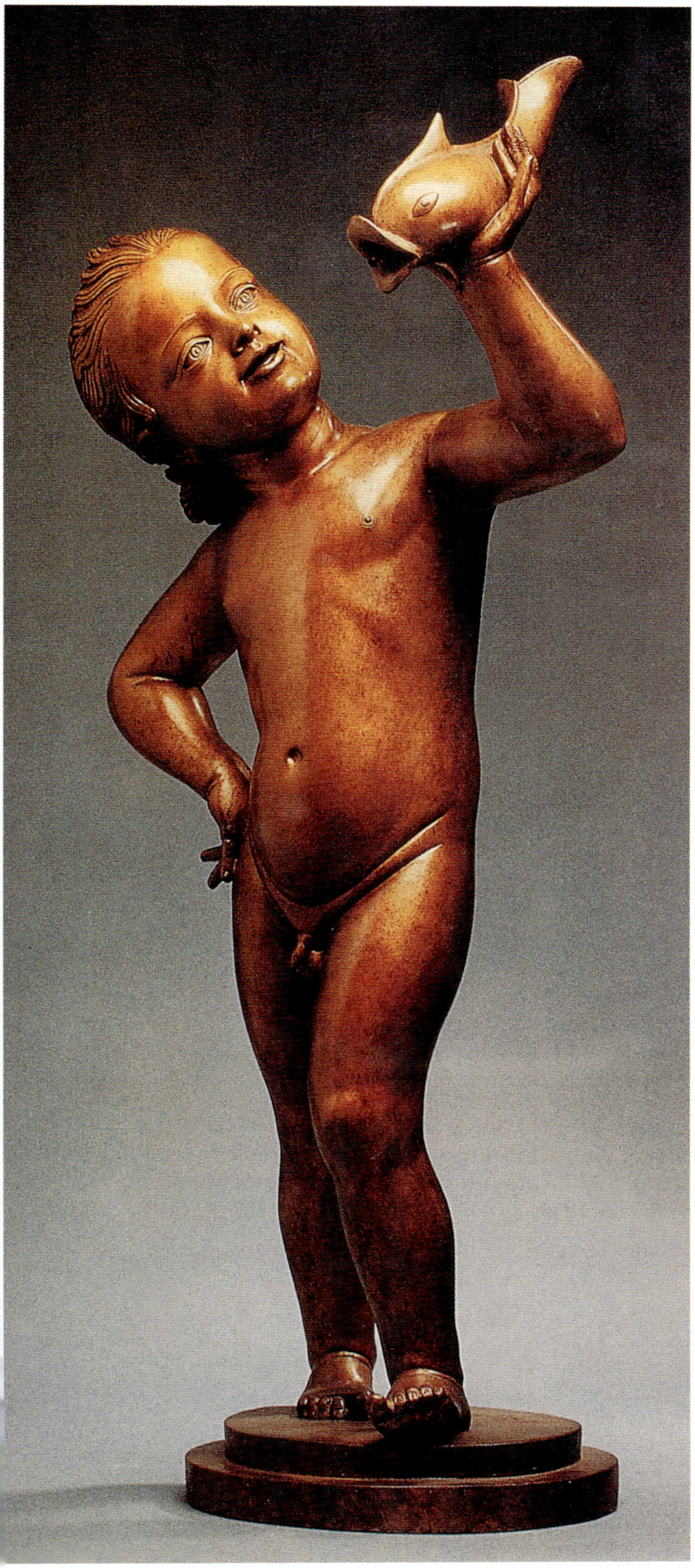

VI-193. "Putto with Dolphin" (fountain figure) by Joseph Arthur Coletti. 1931.

JOSEPH ARTHUR COLETTI (1898-1973)

Joseph Arthur Coletti, who was born in Italy, was educated at Harvard University and was a student of John Singer Sargent; he helped Sargent with the ceiling for the Boston Public Library. A specialist in architectural sculpture, Coletti won a number of awards, as well as a commission from Harvard University. As a sculptor of statuary he produced ecclesiastical works, portrait busts, medals, fountain figures like "Putto with Dolphin" (VI-193) and several animal pieces, including "Klipspringer" (VI-194), which was modeled in 1929 and cast at R.B.W.

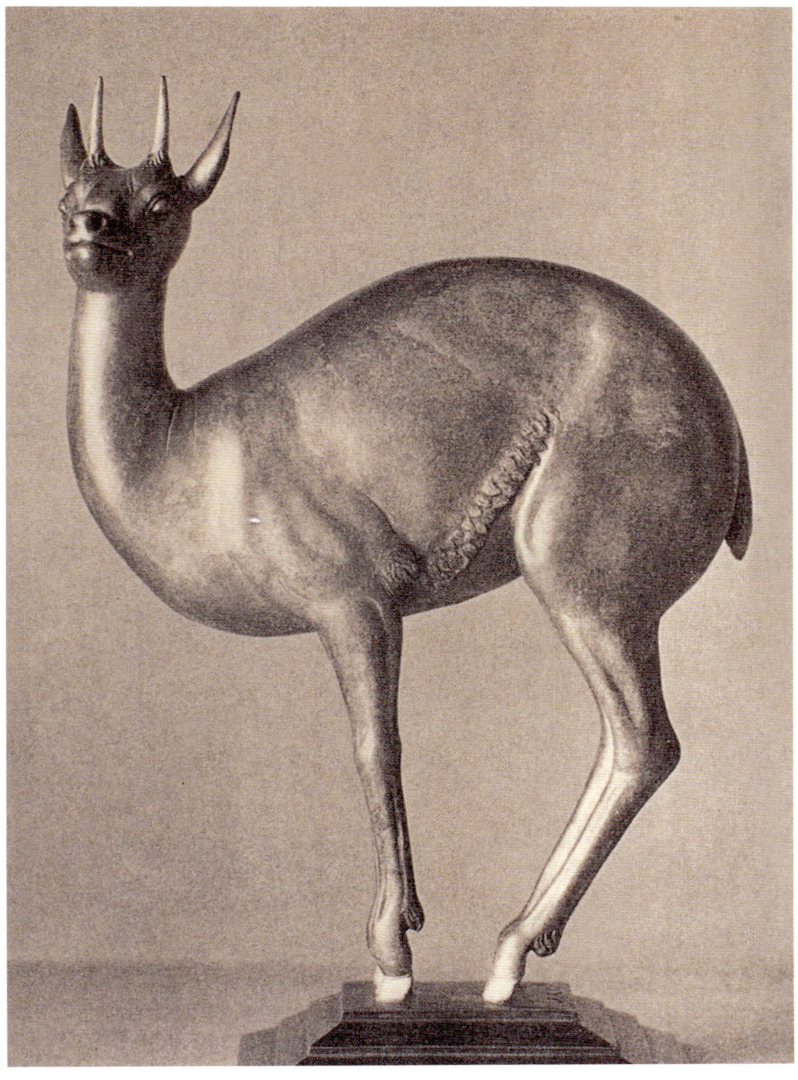

VI-194. "Klipspringer" by Joseph Arthur Coletti. 1929.

POMPEO COPPINI
(1870-1957)

Pompeo Coppini was a sculptor and painter born in Italy and educated at the Academia de Belle Arti in Florence. After arriving in America in 1896, he went on to create thirty-six public monuments in the United States, as well as many portrait statues and a number of painted portraits. One of his monuments was the many-faceted Littlefield, Texas War Memorial, for which Roman Bronze Works cast many separate sections (numbers VI-195 to VI-198). He also cast an "Indian on a Horse" and a "Narcissa" at R.B.W.

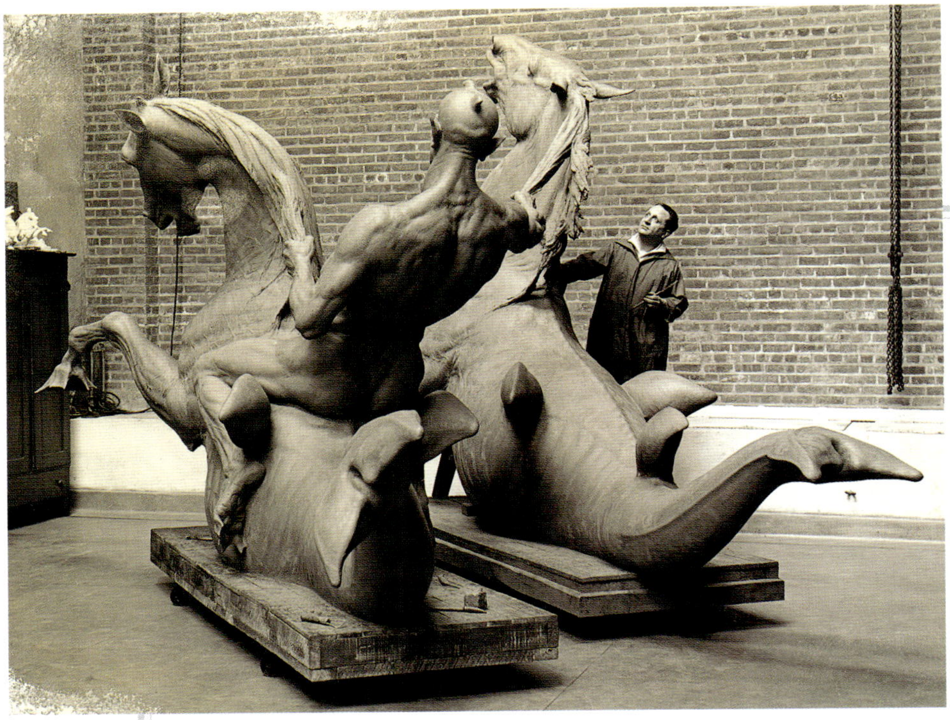

VI-195 to VI-198. "Texas War Memorial," Littlefield, Texas, (details) by Pompeo Coppini. Parts cast beginning in 1910. Completed 1918.

VI-195.

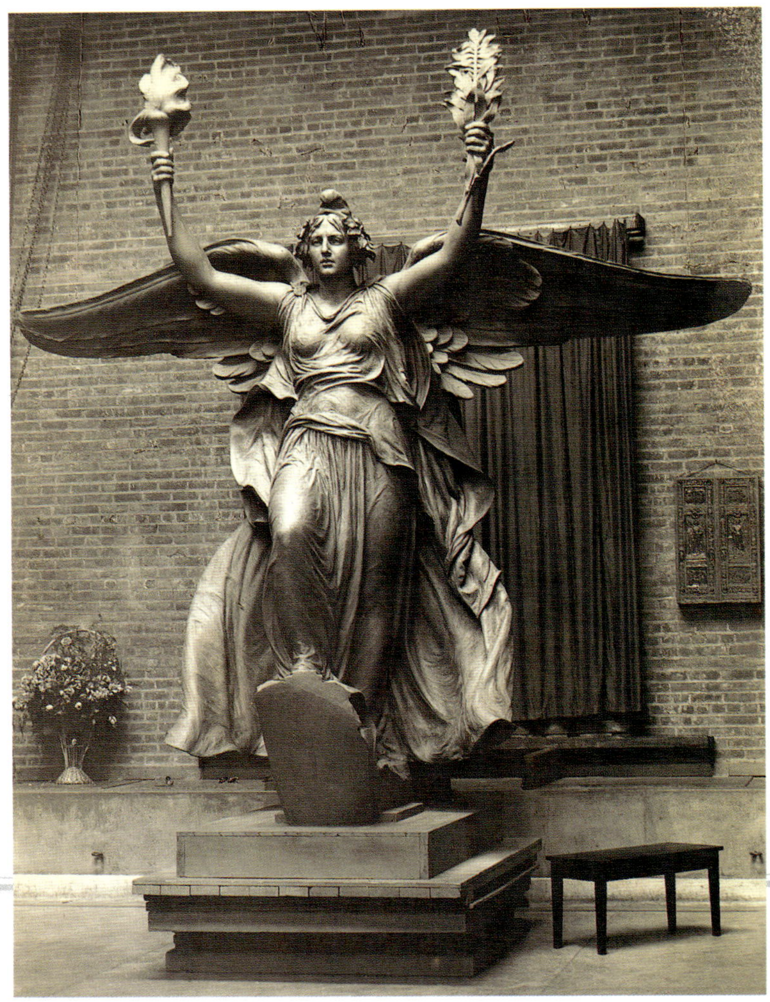

VI-196.

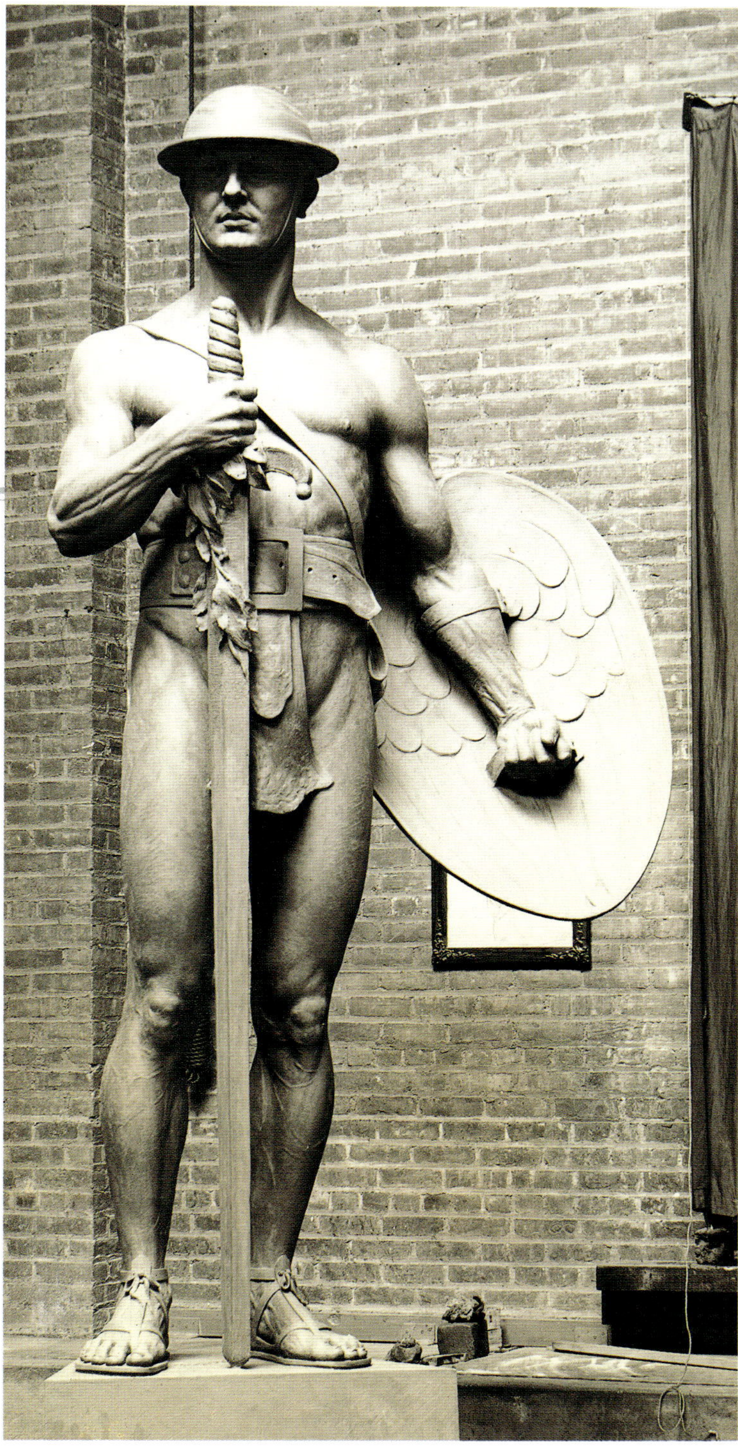

VI-197.

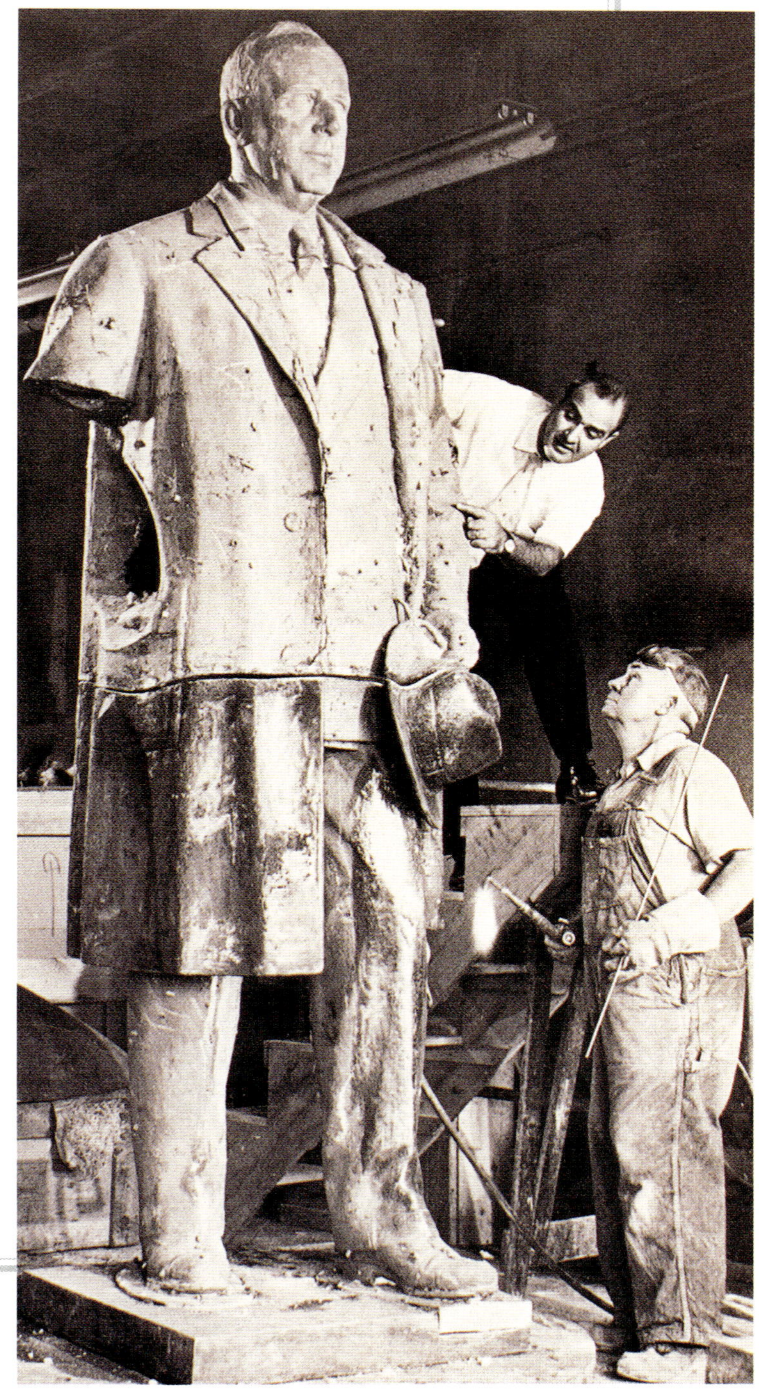

VI-198.

HUNT DIEDERICH (1884-1953)

Hunt Diederich, who was born in Hungary, was a student of Paul Manship. For many years he lived in France and traveled widely. He is best known for his wrought iron sculptures and his silhouettes, but he also made a number of bronzes, including several with animal themes, such as "Two Cats" (VI-199).

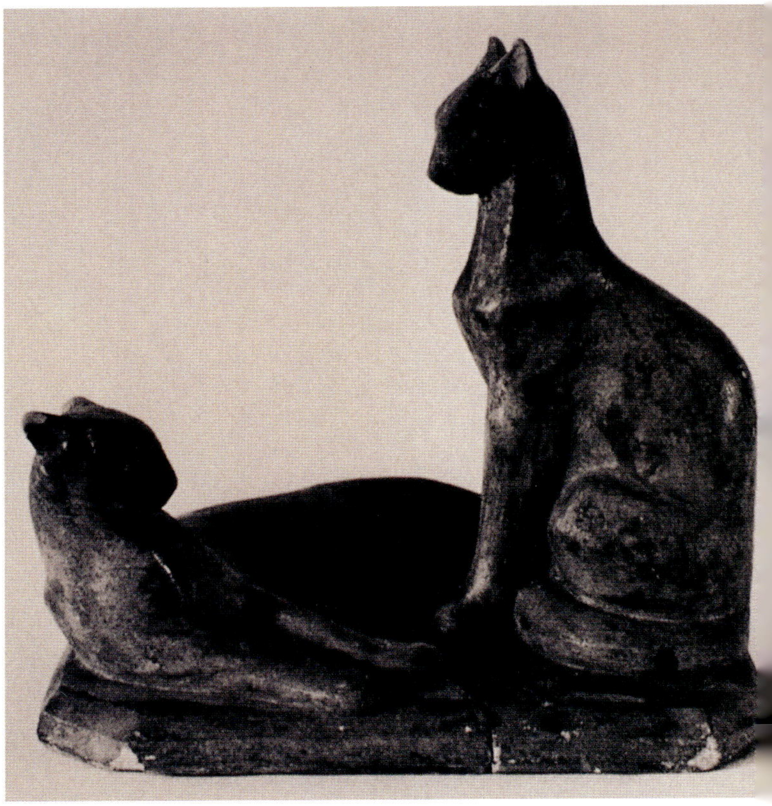

VI-199. "Two Cats" by Hunt Diederich.

ABSTENIA ST. LEGER EBERLE (1878-1942)

Abstenia St. Leger Eberle was a genre artist who specialized in sculptural portraits of people living on the Lower East Side of New York City. She combined a careful naturalism with an interest in making art with social significance. Her images of immigrants and their children in the teeming streets of the city had a rhythmic intensity; of particular note was "The Windy Doorstep" (VI-200) which was modeled in 1910 and later recast by Roman Bronze Works. She also cast a "Salome" at R.B.W. in 1908.

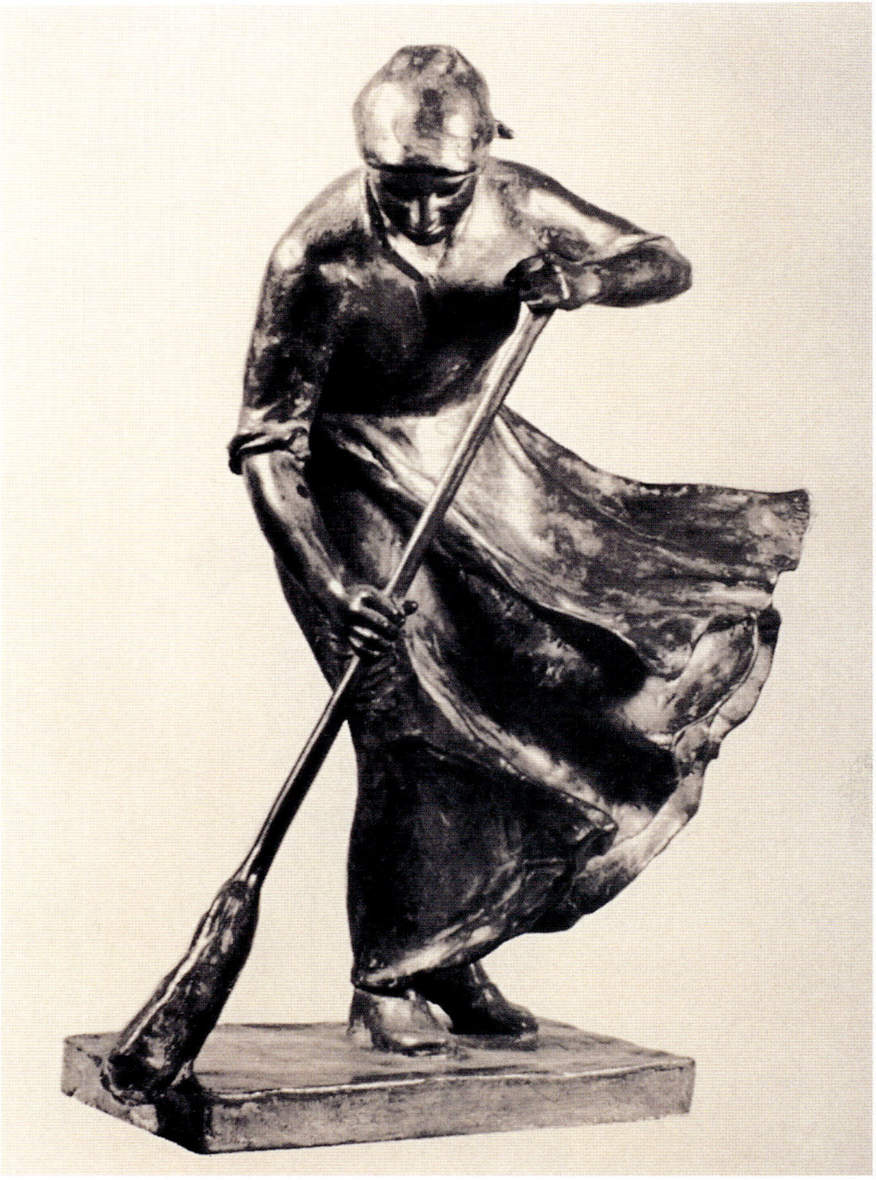

VI-200. "The Windy Doorstep" by Abstenia St. Leger Eberle. Modeled 1910.

MAX KALISH
(1891-1945)

Max Kalish was a specialist in genre sculpture that depicted the American worker, Max Kalish was a part of the socially conscious art movement known as Social Realism. His sculptures of workers, like those of Mahonri Young, emphasized the dignity of work; his subjects were laborers in the industrial age, like his "Steel Into The Sky" (VI-201). Kalish was quoted as saying: "In action, an American worker is naturally graceful and makes a subject that even the Greeks might have envied." Kalish cast only a few works at R.B.W. beginning in 1930.

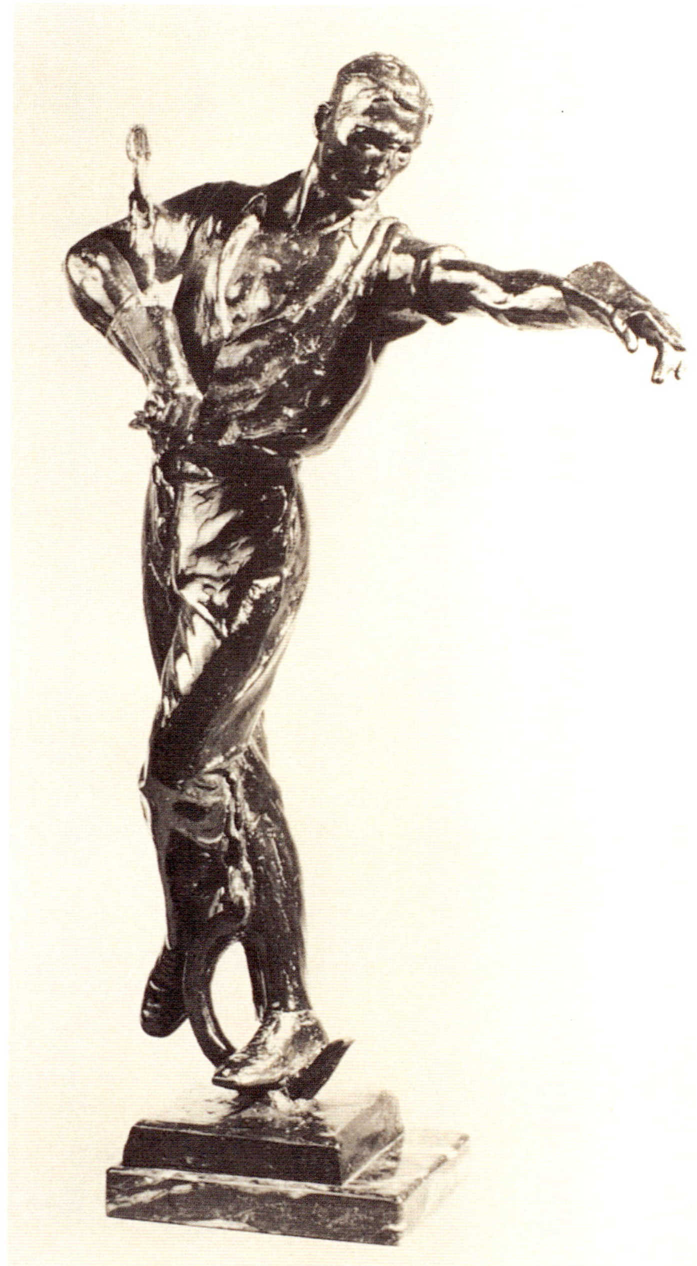

VI-201"Steel Into the Sky" by Max Kalish.

NATHANIEL KAZ
(1917-)

Nathaniel Kaz was born in New York and studied at the Art Student's League. He worked in a variety of media, including wood, limestone and bronze, and was a longtime instructor of sculpture at the Art Student's League. His "Danse Espagnole" (VI-202) was cast at R.B.W. in 1952. Kaz also cast images of "Cyrano" and a guitar player there.

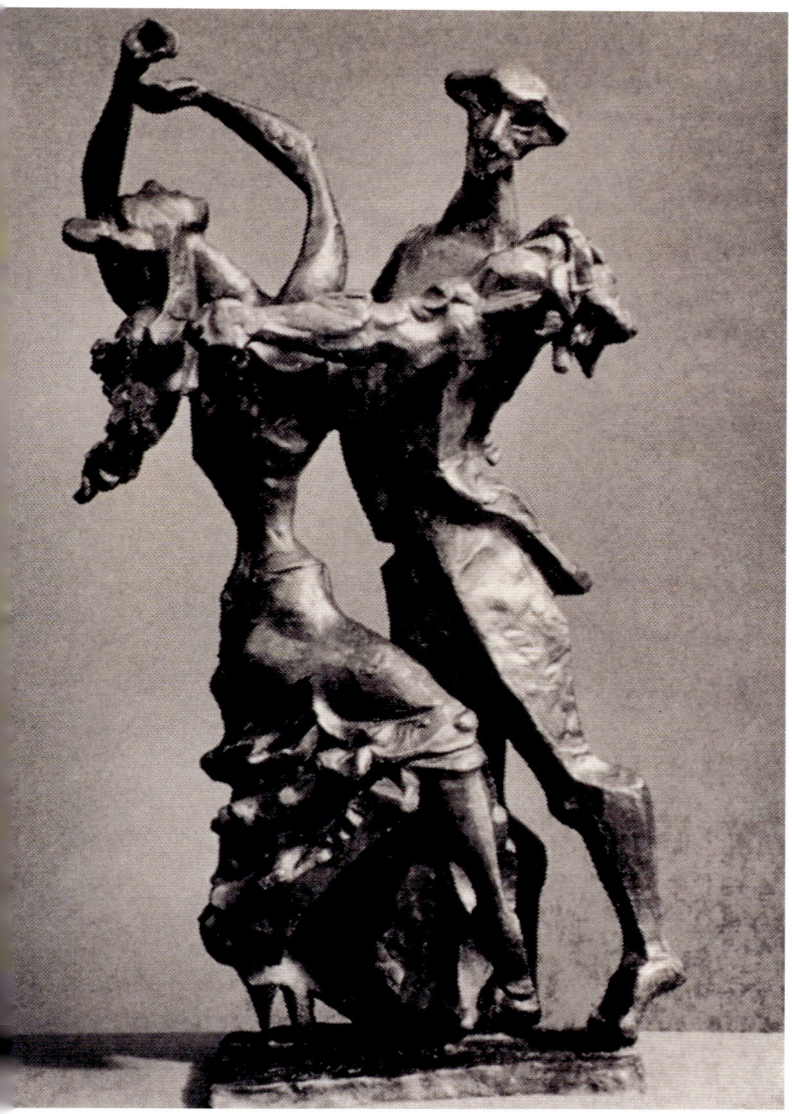

VI-202. "Danse Espagnole" by Nathaniel Kaz. 1952.

LEO LENTELLI
(1879-1961)

Leo Lentelli was born in Bologna, Italy. He was best known for his carvings at the Cathedral of St. John the Divine in New York. He was an assistant to several notable American sculptors after his arrival in the United States in 1903; with Alexander Stirling Calder he developed his own decorative style. He produced a number of ornamental sculptures for architecture, and several memorial sculptures, including an equestrian of General Lee, a bust of Jefferson Davis, and a statue of "James Cardinal Gibbons" (VI-203), which was cast at R.B.W. in 1932.

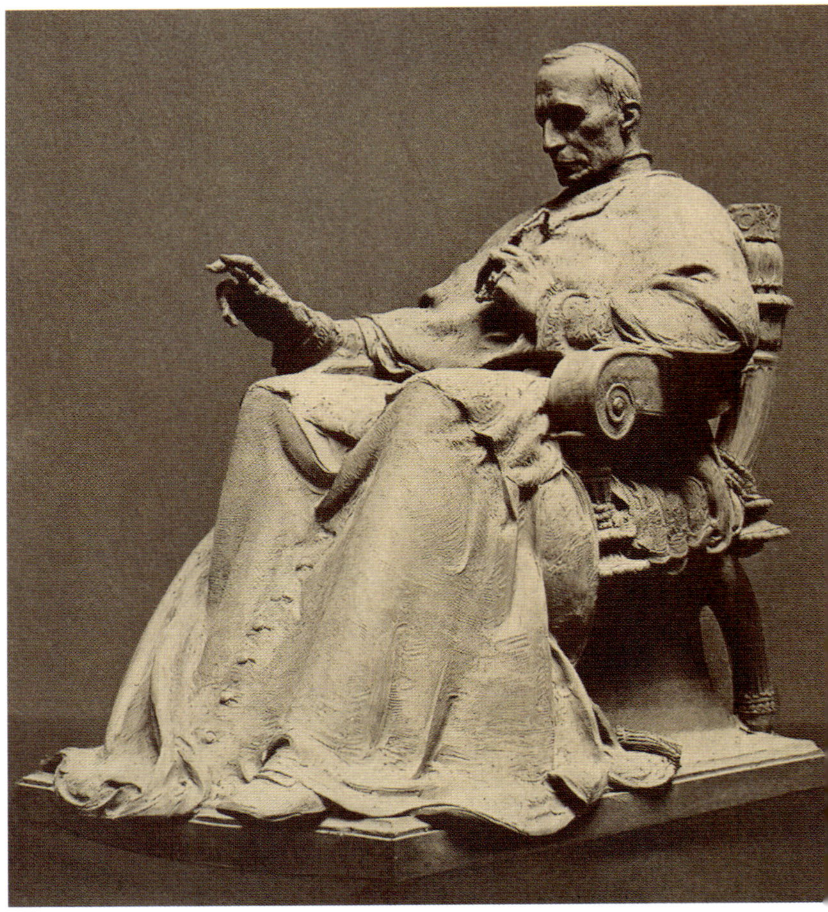

VI-203. "James Cardinal Gibbons" by Leo Lentelli.1931.

HENRY LINDER
(1854-1910)

Henry Linder, a native of Brooklyn, New York, was trained in Munich and Rome. He was known for creating and designing ornamental, decorative works such as andirons (VI-204 and 205) and silverware, as well as several commemorative statues and allegorical works.

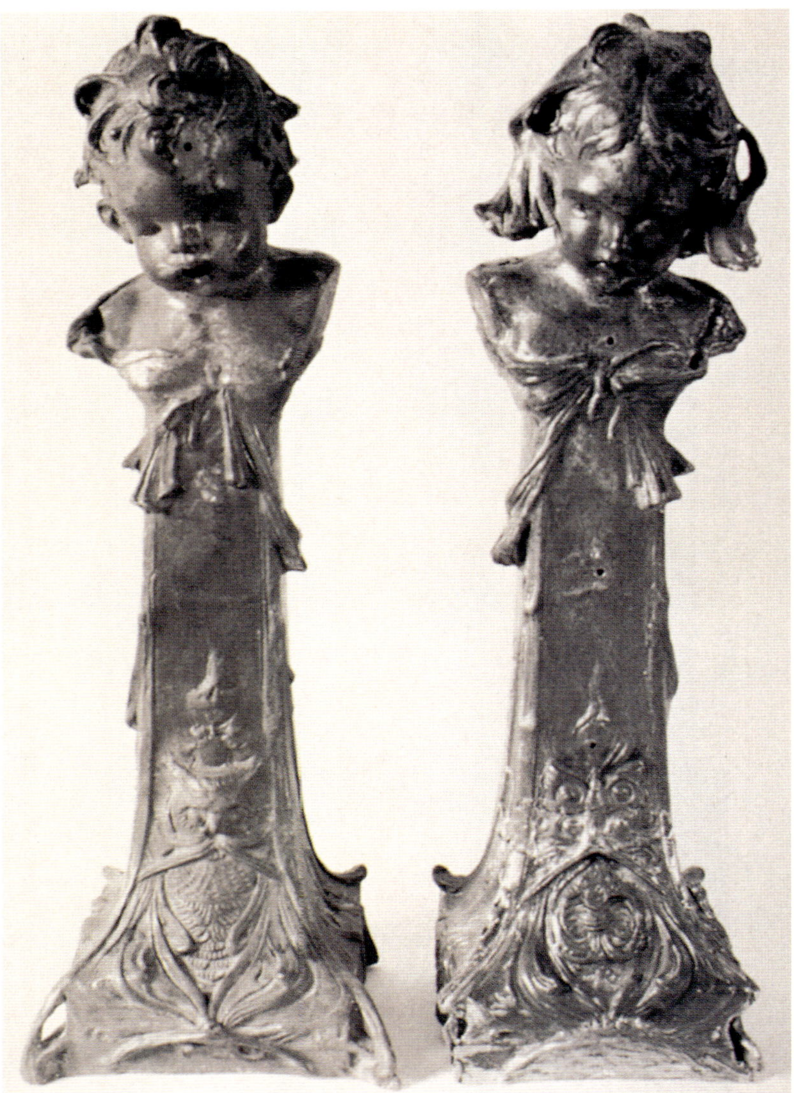

Left:
VI-204 "Pair of Andirons in Art Nouveau Style" by Henry Linder. Modeled 1896.

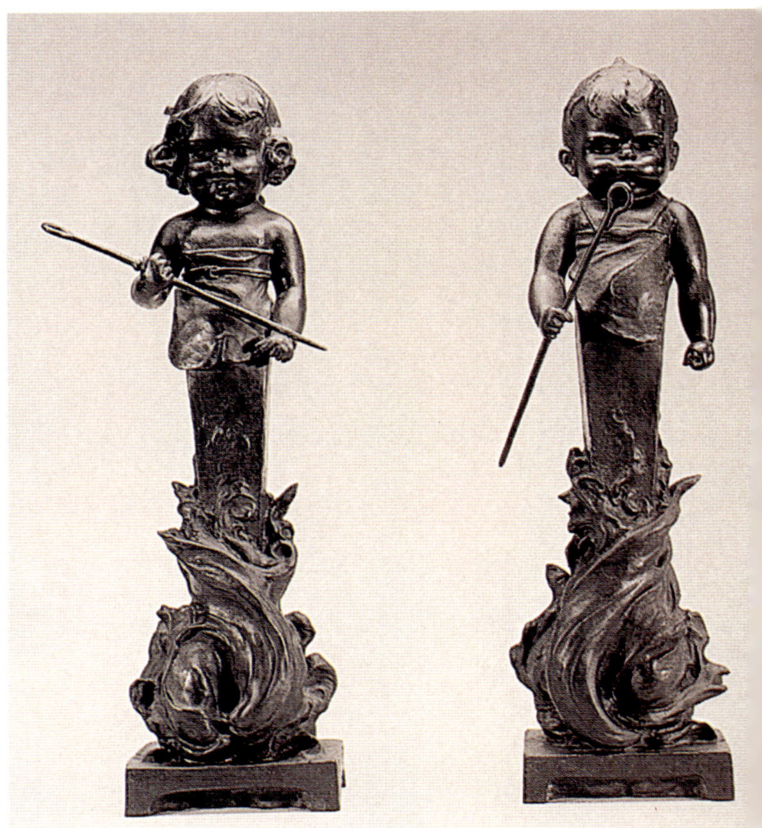

VI-205. Pair of Bronze Andirons of "A Boy and a Girl" by Henry Linder. Modeled 1896.

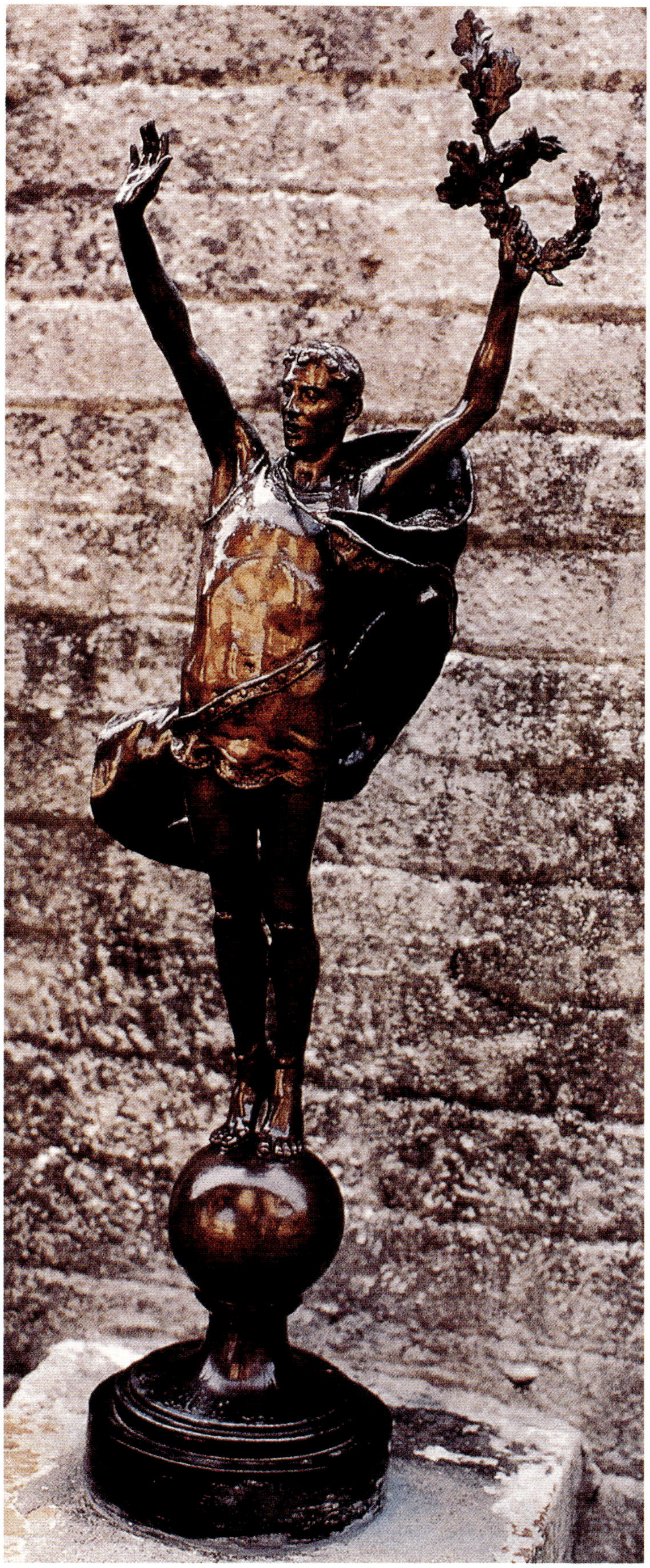

VI-206. "Victory" by Evelyn Beatrice Longman. 1903.

EVELYN BEATRICE LONGMAN (1874-1954)

Evelyn Beatrice Longman was the only woman assistant to Daniel Chester French, and the first women elected a full member of the National Academy of Design. She had a long and illustrious career. A student of Lorado Taft and an assistant to Hermon MacNeil and Isidore Konti for the Pan American Exposition in 1901, Longman went on to work in collaboration with the architect Henry Bacon in a classical, allegorical style, and to produce many public monuments. Her most notable achievements included her "Victory" (VI-206) for the Saint Louis Exposition, and her "Spirit of Electricity" (VI-207) to surmount the American Telephone and Telegraph Building.

VI-207. "The Spirit of Electricity" by Evelyn Beatrice Longman. Based on "Victory" this statue was installed atop the A.T.and T. building in 1916.

BORIS LOVETT-LORSKI (1894-1973)

Boris Lovett-Lorski, who was born in Lithuania and trained in St. Petersburg, had a successful career both in America and in France, where he was chosen to create a bronze head of Charles DeGaulle. He was also a mosaicist. He specialized in busts of statesmen and of other artists, and in genre and allegorical figures, such as his 1920-1930 "Catman Doing a Split" (VI-208), cast at Roman Bronze Works.

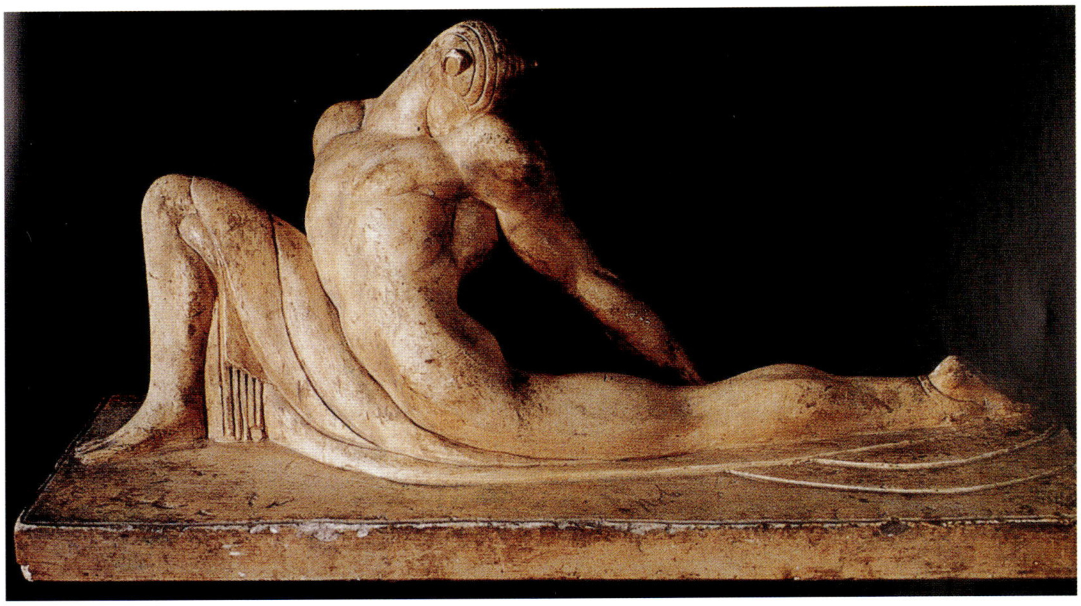

VI-208. "Catman Doing a Split" by Boris Lovett-Lorski. 1920-30.

ELEANOR MELLON (1894-1979)

Eleanor Mellon produced both portrait and figure sculpture, working on a number of ecclesiastical figures, and winning several major commissions. Among them were figure sculptures of Saint Francis, and Saint John in the Wilderness. Among her heroic sized monuments were her angel "Draped Winged Female" (VI-209), her "Man in Armor" (VI-210) and her "Angel of the Nativity" (VI-211), all seen here in plaster models. Her Saint Michael is in the Cathedral of Saint John the Divine in New York. Mellon was a student of several well-known sculptors, including A.A.Weinman, Robert Aitken, Charles Grafly, and Edward McCartan.

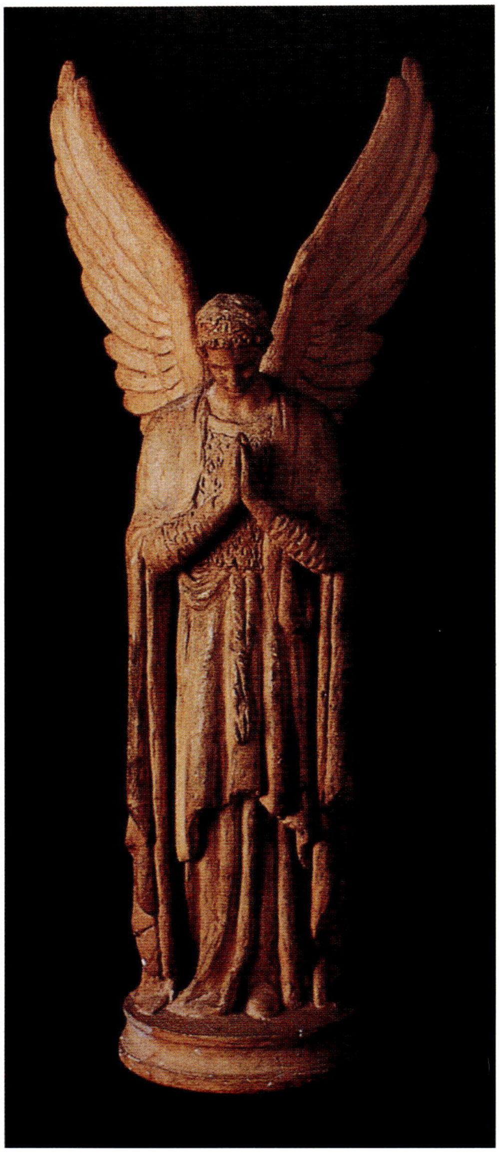

VI-209. "Draped Winged Female" (plaster cast) by Eleanor Mellon.

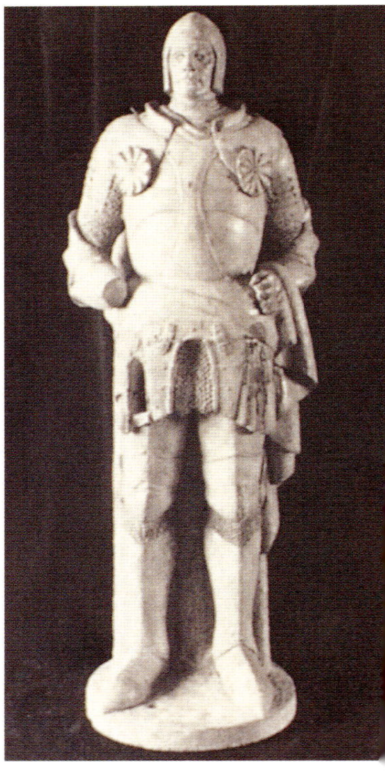

VI-210. "Man in Armor" (plaster cast) by Eleanor Mellon.

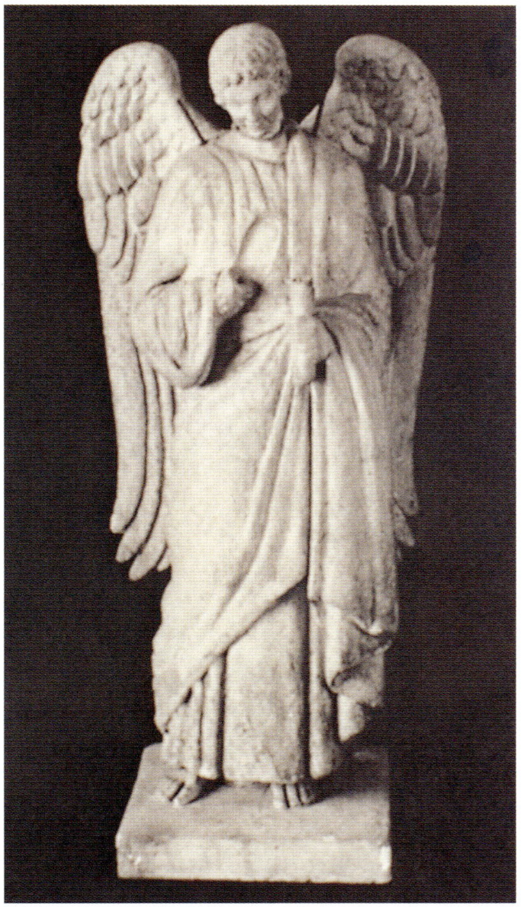

VI-211. "Angel of the Nativity" (plaster cast) by Eleanor Mellon. 1946.

ATTILIO PICCIRILLI (1866-1945)

Attilio Piccirilli was one of six brothers, all trained as marble workers in Italy. Attilio Piccirilli became a sculptor of monuments in the United States. While the entire clan produced much of the marble work of major buildings and commemorative statuary by other artists, Attilio was noted for his own artistic contributions, including the Marconi Memorial (VI-212) in Washington, and the Maine Monument in Central Park, New York City. Piccirilli cast a few works at R.B.W. in the 1930s and 1940s.

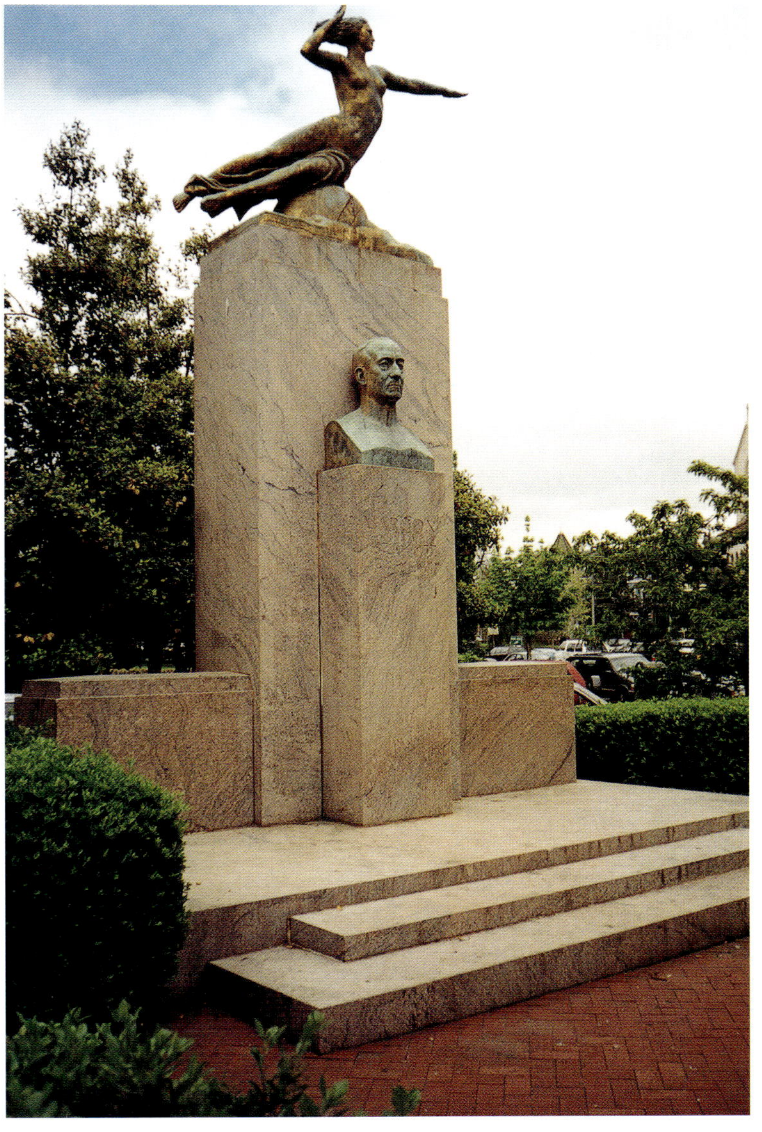

VI-212. "Marconi Memorial" by Attillio Piccirilli. 1941.

JOSEPH POLLIA
(1893-1954)

Joseph Pollia was born in Italy. He made many statues and memorial figures in New York and Massachusetts, and was particularly noted for his "Doughboy," honoring World War I veterans, his monument to General Pulaski, and his standing portrait of General Philip Henry Sheridan (VI-213), the Civil War general. It stands in the small park adjoining Sheridan Square in New York.

BRENDA PUTNAM
(1890-1975)

Brenda Putnam was a sculptor noted for her images of merry children and people making music, as well as decorative sundials and fountains. She is also known for her portrait image of Abraham Lincoln in Chicago. Her "Fountain Dancers" (VI-214) was made in 1919.

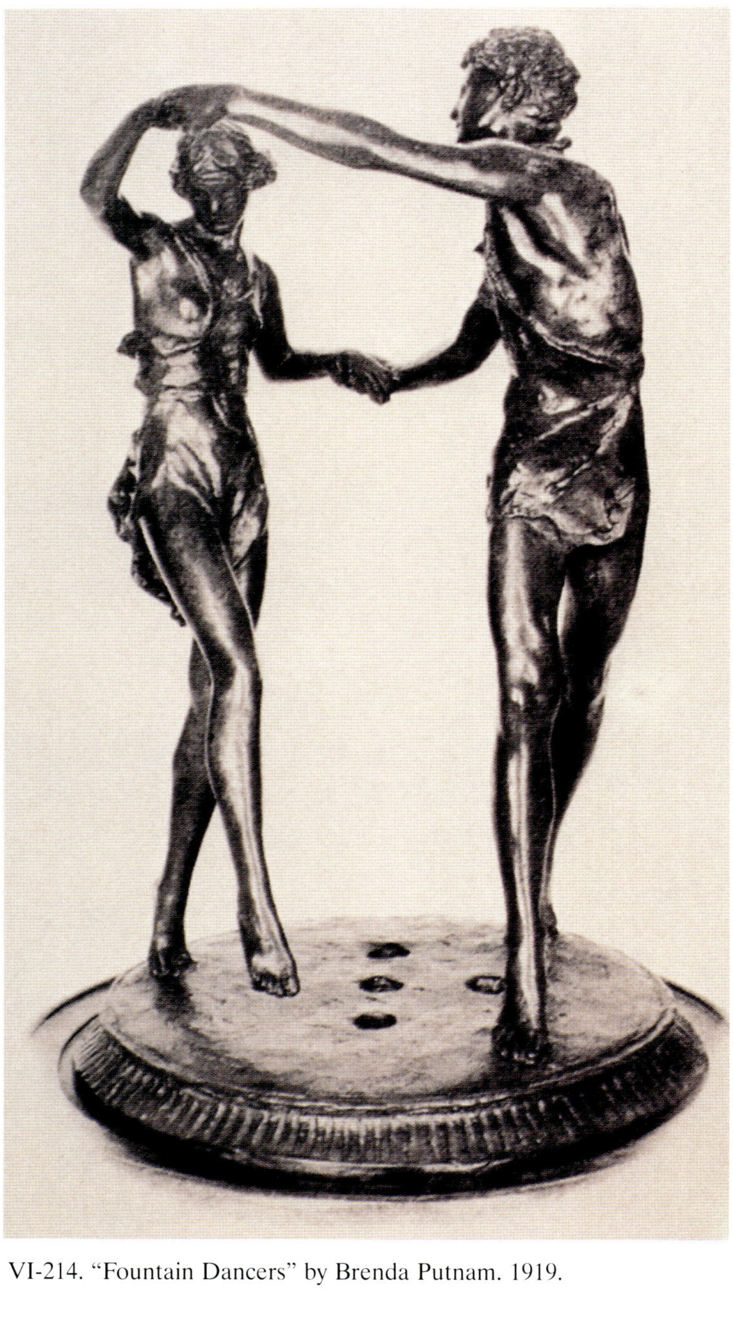

VI-214. "Fountain Dancers" by Brenda Putnam. 1919.

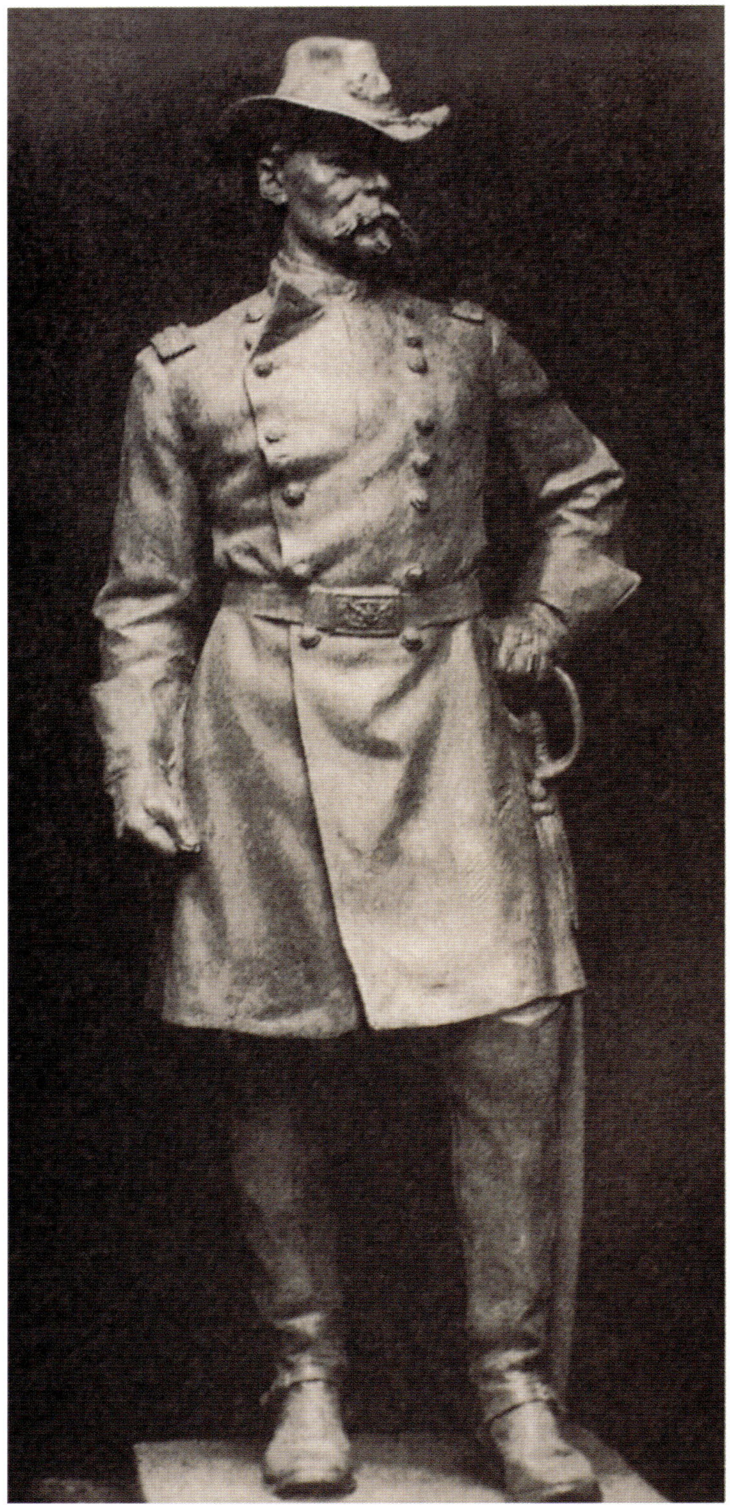

VI-213. "General Philip Henry Sheridan" by Joseph Pollia. c. 1936.

Among other notable sculptors who cast occasional pieces at R.B.W. were: John Angel ("Alexander Hamilton" and "Christ the Teacher"), Leo Friedlander (with James Earle Fraser the sculptor of the equestrian groups at the Arlington Memorial Bridge in Washington), Paul Wayland Bartlett (Lafayette statue in Hartford, Connecticut), Henri Crenier ("Girl and Butterfly"), Douglas Volk (Lincoln masks), Trygve Rovelstad (East front of the United States Capitol in Washington), Louis Saint-Gaudens ("The Pipes of Pan"), and Agop Agopoff (portraits of Will Rogers and John Henry, and an "Abyssinian Pirate"), and George Lober ("Girl with Grapes"). Canadian A. Suzor Cote cast a number of genre pieces at R.B.W.

VII. TO THE PRESENT: R.B.W.'s CURRENT SCULPTORS

\MARVIN BELL, SANDRO CHIA, DANTE V. DE FLORIO, ARTURO DI MODICA, LEONDA FINKE, BRIAN HUNT, BEATRICE JABERG, BRUNO LUCCHESI, LAWRENCE M. LUDKE, GREGORY PERILLO, ALFREDO SASSO, ADOLF SEHRING, RHODA SHERBELL, ELEANOR STILLMAN, AND BISMARCK VICTORIA

In the last few years Roman Bronze Works, under the direction of Philip Schiavo, has moved from Corona, Queens to its new home in Copiague, Long Island to continue its work of casting. Indicative of today's art world, sculptors work in a great variety of styles, and the foundry's clients typify that range of concept and stylistic direction.

Among sculptures cast at R.B.W. in recent decades are the works of Marvin Bell (VII-1 and 2), Sandro Chia (VII-3), Dante V. de Florio (VII-4), Arturo di Modica (VII-5), Leonda Finke (VII-6), Brian Hunt (VII-7), Beatrice Jaberg (VII-8), Bruno Lucchesi (VII-9), Lawrence M. Ludke (VII-10 and 11), Gregory Perillo (VII-12 and 13), Alfredo Sasso (VII-14), Adolf Sehring (VII-15), Rhoda Sherbell (VII-16), Eleanor Stillman (VII-17), and Bismarck Victoria (VII-18).

MARVIN BELL

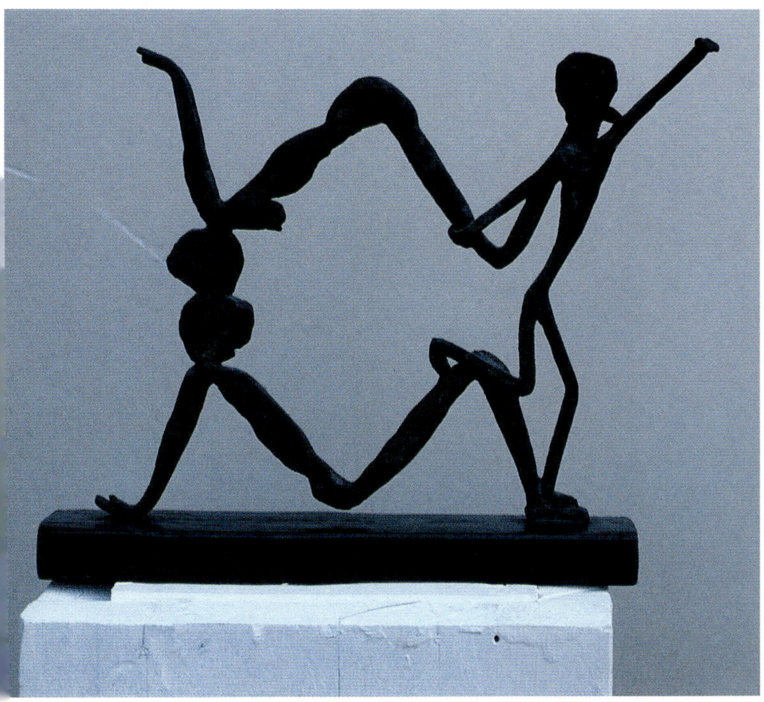

VII-1. "Family at Play" by Marvin Bell. 1985.

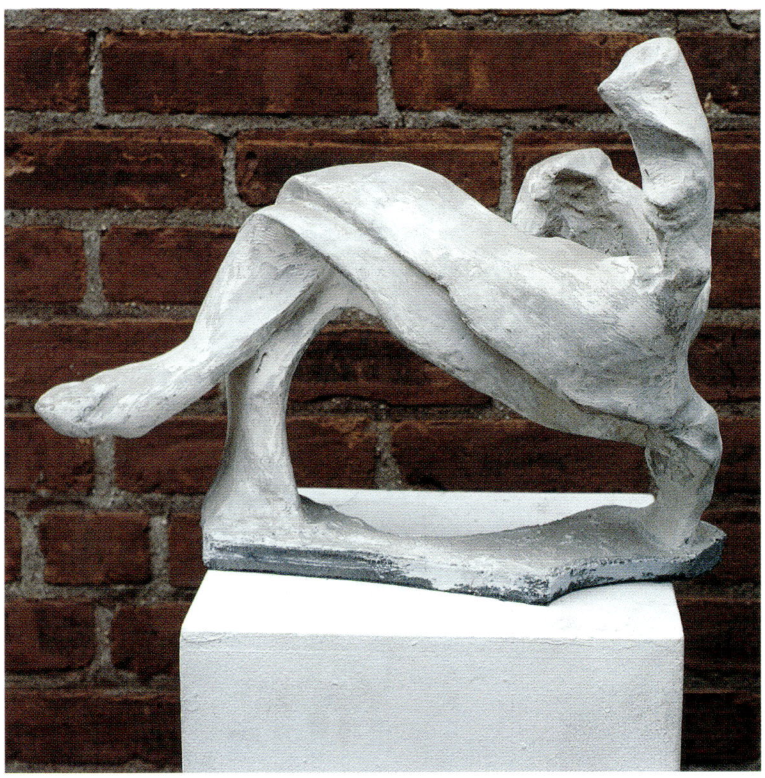

VII-2. "Free" by Marvin Bell.

SANDRO CHIA

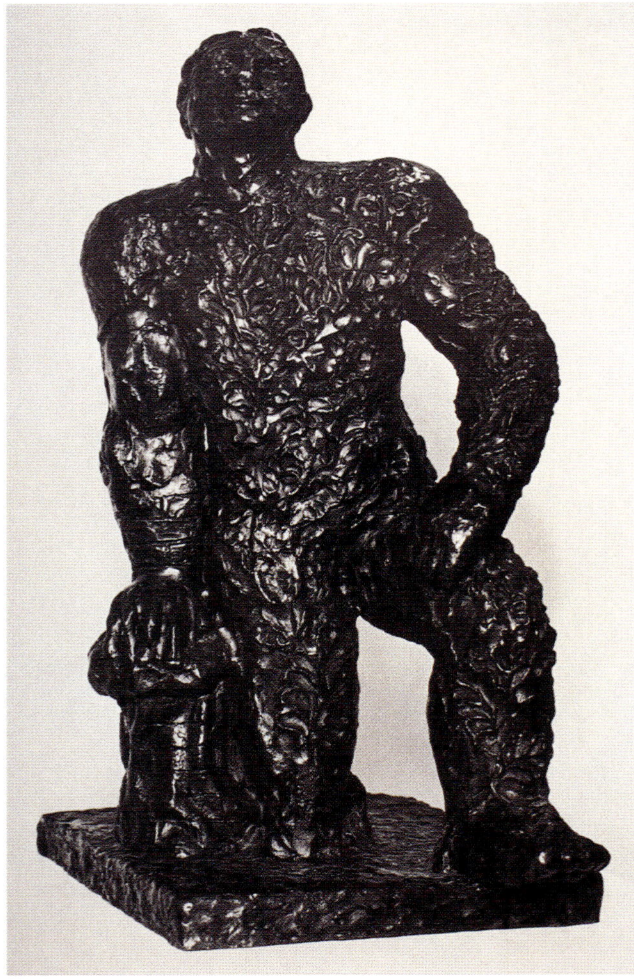

VII-3. "Kneeling Man" by Sandro Chia. 1982.

DANTE V. DE FLORIO

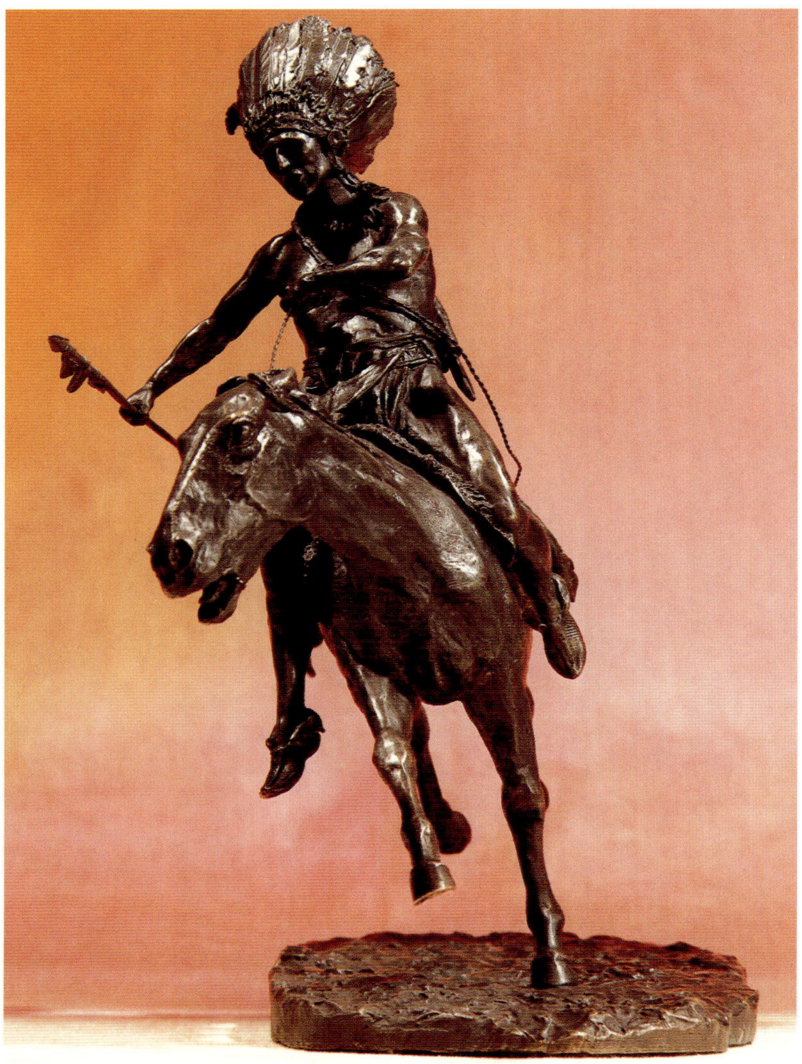

VII-4. "The Warrior" by Dante V. deFlorio.

ARTURO DI MODICA

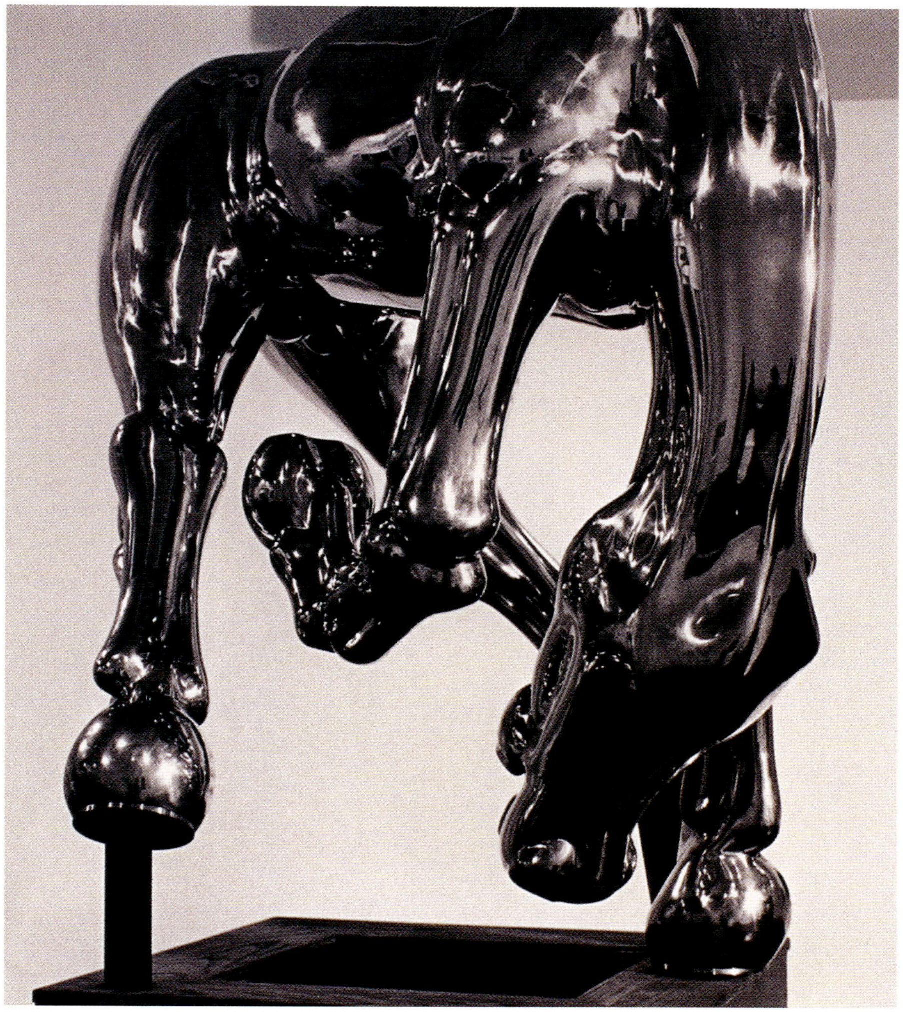

VII-5. "Cavallo" by Arturo diModico. 1981-84.

LEONDA FINKE

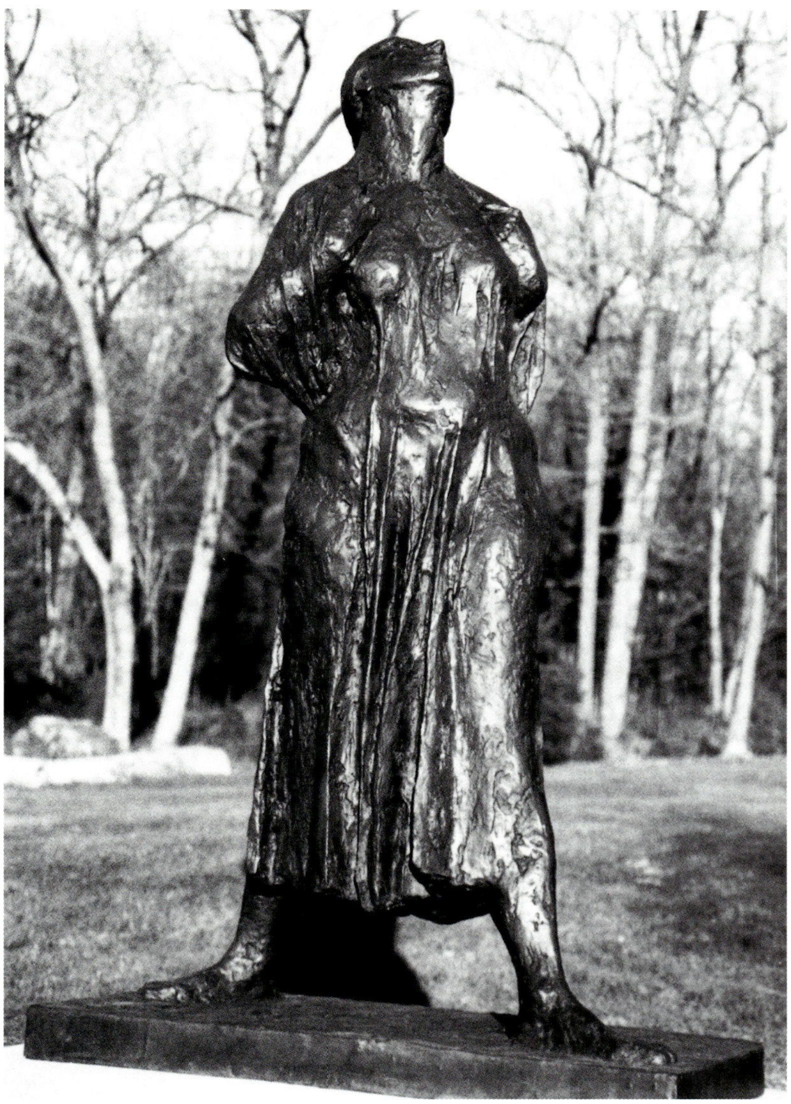

VII-6. "Standing Woman #3" by Leonda Finke.

BRIAN HUNT

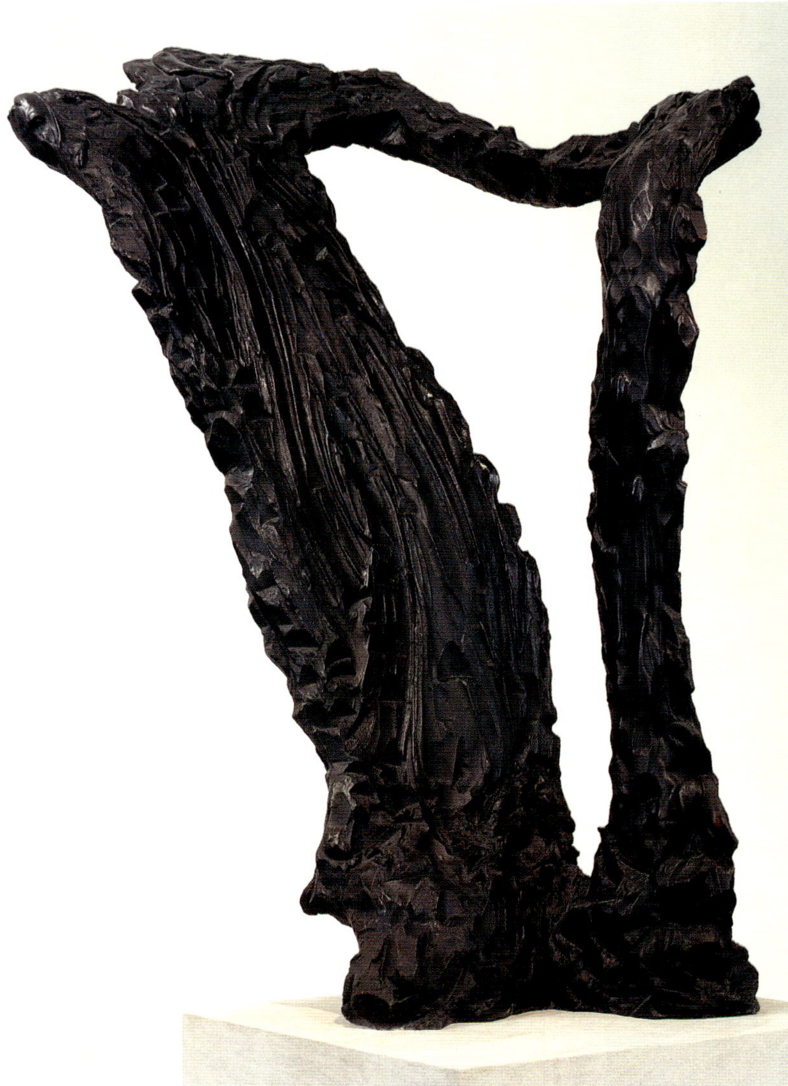

VII-7. "The Cloak of Lorenzo" by Brian Hunt. 1980-81.

BEATRICE JABERG

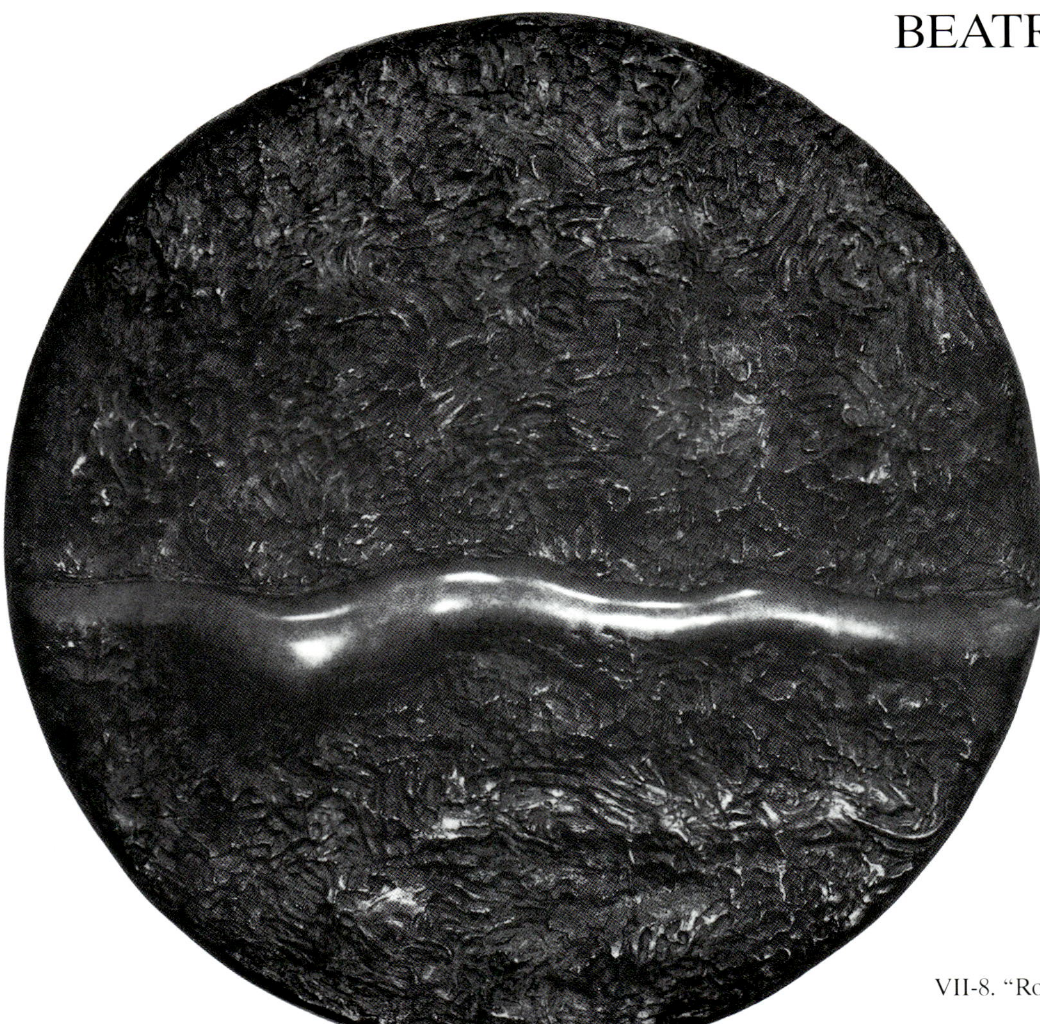

VII-8. "Round Relief" by Beatrice Jaberg. 1978.

BRUNO LUCCHESI

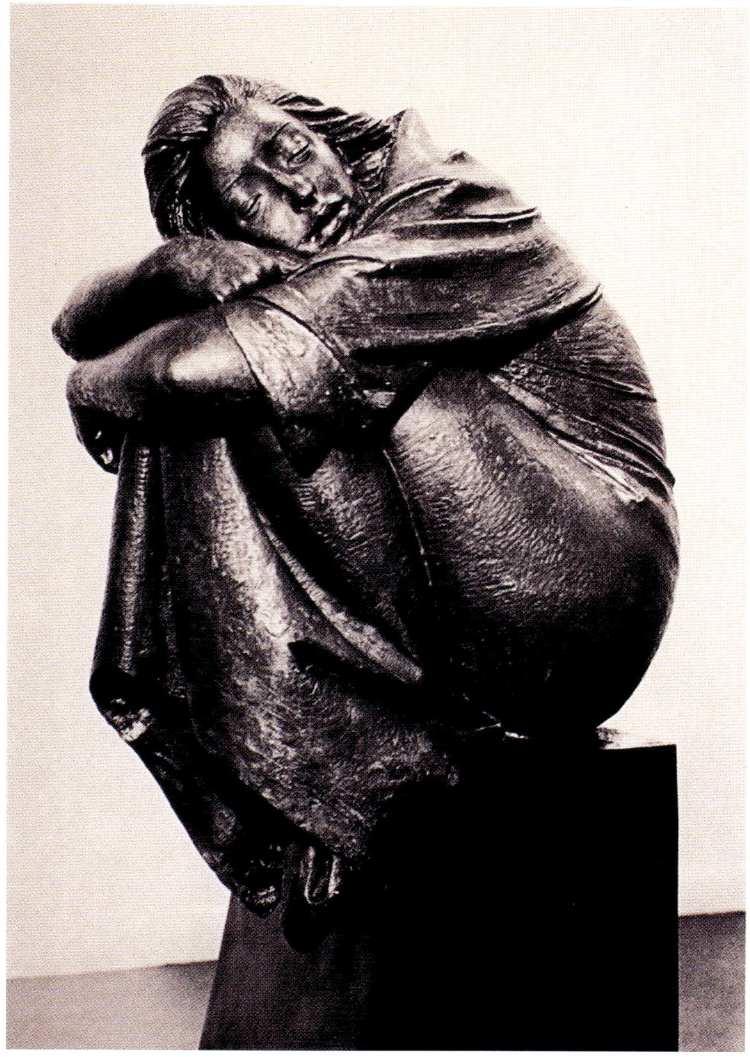

VII-9. "Muse" by Bruno Lucchesi. 1981.

LAWRENCE M. LUDKE

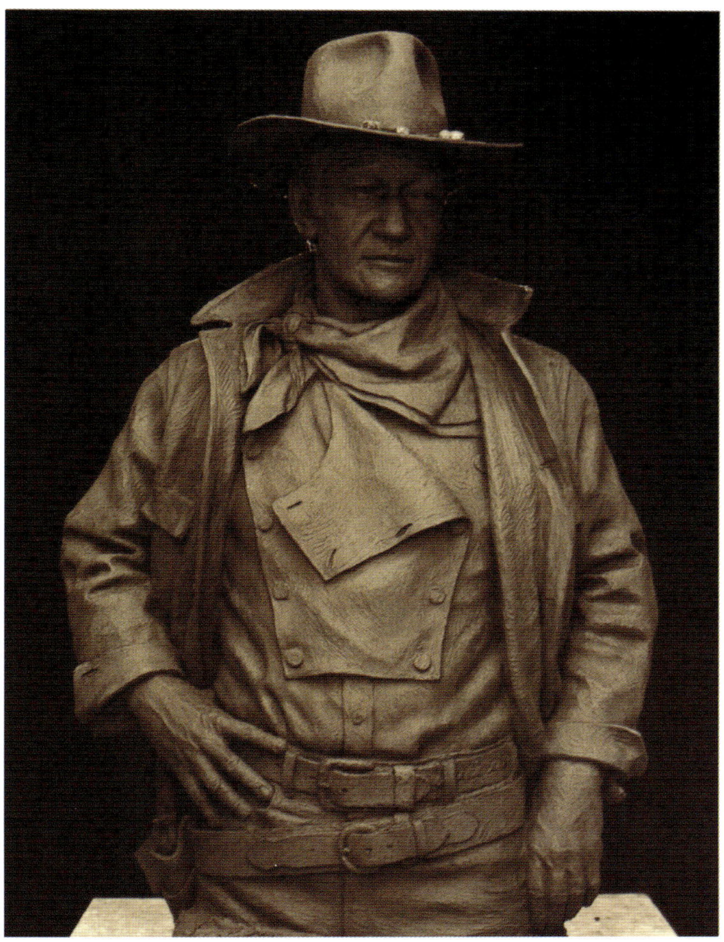

VII-10. "John Wayne" (clay model for bronze), by Lawrence M. Ludke.

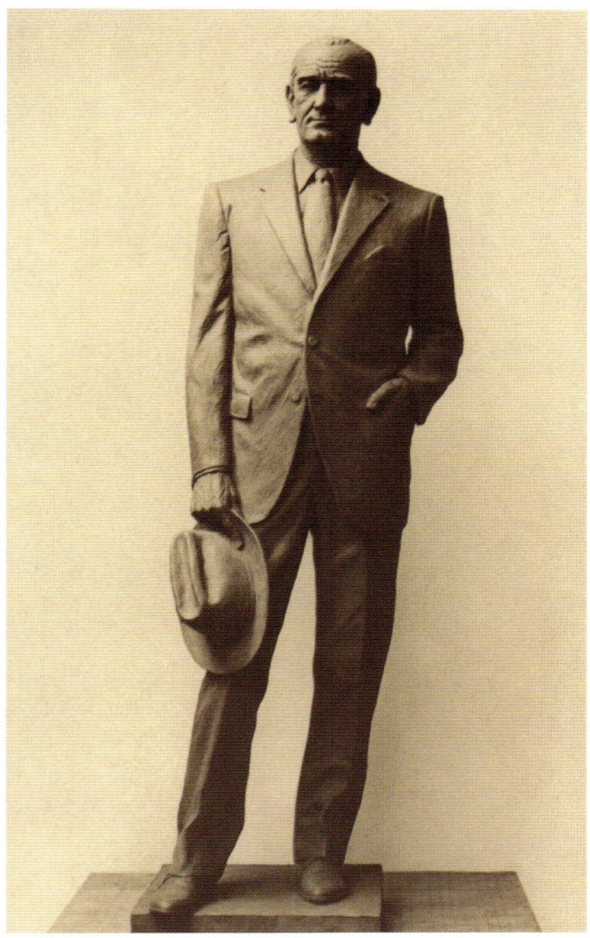

VII-11. "President Lyndon Johnson" (clay model for life-size bronze), by Lawrence M. Ludke.

GREGORY PERILLO

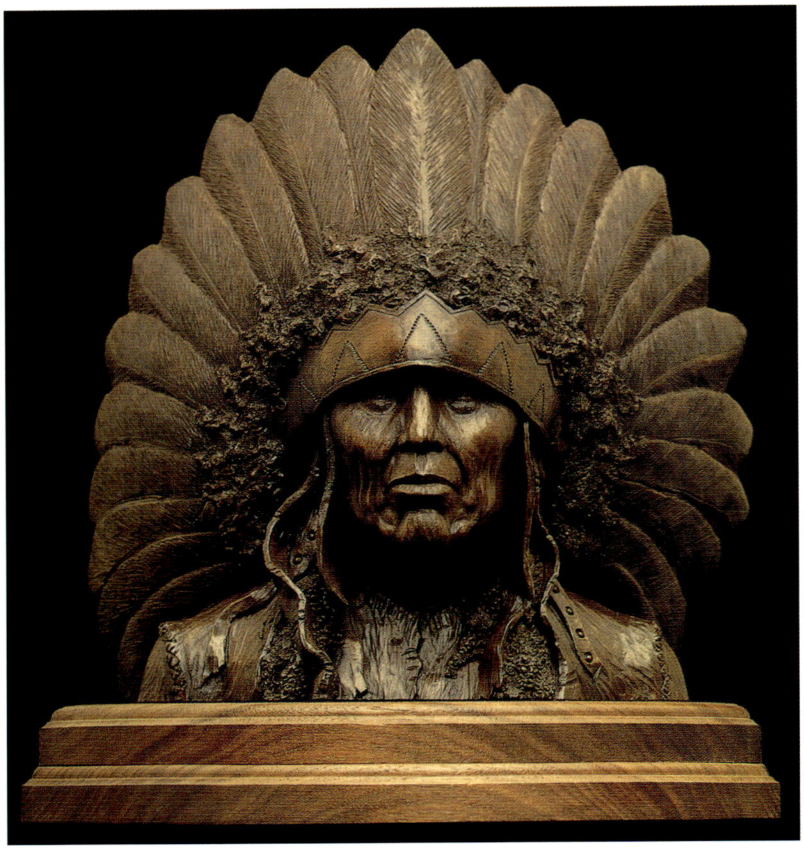

VII-12. "Chief Red Cloud" by Gregory Perillo.

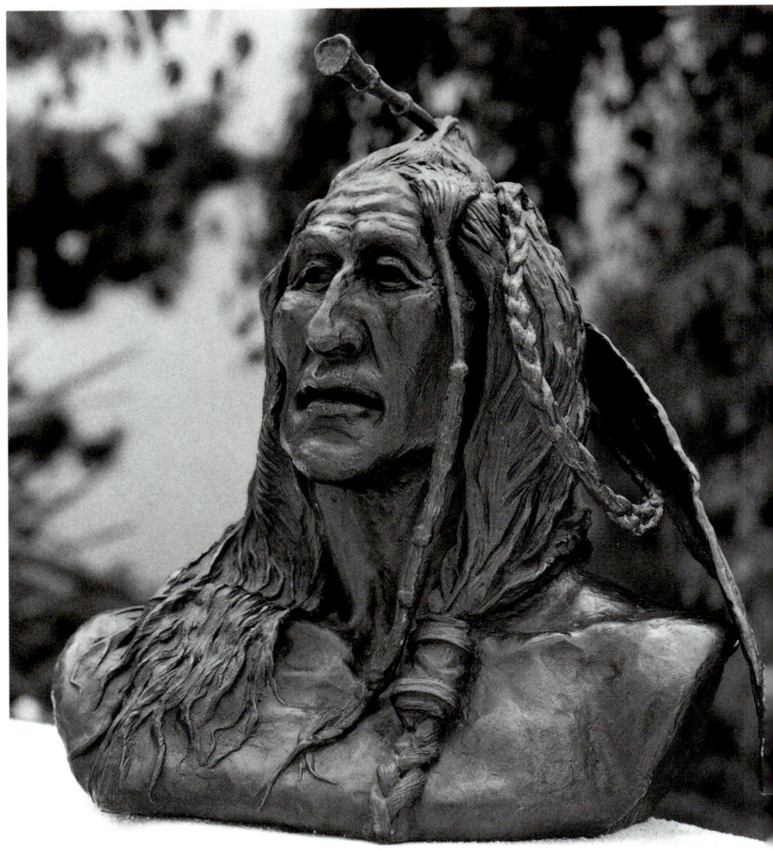

VII-13. "Chief Black Kettle" by Gregory Perillo.

ALFREDO SASSO

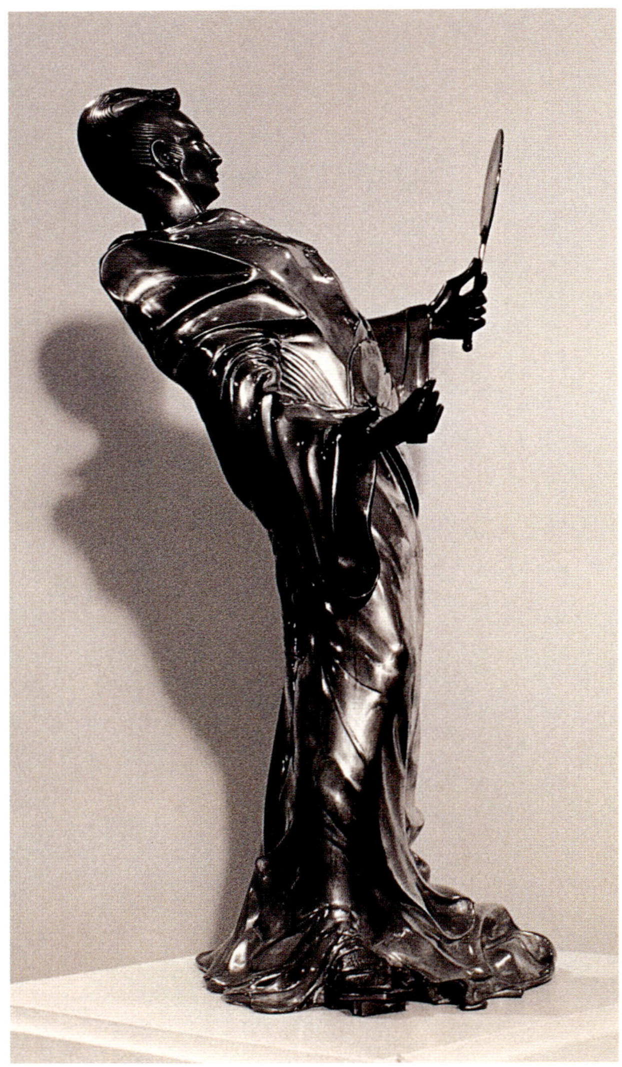

VII-14. "Samurai" by Alfredo Sasso. 1979.

ADOLF SEHRING

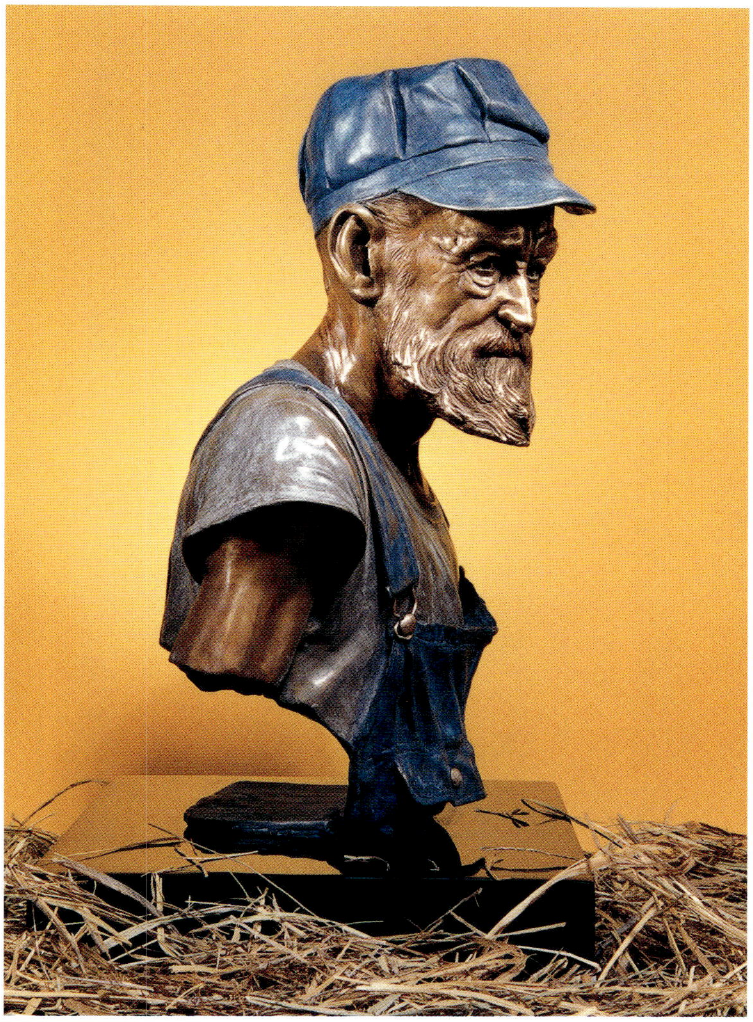

VII-15. "Leroy" by Adolf Sehring.

RHODA SHERBELL

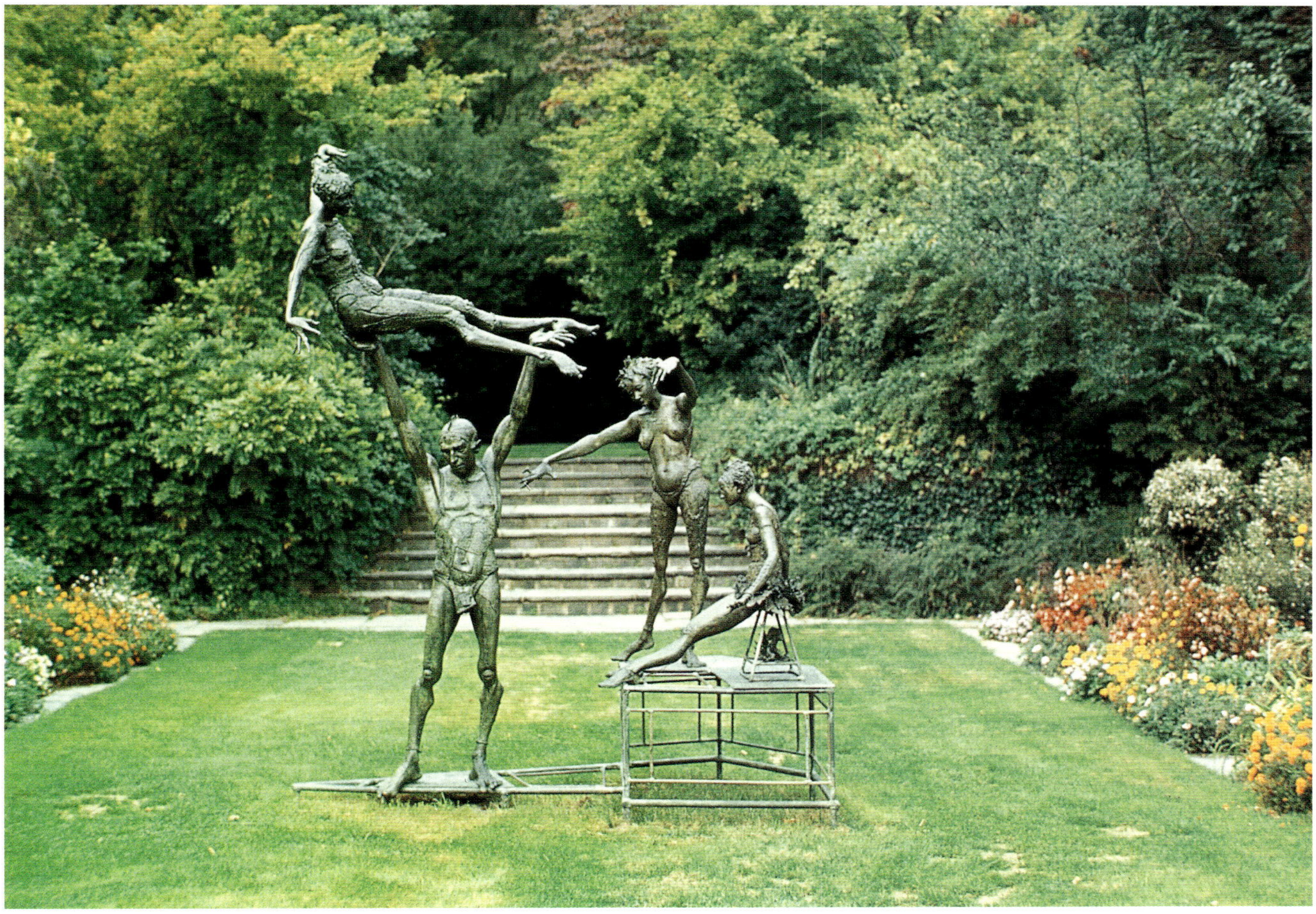

VII-16. "The Performers" by Rhoda Sherbell.

ELEANOR STILLMAN

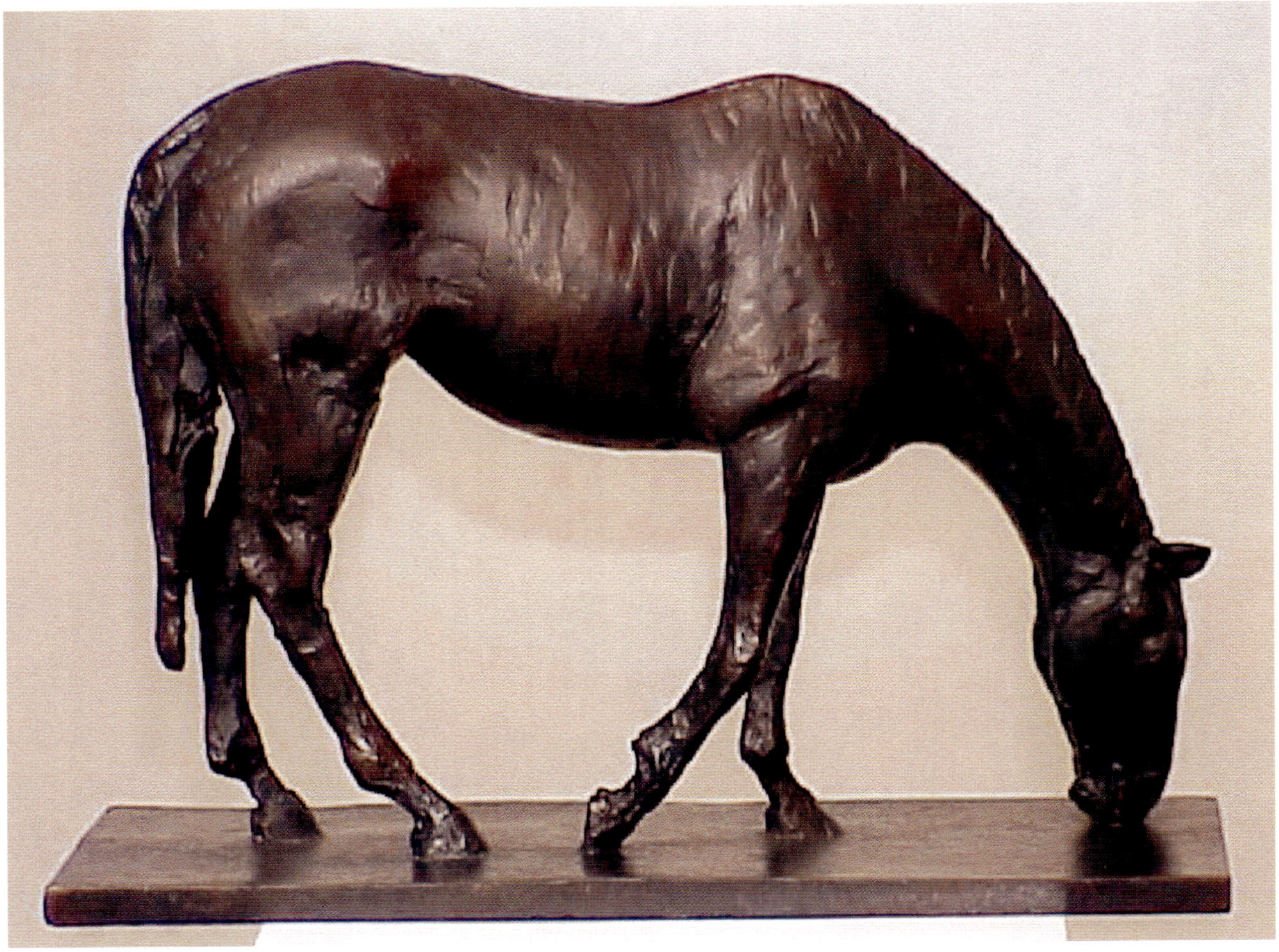

VII-17. "Retired" by Eleanor Stillman.

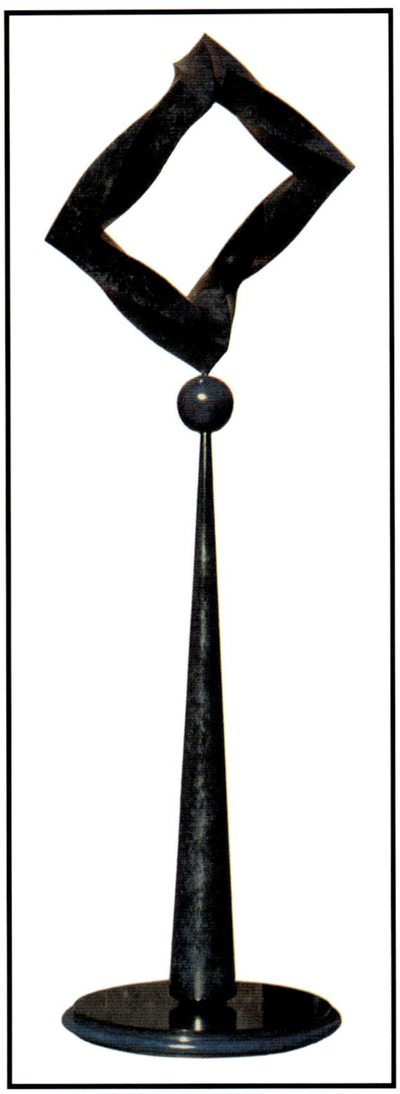

BISMARCK VICTORIA

VII-18. "Forgotten Alchemy" by Bismarck Victoria. 1977.

PICTURE CREDITS

I would like to extend my sincere thanks to everyone listed below for providing the photographs for this book. All of the pictures included in Chapters I through VI are the property of Roman Bronze Works or the National Sculpture Society in New York, except for those listed individually below. Additional outdoor photos were taken by the author. All pictures in Chapter VII are courtesy of the artists. The following pictures were graciously supplied by the organizations listed below:

III-7: Collection of the Cornish Colony and Museum, Cornish, New Hampshire; III-9: Collection of Brookgreen Gardens, Pawleys Island, South Carolina; III-10: Chesterwood, a National Trust Historic Site, Stockbridge, Massachusetts, photograph by A.S. Currier, Currier Studio, Exeter, N.H.; III-24: Courtesy of the U.S. Department of the Interior, National Park Service, Saint-Gaudens National Historic Site, Cornish, New Hampshire; III-26: photograph courtesy of Hirschl and Adler Galleries, Inc.; III-51: Collection of Brookgreen Gardens, Pawleys Island, South Carolina.

IV-78: Collection of Brookgreen Gardens, Pawleys Island, South Carolina; IV-119: Collection of Brookgreen Gardens, Pawleys Island, South Carolina; IV-123: Collection of Brookgreen Gardens, Pawleys Island, South Carolina.

V-21: Courtesy of the Aaron I. Fleischman Collection; V-34: Courtesy of Tishman Speyer Properties, New York; V-42: Private Collection; V-47: Collection of the Westmoreland Museum of American Art, Greensburg, Pennsylvania/ Gift of the Henry L. Hillman Foundation, Pittsburgh, Pennsylvania; V-51: Private Collection, photograph courtesy of Hirschl and Adler Galleries, Inc.; V-53: Private Collection, Frederick Koch, Butler, Pennsylvania; V-59: Courtesy of Tishman Speyer Properties, New York; V-91 through V-96: Courtesy Paul Nakian.

VI-8: Collection of Hirschl and Adler Galleries, Inc.; VI-31: Courtesy of the Shepherd Gallery, New York; VI-36: Collection of Brookgreen Gardens, Pawleys Island, South Carolina; VI-46: Christie's Images, New York 2000; VI-51: Post Road Gallery, Larchmont, New York; VI-57: Courtesy of the Shepherd Gallery, New York; VI-81: Collection of Brookgreen Gardens, Pawleys Island, South Carolina; VI-121: Collection of Brookgreen Gardens, Pawleys Island, South Carolina; VI-125: Collection of Brookgreen Gardens, Pawleys Island, South Carolina; VI-144: Collection of Brookgreen Gardens, Pawleys Island, South Carolina; VI-150: Collection of Brookgreen Gardens, Pawleys Island, South Carolina; VI-152: Courtesy of Tishman Speyer Properties, New York; VI-153: Courtesy of Tishman Speyer Properties, New York; V-171: Collection of Brookgreen Gardens, Pawleys Island, South Carolina.

INDEX of ARTISTS

Adams, Herbert, 38
Agopoff, Agop, 278
Aitken, Robert Ingersoll, 45
Akeley, Carl E., 134
Alston, Charles, 183
Angel, John, 278
Archipenko, Alexander, 140
Baker, Bryant, 64
Barnard, George Grey, 65
Bartlett, Paul Wayland, 278
Barye, Antoine Louis, 134
Beach, Chester, 265
Beckmann, Max, 183
Bell, Marvin, 279
Berge, Edward, 266
Bitter, Karl, 37
Bohland, Gustav, 268
Bonheur, Isidore Jules, 134
Borglum, Gutzon, 42
Borglum, Solon, 88
Calder, Alexander Stirling, 40
Calder, Alexander, 172
Callery, Mary, 184
Chambellan, Réne Paul, 250
Chia, Sandro, 280
Clark, Alan, 252
Clark, James Lippitt, 122
Colleti, Joseph Arthur, 269
Coppini, Pompeo, 270
Cote, A. Suzor, 278
Crenier, Henri, 278
Cressson, Margaret French, 62
Dallin, Cyrus E., 98
Davidson, Jo, 48
DeCreeft, José, 184
DeFlorio, Dante V., 280
DeFrancisi, Anthony, 242
Deming, Edwin W., 120
Derujinsky, Gleb, 246
Diederich, Hunt, 272
DiModica, Arturo, 282
Eakins, Thomas, 187
Eberle, Abstenia St. Leger, 272
Evans, Rudolph, 53
Fairbanks, Avard, 256
Farnham, Sally James, 93
Finke, Leonda, 281
Flanagan, John, 65
Fraser, James Earle, 217
French, Daniel Chester, 22
Friedlander, Leo, 278
Frishmuth, Harriet, 226
Giacomantonio, Archimedes, 262
Grafly, Charles, 55
Gregory, John, 224
Hammer, Trygve, 68
Hansen, Oscar J.W., 68
Hartley, Jonathan Scott, 68
Harvey, Eli, 135
Hering, Henry, 68
Hoffman, Malvina, 234
Humphriss, Charles, 96
Hunt, Brian, 282
Huntington, Anna Hyatt, 113
Jaberg, Beatrice, 282
Jennewein, Paul, 248
Jonas, Louis Paul, 136
Kalish, Max, 273
Kaz, Nathaniel, 273
Keck, Charles, 54
Kiselewski, Joseph, 260
Konti, Isidore, 201
Korbel, Josef Mario, 229
Lachaise, Gaston, 145
Laessle, Albert, 129
Laurent, Robert, 155
Lawrie, Lee, 165
Lentelli, Leo, 274
Linder, Henry, 274
Lipchitz, Jacques, 150
Lober, George, 278
Longman, Evelyn, 275
Lovett-Lorski, Boris, 276
Lucchesi, Bruno, 283
Ludke, Lawrence M., 283
Lukeman, Augustus, 66
MacMonnies, Frederick William, 204
MacNeil, Hermon Atkins, 101
Manship, Paul, 156
Martiny, Philip, 66
McCartan, Edward, 219
McKenzie, Robert Tait, 211
Mellon, Eleanor, 276
Milles, Carl, 184
Murray, Samuel, 67
Nadelman, Elie, 143
Nakian, Reuben, 180
Niehaus, Charles Henry, 67
Noguchi, Isamu, 153
Partridge, William Ordway, 57
Perillo, Gregory, 284
Piccirilli, Attilio, 277
Polasek, Albin, 222
Pollia Joseph, 278
Pope, John Russell, 68
Potter, Louis McClellan, 136
Pratt, Bela Lyon, 68
Proctor, Alexander Phiminster, 109
Putnam, Brenda, 278
Remington, Frederic, 70
Rhind, John Massey, 58
Robus, Hugo, 169
Rogers, John, 186
Roth, Frederic, 131
Rovelstad, Trygve, 278
Russell, Charles Marion, 79
Saint-Gaudens, Augustus, 29
Saint-Gaudens, Louis, 278
Sasso, Alfredo, 284
Schreyvogel, Charles, 138
Sehring, Adolf, 285
Sherbell, Rhoda, 285
Shrady, Henry Merwin, 125
Sterne, Maurice, 184
Stillman, Eleanor, 286
Storrs, John, 183
Taft, Lorado, 68
Troubetskoy, Paul, 208
Umlauf, Charles, 138
Victoria, Bismarck, 286
Volk, Douglas, 278
Vonnoh, Bessie Potter, 190
Ward, John Quincy Adams, 36
Weber, Max, 151
Weinman, Adolph Alexander, 59, 105
Whitney, Gertrude Vanderbilt, 216
Williams, Wheeler, 258
Ximenes, Ettore, 68
Young, Mahonri, 107, 199
Zorach, William, 167